SOLUTIONS MANUAL

PRENTICE HALL

Algebra 1

Jan Fair

Sadie Bragg

PRENTICE HALL
Englewood Cliffs, New Jersey
Needham, Massachusetts

CONTENTS

Printed in the United States of America.

ISBN 0-13-026493-8

1 2 3 4 5 6 7 8 9 10   96 95 94 93 92

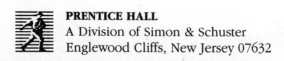

PRENTICE HALL
A Division of Simon & Schuster
Englewood Cliffs, New Jersey 07632

Chapter 1   Real Numbers

page 2   Capsule Review

1. $6\left(\dfrac{3}{4}\right) = \dfrac{18}{4} = \$4.50$

2. $6\left(3\dfrac{1}{2}\right) = \dfrac{42}{2} = \$21$

3. $6(6.8) = \$40.80$

4. $6(10) = \$60$

5. change to $8 \times a$

page 5   Class Exercises

1. x; multiplication

2. a; subtraction

3. m; addition

4. g; division

5. $1 + 89 = 90$

6. $5.4 - 0.5 = 4.9$

7. $5.4 \div 0.5 = 10.8$

8. $(0.5)(1)(5.4) = 2.7$

pages 5–6   Practice Exercises

A 1. $12 + 15 = 27$

2. $30 - 12 = 18$

3. $\dfrac{1}{2}(12) = 6$

4. $\dfrac{5}{6}(12) = 10$

5. $12 \div 6 = 2$

6. $\dfrac{48}{12} = 4$

7. $12 + 12 = 24$

8. $0 + 12 = 12$

9. $\dfrac{(15)(6)}{2(3)} = 15$

10. $5(15)(6)(3) = 1350$

11. $\dfrac{3}{10(15)(6)} = \dfrac{1}{300}$

12. $\dfrac{(15)(6)(3)}{25} = \dfrac{54}{5}$

13. $\dfrac{1}{3}(15 + 3) = 6$

14. $\dfrac{1}{5}(6 + 3) = \dfrac{9}{5}$

15. $\dfrac{2}{3}(3) + 5 = 7$

16. $\dfrac{5}{3}(15) + 1 = 26$

17. $3 + x$

18. $\dfrac{3}{2} + y$

19. $t - 1.07$

20. $m - 99$

B 21. $\dfrac{3}{4} \div \dfrac{2}{3} = \dfrac{9}{8}$

22. $\dfrac{3}{4} - \dfrac{2}{3} = \dfrac{1}{12}$

23. $\dfrac{3}{4} - \dfrac{1}{2} = \dfrac{1}{4}$

24. $\dfrac{\left(\dfrac{2}{3}\right)\left(\dfrac{3}{4}\right)}{\dfrac{1}{2}} = 1$

25. $4\left(\dfrac{2}{3}\right)\left(\dfrac{3}{4}\right)\left(\dfrac{1}{2}\right) = 1$

26. $\dfrac{2}{3} - \dfrac{1}{2} = \dfrac{1}{6}$

27. $\dfrac{1}{2} - \dfrac{1}{2} = 0$

28. $\dfrac{2}{3} + \dfrac{3}{4} + \dfrac{1}{2} = \dfrac{23}{12}$

29. $\dfrac{\dfrac{2}{3} + \dfrac{3}{4}}{\dfrac{1}{2}} = \dfrac{\dfrac{17}{12}}{\dfrac{1}{2}} = \dfrac{17}{6}$

30. $\dfrac{2\left(\frac{3}{4}\right) - \frac{2}{3}}{2\left(\frac{1}{2}\right)} = \dfrac{\frac{3}{2} - \frac{2}{3}}{1} = \dfrac{5}{6}$

31. $3\left(\frac{2}{3}\right) + 2\left(\frac{3}{4}\right) + 2\left(\frac{1}{2}\right) = \dfrac{9}{2}$

32. $\dfrac{3}{2}\left(\frac{2}{3}\right) + \dfrac{3}{4} + 4\left(\frac{1}{2}\right) = \dfrac{15}{4}$

33. $\dfrac{3}{8}j$

34. $z - 5$

35. $\dfrac{6.4}{b}$

36. $x + 10$

C 37. $\left(3\frac{5}{8} - \frac{3}{4}\right)(8) = \dfrac{23}{8}(8) = 23$

38. $\left(\frac{3}{4} + 6\right)\left(\frac{1}{3}\right) = \dfrac{9}{4}$

39. $\dfrac{6 + 4}{6 - 4} = \dfrac{10}{2} = 5$

40. $\dfrac{6 - 1 + 4}{6 - \frac{3}{4}} = \dfrac{9}{\frac{21}{4}} = \dfrac{12}{7}$

41. $(8)\left(\frac{3}{4}\right) + \left(\frac{2}{3}\right)(6) - \left(\frac{1}{2}\right)(4) = 6 + 4 - 2 = 8$

42. $\left(\frac{7}{10}\right)(6 + 4) - (4)\left(\frac{3}{4}\right) = \dfrac{7}{10}(10) - 3 = 4$

43. $0.6(220 - 25) = 117$; 117 beats/min; $0.6(220 - 50) = 102$; 102 beats/min

44. $\dfrac{m}{g}$

45. $h - l$

46. Answers may vary. An example: 1. Substitute values for a, b, c, and d. 2. Add the values in parentheses. 3. Multiply the result in Step 2 by the value of a. 4. Multiply the value of d by 2. 5. Divide the result in Step 3 by the result in Step 4.

page 7   Capsule Review

1. 2

2. -5

3. 3

4. 0

5. E

6. K

7. B

8. F

page 9   Class Exercises

1. $\dfrac{1}{2}$

2. $\dfrac{3}{4}$

3. $1\dfrac{1}{4}$

4. $2\dfrac{1}{4}$

5. E

6. J

7. B

8. L

9.

10.

11.

12.

pages 10–11 Practice Exercises

A **1.** -1.0 **2.** -0.5 **3.** 0.2 **4.** 0.8

5. -1.2 **6.** -0.8 **7.** 1.2 **8.** 1.6

9. G **10.** K **11.** A **12.** E

13. I **14.** H **15.** D **16.** F

17. $\dfrac{7}{1}$ **18.** $\dfrac{-5}{1}$ **19.** $\dfrac{-3012}{1000}$ **20.** $\dfrac{2618}{1000}$

21. $\dfrac{47}{4}$ **22.** $\dfrac{-41}{6}$ **23.** $\dfrac{-1}{1}$ **24.** $\dfrac{0}{1}$

B **25.** ◄—┼—┼●┼●┼●┼◆┼◆┼—►
$-4\ -2\ \ 0\ \ 2\ \ 4\ \ 6$

26. ◄—┼●┼●┼—◆┼—◆┼—┼—►
$-4\ -2\ \ 0\ \ 2\ \ 4\ \ 6$

27. ◄—◆┼—┼●┼●┼—●●┼—┼—►
$-4\ -2\ \ 0\ \ 2\ \ 4\ \ 6$

28. ◄—◆┼—●┼—┼—●┼●●┼—┼—►
$-6\ -3\ \ 0\ \ 3\ \ 6\ \ 9$

29. ◄—◆┼—◆┼—┼—┼●●┼—┼—►
$-1\ \ \ \ \ 0\ \ \ \ \ 1$

30. ◄—◆┼—┼—┼●●┼—┼●┼┼—►
$-2\ -1\ \ 0\ \ 1\ \ 2\ \ 3$

		Natural Numbers	Whole Numbers	Integers	Rational Numbers	Irrational Numbers	Real Numbers
31.	8	✔	✔	✔	✔		✔
32.	-11			✔	✔		✔
33.	$\dfrac{0}{10}$		✔	✔	✔		✔
34.	16.2				✔		✔
35.	$-3\dfrac{1}{6}$				✔		✔
36.	$-\sqrt{3}$					✔	✔

C **37.** Yes; yes; every real number corresponds to a point on the number line.

38. The graph of the real numbers is a continuous line because between every two real numbers there exists another real number.

39–40.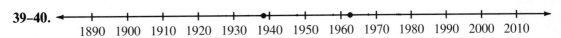
$1890\quad 1900\quad 1910\quad 1920\quad 1930\quad 1940\quad 1950\quad 1960\quad 1970\quad 1980\quad 1990\quad 2000\quad 2010$

41. Answers may vary. **42.** Answers may vary.

43. $5 \div 8 = 0.625$ **44.** $1 + 4 \div 5 = 1.8$

45. $0.25 \cdot 100 = 25\%$ **46.** $4.6 \cdot 100 = 460\%$

page 12 Capsule Review

1. $6 > -5$ **2.** $-4 < 3$ **3.** $-1.5 < 0$ **4.** $-2 = -2$

5. $-\frac{1}{2} > -3$ **6.** $-1\frac{1}{2} < -1$ **7.** $\frac{2}{3} < \frac{3}{4}$ **8.** $-1.2 > -1.25$

pages 14–15 Class Exercises

1. infinitely many **2.** -19 **3.** 5 **4.** $\frac{2}{3}$

5. $-\frac{11}{8}$ **6.** true **7.** true **8.** false

9. false **10.** false **11.** false **12.** false **13.** true

14. Yes. If a number is positive or zero, then its absolute value is equal to the number itself. If a number is negative, then its absolute value is equal to the opposite of the number.

pages 15–16 Practice Exercises

A 1. $-4, -1, 0, 2, 3$

2. $-3, -2, 1, 3, 5$

3. $-2\frac{1}{2}, -\frac{1}{2}, 0, 2, 2\frac{1}{2}$

4. $-1.5, -1, -0.5, 0, 1.5$

5. $-1 < 0$ **6.** $-3 > -4$ **7.** $1.5 > 1.25$ **8.** $3.3 < 3.33$

9. $\frac{3}{9} = \frac{2}{6}$ **10.** $-\frac{1}{4} < -\frac{1}{8}$ **11.** $-\frac{1}{3} > -\frac{1}{2}$ **12.** $-\frac{3}{5} > -\frac{3}{4}$

13. 3.1 **14.** 5.34 **15.** $-\frac{2}{5}$ **16.** $-\frac{1}{4}$

17. 6.2 **18.** $\frac{1}{2}$ **19.** $-3\frac{1}{5}$ **20.** $-\frac{5}{7}$

21. 13 **22.** 44 **23.** 5 **24.** 12

B 25. $\left\{\frac{1}{7}, \frac{1}{6}, \frac{1}{5}\right\}$ **26.** $\left\{-\frac{3}{4}, -\frac{3}{5}, -\frac{3}{8}\right\}$ **27.** $\left\{-1\frac{3}{4}, -1\frac{2}{3}, -\frac{5}{4}\right\}$

28. $\left\{\frac{5}{3}, 1\frac{5}{6}, \frac{5}{2}\right\}$ **29.** $\{-0.1041, -0.104, -0.1\}$ **30.** $\{7.88, 7.8809, 7.885\}$

31. -14 **32.** $4\frac{1}{5}$ **33.** $-\left|\frac{3}{8}\right| = -\frac{3}{8}$ **34.** $-[5.23] = -5.23$ **35.** -21

36. 50 **37.** $3 + 9 = 12$ **38.** $5 - 4 = 1$ **39.** $2 + |-5| = 2 + 5 = 7$

C 40. $|-6.1| = 6.1$ **41.** $-|7| = -7$ **42.** $|-(-6.1)| = |6.1| = 6.1$

43. $-|-6.1| = -(6.1) = -6.1$ **44.** $\left|7 + 2\frac{3}{4}\right| = 9\frac{3}{4}$ **45.** $|-6.1| - \left|2\frac{3}{4}\right| =$

$6.1 - 2.75 = 3.35$ **46.** true   **47.** false   **48.** true   **49.** true

50. $+16$; 16°C above freezing   **51.** -20; 20°C below freezing   **52.** -6.5;

loss of 6.5 lb   **53.** $+3\frac{1}{4}$; $3\frac{1}{4}$ lb weight gain   **54.** -120; 120 ft below sea

level   **55.** $+55$; 55 ft above sea level   **56.** $+2560.38$; $2560.38 credit

57. -785.95; $785.95 loss   **58.** $-1\frac{1}{2}$; loss of $1\frac{1}{2}$ points   **59.** $+\frac{3}{4}$; $\frac{3}{4}$ point

gain   **60.** $INT(17/2) = 8$ **61.** $INT(-\pi) = -4$ **62.** $INT(\sqrt{35}) = 5$

63. $INT(98/99) = 0$ **64.** $INT(-98/99) = -1$ **65.** $INT(-6.6 - 3.3) = -10$

66. $INT(X/2) = X$ when $X = -1$ or 0.   **67.** Answers may vary. A possible

answer is 2.365.   **68.** Answers may vary. A possible answer is 9.495.

69. 7 and 8   **70.** 1 and 18   **71.** 2, 3, 5, 7, 11, 13, 17, 19

72. $2 \cdot 2 \cdot 3 \cdot 5$ **73.** 31, 37, 41, 43, 47

page 17   Capsule Review

1. 3 units   **2.** 1 unit   **3.** 3 units   **4.** 10 units

5. 1 unit   **6.** 4 units   **7.** 8 units   **8.** 4 units

pages 19–20   Class Exercises

1. $3 + (-6) = -3$ **2.** $-2 + 4 = 2$ **3.** $-4 + (-1) = -5$

4. $3 + 3 = 6$ **5.** $6 + (-4) = 2$ **6.** $-3 + 8 = 5$

7. neg.   **8.** neg.   **9.** pos.   **10.** zero   **11.** pos.   **12.** neg.

pages 20–21   Practice Exercises

A 1. -5 **2.** -7 **3.** -6 **4.** -21

5. 1 **6.** -3 **7.** -7 **8.** -17

9. -8 **10.** -6 **11.** -3.5 **12.** -4.3

13. -8.0 **14.** -1 **15.** -6 **16.** -9

17. 1   **18.** −11   **19.** 6   **20.** 2   **21.** −6

B 22. 1   **23.** −17   **24.** −23   **25.** −6.5   **26.** 2.5

27. $-2\frac{1}{2}$ **28.** $2\frac{3}{4}$ **29.** $\frac{7}{3}$ **30.** $-\frac{11}{2}$

31. 2.75   **32.** 2.333   **33.** −5.5

C 34. −16.5 < − 16.3   **35.** 15.35 > − 15.35   **36.** $\left|-\frac{11}{56}\right| < \left|-\frac{19}{56}\right|$ **37.** $\left|-\frac{7}{9}\right| = \left|\frac{7}{9}\right|$

38. 5 + (−17) = −12; −12°C

39. (−3) + (−3) + (−3) + (−3) = −12; 12 yd loss

40. 25 + (−12) + 10 = 23; $23.00 left   **41.** $-3\frac{1}{8} + 4\frac{5}{8} = 1\frac{4}{8} = 1\frac{1}{2}$; up $1\frac{1}{2}$ points

page 21   Biography: Carl Friedrich Gauss

1. 1 + 2 + 3 + ⋯ + 48 + 49 + 50
50 + 49 + 48 + ⋯ + 3 + 2 + 1
$\overline{51 + 51 + 51 + ⋯ + 51 + 51 + 51}$

Adding the pairs of numbers in each column, you get a sum of 51. There are 50 such sums. (50)(51) is two times the correct answer.
$\frac{(50)(51)}{2}$ = (25)(51) = 1275

2. 31 + 32 + 33 + ⋯ + 38 + 39 + 40
40 + 39 + 38 + ⋯ + 33 + 32 + 31
$\overline{71 + 71 + 71 + ⋯ + 71 + 71 + 71}$

Adding the pairs of numbers in each column, you get a sum of 71. There are 10 such sums. (10)(71) is two times the correct answer.
$\frac{(10)(71)}{2}$ = 355

page 22   Capsule Review

1. $|-38|$ **2.** $|-6.2|$ **3.** $|0.56|$

page 24   Class Exercises

1. 21   **2.** −4   **3.** 0   **4.** $-8\frac{5}{8}$

5. Yes, but it can be cumbersome if the numbers are very large and/or nonintegral.

6. Rules help simplify the process.

pages 24–26   Practice Exercises

A 1. −500   **2.** −213   **3.** −453   **4.** 853   **5.** 4.61

6. 49.76   **7.** −1.709   **8.** −0.836   **9.** $-\frac{4}{7}$ **10.** $\frac{6}{9} = \frac{2}{3}$

6   Chapter 1: Real Numbers

11. $-\dfrac{3}{8} + \left(-\dfrac{2}{8}\right) = -\dfrac{5}{8}$

12. $-\dfrac{5}{9} + \left(-\dfrac{3}{9}\right) = -\dfrac{8}{9}$

13. $-\dfrac{6}{14} + \dfrac{7}{14} = \dfrac{1}{14}$

14. $\dfrac{25}{30} + \left(-\dfrac{12}{30}\right) = \dfrac{13}{30}$

15. $-27 + 27 = 0$

16. $59 + (-19) = 40$

17. $15.3 + (-6.7) = 8.6$

18. $-27.5 + 4.3 = -23.2$

19. $-5 + \left(-2\dfrac{1}{2}\right) + \left(-\dfrac{1}{2}\right) = -8$

20. $2\dfrac{1}{2} + (-15) + [-(-5)] = -\dfrac{15}{2} = -7\dfrac{1}{2}$

21. $-(-5) + 2\dfrac{1}{2} + \left[-\left(-\dfrac{1}{2}\right)\right] = \dfrac{16}{2} = 8$

22. $-2\dfrac{1}{2} + 15 + (-5) = \dfrac{15}{2} = 7\dfrac{1}{2}$

B 23. $-\dfrac{9}{24} + \dfrac{14}{24} + \dfrac{12}{24} = \dfrac{17}{24}$

24. $\dfrac{15}{18} + \left(-\dfrac{4}{18}\right) + \left(-\dfrac{6}{18}\right) = \dfrac{5}{18}$

25. $\dfrac{159}{75} + \left(-\dfrac{35}{75}\right) = \dfrac{124}{75} = 1\dfrac{49}{75}$

26. $\dfrac{16}{60} + \left(-\dfrac{205}{60}\right) = -\dfrac{189}{60} - 3\dfrac{9}{60} = -3\dfrac{3}{20}$

27. $27.48 + (-3.5) = 23.98$

28. $-40.1 + 15.22 = -24.88$

29. $-15.76 + 4.9 = -10.86$

30. $-(-1.125) + 0.375 = 1.5$

31. $-(8) + [-5] = -8 + (-5) = -13$

32. $-[22] + (6) = -22 + 6 = -16$

33. $-|5 + (-30)| = -|-25| = -25$

34. $-|13 + (-22)| = -|-9| = -9$

C 35. $82 + (-|-23|) = 82 + (-23) = 59$

36. $-|11| + (-15) = -11 + (-15) = -26$

37. $|-4| + |-11| = 4 + 11 = 15$

38. $-|-16| + |-14| = -16 + 14 = -2$

39. $|-(-22)| + [-|-8 + (-7)|] = 22 + [-15] = 7$

40. $-|7 + [-(-8)]| + |-22| = -|15| + 22 = -15 + 22 = 7$

41. $-|-8 + [-(-22)]| + |7 + (-22)| = -|-14| + |-15| = -14 + 15 = 1$

42. $|-7 + [-(-22)]| + \{-|-22 + [-(-8)]|\} = |15| + [-|-14|] = 15 - 14 = 1$

43. $19 + 6 + (-8) = 17$; 17°C

44. $7 + (-11) + (-6) + 15 = 5$; 5 yd gain

45. $1000 + (-60) + 2200 + (-200) + 1700 = 4640 = 4640$ m

46. $300.30 + (-35) + (-106.15) + 20 + 43.75 = 222.90$; \$222.90

47. 15

48. 63

49. 5.07

50. -98.86

51. 72

52. -23.989

53. always true   **54.** never true   **55.** sometimes true

56. always true   **57.** never true

page 27   Capsule Review

1. 6   **2.** -15 **3.** $\dfrac{1}{4}$ **4.** $4\dfrac{1}{3}$ **5.** -13.09

page 29   Class Exercises

1. $25 + (-8) = 17$

2. $9 + (-15) = -6$

3. $-6 + (-19) = -25$

4. $17 + 8 = 25$

5. $-6 + (-8) + (-15) = -29$

6. $-14 + (-7) + (-5) + (-6) = -32$

7. $-2 + (8) + (-1) = 5$

pages 29–30   Practice Exercises

A **1.** 4.1   **2.** $\dfrac{3}{4}$ **3.** -0.75 **4.** $-1\dfrac{2}{3}$

5. -8 **6.** -15 **7.** -13 **8.** -7

9. $1.4 + 2.3 = 3.7$ **10.** $2.5 + 7.4 = 9.9$ **11.** $1.5 + 8 = 9.5$ **12.** $4.2 + 1.5 = 5.7$

13. $-(-5) + 12 = 5 + 12 = 17$

14. $-7 - (-22) = -7 + 22 = 15$

15. $-(-5) + 13 = 5 + 13 = 18$

16. $-(-77) + (-10) + (-24) = 77 + (-34) = 43$

17. $-\dfrac{1}{9} + \dfrac{8}{9} = \dfrac{7}{9}$

18. $\dfrac{2}{3} + \dfrac{1}{3} = \dfrac{3}{3} = 1$

19. $13 - (-33) = 13 + 33 = 46$

20. $-91 - (-128) = -91 + 128 = 37$

21. $-(-63) - 36 = 63 - 36 = 27$

B **22.** $-6.7 - (11.5 - 10) = -6.7 - 1.5 = -8.2$

23. $11.5 - (15.2 - 4) = 11.5 - 11.2 = 0.3$

24. $[-4.9 - (-6.7)] - (15.2 - 11.5) = (-4.9 + 6.7) - (15.2 - 11.5) =$
$1.8 - 3.7 = -1.9$

25. $[-6.7 - (-4.9)] - (11.5 - 15.2) = (-6.7 + 4.9) - (11.5 - 15.2) =$
$-1.8 - (-3.7) = -1.8 + 3.7 = 1.9$

26. $[-11.5 - (-15.2)] - 15.2 = 3.7 - 15.2 = -11.5$

27. $-(-4.9) - [-6.7 - (-15.2)] = 4.9 - [-6.7 + 15.2] = 4.9 - 8.5 = -3.6$

28. $-\dfrac{3}{6} - \dfrac{2}{6} + \dfrac{2}{6} = -\dfrac{3}{6} = -\dfrac{1}{2}$

29. $-\dfrac{6}{8} - \dfrac{2}{8} - \dfrac{5}{8} = -\dfrac{13}{8}$

30. $-0.26 - 5.3 + 0.87 = -4.69$

31. $-1.9 - 4 + 0.25 = -5.65$

32. $-\dfrac{11}{2} - \left[\dfrac{12}{2} + \dfrac{3}{2}\right] = -\dfrac{11}{2} - \dfrac{15}{2} = -\dfrac{26}{2} = -13$

33. $\left(\dfrac{22}{3} - \dfrac{17}{3}\right) - \dfrac{48}{3} = \dfrac{5}{3} - \dfrac{48}{3} = -\dfrac{43}{3} = -14\dfrac{1}{3}$

34. $-(7.75) - (-2.25) = -7.75 + 2.25 = -5.5$

35. $26.1 - (-5.7) = 26.1 + 5.7 = 31.8$

C 36. $\{-7 - [3 - (-35)]\} = -7 - (38) = -45$

37. $-41 - \{-[6 - (-18)]\} = -41 - [-(24)] = -41 + 24 = -17$

38. $|-8 - 3| - |15| = |-11| - 15 = 11 - 15 = -4$

39. $|44| + |-20| = 44 + 20 = 64$

40. $3 - [(7.8 - (-3.2)) - (-4.7 + 1.4)] = 3 - [11 - (-3.3)] = 3 - [14.3] = -11.3$

41. $-14.2 - [(-4.7 - (-7.8) + (3.8 - (-3.2)))] = -14.2 - [-4.7 + 7.8 + 7.0] =$
$-14.2 - [10.1] = -24.3$

42. $-4.7 - |[1.2 - (-3.2)] - [-(8.5 - 7.8)]| = -4.7 - |4.4 - (-0.7)| =$
$-4.7 - |5.1| = -4.7 - 5.1 = -9.8$

43. $-4.7 - |(-3.2 + 11) - (13.1 - 7.8)| = -4.7 - |7.8 - 5.3| = -4.7 - |2.5| =$
$-4.7 - 2.5 = -7.2$

44. $\text{ABS}(-3) = 3$ **45.** $\text{ABS}(6 - 9) = 3$ **46.** $\text{ABS}(9 - 6) = 3$ **47.** $-\text{ABS}(-5) = -5$

48. $\text{ABS}(-6.1 - (-5.7)) = 0.4$ **49.** $\text{ABS}(-6.1) - \text{ABS}(-5.7) = 0.4$

page 30   Test Yourself

1. $21.9 + 0.4 = 22.3$

2. $(2)(0.4)(8) = 6.4$

3. $10(8 - 0.4) =$
$10(7.6) = 76$

4.
$-4\ -2\ \ 0\ \ 2\ \ 4\ \ 6$

5.
$-4\ -3\ -2\ -1\ \ 0\ \ 1$

6. $-27 > -28$

7. $\dfrac{5}{8} > \dfrac{5}{9}$

8. $-0.42 > -0.425$

9. -1

10. $-1\dfrac{3}{4}$

11. $-10\dfrac{1}{2} + 15 = 4\dfrac{1}{2}$

12. $-27 + 30 = 3$

13. -4.42

14. $[59 + 23] + 25 = 82 + 24 = 106$

1. 1176   **2.** 18, 700   **3.** 6.15   **4.** 0.98   **5.** 40   **6.** $\dfrac{2}{6} = \dfrac{1}{3}$

7. $7 \cdot \dfrac{8}{7} = \dfrac{8}{1} = 8$ **8.** $\dfrac{2}{3} \cdot \dfrac{8}{3} = \dfrac{16}{9}$ **9.** $\dfrac{5}{8} \cdot \dfrac{32}{6} = \dfrac{160}{48} = \dfrac{10}{3}$

10. $\dfrac{5}{2} \cdot \dfrac{6}{9} = \dfrac{30}{18} = \dfrac{5}{3}$ **11.** $\dfrac{18}{7} \cdot \dfrac{14}{3} = 6 \cdot 2 = 12$ **12.** $\dfrac{27}{4} \cdot \dfrac{8}{3} = 9 \cdot 2 = 18$

page 33 Class Exercises

1. -2 **2.** -16 **3.** -0.3 **4.** 0

5. $\dfrac{5}{18}$ **6.** $-\dfrac{1}{2}$ **7.** $-\dfrac{2}{27}$ **8.** 8

pages 33–34 Practice Exercises

A 1. -30 **2.** -120 **3.** -0.468 **4.** 0

5. 65   **6.** 99   **7.** 2.22   **8.** 0.0016

9. 30   **10.** -48 **11.** -45 **12.** 0

13. $(-4)(3)(-10) = 120$ **14.** $(-10)(5)(3) = -150$

15. $(-10)(4)(-5) = 200$ **16.** $(-3)(4)(-5) = 60$

17. $(4)(5)(10)(-6) = -1200$ **18.** $(3)(4)(10)(-5) = -600$

19. $(-10)(-3)(-6)(5) = -900$ **20.** $(6)(-3)(-5) = 90$

B 21. -18 **22.** 45   **23.** -6 **24.** -400

25. 55,555.5   **26.** 0.444444   **27.** $-11,111.1111$

28. 7777.77   **29.** $-1574.5\overline{3}$ **30.** 28.88028

31. $[(-6)(-2)] - (-3) = 12 + 3 = 15$ **32.** $10 + \left[(-2)\left(-\dfrac{1}{2}\right)\right] = 10 + 1 = 11$

33. $[(-6)(-2)] - \left[(10)\left(-\dfrac{1}{2}\right)\right] =$ **34.** $[(-3)(-6)] - [(10)(-2)] =$

$12 - [-5] = 17$ $\qquad 18 - [-20] = 38$

35. $- (-3)(-6 - (-2)) = 3(-4) = -12$ **36.** $[-2 - (-3)](10)\left(-\dfrac{1}{2}\right) = (1)(-5) = -5$

C 37. $-\left|(-6)\left(-\dfrac{1}{2}\right)\right| \cdot (-2)(-3) = -|3| \cdot 6 = -3 \cdot 6 = -18$

38. $(-6)\left|-\dfrac{1}{2} - 10\right| = (-6)\left|-\dfrac{21}{2}\right| = (-6)\left(\dfrac{21}{2}\right) = -63$

39. $\left|(-6)\left(-\dfrac{1}{2}\right)\right| - |(-2)(-3)| = |3| - |6| = 3 - 6 = -3$

40. $\left(-\dfrac{1}{2}\right)(10)|-(-3) - (-2)| = (-5)|3 + 2| = (-5)|5| = -25$

41. $\left|\left[\left(-\dfrac{1}{2}\right)\{10 + (-2)\}\right]\right| = \left|-\dfrac{1}{2}(8)\right| = |-4| = 4$

42. $[(10)(-2)]\left|(-6 - 10)\left[-\left(-\dfrac{1}{2}\right)\right]\right| = [-20]\left|(-16)\left(\dfrac{1}{2}\right)\right| = [-20]|-8| = -20 \cdot 8 = -160$

43. $-7 \times 2 = -14$; $-14°$ drop   **44.** $(3)(-4.5) = -13.5$; lost \$13.5 million

page 34   Algebra in Meteorology

1. $20 - (-10) = 30$; $30°F$; $-15 - (59) = 44$; $44°F$

page 35   Capsule Review

1. positive   **2.** positive   **3.** negative   **4.** zero

5. negative   **6.** positive   **7.** positive   **8.** zero

page 37   Class Exercises

1. negative   **2.** positive   **3.** positive   **4.** positive

5. $\left(\dfrac{3}{4}\right)\left(\dfrac{8}{7}\right)$ **6.** $\left(-\dfrac{24}{1}\right)\left(\dfrac{5}{4}\right) = (-24)\left(\dfrac{5}{4}\right)$

7. $\left(\dfrac{132}{1}\right)\left(-\dfrac{1}{11}\right) = (132)\left(-\dfrac{1}{11}\right)$ **8.** $\left(-\dfrac{15}{1}\right)\left(-\dfrac{4}{1}\right) = (-15)(-4)$

9. $(0)\left(-\dfrac{1}{10}\right)$ **10.** $\left(-\dfrac{2}{3}\right)\left(\dfrac{3}{2}\right)$

11. $(0.8)\left(\dfrac{1}{5}\right) = \left(\dfrac{8}{10}\right)\left(\dfrac{1}{5}\right)$ **12.** $\left(1\dfrac{1}{3}\right)(-2) = \left(\dfrac{4}{3}\right)\left(-\dfrac{2}{1}\right)$

pages 38–40   Practice Exercises

A 1. $\dfrac{1}{5}$ **2.** $\dfrac{1}{8}$ **3.** $\dfrac{1}{75}$ **4.** $\dfrac{1}{10}$

5. -5 **6.** -7 **7.** $-\dfrac{10}{9}$ **8.** $-\dfrac{10}{6} = -\dfrac{5}{3}$

9. -7 **10.** -4 **11.** -12 **12.** -3

13. $\left(-\dfrac{5}{3}\right)\left(-\dfrac{9}{2}\right) = \dfrac{15}{2}$ **14.** $\left(-\dfrac{1}{2}\right)\left(-\dfrac{3}{2}\right) = \dfrac{3}{4}$ **15.** $\left(-\dfrac{7}{12}\right)\left(\dfrac{6}{7}\right) = -\dfrac{1}{2}$

16. $\left(-\dfrac{1}{3}\right)\left(\dfrac{6}{5}\right) = -\dfrac{2}{5}$

17. 0

18. not possible

19. not possible

20. 0

21. $\dfrac{-8}{-4} = 2$

22. $-4 \div \dfrac{1}{2} = (-4)(2) = -8$

23. $\dfrac{1}{2} \div (-3) = \left(\dfrac{1}{2}\right)\left(-\dfrac{1}{3}\right) = -\dfrac{1}{6}$

24. $\dfrac{1}{2} \div (-8) = \left(\dfrac{1}{2}\right)\left(-\dfrac{1}{8}\right) = -\dfrac{1}{16}$

B 25. $\left[\left(\dfrac{1}{2}\right)\left(\dfrac{1}{2}\right)\right] \div (-4) = \dfrac{1}{4} \div (-4) = \left(\dfrac{1}{4}\right)\left(-\dfrac{1}{4}\right) = -\dfrac{1}{16}$

26. $\left[\left(-\dfrac{3}{2}\right)(-8)\right] \div (-3) = 12 \div (-3) = (12)\left(-\dfrac{1}{3}\right) = -4$

27. $\left(-3 - \dfrac{1}{2}\right) \div \dfrac{1}{2} = -\dfrac{7}{2} \div \dfrac{1}{2} = \left(-\dfrac{7}{2}\right)(2) = -7$

28. $\dfrac{1}{2} \div [-4 - (-8)] = \dfrac{1}{2} \div 4 = \left(\dfrac{1}{2}\right)\left(\dfrac{1}{4}\right) = \dfrac{1}{8}$

29. $[(-4)(-3)] \div \left[(8)\left(\dfrac{1}{2}\right)\right] = 12 \div 4 = (12)\left(\dfrac{1}{4}\right) = 3$

30. $\left[\left(-\dfrac{9}{2}\right)(-4)\right] \div (3)(-3) = 18 \div (-9) = -2$

31. $\dfrac{-2}{2} = -1$

32. $\dfrac{18}{-9} = -2$

33. $(-16) \div (-8) = 2$

34. $84 \div (-12) = -7$

35. $\dfrac{-21}{-7} = 3$

36. $\dfrac{-36}{-3} = 12$

37. -2.909088

38. 1518.5

39. $\dfrac{-158.5494}{0.27} = -587.22$

40. $\dfrac{-90.81}{-0.15625} = 581.184$

41. $\dfrac{633.6}{0.352} = 1800$

42. $\dfrac{-26.064}{0.705} = -36.97021277$

C 43. $\left(\dfrac{3}{1} - \dfrac{13}{3}\right) \div \left(-\dfrac{2}{3} + \dfrac{5}{6}\right) = \left(\dfrac{9}{3} - \dfrac{13}{3}\right) \div \left(-\dfrac{4}{6} + \dfrac{5}{6}\right) = -\dfrac{4}{3} \div \dfrac{1}{6} = -\dfrac{4}{3} \cdot \dfrac{6}{1} = -8$

44. $\left(\dfrac{3}{1} + \dfrac{5}{4}\right) \div \left(-\dfrac{5}{1} + \dfrac{3}{4}\right) = \left(\dfrac{12}{4} + \dfrac{5}{4}\right) \div \left(-\dfrac{20}{4} + \dfrac{3}{4}\right) = \dfrac{17}{4} \div \left(-\dfrac{17}{4}\right) = -1$

45. $\dfrac{|-30|}{-4} = \dfrac{30}{-4} = -7\dfrac{1}{2}$

46. $\dfrac{21}{|11 - 4|} = \dfrac{21}{|7|} = \dfrac{21}{7} = 3$

47. $\left(a \cdot \dfrac{1}{b}\right) \cdot \dfrac{1}{c} = \left(\dfrac{a}{b}\right)\left(\dfrac{1}{c}\right) = \dfrac{a}{bc};\ \left(a \cdot \dfrac{1}{c}\right) \cdot \dfrac{1}{b} = \left(\dfrac{a}{c}\right)\left(\dfrac{1}{b}\right) = \dfrac{a}{cb};\ \dfrac{a}{bc} = \dfrac{a}{cb};$ true

48. $\left(a \cdot \dfrac{1}{b}\right) \cdot \dfrac{1}{c} = \left(\dfrac{a}{b}\right)\left(\dfrac{1}{c}\right) = \dfrac{a}{bc}$; true   **49.** true

50. $(a - b) \div (c - d) = (a - b) \div (-1)(d - c) = (-1)(a - b) \div (d - c) =$

   $(b - a) \div (d - c) = \dfrac{b - a}{d - c}$; true

51. $\dfrac{83 + 79 + 73 + 84}{4} = \dfrac{319}{4} = 79.75$; 80   **52.** $\dfrac{76 + 58 + 87 + 80 + 82}{5} = 76.6$; 77

53. $\dfrac{178 + 185 + 193 + 201}{4} = \dfrac{757}{4} = 189.25$; 190 cm

54. $\dfrac{28.90 + 29.10 + 30.20 + 29.60 + 30.10 + 29.90 + 30.10}{7} = \dfrac{207.9}{7}$; 29.70

55. $\dfrac{1.6 + 0.8 + 1.2 + 2.4}{4} = \dfrac{6.0}{4}$; 1.5 km; no

56. $\dfrac{62\frac{1}{4} + 61\frac{1}{2} + 60\frac{3}{4} + 61 + 62\frac{1}{2}}{5} = \dfrac{62\frac{1}{4} + 61\frac{2}{4} + 60\frac{3}{4} + 61 + 62\frac{3}{4}}{5} = \dfrac{308}{5}$; $61\frac{3}{5}$

57. $\dfrac{82.4 + 94.1 + 84.5 + 89.5 + 92.6 + 90.8 + 87.9}{7} = \dfrac{621.8}{7} = 88.82$; 88.8°F

58. 0.05882   **59.** 0.01   **60.** −0.05263   **61.** −2.5

62. 110 cm²   **63.** 117 in.²   **64.** 90 cm²   **65.** $16\frac{2}{3}$ ft²

66. 90 cm²   **67.** 50.37 cm²   **68.** 32 cm   **69.** 135 in.²

70. $(12)(5)(4) = 240$; 240 cm³   **71.** 81, 243, 729 (mult. preceding term by 3)

72. 0.05, 0.005, 0.0005; (divide preceding term by 10)

73. 16, 22, 29; (difference between terms is 1 more than preceding difference)

74. 36, 49, 64; (terms are the squares of the natural numbers)

75. If b were 0, then $a \times b = 0$, which is true for more than one value of a.

76. More than one value of a makes $a \times 0 = 0$ true.

page 41   Integrating Algebra: Meteorology

1. $8.2 - (-0.9) = 9.1$; 9.1°F   **2.** $-0.9 - (-28) = 1.9$; 1.9°F

3. $\dfrac{6.2 + (-1.8) + 3.6 + 0 + (-0.5)}{5} = \dfrac{7.5}{5} = 1.5$; 1.5°F

4. $\dfrac{(-2.3) + 0.6 + 5.2 + (-4.7) + 3.2}{5} = \dfrac{2}{5} = 0.4$; 0.4°F

5. Answers may vary.

page 44   Class Exercises

1. Let x represent the unknown number.

2. Let x represent the unknown number; $|x|$ represents the absolute value.

3. Let x represent the unknown number; $x < 0$ represents the number less than zero.

4. Let a represent one number, and b represent the second number; ab represents the product.

5. Let a represent one number, b represent the second number, and c represent the third number; $a + b + c$ represents the sum.

6. Let m represent one number, and n represent the second number; $|m - n|$ represents the absolute value of the difference.

7. Let r represent one number, and s represent the second number; $2(r + s)$ represents two times the sum of the numbers.

pages 44–45   Practice Exercises

A   **1.** Let m represent Mary's age now; then Mary's age two years ago is $m - 2$.

2. Let p represent the current price; then one-half the current price is $\frac{1}{2}p$.

3. Let s represent John's salary; then John's salary plus $500 in commission is $s + 500$.

4. Let a represent Martha's annual salary; then Martha's annual salary less $4500 in deductions is $a - 4500$.

5. Let p represent the price of an item; then the price plus 6% sales tax is $p + 0.06p$.

6. Let b represent the number of boys; then 25% of the number of boys is $0.25b$.

7. $\dfrac{a + b + c}{3}$ is the average salary.

8. 3, 6, 9, ...   Let x represent a number; then $3x$ is a multiple of 3.

9. 2, 4, 6, 8, 10, 12, 14, 16, 18, 20; odd; $2x - 1$

B **10.** Five times the number of pears

11. Yes, they can be equal to one another since one can be a simplified form of the other.

12. Let x represent an integer; $-|2x|$ represents any negative even integer.

13. Let x represent an integer; $-|2x - 1|$ represents any negative odd integer.

C 14. True; let $x - y = a$, so $y - x = -a$, $|a| = |-a|$, are the same.

15. No, if $a = c$ and $a + b = c$, then $b = 0$, and if $b = c$ and $a + b = c$, then $a = 0$. This cannot be true since both a and b are nonzero.

page 45 Project

The magic constant is $3x$ (the sum of any row, column, or diagonal).

page 45 Test Yourself

1. -34

2. 4.7

3. 8

4. $-\dfrac{8}{12} + \dfrac{3}{12} = -\dfrac{5}{12}$

5. $-12 + 12 = 0$

6. $-\dfrac{1}{4} + \dfrac{3}{5} = -\dfrac{5}{20} + \dfrac{12}{20} = \dfrac{7}{20}$

7. -6

8. $3.2 + 4.1 = 7.3$

9. $-(-5) - (-3) = 5 + 3 = 8$

10. -150

11. $-\dfrac{1}{9}$

12. 84

13. -9

14. $-\dfrac{7}{3} \div \left(-\dfrac{1}{2}\right) = \left(-\dfrac{7}{3}\right)\left(-\dfrac{2}{1}\right) = \dfrac{14}{3}$

15. $\left(\dfrac{3}{4}\right)\left(\dfrac{4}{2}\right) = \dfrac{3}{2}$

16. Let $m =$ the price of the meal; $m + 0.15m$ or $1.15m$

pages 46–47 Summary and Review

1. $4 - 0.5 = 3.5$

2. $\dfrac{16}{4} = 4$

3. $\dfrac{3}{4}(16) = 12$

4. $4 \div \dfrac{1}{2} = (4)(2) = 8$

5. $(3)(16)\left(\dfrac{2}{3}\right) = 32$

6. $4(4 - 0.5) = 4(3.5) = 14$

7. $\dfrac{8}{(4)(16)} = \dfrac{8}{64} = \dfrac{1}{8}$

8. $16 - [(4)(0.5)] = 16 - 2 = 14$

9. $x + 18$

10. $22n$

11. $a - b$

12. $m \div \dfrac{2}{3}$

13.
$\quad -4\ -2\ \ 0\ \ 2\ \ 4\ \ 6$

14.
$\quad -2\ -1\ \ 0\ \ 1\ \ 2\ \ 3$

15.
$\quad -2\ -1\ \ 0\ \ 1\ \ 2\ \ 3$

16. $-11 < 10$

17. $0 > -3.5$

18. $-1\dfrac{1}{3} > -1.5$

19. -4.31

20. 8

21. $-|9| = -9$

22. $-\left(\dfrac{8}{9}\right) = -\dfrac{8}{9}$

23. -1.2

24. $-\dfrac{4}{3} = -1\dfrac{1}{3}$

25. $-14.6 + 2.9 = -11.7$ **26.** $-18 + 11 = -7$ **27.** -5.92

28. $[-48 + 48] + 9 = 0 + 9 = 9$ **29.** 150

30. 0 **31.** $\dfrac{2}{3}(-180) = -120$ **32.** -6

33. $(0)(-3) = 0$ **34.** 2.04

35. $\dfrac{6 + (-10) + 4 + 10.5 + (-0.5) + (-9) + (-6) + 25}{8} = \dfrac{20}{8} = 2.5$ yd

36. Let h represent the number of hours; then $20 + 25(h)$ represents the amount charged by one carpenter, and $30h$ represents the amount charged by the second carpenter. Make a table. It takes 4 h to charge the same amount.

h	$20 + 25h$	$30h$
1	45	30
2	70	60
3	95	90
4	120	120

page 48   Chapter Test

1. $0.9 + 1.4 = 2.3$

2. $(1.4)(25)\left(\dfrac{1}{5}\right) = 7$

3. $5\left(\dfrac{1}{5}\right) - (1.4)(25) = 1 - 35 = -34$

4. $p - 0.85$ **5.** $\dfrac{b}{27}$ **6.** $x - \dfrac{2}{3}$ **7.** wz

8.
 $-1\ 0\ 1\ 2\ 3\ 4$

9.
 $-2 -1\ 0\ 1\ 2\ 3$

10. $\{-2.25, -2, -1.5, 0, 2.25, 2.5\}$

11. $\left\{-3, -1\dfrac{1}{2}, -1\dfrac{1}{3}, -\dfrac{1}{3}, 0, \dfrac{1}{2}\right\}$

12. 0.3 **13.** $-\dfrac{2}{5}$ **14.** $-[-9] = 9$ **15.** -177

16. -20 **17.** $3.2 + 4.07 = 7.27$ **18.** $-11 + (-11) = -22$

19. 0 **20.** 600 **21.** $-96\left(-\dfrac{3}{2}\right) = 144$ **22.** not possible

23. a. $\left[\left(\dfrac{2}{3}\right)\left(\dfrac{3}{4}\right)\right] + \left(-\dfrac{1}{2}\right) = \dfrac{1}{2} + \left(-\dfrac{1}{2}\right) = \dfrac{1}{2} - \dfrac{1}{2} = 0$

b. $-\left(\dfrac{2}{3} \cdot \dfrac{3}{4}\right) - \left(-\dfrac{1}{2}\right) = -\dfrac{1}{2} - \left(-\dfrac{1}{2}\right) = -\dfrac{1}{2} + \dfrac{1}{2} = 0$

c. $\dfrac{-\dfrac{1}{2} + \dfrac{2}{3} + \dfrac{3}{4}}{3} = \dfrac{-\dfrac{6}{12} + \dfrac{8}{12} + \dfrac{9}{12}}{3} = \dfrac{\dfrac{11}{12}}{3} = \left(\dfrac{11}{12}\right)\left(\dfrac{1}{3}\right) = \dfrac{11}{36}$

24. Let h represent the number of hours; then $15 + 20h$ represents the amount one painter charges, and $25h$ represents the amount the second painter charges.

h	$15 + 20h$	$25h$
1	35	25
2	55	50
3	75	75

The job takes 3 hours for the painters to charge the same amount.

Challenge   Make a table as follows:

Time	10:15	11:15	12:15	1:15	2:15
Train 1	0	40	80	120	160
Train 2	0	0	0	35	70
Distance	0	40	80	155	230

The trains will be 230 miles apart at 2:15 P.M.

page 49   Preparing for Standardized Tests

1. C; $2a + 3b - c = 2(3) + 3(0) - 2 = 6 + 0 - 2 = 4$

2. A; $(+9) + (-7) = 2;\ (+4) + (-5) = -1;\ 2 - (-1) = 3$

3. B; $|7 + (3 - 4)| = |7 - 1| = 6$

4. B; Using $1\frac{1}{3} < k$ and $k < 1\frac{1}{2}$, then only the fractional parts of the numbers in mixed form need to be compared.

   $\frac{11}{9} = 1\frac{2}{9}; \frac{2}{9} < \frac{1}{3}; k \neq \frac{11}{9}$

   $\frac{11}{8} = 1\frac{3}{8}; \frac{3}{8} > \frac{1}{3}$ and $\frac{3}{8} < \frac{1}{2}; \frac{1}{3} < \frac{11}{8} < \frac{1}{2}; k = \frac{11}{8}$

   $\frac{11}{7} = 1\frac{4}{7}; \frac{4}{7} > \frac{1}{2}; k \neq \frac{11}{7}$

5. D; Use the method of casting out 9's to check the choices. In choice D, $6 + 8 = 14$, $1 + 4 = 5$, so 608 is not a multiple of 9.

6. B; Since $\left(-\frac{3}{5}\right) \cdot \left(-\frac{5}{3}\right) = 1$ and $\left(\frac{2}{3}\right) \cdot \left(-\frac{3}{2}\right) = -1$, the problem becomes

   $1(-1)\left(\frac{4}{5}\right) = -\frac{4}{5}.$

7. D; Since 18 is $\frac{2}{3}$ of the number, 9 is $\frac{1}{3}$ of it. Thus, the number must be 27.

8. E; Compare each fraction to a form of $\frac{3}{4}$.

$$\frac{3}{4} = \frac{48}{64}; \frac{49}{64} > \frac{3}{4}$$

$$\frac{45}{60} = \frac{3}{4}$$

$$\frac{3}{4} = \frac{105}{140}; \frac{53}{70} = \frac{106}{140}; \frac{3}{4} < \frac{53}{70}$$

$$\frac{3}{4} = \frac{51}{68}; \frac{51}{68} > \frac{3}{4}$$

9. A; In choice A, $2a + 3b = 6 + 2 = 8$, while the other choices give values of $\frac{22}{3}$, 5, 7, and 6.

10. E; 19 dimes   = $1.90
        18 pennies =   0.18
        17 nickels =   0.85
        16 quarters =   4.00
                      —————
                      $6.93

11. D; 1st ounce requires   56¢

$2 \left(\frac{1}{2} \text{ - ounces} \right)$ 34¢

        0.3 ounce   17¢
                          —————
                          107¢ = $1.07

page 50   Mixed Review

1. 3.3147	**2.** 47.55	**3.** 42.807	**4.** 61.1	**5.** 0.9
6. 15	**7.** 3.7	**8.** 540	**9.** 14.94	**10.** 137.17
11. 280.5	**12.** 0.928	**13.** 2.8	**14.** 3.5	**15.** 8.2
16. 9.5	**17.** 0.452	**18.** 0.01808	**19.** −7	**20.** −3
21. −28	**22.** −7	**23.** 14	**24.** 48	**25.** −15
26. −8	**27.** −4	**28.** 6	**29.** −42	**30.** 4

31. 38×12; Dominic travels a total distance of 456 mi.

32. $249 + 5(7.98) = 249 + 39.9 = 288.90$; Sandi spent a total amount of $288.90.

33. $780 \div 12 = 65$; Peter packed a total of 65 packages in a box.

Chapter 2 Algebraic Expressions

page 52 Capsule Review

1. $2 + [-(-1)] + 13 = 16$

2. $(13)(-4)(-1) = 52$

3. $2 \div (-1) = -2$

4. $\dfrac{-(-1)}{-(-4)} = \dfrac{1}{4}$

page 53 Class Exercises

1. $2 + 48 \div 4 = 2 + 12 = 14$

2. $(2)(2) + 6 = 10$

3. $6 + 4 - 3 = 7$

4. $(2)(3) + 1 = 7$

5. $(3)(-4) + 1 = -11$

6. $\dfrac{1}{6-3} = \dfrac{1}{3}$

7. $\dfrac{-4+1}{3} = -\dfrac{3}{3} = -1$

pages 53–54 Practice Exercises

A 1. $16 - 22 + 1 = -5$

2. $10 - 39 + 1 = -28$

3. $21 + 112 - 3 = 130$

4. $42 + 189 - 2 = 229$

5. $\{30 - [12 + 8] \cdot 4\} = \{30 - 48 - 32\} = -50$

6. $\{40 - [24 + 18] \cdot 7\} = \{40 - 168 - 126\} = -254$

7. $4(5) + 3 = 23$

8. $2(2) - (-1) = 5$

9. $5 - \dfrac{2}{-1} = 5 + 2 = 7$

10. $2 - \dfrac{0}{-1} = 2 - 0 = 2$

B 11. $2(5) - 3(-1) + 4(0) = 10 + 3 = 13$

12. $-4(2) + 2(-1) - 3(5) + 0 = -8 - 2 - 15 = -25$

13. $\dfrac{7(5)(-1) - 2(2)}{6(5) - (-1)} = \dfrac{-35 - 4}{30 + 1} = \dfrac{-39}{31}$

14. $\dfrac{1 - 16(-1)(2)}{15(5) + 1} = \dfrac{1 + 32}{75 + 1} = \dfrac{33}{76}$

15. $8[5 + (-1)] - 11(-1)(2) = 8(4) + 22 = 54$

16. $-5[2(5) + 2] + 6(-1) = -5(10 + 2) + (-6) = -5(12) + (-6) = -60 + (-6) = -66$

17. $6 - \{9 + 12(10) \div 4\} = 6 - \{9 + 30\} = 6 - \{39\} = -33$

18. $5 - \{-4[8 - 11(-10)]\} = 5 - \{-4[8 + 110]\} = 5 - \{-4[118]\} = 5 - \{-472\} = 477$

C 19. $6 + \dfrac{6(-2) + 18}{9(-3) - 12} = 6 + \dfrac{-12 + 18}{-27 - 12} = 6 + \dfrac{6}{-39} = 6 - \dfrac{2}{13} = \dfrac{76}{13}$

20. $\dfrac{-(-3) + 7(-2)}{11 - (-3)(9)} = \dfrac{3 - 14}{11 + 27} = -\dfrac{11}{38}$

21. $6 - [-(-2)][-3 - (-9)][6 - (-2)] = 6 - 2[-3 + 9][6 + 2] =$
$6 - (12)(8) = -90$

22. $[18 + (-3)][-2 - (-6)] - 6(-2) = 15[-2 + 6] + 12 = 15(4) + 12 =$
$60 + 12 = 72$

23. 96   **24.** 53.424   **25.** 478.7 cm²

26. $-100 > -101$ **27.** $-2.4 < -2.3$ **28.** $-\dfrac{5}{2} < -\dfrac{5}{3}$

page 55   Capsule Review

1. $3 \cdot 3 \cdot 3 \cdot 3 = 81$ **2.** $(-2)(-2)(-2)(-2) = 16$

3. $(-4)(-2)(-2)(-2) \cdot 3 \cdot 3 = 288$

page 57   Class Exercises

1. 100   **2.** $\dfrac{8}{27}$ **3.** -4 **4.** $(2)(25) = 50$

5. $-(-5)^3 = -(-125) = 125$ **6.** $(2)(2)(3) = 12$ **7.** $-(2)(2)(3) = -12$

8. $2^3 + 3^2 = 8 + 9 = 17$ **9.** $[(2)(3)]^2 = 6^2 = 36$ **10.** $-(2 - 3)^2 = -(-1)^2 = -1$

pages 57–58   Practice Exercises

A   **1.** 25   **2.** 49   **3.** $\dfrac{1}{4}$ **4.** $\dfrac{1}{125}$

5. -16 **6.** -36 **7.** 16   **8.** 36

9. -16 **10.** -36 **11.** -16 **12.** -36

13. $2(7)^2 = 2(49) = 98$ **14.** $3(5)^2 = 3(25) = 75$

15. $(2 - 49)^2 = (-47)^2 = 2209$ **16.** $(3 - 125)^2 = (-122)^2 = 14884$

17. $3(11)^2 = 3(121) = 363$ **18.** $4(9)^2 = 4(81) = 324$

19. $(4 - 216)^2 = (-212)^2 = 44944$ **20.** $(7 - 9)^2 = (-2)^2 = 4$

21. $2(2)^2 (3)^3 = 2(4)(27) = 216$ **22.** $3(2)^3 (3)^2 = 3(8)(9) = 216$

23. $3(-2)^3 - (-3)^2 = 3(-8) - 9 = -24 - 9 = -33$

24. $6(-2)^2 - (-3)^3 = 6(4) - (-27) = 24 + 27 = 51$

B   **25.** 0.09   **26.** 0.25   **27.** 0.000000001   **28.** 0.000001

29. $4(5)^2 = 4(25) = 100$ **30.** $9(6)^2 = 9(36) = 324$

31. $25(-2)^2 = 25(4) = 100$

32. $16(-4)^2 = 16(16) = 256$

33. $-4 + 3^2 = -4 + 9 = 5$

34. $-3 + (-2)^2 = -3 + 4 = 1$

35. $3(4)^2 - (-2)^3 = 3(16) - (-8) = 48 + 8 = 56$

36. $(-2)^3 - 2(4)^2 = -8 - 2(16) = -8 - 32 = -40$

37. $4^3[3 + (-2)]^2 = 64(1)^2 = 64$

38. $(-2 - 3)^2 (4)^2 = (-5)^2 (4)^2 = (25)(16) = 400$

39. $[2(4) + 4(3)]^2 = [8 + 12]^2 = [20]^2 = 400$

40. $[-2 - 2(3)]^3 = [-2 - 6]^3 = [-8]^3 = -512$

C 41. $\dfrac{(-1)^4 - (-2)^4}{[-1 - (-2)]^4} = \dfrac{1 - 16}{(-1 + 2)^4} = -\dfrac{15}{1} = -15$

42. $\dfrac{1^3 + (-1)^3}{[1 - (-1)]^3} = \dfrac{1 + (-1)}{[1 + 1]^3} = \dfrac{0}{[2]^3} = 0$

43. $(-3)^3 \div [4^2 - 3^2 - (-2)^2] = -27 \div [16 - 9 - 4] = -27 \div 3 = -9$

44. $[3^3 - (-2)^3 + (-1)^3](4^3)] = [27 - (-8) + (-1)](64) = [27 + 8 - 1](64) = [34](64) = 2176$

45. $-(-1)^4 + 5^2[0^2 - 2(3)^2] \div 2 = -1 + 25[0 - 18] \div 2 = -1 - 450 \div 2 = -1 - 225 = -226$

46. 6561   **47.** 7776   **48.** 128,100.2839   **49.** -231.3441

50. $a = \dfrac{25^2}{15} = 41.\overline{6}$; approximately 41.7 m/s^2

page 58   Biography

Ramanujan is called the "formula man" because he discovered many important algebra formulas.

page 59   Capsule Review

1. $2 = 2$, true

2. $-4 = 4$, false

3. $-4 + 21 = 8 + 9$, $17 = 17$, true

4. $6 - (-13) = 4 - 15$, $19 = -11$, false

page 61   Class Exercises

1. Associative property for addition

2. Commutative property for multiplication

3. Distributive property

4. Identity property for addition

5. Commutative property for multiplication

6. Associative property for addition

7. no; For example: $3 - 2 \neq 2 - 3$

8. no; For example: $8 \div (4 \div 2) \neq (8 \div 4) \div 2$

page 61 Practice Exercises

A 1. $13 + 55 = 68$

2. $23 + 32 = 55$

3. $98 + 36 = 134$

4. $53 + 59 = 112$

5. $\dfrac{2}{3}\left(\dfrac{3}{14}\right) = \dfrac{1}{7}$

6. $\dfrac{7}{8}\left(\dfrac{2}{7}\right) = \dfrac{7}{28} = \dfrac{1}{4}$

7. $(10)(500) + (10)(13) = 5000 + 130 = 5130$

8. $(12)(200) + (12)(10) = 2400 + 120 = 2520$

9. $(25)(98) - (25)(5) = 2450 - 125 = 2325$

10. $(36)(53) - (36)(12) = 1908 - 432 = 1476$

11. $(1)(0.25) - (100)(0.25) = 0.25 - 25 = -24.75$

12. $(1000)(1.5) + (30)(1.5) = 1500 + 45 = 1545$

13. $\left(\dfrac{2}{3}\right)(7) - \left(\dfrac{8}{21}\right)(7) = \dfrac{14}{3} - \dfrac{8}{3} = \dfrac{14}{3} - \dfrac{8}{3} = \dfrac{6}{3} = 2$

14. $\left(\dfrac{2}{5}\right)(5) - \left(\dfrac{7}{8}\right)(5) = 2 - \dfrac{35}{8} = \dfrac{16}{80} - \dfrac{35}{8} = -\dfrac{19}{8}$

B 15. Distributive property

16. Distributive property

17. Commutative property for multiplication

18. Associative property for multiplication

19. Inverse prop. for add.

20. Identity prop. for mult.

21. $(2)(3x) - (2)(11)$

22. $6(7) + 2y(7)$

23. $\left(-\dfrac{3}{4}\right)\left(\dfrac{2}{3}x\right) - \left(-\dfrac{3}{4}\right)(8)$

24. $(-5j)\left(-\dfrac{3}{5}\right) + 20\left(-\dfrac{3}{5}\right)$

25. false; For example: $(2 - 3) - 4 \neq 2 - (3 - 4)$

26. false; For example: $2 \div 4 \neq 4 \div 2$

C 27. false; For example: $12 \div (6 \cdot 2) \neq (12 \div 6)(12 \div 2)$

28. false; For example: $2(8 \div 4) \neq (2 \cdot 8) \div (2)(4)$

29. false; For example: $2 - (10 \div 5) \neq (2 - 10) \div (2 - 5)$

30. false; For example: $3 \cdot 7 \cdot 5 \neq (3 \cdot 7)(3 \cdot 5)$

31. $3(0.99 + 2.19 + 1.31) = 3(4.49) = 13.47$; The total cost is $13.47.

32. $(10)(0.89 + 1.59 + 2.59) = (10)(5.07) = 50.70$; The total cost is \$50.70.

33. Think of 206 as $200 + 6$; then $25(200 + 6) = 5000 + 150 = 5150$.

34. Answers may vary. An example: A number and its *additive inverse* have the same *absolute value* because they are equidistant from *zero*.

35. Answers may vary. An example: Parentheses can be used as a *grouping symbol* or to indicate the *operation* of *multiplication*.

36. Answers may vary. An example: To *evaluate* an algebraic expression, substitute the given value for each *variable* and *simplify* the resulting numerical expression.

37. Answers may vary. An example: The *multiplicative inverse* of every *integer* except zero is a *real number*.

38. Answers may vary. An example: An *exponent* indicates how many times the *base* is used as a *factor*.

page 63   Capsule Review

1. $3x^2 + 4x^2$ **2.** $(4 + 3)x^2$ **3.** $9 \cdot 3x - 9 \cdot 7$ **4.** $(3x - 7)9$

page 64   Class Exercises

1. 2; $3x$, y; 3, 1 **2.** 1; $-4xy^3$; -4 **3.** 4; $3a$, a, $5a$, 4; 3, 1, 5

4. 2; xy^2, x^2y; 1, 1 **5.** unlike **6.** like

7. like **8.** unlike **9.** $11x$

10. $9y$ **11.** $7a^2$ **12.** $13m^2$

pages 64–66   Practice Exercises

A **1.** like **2.** like **3.** unlike **4.** unlike

5. like **6.** like **7.** unlike **8.** unlike

9. unlike **10.** unlike **11.** like **12.** like

13. $8x$ **14.** $10y$ **15.** $-12t$ **16.** $-7q$

17. $2t + 15$ **18.** $3x + 15$ **19.** $12g^2$ **20.** $3m^3$

21. $7x^2 + 7x$ **22.** $5y^3 + 7y$ **23.** $12g - h + 2$ **24.** $-8x + 2$

25. $9ab$ **26.** $23rs$ **27.** $5ab^2$ **28.** $9m^2n$

29. $1 + 6m + 12 - 2m = 4m + 13$ **30.** $-2 + 6b - 24 + 4b = 10b - 26$

B 31. $7y - 6x - 4$ **32.** $10m - 22n + 8$

33. $3x + 6 + 5 = 3x + 11$ **34.** $2x + 2 + 3 = 2x + 5$

35. $3r + 2r - 2 = 5r - 2$ **36.** $6s + 3s - 6 = 9s - 6$

37. $4 + 6x + 3 = 6x + 7$ **38.** $3 + 12x + 8 = 12x + 11$

39. $15c - 15c^2 + c = 16c - 15c^2$ **40.** $9b^2 + 20b - 8b^2 = b^2 + 20b$

41. $8x^3 + 4x^2 - 5x^2 = 8x^3 - x^2$ **42.** $-8b^3 + 5b^2 - 20b^3 = -28b^3 + 5b^2$

43. $130 + 65r + 165 + 90r = 295 + 155r$ **44.** $30p - 110 + 28p - 100 = 58p - 210$

45. $4m + 4n + 3m + 3n = 7m + 7n$ **46.** $6a + 6b + 7a + 7b = 13a + 13b$

47. $2x^2 - 2y + 4x^2 - 4y = 6x^2 - 6y$ **48.** $7g^3 - 7h + 3g^3 - 3h = 10g^3 - 10h$

49. $2t - 2t^2 + 5t + 5t^2 = 7t + 3t^2$ **50.** $5b^2 + 5b + 8b^2 - 8b = 13b^2 - 3b$

C 51. $3x^2y + 3xy^2 + 6x^2y^2 - 12xy^2 = 3x^2y - 9xy^2 + 6x^2y^2$

52. $21mn - 7mn^2 + 5mn^2 - 5m^2n = 21mn - 2mn^2 - 5m^2n$

53. $2a + b + 2a - b = 4a$ **54.** $2v + w + 5w - 6v = 6w - 4v$

55. $\frac{1}{4}y - \frac{1}{8} + \frac{3}{8}y + \frac{1}{8} = \frac{5}{8}y$

56. $\frac{3}{10}a - 3 + \frac{6}{20}a - \frac{12}{5} = \frac{6a}{20} - \frac{15}{5} + \frac{6a}{20} - \frac{12}{5} = \frac{12a}{20} - \frac{27}{5} = \frac{3}{5}a - \frac{27}{5}$

57. $6[5c + 4d - 8c] - 5c = 30c + 24d - 48c - 5c = -23c + 24d$

58. $4[2a + 6b - 3a] + 8b = 8a + 24b - 12a + 8b = -4a + 32b$

59. $0.69[b + 1.2b - 1] = 0.69b + 0.828b - 0.69 = 1.518b - 0.69$

60. $0.4[2.54a + 0.508 - a] = 1.016a + 0.2032 - 0.4a = 0.616a + 0.2032$

61. $(7w)(5) + (4w)(2) = 35w + 8w = 43w$

62. $7(x + 8) + 3[(x + 10) - 7] = 7(x + 8) + 3(x + 3) = 7x + 56 + 3x + 9 = 10x + 65$

page 66   Test Yourself

1. $9 - 18 = -9$ **2.** $-1 + 7(8) = -1 + 56 = 55$

3. $\dfrac{8}{16}(2) = \dfrac{1}{2}(2) = 1$

4. $5(-2) + 3 = -10 + 3 = -7$

5. $2(-2) - 3(0) + 4(3) = -4 - 0 + 12 = 8$

6. $-2 + \dfrac{0+3}{0-3} = -2 + -\dfrac{3}{3} = -2 + (-1) = -3$

7. $3(16)(4) = 192$

8. $-(-3)^2 + (-2) = -9 + (-2) = -11$

9. $s = (0.5)(32)(5)^2 = (0.5)(32)(25) = 400$

10. Commutative property for addition

11. Associative property for multiplication

12. Distributive property

13. Commutative property for addition

14. $-y^2 + y$

15. $11hk - 5hk^2 - 5h^2k$

16. $3a^2 - 3 + 4a^2 - 8 = 7a^2 - 11$

17. $5x^2 + 10 - 3x^2 - 3 = 2x^2 + 7$

page 67   Capsule Review

1. $(2)(-3) = -6$

2. $2(2)(-3) = -12$

3. $2(2) + (-3) = 4 + (-3) = 1$

4. $2 + 2(-3) = 2 + (-6) = -4$

5. $2 - (-3) = 5$

6. $-3 - 2 = -5$

7. $3(2) - (-3) = 6 + 3 = 9$

8. $2(-3) - 2(2) = -6 - 4 = -10$

page 69   Class Exercises

1. $-x - 3$

2. $-y + 7$

3. $a - 4$

4. $b + 3$

5. $-2m^2 - 3$

6. $-5n^2 + 6$

7. $f^2 - f$

8. $2g^2 + g$

9. $-ab^2 - ab^2 = -2ab^2$

10. $-2rs - rs = -3rs$

11. $2x - 4 - x + 5 = x + 1$

12. $y - 7 - 4 - 2y = -y - 11$

13. $-b + a^2;\ -(-5) + 6^2 = 41$

14. $-4[2b + 6 - a] = -8b - 24 + 4a;\ -8(-5) - 24 + 4(6) = 40 - 24 + 24 = 40$

pages 69–71   Practice Exercises

A **1.** $2g - 4 - g - 4 = g - 8$

2. $4 + 5m - 3 - m = 4m + 1$

3. $3a + 4 - 5a + 9 = 13 - 2a$

4. $2h - 5 - 7h - 5 = -5h - 10$

5. $5t^2 - 2 - 2t^2 - 7 = 3t^2 - 9$

6. $9p^2 - 1 - 3p^2 - 2 = 6p^2 - 3$

7. $x + 5 - 3x - 6 = -2x - 1$

8. $y + 3 - 2y - 14 = -y - 11$

9. $-ab^2c - 3abc^2 + 2ab^2c = ab^2c - 3abc^2$

10. $-a^2bc + a^2bc^2 + a^2bc = a^2bc^2$

11. $-a - b^2;\ -2 - (-5)^2 = -2 - 25 = -27$

12. $-m - n^2;\ -3 - (-7)^2 = -3 - 49 = -52$

13. $-x + y^2;\ -4 + (-1)^2 = -4 + 1 = -3$

14. $-r + s^2$; $-1 + (-5)^2 = -1 + 25 = 24$

15. $8 + 3[-2p + 6 + 2p] = 8 - 6p + 18 + 6p = 26$

16. $5 + 2[-z + 7 + 4z] = 5 + 2[3z + 7] = 5 + 6z + 14 = 6z + 19$;
$6(0.2) + 19 = 1.2 + 19 = 20.2$

17. $13 - 4[3q + 3 - 2q] = 13 - 4[q + 3] = 13 - 4q - 12 = 1 - 4q$;
$1 - 4(-1) = 5$

18. $10 - 3[4a + 8 - 3a] = 10 - 3[a + 8] = 10 - 3a - 24 = -3a - 14$;
$-3(-3) - 14 = 9 - 14 = -5$

B 19. $-m + n^2$; $-4 + (5)^2 = -4 + 25 = 21$ **20.** $-y + z^2$; $-(-6) + (7)^2 = 6 + 49 = 55$

21. $5 - 6[10x^2 + 20x + 3x] = 5 - 6[10x^2 + 23x] = 5 - 60x^2 - 138x$;
$5 - 60(-2)^2 - 138(-2) = 5 - 240 + 276 = 41$

22. $11 - [4y + 3y^2 - 15y]8 = 11 - [-11y + 3y^2]8 = 11 + 88y - 24y^2$;
$11 + 88(2) - 24(2)^2 = 11 + 176 - 96 = 91$

23. $x^2yz + x^2yz^2 + x^2yz = 2x^2yz + x^2yz^2$

24. $x^2yz^2 + xyz^2 + x^2yz^2 = 2x^2yz^2 + xyz^2$

25. $2 + 2[-3x + 6] = 2 - 6x + 12 = -6x + 14$

26. $7 + 3[-2x + 2] = 7 - 6x + 6 = -6x + 13$

27. $5r - 9 + 16r - 144 = 21r - 153$ **28.** $12x - 132 + x + 13 = 13x - 119$

C 29. $4 - \{m + 5m + 5 - m\} = 4 - \{5m + 5\} = 4 - 5m - 5 = -1 - 5m$

30. $2 - \{g + 3g - 3 + g\} = 2 - \{5g - 3\} = 2 - 5g + 3 = -5g + 5$

31. $1 - \{a - [a + a - 3]\} = 1 - \{a - [2a - 3]\} = 1 - \{-a + 3\} =$
$1 + a - 3 = a - 2$

32. $1 - \{q - [q - q - 4]\} = 1 - \{q - [-4]\} = 1 - \{q + 4\} = 1 - q - 4 =$
$-q - 3$

33. $5 - 4\{13 + 6x - 21 + 3x\} = 5 - 4\{9x - 8\} = 5 - 36x + 32 = 37 - 36x$

34. $7 + 9\{11 - 12z - 24 + 4z\} = 7 + 9\{-13 - 8z\} = 7 - 117 - 72z =$
$-110 - 72z$

35. $A = \frac{1}{2}(3)(5 + 7)$, $A = \frac{1}{2}(3)(12)$, $A = 18$ cm^2

36. $A = \frac{1}{2}(14)(10 + 12)$, $A = \frac{1}{2}(14)(22)$, $A = 154$ ft^2

37. $4(3a) + 2(a + 6)$; $12a + 2a + 12 = 14a + 12$

38. $2(x + 2) + 2(2) + 2(x + 4) + x$; $2x + 4 + 4 + 2x + 8 + x = 5x + 16$

39. 13.19   **40.** 21.81   **41.** 5

42. 100   **43.** 40   **44.** $-\dfrac{10}{3}$

45. $-(1) + (5) = 4$ **46.** $-(-4) + 10 = 4 + 10 = 14$

47. $(-12) - (-7) = -12 + 7 = -5$

pages 72–73   Technology: Introduction to Spreadsheets

1. C3 + C4 + C5 + C6   **2.** C12 + C13 + C14 + C15 + C16

3. C8 − C18   **4.** D4 + D5 + D6

5. D12 + D13 + D14 + D15 + D16   **6.** D8 − D18

7. E4 + E5 + E6   **8.** E12 + E13 + E14 + E15 + E16

9. E8 − E18   **10.** Answers may vary. Possible answer is C21 ∗ 4.

11. Answers may vary.   **12.** Answers may vary.

page 74   Capsule Review

1. $a + b$ **2.** $x - d$ **3.** ef **4.** $\dfrac{g}{h}$ **5.** $j - i$

6. $l + k$ **7.** $\dfrac{m}{n}$ **8.** $o - p$ **9.** $2 - y$ **10.** $4r$

pages 75–76   Class Exercises

1. $n + 6$ **2.** $n - 2$ **3.** $\dfrac{1}{2}n$ **4.** $\dfrac{n}{6}$ **5.** $5n - 2$

6. $2n + 5$ **7.** $4n^2$ **8.** $\dfrac{n^2}{8}$ **9.** $4(n + 9)$ **10.** $2(n - 7)$

11. Two times a number plus three; three more than two times a number.

12. Nine minus three times a number; the difference of nine and three times a number.

13. A number squared plus one; the sum of a number squared and one.

14. Ten times a number squared plus sixty; sixty more than 10 times a number squared.

pages 76–77   Practice Exercises

A **1.** $n - 7$ **2.** $n - 6$ **3.** $n + 5$ **4.** $n + 6$

    5. $8n - 1$ **6.** $n + 9$ **7.** $5n^2$ **8.** $6n^2$

    9. $\dfrac{n^2}{3}$ **10.** $\dfrac{n^2}{2}$ **11.** $6(n + 3)$ **12.** $2(n + 3)$

13. Five less than three times a number; the difference of three times a number and 5.

14. The sum of seven times a number and six; six more than seven times a number.

15. A number times the sum of that number and three; the product of a number and that number plus three.

B 16. a. $s - 4$ **b.** $s + 2$ **c.** $s - 2$ **17. a.** $m - 4$ **b.** $m - 4$ **c.** $m - 8$

18. a. w **b.** $w + 4$ **c.** $w(w + 4) = w^2 + 4w$
     d. $2(w + 4) + 2w = 2w + 8 + 2w = 4w + 8$

19. a. l **b.** $l - 7$ **c.** $l(l - 7) = l^2 - 7l$
     d. $2l + 2(l - 7) = 2l + 2l - 14 = 4l - 14$

C 20. $\dfrac{b}{5} - \dfrac{a}{3}$ assuming $3b > 5a$

21. $4t + 3(t + 12) + 2t = 4t + 3t + 36 + 2t = 9t + 36$ dollars;
     $11t - (9t + 36) = 2t - 36$ dollars

22. $1.05 + 0.25(m - 3) = 1.05 + 0.25m - 0.75 = 0.25m + 0.30$

23. $0.14(2000) + 0.155(d - 2000) = 280 + 0.155d - 310 = 0.155d - 30$

page 77 Historical Note

1. Answers may vary. Possible answers are: **2.** Answers may vary.

$$1038 =$$
$$54{,}174 =$$
$$310{,}651 =$$

page 78 Capsule Review

1. $A = (8)(6) = 48$ sq. units   **2.** $A = 55.5$ sq. units   **3.** $A = \left(\dfrac{1}{2}\right)(7) = \dfrac{7}{2}$ sq. units

4. $d = (10)(2) = 20$ **5.** $d = (50)(3) = 150$ **6.** $d = (26)\left(\dfrac{1}{2}\right) = 13$

page 80 Class Exercises

1. $12 - 8 = 10$, $4 = 10$, false   **2.** $-30 < -7$, true   **3.** $-1 \neq -1$, false

4. Ø   **5.** $\{-1\}$ **6.** $\{-1, 0\}$ **7.** $\{1\}$ **8.** $\{0, 1\}$ **9.** Ø

A 1. $\{1\}$ 2. $\{-1\}$ 3. $\{0\}$ 4. $\{-1\}$ 5. $\{-1, 0, 1\}$

6. $\{-1, 0\ 1\}$ 7. $\{-1, 1\}$ 8. $\{-1, 0\}$ 9. $\{0\}$ 10. $\{1\}$

11. $\{-1, 0, 1\}$ 12. $\{0, 1\}$ 13. $\{-1\}$ 14. $\{1\}$ 15. $\{-1, 0\}$

16. $\left\{-2, -1, 0, \dfrac{1}{2}\right\}$ 17. $\varnothing$ 18. $\varnothing$

B 19. $\left\{\dfrac{1}{2}\right\}$ 20. $\{-2\}$ 21. $\{-1\}$ 22. $\{0\}$ 23. $\{-1\}$

24. $\varnothing$ 25. $\{-1\}$ 26. $\left\{-2, -1, 0, \dfrac{1}{2}\right\}$ 27. $\varnothing$

For Exercises 28–31 answers may vary. Possible answers are shown.

C 28. $x - 1 = 0;\ 3x = 3$ 29. $x^2 - 4 = 0;\ 3x^2 = 12$ 30. $x < 3;\ 2x = 2x$

31. $x^2 - 2x = 0;\ 3x^2 = 6x$ 32. $3x = 7,\ x = \dfrac{7}{3}$ 33. any negative number

34. There is no such number; the left and right sides are always equal.

35. $\dfrac{3 + 2 + 3 + x}{4} \geq 3,\ \dfrac{8 + x}{4} \geq 3$; possible scores: 4 or 5

36. $\dfrac{5 + 2 + 3 + x}{4} \geq 3,\ \dfrac{10 + x}{4} \geq 3$; possible scores: 2, 3, 4, or 5

37. $\dfrac{5 + 2 + 2 + x}{4} \geq 3,\ \dfrac{9 + x}{4} \geq 3$; possible scores: 3, 4, or 5

38. $\dfrac{4 + 5 + 2 + x}{4} \geq 3,\ \dfrac{11 + x}{4} \geq 3$; possible scores: 1, 2, 3, 4, or 5

39. $3(8) + 5(-2) = 24 - 10 = 14$ 40. $6\left(-\dfrac{1}{2}\right) - 6\left(\dfrac{2}{3}\right) = -3 - 4 = -7$

41. $[2(-2) + (12)] - \left[(8)\left(-\dfrac{1}{2}\right) + 1\right] = -4 + 12 - (-4 + 1) = -4 + 12 + 4 - 1 = 11$

42. $\dfrac{(8)(-2)}{(12)\left(\dfrac{2}{3}\right)} = \dfrac{-16}{8} = -2$ 43. $\dfrac{1 - 2(-2)(12)}{4(-2) + 1} = \dfrac{1 - (-48)}{-8 + 1} = -\dfrac{49}{7} = -7$

44. $\dfrac{(12)\left(-\dfrac{1}{2}\right) - (-2)}{10 + 3(-2)} = \dfrac{-6 + 2}{10 - 6} = \dfrac{-4}{4} = -1$ 45. $(8)^2 + (-2)^3 = 64 + (-8) = 56$

46. $3\left(-\dfrac{1}{2}\right)^3\left(\dfrac{2}{3}\right)^2 = 3\left(-\dfrac{1}{8}\right)\left(\dfrac{4}{9}\right) = -\dfrac{12}{72} = -\dfrac{1}{6}$ 47. $3\left(-\dfrac{1}{2}\right)^2\left(\dfrac{2}{3}\right)^3 = 3\left(\dfrac{1}{4}\right)\left(\dfrac{8}{27}\right) = \dfrac{24}{108} = \dfrac{2}{9}$

48. $2x - 8$ 49. $x - 9$ 50. $x + 15$ 51. $7x - 2$ 52. $2(x + 5)^2$

page 82   Capsule Review

1. $x + 2x$
  2. $\dfrac{j}{5j}$
  3. $50n$
  4. $z - 9$

5. two more than a number r
  6. thirty decreased by a number x

7. the product of l and w

8. a number p increased by 8 times that number

page 84   Class Exercises

1. $2n = 14$
  2. $n - 12 = 7$
  3. $\dfrac{1}{2}n = 34$
  4. $\dfrac{n}{6} = \dfrac{2}{3}$

5. $2n - 6 = 5$
  6. $5n^2 = 45$

pages 84–86   Practice Exercises

A 1. $x - 3 = 18$
  2. $x - 12 = 4$
  3. $x + 4 = 27$
  4. $x + (-7) = 2$

5. $5x - 2 = 18$
  6. $2x + 1 = -13$
  7. $8x^2 = 56$
  8. $7(8x) = 79$

9. $\dfrac{5}{3x} = 10$
  10. $\dfrac{5}{x^2} = 100$
  11. $6(x + 9) = 132$

12. $3(x + 3) = -18$

13. a. $b - 6$
  b. $b - 1$
  c. $(b - 1) - 6 = b - 7$
  d. $(b - 6) + (b - 7) = 35$

14. a. $j + 9$
  b. $j - 15$
  c. $(j - 15) + 9 = j - 6$
  d. $j + 9 = 2.5\,(j - 6)$

B 15. a. $45n$
  b. $45n + 35$
  c. $45n + 35 = 485$

16. a. $55n$
  b. $55n + 65$
  c. $55n + 65 = 670$

17. Three times a number decreased by seven is thirty-two.

18. The difference of a number and three is four times that number.

19. A number times the sum of that number and six is forty-eight.

20. Twice the difference of a number and five increased by two times the number is one hundred and twenty-eight.

C 21. $3j + 5 = 2(j + 5)$
  22. $2p - 14 = 4(p - 14)$

23. $25e + 10(e + 6) = 235$
  24. $25q + 5(q + 11) = 265$

25. $2x + 150 = 13{,}250$; What was the cost of the old car?

26. $w(2w - 10) = 208$; What is the length and width of the rectangle?

27. $5x + 35 = 110$; What is her monthly payment?

28. $2w + 2(3w + 25) = 130$; What is the length and width of the rectangle?

29. $s = 2(1200) + 100$; What is Sheila's tuition?

30. $12x + 65 = 245$; What is his monthly payment?

31. $\frac{1}{2}(b)(4b + 2) = 55$; What are the height and base of the triangle?

32. $2s + 3s - 12 = 23$; What is the length of each side of the triangle?

page 86 Test Yourself

1. $2r - 4r - 12 = -2r - 12$ **2.** $y - 6 - 2y - 4 = -y - 10$

3. $-m^2 + n$; $-(-3)^2 + (-2) = -9 + (-2) = -11$

4. $7(n + 3)$ **5.** $\frac{n}{9} - 8$

6. Ten times a number decreased by six; the difference of ten times a number and six.

7. The quotient of a number and four increased by six; six added to a number divided by four.

8. $\{-1\}$ **9.** $\varnothing$ **10.** $\{-2, 0, 1, 2\}$

11. $n - (-2) = -11$ **12.** $5x + 2 = 25$

page 88 Class Exercises

1. *Finite* means "a countable number."

2. An open sentence is either true or false.

3. A solution is a value which makes an open sentence true.

4. *Infinite* means "unending" or "uncountable."

pages 88–89 Practice Exercises

A 1. $\varnothing$ **2.** $\{5\}$ **3.** $\{0\}$ **4.** $\{-4\}$ **5.** $\{-2\}$

6. $\{2\}$ **7.** $\{-3\}$ **8.** $\{0\}$ **9.** $\{1\}$

10. The solution to the equation must be greater than 5, since
$10(5 + 6) = 110.$
$10(6 + 6) = 120$; No.
$10(7 + 6) = 130$; No.
$10(8 + 6) = 140$; Yes; The solution is 8; Lois has 8 dimes.

11. The solution to the equation must be greater than 10, since
$2(10) + 4 = 24.$ $2(13) + 4 = 30$; No.
$2(11) + 4 = 26$; No. $2(14) + 4 = 32$; No.
$2(12) + 4 = 28$; No. $2(15) + 4 = 34$; Yes;
The solution is 15; Bob is 15 years old.

12. The solution to the equation must be greater than 48, since

$\frac{48}{2} - 1 = 23.$ $\frac{49}{2} - 1 = 23\frac{1}{2};$ No. $\frac{50}{2} - 1 = 24;$ Yes;

The solution is 50; Ken worked 50 weeks.

13. The solution to the equation must be greater than 30, since

$5 + \frac{1}{3}(30) = 15.$ $5 + \frac{1}{3}(32) = 15\frac{2}{3};$ No.

$5 + \frac{1}{3}(31) = 15\frac{1}{3};$ No. $5 + \frac{1}{3}(33) = 16;$ Yes;

The solution is 33; Suzanne edited 33 pages.

14. $x + (x + 9) = 229;$ The solution to the equation must be greater than 105, since

105 + 114 = 219.   108 + 117 = 225; No.
106 + 115 = 221; No.   109 + 118 = 227; No.
107 + 116 = 223; No.   110 + 119 = 229; Yes;

The solution is 110; Helen weighs 110 lb and her sister weighs 119 lb.

B 15. $m + 50 = 2m;$ The solution to the equation must be greater than 45, since

45 + 50 $\neq$ 2(45).   48 + 50 = 2(48); No.
46 + 50 = 2(46); No.   49 + 50 = 2(49); No.
47 + 50 = 2(47); No.   50 + 50 = 2(50); Yes;

The solution is 50; Carmen has $100 and Bill has $50.

16. $25q + 25(q + 5) = 475;$ The solution must be greater than 5, since 25(5) + 25(10) = 375

25(6) + 25(11) = 425; No.
25(7) + 25(12) = 475; Yes; The solution is 7. Beth has 7 quarters, and Harry has 12 quarters.

17. $n + (n + 6) = -8;$ The solution to the equation must be greater than -10, since

$-10 + (-4) = -14.$
$-9 + (-3) = -12;$ No.
$-8 + (-2) = -10;$ No.
$-7 + (-1) = -8;$ Yes; The solution is -7; the number is -7.

C 18. Yes, it is possible to list the solution to $x + 4 < -3;$ $-100 < x < -7.$

19. $2(l + w) = 64,$ $l + w = 32,$ $l = 32 - w;$ $(l)(w) = 252,$ $w(32 - w) = 252,$
12(32 − 12) = 240; No.
13(32 − 13) = 247; No.
14(32 − 14) = 252; Yes, then $l = 32 - 14 = 18;$ The dimensions of the rectangle are 18 ft x 14 ft.

1. $419 - 212 = 207$; The final depth of the submarine was 207 ft.

2. a. $x + 2 = -1, \{-3\}$ **b.** $2x - (-3) = 7, 2x + 3 = 7, 2x = 4, \{2\}$

c. $\dfrac{9}{x} = -3, \{-3\}$ **d.** $x - 16 = -1$, there is no solution to this equation using the given replacement set.

page 89   **Project**

$$
\begin{array}{r}
5\,6 \\
\times\ 4\,7 \\
\hline
3\,9\,2 \\
2\,2\,4 \\
\hline
2\,6\,3\,2
\end{array}
\qquad
\begin{array}{r}
9\,3 \\
2\,8\,)\overline{2\,6\,0\,4} \\
2\,5\,2 \\
\hline
8\,4 \\
8\,4 \\
\hline
0
\end{array}
\qquad
\begin{array}{r}
3\,5\,9 \\
\times\ 2\,8 \\
\hline
2\,8\,7\,2 \\
7\,1\,8 \\
\hline
1\,0\,,0\,5\,2
\end{array}
$$

pages 90–91   **Summary and Review**

1. $-2(-3) + 6 = 6 + 6 = 12$

2. $4\left(\dfrac{1}{16}\right) = \dfrac{1}{4}$

3. -0.0001

4. $5(16 - 27) = 5(-11) = -55$

5. $2 + \dfrac{4 - (-1)}{2 + (-1)} = 2 + \dfrac{5}{1} = 2 + 5 = 7$

6. $(2 - 4)^3 = (-2)^3 = -8$

7. Distributive property

8. Commutative property for multiplication

9. Associative property for addition

10. Associative property for addition

11. Identity property for addition

12. Distributive property

13. Commutative property for multiplication

14. Inverse property for mult.

15. $-2c + 5 + 4c^2$

16. $-8x + 3x - 15 = -5x - 15$

17. $4mn^2 - 3mn$

18. $20 - 12r - 10r = 20 - 22r$

19. $2h^2 - h + h = 2h^2$

20. $8t^3 - 4t^2 + t^3 = 9t^3 - 4t^2$

21. $-3[8 - 10p + p] - p = -3[8 - 9p] - p = -24 + 27p - p = -24 + 26p$;
$-24 + 26(2) = -24 + 52 = 28$

22. $\dfrac{11}{n}$

23. $2[n + (-2)]$

24. $12n - 8$

25. $\{0\}$

26. $\varnothing$

27. $\{-2, -1, 1, 2\}$

28. $2x = -15$

29. $2x - 2 = 1$

30. $5x + 5(x - 4) = 160$

31. $y + (y + 5) = 31$; The solution to the equation must be greater than 10, since

$10 + 15 = 25$.

$11 + 16 = 27$; No.

$12 + 17 = 29$; No.

$13 + 18 = 31$; Yes; The solution is 13; Jenna is 13 years old and Kim is 18 years old.

page 92   Chapter Test

1. $-6 + 27 = 21$

2. $24 \div 3 + (-2) = 8 + (-2) = 6$

3. 16

4. $9\left(\dfrac{1}{27}\right) = \dfrac{1}{3}$

5. $\dfrac{-1 + 4}{-2} = -\dfrac{3}{2}$

6. $\dfrac{4 - 2(-2)}{-1 + 4} = \dfrac{4 - (-4)}{3} = \dfrac{8}{3}$

7. Distributive property

8. Distributive property

9. Associative property for multiplication

10. Commutative property for addition

11. $-8q + 5 + q^2$

12. $-10rt + r^2t + 2rt^2$

13. $4x - 8 + x = 5x - 8$

14. $7 - 3 + 2j = 4 + 2j$

15. $-5[2 - 6t - 21 - t] = -5[-19 - 7t] = 95 + 35t$

16. $y - [2y + 3y - 3] = y - [5y - 3] = y - 5y + 3 = -4y + 3$

17. $a - 5$

18. $4l$

19. $\{1\}$

20. $\{-1, 1\}$

21. $\{-2, -1, 0, 1\}$

22. $\varnothing$

23. $\{0, 1\}$

24. $\varnothing$

25. $\dfrac{1}{2}x + 3 = 25$

26. $2b - 11 = 3(b - 11)$

Challenge

1. $3 - \{x - [x - 3x + 9]\} = 3 - \{x - [-2x + 9]\} = 3 - \{x + 2x - 9\} =$ $3 - 3x + 9 = -3x + 12$

2. Let b = Billy's age, then $b - 14$ = Annie's age; $b - 14 = \dfrac{2}{3}b$

page 93   Preparing for Standardized Tests

1. D; $3a + b - 2c + a - 2b + c = 4a - b - c$

2. B; $a^2 + b^2 = 3^2 + 4^2 = 9 + 16 = 25$

3. A; Seven less than twice a certain number is $2x - 7$.

4. E; Write the estimate as 7000×20, which becomes 140,000.

5. E; Check only the last two digits of each number for divisibility by 4.

  A. _____ 74 no   B. _____ 15 no

  C. _____ 82 no   D. _____ 10 no

  E. _____ 56 yes, 56 = 4 × 14

6. B; $2[3(x^2 - 4) + 5y] = 2[3(3^2 - 4) + 5 \cdot 1] = 2[3(5) + 5] = 2[20] = 40$

7. D; Use decimal values or their approximations (estimated) for each

  fraction. It can be seen that $\frac{4}{9} = 0.444\ldots$ is the only one which satisfies

  the requirements.

8. B; Since $x + 7 = 12$, by inspection, x must be 5.

9. E; If $2k < 60$, then $k < 30$, and all choices are possible except 32.

10. B; $\dfrac{96 + 85 + 79 + 91 + 89}{5} = \dfrac{440}{5} = 88$, the average Alex would like to

  maintain. Thus, he needs a score of 88.

11. B; Since $14.70 = 2 \times 7.35$, Betty must have 3×7.35 or $22.05.

page 94   Cumulative Review

1. $-1.5, -1, 0, 2, 2.5$ **2.** $-2, -1, -\frac{1}{2}, 1, 1\frac{1}{2}$ **3.**
$$\begin{array}{c} \leftarrow\!\!+\!\!\leftbullet\!\!+\!\!\bullet\!\bullet\!\!+\!\!\bullet\!\!+\!\!+\!\!\rightarrow \\ {\scriptstyle -2\,-1\ 0\ 1\ 2\ 3} \end{array}$$

4. -4 **5.** 0.8 **6.** -63

7. $\dfrac{6}{4} = 1\dfrac{1}{2}$ **8.** 9 **9.** 14

10. $\dfrac{1}{2}\left(-\dfrac{4}{3}\right) = -\dfrac{2}{3}$ **11.** $-5 + 0.9 = -4.1$ **12.** $-\dfrac{1}{2}$

13. -4 **14.** $-5.4\left(\dfrac{1}{9}\right) = -0.6$ **15.** $\dfrac{2}{8} = \dfrac{1}{4}$

16. $-\left|\dfrac{3}{4}\right| = -\dfrac{3}{4}$ **17.** $9 + 7 = 16$

18. $6 + 24 \div 6 = 6 + 4 = 10$

19. $-2[12 - 3(7)] = -2[12 - 21] = -2[-9] = 18$

20. $[6 - 4]^3 = 2^3 = 8$ **21.** $6x - 3y$

22. $-m^2 + m$ **23.** $3a + 6 - 5a = -2a + 6$

24. $p - 5 + 2p = 3p - 5$ **25.** $5c + 2 - 3c + 6 = 2c + 8$

26. $-2st^2 + 3st^2 - 1 = st^2 - 1$ **27.** $3\left(\dfrac{2}{3}\right) - 5 = 2 - 5 = -3$

28. $3\left(\dfrac{2}{3}\right) - 3(4) = 2 - 12 = -10$ **29.** $3\left(\dfrac{2}{3}\right)(4) = 8$

30. $|3 - 2(4)| = |3 - 8| = |-5| = 5$

31. $\left[5(3) - 6\left(\dfrac{2}{3}\right)\right](4) = [15 - 4](4) = 11(4) = 44$

32. $\dfrac{2(3) - 4}{\dfrac{2}{3}} = \dfrac{6 - 4}{\dfrac{2}{3}} = \dfrac{2}{\dfrac{2}{3}} = 2\left(\dfrac{3}{2}\right) = 3$

33. $5\left\{\dfrac{2}{3} \div [2(3) + 4]\right\} = 5\left\{\dfrac{2}{3} \div [6 + 4]\right\} = 5\left\{\dfrac{2}{3} \div 10\right\} = 5\left\{\dfrac{2}{3}\left(\dfrac{1}{10}\right)\right\} = 5\left\{\dfrac{1}{15}\right\} = \dfrac{1}{3}$

34. $\dfrac{3(3) + 1}{\dfrac{2}{3} + \dfrac{4}{3}} = \dfrac{10}{\dfrac{6}{3}} = \dfrac{10}{2} = 5$ **35.** $2(3)^2 - 4 = 18 - 4 = 14$ **36.** $x + 5$

37. $2n - 3$ **38.** $5(n + 3)$ **39.** $l = 2w + 1$ **40.** $3w + 14$

Chapter 3   Equations in One Variable

page 96   Capsule Review

1. 0

2. 0

3. y

4. x

5. subtract 6

6. add 21

7. subtract $\dfrac{4}{5}$

8. subtract (-2.1)

page 98   Class Exercises

1. subtract 4

2. add 2

3. add 1

4. add 4

5. subtract 15

6. add 11

7. subtract 24

8. add 35

9. $a - 7 = 24$, $a - 7 + 7 = 24 + 7$, $a + 0 = 31$; $a = 31$

10. $x + 21 = -25$, $x + 21 - 21 = -25 - 21$, $x + 0 = -46$; $x = -46$

11. $12 + d = -15$, $12 + d - 12 = -15 - 12$, $d + 0 = -27$; $d = -27$

12. $r - (-7) = 13$, $r + 7 = 13$, $r + 7 - 7 = 13 - 7$, $r + 0 = 6$; $r = 6$

13. $-2 = -15 + n$, $-2 + 15 = -15 + n + 15$, $13 = n + 0$; $n = 13$

14. $33 = -14 + c$, $33 + 14 = -14 + 14 + c$, $47 = c + 0$; $c = 47$

pages 98–99   Practice Exercises

A 1. $x - 3.4 = 9.61$, $x - 3.4 + 3.4 = 9.61 + 3.4$, $x + 0 = 13.01$; $x = 13.01$

2. $a - 12.5 = 13.9$, $a - 12.5 + 12.5 = 13.9 + 12.5$, $a + 0 = 26.4$; $a = 26.4$

3. $y + 1.9 = 10.2$, $y + 1.9 - 1.9 = 10.2 - 1.9$, $y + 0 = 8.3$; $y = 8.3$

4. $z + 2.4 = 5.3$, $z + 2.4 - 2.4 = 5.3 - 2.4$, $z + 0 = 2.9$; $z = 2.9$

5. $m - 3.6 = 4.5$, $m - 3.6 + 3.6 = 4.5 + 3.6$, $m + 0 = 8.1$; $m = 8.1$

6. $t - 5.3 = 2.3$, $t - 5.3 + 5.3 = 2.3 + 5.3$, $t + 0 = 7.6$; $t = 7.6$

7. $9 = x + 2\frac{1}{3}$, $9 - 2\frac{1}{3} = x + 2\frac{1}{3} - 2\frac{1}{3}$, $6\frac{2}{3} = x + 0$; $x = 6\frac{2}{3}$

8. $11 = z + 3\frac{2}{5}$, $11 - 3\frac{2}{5} = z + 3\frac{2}{5} + 3\frac{2}{5}$, $7\frac{3}{5} = z + 0$; $z = 7\frac{3}{5}$

9. $4 = m + 1\frac{1}{3}$, $4 - 1\frac{1}{3} = m + 1\frac{1}{3} - 1\frac{1}{3}$, $2\frac{2}{3} = m + 0$; $m = 2\frac{2}{3} = \frac{8}{3}$

10. $8 = s + 4\frac{1}{2}$, $8 - 4\frac{1}{2} = s + 4\frac{1}{2} - 4\frac{1}{2}$, $3\frac{1}{2} = s + 0$; $s = 3\frac{1}{2}$

11. $3 = t - 1\frac{2}{3}$, $3 + 1\frac{2}{3} = t - 1\frac{2}{3} + 1\frac{2}{3}$, $4\frac{2}{3} = t + 0$; $t = 4\frac{2}{3}$

12. $5 = y - 1\frac{1}{4}$, $5 + 1\frac{1}{4} = y - 1\frac{1}{4} + 1\frac{1}{4}$, $6\frac{1}{4} = y + 0$; $y = 6\frac{1}{4}$

13. $a - (-60) = 30$, $a + 60 = 30$, $a + 60 - 60 = 30 - 60$, $a + 0 = -30$; $a = -30$

14. $b - (-25) = 24$, $b + 25 = 24$, $b + 25 - 25 = 24 - 25$, $b + 0 = -1$; $b = -1$

15. $x - (-5) = 10$, $x + 5 = 10$, $x + 5 - 5 = 10 - 5$, $x + 0 = 5$; $x = 5$

16. $m - (-2) = 12$, $m + 2 = 12$, $m + 2 - 2 = 12 - 2$, $m + 0 = 10$; $m = 10$

17. $z - (-6) = 50$, $z + 6 = 50$, $z + 6 - 6 = 50 - 6$, $z + 0 = 44$; $z = 44$

18. $t - (-9) = 41$, $t + 9 = 41$, $t + 9 - 9 = 41 - 9$, $t + 0 = 32$; $t = 32$

19. $-2.3 + x = -5.9$, $-2.3 + x + 2.3 = -5.9 + 2.3$, $x + 0 = -3.6$; $x = -3.6$

20. $-5.3 + m = 10.2$, $-5.3 + m + 5.3 = 10.2 + 5.3$, $m + 0 = 15.5$; $m = 15.5$

21. $-3.4 + s = -9.5$, $-3.4 + s + 3.4 = -9.5 + 3.4$, $s + 0 = -6.1$; $s = -6.1$

22. $-8.2 + t = -12.4$, $-8.2 + t + 8.2 = -12.4 + 8.2$, $t + 0 = -4.2$; $t = -4.2$

23. $-5.3 + r = -12.3$, $-5.3 + r + 5.3 = -12.3 + 5.3$, $r + 0 = -7$; $r = -7$

24. $-2.5 + a = -5.5$, $-2.5 + a + 2.5 = -5.5 + 2.5$, $a + 0 = -3$; $a = -3$

B 25. $y - 7.01 = 12.009$, $y - 7.01 + 7.01 = 12.009 + 7.01$, $y + 0 = 19.019$; $y = 19.019$

26. $x - 3.12 = 5.23$, $x - 3.12 + 3.12 = 5.23 + 3.12$, $x + 0 = 8.35$; $x = 8.35$

27. $z - 0.032 = 1.03$, $z - 0.032 + 0.032 = 1.03 + 0.032$, $z + 0 = 1.062$; $z = 1.062$

28. $4 = n + 3\frac{1}{2}$, $4 - 3\frac{1}{2} = n + 3\frac{1}{2} - 3\frac{1}{2}$, $\frac{1}{2} = n + 0$; $n = \frac{1}{2}$

29. $5 = a + 2\frac{1}{3}$, $5 - 2\frac{1}{3} = a + 2\frac{1}{3} - 2\frac{1}{3}$, $2\frac{2}{3} = a + 0$; $a = 2\frac{2}{3}$

30. $8 = x + 3\frac{1}{4}$, $8 - 3\frac{1}{4} = x + 3\frac{1}{4} - 3\frac{1}{4}$, $4\frac{3}{4} = x + 0$; $x = 4\frac{3}{4}$

31. $45 - (-a) = 50$, $45 + a = 50$, $45 + a - 45 = 50 - 45$, $a + 0 = 5$; $a = 5$

32. $17 - (-x) = 22$, $17 + x = 22$, $17 + x - 17 = 22 - 17$, $x + 0 = 5$; $x = 5$

33. $52 = 25 - (-a)$, $52 = 25 + a$, $52 - 25 = 25 + a - 25$, $27 = a + 0$; $a = 27$

34. $-5.8 + 2.75 = a - 2.75 + 2.75$, $-3.05 = a + 0$; $a = -3.05$

35. $-8.3 = x - 2.5$, $-8.3 + 2.5 = x - 2.5 + 2.5$, $-5.8 = x + 0$; $x = -5.8$

36. $-3.9 = b - 5.1$, $-3.9 + 5.1 = b - 5.1 + 5.1$, $1.2 = b + 0$; $b = 1.2$

37. $-15 + y + 15 = -15 + 15$, $y + 0 = 0$; $y = 0$; no solution

38. $x - 12 = -12$, $x - 12 + 12 = -12 + 12$, $x + 0 = 0$; $x = 0$; no solution

39. $\frac{1}{2} + y = 5\frac{1}{2}$, $\frac{1}{2} + y - \frac{1}{2} = 5\frac{1}{2} - \frac{1}{2}$, $y + 0 = 5$; $y = 5$

40. $-\frac{4}{5} + x = -\frac{8}{5}$, $-\frac{4}{5} + x + \frac{4}{5} = -\frac{8}{5} + \frac{4}{5}$, $x + 0 = -\frac{4}{5}$; $x = -\frac{4}{5}$; no solution

41. $\frac{2}{3} + y = -\frac{1}{2}$, $\frac{2}{3} + y - \frac{2}{3} = -\frac{1}{2} - \frac{2}{3}$, $y + 0 = -\frac{3}{6} - \frac{4}{6}$; $y = -\frac{7}{6}$; no solution

42. $-\frac{3}{4} + x = 1\frac{1}{4}$, $-\frac{3}{4} + x + \frac{3}{4} = 1\frac{1}{4} + \frac{3}{4}$, $x + 0 = 2$; $x = 2$

C 43. $a - 12 = 15$, $a - 12 + 11 = 15 + 11$; $a - 1 = 26$

44. $8 = t + 3$, $8 - 20 = t + 3 - 20$; $-12 = t - 17$

45. $7 - 2x = 8$, $7 - 2x + 4 = 8 + 4$; $11 - 2x = 12$

46. $-3n - 5 = 17$, $-3n - 5 + 10 = 17 + 10$; $-3n + 5 = 27$

47. $x = |3|$; The solution set is {3}.   **48.** $y = |-4|$; The solution set is {4}.

49. $x - 6 = |-3|$, $x - 6 = 3$, $x - 6 + 6 = 3 + 6$, $x + 0 = 9$; $x = 9$; {9}

50. $|z| = |-4| - |-6|$, $|z| = 4 - 6$, $|z| = -2$. The solution set is Ø, since the absolute value of any real number is greater than or equal to 0.

51. Go 1 block east, 2 blocks north, 3 blocks west, and 7 blocks north.

52. Drive 2 blocks east to traffic light, make a right turn and drive 2 miles south on Rt. 9, and 6 miles east on Rt. 117.

53. Answers may vary.

54. Answers may vary. An example: For a replacement set of positive integers, there is no solution; for a replacement set of negative integers, the solution set is {−6}.

55. Fill the 3-qt measure and empty its contents into the 5-qt measure. Fill the 3-qt measure again and fill the 5-qt measure from it; this leaves 1 qt in the 3-qt measure. Empty the 5-qt measure. Pour in the 1 qt from the 3-qt measure. Fill the 3-qt measure and empty its contents into the 5-qt measure.

56. Everett and Alice do not play in the orchestra. Enid and Toni are not on the cycling team. Everett is not a cheerleader or on the cycling team. Enid is not a cheerleader. Use that information to make this chart.

	Orchestra	Cheerleader	Play	Cycling Team
Everett			✔	
Alice		✔	✔	✔
Toni	✔	✔	✔	
Enid	✔		✔	

Since the chart shows Everett is in the play, Enid must be in the orchestra, Toni must be the cheerleader, and Alice is on the cycling team.

page 101   Capsule Review

1. $\dfrac{1}{4} \cdot 4 = \dfrac{4}{4} = 1$

2. $\dfrac{-5}{-5} = 1$

3. $\dfrac{r}{5} \cdot 5 = r \cdot \dfrac{5}{5} = r(1) = r$

4. $-\dfrac{1}{8}a \cdot (-8) = \dfrac{-8}{-8}a = (1)a = a$

5. mult. by 9   **6.** div. by -15 **7.** mult. by -3 **8.** div. by 2.9

page 104   Class Exercises

1. div. by 5; $\dfrac{5x}{5} = \dfrac{30}{5}$; $x = 6$

2. div. by -6; $\dfrac{-6y}{-6} = \dfrac{48}{-6}$; $y = -8$

3. multiply by 5; $\dfrac{a}{5} = -4$, $5 \cdot \dfrac{a}{5} = 5 \cdot (-4)$; $a = -20$

4. multiply by -3; $6 = -\dfrac{1}{3}t$, $6(-3) = -3 \cdot -\dfrac{1}{3}t$; $t = -18$

5. divide by -1; $-24 = -y$, $\dfrac{-24}{-1} = \dfrac{-y}{-1}$; $y = 24$

6. multiply by $\dfrac{3}{2}$; $\dfrac{2}{3}x = 1$, $\dfrac{3}{2} \cdot \dfrac{2}{3}x = \dfrac{3}{2} \cdot 1$; $x = \dfrac{3}{2}$

7. divide by -1.01; $-1.01b = 6.06$, $\dfrac{-1.01b}{-1.01} = \dfrac{6.06}{-1.01}$; $b = -6$

8. multiply by -6; $\dfrac{x}{-6} = 5$, $-6 \cdot \dfrac{x}{-6} = -6(5)$; $x = -30$

A 1. $-9t = 72, \dfrac{-9t}{-9} = \dfrac{72}{-9}; t = -8$

2. $-17t = 51, \dfrac{-17t}{-17} = \dfrac{51}{-17}; t = -3$

3. $8a = 56, \dfrac{8a}{8} = \dfrac{56}{8}; a = 7$

4. $2x = 12, \dfrac{2x}{2} = \dfrac{12}{2}; x = 6$

5. $-24r = 120, \dfrac{-24r}{-24} = \dfrac{120}{-24}; r = -5$

6. $-5p = 75, \dfrac{-5p}{-5} = \dfrac{75}{-5}; p = -15$

7. $3 = -x, \dfrac{3}{-1} = \dfrac{-x}{-1}; x = -3$

8. $15 = -z, \dfrac{15}{-1} = \dfrac{-z}{-1}; z = -15$

9. $7 = -x, \dfrac{7}{-1} = \dfrac{-x}{-1}; x = -7$

10. $\dfrac{c}{25} = -1, 25 \cdot \dfrac{c}{25} = 25 \cdot -1; c = -25$

11. $\dfrac{x}{7} = -3, 7 \cdot \dfrac{x}{7} = 7 \cdot -3; x = -21$

12. $\dfrac{t}{3} = -12, 3 \cdot \dfrac{t}{3} = 3 \cdot -12; t = -36$

13. $-\dfrac{a}{4} = -6, -4 \cdot -\dfrac{a}{4} = -4 \cdot -6; a = 24$

14. $-\dfrac{m}{2} = -10, -2 \cdot -\dfrac{m}{2} = -2 \cdot -10; m = 20$

15. $\dfrac{s}{5} = 11, 5 \cdot \dfrac{s}{5} = 5 \cdot 11; s = 55$

16. $\dfrac{1}{2}e = 2, 2 \cdot \dfrac{1}{2}e = 2 \cdot 2; e = 4$

17. $\dfrac{3}{5}a = 12, \dfrac{5}{3} \cdot \dfrac{3}{5}a = \dfrac{5}{3} \cdot 12; a = 20$

18. $\dfrac{1}{4}x = 9, 4 \cdot \dfrac{1}{4}x = 4 \cdot 9; x = 36$

19. $\dfrac{1}{8}n = 3, 8 \cdot \dfrac{1}{8}n = 8 \cdot 3; n = 24$

20. $-\dfrac{2}{3}m = 4, -\dfrac{3}{2} \cdot -\dfrac{2}{3}m = -\dfrac{3}{2} \cdot 4; m = -6$

21. $-\dfrac{3}{7}t = 15, -\dfrac{7}{3} \cdot -\dfrac{3}{7}t = -\dfrac{7}{3} \cdot 15; t = -35$

22. $\dfrac{a}{12} = 11.5, 12 \cdot \dfrac{a}{12} = 12 \cdot 11.5; a = 12 \cdot 11.5$

23. $\dfrac{m}{15} = 10.5, 15 \cdot \dfrac{m}{15} = 15 \cdot 10.5; m = 15 \cdot 10.5$

24. $\dfrac{2x}{7} = -21.2, \dfrac{7}{2} \cdot \dfrac{2x}{7} = \dfrac{7}{2} \cdot -21.2; x = \dfrac{7 \cdot -21.2}{2}$

25. $\dfrac{5t}{11} = -55.5, \dfrac{11}{5} \cdot \dfrac{5}{11}t = \dfrac{11}{5} \cdot -55.5; t = \dfrac{11 \cdot -55.5}{5}$

26. $\dfrac{-2s}{3} = 33.5, \dfrac{3}{-2} \cdot \dfrac{-2s}{3} = \dfrac{3}{-2} \cdot 33.5; \; s = \dfrac{3 \cdot 33.5}{-2}$

27. $\dfrac{-5r}{6} = 24.6, \dfrac{6}{-5} \cdot \dfrac{-5r}{6} = \dfrac{6}{-5} \cdot 24.6; \; r = \dfrac{6 \cdot 24.6}{-5}$

B 28. $\dfrac{x}{5} = -1\dfrac{1}{10}, \dfrac{x}{5} = -\dfrac{11}{10}, 5 \cdot \dfrac{x}{5} = 5 \cdot -\dfrac{11}{10}; \; x = -\dfrac{11}{2} = -5\dfrac{1}{2}$

29. $\dfrac{m}{8} = -2\dfrac{1}{4}, \dfrac{m}{8} = -\dfrac{9}{4}, 8 \cdot \dfrac{m}{8} = 8 \cdot -\dfrac{9}{4}; \; m = -18$

30. $\dfrac{y}{9} = -\dfrac{5}{3}, 9 \cdot \dfrac{y}{9} = 9 \cdot -\dfrac{5}{3}; \; y = -15$

31. $\dfrac{z}{3} = -\dfrac{9}{3}, 3 \cdot \dfrac{z}{3} = 3 \cdot -\dfrac{9}{3}; \; z = -9$

32. $\dfrac{7}{9}d = \dfrac{14}{3}, \dfrac{9}{7} \cdot \dfrac{7}{9}d = \dfrac{9}{7} \cdot \dfrac{14}{3}; \; d = 6$

33. $\dfrac{2}{3}a = \dfrac{4}{9}, \dfrac{3}{2} \cdot \dfrac{2}{3}a = \dfrac{3}{2} \cdot \dfrac{4}{9}; \; a = \dfrac{2}{3}$

34. $-4.027m = 50.1, \dfrac{-4.027m}{-4.027} = \dfrac{50.1}{-4.027}; \; m = -12.441$

35. $-7.850n = 0.929, \dfrac{-7.850n}{-7.850} = \dfrac{0.929}{-7.850}; \; n = -0.118$

36. $-\dfrac{x}{5} = -2.5, -5 \cdot -\dfrac{x}{5} = -5(-2.5); \; x = 12.5$

37. $-\dfrac{z}{2} = -4.6, -2 \cdot -\dfrac{z}{2} = -2(-4.6); \; z = 9.2$

C 38. $3a = -11, 4 \cdot 3a = 4(-11); \; 12a = -44$

39. $2c = -5, -4 \cdot 2c = -4\,(-5); \; -8c = 20$

40. $|a| = 30, a = 30 \;$ or $\; a = -30; \{30, -30\}$

41. $-12|x| = -144, \dfrac{-12\,|x|}{-12} = \dfrac{-144}{-12}, |x| = 12, x = 12 \;$ or $\; x = -12; \{12, -12\}$

42. $\dfrac{|c|}{3} = 15, 3 \cdot \dfrac{|c|}{3} = 3 \cdot 15, |c| = 45, c = 45 \;$ or $\; c = -45; \{45, -45\}$

43. $-6|2t| = -8, \dfrac{-6|2t|}{-6} = \dfrac{-8}{-6}, |2t| = \dfrac{4}{3}, 2t = \dfrac{4}{3} \;$ or $\; 2t = -\dfrac{4}{3}, t = \dfrac{2}{3} \;$ or

$t = -\dfrac{2}{3}; \left\{\dfrac{2}{3}, -\dfrac{2}{3}\right\}$

44. $\dfrac{m}{4.5} = 33, 4.5 \cdot \dfrac{m}{4.5} = 4.5 \cdot 33, m = 4.5 \cdot 33; \; m = 148.5$

45. $\dfrac{r}{29} = 1.1, 29 \cdot \dfrac{r}{29} = 29 \cdot 1.1, r = 29 \cdot 1.1; \; r = 31.9$

46. $0.89 = \frac{1}{6}y$, $6 \cdot 0.89 = 6 \cdot \frac{1}{6}y$, $y = 6 \cdot 0.89$; $y = 5.34$

47. $17.25x \div 17.25 = 50.59 \div 17.25$, $x = 50.59 \div 17.25$; $x = 2.93$

48. $81.23m \div 81.23 = 62.4 \div 81.23$, $m = 62.4 \div 81.23$; $m = 0.77$

49. $-2.98 = \frac{x}{5.023}$, $5.023\,(-2.98) = 5.023 \cdot \frac{x}{5.023}$, $x = 5.023\,(-2.98)$; $x = -14.97$

50. $\frac{28}{56.7} = 8.14c$, $\frac{1}{8.14} \cdot \frac{28}{56.7} = \frac{1}{8.14} \cdot 8.14c$, $c = \frac{28}{8.14 \cdot 56.7}$; $c = 0.06$

51. $\frac{5}{9.2} = \frac{n}{36.74}$, $36.74 \cdot \frac{5}{9.2} = 36.74 \cdot \frac{n}{36.74}$, $n = \frac{36.74 \cdot 5}{9.2}$; $n = 19.97$

52. 25

53. $(2 - 9)^2 = (-7)^2 = 49$

54. $16(-2)^3 = 16(-8) = -128$

55. $\frac{n^2}{4} = 25$

56. One less than three times a number is 23.

57. $4(-2)^3(5)^2 = 4(-8)(25) = -800$

58. $-[5 - (-3)^2] = -(5 - 9) = -(-4) = 4$

59. $\frac{(5)^2(4)}{2(-2) - 3(-3)} = \frac{(25)(4)}{(-4) + 9} = \frac{100}{5} = 20$

60. $\frac{(-2)^4 - (-2)^3}{(-2)^2 - (-2)} = \frac{16 - (-8)}{4 - (-2)} = \frac{24}{6} = 4$

61. Dividing both sides of an equation by c, $c \ne 0$, has the same effect as multiplying both sides by $\frac{1}{c}$.

62. Answers may vary. An example: find the value of the variable that makes the equation true.

page 106   Capsule Review

1. subtract 2; divide by 3

2. subtract 24; divide by -4

3. add 3; multiply by 4

4. add 18; multiply by $\frac{3}{2}$

page 108   Class Exercises

1. subtract 8   **2.** add 15

3. add 19   **4.** subtract 11

5. add 3; divide by 2

6. subtract 7; divide by 3

7. subtract 11; divide by -4

8. subtract 15; divide by -1.3

9. subtract 28; divide by 4

10. add 4.7; multiply by $\frac{4}{3}$

11. add 15; divide by 3   **12.** add 20; multiply by 2   **13.** subtract 2; divide by -1

14. $3x + 2 = 20$, $3x + 2 - 2 = 20 - 2$, $3x = 18$, $\dfrac{3x}{3} = \dfrac{18}{3}$; $x = 6$

15. $7m - 1 = 34$, $7m - 1 + 1 = 34 + 1$, $7m = 35$, $\dfrac{7m}{7} = \dfrac{35}{7}$; $m = 5$

16. $31 = 3 - 4h$, $31 - 3 = 3 - 4h - 3$, $28 = -4h$, $\dfrac{28}{-4} = \dfrac{-4h}{-4}$; $h = -7$

17. $2.5 = 1.3 - r$, $2.5 - 1.3 = 1.3 - r - 1.3$, $1.2 = -r$, $\dfrac{1.2}{-1} = \dfrac{-r}{-1}$; $r = -1.2$

18. $\dfrac{1}{2}y + 5 = -12$, $\dfrac{1}{2}y + 5 - 5 = -12 - 5$, $\dfrac{1}{2}y = -17$, $\dfrac{2}{1} \cdot \dfrac{1}{2}y = \dfrac{2}{1}(-17)$; $y = -34$

19. $\dfrac{2}{3}x - 9 = -21$, $\dfrac{2}{3}x - 9 + 9 = -21 + 9$, $\dfrac{2}{3}x = -12$, $\dfrac{3}{2} \cdot \dfrac{2}{3}x = \dfrac{3}{2} \cdot -12$; $x = -18$

20. $-11h - 7 = -18$, $-11h - 7 + 7 = -18 + 7$, $-11h = -11$, $\dfrac{-11h}{-11} = \dfrac{-11}{-11}$; $h = 1$

21. $\dfrac{x}{5} + 15 = 0$, $\dfrac{x}{5} + 15 - 15 = 0 - 15$, $\dfrac{x}{5} = -15$, $5 \cdot \dfrac{x}{5} = 5(-15)$; $x = -75$

22. $-8 - x = 11$, $-8 - x + 8 = 11 + 8$, $-x = 19$, $\dfrac{-x}{-1} = \dfrac{19}{-1}$; $x = -19$

pages 108–109   Practice Exercises

A   **1.** $3a - 1 = 7$, $3a - 1 + 1 = 7 + 1$, $3a = 8$, $\dfrac{3a}{3} = \dfrac{8}{3}$; $a = \dfrac{8}{3}$

2. $2y - 18 = 44$, $2y - 18 + 18 = 44 + 18$, $2y = 62$, $\dfrac{2y}{2} = \dfrac{62}{2}$; $y = 31$

3. $3x - 1 = 8$, $3x - 1 + 1 = 8 + 1$, $3x = 9$, $\dfrac{3x}{3} = \dfrac{9}{3}$; $x = 3$

4. $23 = 4n + 11$, $23 - 11 = 4n + 11 - 11$, $12 = 4n$, $\dfrac{12}{4} = \dfrac{4n}{4}$; $n = 3$

5. $-14 = 10 + 6q$, $-14 - 10 = 10 + 6q - 10$, $-24 = 6q$, $\dfrac{-24}{6} = \dfrac{6q}{6}$; $q = -4$

6. $0 = -9t - 27$, $0 + 27 = -9t - 27 + 27$, $27 = -9t$, $\dfrac{27}{-9} = \dfrac{-9t}{-9}$; $t = -3$

7. $20 - 5y = 45$, $20 - 5y - 20 = 45 - 20$, $-5y = 25$, $\dfrac{-5y}{-5} = \dfrac{25}{-5}$; $y = -5$

8. $25 - 3c = 36$, $25 - 3c - 25 = 36 - 25$, $-3c = 11$, $\dfrac{-3c}{-3} = \dfrac{11}{-3}$;

$c = -\dfrac{11}{3}$

9. $14 - 2x = 18$, $14 - 2x - 14 = 18 - 14$, $-2x = 4$, $\dfrac{-2x}{-2} = \dfrac{4}{-2}$; $x = -2$

10. $34 = 13 - 4p$, $34 - 13 = 13 - 4p - 13$, $21 = -4p$, $\dfrac{21}{-4} = \dfrac{-4p}{-4}$; $p = -\dfrac{21}{4}$

11. $-7 = 11 + 3b$, $-7 - 11 = 11 + 3b - 11$, $-18 = 3b$, $\dfrac{-18}{3} = \dfrac{3b}{3}$; $b = -6$

12. $15 = 25 + \dfrac{w}{5}$, $15 - 25 = 25 + \dfrac{w}{5} - 25$, $-10 = \dfrac{w}{5}$, $-10 \cdot 5 = \dfrac{w}{5} \cdot 5$;

$w = -50$

13. $\dfrac{4}{9}x - 13 = -5$, $\dfrac{4}{9}x - 13 + 13 = -5 + 13$, $\dfrac{4}{9}x = 8$, $\dfrac{9}{4} \cdot \dfrac{4}{9}x = \dfrac{9}{4} \cdot 8$; $x = 18$

14. $21 + \dfrac{a}{3} = 10$, $21 + \dfrac{a}{3} - 21 = 10 - 21$, $\dfrac{a}{3} = -11$, $3 \cdot \dfrac{a}{3} = 3(-11)$; $a = -33$

15. $17 = 9 - \dfrac{6}{5}y$, $17 - 9 = 9 - \dfrac{6}{5}y - 9$, $8 = -\dfrac{6}{5}y$, $-\dfrac{5}{6} \cdot 8 = -\dfrac{5}{6}\left(-\dfrac{6}{5}y\right)$;

$y = -\dfrac{20}{3}$

16. $2(x - 4) = 26$, $2x - 8 = 26$, $2x - 8 + 8 = 26 + 8$, $2x = 34$, $\dfrac{2x}{2} = \dfrac{34}{2}$;

$x = 17$

17. $5(3 - x) = 40$, $15 - 5x = 40$, $15 - 5x - 15 = 40 - 15$, $-5x = 25$,

$\dfrac{-5x}{-5} = \dfrac{25}{-5}$; $x = -5$

18. $\dfrac{1}{4}(x - 24) = 13$, $\dfrac{1}{4}x - 6 = 13$, $\dfrac{1}{4}x - 6 + 6 = 13 + 6$, $\dfrac{1}{4}x = 19$, $\dfrac{4}{1} \cdot \dfrac{1}{4}x =$

$\dfrac{4}{1} \cdot 19$; $x = 76$

19. $\dfrac{1}{3}(x + 27) = 4$, $\dfrac{1}{3}x + 9 = 4$, $\dfrac{1}{3}x + 9 - 9 = 4 - 9$, $\dfrac{1}{3}x = -5$, $\dfrac{3}{1} \cdot \dfrac{1}{3}x =$

$\dfrac{3}{1}(-5)$; $x = -15$

20. $-7(3 + 2x) = 84$, $-21 - 14x = 84$, $-21 - 14x + 21 = 84 + 21$, $-14x =$

105, $\dfrac{-14x}{-14} = \dfrac{105}{-24}$; $x = -\dfrac{15}{2}$

21. $6(5 - 3x) = 84$, $30 - 18x = 84$, $30 - 18x - 30 = 84 - 30$, $-18x = 54$,

$\dfrac{-18x}{-18} = \dfrac{54}{-18}$; $x = -3$

22. $-45 = 3(1 - 4d)$, $-45 = 3 - 12d$, $-45 - 3 = 3 - 12d - 3$, $-48 = -12d$, $\dfrac{-48}{-12} = \dfrac{-12d}{-12}$; $d = 4$

23. $-1 = \dfrac{1}{4}(x - 20)$, $-1 = \dfrac{1}{4}x - 5$, $-1 + 5 = \dfrac{1}{4}x - 5 + 5$, $4 = \dfrac{1}{4}x$, $4 \cdot 4 = 4 \cdot \dfrac{1}{4}x$; $x = 16$

24. $12 = \dfrac{1}{2}(18 - g)$, $12 = 9 - \dfrac{1}{2}g$, $12 - 9 = 9 - \dfrac{1}{2}g - 9$, $3 = -\dfrac{1}{2}g$, $-\dfrac{2}{1} \cdot 3 = -\dfrac{2}{1}\left(-\dfrac{1}{2}g\right)$; $g = -6$

B 25. $-\dfrac{3}{4} = \dfrac{12}{4} - \dfrac{1}{2}x$, $-\dfrac{3}{4} - \dfrac{12}{4} = \dfrac{12}{4} - \dfrac{1}{2}x - \dfrac{12}{4}$, $-\dfrac{15}{4} = -\dfrac{1}{2}x$, $-2\left(-\dfrac{15}{4}\right) = -2\left(-\dfrac{1}{2}x\right)$; $x = \dfrac{15}{2}$

26. $-\dfrac{5}{6} = \dfrac{5}{6} - \dfrac{2}{3}x$, $-\dfrac{5}{6} - \dfrac{5}{6} = \dfrac{5}{6} - \dfrac{2}{3}x - \dfrac{5}{6}$, $-\dfrac{5}{3} = -\dfrac{2}{3}x$, $-\dfrac{3}{2}\left(-\dfrac{5}{3}\right) = -\dfrac{3}{2}\left(-\dfrac{2}{3}x\right)$; $x = \dfrac{5}{2}$

27. $\dfrac{2}{9} = \dfrac{1}{3} - \dfrac{4}{9}x$, $\dfrac{2}{9} - \dfrac{1}{3} = \dfrac{1}{3} - \dfrac{4}{9}x - \dfrac{1}{3}$, $\dfrac{2}{9} - \dfrac{3}{9} = -\dfrac{4}{9}x$, $-\dfrac{1}{9} = -\dfrac{4}{9}x$, $-\dfrac{9}{4}\left(-\dfrac{1}{9}\right) = -\dfrac{9}{4}\left(-\dfrac{4}{9}x\right)$; $x = \dfrac{1}{4}$

28. $\dfrac{1}{2} = \dfrac{5}{4} - \dfrac{3}{2}x$, $\dfrac{1}{2} - \dfrac{5}{4} = \dfrac{5}{4} - \dfrac{3}{2}x - \dfrac{5}{4}$, $\dfrac{2}{4} - \dfrac{5}{4} = -\dfrac{3x}{2}$, $-\dfrac{3}{4} = -\dfrac{3x}{2}$, $-\dfrac{2}{3}\left(-\dfrac{3}{4}\right) = -\dfrac{2}{3}\left(-\dfrac{3}{2}x\right)$; $x = \dfrac{1}{2}$

29. $\dfrac{1}{5} = \dfrac{2}{3} + \dfrac{3}{5}x$, $\dfrac{1}{5} - \dfrac{2}{3} = \dfrac{2}{3} + \dfrac{3}{5}x - \dfrac{2}{3}$, $\dfrac{3}{15} - \dfrac{10}{15} = \dfrac{3}{5}x$, $-\dfrac{7}{15} = \dfrac{3}{5}x$, $\dfrac{5}{3}\left(-\dfrac{7}{15}\right) = \dfrac{5}{3}\left(\dfrac{3}{5}x\right)$; $x = -\dfrac{7}{9}$

30. $\dfrac{1}{8} = \dfrac{3}{4} + \dfrac{1}{8}x$, $\dfrac{1}{8} - \dfrac{3}{4} = \dfrac{3}{4} + \dfrac{1}{8}x - \dfrac{3}{4}$, $\dfrac{1}{8} - \dfrac{6}{8} = \dfrac{1}{8}x$, $-\dfrac{5}{8} = \dfrac{1}{8}x$, $8\left(-\dfrac{5}{8}\right) = 8\left(\dfrac{1}{8}x\right)$; $x = -5$

31. $49w - 186 = 5351$, $49w - 186 + 186 = 5351 + 186$, $49w = 5537$, $\dfrac{49w}{49} = \dfrac{5537}{49}$; $w = 113$

32. $44x - 728 = 1736$, $44x - 728 + 728 = 1736 + 728$, $44x = 2454$, $\dfrac{44x}{44} =$

$\dfrac{2464}{44}$; $x = 56$

33. $\dfrac{x}{3} + 3 = 42$, $\dfrac{x}{3} + 3 - 3 = 42 - 3$, $\dfrac{x}{3} = 39$, $3\left(\dfrac{x}{3}\right) = 3(39)$; $x = 117$

34. $\dfrac{f}{34} + 16 = 35$, $\dfrac{f}{34} + 16 - 16 = 35 - 16$, $\dfrac{f}{34} = 19$, $34\left(\dfrac{f}{34}\right) = 34(19)$;

$f = 646$

35. $-\dfrac{7}{12} = \dfrac{5}{12} - \dfrac{2}{3}x$, $-\dfrac{7}{12} - \dfrac{5}{12} = \dfrac{5}{12} - \dfrac{2}{3}x - \dfrac{5}{12}$, $-\dfrac{12}{12} = -\dfrac{2}{3}x$, $-1 = -\dfrac{2}{3}x$,

$-\dfrac{3}{2}(-1) = -\dfrac{3}{2}\left(-\dfrac{2}{3}x\right)$; $x = \dfrac{3}{2}$

36. $-\dfrac{3}{24} = \dfrac{5}{24} - \dfrac{4}{12}x$, $-\dfrac{3}{24} - \dfrac{5}{24} = \dfrac{5}{24} - \dfrac{4}{12}x - \dfrac{5}{24}$, $-\dfrac{8}{24} = -\dfrac{4}{12}x$, $-\dfrac{1}{3} = -\dfrac{1}{3}x$,

$-3\left(-\dfrac{1}{3}\right) = -3\left(-\dfrac{1}{3}x\right)$; $x = 1$

C 37. $2|x| - 7 = 1$, $2|x| - 7 + 7 = 1 + 7$, $2|x| = 8$, $\dfrac{1}{2} \cdot 2|x| = \dfrac{1}{2} \cdot 8$, $|x| = 4$; $x =$

4 or $x = -4$

38. $3|n| + 6 = -3$, $3|n| + 6 - 6 = -3 - 6$, $3|n| = -9$, $\dfrac{1}{3} \cdot 3|n| = \dfrac{1}{3}(-9)$, $|n| =$

-3; no solution

39. $5|t| - 10 = 0$, $5|t| - 10 + 10 = 0 + 10$, $5|t| = 10$, $\dfrac{1}{5} \cdot 5|t| = \dfrac{1}{5}(10)$, $|t| = 2$;

$t = 2$ or $t = -2$

40. $9|x| - 7 = 7$, $9|x| - 7 + 7 = 7 + 7$, $9|x| = 14$, $\dfrac{1}{9} \cdot 9|x| = \dfrac{1}{9}(14)$, $|x| = \dfrac{14}{9}$;

$x = \dfrac{14}{9}$ or $x = -\dfrac{14}{9}$

41. $-5|x| + 7 = 2$, $-5|x| + 7 - 7 = 2 - 7$, $-5|x| = -5$, $\dfrac{-5|x|}{-5} = \dfrac{-5}{-5}$, $|x| = 1$;

$x = 1$ or $x = -1$

42. $-7|r| - 8 = -1$, $-7|r| - 8 + 8 = -1 + 8$, $-7|r| = 7$, $\dfrac{-7|r|}{-7} = \dfrac{7}{-7}$, $|r| = -1$;

no solution

43. $4x + 2 = 14$, $\dfrac{1}{2}(4x + 2) = \dfrac{1}{2} \cdot 14$, $2x + 1 = 7$

44. $3x - 1 = 17$, $2(3x - 1) = 2 \cdot 17$, $6x - 2 = 34$

45. $x + 7 = 8$, $x + 7 - 14 = 8 - 14$, $x - 7 = -6$, $(-1)(x - 7) = (-1)(-6)$, $-x + 7 = 6$, $7 - x = 6$

46. $2x + 5 = 21$, $2x + 5 - 5 = 21 - 5$, $2x = 16$, $4(2x) = 4 \cdot 16$, $8x = 64$

47. $583r + 23.58 = 2.79$, $583r + 23.58 - 23.58 = 2.79 - 23.58$, $583r = 2.79 - 23.58$, $r = \dfrac{2.79 - 23.58}{583} \approx -0.036$; -0.04

48. $-51.5 = 29m - 4.06$, $-51.5 + 4.06 = 29m - 4.06 + 4.06$, $-51.5 + 4.06 = 29m$, $m = \dfrac{-51.5 + 4.06}{29} \approx -1.635$; -1.64

49. $\dfrac{x}{0.24} - 0.03 = -0.14$, $\dfrac{x}{0.24} - 0.03 + 0.03 = -0.14 + 0.03$, $\dfrac{x}{0.24} = -0.14 + 0.03$, $x = (-0.14 + 0.03)\,0.24 \approx -0.0264$; -0.03

50. $\dfrac{a}{2.5} + 11.9 = 0.02$, $\dfrac{a}{2.5} + 11.9 - 11.9 = 0.02 - 11.9$, $\dfrac{a}{2.5} = 0.02 - 11.9$, $a = (0.02 - 11.9)\,2.5 \approx -29.7$; -29.70

page 109   Test Yourself

1. $y + 11 = 4$, $y + 11 - 11 = 4 - 11$; $y = -7$

2. $-7.6 = x - 1.4$, $-7.6 + 1.4 = x - 1.4 + 1.4$; $x = -6.2$

3. $-\dfrac{3}{4} + z = \dfrac{5}{2}$, $-\dfrac{3}{4} + z + \dfrac{3}{4} = \dfrac{5}{2} + \dfrac{3}{4}$, $z = \dfrac{10}{4} + \dfrac{3}{4}$; $z = \dfrac{13}{4}$

4. $9t = 30$, $\dfrac{9t}{9} = \dfrac{30}{9}$; $t = \dfrac{10}{3}$

5. $24 = -\dfrac{2}{5}m$, $-\dfrac{5}{2} \cdot 24 = -\dfrac{5}{2}\left(-\dfrac{2}{5}m\right)$; $m = -60$

6. $\dfrac{x}{7} = -3.15$, $7 \cdot \dfrac{x}{7} = 7(-3.15)$; $x = -22.05$

7. $6k + 5 = 23$, $6k + 5 - 5 = 23 - 5$, $6k = 18$, $\dfrac{6k}{6} = \dfrac{18}{6}$; $k = 3$

8. $5 - \dfrac{j}{12} = 1$, $5 - \dfrac{j}{12} - 5 = 1 - 5$, $-\dfrac{j}{12} = -4$, $-12\left(-\dfrac{j}{12}\right) = -12(-4)$; $j = 48$

9. $-7.13 = 0.15p - 7.13$, $-7.13 + 7.13 = 0.15p - 7.13 + 7.13$; $0 = 0.15p$, $\dfrac{0}{0.15} = \dfrac{0.15p}{0.15}$; $p = 0$

   1. $64 - n$

   2. Let l represent the length, w represent the width, and A represent the area of a rectangular floor, then $A = lw$.

   3. line segment   **4.** $95d$ **5.** $d = rt$ or $d = 35t$

pages 111–112   Practice Exercises

A **1.** Let x represent the number of records bought, then $7x + 3 = 24$.

   $7x + 3 - 3 = 24 - 3$, $7x = 21$, $\dfrac{7x}{7} = \dfrac{21}{7}$; $x = 3$; Jane bought 3 records.

   2. Let x represent the number of guests at each table, then $4x = 24$. $\dfrac{4x}{4} = \dfrac{24}{4}$; $x = 6$; Each table will accomodate 6 guests.

   3. Let x represent the number of books packed into each box, then $5x = 40$.

   $\dfrac{5x}{5} = \dfrac{40}{5}$; $x = 8$; The number of books packed into each box is 8, with 0 books left over.

B **4.** Let p represent the number of people seated at each of the other tables,

   then $8p + 3 = 35$. $8p + 3 - 3 = 35 - 3$, $8p = 32$, $\dfrac{8p}{8} = \dfrac{32}{8}$; $p = 4$; There were 4 people seated per table.

   5. Let e represent how much Lois earned per hour, then $4e + 5 = 25$.

   $4e + 5 - 5 = 25 - 5$, $4e = 20$, $\dfrac{4e}{4} = \dfrac{20}{4}$; $e = 5$; Lois earns \$5 per hour.

   6. Let n represent the number, then $7n + 5 = 33$. $7n + 5 - 5 = 33 - 5$,

   $7n = 28$, $\dfrac{7n}{7} = \dfrac{28}{7}$; $n = 4$; The number is 4.

   7. Let x represent the original price of each ticket, then $15x - 25 = 155$.

   $15x - 25 + 25 = 155 + 25$, $15x = 180$, $\dfrac{15x}{15} = \dfrac{180}{15}$; $x = 12$; The original price of each ticket was \$12.

C For Exercises 8–11, answers may vary. Possible answers are:

   8. The product of one half and b equals eight.

   9. Five times y less fourteen equals twenty-seven.

   10. a less three is less than seven.

   11. One seventh times m plus one is greater than nineteen.

1. Total gain $= 400 + (-700) + 1000 + 200 + (-500) + 100 = 500$; the total gain is \$500; the average gain is $500 \div 6 = 83.33$ or \$83.

2. a. $2n + 5 = 11, 2n + 5 - 5 = 11 - 5, 2n = 6, \dfrac{2n}{2} = \dfrac{6}{2}; n = 3$

  b. $n - 8 = 2; n = 10$

  c. $\dfrac{20}{a} = 4, 20 = 4a, \dfrac{20}{4} = \dfrac{4a}{4}; a = 5$

  d. $b + 9b = 10, 10b = 10, \dfrac{10b}{10} = \dfrac{10}{10}; b = 1$

page 112   Project

Answers may vary.

page 113   Capsule Review

1. Comm. prop. for add.   **2.** Distrib. prop.   **3.** Assoc. prop. for add.

4. Ident. prop. for mult.   **5.** Ident. prop. for add.   **6.** Distrib. prop.

page 115   Class Exercises

1. Assoc. prop. for add.   **2.** 0; Inv. prop. for add.   **3.** Ident. prop. for add.

pages 115–116   Practice Exercises

A   **1.**

Statements	Reasons
1. $a = b, c \neq 0$	1. Given
2. $\dfrac{a}{c} = \dfrac{a}{c}$	2. Reflexive property
3. $\dfrac{a}{c} = \dfrac{b}{c}$	3. Substitution property

2.

Statements	Reasons
1. $-(x - y) + (y - x)$ $= -[x + (-y)] + (y - x)$	1. Definition of subtraction
2. $= -1[x + (-y)] + (y - x)$	2. Property of -1 for multiplication
3. $= (-1)(x) + (-1)(-y) + (y - x)$	3. Distributive property
4. $= -x + y + (y - x)$	4. Property of -1 for multiplication
5. $= y + y - x - x$	5. Commutative property for addition
6. $= 2y - 2x$	6. Combine like terms
7. $2(y - x)$	7. Distributive property

3.

Statements	Reasons
1. $a = b$	1. Given
2. $a - c = a - c$	2. Reflexive prop.
3. $a - c = b - c$	3. Substitution prop.

4.

Statements	Reasons
1. $x = y$	1. Given
2. $z - x = z - x$	2. Reflexive prop.
3. $z - x = z - y$	3. Substitution prop.

B 5.

Statements	Reasons
1. $a + [b + (-a)] = a + [(-a) + b]$	1. Commutative property for addition
2. $\qquad = [a + (-a)] + b$	2. Associative property for addition
3. $\qquad = 0 + b$	3. Inverse property for addition
4. $\qquad = b$	4. Identity property for addition

6.

Statements	Reasons
1. $x + [-(x + y)] = x + [(-x) + (-y)]$	1. Distributive property
2. $\qquad = [x + (-x)] + (-y)$	2. Associative property for addition
3. $\qquad = 0 + (-y)$	3. Inverse property for addition
4. $\qquad = -y$	4. Identity property for addition

C 7.

Statements	Reasons
1. $a = b$	1. Given
2. $a - d = a - d$	2. Reflexive property
3. $\qquad = b - d$	3. Substitution property
4. $\qquad = b + (-d)$	4. Definition of subtraction
5. $\qquad = -d + b$	5. Commutative property for addition
6. $\qquad = -d - (-b)$	6. Definition of subtraction
7. $\qquad = (-1)(d) - (-1)(b)$	7. Property of -1 for multiplication
8. $\qquad = (-1)(d - b)$	8. Distributive property
9. $\qquad = -(d - b)$	9. Property of -1 for multiplication

8.

Statements	Reasons
1. a is a real number	1. Given
2. $a + 0 = a$	2. Identity property for addition
3. $a(a + 0) = (a)(a)$	3. Multiplication property
4. $(a)(a) + (a)(0) = (a)(a)$	4. Distributive property
5. $(a)(a) + (a)(0) - (a)(a) = 0$	5. Subtraction property
6. $(a)(a) + (a)(0) + [-(a)(a)] = 0$	6. Definition of subtraction
7. $(a)(a) + [-(a)(a)] + (a)(0) = 0$	7. Commutative property for addition
8. $0 + (a)(0) = 0$	8. Inverse property for addition
9. $(a)(0) = 0$	9. Identity property for addition

9. transitive   **10.** transitive   **11.** symmetric   **12.** symmetric

13. The fewest number of moves is 6.

page 117   Capsule Review

1. $-49 = 7x, \dfrac{-49}{7} = \dfrac{7x}{7}; x = -7$

2. $\dfrac{1}{2}x = -35, 2 \cdot \dfrac{1}{2}x = 2(-35); x = -70$

3. $\dfrac{x}{5} + 6 - 6 = 11 - 6, \dfrac{x}{5} = 5; x = 25$

4. $-4x - 9 = 31, -4x - 9 + 9 = 31 + 9, -4x = 40, \dfrac{-4x}{-4} = \dfrac{40}{-4}; x = -10$

5. $13 - 13x = 0, 13 - 13x - 13 = 0 - 13, -13x = -13, \dfrac{-13x}{-13} = \dfrac{-13}{-13}; x = 1$

6. $\dfrac{1}{3}x - 9 = -6, \dfrac{1}{3}x - 9 + 9 = -6 + 9, \dfrac{1}{3}x = 3, 3 \cdot \dfrac{1}{3}x = 3(3); x = 9$

7. $16.2 = 2x + 6.2, 16.2 - 6.2 = 2x + 6.2 - 6.2, 10 = 2x, \dfrac{10}{2} = \dfrac{2x}{2}; x = 5$

8. $0.7x - 5.6 = 0, 0.7x - 5.6 + 5.6 = 0 + 5.6, 0.7x = 5.6, \dfrac{0.7}{0.7}x = \dfrac{5.6}{0.7}; x = 8$

page 118   Class Exercises

1. $A = \dfrac{1}{2}bh, 16 = \dfrac{1}{2}(4)h, 16 = 2h; 8 = h$; The height of the triangle is 8 units.

2. $A = p + prt, A = 800 + 800(0.04)\left(\dfrac{1}{4}\right), A = 800 + 8, A = 808$; The amount is $808.00.

3. $S = \dfrac{n}{2}(a + l), 52 = \dfrac{n}{2}(-4 + 12), 104 = n(8); 13 = n$; There are 13 terms.

4. $V = lwh, 3.6 = l(0.8)(3), 3.6 = 2.4l; 1.5 = l$; The length is 1.5 units.

5. $F = \dfrac{9}{5}C + 32, -31 = \dfrac{9}{5}C + 32, -63 = \dfrac{9}{5}C; -35 = C$; The temperature is $-35°$C.

pages 118–119   Practice Exercises

A **1.** $P = 2l + 2w, 15.5 = 2l + 2(4.25), 15.5 = 2l + 8.5, 7 = 2l; 3.5 = l$; The length of the rectangle is 3.5 units.

2. $P = 2l + 2w, 17.6 = 2(3.3) + 2w, 17.6 = 6.6 + 2w, 11.0 = 2w; 5.5 = w$; The width of the rectangle is 5.5 units.

3. $V = \pi r^2 h$, $168 = 3.14(3^2)h$, $168 = 3.14(9)h$, $168 = 28.26h$; $5.9 = h$; The height is 5.9 units.

4. $V = \pi r^2 h$, $235.5 = 3.14(5^2)h$, $235.5 = 3.14(25)h$, $235.5 = 78.5h$; $3 = h$; The height is 3 units.

B 5. $d = rt$, $28 = (r)(4)$; $r = 7$; rate: 7.   6. $d = rt$, $45 = 5t$; $t = 9$; time: 9.

7. $V = lwh$, $36 = 4(3)h$, $36 = 12h$; $h = 3$; The height is 3 units.

8. $V = lwh$, $120 = (5)(w)(3)$, $120 = 15w$; $8 = w$; The width is 8 units.

9. $P = 2l + 2w$, $240 = 2(10) + 2w$, $240 = 20 + 2w$, $220 = 2w$; $w = 110$; The width is 110 units.

10. $P = 2l + 2w$, $360 = 2l + 2(6)$, $360 = 2l + 12$, $348 = 2l$; $l = 174$; The length is 174 units.

C 11. $A = \dfrac{b}{2} + i - 1$, $15 = \dfrac{b}{2} + 10 - 1$, $15 = \dfrac{b}{2} + 9$, $15 - 9 = \dfrac{b}{2} + 9 - 9$, $6 = \dfrac{b}{2}$,

    $6 \cdot 2 = 2 \cdot \dfrac{b}{2}$; $b = 12$; The value of b is 12.

12. $V = \pi r^2 h$, $420 = 3.14(3^2)h$, $420 = 3.14(9)h$, $420 = 28.26h$; $h \approx 14.9$; The height is approximately 14.9 units.

13. $d = rt$, $334.25 = r\left(3\dfrac{1}{2}\right)$, $334.25 = \dfrac{7}{2}r$; $r = 95.5$; average speed: 95.5 km/h.

14. $C = 2\pi r$, $\pi \approx 3.14$, $219.8 = 2(3.14)r$, $219.8 = 6.28r$; $35 = r$; The length of the radius is approximately 35 cm.

15. $18 - 3 \cdot 4 + 10 \div 2 = 18 - 12 + 5 = 11$

16. $15 + 6(1 - 4) = 15 + 6(-3) = 15 - 18 = -3$

17. $5^2 - 3(9 - 11)^3 = 25 - 3(-2)^3 = 25 - 3(-8) = 25 + 24 = 49$

18. $\dfrac{6 - 3(-9)}{2^2 + 7} = \dfrac{6 + 27}{4 + 7} = \dfrac{33}{11} = 3$ 19. $\dfrac{(2^2)(5^2)}{6^2 + 8^2} = \dfrac{(4)(25)}{36 + 64} = \dfrac{100}{100} = 1$

20. $\dfrac{(15 - 4^2)^5}{-3^2} = \dfrac{(15 - 16)^5}{-9} = \dfrac{(-1)^5}{-9} = \dfrac{-1}{-9} = \dfrac{1}{9}$

21. $5(2n - 1) + 3(6 - n) = 10n - 5 + 18 - 3n = 7n + 13$

22. $8 - 6x - (6x + 4) = 8 - 6x - 6x - 4 = -12x + 4$

23. $9(5y + 1) - (10 - y) = 45y + 9 - 10 + y = 46y - 1$

24. Since the area and length are known, and $A = l \cdot w$, the width is $\dfrac{A}{l} = \dfrac{x}{10}$ cm.

    The perimeter can then be found by adding two lengths and two widths:

    $10 + 10 + \dfrac{x}{10} + \dfrac{x}{10}$. The perimeter is $20 + \dfrac{x}{5}$ cm.

25. Answers may vary.

1. $n - 59$, $n - 59$ **2.** 564   **3.** $1.59n$ **4.** 148

5. Let n represent the number. $n + (-32) = 79$, $n - 32 = 79$, $n - 32 + 32 = 79 + 32$; $n = 111$; The number is 111.

6. Let x represent the amount in Charles' savings account. $3x + 14.95 = 1809.46$, $3x = 1794.51$; $x = 598.17$; Charles has $598.17 in his account.

7. An equation and an inequality are two types of open sentences.

8. A solution makes an open sentence true.   **9.** The word "is" implies equality.

10. An equation is the mathematical term for a relationship of equality.

11. Yes, they express relationships and display information algebraically.

pages 122–124   Practice Exercises

A   **1.** Let x represent the money Jason earned in one week, then $\frac{1}{3}x = 10$.

2. Let x represent the money Mary earned, then $\frac{1}{4}x = 25$.

3. Let w represent the weight of the elephant, then $2w + 3 = 7$.

4. Let w represent the weight of the Shot Put, then $3w + 2 = 18$.

5. Let n represent the number; $n - 92 = -28$, $n - 92 + 92 = -28 + 92$; $n = 64$; The number is 64.

6. Let n represent the number; $n - 47 = -15$, $n - 47 + 47 = -15 + 47$; $n = 32$; The number is 32.

7. Let n represent the number; $11n = -165$, $\frac{11n}{11} = \frac{-165}{11}$; $n = -15$; -15.

8. Let n represent the number; $15n = -240$, $\frac{15n}{15} = \frac{-240}{15}$; $n = -16$; -16.

9. Let n represent a number; $n + 7.13 = 2.09$, $n + 7.13 - 7.13 = 2.09 - 7.13$; $n = -5.04$; The number is -5.04.

10. Let n represent a number; $n + 4.12 = 3.15$, $n + 4.12 - 4.12 = 3.15 - 4.12$; $n = -0.97$; The number is -0.97.

11. Let x represent the age of Tom's mother. Tom's age of 16 years is $\frac{2}{5}$ of his mother's age; $16 = \frac{2}{5}x$; $\frac{5}{2} \cdot 16 = \frac{5}{2} \cdot \frac{2}{5}x$; $40 = x$; Tom's mother is 40 yr old.

12. Let x represent the father's age, the brother's age of 22 years is $\frac{1}{2}$ his

father's age. $22 = \frac{1}{2}x$, $2(22) = 2 \cdot \frac{1}{2}x$; $44 = x$; The father is 44 yr. old.

B 13. Let n represent a number; $3n + 11 = 50$, $3n + 11 - 11 = 50 - 11$,

$3n = 39$, $\frac{3n}{3} = \frac{39}{3}$; $n = 13$; The number is 13.

14. Let n represent a number; $5n + 14 = 129$, $5n + 14 - 14 = 129 - 14$,

$5n = 115$, $\frac{5n}{5} = \frac{115}{5}$; $n = 23$; The number is 23.

15. Let l represent the length of the rectangle, then $P = 2l + 2w$.

$54 = 2l + 2(12)$, $54 = 2l + 24$, $54 - 24 = 2l + 24 - 24$, $30 = 2l$, $\frac{30}{2} = \frac{2l}{2}$;

$15 = l$; The length of the rectangle is 15 cm.

16. Let x represent the length of each equal side of an isosceles triangle, then
$P = x + x + c$. $56 = x + x + 18$, $56 = 2x + 18$, $56 - 18 = 2x + 18 -$

18, $38 = 2x$, $\frac{38}{2} = \frac{2x}{2}$; $19 = x$; The length of each equal side is 19 mm.

17. Let x represent the original price of the VCR. The cost of the VCR was $\frac{3}{5}$

of the original price. $245.97 = \frac{3}{5}x$, $\frac{5}{3}(245.97) = \frac{5}{3} \cdot \frac{3}{5}x$; $409.95 = x$; The
original price of the VCR was \$409.95.

18. Let r represent the regular price of a sweater. The cost of a sweater was $\frac{1}{2}$

of the regular price. $25 = \frac{1}{2}r$, $2(25) = 2 \cdot \frac{1}{2}r$; $50 = r$; The regular price of
the sweater was \$50.

C 19. Let x represent the number of adult tickets sold, adults' ticket sales +
children's ticket sales = total sales; $4.50x + 2.50(117) = 733.50$, $4.50x +$
$292.50 = 733.50$, $4.50x + 292.50 - 292.50 = 733.50 - 292.50$, $4.5x =$

441, $\frac{4.5x}{4.50} = \frac{441}{4.50}$; $x = 98$; 98 adult tickets were sold.

20. Use the formula for distance: $d = rt$. $2.1 = 70t$, $\frac{2.1}{70} = \frac{70t}{70}$; $0.03 = t$.

t is expressed in hours; since 1 hr = 60 min, 0.03 hr = 0.03(60 min) = 1.8
min. The cheetah sprinted for 1.8 min.

21. Let x represent the total amount of money. Amount invested in land + amount invested in stock + amount invested in machinery + amount in savings account = total amount of money saved or invested.

$\frac{1}{2}x + \frac{1}{10}x + \frac{1}{20}x + 35,000 = x$, $\frac{10}{20}x + \frac{2}{20}x + \frac{1}{20}x + 35,000 = x$,

$\frac{13}{20}x + 35,000 = x$, $\frac{13}{20}x + 35,000 - \frac{13x}{20} = x - \frac{13x}{20}$, $35,000 = \frac{20x}{20} - \frac{13x}{20}$,

$35,000 = \frac{7x}{20}$, $\frac{20}{7}(35,000) = \frac{20}{7} \cdot \frac{7x}{20}$; $100,000 = x$; The total amount of money saved or invested is $100,000.

page 124   Mixed Problem Solving Review

1. The net charge in stock price for the week is equal to the sum of the charges for each day of the week.

Net charge $= 3\frac{1}{8} + 2\frac{1}{4} - 3 - 3\frac{3}{8} + 1\frac{1}{4}$

$= \left(3\frac{1}{8} - 3\frac{3}{8}\right) + \left(2\frac{1}{4} + 1\frac{1}{4}\right) - 3 = -\frac{2}{8} + 3\frac{2}{4} - 3$

$= \left(-\frac{1}{4} + 3\frac{2}{4}\right) - 3 = 3\frac{1}{4} - 3 = +\frac{1}{4}$

2. a. Since the mat is a square, the outer perimeter p of the mat is equal to 4 times the length of a side s. The length of each side s is 2 in. + 12 in. + 2 in. = 16 in. Thus $P = 4s = 4(16) = 64$. The perimeter is 64 in.
b. $A = 16^2$; $A = 256$. The area of the mat = $16^2 - 12^2 = 256 - 144 = 112$ in.^{2}

3. Eduardo's mother's age is 1 year more than 3 times Eduardo's age; $3a + 1$.

4. $P = \frac{1}{2}\pi x + 2y + \frac{1}{2}\pi x$; $\frac{1}{2}x = 11$ ft, $y = 28$ ft; $P = 125.1$; the perimeter is 125.1 ft.

5. $r + s = 24$, The maximum value of rs occurs when $r = s$, $12 + 12 = 24$, therefore, the maximum value of $rs = 12(12) = 144$.

page 124   Project

Answers may vary.

pages 125–126   Technology: Using Spreadsheets

1. @SUM (D6 . . . D10)   **2.** percentage grade

3. Yes, at least a 95 on Test 2, if a grade of an A starts at an average of 90.

4. +G4 + G5

5. @SUM (G10 . . . G15)

6. yes

7. Check students' work.

pages 127–128 Class Exercises

1. Let n represent the number; $8n + 12 = -84$, $8n + 12 - 12 = -84 - 12$, $8n = -96$, $\dfrac{8n}{8} = \dfrac{-96}{8}$; $n = -12$; The number is -12.

2. Let n represent a number; $\dfrac{2}{3}n - 7 = 27$, $\dfrac{2}{3}n - 7 + 7 = 27 + 7$, $\dfrac{2}{3}n = 34$, $\dfrac{3}{2} \cdot \dfrac{2}{3}n = \dfrac{3}{2} \cdot 34$; $n = 51$; The number is 51.

3. Let s represent the side of a square; $4s = 196$, $\dfrac{4s}{4} = \dfrac{196}{4}$; $s = 49$; The length of a side is 49 m.

4. Let x represent the number of students in the freshmen class; $\dfrac{8}{9}x = 160$, $\dfrac{9}{8} \cdot \dfrac{8}{9}x = \dfrac{9}{8} \cdot 160$; $x = 180$; There are 180 freshmen students.

pages 128–129 Practice Exercises

A **1.** Let n represent the number; $12 + 3n = 21$, $12 + 3n - 12 = 21 - 12$, $3n = 9$, $\dfrac{3n}{3} = \dfrac{9}{3}$; $n = 3$; The number is 3.

2. Let n represent the number; $15 + 3n = 54$, $15 + 3n - 15 = 54 - 15$, $3n = 39$, $\dfrac{3n}{3} = \dfrac{39}{3}$; $n = 13$; The number is 13.

3. Let n represent the number; $5n - 1 = -26$, $5n - 1 + 1 = -26 + 1$, $5n = -25$, $\dfrac{5n}{5} = \dfrac{-25}{5}$; $n = -5$; The number is -5.

4. Let n represent the number; $3n - 3 = -18$, $3n - 3 + 3 = -18 + 3$, $3n = -15$, $\dfrac{3n}{3} = \dfrac{-15}{3}$; $n = -5$; The number is -5.

5. Let n represent the number; $\dfrac{0.5n}{3} = 1.5$, $\dfrac{3}{0.5} \cdot \dfrac{0.5}{3}n = \dfrac{3}{0.5}(1.5)$; $n = 9$; The number is 9.

6. Let n represent the number; $\dfrac{2.4n}{6} = 12$, $\dfrac{6}{2.4} \cdot \dfrac{2.4}{6}n = \dfrac{6}{2.4}(12)$; $n = 30$; The number is 30.

7. Let n represent the number; $\frac{1}{5}n - 8 = 7$, $\frac{1}{5}n - 8 + 8 = 7 + 8$, $\frac{1}{5}n = 15$,

$5 \cdot \frac{1}{5}n = 5 \cdot 15$; $n = 75$; The number is 75.

8. Let n represent the number; $\frac{1}{4}n - 5 = 10$, $\frac{1}{4}n - 5 + 5 = 10 + 5$, $\frac{1}{4}n = 15$,

$4 \cdot \frac{1}{4}n = 4(15)$; $n = 60$; The number is 60.

9. Let x represent the original price of the sweater; $16 = x - 5.98$, $16 +$ $5.98 = x - 5.98 + 5.98$; $21.98 = x$; The original price of the sweater was $21.98.

10. Let x represent the cost of the shirt; $x + 0.87 = 16.86$, $x + 0.87 - 0.87 = 16.86 - 0.87$; $x = 15.99$; The cost of the shirt is $15.99.

11. Let x represent Lisa's age; $2x - 8 = 36$, $2x - 8 + 8 = 36 + 8$, $2x = 44$,

$\frac{2x}{2} = \frac{44}{2}$; $x = 22$; Lisa is 22 years old.

12. Let x represent Tony's age; $3x - 10 = 50$, $3x - 10 + 10 = 50 + 10$, $3x = 60$, $\frac{3x}{3} = \frac{60}{3}$; $x = 20$; Tony is 20 years old.

13. Let n represent the number; $\frac{n}{5} - 15 = -100$, $\frac{n}{5} - 15 + 15 = -100 + 15$,

$\frac{n}{5} = -85$, $5 \cdot \frac{n}{5} = 5(-85)$; $n = -425$; The number is -425.

14. Let n represent the number; $\frac{n}{7} - 14 = -150$, $\frac{n}{7} - 14 + 14 = -150 + 14$;

$\frac{n}{7} = -136$, $7 \cdot \frac{n}{7} = 7(-136)$; $n = -952$; The number is -952.

B 15. Let n represent the number; $6(n - 7) = 4.2$, $\frac{1}{6} \cdot 6(n - 7) = \frac{1}{6}(4.2)$,

$n - 7 = 0.7$, $n - 7 + 7 = 0.7 + 7$; $n = 7.7$; The number is 7.7.

16. Let n represent the number; $8(n - 12) = 9.2$, $\frac{1}{8} \cdot 8(n - 12) = \frac{1}{8}(9.2)$,

$n - 12 = 1.15$, $n - 12 + 12 = 1.15 + 12$; $n = 13.15$; The number is 13.15.

17. Let P represent the perimeter, w the width, and l the length; the perimeter of the rectangle is 64 m. $P = 2l + 2w$, $64 = 2(14) + 2w$, $64 = 28 + 2w$,

$64 - 28 = 28 + 2w - 28$, $36 = 2w$, $\frac{36}{2} = \frac{2w}{2}$; $18 = w$; The width of the rectangle is 18 m.

18. Let P represent the perimeter, w the width, and l the length of the rectangle; the perimeter of the rectangle is 96 m. $P = 2l + 2w$, $96 = 2l + 2(12)$, $96 = 2l + 24$, $96 - 24 = 2l + 24 - 24$, $72 = 2l$, $\dfrac{72}{2} = \dfrac{2l}{2}$; $36 = l$; The length of the rectangle is 36 m.

19. $d = rt$, $40 = 12.5t$; $3.2 = t$; The time is 3.2 hours.

20. $d = rt$, $2.5 = (20 \text{ min}) \, r$, $2.5 = \left(\dfrac{1}{3}\right)r$, $\dfrac{3}{1} \cdot 2.5 = \left(\dfrac{3}{1}\right)\left(\dfrac{1}{3}\right)r$; $r = 7.5$; The runner's rate is 7.5 mi/h.

21. Let x represent the equal sides, then $x + x + 12 + 9 + 13 = 56$, $2x + 34 = 56$, $2x + 34 - 34 = 56 - 34$, $2x = 22$, $\dfrac{1}{2} \cdot 2x = \dfrac{1}{2} \cdot 22$; $x = 11$; The equal sides measure 11 in.

22. Let x represent the equal sides, then $x + x + x + 15 + 20 + 25 = 150$, $3x + 60 = 150$, $3x + 60 - 60 = 150 - 60$, $3x = 90$, $\dfrac{1}{3} \cdot 3x = \dfrac{1}{3} \cdot 90$; $x = 30$; The equal sides measure 30 cm.

C 23. Let x represent the amount invested in Gateway Airline stock, then $\dfrac{1}{2}x$ represents the amount invested in National Computing stock; $\dfrac{1}{2}x + x = 4500$, $\dfrac{3}{2}x = 4500$, $\dfrac{2}{3} \cdot \dfrac{3}{2}x = \dfrac{2}{3}(4500)$; $x = 3000$; The amount invested in Gateway Airline stock was $3000.

24. Let a equal the number of A's, then $a + 5$ is the number of B's, and $2(a + 5)$ is the number of C's; $a + a + 5 + 2(a + 5) = 35$, $4a + 15 = 35$, $4a + 15 - 15 = 35 - 15$, $4a = 20$, $a = 5$; Therefore, there are 5 A's, 5 + 5 or 10 B's, and 2(5 + 5) or 20 C's.

25. If $C = -20$, then $F = \dfrac{9}{5}(-20) + 32$, $F = -36 + 32$, $F = -4$; The temperature is $-4°F$.

26. If $F = 59$, then $C = \dfrac{5}{9}(59 - 32)$, $C = \dfrac{5}{9}(27)$, $C = 15$; The temperature is $15°C$.

page 129   Test Yourself

1. Let n represent the number; $\dfrac{n}{-6} = 36$, $-6\left(\dfrac{n}{-6}\right) = -6(36)$; $n = -216$; The number is -216.

2. Let x represent the amount Barbara paid; $360 = 3x - 30$, $360 + 30 =$ $3x - 30 + 30$, $390 = 3x$, $\dfrac{390}{3} = \dfrac{3x}{3}$; $130 = x$; Barbara paid $130.

3.

Statements	Reasons
1. $m = r$; $n \neq 0$	1. Given
2. $\dfrac{m}{n} = \dfrac{m}{n}$	2. Reflexive property
3. $\dfrac{m}{n} = \dfrac{r}{n}$	3. Substitution property

4. $P = 4s$, $\dfrac{1}{4} = 4s$, $\dfrac{1}{4}\left(\dfrac{1}{4}\right) = \dfrac{1}{4} \cdot 4s$; $s = \dfrac{1}{16}$

5. $A = \dfrac{1}{2}h(a + b)$, $40 = \dfrac{1}{2}h(5.5 + 2.5)$, $40 = \dfrac{1}{2}h(8)$, $40 = 4h$, $\dfrac{1}{4}(40) = \dfrac{1}{4} \cdot 4h$; $h = 10$

pages 130–131   Summary and Review

1. $x + 7 = 28$, $x + 7 - 7 = 28 - 7$; $x = 21$

2. $y - 11 = -38$, $y - 11 + 11 = -38 + 11$; $y = -27$

3. $-27 = a + 5$, $-27 - 5 = a + 5 - 5$; $a = -32$

4. $3.1 = -5.8 + b$, $3.1 + 5.8 = -5.8 + b + 5.8$; $b = 8.9$

5. $9a = 45$, $\dfrac{9a}{9} = \dfrac{45}{9}$; $a = 5$

6. $\dfrac{t}{3} = -15$, $3 \cdot \dfrac{t}{3} = 3(-15)$; $t = -45$

7. $-18 = -\dfrac{5x}{2}$, $-\dfrac{2}{5}(-18) = -\dfrac{2}{5} \cdot -\dfrac{5x}{2}$; $x = \dfrac{36}{5}$

8. $-22y = -121$, $\dfrac{-22y}{-22} = \dfrac{-121}{-22}$; $y = 5.5$

9. $3a + 4 = 19$, $3a + 4 - 4 = 19 - 4$, $3a = 15$, $\dfrac{3a}{3} = \dfrac{15}{3}$; $a = 5$

10. $8 = 5x - 12$, $8 + 12 = 5x - 12 + 12$, $20 = 5x$, $\dfrac{20}{5} = \dfrac{5x}{5}$; $x = 4$

11. $-5(y + 7) = 25$, $-\dfrac{1}{5} \cdot -5(y + 7) = -\dfrac{1}{5}(25)$, $y + 7 = -5$, $y + 7 - 7 = -5 - 7$; $y = -12$

12. $13a + 7 = 7$, $13a + 7 - 7 = 7 - 7$, $13a = 0$, $\dfrac{13a}{13} = \dfrac{0}{13}$; $a = 0$

13. $15.8 = 16 - 0.2d$, $15.8 - 16 = 16 - 0.2d - 16$, $-0.2 = -0.2d$, $\dfrac{-0.2}{-0.2} = \dfrac{-0.2d}{-0.2}$; $d = 1$

14. $4.9w - 18.6 = 535.1$, $4.9w - 18.6 + 18.6 = 535.1 + 18.6$, $4.9w = 553.7$, $\dfrac{4.9w}{4.9} = \dfrac{553.7}{4.9}$; $w = 113$

15. $0 = 1\frac{3}{5} - \frac{1}{5}y$, $0 - 1\frac{3}{5} = 1\frac{3}{5} - \frac{1}{5}y - 1\frac{3}{5}$, $-1\frac{3}{5} = -\frac{1}{5}y$, $-\frac{8}{5} = -\frac{1}{5}y$, $-5\left(-\frac{8}{5}\right) = -5 \cdot -\frac{1}{5}y$; $y = 8$

16. $\frac{1}{2} + \frac{3}{4}p = \frac{1}{4}$, $\frac{1}{2} + \frac{3}{4}p - \frac{1}{2} = \frac{1}{4} - \frac{1}{2}$, $\frac{3}{4}p = -\frac{1}{4}$, $\frac{4}{3} \cdot \frac{3}{4}p = \frac{4}{3}\left(-\frac{1}{4}\right)$; $p = -\frac{1}{3}$

17. $P = 2l + 2w$, $72 = 2(28) + 2w$, $72 = 56 + 2w$, $72 - 56 = 56 + 2w - 56$, $16 = 2w$, $\dfrac{16}{2} = \dfrac{2w}{2}$; $w = 8$

18. $A = \frac{1}{2}h(a + b)$, $16 = \frac{1}{2}h\left(3\frac{1}{4} + 2\right)$, $16 = \frac{1}{2}h\left(5\frac{1}{4}\right)$, $16 = \frac{1}{2}h\left(\frac{21}{4}\right)$, $16 = \frac{21}{8}h$, $\frac{8}{21}(16) = \frac{8}{21} \cdot \frac{21}{8}h$; $h = 6\frac{2}{21}$

19. $A = p + prt$, $896 = 800 + 800r(3)$, $896 = 800 + 2400r$, $896 - 800 = 800 + 2400r - 800$, $96 = 2400r$, $\dfrac{96}{2400} = \dfrac{2400r}{2400}$; $r = 0.04$

20. $C = 2\pi r$, $16.956 = 2(3.14)r$, $16.956 = 6.28r$, $\dfrac{16.956}{6.28} = \dfrac{6.28r}{6.28}$; $r = 2.7$

21. For all real numbers x, y, and z, if $x = y$, then $x + z = y + z$.

22.

Statements	Reasons
1. $x = y$	1. Given
2. $x - z = x - z$	2. Reflexive property
3. $x - z = y - z$	3. Substitution property

23. Let x represent the weight of the truck; $3x + 500 = 12{,}200$, $3x + 500 - 500 = 12{,}200 - 500$, $3x = 11{,}700$, $\dfrac{3x}{3} = \dfrac{11{,}700}{3}$; $x = 3900$; The weight of the truck is 3900 lb.

24. Let n represent the number; $-\frac{3}{8}n = -24$, $-\frac{8}{3} \cdot -\frac{3}{8}n = -\frac{8}{3}(-24)$; $n = 64$; the number is 64.

25. Let x represent the amount Eric has in his account; $809.45 = 2x + 14.95$,

$809.45 - 14.95 = 2x + 14.95 - 14.95$, $794.50 = 2x$, $\dfrac{794.50}{2} = \dfrac{2x}{2}$;

$x = 397.25$; The amount in Eric's account is $397.25.

26. Let x represent the age of Roberto's brother; $45 = \dfrac{3}{4}x$, $\dfrac{4}{3}(45) = \dfrac{4}{3} \cdot \dfrac{3}{4}x$;

$60 = x$; Roberto's brother is 60 years old.

27. $P = 2l + 2w$, $54 = 2l + 2(12)$, $54 = 2l + 24$, $54 - 24 = 2l + 24 - 24$,

$30 = 2l$, $\dfrac{30}{2} = \dfrac{2l}{2}$; $15 = l$; The length of the rectangle is 15 cm.

28. $d = rt$, $93.75 = 37.5t$, $\dfrac{93.75}{37.5} = \dfrac{37.5t}{37.5}$; $2.5 = t$; It will take Eli 2.5 h to drive

the distance of 93.75 mi.

29. Let x represent Brenda's weight; $210 = 2x$, $\dfrac{210}{2} = \dfrac{2x}{2}$; $105 = x$; Brenda

weighs 105 lb.

30. $d = rt$, $5.25 = 3.5t$, $\dfrac{5.25}{3.5} = \dfrac{3.5t}{3.5}$; $1.5 = t$; It will take Greg 1.5 h to walk

5.25 mi.

page 132   Chapter Test

1. $b - 2.7 = 2.5$, $b - 2.7 + 2.7 = 2.5 + 2.7$; $b = 5.2$

2. $7 = y + 3\dfrac{2}{5}$, $7 - 3\dfrac{2}{5} = y + 3\dfrac{2}{5} - 3\dfrac{2}{5}$; $y = 3\dfrac{3}{5}$

3. $-\dfrac{1}{2}(y - 8) = 7$, $-2 \cdot -\dfrac{1}{2}(y - 8) = -2(7)$, $y - 8 = -14$, $y - 8 + 8 =$

$-14 + 8$; $y = -6$

4. $3t + 13 = -11$, $3t + 13 - 13 = -11 - 13$, $3t = -24$, $\dfrac{3t}{3} = \dfrac{-24}{3}$; $t = -8$

5. $-\dfrac{5}{7}x = 35$, $-\dfrac{7}{5} \cdot -\dfrac{5}{7}x = -\dfrac{7}{5}(35)$; $x = -49$

6. $-11 = a + 2$, $-11 - 2 = a + 2 - 2$; $a = -13$

7. $6d + \dfrac{3}{4} = -\dfrac{3}{4}$, $6d + \dfrac{3}{4} - \dfrac{3}{4} = -\dfrac{3}{4} - \dfrac{3}{4}$, $6d = -\dfrac{3}{2}$, $\dfrac{1}{6} \cdot 6d = \dfrac{1}{6}\left(-\dfrac{3}{2}\right)$; $d = -\dfrac{1}{4}$

8. $\dfrac{3}{4}(t + 8) = -6$, $\dfrac{4}{3} \cdot \dfrac{3}{4}(t + 8) = \dfrac{4}{3}(-6)$, $t + 8 = -8$, $t + 8 - 8 = -8 - 8$;

$t = -16$

9. $P = 2l + 2w$, $80 = 2(15) + 2w$, $80 = 30 + 2w$, $80 - 30 = 30 + 2w - 30$,

$50 = 2w$, $\dfrac{50}{2} = \dfrac{2w}{2}$; $w = 25$

10. $d = rt$, $1140 = r\left(4\dfrac{3}{4}\right)$, $1140 = \dfrac{19}{4}r$, $\dfrac{4}{19}(1140) = \dfrac{4}{19}(19r)$; $r = 240$

11.

Statements	Reasons
1. $x = y$	1. Given
2. $x \cdot z = x \cdot z$	2. Reflexive property
3. $x \cdot z = y \cdot z$	3. Substitution property

12.

Statements	Reasons
1. $ab + cb + d = (a + c)b + d$	1. Distributive property
2. $\quad\quad\quad\quad = d + (a + c)b$	2. Commutative property for addition
3. $\quad\quad\quad\quad = d + b(a + c)$	3. Commutative property for multiplication

13. Let n represent the number of shirts Anne Marie bought. Cost of the shirts + cost of the socks = the total cost. $8n + 3(1) = 27$, $8n + 3 = 27$,

$8n + 3 - 3 = 27 - 3$, $8n = 24$, $\dfrac{8n}{8} = \dfrac{24}{8}$; $n = 3$; Anne Marie bought 3 shirts.

14. Let n represent the number; $-3n = 87$, $\dfrac{-3n}{-3} = \dfrac{-87}{-3}$; $n = 29$; The number is 29.

15. Let p represent the price of one individual bagel; $3.50 = 12p - 0.34$,

$350 = 12p - 34$, $350 + 34 = 12p - 34 + 34$, $384 = 12p$, $\dfrac{384}{12} = \dfrac{12p}{12}$;

$p = 32$; The price of one individual bagel is 32¢.

Challenge

1. Let l represent the length of the shorter piece, then $l + 9\dfrac{1}{2}$ represents the

length of the longer piece; $l + \left(l + 9\dfrac{1}{2}\right) = 35$.

2. Let k represent the cost of the keyboard, then $k + 300$ represents the cost of the monitor and $(k + 300) + 600$ or $k + 900$ represents the cost of the computer; $1800 = k + (k + 300) + (k + 900)$.

page 133 Preparing for Standardized Tests

1. B; since $\dfrac{17}{25} = 0.68$, $0.68 > 0.63$.

2. A; $-\dfrac{2}{7} > -\dfrac{2}{5}$

3. C; since $\dfrac{0}{-1} = 0$ and $\dfrac{0}{1} = 0$.

4. A; Using the rules for order of operations, $10 + 4 \div 2 = 10 + 2 = 12$,
$5 \cdot 6 \div 3 = 30 \div 3 = 10$; $12 > 10$.

5. C; $\dfrac{1}{5} \times \$640 = \$128$, $\dfrac{1}{3} \times \$384 = \$128$

6. B; $P = 2(10) + 2(5) = 30$, $P = 2(12) + 2(4) = 32$; $32 > 30$

7. C; $-(a - b) = -a + b = b - a$

8. D; Not enough information is given concerning a and b.

9. A; $2(2^2 + 3^2) = 2(4 + 9) = 26$, $(2 + 3)^2 = 5^2 = 25$; $26 > 25$

10. A;
$$
\begin{aligned}
5x - 8 &= 12 & 15 - 4y &= 7 \\
5x &= 20 & -4y &= -8 \\
x &= 4 & y &= 2
\end{aligned}
$$
$$x > y$$

11. B; The sum of 2 times a number and 3 is $2n + 3$ and 2 times the sum of a number and 3 is $2(n + 3)$ or $2n + 6$; $2n + 6 > 2n + 3$.

12. B; $\dfrac{55}{154} \times \dfrac{42}{70} = \dfrac{5 \cdot 11}{2 \cdot 7 \cdot 11} \times \dfrac{2 \cdot 3 \cdot 7}{2 \cdot 5 \cdot 7} = \dfrac{3}{14}$

$\dfrac{65}{78} \times \dfrac{54}{126} = \dfrac{5 \cdot 13}{2 \cdot 3 \cdot 13} \times \dfrac{2 \cdot 3 \cdot 9}{2 \cdot 7 \cdot 9} = \dfrac{5}{14}$; $\dfrac{5}{14} > \dfrac{3}{14}$

13. A;
$$
\begin{aligned}
\tfrac{1}{2} \text{ of father's age} &= 24 & \tfrac{2}{3} \text{ of mother's age} &= 24 \\
\text{father's age} &= 24 \cdot \tfrac{2}{1} & \text{mother's age} &= 24 \cdot \tfrac{3}{2} \\
&= 48 & &= 36
\end{aligned}
$$
father's age > mother's age

14. B; $36 > 30$

15. C; $48 = 48$

page 134   Mixed Review

1. $7(b + 5) - 9b = 7b + 35 - 9b = -2b + 35$

2. $6x + 5(x - 2) + 4 = 6x + 5x - 10 + 4 = 11x - 6$

3. $5y + 3 - (4y - 2) = 5y + 3 - 4y + 2 = y + 5$

4. $-2(3c + 9) + 4c = -6c - 18 + 4c = -2c - 18$

5. $m - 5 + (-5m + 9) = m - 5 - 5m + 9 = -4m + 4$

6. $-(8 - 3n) + 6(2n - 12) = -8 + 3n + 12n - 72 = 15n - 80$

7. $20\% = 0.20$

8. $42\% = 0.42$

9. $6\% = 0.06$

10. $12.5\% = 0.125$

11. $10\% = 0.10$

12. $30.3\% = 0.303$

13. $0.75 = 75\%$

14. $0.18 = 18\%$

15. $0.07 = 7\%$

16. $0.01 = 1\%$

17. $0.385 = 38.5\%$

18. $0.065 = 6.5\%$

19. $\frac{1}{4} \rightarrow 4\overline{)1.00}^{\,0.25}, 0.25 \rightarrow 25\%$

20. $\frac{2}{5} \rightarrow 5\overline{)2.00}^{\,0.40}, 0.40 \rightarrow 40\%$

21. $\frac{7}{10} \rightarrow 10\overline{)7.00}^{\,0.70}, 0.70 \rightarrow 70\%$

22. $\frac{1}{2} \rightarrow 2\overline{)1.00}^{\,0.50}, 0.50 \rightarrow 50\%$

23. $\frac{3}{8} \rightarrow 8\overline{)3.000}^{\,0.375}, 0.375 \rightarrow 37.5\%$

24. $\frac{2}{3} \rightarrow 3\overline{)2.00}^{\,0.66\frac{2}{3}}, 0.66\frac{2}{3} \rightarrow 66\frac{2}{3}\%$

25. 35% means $\frac{35}{100} = \frac{7}{20}$

26. 80% means $\frac{80}{100} = \frac{4}{5}$

27. $37\frac{1}{2}\%$ means $\frac{37\frac{1}{2}}{100} = \frac{3}{8}$

28. 75% means $\frac{75}{100} = \frac{3}{4}$

29. $8\frac{1}{3}\%$ means $\frac{8\frac{1}{3}}{100} = \frac{1}{12}$

30. 36% means $\frac{36}{100} = \frac{9}{25}$

31. $d = rt$, $75 = 50t$, $\frac{75}{50} = \frac{50t}{50}$, $\frac{3}{2} = t$; $t = 1\frac{1}{2}$; It would take Miriam $1\frac{1}{2}$ h to drive the distance.

32. $P = 2l + 2w$, $42 = 2(12) + 2w$, $42 = 24 + 2w$, $42 - 24 = 24 + 2w - 24$, $18 = 2w$, $\frac{18}{2} = \frac{2w}{2}$; $9 = w$; The width of the garden is 9m.

33. $F = \frac{9}{5}C + 32$, $F = \frac{9}{5}(35) + 32$, $F = 63 + 32$; $F = 95$; The temperature is $95°F$.

Chapter 4   More Equations in One Variable

page 136   **Capsule Review**

1. $8n - 2$ **2.** $5a^2 + 7b^2$ **3.** $-11n - 1$

4. $-11 + 8x$ **5.** $-6b - 8b^2$ **6.** $-3w + 3$

page 138   **Class Exercises**

1. $3a + 6$ **2.** $5 - 10b$ **3.** $-12y - 32$

4. $21 + 12x$ **5.** $2.8m - 1.4$ **6.** $1 - 3a$

7. $4a = 8, a = 2$ **8.** $8x = 30, x = \dfrac{30}{8}, x = \dfrac{15}{4}$

9. $2x - 6 = 6, 2x = 12, x = 6$ **10.** $2x - 5 = 4, 2x = 9, x = \dfrac{9}{2}$

11. $5 - 2t = -1, -2t = -6, t = 3$ **12.** $3b + 15 = 21, 3b = 6, b = 2$

13. $4y - 8 = 16, 4y = 24, y = 6$

14. $18 + 15m = -12, 15m = -30, m = -2$

15. $7x - 3 - 4x - 2 = 13, 3x - 5 = 13, 3x = 18, x = 6$

16. $5y + 4 + 3y - 9 = 11, 8y - 5 = 11, 8y = 16, y = 2$

pages 138–139   **Practice Exercises**

A   **1.** $8y = 16, y = 2$ **2.** $2x = 18, x = 9$

    3. $7a - 7 = 21, 7a = 28, a = 4$ **4.** $5b = 30, b = 6$

    5. $3 - 11t = -19, -11t = -22, t = 2$ **6.** $5 - 10x = -25, -10x = -30, x = 3$

    7. $-13 = x - 10, x = -3$ **8.** $-8x + 4 = 36, -8x = 32, x = -4$

    9. $1 - 10y = 1, -10y = 0, y = 0$ **10.** $-5 = -5 - 2s, -2s = 0, s = 0$

    11. $3c + 6 = 27, 3c = 21, c = 7$

    12. $2y + 5 - y - 3 + 7y - 3y = 17, 5y + 2 = 17, 5y = 15, y = 3$

    13. $4x + 12 = 40, 4x = 28, x = 7$ **14.** $4x + 15 = 45, 4x = 30, x = \dfrac{30}{4} = 7\dfrac{1}{2}$

    15. $2n - 6 = 12, 2n = 18, n = 9$ **16.** $3b + 12 = 24, 3b = 12, b = 4$

    17. $8x - 8 = -24, 8x = -16, x = -2$ **18.** $15 - 3r = 18, -3r = 3, r = -1$

19. $-6 + 2m = 14$, $2m = 20$, $m = 10$

20. $-15 + 5d = 25$, $5d = 40$, $d = 8$

21. $42 - 12t = 30$, $-12t = -12$, $t = 1$

22. $20x - 20 = 0$, $20x = 20$, $x = 1$

23. $-6x - 10 = 2$, $-6x = 12$, $x = -2$

B 24. $4x - 7x + 63 = 42$, $-3x + 63 = 42$, $-3x = -21$, $x = 7$

25. $3n + 4n - 36 = -78$, $7n = -42$, $n = -6$

26. $-5t + 35 + 5 = -5$, $-5t + 40 = -5$, $-5t = -45$, $t = 9$

27. $14 - 2a - 4 = 0$, $10 - 2a = 0$, $-2a = -10$, $a = 5$

28. $-0.2d = 0.8$, $d = -4$

29. $9c = 1.8$, $c = 0.2$

30. $5z = 3.2$, $z = 0.64$

31. $-4.5 = 3n$, $n = -1.5$

32. $3a - 15 = 4$, $3a = 19$, $a = \dfrac{19}{3}$

33. $-2b - 8 = 11$, $-2b = 19$, $b = -\dfrac{19}{2}$

34. $x - 18 + 3x = -14$, $4x - 18 = -14$, $4x = 4$, $x = 1$

35. $6 - 22p - 132 = 2$, $-22p - 126 = 2$, $-22p = 128$, $p = -\dfrac{128}{22} = -\dfrac{64}{11}$

36. $-4 - 7 - 3y = 7$, $-3y - 11 = 7$, $-3y = 18$, $y = -6$

37. $-9 - 8 + 5t = 18$, $-17 + 5t = 18$, $5t = 35$, $t = 7$

38. $\dfrac{1}{3}x - 3 = -6$, $\dfrac{1}{3}x = -3$, $x = 3(-3)$, $x = -9$

39. $\dfrac{2}{5}y + 4 = 0$, $\dfrac{2}{5}y = -4$, $y = \dfrac{5}{2}(-4)$, $y = -10$

C 40. $-\dfrac{3}{4}a - 8 = 2$, $-\dfrac{3}{4}a = 10$, $a = -\dfrac{4}{3}(10)$, $a = -\dfrac{40}{3}$

41. $\dfrac{1}{5}b + 5 = 1$, $\dfrac{1}{5}b = -4$, $b = 5(-4)$, $b = -20$

42. $\dfrac{2}{3} + 6 - 3m = -2$, $\dfrac{2}{3} + \dfrac{18}{3} - 3m = -\dfrac{6}{3}$, $\dfrac{20}{3} - 3m = -\dfrac{6}{3}$, $-3m = -\dfrac{26}{3}$, $m = \left(-\dfrac{1}{3}\right)\left(-\dfrac{26}{3}\right)$, $m = \dfrac{26}{9}$

43. $-8 - 2n - \dfrac{5}{8} = \dfrac{7}{8}$, $-\dfrac{64}{8} - 2n - \dfrac{5}{8} = \dfrac{7}{8}$, $-2n = \dfrac{69}{8} + \dfrac{7}{8}$, $-2n = \dfrac{76}{8}$, $n = \left(-\dfrac{1}{2}\right)\left(\dfrac{76}{8}\right)$, $n = -\dfrac{76}{18} = -\dfrac{19}{4}$

44. $1.07y + 8.42 = 2$, $1.07y = -6.42$, $y = -6$

45. $-0.35x + 0.02 = -0.68$, $-0.35x = -0.70$, $x = 2$

46. $2[x + 3x - 3] = 18$, $2[4x - 3] = 18$, $8x - 6 = 18$, $8x = 24$, $x = 3$

47. $8k - 28 + 3k - 3 = 46$, $11k - 31 = 46$, $11k = 77$, $k = 7$

48. Let n represent the number; $3n - 5n = 20$, $-2n = 20$, $n = -10$; The number is -10.

49. Let w represent the width; $2(2w - 10) + 2w = 118$, $4w - 20 + 2w = 118$, $6w - 20 = 118$, $6w = 138$, $w = 23$; $l = 2(23) - 10$, $l = 36$; The length is 36 ft and the width is 23 ft.

50. Let r represent the number of records Jocelyn has; $3r + r + (2r + 2) = 56$, $6r + 2 = 56$, $6r = 54$, $r = 9$; Jasmine has 20 records, Jane has 27 records, and Jocelyn has 9 records.

51. $x + y = xy$; $x = 0$, $y = 0$ or $x = 2$, $y = 2$

52. $r - s = r + s$; $s = 0$, $r = $ any real number

53. $\dfrac{c}{d} = cd$; $c = 1$, $d = 1$ or $c = 0$, $d = $ any real number except 0

54. $m + \dfrac{1}{m} = 2$; $m = 1$

55. Answers may vary. An example: Henry bought pens priced at $3.50 each and $2.25 each. If he bought 5 more $2.25 pens than $3.50 pens, and paid a total of $80.25, how many of each kind did he buy?

page 140   Capsule Review

1. $2x = -12$, $x = -6$

2. $24 = -3y$, $y = -8$

3. $2y = 16$, $y = 8$

4. $a - 10 = -19$, $a = -9$

5. $2n - 6 = 14$, $2n = 20$, $n = 10$

6. $20 - 5b - 35 = 0$, $-15 - 5b = 0$, $-5b = 15$, $b = -3$

page 142   Class Exercises

1. Subtract m from each side of the equation.

2. Subtract x and subtract 28 from each side of the equation.

3. Add z to each side of the equation.

4. Distribute 4 on the right side and subtract $4n$ from each side of the equation.

5. Distribute 3 on the right side and add $9b$ to each side of the equation.

6. Distribute 2 on the left side and subtract $2y$ from each side of the equation.

7. $6t = 0$, $t = 0$

8. $8a = 8$, $a = 1$

9. $4x + 28 = 7 + x$, $3x = -21$, $x = -7$

10. You would do that for Exercises 3 and 6 in order to avoid obtaining variables with a negative coefficient and dividing by a negative.

11. Some possible answers are absolute value equations, rational equations, or equations which involve division by zero. An example: $|x| = -2$

pages 142–143   Practice Exercises

A 1. $3y = -9$, $y = -3$ **2.** $4m = -12$, $m = -3$ **3.** $2d = 2$, $d = 1$

4. $3a = -9$, $a = -3$ **5.** $8m = 0$, $m = 0$ **6.** $10j = 0$, $j = 0$

7. $2n = 20$, $n = 10$ **8.** $2e = 14$, $e = 7$ **9.** $3d = 10$, $d = \dfrac{10}{3}$

10. $6j = -12$, $j = -2$ **11.** $-22 - 3d = -4d$, $-22 = -d$, $d = 22$

12. $-42 + 4c = c$, $-42 = -3c$, $c = 14$ **13.** $8y + 20 = -2y$, $20 = -10y$, $y = -2$

14. $4b - 10 = 2b$, $-10 = -2b$, $b = 5$

15. $-5f = -36 + 4f$, $-9f = -36$, $f = \dfrac{-36}{-9} = 4$

16. $2a = -18 + 8a$, $-6a = -18$, $a = 3$

17. $9y + 2 = 3y + 12$, $6y = 10$, $y = \dfrac{10}{6} = \dfrac{5}{3}$

18. $3x - 2 = 4x - 8$, $-x = -6$, $x = 6$

19. $3y + 2y - 10 = -3y + 35 - y$, $5y - 10 = -4y + 35$, $9y = 45$, $y = 5$

20. $4x + 3x - 6 = -5x + 20 - x$, $7x - 6 = -6x + 20$, $13x = 26$, $x = 2$

21. $5y - 2y - 10 = -2y - 15 + y$, $3y - 10 = -y - 15$, $4y = -5$, $y = -\dfrac{5}{4}$

22. $2z - 3z - 3 = -5z - 3 + z$, $-z - 3 = -4z - 3$, $3z = 0$, $z = 0$

23. $6m + 3m + 6 = -2m - 7 + m$, $9m + 6 = -m - 7$, $10m = -13$, $m = -\dfrac{13}{10}$

24. $9t + 5t + 15 = -t - 13 + t$, $14t + 15 = -13$, $14t = -28$, $t = -2$

B 25. $5f = 4f + \dfrac{3}{4} + \dfrac{1}{2}, f = \dfrac{3}{4} + \dfrac{2}{4}, f = \dfrac{5}{4}$

26. $2g = g + \dfrac{5}{8} + \dfrac{5}{8}, g = \dfrac{10}{8} = \dfrac{5}{4}$

27. $14d - 7 + 10 - 15d = 2d, -d + 3 = 2d, 3 = 3d, d = 1$

28. $18e - 6 + 4 - 20e = e, -2e - 2 = e, -2 = 3e, e = -\dfrac{2}{3}$

29. $15 - 20y + 14y = 14 - 35y, 15 - 6y = 14 - 35y, 29y = -1, y = -\dfrac{1}{29}$

30. $10 - 8x + x = 9x - 33, 10 - 7x = 9x - 33, 16x = 43, x = \dfrac{43}{16}$

31. $12n - 30 = -21 + 9n + 2n, 12n - 30 = -21 + 11n, n = 9$

32. $33 - 22q = 27 - 21q + q, 33 - 22q = 27 - 20q, -2q = -6, q = 3$

33. $5x - 1 = -8 + 6x + 10, 5x - 1 = 2 + 6x, x = -3$

34. $14t - 1 = -3 + 6t + 12, 14t - 1 = 9 + 6t, 8t = 10, t = \dfrac{10}{8} = \dfrac{5}{4}$

35. $-7y + 3 = -3y - 5 - 2, -7y + 3 = -3y - 7, 4y = 10, y = \dfrac{10}{4} = \dfrac{5}{2}$

36. $-9m + 5 = -2m - 8 - 6, -9m + 5 = -2m - 14, -7m = -19, m = \dfrac{19}{7}$

37. $-15x + 10 + 12 - 12x = 3x, -27x + 22 = 3x, 30x = 22, x = \dfrac{22}{30} = \dfrac{11}{15}$

38. $-36y + 24 + 15 - 35y = -68y, -71y + 39 = -68y, 3y = 39, y = 13$

C 39. $-6t - [4 - 2 + 3t] = 4t + 4, -6t - [2 + 3t] = 4t + 4, -6t - 2 - 3t =$
$4t + 4, -9t - 2 = 4t + 4, 13t = -6, t = -\dfrac{6}{13}$

40. $-8b - [22 - 4b + 4] = 9b, -8b - [26 - 4b] = 9b, -8b - 26 + 4b = 9b,$
$-4b - 26 = 9b, -13b = 26, b = -2$

41. $4[5y - 4y + 4] = 3[4y + 4], 4[y + 4] = 3[4y + 4], 4y + 16 = 12y + 12,$
$8y = 4, y = \dfrac{1}{2}$

42. $-2[10j - 30 - 7j] = 5[3j - 6], -2[3j - 30] = 5[3j - 6], -6j + 60 =$
$15j - 30, 21j = 90, j = \dfrac{90}{21} = \dfrac{30}{7}$

43. $\frac{1}{2}\left[\frac{8}{3}d - \frac{2}{3}\right] - 3d + 2 = \frac{5}{9}d,\ \frac{8}{6}d - \frac{1}{3} - 3d + 2 = \frac{5}{9}d,\ \frac{8}{6}d - 3d - \frac{5}{9}d = \frac{1}{3} - 2,$

$\frac{24}{18}d - \frac{54}{18}d - \frac{10}{18}d = -\frac{5}{3},\ -\frac{40}{18}d = -\frac{5}{3},\ d = -\frac{5}{3}\left(-\frac{18}{40}\right),\ d = \frac{3}{4}$

44. $4 - \frac{1}{2}c - \frac{1}{3}\left(\frac{3}{4} - \frac{3}{2}c\right) = \frac{1}{12}c,\ 4 - \frac{1}{2}c - \frac{1}{4} + \frac{1}{2}c = \frac{1}{12}c,\ \frac{15}{4} = \frac{1}{12}c,\ c = 12\left(\frac{15}{4}\right),$

$c = 45$

45. $-3[4m + 2m - 8] = -5[2m + 10],\ -3[6m - 8] = -5[2m + 10],$

$-18m + 24 = -10m - 50,\ 8m = 74,\ m = \frac{74}{8} = \frac{37}{4} = 9\frac{1}{4}$

46. $-2[3k + 3k - 24] = -2[3k + 21],\ -2[6k - 24] = -2[3k + 21],\ -12k +$

$48 = -6k - 42,\ 6k = 90,\ k = 15$

47. Let A represent the area of a rectangle; $A + 4 = 36 - 5A,\ 6A = 32,$

$A = \frac{32}{6} = \frac{16}{3}$

48. $x + x + 10,000 = 3x - 15,000,\ x = 25,000;$ combined income = $60,000

page 143 Algebra in Transportation

Let x represent the rate of the SST, $2x = 15(x - 1300),\ 2x = 15x - 19,500,$ $13x = 19,500,\ x = 1500,$ the rate of the propeller plane $= x - 1300 =$ $1500 - 1300 = 200;$ The propeller plane flies at 200 mi/h.

page 146 Class Exercises

1. 3, 4, 5; 4, 6, 8; cannot be done **2.** 1, 2, 3; 2, 4, 6; cannot be done

3. $-108, -107, -106;$ cannot be done; $-107, -105, -103$

pages 146–147 Practice Exercises

A 1. Let $x =$ the first integer, then $x + 1 =$ the second integer, and $x + 2 =$ the third integer. The equation is: $x + (x + 1) + (x + 2) = 99;$ solve $3x + 3 = 99,\ 3x = 96,\ x = 32.$ Then $x + 1 = 33,$ and $x + 2 = 34.$ The three consecutive integers are 32, 33, 34.

2. Let $x =$ the first integer, then $x + 1 =$ the second integer, $x + 2 =$ the third integer, and $x + 3 =$ the fourth integer. The equation is: $x + (x + 1) + (x + 2) + (x + 3) = 26;$ solve $4x + 6 = 26,\ 4x = 20,\ x = 5.$ Then $x + 1 = 6,\ x + 2 = 7,$ and $x + 3 = 8.$ The four consecutive integers are 5, 6, 7, 8.

3. Let x = the first even integer, then $x + 2$ = the second even integer, $x + 4$ = the third even integer, and $x + 6$ = the fourth even integer. The equation is: $x + (x + 2) + (x + 4) + (x + 6) = -124$; solve $4x + 12 = -124$, $4x = -136$, $x = -34$. Then $x + 2 = -32$, $x + 4 = -30$, and $x + 6 = -28$. The four consecutive even integers are $-34, -32, -30, -28$.

4. Let x = the first even integer, then $x + 2$ = the second even integer, $x + 4$ = the third even integer, and $x + 6$ = the fourth even integer. The equation is: $x + (x + 2) + (x + 4) + (x + 6) = -36$; solve $4x + 12 = -36$, $4x = -48$, $x = -12$. Then $x + 2 = -10$, $x + 4 = -8$, and $x + 6 = -6$. The four consecutive even integers are $-12, -10, -8, -6$.

5. Let x = the first even integer, then $x + 2$ = the second even integer, $x + 4$ = the third even integer, $x + 6$ = the fourth even integer, and $x + 8$ = the fifth even integer. The equation is: $x + (x + 8) = 3(x + 6) - 2$; solve $2x + 8 = 3x + 18 - 2$, $2x + 8 = 3x + 16$, $-8 = x$. Then $x + 2 = -6$, $x + 4 = -4$, $x + 6 = -2$, and $x + 8 = 0$. The five consecutive even integers are $-8, -6, -4, -2, 0$.

6. Let x = the first odd integer, then $x + 2$ = the second odd integer, $x + 4$ = the third odd integer, and $x + 6$ = the fourth odd integer. The equation is: $x + (x + 6) = 3(x + 2) - 3$; solve $2x + 6 = 3x + 6 - 3$, $2x + 6 = 3x + 3$, $3 = x$. Then $x + 2 = 5$, $x + 4 = 7$, and $x + 6 = 9$. The four consecutive odd integers are $3, 5, 7, 9$.

7. Let x = the first integer, then $x + 1$ = the second integer. The equation is: $x + (x + 1) = 105$, solve $2x + 1 = 105$; $2x = 104$, $x = 52$. Then $x + 1 = 53$. The two consecutive integers are $52, 53$.

8. Let x = the first integer, then $x + 1$ = the second integer. The equation is: $x + (x + 1) = -35$; solve $2x + 1 = -35$, $2x = -36$, $x = -18$. Then $x + 1 = -17$. The two consecutive integers are $-18, -17$.

9. Let x = the first integer, then $x + 1$ = the second integer, and $x + 2$ = the third integer. The equation is: $x + (x + 1) + (x + 2) = -354$; solve $3x + 3 = -354$, $3x = -357$, $x = -119$. Then $x + 1 = -118$ and $x + 2 = -117$. The three consecutive integers are $-119, -118, -117$.

10. Let x = the first integer, then $x + 1$ = the second integer, $x + 2$ = the third integer, and $x + 3$ = the fourth integer. The equation is: $x + (x + 1) + (x + 2) + (x + 3) = 50$; solve $4x + 6 = 50$, $4x = 44$, $x = 11$. Then $x + 1 = 12$, $x + 2 = 13$, and $x + 3 = 14$. The four consecutive integers are $11, 12, 13, 14$.

11. Let x = the first even integer, then $x + 2$ = the second even integer. The equation is: $x + (x + 2) = -54$; solve $2x + 2 = -54$, $2x = -56$, $x = -28$. Then $x + 2 = -26$. The two consecutive even integers are $-28, -26$.

12. Let x = the first even integer, then $x + 2$ = the second even integer, and $x + 4$ = the third even integer. The equation is: $x + (x + 2) + (x + 4) =$ 312; solve $3x + 6 = 312$, $3x = 306$, $x = 102$. Then $x + 2 = 104$ and $x + 4 = 106$. The three consecutive even integers are 102, 104, 106.

13. Let x = the first odd integer, then $x + 2$ = the second odd integer, and $x + 4$ = the third odd integer. The equation is: $x + (x + 2) + (x + 4) =$ -45; solve $3x + 6 = -45$, $3x = -51$, $x = -17$. Then $x + 2 = -15$ and $x + 4 = -13$. The three consecutive odd integers are $-17, -15, -13$.

14. Let x = the first even integer, then $x + 2$ = the second even integer, $x + 4$ = the third even integer, and $x + 6$ = the fourth even integer. The equation is: $x + (x + 2) + (x + 4) + (x + 6) = 180$; solve $4x + 12 = 180$, $4x = 168$, $x = 42$. Then $x + 2 = 44$, $x + 4 = 46$, and $x + 6 = 48$. The four consecutive even integers are 42, 44, 46, 48.

15. Let x = the first integer, then $x + 1$ = the second integer, $x + 2$ = the third integer, and $x + 3$ = the fourth integer. The equation is: $(x + 1) + (x + 3) = 48$; solve $2x + 4 = 48$, $2x = 44$, $x = 22$. Then $x + 1 = 23$, $x + 2 = 24$, and $x + 3 = 25$. The four consecutive integers are 22, 23, 24, 25.

16. Let x = the first integer, then $x + 1$ = the second integer, and $x + 2 =$ the third integer. The equation is: $x + (x + 2) = -34$; solve $2x + 2 =$ -34, $2x = -36$, $x = -18$. Then $x + 1 = -17$ and $x + 2 = -16$. The three consecutive integers are $-18, -17, -16$.

B 17. Let x = the first odd integer, then $x + 2$ = the second odd integer. The equation is: $2(x + 2) + x = 85$; solve $2x + 4 + x = 85$, $3x + 4 = 85$, $3x =$ 81, $x = 27$. Then $x + 2 = 29$. The two consecutive odd integers are 27, 29.

18. Let x = the first even integer, then $x + 2$ = the second even integer, and $x + 4$ = the third even integer. The equation is: $[x + (x + 2) + (x + 4)] -$ $(x + 4) = -22$; solve $3x + 6 - (x + 4) = -22$, $3x + 6 - x - 4 = -22$, $2x + 2 = -22$, $2x = -24$, $x = -12$. Then $x + 2 = -10$ and $x + 4 = -8$. The three consecutive even integers are $-12, -10, -8$.

19. Let x = the age of the first brother, then $x + 1$ = the age of the second brother, and $x + 2$ = the age of the third brother. The equation is: $x +$ $(x + 1) + (x + 2) = 39 - x$; solve $3x + 3 = 39 - x$, $4x + 3 = 39$, $4x =$ 36, $x = 9$. Then $x + 1 = 10$ and $x + 2 = 11$. The ages of the three brothers are 9, 10, 11.

20. Let x = the first even integer, then $x + 2$ = the second even integer, and $x + 4$ = the third even integer. The equation is: $x + (x + 2) + (x + 4) =$ $50 + (x + 4)$; solve $3x + 6 = 50 + x + 4$, $3x + 6 = 54 + x$, $2x = 48$, $x = 24$. Then $x + 2 = 26$ and $x + 4 = 28$. The three consecutive even integers are 24, 26, 28.

21. Let x = the first integer, then $x + 1$ = the second integer, and $x + 2 =$ the third integer. The equation is: $x + (x + 1) + (x + 2) = 4x - 9$; solve $3x + 3 = 4x - 9$, $12 = x$. Then $x + 1 = 13$ and $x + 2 = 14$. The three consecutive integers are 12, 13, 14.

22. Let x = the first even integer, then $x + 2$ = the second even integer, $x + 4$ = the third even integer, and $x + 6$ = the fourth even integer. The equation is: $\dfrac{x + (x + 2) + (x + 4) + (x + 6)}{7} = 4$; solve $\dfrac{4x + 12}{7} = 4$, $4x + 12 = (4)(7)$, $4x + 12 = 28$, $4x = 16$, $x = 4$. Then $x + 2 = 6$, $x + 4 = 8$, and $x + 6 = 10$. The four consecutive even integers are 4, 6, 8, 10.

C 23. Let x = the first multiple of five, then $x + 5$ = the second multiple of five, $x + 10$ = the third multiple of five, and $x + 15$ = the fourth multiple of five. The equation is: $x + (x + 5) + (x + 10) + (x + 15) = 90$; solve $4x + 30 = 90$, $4x = 60$, $x = 15$. Then $x + 5 = 20$, $x + 10 = 25$, and $x + 15 = 30$. Four consecutive multiples of five are 15, 20, 25, 30.

24. Let x = the first multiple of 3, then $x + 3$ = the second multiple of 3, and $x + 6$ = the third multiple of 3. The equation is: $x + (x + 6) = 12$; solve $2x + 6 = 12$, $2x = 6$, $x = 3$. Then $x + 3 = 6$ and $x + 6 = 9$. The three consecutive multiples of three are 3, 6, 9.

25. Let x = the first integer, then $x + 1$ = the second integer, and $x + 2 =$ the third integer. The equation is: $2(x + 1) = x + (x + 2)$; solve $2x + 2 = 2x + 2$. Then any integer will make this sentence true. Any three consecutive integers is the solution.

26. Let x = the first odd integer, then $x + 2$ = the second odd integer. The equation is: $x + (x + 2) = 0$; solve $2x + 2 = 0$, $2x = -2$, $x = -1$. Then $x + 2 = 1$. The two consecutive odd integers are -1, 1.

27. Let x = the first integer, then $x + 1$ = the second integer, $x + 2 =$ the third integer, and $x + 3$ = the fourth integer. The equation is: $3[x + (x + 1)] - [(x + 2) + (x + 3)] = 70$; solve $3[2x + 1] - [2x + 5] = 70$, $6x + 3 - [2x + 5] = 70$, $4x - 2 = 70$, $4x = 72$, $x = 18$. Then $x + 1 = 19$, $x + 2 = 20$, and $x + 3 = 21$. The four consecutive integers are 18, 19, 20, 21.

28. Let x = the first odd integer, then $x + 2$ = the second odd integer, $x + 4 =$ the third odd integer, and $x + 6$ = the fourth odd integer. The equation is: $[x + (x + 2)] + 4(x + 6) = 92$; solve $2x + 2 + 4x + 24 = 92$, $6x + 26 = 92$, $6x = 66$, $x = 11$. Then $x + 2 = 13$, $x + 4 = 15$, and $x + 6 = 17$. The four consecutive odd integers are 11, 13, 15, 17.

29. Commutative property of multiplication

30. Associative property of addition

31. Distributive property

32. Identity property of multiplication

33. Inverse property of addition

34. Commutative property of addition

35. Identity property of addition

36. Inverse property of multiplication

37. $9x - 12 = 33$, $9x = 45$, $x = 5$

38. $26 + 11y = -7$, $11y = -33$, $y = -3$

39. $8 = 20 + \frac{1}{3}q$, $-12 = \frac{1}{3}q$, $q = -36$

40. $\frac{3}{4} = \frac{1}{2} - \frac{2}{3}w$, $\frac{1}{4} = -\frac{2}{3}w$, $w = -\frac{3}{8}$

page 148   Capsule Review

1. 428

2. 875

3. -635

4. 70,010

5. $41 \div 0.041 = 1000 = 10^3$

6. $3208 \div 32.08 = 100 = 10^2$

7. $32 \div 0.00032 = 100{,}000 = 10^5$

pages 149–150   Class Exercises

1. Divide each side of the equation by 4; $3 - x = 4$.

2. Divide each side of the equation by -6; $-3 = y + 5$.

3. Add $4a$ to each side of the equation; $33 = 11a$.

4. Subtract 5 from each side of the equation; $-7(b + 4) = 0$.

5. Multiply each side of the equation by 10; $3 - c = 15$.

6. Multiply each side of the equation by 100; $210 + 30d = 33d$.

7. Divide each side of the equation by 5; $2m - 3 = 3m$.

8. Add 8 to each side of the equation; $5(p - 6) = 45$.

9. $4(0.2q + 5.1) = -72$

10. First add 8 to each side of the equation, then divide each side of the equation by 5; $p - 6 = 9$.

pages 150–151   Practice Exercises

A 1. $2y - 4 = 15$, $2y = 19$, $y = 9.5$ **2.** $8x + 3 = 597$, $8x = 594$, $x = 74.25$

3. $180 - 2y = 132$, $2j = 48$, $j = 24$

4. $321 - 427k = 748$, $-427k = 427$, $k = -1$

5. $2340 - 100d = 25d$, $2340 = 125d$, $d = 18.72$

6. $8260 - 100e = 18e$, $8260 = 118e$, $e = 70$

7. $3a = (a + 3)$, $2a = 3$, $a = \dfrac{3}{2}$ **8.** $3x = (x + 4)$, $2x = 4$, $x = 2$

9. $5 - a = 2$, $a = 3$ **10.** $-2 = t - 4$, $t = 2$

11. $b + 7 = -3$, $b = -10$ **12.** $5 + c = -5$, $c = -10$

13. $5 = -25r$, $r = -\dfrac{1}{5}$ **14.** $-8t = 24$, $t = -3$

15. $-8 = 3f - 11$, $3f = 3$, $f = 1$ **16.** $-13 = 4g - 23$, $4g = 10$, $g = \dfrac{10}{4} = \dfrac{5}{2}$

17. $2x - 7 = -3(3x - 5)$, $2x - 7 = -9x + 15$, $11x = 22$, $x = 2$

18. $4y - 6 = -2(2y - 5)$, $4y - 6 = -4y + 10$, $8y = 16$, $y = 2$

19. $7(14 - 2r) = 8 + r$, $98 - 14r = 8 + r$, $15r = 90$, $r = 6$

20. $6(7 - 3t) = 2 + 2t$, $42 - 18t = 2 + 2t$, $20t = 40$, $t = 2$

21. $3(2y - 1) = -3$, $2y - 1 = -1$, $2y = 0$, $y = 0$

22. $4(3x - 5) = 16$, $3x - 5 = 4$, $3x = 9$, $x = 3$

23. $1.5a = 157.5$, $a = 105$ **24.** $2.5b = -131.5$, $b = -52.6$

B 25. $45p = 60(2000 - p) + 27000$, $45p = 120{,}000 - 60p + 27000$, $105p = 147{,}000$, $p = 1400$

26. $75v = 30(50 - v) + 3000$, $75v = 1500 - 30v + 3000$, $105v = 4500$, $v = 42.857$

27. $20q + 7q + 28 = q$, $28 = -26q$, $q = -1.0769$

28. $80r + 9(r + 3) = 8r$, $80r + 9r + 27 = 8r$, $81r = -27$, $r = -\dfrac{27}{81}$, $r = -0.33\overline{3}$

29. $10a + 5(2 - a) = 8$, $10a + 10 - 5a = 8$, $5a = -2$, $a = -0.4$

30. $70b + 15(5 - b) = 20$, $70b + 75 - 15b = 20$, $55b = -55$, $b = -1$

31. $0.02n = 19.2$, $2n = 1920$, $n = 960$ **32.** $5.36m = 26.8$, $m = 5$

33. $3x - 4 + 2(x + 3) = -14$, $3x - 4 + 2x + 6 = -14$, $5x + 2 = -14$, $5x = -16$, $x = -\dfrac{16}{5}$

34. $2y + 3 + 3(y + 2) = -4$, $2y + 3 + 3y + 6 = -4$, $5y + 9 = -4$, $5y = -13$, $y = -\dfrac{13}{5}$

35. $3 - 4f - 2(6 - f) = 2$, $3 - 4f - 12 + 2f = 2$, $-2f - 9 = 2$, $-2f = 11$, $f = -\dfrac{11}{2}$

36. $2(2 - g) - (1 - 3g) = 6$, $4 - 2g - 1 + 3g = 6$, $3 + g = 6$, $g = 3$

37. $0.2b = 0$, $b = 0$ **38.** $3.4c = 0$, $c = 0$

39. $2a - (2a - 3) + 3(3 - a) = 4$, $2a - 2a + 3 + 9 - 3a = 4$, $-3a + 12 = 4$, $-3a = -8$, $a = \dfrac{8}{3} = 2\dfrac{2}{3}$

40. $-b - 3(1 - 2b) + 6(b - 2) = 12$, $-b - 3 + 6b + 6b - 12 = 12$, $11b - 15 = 12$, $11b = 27$, $b = \dfrac{27}{11} = 2\dfrac{5}{11}$

41. $-5(y + 2) = 3(3 - 2y)$, $-5y - 10 = 9 - 6y$, $y = 19$

42. $-54(2x + 5) = 18(6 + 3x)$, $-3(2x + 5) = 6 + 3x$, $-6x - 15 = 6 + 3x$, $9x = -21$, $x = -\dfrac{21}{9} = -2.3\overline{3}$

43. $310t - 100(0.02 - 0.2t) = 64$, $310t - 2 + 20t = 64$, $330t = 66$, $t = 0.2$

44. $73w - 30(0.15 - 0.3w) = -578.5$, $73w - 4.5 + 9w = -578.5$, $82w = -574$, $w = -7$

45. $-6x - [4 - (2 - 3x)] = 4(x + 1)$, $-6x - [4 - 2 + 3x] = 4x + 4$, $-6x - 2 - 3x = 4x + 4$, $-9x - 2 = 4x + 4$, $13x = -6$, $x = -\dfrac{6}{13} \approx -0.46$

46. $3.2y - 0.5[7 - (4 - 5y)] = 0.6(2y + 2)$, $32y - 5[7 - 4 + 5y] = 6(2y + 2)$, $32y - 5[3 + 5y] = 12y + 12$, $32y - 15 - 25y = 12y + 12$, $7y - 15 = 12y + 12$, $5y = -27$, $y = -\dfrac{27}{5} = -5.4$

47. $7.03(2 - 5v) = 0.2[2v - (1 - v)] + 3.2v$, $703(2 - 5v) = 20[2v - 1 + v] + 320v$, $1406 - 3515v = 20[3v - 1] + 320v$, $1406 - 3515v = 60v - 20 + 320v$, $1406 - 3515v = 380v - 20$, $3895v = 1426$, $v \approx 0.37$

48. $-2.25(3 - 2z) = -0.5[-6z - (1 - 2z)] + z$, $-225(3 - 2z) = -50[-6z - 1 + 2z] + 100z$, $-675 + 450z = -50[-4z - 1] + 100z$, $-675 + 450z = 200z + 50 + 100z$, $-675 + 450z = 300z + 50$, $150z = 725$, $z = 4.8\overline{333}$

49. $2[2(y + 5)] + 2(2y) = 84$, $2[2y + 10] + 4y = 84$, $4y + 20 + 4y = 84$, $8y = 64$, $y = 8$; $2y = 16$, $2(y + 5) = 2(13) = 26$;   The length is 26 m and the width is 16 m.

50. Let x represent the number of bottles sold; $1.25x + 7.50 = 86.25$, $1.25x = 78.75$, $x = 63$; Charles sold 63 bottles.

51. Let x represent the number of hours worked beyond the initial 3 h; $250 + 75x = 475$, $75x = 225$, $x = 3$; $3 + 3 = 6$; He worked 6 h.

page 151   Test Yourself

1. $3x + 3 = 45$, $x + 1 = 15$, $x = 14$

2. $-9 = 6m - 2 + m$, $-9 = 7m - 2$, $7m = -7$, $m = -1$

3. Let n represent the number; $n + (2n + 2) = 8$, $3n + 2 = 8$, $3n = 6$, $n = 2$; The number is 2.

4. $2y = -4$, $y = -2$

5. $2 - 3r = -3 - 6r$, $3r = -5$, $r = -\dfrac{5}{3}$

6. $-8j + 5 = 44 - 8j$; no solution

7. Let $i = $ the first even integer, then $i + 2 = $ the second even integer, and $i + 4 = $ the third even integer.

8. Let $d = $ the first integer, then $d + 1 = $ the second integer, and $d + 2 = $ the third integer.

9. Let $x = $ the first odd integer, then $x + 2 = $ the second odd integer, $x + 4 = $ the third odd integer, and $x + 6 = $ the fourth odd integer. The

equation is: $x + (x + 2) + (x + 4) + (x + 6) = 56$; solve $4x + 12 = 56$, $4x = 44$, $x = 11$. Then $x + 2 = 13$, $x + 4 = 15$, $x + 6 = 17$. The four consecutive odd integers are 11, 13, 15, 17.

10. $-2t = 18$, $t = -9$ **11.** $1 + 20x = 700$, $20x = 699$, $x = 34.95$

12. $-k = 2(k - 1)$, $-k = 2k - 2$, $3k = 2$, $k = \dfrac{2}{3}$

page 152   Capsule Review

1. 0.29   **2.** 0.031   **3.** 1   **4.** 2.80   **5.** 0.0002

6. 5%   **7.** 13%   **8.** 200%   **9.** 41.15%   **10.** 1500%

page 154   Class Exercises

1. a is 25% of 32; $a = (0.25)(32)$

2. 150 is b% of 200; $150 = b(200)$

3. 30 is 15% of c; $30 = (0.15)c$

4. a is 35% of 400; $a = (0.35)(400)$

5. 16 is b% of 48; $16 = b(48)$

6. 8 is 4% of c; $8 = (0.04)c$

7. Let $x =$ the number.
x is 4% of 150.
$x = (0.04)(150)$

$x = 6$

8. Let $p =$ the percent.
32 is what percent of 40?
$32 = p(40)$

$\dfrac{32}{40} = p$; $p = 0.8$ or 80%

9. Let $x =$ the number.
24 is 150% of x.
$24 = (1.5)x$

$\dfrac{24}{1.5} = x$; $x = 16$

10. Let $x =$ the number.
x is 75% of 200.
$x = (0.75)(200)$

$x = 150$

pages 154–155   Practice Exercises

A   **1.** Let $x =$ the number.
x is 2% of 49.
$x = (0.02)(49) = 0.98$

2. Let $x =$ the number.
x is 65% of 130.
$x = (0.65)(130) = 84.5$

3. Let $x =$ the number.
x is 37.5% of 1000.
$x = (0.375)(1000) = 375$

4. Let $x =$ the number.
x is 6% of 248.
$x = (0.06)(248) = 14.88$

5. Let $p =$ the percent.
5 is what percent of 25?
$5 = p(25)$

$\dfrac{1}{5} = p$; $p = 0.2$ or 20%

6. Let $p =$ the percent.
35 is what percent of 35?
$35 = p(35)$

$1 = p$; $p = 1$ or 100%

7. Let p = the percent.
6 is what percent of 9?
$6 = p(9)$
$\frac{6}{9} = p;\ p = 0.66\overline{6}$ or $66\frac{2}{3}\%$

8. Let p = the percent.
30 is what percent of 80?
$30 = p(80)$
$\frac{30}{80} = p;\ p = 0.375$ or 37.5%

9. Let t = the number.
0.75 is 5% of t.
$0.75 = (0.05)t$
$75 = 5t$
$15 = t$

10. Let t = the number.
8.9 is 100% of t.
$8.9 = 1(t)$
$8.9 = t$

11. 80% of m = 28.
$(0.8)m = 28$
$m = 35$

12. m is $62\frac{1}{2}\%$ of 480.
$m = (0.625)(480)$
$m = 300$

13. x is $87\frac{1}{2}\%$ of 4000.
$x = (0.875)(4000)$
$x = 3500$

14. 60 is 150% of t.
$60 = (1.5)t$
$40 = t$

15. 0.8 is what percent of 32?
$0.8 = p(32)$
$0.025 = p$
$p = 0.025$ or 2.5%

16. 2.4 is what percent of 75?
$2.4 = p(75)$
$0.032 = p$
$p = 0.032$ or 3.2%

17. 36 is $66\frac{2}{3}\%$ of t.
$36 = \left(\frac{2}{3}\right)(t)$
$54 = t$

18. $33\frac{1}{3}\%$ of t is 75.
$\frac{1}{3}t = 75$
$t = 225$

19. 3.5 is what percent of 1.4?
$3.5 = p(1.4)$
$2.5 = p$
$p = 2.5$ or 250%

20. 30 is what percent of 25?
$30 = p(25)$
$1.2 = p$
$p = 1.2$ or 120%

B 21. $16\frac{2}{3}\%$ of m = 1.2.
$(0.166...7)m = 1.2$
$m = 7.2$

22. 12.5% of m = 1.5.
$(0.125)m = 1.5$
$m = 12$

23. Let p = the percent.
12 is what percent of 9?
$12 = p(9)$

$$\frac{12}{9} = p$$

$p = 1.3\overline{3}$ or $133\frac{1}{3}\%$

24. Let p = the percent.
15 is what percent of 5?
$15 = p(5)$

$$\frac{15}{5} = p$$

$p = 3$ or 300%

25. Let x = the amount.

x is $1\frac{1}{2}\%$ of $6000.

$x = (0.015)(6000)$
$x = 90$

26. Let x = the amount.

x is $8\frac{1}{4}\%$ of 720.

$x = (0.0825)(720)$
$x = 59.40$

27. Let x = the number.

x is $\frac{1}{2}\%$ of $5000.

$x = (0.005)\ 5000$
$x = 25$

28. Let p = the percent.

10 is what percent of 4000?

$10 = p(4000)$
$0.0025 = p$
$p = 0.0025$ or 0.25%

C 29. Let p = the percent.
$0.16a$ is what percent of $2a$?
$0.16a = p(2a)$
$0.08 = p$

$p = 0.08$ or 8%

30. Let p = the percent.
$5b$ is what percent of $0.3b$?
$5b = p(0.3b)$
$16.666 = p$

$p = 1666.\overline{666}$ or $1666\frac{2}{3}\%$

31. Let x = the number.
y is $y\%$ of x.

$$y = \left(\frac{y}{100}\right)(x)$$

$100 = x$

32. Let t = the number.
$3x$ is $x\%$ of t.

$$3x = \left(\frac{x}{100}\right)(t)$$

$t = 300$

33. Let x represent the number of people who agreed with the mayor.
$x = 44\%$ of 625.
$x = (0.44)(625)$
$x = 275;$ 275 people agreed with the mayor.

34. Let r represent the amount of interest.

r is $6\frac{3}{4}\%$ of 440.

$r = (0.0675)(440)$
$r = 29.70;$ Ben earned $29.70 in interest.

35. Let r = the annual interest rate.
$129.69 is what percent of $1572?
$129.69 = 1572(1)(r)$
$0.0825 = r$; The annual rate of interest was 8.25%.

36. Let x = the earlier population.
25,868 is 145% of x.
$25{,}868 = 1.45(x)$
$17{,}840 = x$; The earlier population was 17,840 people.

37. $a = 5\%$ of 80

$a = (0.05)(80)$
$a = 4$

38. $b = 4\frac{1}{2}\%$ of 200

$b = (0.045)(200)$
$b = 9$

39. $4 = 200\%$ of c
$4 = 2(c)$
$2 = c$

40. $d = 15\%$ of 20
$d = (0.15)(20)$
$d = 3$

41. $e = \frac{1}{2}\%$ of 1000

$e = (0.005)(1000)$
$e = 5$

42. $f = 33\frac{1}{3}\%$ of 21

$f = (0.33\overline{3})(21)$
$f = 7$

43. $g = 12\%$ of $66\frac{2}{3}$

$g = (0.12)(66.66\overline{6})$
$g = 8$

44. $h = \frac{1}{3}\%$ of 300

$h = (0.00\overline{3})(300)$
$h = 1$

45. $\$90 = i\%$ of $1500
$90 = i(1500)$
$0.06 = i$
$6\% = i$

$a = 4$	$b = 9$	$c = 2$
$d = 3$	$e = 5$	$f = 7$
$g = 8$	$h = 1$	$i = 6$

page 157   Class Exercises

1. Let c = change; $c = 36 - 24$, $c = 12$, $12 (decrease); Let p = percent change; $p = \frac{12}{36}$, $p = \frac{1}{3}$, $33\frac{1}{3}\%$ (decrease)

2. Let n = new price; $n = 72 + 16$, $n = 88$, $88;

Let p = percent change; $p = \frac{16}{72}$, $p = 0.222$, 22.22% (increase)

3. Let n = original price; $n = 64 - 12$, $n = 52$, \$52; Let p = percent change; $p = \dfrac{12}{52}$, $p = 23.08\%$ (increase)

4. Let c = change; $\dfrac{c}{75} = 0.33\overline{3}$, $c = 25$, \$25; Let n = new price; $n = 75 - 25$, $n = 50$, \$50

5. Let r = original price; $\dfrac{15}{r} = (0.10)$, $r = 150$, \$150; Let n = new price; $n = 150 - 15$, $n = 135$, \$135

pages 157–159   Practice Exercises

A **1.** Let n = new price.; $n = 196 - 49$, $n = 147$, \$147; Let p = percent change; $p = \dfrac{49}{196}$, $p = 0.25$, 25% (decrease)

2. Let c = change; $\dfrac{c}{144} = 0.166\overline{6}$, $c = 24$, \$24 (increase); Let n = new price; $n = 144 + 24$, $n = 168$, \$168

3. Let r = original price; $\dfrac{29}{r} = 0.35$, $r = 82.86$, \$82.86; Let n = new price; $n = 82.86 + 29$, $n = 111.86$, \$111.86

4. Let r = original price; $r = 1028 + 64$, $r = 1092$, \$1092; Let p = percent change; $p = \dfrac{64}{1092}$, $p = 0.059$, 5.9% (decrease)

5. Let c = change; $c = 883 - 909$, $c = -26$, \$26 (increase); Let p = percent change; $p = \dfrac{26}{883}$, $p = 0.029$, 2.9% (increase)

6. Let n = new price; $n - 684 = 36$, $n = 720$, \$720; Let p = percent change; $p = \dfrac{36}{684}$, $p = 0.053$, 5.3% (decrease)

7. Let c = change; $\dfrac{c}{84} = 0.0625$, $c = 5.25$, \$5.25 (decrease); Let n = new price; $84 - n = 5.25$, $n = 78.75$, \$78.75

8. Let r = original price; $514 - r = 18$, $r = 496$, \$496; Let p = percent change; $p = \dfrac{18}{496}$, $p = 0.036$, 3.6% (increase)

9. Let c = increase; $c = 19{,}500(0.15)$, $c = 2925$, \$2925 (increase); Let n = new salary; $n = 19{,}500 + 29.25$, $n = 22{,}425$, Mary could hope to earn \$22,425 after six months.

10. Let c = change in price; $c = 48(0.125)$, $c = 6$, $6; Let n = new price; $n = 48 - 6$, $n = 42$, The price of oil will be $42 if prices decline by $12\frac{1}{2}\%$

11. Let c = change; $c = 810, 153 - 708, 417$; $c = 101, 736$ people;

Let p = percent of decrease; $p = \dfrac{101,736}{810,153}$, $p = 0.126$; 12.6%. This represents a decrease in subscribers.

12. Let c = dollar change; $c = 84.95 - 78.90$, $c = 6.05$, $6.05; Let p = percent change; $p = \dfrac{6.05}{78.90}$, $p = 0.77$, MARTCO stock increased by 7.7%.

13. Let p = percent change; $p = \dfrac{1084}{9785}$, $p = 0.111$, Chowan's population has increased by 11.1%.

B 14. Let p = percent loss; $p = \dfrac{9}{135}$, $p = 0.67$, Razi must decrease his weight by 6.7%.

15. Let r = original number; $r = 7826 + 115$, $r = 7941$; Let p = percent decrease; $p = \dfrac{115}{7941}$, $p = 0.014$, There is a 1.4% decrease in the number of registered voters.

16. Let r = original price; $r = 48.50 - 6.75$, $r = 41.75$, $41.75; Let p = percent increase; $p = \dfrac{6.75}{41.75}$, $p = 0.162$, The price of sneakers increased by 16.2%.

17. Let x = old price; $\dfrac{10}{x} = 0.16\overline{6}$, $x = 60$, The old price of a cheese stick is 60¢. Let n = new price; $n = 60 + 10$, $n = 70$, The new price of a cheese stick is 70¢.

18. Let x = last year's number; $\dfrac{7}{x} = 0.14$, $x = 50$, Last year 50 students tried out for Lesterville High School's football team. Let n = this year's number; $n = 50 - 7$, $n = 43$, This year 43 students tried out for Lesterville High School's football team.

19. Let x = amount saved; $x = 0.20 (12,998)$, $x = 2599.60$, $2599.60; Let n = actual price; $n = 12,998 - 2599.60$, $n = 10,398.40$, Mr. D'Andrea actually paid $10,398.40 for his new automobile.

20. Let x = dollar change; $x = 0.15(18.50)$, $x = 2.78$, $2.78; Let n = new price; $n = 18.50 + 2.78$, $n = 21.28$, The new price for a women's haircut is $21.28.

21. Let c = price change; $c = 1.19 - 0.89$, $c = 0.30$, 30¢; Let p = percent change; $p = \dfrac{30}{89}$, $p = 0.337$, The price of tomatoes increased by 33.7%.

22. Let c = price change; $c = 21.95 \,(0.0833...)$, $c = 1.83$, $1.83; Let n = new price; $n = 21.95 + 1.83$, $n = 23.78$, This year Julio's dance lessons will cost $23.78.

23. Let c = price change; $c = 0.25\,(88)$, $c = 22$, 22¢; Let n = new price; $n = 88 - 22$, $n = 66$, The promotion price for Bagel King's new bagels is 66¢.

24. Let r = original price; $r = 189{,}000 - 19{,}000$, $r = 170{,}000$, $170,000; Let p = percent change; $p = \dfrac{19{,}000}{170{,}000}$, $p = 0.112$, This year's assessment of the Mehtas' home is an 11.2% increase.

25. Let r = original price; $r = 8990 + 1850$, $r = 10{,}840$, $10,840; Let p = percent decrease; $p = \dfrac{1850}{10{,}840}$, $p = 0.171$, The value of Ari's car decreased by 17.1%.

26. Let r = original price; $\dfrac{1.10}{x} = 0.20$, $x = 5.50$, $5.50; Let n = current price; $n = 5.50 - 1.10$, $n = 4.40$, The current price for a record by Girl Jill is $4.40.

27. Let r = original price; $\dfrac{20}{r} = 0.26$, $r = 76.9$, 76.9¢/gal; Let n = super unleaded price; $n = 76.9 + 20$, $n = 96.9$, Super unleaded gasoline costs 96.9¢/gal.

C 28. Let x = original price; $\dfrac{x - 2.75}{x} = 0.30$, $0.30x = x - 2.75$, $0.70x = 2.75$, $x = 3.93$, The price of angelfish was $3.93.

29. Let x = number of students who went last year; $\dfrac{99 - x}{x} = 0.125$, $0.125x = 99 - x$, $1.125x = 99$, $x = 88$, Last year 88 students signed up for a museum tour.

30. Let x = number of shades last year; $\dfrac{24 - x}{x} = 0.20$, $0.20x = 24 - x$, $1.2x = 24$, $x = 20$; Last year Breathless Cosmetics offered 20 shades of eye shadows.

31. Let r = original price; $\dfrac{r - 13.98}{r} = 0.39$, $r - 13.98 = 0.39r$, $0.61r = 13.98$, $r = 22.92$, The earlier price of T-shirt dresses was $22.92.

1. 1(C),   2(B),   3(A),   4(D),   5(E)

2. Let x = the increase; x = 4.25% of 36.5, x = (0.0425)(36.5), x = 1.6; Let r = the new mileage; r = 36.5 + 1.6 = 38.1, The new mileage is 38.1 mi/gal.

3. Let x = the decrease; x = 4.5% of 28.0, x = (0.045)(28.0), x = 1.26; Let r = the gas mileage with luggage; r = 28.0 − 1.26 = 26.74, The gas mileage with luggage was 26.7 mi/gal.

4. Let x = the reduction in mileage; x = 9% of 43, x = (0.09)(43), x = 3.87; Let r = the mileage with the head wind; r = 43.0 − 3.87 = 39.13, The mileage with the head wind was 39.13 mi/gal.

1. Let c = cost of service and e_2 = earnings for year 2, then e_2 = 55% of c, e_2 = 0.55c.

2. Let c = cost of service and e_3 = earnings for year 3, then e_3 = 60% of c, e_3 = 0.60c.

3. Let e = earnings for the day; e = 0.55(228), e = 125.40, Jack earned $125.40 for the day.

4. Let e_w = earnings for the week and e = earnings for the day. e = 0.6(240), e = 144, $144, e_w = 5(144), e_w = 720, Ann earned $720.

5. Let e = earnings for the day; e = 0.55c, 90 = 0.55c, $c = \dfrac{90}{0.55}$, c = 163.64, The cost of service was $163.64 for one day.

6. Let e_w = earnings for the week; e_w = 0.60c, 132 = 0.6c, c = 220, The cost of service was $220 for one week.

1. 150 mL   2. 0.35(250) oz   3. 540 g   4. 0.15(185) g

A 1. Let x = the percent of antifreeze concentration in the new solution.

	Substance	Total No. of mL	mL Pure Antifreeze
Start with	20% antifreeze concentration	125 mL	(0.20)(125)
Add	water	25 mL	0
Finish with	x% antifreeze solution	150	150x

$150x = (0.20)(125)$, $150x = 25$, $x = \dfrac{25}{150}$, $x = \dfrac{1}{6}$, $x = 0.1666$;

The solution will have an antifreeze concentration of $16\frac{2}{3}\%$.

2. Let x = the percent of acid concentration in the new solution.

	Substance	Total No. of L	No. of L Pure Nitric Acid
Start with	48% nitric acid solution	8 L	(0.48)(8)
Add	water	2 L	0
Finish with	x% nitric acid solution	10 L	$10x$

$10x = (0.48)(8)$, $10x = 3.84$, $x = \dfrac{3.84}{10}$, $x = 0.384$;

The solution will have a nitric acid concentration of 38.4%.

3. Let x = the percent of Costa Rican coffee in the new mixture.

	Substance	Total No. of g	g of Pure Costa Rican Coffee
Start with	5% Costa Rican coffee mixture	210 g	(0.05)(210)
Add	100% pure Costa Rican coffee	15 g	15
Finish with	x% Costa Rican coffee mixture	225 g	$225x$

$225x = (0.05)(210) + 15$, $225x = 25.5$, $x = \dfrac{25.5}{225}$, $x = 0.11333$;

The mixture will be $11.\overline{3}\%$ Costa Rican coffee.

B 4. Let x = the percent of orange juice concentrate in the new solution.

	Substance	Total No. of qt	No. of qt Pure Concentrate
Start with	20% concentrate sol.	4 qt	(0.20)(4)
Add	100% concentrate	1 qt	1.0
Finish with	x% concentrate sol.	5 qt	$5x$

$5x = (0.20)(4) + 1$, $5x = 1.8$, $x = \dfrac{1.8}{5}$, $x = 0.36$;

The solution will have an orange juice concentration of 36%.

5. Let x = the number of mL of pure acid to be added.

	Substance	Total No. of mL	No. of mL Pure Acid
Start with	15% acid solution	50 mL	(0.15)(50)
Add	100% pure acid	x	1.00(x) or x
Finish with	25% acid solution	50 + x	0.25(50 + x)

$0.25(50 + x) = (0.15)(50) + x$, $0.25(50 + x) = 7.5 + x$, $12.5 + 0.25x = 7.5 + x$, $5 = 0.75x$, $x = \dfrac{5}{0.75}$, $x = 7.7 = 6\dfrac{2}{3}$;

Add $6\dfrac{2}{3}$ mL of pure acid to raise the concentration.

6. Let x = the number of ounces of meat to be added.

	Substance	Total No. of oz	No. of oz Pure Cereal
Start with	45% cereal mixture	125 oz	(0.45)(125)
Add	0% cereal	x	0
Finish with	30% cereal mixture	125 + x	0.30(125 + x)

$0.30(125 + x) = (0.45)(125)$, $0.30(125 + x) = 56.25$, $37.5 + 0.3x = 56.25$, $0.3x = 18.75$, $x = \dfrac{18.75}{0.3}$, $x = 62.5$;

Add 62.5 oz of meat to the mixture to lower the concentration of cereal.

C 7. Let x = the number of mL of pure salt to be added.

	Substance	Total No. of mL	No. of mL Pure Salt
Start with	5% salt solution	25 g	(0.05)(25)
Add	100% pure salt	x	x
Finish with	15% salt solution	25 + x	0.15(25 + x)

$0.15(25 + x) = (0.05)(25) + x$, $0.15(25 + x) = 125 + x$, $3.75 + 0.15x = 1.25 + x$, $2.5 = 0.85x$, $x = \dfrac{2.5}{0.85}$, $x \approx 2.9$;

Add 2.9 g of salt to the solution to increase the concentration.

8. Let x = the number of gallons of butterfat milk to be added.

	Substance	Total No. gal	No. of gal Pure Butter Fat
Start with	1% butterfat milk	80 gal	(0.01)(80)
Add	$3\frac{1}{2}$% butterfat	x	0.035x
Finish with	2% butterfat milk	80 + x	0.02(80 + x)

$0.02(80 + x) = (0.01)(80) + (0.35)x$, $0.02(80 + x) = 0.8 + 0.035x$, $1.6 +$
$0.02x = 0.8 + 0.035x$, $0.8 = 0.015x$, $x = \dfrac{0.8}{0.015}$, $x = 53.333$; add $53\frac{1}{3}$ gal.

9. Area = (base)(height) ÷ 2 = (10.41)(6.62) ÷ 2 = 34.4571; 34.4571 in.^{2}

10. Area = (π)(radius)^2 = $\pi(2.8)^2 \approx 24.63$; 24.63 m^2

11. Area = Area of square + Area of semicircle = (side)^2 + (π)(radius)^2 ÷ 2 =
$15^2 + (\pi)(7.5)^2 \div 2 \approx 225 + 88.357 = 313.357$; 313.3573 mm^2

12. $9rs - (7 + 10rs) = 9rs - 7 - 10rs = -rs - 7$

13. $3(12 - 9pq) - (5pg - 6) = 36 - 27pq - 5pq + 6 = -32pq + 42$

14. $x + 4[3x - 8(1 - 2x)] = x + 4[3x - 8 + 16x] = x + 4(19x - 8) =$
$x + 76x - 32 = 77x - 32$

15. $-2(5m + 16) - [9 + 7(3m - 4)] = -10m - 32 - [9 + 21m - 28] =$
$-10m - 32 - (-19 + 21m) = -10m - 32 + 19 - 21m = -31m - 13$

16. $u^2v - (-3uv - u^2v) = u^2v + 3uv + u^2v = 2u^2v + 3uv$

17. $x^3y^2 - 5(2x^2y^3 + x^3y^2) - x^2y^3 = x^3y^2 - 10x^2y^3 - 5x^3y^2 - x^2y^3 =$
$-4x^3y^2 - 11x^2y^3$

page 164 Capsule Review

1. Subtraction property for equations

2. Division property for equations

3. Addition property for equations

4. Addition property for equations

5. Division property for equations

6. Mult. prop. for equations

7. Subtraction prop. for equations

8. Subtraction prop. for equations

page 166 Class Exercises

1. $x = y + 7$

2. $q = -\dfrac{12}{p}$

3. $r = \dfrac{4}{3}st$

4. $1 - n = 2m$, $m = \dfrac{1 - n}{2}$

5. $3a - 2c = 2b$, $b = \dfrac{3a - 2c}{2}$

6. $7d + 14 = 7e$, $7d = 7e - 14$, $d = \dfrac{7e - 14}{7}$, $d = e - 2$

A 1. $x = y - 10$ 2. $a = b - 13$ 3. $c = d + 3$

4. $z = -8 - y$ 5. $d = \dfrac{2y}{a}$ 6. $x = \dfrac{c}{a}$

7. $n = \dfrac{8}{3}k$ 8. $1 = -\dfrac{8}{3}m$ 9. $m = \dfrac{2j}{kl}$

10. $p = \dfrac{4t}{3sr}$ 11. $4y = 12 - 3x, \; y = \dfrac{12 - 3x}{4}$

12. $j = -2 - 3m$ 13. $y = 4t$ 14. $z = 6 + 7p$

15. $-7 - 11b = 3a, \; a = \dfrac{-7 - 11b}{3}$ 16. $5p = r + b, \; p = \dfrac{r + b}{5}$

17. $h = 2l - 4m, \; 4m = 2l - h, \; m = \dfrac{2l - h}{4}$

18. $-6a + 3b = c, \; 3b = c + 6a, \; b = \dfrac{c + 6a}{3}$

19. $w = \dfrac{A}{l}; \; w = \dfrac{63}{9} = 7$ 20. $h = \dfrac{V}{lw}; \; h = \dfrac{64}{10(2)}, \; h = \dfrac{64}{20} = 3.2$

21. $b = p - 2a; \; b = 74 - 2(27), \; b = 74 - 54, \; b = 20$

22. $l = \dfrac{P - 2w}{2}; \; l = \dfrac{30 - 2(6)}{2}, \; l = \dfrac{30 - 12}{2}, \; l = \dfrac{18}{2} = 9$

23. $b = \dfrac{V}{h}; \; b = \dfrac{260}{3}, \; b = 260\left(\dfrac{4}{3}\right), \; b = \dfrac{1040}{3} = 346\dfrac{2}{3}$

24. $r = \dfrac{C}{2\pi}; \; r = \dfrac{81.64}{2(3.14)}, \; r = \dfrac{81.64}{6.28}, \; r = 13$

B 25. $C = \dfrac{5}{9}(F - 32); \; C = \dfrac{5}{9}(-4 - 32), \; C = \dfrac{5}{9}(-36), \; C = -20$

26. $s = \dfrac{3t - G}{3}; \; s = \dfrac{3(8) - 18}{3}, \; s = \dfrac{24 - 18}{3}, \; s = \dfrac{6}{3}, \; s = 2$

27. $-7x - 14z = 4y, \; \dfrac{-7x - 14z}{4} = y$

28. $3b = -3a - 9, \; b = \dfrac{-3a - 9}{3}, \; b = -a - 3$

29. $2.1r + 0.2r = 3.16 - 0.02q, \; 2.3r = 3.16 - 0.02q, \; r = \dfrac{3.16 - 0.02q}{2.3},$

$r = \dfrac{316 - 2q}{230}, \; r = \dfrac{2(158 - q)}{230}, \; r = \dfrac{158 - q}{115}$

30. $5.12t - 0.12t = 4.5 - 3.5s$, $5t = 4.5 - 3.5s$, $t = \dfrac{4.5 - 3.5s}{5}$, $t = \dfrac{45 - 35s}{50}$,

$t = \dfrac{5(9 - 7s)}{50}$, $t = \dfrac{9 - 7s}{10}$

31. $-12t + 3w = 4t$, $3w = 16t$, $w = \dfrac{16t}{3}$ **32.** $-a = 7j - 21a$, $20a = 7j$, $j = \dfrac{20a}{7}$

33. $\dfrac{1}{3}v - \dfrac{2}{3}q = 2v + 3$, $-\dfrac{2}{3}q = \dfrac{5}{3}v + 3$, $q = -\dfrac{3}{2}\left(\dfrac{5}{3}v + 3\right)$, $q = -\dfrac{5}{2}v - \dfrac{9}{2}$,

$q = \dfrac{-5v - 9}{2}$

34. $\dfrac{3}{2}p - \dfrac{t}{2} = t - 8$, $\dfrac{3}{2}p = \dfrac{3}{2}t - 8$, $p = \dfrac{2}{3}\left(\dfrac{3}{2}t - 8\right)$, $p = t - \dfrac{16}{3}$, $p = \dfrac{3t - 16}{3}$

35. $\dfrac{3}{4}l - \dfrac{3}{4}h = l - 4j$, $-\dfrac{3}{4}h = \dfrac{1}{4}l - 4j$, $h = -\dfrac{4}{3}\left(\dfrac{1}{4}l - 4j\right)$, $h = -\dfrac{1}{3}l + \dfrac{16}{3}j$,

$h = \dfrac{-l + 16j}{3}$

36. $6 - 5m = \dfrac{3}{2}n + 3$, $\dfrac{3}{2}n = 3 - 5m$, $n = \dfrac{2}{3}(3 - 5m)$, $n = 2 - \dfrac{10}{3}m$,

$n = \dfrac{6 - 10m}{3}$

37. $prt = A - p$, $t = \dfrac{A - p}{pr}$

38. $h = \dfrac{V}{\pi r^2}$

C 39. $L = a + dn - d$, $dn = L - a + d$, $n = \dfrac{L - a + d}{d}$

40. $A = \dfrac{1}{2}ha + \dfrac{1}{2}bh$, $A - \dfrac{1}{2}ha = \dfrac{1}{2}bh$, $b = \dfrac{2}{h}\left(A - \dfrac{1}{2}ha\right)$, $b = \dfrac{2A}{h} - \dfrac{ha}{h}$,

$b = \dfrac{2A - ha}{h}$

41. $S = \dfrac{1}{2}na + \dfrac{1}{2}nl$, $\dfrac{1}{2}nl = S - \dfrac{1}{2}na$, $l = \dfrac{2}{n}\left(S - \dfrac{1}{2}na\right)$, $l = \dfrac{2S}{n} - \dfrac{na}{n}$, $l = \dfrac{2S - na}{n}$

42. $F - 32 = \dfrac{9}{5}C$, $F = \dfrac{9}{5}C + 32$ **43.** $12bc = a + 2b$, $c = \dfrac{a + 2b}{12b}$

44. $\dfrac{3}{2}x = 5 + y$, $x = \dfrac{2}{3}(5 + y)$, $x = \dfrac{10 + 2y}{3}$

45. $d = rt$, $r = \dfrac{d}{t}$; $r = \dfrac{18}{3.5}$, $r = 5.14$;   The rate is 5.14 km/h.

46. Let l = length and w = width. $P = 2l + 2w$, $2w = P - 2l$, $w = \dfrac{P}{2} - l$;

$w = \dfrac{42}{2} - 14$, $w = 21 - 14$, $w = 7$; The width is 7 ft.

page 167 Algebra in Aviation

1. $A = \dfrac{s^2}{R}$

2. $A = \dfrac{s^2}{R}$; $A = \dfrac{(9)^2}{3}$, $A = \dfrac{81}{3}$, $A = 27$ ft^2

3. $R = \dfrac{s^2}{A}$; $R = \dfrac{(12.2)^2}{30.5}$, $R = \dfrac{148.84}{30.5}$, $R = 4.88$

4. $s^2 = AR$, $s = \sqrt{AR}$; $s = \sqrt{32(4.5)}$, $s = \sqrt{144}$, $s = 12$ ft

page 169 Class Exercises

1.

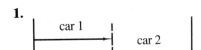

2.
$$\text{plane 1} \quad \text{plane 2}$$
$$\text{west} \quad\quad \text{east}$$
$$\overset{\longleftarrow}{d} \quad\quad \overset{\longrightarrow}{d}$$

3.
going ————→ N.Y.
returning ←———— N.Y.

4. $t + 1$

5. $t - \dfrac{1}{2}$

6. $45t = 55(t + 1)$

pages 170–171 Practice Exercises

A 1. $45t$; $55(t - 1)$

2. $65(t + 1)$; $70t$

3. $550(t + 2)$; $625t$

4. $40(t - 1)$; $30t$; $40(t - 1) = 30t$

5. $3(r + 20)$; $4r$; $3(r + 20) = 4r$

6. $20t$; $35\left(t - \dfrac{1}{2}\right)$; $20t = 35\left(t - \dfrac{1}{2}\right)$

7. $45t = 55t - 55$, $55 = 10t$, $t = \dfrac{55}{10} = 5\dfrac{1}{2}$

8. $65t + 65 = 70t$, $65 = 5t$, $t = 13$

9. $550t + 1100 = 625t$, $75t = 1100$, $t = 14\dfrac{2}{3}$

10. $40t - 40 = 30t$, $10t = 40$, $t = 4$

11. $3r + 60 = 4r$, $r = 60$

12. $20t = 35t - \dfrac{35}{2}$, $15t = \dfrac{35}{2}$, $t = \dfrac{35}{30} = 1\dfrac{1}{6}$

B 13.

	rate	× time	= distance
car 1	40	$t + 1$	$40(t + 1)$
car 2	55	t	$55t$

14.

	rate	× time	= distance
hiker 1	$r + 3$	2	$2(r + 3)$
hiker 2	r	2	$2r$

15.

rate × time = distance			
going	50	$t + \dfrac{1}{2}$	$50\left(t + \dfrac{1}{2}\right)$
returning	55	t	$55t$

16. $40(t + 1) = 55t$; $2(r + 3) + 2r = 22$; $50\left(t + \dfrac{1}{2}\right) = 55t$

17. $40t + 40 = 55t$, $15t = 40$, $t = \dfrac{40}{15} = \dfrac{8}{3}$ h;   The second car will overtake the first car in $2\dfrac{2}{3}$ h. $2r + 6 + 2r = 22$, $4r + 6 = 22$, $4r = 16$, $r = 4$;   The first hiker is walking at a rate of 7 mi/h, the second hiker is walking at a rate of 4 mi/h. $55t = 50t + \dfrac{50}{2}$, $55t = 50t + 25$, $5t = 25$, $t = 5$; $d = 55(5)$, $d = 275$;   The beach is 275 mi from their home.

C 18. Let $t =$ the time for the faster train and $t + \dfrac{30}{60} =$ the time for the slower train. $55(t + 0.5) = 65t$, $55t + 27.5 = 65t$, $10t = 27.5$, $t = 2.75$;   The faster train will catch the slower train in $2\dfrac{3}{4}$ h.

page 171   Mixed Problem Solving Review

1. Let x represent the number, then $4x^2 = 64$.

2. Let x represent the number, then $x - 5 = 2(6)$; $x - 5 = 12$.

3. Let x represent the number, then $5x - 1 = -21$, $5x = -20$, $x = -4$.

4. Let d represent the discount amount of the sneakers, then $d = (25\%)(60)$, $d = (0.25)(60)$, $d = 15$;   The sneakers are discounted by $15.00.
Let r represent the price of the sneakers before the sales tax, then $r = 60 - 15 = 45$;   The sneakers cost $45.00.
Let t represent the sales tax, then $t = (5\%)(45)$, $t = (0.05)(45)$, $t = 2.25$;   The tax is $2.25.
Let x represent the price Ted paid for the sneakers, then $x = 45 + 2.25 = 47.25$;   Ted paid a total of $47.25 for the sneakers.

page 171   Project

Answers may vary.

1.

rate x time = distance			
Smiths	50	t + 3	50(t + 3)
Duprees	70	t	70t

Duprees overtake Smiths

same direction; $50(t + 3) = 70t$

2. southbound | northbound

2000 mi

	rate x time = distance		
north–bound	r	5	5r
south–bound	$r - 100$	5	$5(r - 100)$

opposite directions; $5r + 5(r - 100) = 2000$

3.
going

home   office

returning

home   office

	rate x time = distance		
going	25	t	25t
returning	30	$t - \frac{1}{2}$	$30(t - \frac{1}{2})$

roundtrip; $25t = 30\left(t - \frac{1}{2}\right)$

4.
southbound | northbound

399 mi

	rate x time = distance		
north–bound	$r + 5$	$3\frac{1}{2}$	d
south–bound	r	$3\frac{1}{2}$	d

opposite directions; $3\frac{1}{2}r + 3\frac{1}{2}(r + 5) = 399$

pages 175–177   Practice Exercises

A 1. Let t = the time for the second ship, then $t + 3$ = the time for the first ship.
$24(t - 3) = 32(t)$, $24t - 72 = 32t$, $8t = 72$, $t = 9$;   The two ships will pass in 9 h.

2. Let t = the time for Jonathan, then $t + 1$ = the time for his little sister.
$8(t) = 5(t + 1)$, $8t = 5t + 5$, $3t = 5$, $t = \frac{5}{3} = 1\frac{2}{3}$;   Jonathan will overtake his sister in $1\frac{2}{3}$ h.

3. Let t = the time it takes for the two trains to pass.
$43t + 38t = 324$, $81t = 324$, $t = 4$;   One train will pass the other in 4 h.

4. Let t = the time it will take the two trains to be 550 km apart.
$62t + 48t = 550$, $110t = 550$, $t = 5$;   The trains will be 550 km after 5 h.

5. Let t = the time it will take the two cyclists to be 175 km apart.
$22t + 28t = 175$, $50t = 175$, $t = 3.5$;   The cyclists will be 175 km apart
after $3\frac{1}{2}$ h.

6. Let r = the rate of the westbound plane, then $r + 120$ = the rate of the
eastbound plane. $\frac{3}{4}(r) + \frac{3}{4}(r + 120) = 870$, $\frac{3}{4}r + \frac{3}{4}r + 90 = 870$, $\frac{6}{4}r = 780$,
$r = 780\left(\frac{4}{6}\right)$, $r = 520$, then $r + 120 = 640$;   The westbound train is
traveling at 520 mi/h and the eastbound train is traveling at 640 mi/h.

plane W
plane (E)

Redo

7. Let r = the rate of the second bus, then $r + 5$ = the rate of the first bus.
$r\left(1\frac{1}{2}\right) + (r + 5)\left(1\frac{1}{2}\right) = 142.5$, $1\frac{1}{2}r + 1\frac{1}{2}r + \frac{15}{2} = 142.5$, $\frac{6}{2}r + \frac{15}{2} = 142.5$,
$3r = 142.5 - 7.5$, $3r = 135$, $r = 45$;   The rate of the first bus is 50 mi/h,
and the rate of the second bus is 45 mi/h.

8. Let t = the time returning, then $13\frac{1}{2} - t$ = the time going.
$24\left(13\frac{1}{2} - t\right) = 30t$, $324 - 24t = 30t$, $324 = 54t$, $t = 6$, then $13\frac{1}{2} - t = 7\frac{1}{2}$;
$r \times t = d$, $24 \times 7\frac{1}{2} = 180$ and $30 \times 6 = 180$;   The distance is 180 mi.

9. Let t = the time returning, then $7\frac{20}{60} - t$ = the time going.
$15\left(7\frac{1}{3} - t\right) = 18t$, $110 - 15t = 18t$, $110 = 33t$, $t = 3.333$, $t = 3\frac{1}{3}$ then
$7\frac{1}{3} - t = 4$, $r \times t = d$, $(15)(4) = 60$ and $(18)\left(3\frac{1}{3}\right) = 60$;   The distance is
60 km.

10. Let r = the rate of one snail, then $r - 2$ = the rate of another snail.
$r(27) + 27(r - 2) = 432$, $27r + 27r - 54 = 432$, $54r = 486$, $r = 9$ then
$r - 2 = 7$;   The rate of one snail is 9 cm/min, and the other is 7 cm/min.

11. Let t = the time on his skateboard, then $1 - t$ = the time walking.
$7(t) + 4(1 - t) = 6$, $7t + 4 - 4t = 6$, $3t = 2$, $t = \frac{2}{3}$, then $1 - t = \frac{1}{3}$;
$r \times t = d$, $7\left(\frac{2}{3}\right) = \frac{14}{3} = 4\frac{2}{3}$;   Paul traveled $4\frac{2}{3}$ mi on his skateboard.

12. Let t = the time driving, then $2 - t$ = the time walking.

$44t + 4(2 - t) = 68$, $44t + 8 - 4t = 68$, $40t = 60$, $t = \dfrac{60}{40} = \dfrac{3}{2}$, then

$2 - t = \dfrac{1}{2}$; $r \times t = d$, $44\left(\dfrac{3}{2}\right) = 66$;   Daniel drove Erik 66 mi.

B 13. Let r = the rate of the bus, then $r + 12$ = the rate of the car.

$\dfrac{40}{60}(r + 12) = r(1)$, $\dfrac{2}{3}(r + 12) = r$, $\dfrac{2}{3}r + 8 = r$, $8 = \dfrac{1}{3}r$, $r = 24$, then $r + 12 = 36$;

The bus was traveling at 24 mi/h, the car at a rate of 36 mi/h.

14. Let r = the rate of the second part of the trip, then $\dfrac{r}{2}$ = the rate of the

first part of the trip. $2\left(\dfrac{r}{2}\right) + 4r = 70$, $r + 4r = 70$, $5r = 70$, $r = 14$, then

$\dfrac{r}{2} = 7$;   Lila's speed on the first part of her trip was 7 mi/h.

15. Let t = the time for the plane, then $t + 2$ = the time for the train.
$48(t + 2) + 240(t) = 1200$, $48t + 96 + 240t = 1200$, $288t + 96 = 1200$,

$288t = 1104$, $t = 3\dfrac{5}{6}$, then $t + 2 = 5\dfrac{5}{6}$;   The entire trip took $3\dfrac{5}{6} + 5\dfrac{5}{6}$ or $9\dfrac{2}{3}$ h.

16. Let r = the rate of the freight train, then $r + 10$ = the rate of the passenger train. $8r + 5(r + 10) = 635$, $8r + 5r + 50 = 635$, $13r + 50 = 635$, $13r = 585$, $r = 45$, then $r + 10 = 55$;   The freight train was traveling at a rate of 45 mi/h, and the passenger train was traveling at 55 mi/h.

17. Let t = the time for the eastbound train, then $t - 2$ = the time for the westbound train. $40t + 35(t - 2) = 230$, $40t + 35t - 70 = 230$, $75t = 300$, $t = 4$; 10:15 PM + 4h = 2:15 AM;   The two trains will be 230 mi apart at 2:15 AM.

18. Let r = the rate of the slower bus, then $2r - 3$ = the rate of the faster bus. $8r + 8(2r - 3) = 360$, $8r + 16r - 24 = 360$, $24r = 384$, $r = 16$, then $2r - 2 = 29$;   The slower bus traveled at a rate of 16 km/h, and the faster bus at 29 km/h.

19. Let t = the time for the freight train, then $t - 1\dfrac{1}{2}$ = the time for the

passenger train. $40(t) = 64\left(t - 1\dfrac{1}{2}\right)$, $40t = 64t - 96$, $96 = 24t$, $t = 4$;

6:00 AM + 4h = 10:00 AM;   The passenger train will overtake the freight train at 10:00 AM.

20. Let t = the time for the cyclists. $25t + 30t = 68.75$, $55t = 68.75$, $t =$ 1.25, 7:45 AM + $1\frac{1}{4}$h = 9:00 AM;   The two cyclists will meet at 9:00 AM.

21. Let t = the time for Chris, then $t - \frac{1}{2}$ = the time for Benigno.

$16\left(t - \frac{1}{2}\right) + 12(t) = 96$, $16t - 8 + 12t = 96$, $28t = 104$, $t = 3.714 = 3\frac{5}{7}$;

They will meet after $3\frac{5}{7}$ h.

C 22. Let t = the time going, then $t + \frac{15}{60}$ = the time returning.

$50(t) = 40\left(t + \frac{1}{4}\right) - 5$, $50t = 40t + 10 - 5$, $10t = 5$, $t = \frac{1}{2}$, then $t + \frac{1}{4} =$

$\frac{3}{4}$; $r \times t = d$, $50\left(\frac{1}{2}\right) = 25$, and $40\left(\frac{3}{4}\right) = 30$;   The distance going was 25 mi, and returning was 30 mi.

23. Let r = the rate of Tina's sister, then $r + 2$ = the rate of Tina. $125(r) = 100(r + 2)$, $125r = 100r + 200$, $25r = 200$, $r = 8$, then $r + 2 = 20$; $r \times t = d$, $(10)(100) = 1000$;   The length of one lap is 1000 m.

24. Let t_1 = the time for Hendrik, and t_2 = the time for Esteban. $375(t_1) = 1500$, $t_1 = 4$; $350(t_2) = 1500$, $t_2 = 4.28$; $t_2 - t_1 = 4.28 - 4 = 0.28$; $(0.28)(60) = 16.8 \approx 17$;   The length of the headstart should be approximately 17 s.

page 177   Test Yourself

1. Let n = the number.
n is 2.5% of 30.
$n = (0.025)(30)$
$n = 0.75$

2. Let m = the number.
15% *of m* is 120.
$(0.15)(m) = 120$
$m = 800$

3. Let p = the percent.
84 is what percent of 28?
$84 = p(28)$
$p = \frac{84}{28}$
$p = 3$ or 300%

4. Let n = the number.
n is 30% of $198.
$n = (0.30)(198)$
$n = 59.4$; $59.40

5. Let n = the change in price; $n = 8400 - 7980$; $n = 420$; Let p = the percent of change; $p = \frac{420}{8400} = 0.05$, $(0.05)(100) = 5$; 5% decrease

6. Let n = the amount of increase, $n = (28)(20\%) = (28)(0.20) = 5.6$.
   Let x = the new price, $x = 28 + 5.6$, $x = 33.60$; \$33.60.

7. Let x = the percent of acid in the new solution.

	Substance	No. of mL	No. of mL pure acid
Start with	10% acid solution	25 mL	(0.10)(25)
Add	water	15 mL	0
Finish with	x% acid solution	40	$40x$

$40x = (0.10)(25)$, $40x = 2.5$, $x = \dfrac{2.5}{40}$, $x = 0.0625$

The new solution will have a 6.25% concentration of acid.

8. $y = -x$

9. $5a = 15 + 10b$, $a = \dfrac{15 + 10b}{5}$; $a = 3 + 2b$

10. $M = -2j + 2k$, $M + 2j = 2k$, $\dfrac{M + 2j}{2} = k$

11. Let t = the time for the cyclists; $10(t) + 12(t) = 66$, $22t = 66$, $t = 3$; The two cyclists will be 66 mi apart after 3 h.

pages 178–179   Summary and Review

1. $27 + 2d = 9$, $2d = -18$, $d = -9$

2. $5t - 8t + 12 = 0$, $-3t + 12 = 0$, $-3t = -12$, $t = 4$

3. $-9 = 3\frac{3}{4} + c$, $-9 - \dfrac{15}{4} = c$, $-\dfrac{36}{4} - \dfrac{15}{4} = c$, $c = -\dfrac{51}{4} = -12\frac{3}{4}$

4. $-12 + 6r + r = 2r$, $-12 + 7r = 2r$, $5r = 12$, $r = \dfrac{12}{5}$

5. $50t + 50(t + 6) = 18$, $50t + 50t + 300 = 18$, $100t = -282$, $t = -2.82$

6. $3p - 7 = -6$, $3p = 1$, $p = \dfrac{1}{3}$

7. Let z represent an even integer; then $z + 8$, $z + 10$, $z + 12$, and $z + 14$ follow the even integer $z + 6$.

8. Let x = the first integer, then $x + 1$ = the second integer and $x + 2$ = the third integer.
   The equation is: $x + (x + 1) + (x + 2) = -9$; solve $3x = 3 = -9$, $3x = -12$, $x = -4$. Then $x + 1 = -3$ and $x + 2 = -2$. The three consecutive integers are $-4, -3, -2$.

9. Let x = the number.

x is 5% of 57.

$x = (0.05)(57)$

$x = 2.85$

10. Let p = the percent

4.5 is what percent of 150?

$4.5 = p(150)$

$p = \dfrac{4.5}{150}$

$p = 0.03 \ or \ 3\%$

11. Let x = the investment.

64.80 is 4.8% of x for 1 year.

$64.80 = (0.048)(x)(1)$

$64.80 = (0.048)(x)$

$x = 1350; \ \$1350$

12. Let p = the percent of defective machines.

$p = \dfrac{4}{128}$

$p = 0.03125; \ 3\frac{1}{8}\%$

13. Let x = number of increase; x is 20% of 90. $x = (0.20)(90)$, $x = 18$; Let n = number of students in the course. $n = 90 + 18$, $n = 108$; There are 108 students taking the course.

14. Let x = the percent of decrease. 50 is what percent of 60? $50 = x(60)$,

$x = 0.8\overline{3}; \ 83\frac{1}{3}\%$

15. Let x = the number of mL of water added.

	Substance	No. of mL	No. of mL Pure Acid
Start with	15% acid solution	75 mL	(0.15)(75)
Add	water	x	0
Finish with	10% acid solution	75 + x	0.10(75 + x)

$0.10(75 + x) = (0.15)(75)$, $0.10(75 + x) = 11.25$, $7.5 + 0.1x = 11.25$, $0.1x = 3.75$, $x = 37.5$

Add 37.5 mL water to the solution to lower the concentration of acid.

16. Let x = the percent of antifreeze in the new solution.

	Substance	No. of qt	Qt of Pure Antifreeze
Start with	40% antifreeze sol.	9 qt	(0.40)(9)
Add	100% antifreeze	3 qt	3
Finish with	x% antifreeze sol.	12 qt	12x

$12x = (0.40)(9) + 3$, $12x = 6.6$, $x = \dfrac{6.6}{12}$, $x = 0.55$;

The solution will have an antifreeze concentration of 55%.

17. $15m - 5n = p$, $15m - p = 5n$; $n = \dfrac{15m - p}{5}$

18. $-5d - 7d = -4bx + 5bx$, $-12d = bx$, $x = -\dfrac{12d}{b}$

19. $2a = p - b$, $a = \dfrac{p - b}{2}$; $a = \dfrac{15 - 4}{2}$, $a = \dfrac{11}{2}$

20. $r = \dfrac{c}{2\pi}$, $r = \dfrac{157}{2(3.14)}$, $r = \dfrac{157}{6.28} = 25$

21.

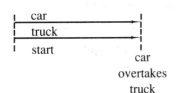

	rate	x time	= distance
north-bound	250	3	3(250)
south-bound	300	3	3(300)

$250(3) + 300(3) = 750 + 900 = $ **1650 mi apart**

22.

	rate	x time	= distance
truck	72	$t + 1$	$72(t + 1)$
car	80	t	$80t$

$72(t + 1) = 80t$, $72t + 72 = 80t$, $8t = 72$, $t = 9$;   The car will overtake the truck in 9 h.

page 180   Chapter Test

1. $-15t + 9 = -31$, $-15t = -40$, $t = \dfrac{40}{15} = 2\dfrac{2}{3}$

2. $-5 - 6 - 3y = 7$, $-11 - 3y = 7$, $-3y = 18$, $y = -6$

3. $3d - 2 = 1\dfrac{3}{4}$, $3d = 1\dfrac{3}{4} + 2$, $3d = \dfrac{15}{4}$, $d = \dfrac{15}{12} = \dfrac{5}{4}$

4. $2p + 2 - 12p + 6 = 6p$, $-10p + 8 = 6p$, $16p = 8$, $p = \dfrac{1}{2}$

5. $\dfrac{2}{3}y - 3 = -18$, $\dfrac{2}{3}y = -15$, $y = -15\left(\dfrac{3}{2}\right)$, $y = -\dfrac{45}{2} = -22\dfrac{1}{2}$

6. $5x - 2(8000 - x) = 8500$; $5x - 16{,}000 + 2x = 8500$; $7x = 24{,}500$; $x = 3500$

7. $x = -10c + c + 8$, $x = -9c + 8$

8. $h = \dfrac{2A}{b}$

9. $2w = P - 21$, $w = \dfrac{P - 21}{2}$; $w = \dfrac{15.5 - (2)(3.5)}{2}$, $w = \dfrac{8.5}{2} = 4.25$

10. Let p = the percent.
    5 is what percent of 125?
    $5 = p(125)$
    $p = \dfrac{1}{25}$
    $p = 0.4$; 4%

11. Let n = the number.
    5.6 is 25% of n.
    $5.6 = (0.25)(n)$
    $n = 22.4$

12. Let m = the amount.
    m is $33\frac{1}{3}$% of $6.90.
    $m = (0.3\overline{3})(6.90)$
    $m = 2.3$; $2.30

13. Let x = the amount of tax.
    x is 7.5% of $24.98.
    $x = (0.075)(24.98)$
    $x = 1.87$; $1.87

14. Let x = the first even integer; then $x + 2$ = the second even integer and $x + 4$ = the third even integer. The equation is: $x + (x + 2) + (x + 4) = -42$; solve $3x + 6 = -42$, $3x = -48$, $x = -16$; then $x + 2 = -14$ and $x + 4 = -12$. The three consecutive even integers are $-16, -14, -12$.

15. Let x = the change in population; $x = 3.0 - 2.5$, $x = 0.5$; 0.5 million.
    Let p = the percent of increase; $p = \dfrac{0.5}{2.5}$, $p = 0.2$ or 20%.

16. Let x = the percent of sugar in the new solution.

	Substance	No. of g	No. of g pure sugar
Start with	5% sugar solution	150 g	(0.05)(150)
Add	100% pure sugar	25 g	25
Finish with	x% sugar solution	175	175x

$175x = (0.05)(150) + 25$, $175x = 7.5 + 25$, $175x = 32.5$, $x = \dfrac{32.5}{175}$,

$x \approx 0.186$
The new solution has a concentration of 18.6%.

17. Let t = the time of the cars. $55t + 45t = 240$, $100t = 240$, $t = 2.4$;
    $d = (55)(2.4) = 132$ mi; $(45)(2.4) = 108$ mi

Challenge Let x = the total amount invested. Then $\left(\frac{1}{2}\right)(10\%)x + \left(\frac{1}{3}\right)(9\%)x +$ $\left(\frac{1}{6}\right)(12\%)x = 1110$, $\left(\frac{1}{2}\right)(0.1)x + \left(\frac{1}{3}\right)(0.9)x + \left(\frac{1}{6}\right)(0.12)x = 1110$, $0.05x +$ $0.03x + 0.02x = 1110$, $0.1x = 1110$, $x = 11{,}100$. The total amount invested was \$11,100. Then $\frac{1}{2}(\$11{,}100) = \$5550$ at 10%; $\frac{1}{3}(\$11{,}100) = \$3700$ at 9%; $\frac{1}{6}(\$11{,}100) = \$1850$ at 12%.

page 181 Preparing for Standardized Tests

1. $2x + 5 = x + 9$, $x = 4$

2. $0.025 = \dfrac{25}{1000} = \dfrac{1}{40}$

3. $\dfrac{2}{1000} + \dfrac{4}{10} + \dfrac{6}{100} = 0.002 + 0.4 + 0.06 = 0.462$

4. $5x + 5 = 3x - 13$, $2x + 5 = -13$, $2x = -18$, $x = -9$

5. $3x(x + y) - 2xy^2 = 3(-1)(-1 + 3) - 2(-1)(3)^2 = -3(2) - 2(-1)(9) =$
$- 6 + 18 = 12$

6. $-\dfrac{4}{9} \div \left(-\dfrac{2}{3}\right) = -\dfrac{4}{9}\left(-\dfrac{3}{2}\right) = \dfrac{2}{3}$

7. $\left|-6 - 4\left(\dfrac{3}{4}\right)\right| = \left|-6 - 3\right| = \left|-9\right| = 9$

8. 39, 45, and 69 are all possible answers.

9. Let t = the time (in hours) it took to go to school, then $t - 0.5$ = the time it took to return home. $25(t) = 50(t - 0.5)$, $25t = 50t - 25$, $-25t = -25$, $t = 1$. Since $d = rt$, $d = 25 \cdot 1$, $d = 25$. It is 25 miles from Greg's home to the school.

10. 30% of \$150 = \$45, leaving \$105 for the sale price of the suit. The 6% sales tax on \$105 is \$6.30, so the total cost is \$111.30.

11. Let x = the number made by Pete, and $2x - 29$ = the number made by Al. Then, $x + (2x - 29) = 88$, $3x - 29 = 88$, $3x = 117$, $x = 39$.

12. $17 + 7 - 9 = 15$; 15°C

13. $400.40 - (45 - 116.24) + (30 + 63.75) = 400.40 - 161.24 + 93.75 = 332.91$; \$332.91

14. Let $3x$ represent Mac's earnings, then $5x$ represents Tom's earnings. $3x = 30.15$, $x = 10.05$; therefore $5x = 50.25$; Tom earned \$50.25.

15. The original value of Lisa's car was \$10,000. $x(10{,}000) = 2000$, $x = \dfrac{1}{5}$, $x = 20\%$. Its value decreased by 20%.

1. $\dfrac{-6}{12} = -\dfrac{1}{2}$

2. $\dfrac{5}{8} - \dfrac{2}{8} = \dfrac{3}{8}$

3. $-\dfrac{3}{10} \cdot -\dfrac{5}{2} = -\dfrac{3}{2} \cdot -\dfrac{1}{2} = \dfrac{3}{4}$

4. $-\dfrac{2}{3} + \dfrac{5}{6} = -\dfrac{4}{6} + \dfrac{5}{6} = \dfrac{1}{6}$

5. $\dfrac{1}{2} \cdot -\dfrac{4}{1} = -\dfrac{4}{2} = -2$

6. $-\dfrac{3}{8} \cdot -\dfrac{4}{5} = \dfrac{3}{10}$

7. $-\dfrac{8}{12} + \dfrac{9}{12} = \dfrac{1}{12}$

8. $\dfrac{1}{3} \cdot \dfrac{5}{3} = \dfrac{5}{9}$

9. $-\dfrac{20}{24} - \dfrac{9}{24} = -\dfrac{29}{24}$

10. $-\dfrac{2}{3} \cdot \dfrac{5}{4} = -\dfrac{5}{6}$

11. $-\dfrac{9}{12} - \dfrac{5}{12} = -\dfrac{14}{12} = -\dfrac{7}{6}$

12. $\dfrac{4}{9} \cdot -\dfrac{3}{8} = -\dfrac{1}{6}$

13. $\dfrac{2}{5} + \dfrac{3}{10} = \dfrac{4}{10} + \dfrac{3}{10} = \dfrac{7}{10}$

14. $-\dfrac{10}{12} + \dfrac{9}{12} = -\dfrac{1}{12}$

15. $-\dfrac{4}{9} \cdot -\dfrac{3}{2} = \dfrac{2}{3}$

16. $\dfrac{5}{6} \cdot \dfrac{4}{15} = \dfrac{2}{9}$

17. $|8 - 15| = |-7| = 7$

18. $24 - 16 = 8$

19. $-5[64 \div (-4)^2] = -5[64 \div 16] = -5[4] = -20$

20. $10a^2 - 12a$

21. $5x - 3x - 6 = 2x - 6$

22. $m^2 + 5$

23. $5h + 4 - 3h + 5 = 2h + 9$

24. $5a^2 + 5b + 2b - 2a^2 = 3a^2 + 7b$

25. $7mn - 2mn - 6 + 5 = 5mn - 1$

26. $-3x^2 + 3y^2 + 5y^2 + 5x^2 = 2x^2 + 8y^2$

27. $8\left(\dfrac{3}{4}\right) + (-6) = 6 - 6 = 0$

28. $\left| -6 - 4\left(\dfrac{3}{4}\right) \right| = |-6 - 3| = |-9| = 9$

29. $\dfrac{5(8) + 2}{\frac{3}{4}} = \dfrac{40 + 2}{\frac{3}{4}} = \dfrac{42}{\frac{3}{4}} = 42\left(\dfrac{4}{3}\right) = 56$

30. $2[8^2 + 4(-6)] = 2[64 - 24] = 2[40] = 80$

31. $\dfrac{[8 + (-6)]^2}{\frac{3}{4} - \frac{1}{2}} = \dfrac{[2]^2}{\frac{3}{4} - \frac{2}{4}} = \dfrac{4}{\frac{1}{4}} = 4(4) = 16$

32. $2(8) + 4\left(-6 + \dfrac{3}{4}\right) = 16 + 4\left(-\dfrac{24}{4} + \dfrac{3}{4}\right) = 16 + 4\left(\dfrac{-21}{4}\right) = 16 - 21 = -5$

33. $(|8| - |-6|)^3 = (8 - 6)^3 = (2)^3 = 8$

34. $8\left\{2 + \dfrac{3}{4}[2 - (-6)]\right\} = 8\left\{2 + \dfrac{3}{4}[8]\right\} = 8\{2 + 6\} = 8\{8\} = 64$

35. $8\left(\dfrac{3}{4} \div -6\right) = 8\left(\dfrac{3}{4} \cdot -\dfrac{1}{6}\right) = 8\left(-\dfrac{1}{8}\right) = -1$

36. $m - 6$

37. $m - 2$

38. $m + 2$

39. $m + (m - 2) = 30$

40. $(m - 3) + (m - 5) = 24$

41. $4 = c - 1, c = 5$

42. $y = 0$

43. $x = -9$

44. $4p = 12, p = 3$

45. $9d = 1, d = \dfrac{1}{9}$

46. $48 = 14n, n = \dfrac{48}{14} = \dfrac{24}{7}$

47. $g = -14$

48. $t = \left(-\dfrac{1}{4}\right)(8), t = -2$

49. $2a = -\dfrac{1}{5}, a = -\dfrac{1}{10}$

50. $0.3r = 1.2, r = 4$

51. $b = -1$

52. $2m = 12, m = 6$

53. $7h = 14, h = 2$

54. $\dfrac{2}{5}r = 2\dfrac{1}{5} - 1, \dfrac{2}{5}r = \dfrac{11}{5} - \dfrac{5}{5}, \dfrac{2}{5}r = \dfrac{6}{5}, r = 3$

55. $9t - 4t - 12 = 8, 5t = 20, t = 4$

56. $8p - 2p + 5 = 23, 6p = 18, p = 3$

57. $S(1 - r) = a, a = 8[1 - (-3)], a = 8[4], a = 32$

58. $prt = A - p, t = \dfrac{A - p}{pr}; t = \dfrac{840 - 600}{(600)(0.08)}; t = \dfrac{240}{48} = 5$

59. $h = \dfrac{V}{\pi r^2}; h = \dfrac{628}{(3.14)(5)^2}, h = \dfrac{628}{78.5}, h = 8$

60. $E = I(R + r); E = 80(12 + 13), E = 80(25), E = 2000$

61. $2w = P - 2l, w = \dfrac{P - 2l}{2}; w = \dfrac{200 - 2(75)}{2}, w = \dfrac{200 - 150}{2}, w = \dfrac{50}{2} = 25$

62. Let $n = $ the number; $4n - 9 = 55, 4n = 64, n = 16$; The number is 16.

63. Let $l = $ the length of the rectangle; $42 = 2l + 2(6), 42 = 2l + 12, 2l = 30,$ $l = 15$; the length is 15 cm.

64. Let $a = $ Frank's age; $2a + 3 = 35, 2a = 32, a = 16$; Frank is 16 years old.

65. Let $n = $ the first integer; then $n + 1 = $ the second integer. The equation is: $n + (n + 1) = 49$; solve $2n + 1 = 49, 2n = 48, n = 24$. Then $n + 1 = 25$. The two consecutive integers are 24, 25.

66. Let $q = $ Gloria's allowance; $3q + q = 48, 4q = 48, q = 12$; Gloria's allowance is $12.

67. Let c = the number.

c is 3% of 80.

$c = (3\%)(80)$

$c = (0.03)(80)$

$c = 2.4$

68. Let x = the number.

x is 7% of 30.

$x = (7\%)(30)$

$x = (0.07)(30)$

$x = 2.1$

69. Let p = the percent.

15 is what percent of 50?

$15 = p(50)$

$p = 0.3$ or 30%

70. Let p = the percent.

10 is what percent of 24?

$10 = p(24)$

$p = 0.416\overline{66}$ or $41\frac{2}{3}\%$

71. Let n = the amount.

n is 25% of \$500.

$n = (25\%)(500)$

$n = (0.25)(500)$

$n = 125; \$125.00$

72. Let n = the amount.

n is 45% of \$650.

$n = (45\%)(650)$

$n = (0.45)(650)$

$n = 292.5; \$292.50$

73. Let x = the number.

5 is 20% of x.

$5 = (20\%)(x)$

$5 = (0.20)(x)$

$x = 25$

74. Let x = the number.

10 is 80% of x.

$10 = (80\%)(x)$

$10 = (0.8)(x)$

$x = 12.5$

75. Let p = the percent.

18 is what percent of 24?

$18 = p(24)$

$p = 0.75$ or 75%

76. Let p = the percent.

24 is what percent of 144?

$24 = p(144)$

$p = 0.1\overline{66}$ or $16\frac{2}{3}\%$

77. Let x = the number.

x is $37\frac{1}{2}\%$ of 96.

$x = \left(37\frac{1}{2}\%\right)(96)$

$x = (0.375)(96)$

$x = 36$

78. Let x = the number.

x is $8\frac{1}{4}\%$ of 120.

$x = \left(8\frac{1}{4}\%\right)(120)$

$x = (0.0825)(120)$

$x = 9.9$

79. Let x = the number.

16 is 5% of x.

$16 = (5\%)(x)$

$16 = (0.05)(x); x = 320$

80. Let x = the number.

36 is 2% of x.

$36 = (0.02)(x)$

$x = 1800$

81. Let x = the number.
   30 is 150% of x.
   $30 = (150\%)(x)$
   $30 = (1.5)(x)$
   $x = 20$

82. Let x = the number.
   50 is 250% of x.
   $50 = (250\%)(x)$
   $50 = (2.5)(x)$
   $x = 20$

83. Let p = the percent.
   4 is what percent of 12?
   $4 = p(12)$
   $p = \dfrac{1}{3}$ or $33\dfrac{1}{3}\%$

84. Let p = the percent.
   17 is what percent of 85?
   $17 = p(85)$
   $p = 0.2$ or 20%

85. Let x = the number.
   x is $\dfrac{1}{2}\%$ of $200.
   $x = \left(\dfrac{1}{2}\%\right)(200)$
   $x = (0.005)(200)$
   $x = 1;\ \$1.00$

86. Let x = the number.
   x is $\dfrac{1}{4}\%$ of $100.
   $x = \left(\dfrac{1}{4}\%\right)(100)$
   $x = (0.0025)(100)$
   $x = 0.25;\ \$0.25$

87. $y = x - 9$

88. $3m = n + 5,\ m = \dfrac{n + 5}{3}$

89. $\dfrac{1}{2}g = h - 3,\ g = 2(h - 3),\ g = 2h - 6$

90. $a = 2b - 6c,\ 2b = a + 6c,\ b = \dfrac{a + 6c}{2}$

91. $12t - p = 4t + 20,\ p = 8t - 20$

92. $\dfrac{1}{2}x = 1\dfrac{1}{2} - y,\ x = 2\left(1\dfrac{1}{2} - y\right),\ x = 3 - 2y$

93. $ac = 3d;\ c = \dfrac{3d}{a}$

94. $j - 5 = \dfrac{1}{3}k,\ k = 3(j - 5),\ k = 3j - 15$

95. $2p - 3q = 400,\ 2p - 400 = 3q,\ q = \dfrac{2p - 400}{3}$

96. Let n = the amount of decrease; n is 25% of $85, $n = (0.25)(85)$, $n =$
   21.25; $21.25; Let x = the new price; $x = 85 - 21.25$, $x = 63.75;\ \$63.75$

97. Let x = the amount of increase; $x = 1002 - 835$, $x = 167$; Let p = the percent of increase; $p = \dfrac{167}{835}$, $p = 0.2$ or 20%

98. Let x = the number of mL of water to be added.

	Substance	No. of mL	No. of mL pure acid
Start with	30% acid solution	200 mL	(0.30)(200)
Add	water	x	0
Finish with	25% acid solution	200 + x	0.25(200 + x)

$0.25(200 + x) = (0.30)(200)$, $0.25(200 + x) = 60$, $50 + 0.25x = 60$, $0.25x = 10$, $x = \dfrac{10}{0.25}$, $x = 40$;

Add 40 mL of water to the solution to decrease the concentration.

99. Let x = the percent of peanuts in the new mixture.

	Substance	No. of lb	No. of lb pure peanuts
Start with	25% peanut mixture	15 lb	(0.25)(15)
Add	100% peanuts	3 lb	3
Finish with	x% peanut mixture	18 lb	18x

$18x = (0.25)(15) + 3$, $18x = 6.75$, $x = \dfrac{6.75}{18}$, $x = 0.375$;

The mixture will be 37.5% peanuts.

100. Let r = the rate of the southbound train, then $r + 6$ = the rate of the northbound train. $r(5) + (r + 6)(5) = 510$, $5r + 5r + 30 = 510$, $10r = 480$, $r = 48$; The rate of the southbound train is 48 mi/h, and the rate of the northbound train is 54 mi/h.

101.

	Rate	× time	= distance
Car 1	40	$t + 2$	40(t + 2)
Car 2	50	t	50t

$40(t + 2) = 50t$, $40t + 80 = 50t$, $80 = 10t$, $t = 8$;
The second car will overtake the first in 8 h.

Chapter 5   Inequalities in One Variable

page 186   Capsule Review

1. $-2\,-1\ 0\ 1\ 2\ 3$

2. $-2\,-1\ 0\ 1\ 2\ 3$

3. $-2\,-1\ 0\ 1\ 2\ 3$

4. $-2\,-1\ 0\ 1\ 2\ 3$

5. $-2\,-1\ 0\ 1\ 2\ 3$

page 188   Class Exercises

1. false   **2.** false   **3.** true   **4.** true

5. $-3\,-2\,-1\ 0\ 1\ 2$

6. $-2\,-1\ 0\ 1\ 2\ 3$

7. $-4\,-3\,-2\,-1\ 0\ 1$

8. $-2\,-1\ 0\ 1\ 2\ 3$

9. $\left\{\dfrac{1}{2}\right\}$

10. {all real numbers greater than 0}

11. {all real numbers less than or equal to -1}

pages 188–189   Practice Exercises

A 1. $4a = a - 9,\ 3a = -9,\ a = -3$ $-3\,-2\,-1\ 0\ 1\ 2$

2. $6x = x - 15,\ 5x = -15,\ x = -3$ $-3\,-2\,-1\ 0\ 1\ 2$

3. $-7m = -m - 6,\ -6m = -6,\ m = 1$ $-2\,-1\ 0\ 1\ 2\ 3$

4. $-z + 6 = -7z,\ 6 = -6z,\ -1 = z$ $-3\,-2\,-1\ 0\ 1\ 2$

5. $2y = 12 - y,\ 3y = 12,\ y = 4$ $-1\ 0\ 1\ 2\ 3\ 4$

6. $4n = 15 - n,\ 5n = 15,\ n = 3$ $-2\,-1\ 0\ 1\ 2\ 3$

7. $-2\,-1\ 0\ 1\ 2\ 3$

8. $0\ 5\ 10\ 15\ 20\ 25$

9. $-2\,-1\ 0\ 1\ 2\ 3$

10. $-3\,-2\,-1\ 0\ 1\ 2$

11. $-2\,-1\ 0\ 1\ 2\ 3$

12. $-1\ 0\ 1\ 2\ 3\ 4$

13. $-2\,-1\ 0\ 1\ 2\ 3$

14. $-2\,-1\ 0\ 1\ 2\ 3$

15. $-2\,-1\ 0\ 1\ 2\ 3$

16. $-2\,-1\ 0\ 1\ 2\ 3$

17. $-2\,-1\ 0\ 1\ 2\ 3$

18. $-2\,-1\ 0\ 1\ 2\ 3$

19. {0}

20. $\left\{\dfrac{1}{2}\right\}$

21. {all real numbers}

22. {all real numbers less than 2}

23. {all real numbers, except −1}

24. {all real numbers greater than or equal to −1}

B 25. $-4x - 26 = -11$, $-4x = 15$, $x = -\dfrac{15}{4}$, $\left\{-\dfrac{15}{4}\right\}$

26. $-6y + 21 = 36$, $-6y = 15$, $y = -\dfrac{5}{2}$, $\left\{-\dfrac{5}{2}\right\}$

27. $6c - 9 = 6$, $6c = 15$, $c = \dfrac{5}{2}$, $\left\{\dfrac{5}{2}\right\}$

28. $-3r - 19 = -19$, $-3r = 0$, $r = 0$, $\{0\}$

29. $12m + 18 = 4 - m + 1$, $12m + 18 = 5 - m$, $13m + 18 = 5$, $13m = -13$, $m = -1$, $\{-1\}$

30. $15 + 3z - 6 = 18 + 2z$, $9 + 3z = 18 + 2z$, $9 + z = 18$, $z = 9$, $\{9\}$

31.

32.

33.

34.

C 35. all nonnegative real numbers

36. all real numbers

37. all nonzero real numbers, except when $a < 0$, and $b > 0$

38. all real numbers where a and b are additive inverses

39. 0, false   **40.** 0, false   **41.** 1, true

42. Answers may vary. An example: Let $x = 5$, $y = 8$, and $z = 10$. $5 < 8$ and $8 < 10$. The property guarantees the conclusion that $5 < 10$. The relative positions of these numbers on the number line illustrate this property.

43. Divide the cubes into 3 equal groups A, B, and C. Each group has 5 cubes.

Comparison #1:   If $A = B$, then C has the lightest cube. If $A > B$ then B has the lightest cube. If $A < B$ then A has the lightest cube.

Comparison #2:   Divide the lightest weight group into D, E, and F. Group D and E have two cubes each, group F has one. If $D = E$ then group F is the lightest cube. If $D > E$ then group E has the lightest cube. If $D < E$ then group D has the lightest cube.

Comparison #3:   Take the lightest group E or D and compare one cube against another.

1. subtr. 15; $m = -11$ **2.** add 12; $v = 7$ **3.** add 8; $q = -15$ **4.** subtr. 17; $c = -11$

page 192   Class Exercises

1. Add 4 to each side.   **2.** Add 4 to each side.

3. Subtract 1 from each side.   **4.** Subtract 4 from each side.

5. Subtract $3x$ from each side; subtract 8 from each side.

6. Subtract $\frac{1}{2}t$ from each side; add 1 to each side.

7. Combine like terms; subtract 3 from each side.

8. Combine like terms; subtract 8 from each side.

9. $x + 4 - 4 > 0 - 4$, $x > -4$

10. $y - 5 + 5 \le -5 + 5$, $y \le 0$

11. $1x - 3 \le -4$, $x - 3 + 3 \le -4 + 3$, $x \le -1$

12. $3\frac{1}{4}x - 2\frac{1}{4}x - 6 > 2\frac{1}{4}x - 2\frac{1}{4}x + 1$, $x - 6 > 1$,
$x - 6 + 6 > 1 + 6$, $x > 7$

13. $7x - 7 - 6x < -8$, $x - 7 < -8$, $x < -1$

14. $6x + 9 \ge 5x + 5$, $x + 9 \ge 5$, $x \ge -4$

15. Answers may vary. For example: $3 > 2$, $4 > -8$, but $3 - 4 \not> 2 - (-8)$.

pages 192–193   Practice Exercises

A   1. $m + 2 - 2 > 0 - 2$, $m > -2$

2. $c + 5 - 5 > 4 - 5$, $c > -1$

3. $3 + a - 3 < 4\frac{1}{2} - 3$, $a < 1\frac{1}{2}$

4. $2 + x - 2 > 3\frac{1}{2} - 2$, $x > 1\frac{1}{2}$

5. $x - 17 + 17 \ge -15 + 17$, $x \ge 2$

6. $m - 25 + 25 \ge -35 + 25$, $m \ge -10$

7. $3\frac{1}{2}y - 2\frac{1}{2}y + 2 \le 2\frac{1}{2}y - 2\frac{1}{2}y + 2,$

$y + 2 \le 2, y + 2 - 2 \le 2 - 2, y \le 0$

8. $4\frac{2}{3}z - 3\frac{2}{3}z - 1 \le 3\frac{2}{3}z - 3\frac{2}{3}z - 1,$

$z - 1 \le -1, z - 1 + 1 \le -1 + 1, z \le 0$

9. $1\frac{1}{2}x - 1\frac{1}{2}x + 6 > 2\frac{1}{2}x - 1\frac{1}{2}x - 6, 6 > x - 6,$

$6 + 6 > x - 6 + 6, x < 12$

10. $3x - 2x + 8 > 7, x + 8 > 7,$
$x + 8 - 8 > 7 - 8, x > -1$

11. $4a - 3a + 6 > 10, a + 6 > 10,$
$a + 6 - 6 > 10 - 6, a > 4$

12. $6c - 5c + 15 > -18, c + 15 > -18,$
$c + 15 - 15 > -18 - 15, c > -23$

13. $z + 7 \ge 7, z + 7 - 7 \ge 7 - 7, z \ge 0$

14. $x + 15 \ge 1, x + 15 - 15 \ge 1 - 15,$
$x \ge -14$

15. $x - 10 < 10, x - 10 + 10 < 10 + 10,$
$x < 20$

B 16. $4z - 3z + 6 < 3z - 3z - 6, z + 6 < -6, z + 6 - 6 < -6 - 6, z < -12;$
all negative integers less than $-12;$ $\{\ldots, -15, -14, -13\}$

17. $5x - 4x + 2 < 4x - 4x - 8, x + 2 < -8, x + 2 - 2 < -8 - 2, x <$
$-10;$ all negative integers less than $-10;$ $\{\ldots, -13, -12, -11\}$

18. $2t - t - 5 < t - t + 2, t - 5 < 2, t - 5 + 5 < 2 + 5, t < 7;$ all negative
integers less than $7;$ $\{\ldots, -3, -2, -1\}$

19. $2x - x - 6 < -5 + x - x, x - 6 < -5, x - 6 + 6 < -5 + 6, x < 1;$ all
negative integers less than $1;$ $\{\ldots, -3, -2, -1\}$

20. $3x - 2x + 6 < 2x - 2x + 6, x + 6 < 6, x + 6 - 6 < 6 - 6, x < 0;$ all
negative integers less than $0;$ $\{\ldots, -3, -2, -1\}$

21. $6y - 6y + 5 < -5 + 7y - 6y, 5 < -5 + y, 5 + 5 < -5 + 5 + y, 10 < y,$
$y > 10;$ all negative integers greater than $10;$ $\emptyset$

22. $6x - 5x + 5 > 2$, $x + 5 > 2$, $x + 5 - 5 > 2 - 5$, $x > -3$; all negative integers greater than -3; $\{-2, -1\}$

23. $6.5 < 2t + 6 - t$, $6.5 < t + 6$, $6.5 - 6 < t$, $0.5 < t$, $t > 0.5$; all negative integers greater than 0.5; Ø

24. $7.2 < 3x + 3 - 2x$, $7.2 < x + 3$, $7.2 - 3 < x + 3 - 3$, $4.2 < x$, $x > 4.2$; all negative integers greater than 4.2; Ø

25. $3t - 2t + 5.8 < 5.6 + 2t - 2t$, $t + 5.8 < 5.6 - 5.8$, $t < -0.2$

26. $2m - m - 0.5 > m - m + 3.5$,
$m - 0.5 + 0.5 > 3.5 + 0.5$, $m > 4$

27. $5x - 4x + \dfrac{1}{4} > 4x - 4x + 10\dfrac{3}{4}$,

$x + \dfrac{1}{4} - \dfrac{1}{4} > 10\dfrac{3}{4} - \dfrac{1}{4}$, $x > 10\dfrac{1}{2}$

28. $2 + 3t \le 8\dfrac{1}{2} + 2t$, $2 + 3t - 2t \le 8\dfrac{1}{2} + 2t - 2t$,

$2 - 2 + t \le 8\dfrac{1}{2} - 2$, $t \le 6\dfrac{1}{2}$

29. $m - 5 + 5 > 1 + 5$, $m > 6$; all positive integers greater than 6; $\{7, 8, 9, \ldots\}$

30. $-3 + 10x - 9x \ge 9x - 9x + 7$, $-3 + x \ge 7$, $-3 + 3 + x \ge 7 + 3$, $x \ge 10$; all positive integers greater than 10; $\{10, 11, 12, \ldots\}$

31. $4z + 3 > 5z + 9$, $4z - 4z + 3 > 5z - 4z + 9$, $3 > z + 9$, $3 - 9 > z + 9 - 9$, $-6 > z$, $z < -6$; all positive integers less than -6; Ø

C 32. Answers may vary. For example: $5 > 2$, and $4 < 7$, then $5 - 4 > 2 - 7$; $1 > -5$ is true.

33. Answers may vary. For example: $5 > 2$, and $3 > -1$, then $5 - 3 > 2 - (-1)$; $2 > 3$ is false.

34. false   **35.** false

36. Let b = the length of the shortest side of the triangle, then $\dfrac{1}{2}(18)(b) \le 36$; $9b \le 36$, $b \le 4$; The length of fencing needed for the shortest side can be no longer than 4 ft.

37. $\dfrac{1}{2}(18)(b) \ge 54$; $9b \ge 54$, $b \ge 6$; The fencing must be at least 6 ft long.

38. $3x + 12 = 9x + 2$, $-6x = -10$, $x = \dfrac{5}{3}$

39. $y - 5y - 3 = -3y - 3 + 2y$, $-4y - 3 = -y - 3$, $-3y = 0$, $y = 0$

40. Let n represent the number. $8(n - 5)$

page 194 Capsule Review

1. Divide by 3. $y = -4$

2. Multiply by -5. $y = -125$

3. Divide by -7. $y = -3$

4. Divide by 6. $y = \dfrac{3}{2}$

page 196 Class Exercises

1. Divide each side by 4.

2. Multiply each side by $-\dfrac{3}{2}$.

3. Divide each side by -2.

4. Multiply each side by $\dfrac{4}{3}$.

5. Divide each side by 0.5.

6. Divide each side by -25.

7. $\dfrac{1}{2}(2x) \geq \dfrac{1}{2}(0)$; $\{x: x \geq 0\}$

8. $\dfrac{5}{4}\left(\dfrac{4}{5}y\right) < \dfrac{5}{4}\left(-\dfrac{4}{1}\right)$; $\{y: y < -5\}$

9. $-\dfrac{1}{3}(-3c) > -\dfrac{1}{3}(9)$; $\{c: c < -3\}$

10. $-2\left(-\dfrac{1}{2}d\right) \leq -2(4)$; $\{d: d \geq -8\}$

11. $-\dfrac{3}{2}\left(-\dfrac{2}{3}m\right) \geq -\dfrac{3}{2}\left(\dfrac{12}{1}\right)$; $\{m: m \leq -18\}$

12. $\dfrac{-1.2}{-1.2}t < \dfrac{-0.6}{-1.2}$; $\{t: t > 0.5\}$

13. $\dfrac{-4.6}{-4.6}k > \dfrac{1.38}{-4.6}$; $\{k: k < -0.3\}$

14. $\dfrac{6}{5}y > -\dfrac{31}{10}$, $\left(\dfrac{5}{6}\right)\left(\dfrac{6}{5}y\right) > \left(\dfrac{5}{6}\right)\left(-\dfrac{31}{10}\right)$; $\left\{y: y > -\dfrac{31}{12}\right\}$

15. Multiplication by 0 is excluded because if $a > b$, then $a \cdot 0 > b \cdot 0$ then is $0 > 0$; which is a false statement.

16. Division by 0 is excluded because if $a > b$, then $\dfrac{a}{0} > \dfrac{b}{0}$ is undefined.

17. Addition or subtraction by 0 is included because if $a > b$, then $a + 0 > b + 0$ is true. Also, if $a > b$ then $a - 0 > b - 0$ is true.

page 197 Practice Exercises

A 1. $\dfrac{4x}{4} < \dfrac{20}{4}$, $x < 5$

2. $\dfrac{7y}{7} < \dfrac{28}{7}$, $y < 4$

3. $\dfrac{8a}{8} < \dfrac{56}{8}$, $a < 7$

4. $\dfrac{5z}{5} < \dfrac{55}{5}$, $z < 11$

5. $\dfrac{6b}{6} > \dfrac{48}{6}$, $b > 8$

6. $\dfrac{9x}{9} > \dfrac{63}{9}$, $x > 7$

7. $\dfrac{-7y}{-7} < \dfrac{42}{-7}$, $y < -6$

8. $\dfrac{-3t}{-3} < \dfrac{36}{-3}$, $t < -12$

9. $\dfrac{28}{-4} < \dfrac{-4a}{-4}$, $-7 < a$

10. $\dfrac{54}{-9} < \dfrac{-9c}{-9}$, $-6 < c$

11. $\dfrac{-2b}{-2} < \dfrac{-64}{-2}$, $b < 32$

12. $\dfrac{-5z}{-5} < \dfrac{-40}{-5}$, $z < 8$

13. $\dfrac{3x}{3} > \dfrac{-6}{3}$, $x > -2$

14. $\dfrac{4a}{4} > \dfrac{-16}{4}$, $a > -4$

15. $2\left(\dfrac{1}{2}y\right) \le 2(3)$, $y \le 6$

16. $\dfrac{5}{4}\left(\dfrac{4}{5}z\right) \le \dfrac{5}{4}\left(\dfrac{12}{1}\right)$, $z \le 15$

17. $4 \le \dfrac{9}{5}x - \dfrac{5}{5}x$, $4 \le \dfrac{4}{5}x$, $\dfrac{5}{4}\left(\dfrac{4}{1}\right) \le \dfrac{5}{4}\left(\dfrac{4}{5}x\right)$, $5 \le x$, $x \ge 5$

18. $3 \le \dfrac{4}{3}y - \dfrac{3}{3}y$, $3 \le \dfrac{1}{3}y$, $3(3) \le \left(\dfrac{3}{1}\right)\left(\dfrac{1}{3}y\right)$, $9 \le y$, $y \ge 9$

B 19. $5 \ge \dfrac{3}{2}z - \dfrac{2}{2}z$, $5 \ge \dfrac{1}{2}z$, $2(5) \ge 2\left(\dfrac{1}{2}\right)z$, $10 \ge z$, $z \le 10$

20. $2 \ge \dfrac{7}{6}b - \dfrac{6}{6}b$, $2 \ge \dfrac{1}{6}b$, $6(2) \ge 6\left(\dfrac{1}{6}b\right)$, $12 \ge b$, $b \le 12$

21. $-3p < -21$, $\dfrac{-3p}{-3} > \dfrac{-21}{-3}$, $p > 7$

22. $-2x > 40$, $\dfrac{-2x}{-2} < \dfrac{40}{-2}$, $x < -20$

23. $-1 > 4y$, $\dfrac{-1}{4} > \dfrac{4y}{4}$, $-\dfrac{1}{4} > y$, $y < -\dfrac{1}{4}$

24. $9 > -2c$, $\dfrac{9}{-2} < \dfrac{-2c}{-2}$, $-\dfrac{9}{2} < c$, $c > -\dfrac{9}{2}$

25. $-\dfrac{8}{5}\left(-\dfrac{5}{8}x\right) \le -\dfrac{8}{5}\left(\dfrac{25}{1}\right),\ x \le -40$

$-80\ -60\ -40\ -20\ 0\ 20$

26. $\dfrac{-1.5}{-1.5}t < \dfrac{-75}{-1.5},\ t < 50$

$-40\ -20\ 0\ 20\ 40\ 60$

27. $\dfrac{-11z}{-11} \ge \dfrac{1.21}{-11},\ z \ge -0.11$

$-3\ -2\ -1\ 0\ 1\ 2$

28. $-\dfrac{5}{3}\left(-\dfrac{3}{5}x\right) \le -\dfrac{5}{3}\left(\dfrac{15}{1}\right),\ x \le -25$

$-30\ -20\ -10\ 0\ 10\ 20$

29. $\dfrac{-0.5m}{-0.5} \ge \dfrac{-1.5}{-0.5},\ m \ge 3$

$-1\ 0\ 1\ 2\ 3\ 4$

30. $\dfrac{-0.6t}{-0.6} < \dfrac{1.08}{-0.6},\ t < -1.8$

$-2\ -1\ 0\ 1\ 2\ 3$

C 31. $\dfrac{-1.3t}{-1.3} < \dfrac{0.6}{-1.3},\ t < -0.462$

$-3\ -2\ -1\ 0\ 1\ 2$

32. $\dfrac{-4h}{-4} > \dfrac{0.25}{-4},\ h > -0.0625$

$-1\ 0\ 1$

33. $-4(-2.4) > -4\left(-\dfrac{1}{4}k\right),\ 9.6 > k,\ k < 9.6$

$-15\ -10\ -5\ 0\ 5\ 10$

34. $-\dfrac{7}{4} \le -\dfrac{5}{2}m,\ -\dfrac{2}{5}\left(-\dfrac{7}{4}\right) \ge -\dfrac{2}{5}\left(-\dfrac{5}{2}m\right),\ \dfrac{7}{10} \ge m,\ m \le \dfrac{7}{10}$

$-3\ -2\ -1\ 0\ 1\ 2$

35. $-22y \le 44,\ \dfrac{-22y}{-22} \ge \dfrac{44}{-22},\ y \ge -2$

$-2\ -1\ 0\ 1\ 2\ 3$

36. $-25 \ge -5w,\ \dfrac{-25}{-5} \le \dfrac{-5w}{-5},\ 5 \le w,\ w \ge 5$

$-15\ -10\ -5\ 0\ 5\ 10$

37. When costs are greater than income, the company operates at a loss. One hundred fifty computers must be sold for the company to break even; 151 or more to make a profit.

38. Let x represent the number of chairs in the row, then $16x \le 25(12)$. $16x \le 300,\ \dfrac{16x}{16} \le \dfrac{300}{16},\ x \le 18.75$; Eighteen chairs will fit in the row.

39. Let x represent the number of bundles of blocks, then $16x \ge 18(13)$.

$16x \ge 234,\ \dfrac{16x}{16} \ge \dfrac{234}{16},\ x \ge 14.625$; She should order at least 15 bundles.

page 198   Capsule Review

1. $8x = 40,\ x = 5$ **2.** $-7y = -42,\ y = 6$ **3.** $3t - 12 + 14 = 16,\ t = \dfrac{14}{3}$

page 200 Class Exercises

1. Add 10 to each side; divide each side by 2.

2. Subtract 1 from each side; multiply each side by -4.

3. Subtract 15 from each side; multiply each side by $\frac{3}{2}$.

4. Distribute -3, combine like terms; subtract 2 from each side; divide each side by -3.

5. $\frac{2}{3}t < 4$, $\frac{3}{2}\left(\frac{2}{3}t\right) < 4\left(\frac{3}{2}\right)$, $t < 6$

6. $-4x > -28$, $-\frac{1}{4}(-4x) < -\frac{1}{4}(-28)$, $x < 7$

7. $2 \le \frac{a}{4}$, $2(4) \le \frac{a}{4}(4)$, $a \ge 8$

8. $-3d < -10\frac{1}{2}$, $-\frac{1}{3}(-3d) > -\frac{1}{3}\left(-10\frac{1}{2}\right)$, $d > 3\frac{1}{2}$

9. $-p + 1 \ge 5$, $-p + 1 - 1 \ge 5 - 1$, $-p \ge 4$, $p \le -4$

10. $2x - 10 + 10 < -8$, $2x < -8$, $x < -4$

pages 200–201 Practice Exercises

A 1. $\frac{1}{3}y > -1$, $3\left(\frac{1}{3}y\right) > 3(-1)$, $y > -3$, $\{y: y > -3\}$

2. $\frac{2}{3}d > 6$, $\frac{3}{2}\left(\frac{2}{3}d\right) > \frac{3}{2}(6)$, $d > 9$, $\{d: d > 9\}$

3. $-2x + 5 \le 9$, $-2x \le 4$, $x \ge -2$, $\{x: x \ge -2\}$

4. $-2y + 7 \le 13$, $-2y \le 6$, $y \ge -3$, $\{y: y \ge -3\}$

5. $3c + 12 \ge 15$, $3c \ge 3$, $c \ge 1$; $\{c: c \ge 1\}$

6. $2y + 2 \ge 15$, $2y \ge 13$, $y \ge \frac{13}{2}$; $\left\{y: y \ge \frac{13}{2}\right\}$

7. $-\frac{2}{3}x < -2$, $x > 3$; all real numbers greater than 3; $\{x: x > 3\}$

8. $-\frac{1}{4}c > 2$, $c < -8$; all real numbers less than -8; $\{c: c < -8\}$

9. $0.8b \le 7.2$, $b \le 9$; all real numbers less than or equal to 9; $\{b: b \le 9\}$

10. $0.4p > 7.6$, $p > 19$; all real numbers greater than 19; $\{p: p > 19\}$

11. $-4y + 2 \geq 22$, $-4y \geq 20$, $y \leq -5$; all real numbers less than or equal to -5; $\{y: y \leq -5\}$

12. $-6n + 5 \geq 13$, $-6n \geq 8$, $n \leq -\frac{4}{3}$; all real numbers less than or equal to $-\frac{4}{3}$; $\left\{n: n \leq -\frac{4}{3}\right\}$

13. $1 + 3a < 3$, $3a < 2$, $a < \frac{2}{3}$; all real numbers less than $\frac{2}{3}$; $\left\{a: a < \frac{2}{3}\right\}$

14. $1 + 10x < 5$, $10x < 4$, $x < \frac{2}{5}$; all real numbers less than $\frac{2}{5}$; $\left\{x: x < \frac{2}{5}\right\}$

15. $4d - 4 \geq 12$, $4d \geq 16$, $d \geq 4$; all real number greater than or equal to 4; $\{d: d \geq 4\}$

16. $8a - 5 > 11$, $8a > 16$, $a > 2$; all real numbers greater than 2; $\{a: a > 2\}$

17. $5y - 2y + 30 < 10$, $3y + 30 < 10$, $3y < -20$, $y < -\frac{20}{3}$; $\left\{y: y < -\frac{20}{3}\right\}$

18. $4t - 2t + 2 < 4$, $2t + 2 < 4$, $2t < 2$, $t < 1$; all real numbers less than 1; $\{t: t < 1\}$

19. $y + 4 > -11$; $y > -15$, $\{y: y > -15\}$ 20. $-\frac{2}{3}z \leq 16$; $z \geq -24$, $\{z: z \geq -24\}$

B 21. $1 - 7k < 5k + 3$; $1 < 12k + 3$, $-2 < 12k$, $-\frac{1}{6} < k$, $\left\{k: k > -\frac{1}{6}\right\}$

22. $6(d + 5) \geq 3 + 6d$; $6d + 30 \geq 3 + 6d$, $30 \geq 3$, $\{d: d$ is any real number$\}$

23. $2n - 3n - 9 \leq 14$, $-n - 9 \leq 14$, $-n \leq 23$, $n \geq -23$, $\{n: n \geq -23\}$

24. $7x - 9x - 1 > -5$, $-2x - 1 > -5$, $-2x > -4$, $x < 2$, $\{x: x < 2\}$

25. $12m - 4 \geq 2m + 6$, $10m - 4 \geq 6$, $10m \geq 10$, $m \geq 1$, $\{m: m \geq 1\}$

26. $17 - 4y + 2 \geq 2y + 6$, $19 - 4y \geq 2y + 6$, $19 - 6y \geq 6$, $-6y \geq -13$, $y \leq \frac{13}{6}$, $\left\{y: y \leq \frac{13}{6}\right\}$

27. $48 - 2w - 11 \geq 17 - 2w$, $-2w + 37 \geq 17 - 2w$, $37 \geq 17$; $\{w: w$ is any real number$\}$

28. $5d - \frac{3}{2}d - 4 \leq -4 + \frac{7}{2}d$, $5d - \frac{3}{2}d - \frac{7}{2}d - 4 \leq -4$, $5d - 5d - 4 \leq -4$, $-4 \leq -4$, $\{d: d$ is any real number$\}$

29. $2 \neq 4 + 8x$, $-2 \neq 8x$, $-\frac{1}{4} \neq x$, $\left\{x: x \neq -\frac{1}{4}\right\}$

30. $-1 + 4z \neq 5$, $4z \neq 6$, $z \neq \frac{3}{2}$, $\left\{z: z \neq \frac{3}{2}\right\}$

31. $\frac{3}{2} - 4t > 20 - 4t$, $\frac{3}{2} > 20$; no solution; $\emptyset$

32. $9x + 6 < 6 + 12x - 3x$, $9x + 6 < 6 + 9x$, $0 < 0$; no solution; $\emptyset$

33. $4 + 6 - 4w \geq 12w - 20 - 6w$, $10 - 4w \geq 6w - 20$, $-10w \geq -30$, $w \leq 3$, $\{w: w \leq 3\}$

34. $5 - 5y - 2y - 14 \leq -10y$, $-9 - 7y \leq -10y$, $-9 \leq -3y$, $3 \geq y$, $\{y: y \leq 3\}$

C 35. $\frac{1}{2} - \frac{1}{2}y < \frac{3}{8}$, $-\frac{1}{2}y < -\frac{1}{8}$, $y > \frac{1}{4}$, $\left\{y: y > \frac{1}{4}\right\}$

36. $-\frac{7}{10}n - \frac{1}{8} > \frac{3}{4}$, $-\frac{7}{10}n > \frac{7}{8}$, $n < -\frac{5}{4}$, $\left\{n: n < -\frac{5}{4}\right\}$

37. $6 - [-35a - 43] \geq a + a - 8$, $35a + 49 \geq 2a - 8$, $33a \geq -57$, $a \geq -\frac{19}{11}$, $\left\{a: a \geq -\frac{19}{11}\right\}$

38. $y - 4 \leq 3 - [-6y - 20]$, $y - 4 \leq 6y + 23$, $-5y \leq 27$, $y \geq -\frac{27}{5}$, $\left\{y: y \geq -\frac{27}{5}\right\}$

39. $8m - 3 + m \leq -\{-[-3 + 9m]\}$, $9m - 3 \leq -\{3 - 9m\}$, $9m - 3 \leq -3 + 9m$, $-3 \leq -3$, $\{m: m$ is any real number$\}$

40. $-2\{-2[-4 + 2z]\} > -16 + 8z$, $-2\{8 - 4z\} > -16 + 8z$, $-16 + 8z > -16 + 8z$, $0 > 0$; no solution; $\emptyset$

41. Let x represent the number of people who must purchase half-day tickets, then $24(18) + 10x \geq 600$. $432 + 10x \geq 600$, $10x \geq 168$, $x \geq 16.8$; At least 17 people must purchase half-day tickets.

42. $26(18) + 10x \geq 600$; $468 + 10x \geq 600$, $10x \geq 132$, $x \geq 13.2$; At least 14 people must purchase half-day tickets.

page 201 Test Yourself

1.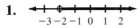

2.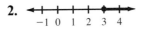

3.

4. all real numbers less than or equal to 1; $\{x: x \leq 1\}$

5. all real numbers greater than -1; $\{x: x > -1\}$

6. $5 > d$, $\{d: d < 5\}$

7. $p < 0.2$, $\{p: p < 0.2\}$

8. $y \geq -3$, $\{y: y \geq -3\}$

9. $a < -12$, $\{a: a < -12\}$

10. $-2t > -2$, $t < 1$, $\{t: t < 1\}$

11. $2a - 12 > 2a - 2$, $-12 > -2$; $\varnothing$

No Solution

page 202 Capsule Review

1. $9 < 5$; false **2.** $6 \geq 6$; true **3.** $-5 > -6$; true

page 204 Class Exercises

1. $-1 < 2 + 3x$, $-3 < 3x$, $-1 < x$; and $2 + 3x < 5$, $3x < 3$, $x < 1$;
$-1 < x$ and $x < 1$, $-1 < x < 1$

2. $y - 1 < -2$, $y < -1$; or $y - 1 \geq 2$, $y \geq 3$; $y < -1$ or $y \geq 3$

page 205 Practice Exercises

A **1.** $34 \leq m \leq 35$ **2.** $21 \leq x \leq 22$

3. $-1 < y < 1$ **4.** $-5 < t < 5$

5. $-2 < r < 3$ **6.** $-4 < z < 5$

7. $3 \leq w < 9$ **8.** $8 \leq t < 15$

9. $a < -4$ or $a > 4$ **10.** $t \leq 0$ or $t > 3$

11. All real numbers **12.** All real numbers

13. $4 < 3 - 2x < 8$, $1 < -2x < 5$, $-\dfrac{5}{2} < x < -\dfrac{1}{2}$

14. $3 < 5 - 3x < 9$, $-2 < -3x < 4$, $-\dfrac{4}{3} < x < \dfrac{2}{3}$

15. $5 < 2z + 1 < 7$, $4 < 2z < 6$, $2 < z < 3$

B **16.** $1 < 3a + 1 < 8$, $0 < 3a < 7$, $0 < a < \dfrac{7}{3}$

17. $-1 \leq 2 - t < 3$, $-3 \leq -t < 1$, $3 \geq t > -1$

18. $-4 < 5 - p < 0$, $-9 < -p < 0$, $9 > p > 0$

19. $-3 \leq 8s + 5 \leq 21$, $-8 \leq 8s \leq 16$, $-1 \leq s \leq 2$

20. $-10 \leq 12r + 2 \leq 14$, $-12 \leq 12r \leq 12$, $-1 \leq r \leq 1$

21. $-4 \leq -4z - 8 \leq 8$, $4 \leq -4z \leq 16$, $-1 \geq x \geq -4$

22. $-2 \le -3m - 9 \le 3$, $7 \le -3m \le 12$, $-\dfrac{7}{3} \ge m \ge -4$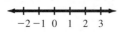

23. $-6 < 2w + 2 < 4$, $-8 < 2w < 2$, $-4 < w < 1$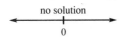

24. $-6 < 3x + 3 < 9$, $-9 < 3x < 6$, $-3 < x < 2$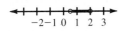

25. $6a > 12$ *and* $-9 < -5a$

   $a > 2$ *and* $\dfrac{9}{5} > a$

   The solution set is { }.

   no solution
   ———————————
   0

26. $3m - 9 > -3$ *or* $m < 3$

   $3m > 6$ *or*

   $m > 2$ *or* $m < 3$

   The solution set is {all real numbers}.

   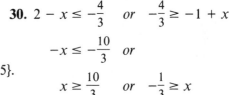

C 27. $4 - 2w < 3w + 1$ *and* $3w + 1 < 7$

   $3 < 5w$ *and* $3w < 6$

   $\dfrac{3}{5} < w$ *and* $w < 2$

   The solution set is $\left\{ w: 2 < w < \dfrac{3}{5} \right\}$.

   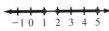

28. $z - 2 \ge 2 - 3z$ *and* $2 - 3z \ge 1 - 4z$

   $4z \ge 4$ *and*

   $z \ge 1$ *and* $z \ge -1$

   The solution set is $\{z: z \ge 1$ and $z \ge -1\}$.

29. $5y < 5$ *or* $-2 < 2y - 6 < 4$

   or $4 < 2y < 10$

   $y < 1$ *or* $2 < y < 5$

   The solution set is $\{y: y < 1$ or $2 < y < 5\}$.

30. $2 - x \le -\dfrac{4}{3}$ *or* $-\dfrac{4}{3} \ge -1 + x$

   $-x \le -\dfrac{10}{3}$ *or*

   $x \ge \dfrac{10}{3}$ *or* $-\dfrac{1}{3} \ge x$

   The solution set is $\left\{ x: x \ge \dfrac{10}{3} \text{ or } x \le -\dfrac{1}{3} \right\}$.

31. $23 < c < 23.5$

32. $142 < M < 148.8$

33. $l \le 107$

page 205 Algebra in Geometry

1. $3r \le -3$ or $\dfrac{1}{2}r \ge 2$, $\{r: r \le -1 \text{ or } r \ge 4\}$; two rays

2. $-2y \ge -1$ and $2y \ge 4$, $\left\{ y: y \le \dfrac{1}{2} \text{ and } y \ge 2 \right\}$; no solution; none

page 206 Capsule Review

1. 5   **2.** 4   **3.** $-6 - 4 = -10$ **4.** $-|0| = 0$

1. $|x - (-9)| = 5, |x + 9| = 5$

2. $|m - (-2)| = 6, |m + 2| = 6$

3. $y + 4 = 11 \quad or \quad y + 4 = -11$
$\qquad y = 7 \quad or \qquad y = -15$
The solution set is $\{-15, 7\}$.

4. $12 - 2t = 9 \quad or \quad 12 - 2t = -9$
$\qquad -2t = -3 \quad or \qquad -2t = -21$
$\qquad t = \dfrac{3}{2} \quad or \qquad t = \dfrac{21}{2}$
The solution set is $\left\{\dfrac{3}{2}, \dfrac{21}{2}\right\}$.

5. $3z - 7 = 5 \quad or \quad 3z - 7 = -5$
$\quad 3z = 12 \quad or \qquad 3z = 2$
$\qquad z = 4 \quad or \qquad z = \dfrac{2}{3}$
The solution set is $\left\{\dfrac{2}{3}, 4\right\}$.

pages 208–209   Practice Exercises

A 1. $x + 1 = 4 \quad or \quad x + 1 = -4$
$\qquad x = 3 \quad or \qquad x = -5$
The solution set is $\{-5, 3\}$.

2. $z + 7 = 8 \quad or \quad z + 7 = -8$
$\qquad z = 1 \quad or \qquad z = -15$
The solution set is $\{-15, 1\}$.

3. No solution; the absolute value of a number cannot be negative.

4. No solution; the absolute value of a number cannot be negative.

5. $x + 6 = 0$
$\quad x = -6$
The solution set is $\{-6\}$.

6. $4m + 1 = 3 \quad or \quad 4m + 1 = -3$
$\quad 4m = 2 \quad or \qquad 4m = -4$
$\quad m = \dfrac{1}{2} \quad or \qquad m = -1$
The solution set is $\left\{-1, \dfrac{1}{2}\right\}$.

7. $9 = 2p + 7 \quad or \quad -9 = 2p + 7$
$\quad 2 = 2p \qquad or \quad -16 = 2p$
$\quad 1 = p \qquad or \quad -8 = p$
The solution set is $\{-8, 1\}$.

8. $11 = 3x + 8 \quad or \quad -11 = 3x + 8$
$\quad 3 = 3x \qquad or \quad -19 = 3x$
$\quad 1 = x \qquad or \quad -\dfrac{19}{3} = x$
The solution set is $\left\{-\dfrac{19}{3}, 1\right\}$.

9. $6z + 4 = 10 \quad or \quad 6z + 4 = -10$
$\quad 6z = 10 \quad or \qquad 6z = -14$
$\quad z = 1 \quad or \qquad z = -\dfrac{14}{6} = -\dfrac{7}{3}$
The solution set is $\left\{-\dfrac{7}{3}, 1\right\}$.

10. $3x - 1 = 5 \quad or \quad 3x - 1 = -5$
$\quad 3x = 6 \quad or \qquad 3x = -4$
$\quad x = 2 \quad or \qquad x = -\dfrac{4}{3}$
The solution set is $\left\{-\dfrac{4}{3}, 2\right\}$.

11. $4y - 2 = 5y$ *or* $4y - 2 = -5y$
   $\quad -2 = y \quad$ *or* $\quad -2 = -9y$
   $$\frac{2}{9} = y$$
   -2 is not a solution since
   $|4(-2) - 2| \neq 5(-2)$.
   The solution set is $\left\{\dfrac{2}{9}\right\}$.

12. $2m - 15 = 3m \quad$ *or* $\quad 2m - 15 = -3m$
   $\quad\quad -15 = m \quad$ *or* $\quad\quad -15 = -5m$
   $$3 = m$$
   -15 is not a solution since
   $|2(-15) - 15| \neq 3(-15)$.
   The solution set is $\{3\}$.

13. $17 = 7h - 4 \quad$ *or* $\quad -17 = 7h - 4$
   $21 = 7h \quad\quad$ *or* $\quad -13 = 7h$
   $\quad 3 = h \quad\quad$ *or* $\quad -\dfrac{13}{7} = h$
   The solution set is $\left\{-\dfrac{13}{7}, 3\right\}$.

14. $\quad 4 = \dfrac{1}{3}a - 2 \quad$ *or* $\quad -4 = \dfrac{1}{3}a - 2$
   $\quad 6 = \dfrac{1}{3}a \quad\quad$ *or* $\quad -2 = \dfrac{1}{3}a$
   $\quad 18 = a \quad\quad$ *or* $\quad -6 = a$
   The solution set is $\{-6, 18\}$.

15. $\dfrac{3}{4}h - 5 = 1 \quad$ *or* $\quad \dfrac{3}{4}h - 5 = -1$
   $\quad \dfrac{3}{4}h = 6 \quad$ *or* $\quad\quad \dfrac{3}{4}h = 4$
   $h = \dfrac{24}{3} = 8 \quad$ *or* $\quad\quad h = \dfrac{16}{3}$
   The solution set is $\left\{\dfrac{16}{3}, 8\right\}$.

16. $|g| + 4 = -4, |g| = -8$;
   No solution, the absolute value of
   a number cannot be negative.

17. $|m| - 4 = -2, |m| = 2$;
   $m = 2$ *or* $m = -2$;
   The solution set is $\{-2, 2\}$.

18. $|p| - 3 = -5, |p| = -2$;
   No solution, the absolute value of
   a number cannot be negative.

19. $|2 - 4m| = 3$;
   $2 - 4m = 3 \quad$ *or* $\quad 2 - 4m = -3$
   $\quad -4m = 1 \quad$ *or* $\quad\quad -4m = -5$
   $\quad m = -\dfrac{1}{4} \quad$ *or* $\quad\quad m = \dfrac{5}{4}$
   The solution set is $\left\{-\dfrac{1}{4}, \dfrac{5}{4}\right\}$.

20. $|-6 + 3m| = 6$;
   $-6 + 3m = 6 \quad$ *or* $\quad -6 + 3m = -6$
   $\quad\quad 3m = 12 \quad$ *or* $\quad\quad\quad 3m = 0$
   $\quad\quad m = 4 \quad$ *or* $\quad\quad\quad m = 0$
   The solution set is $\{0, 4\}$.

21. $|5b + 5| = 10; 5b + 5 = 10 \quad$ *or* $\quad 5b + 5 = -10$
   $\quad\quad\quad\quad\quad\quad 5b = 5 \quad$ *or* $\quad\quad 5b = -15$
   $\quad\quad\quad\quad\quad\quad\ b = 1 \quad$ *or* $\quad\quad\ \ b = -3$ The solution set is $\{-3, 1\}$.

B 22. $|3y - 2y + 4 - 1| = 7, |y + 3| = 7$;
   $y + 3 = 7 \quad$ *or* $\quad y + 3 = -7$
   $\quad y = 4 \quad$ *or* $\quad\quad y = -10$
   The solution set is $\{-10, 4\}$.

23. $|4x - 3x + 3 - 2| = 9$,
   $|x + 1| = 9$;
   $x + 1 = 9 \quad$ *or* $\quad x + 1 = -9$
   $\quad x = 8 \quad$ *or* $\quad\quad x = -10$
   The solution set is $\{-10, 8\}$.

24. $|13 - k - 2| = 1, |11 - k| = 1;$
$$11 - k = \quad 1 \quad or \quad 11 - k = -1$$
$$-k = -10 \quad or \quad -k = -12$$
$$k = \quad 10 \quad or \quad k = 12$$
The solution set is $\{10, 12\}$.

25. $|y - 2y - 1| = 3, |-y - 1| = 3;$
$$-y - 1 = \quad 3 \quad or \quad -y - 1 = -3$$
$$-y = \quad 4 \quad or \quad -y = -2$$
$$y = -4 \quad or \quad y = 2$$
The solution set is $\{-4, 2\}$.

26. $|4d| = 6;$
$$4d = 6 \quad or \quad 4d = -6$$
$$d = \frac{3}{2} \quad or \quad d = -\frac{3}{2}$$
The solution set is $\left\{-\frac{3}{2}, \frac{3}{2}\right\}$.

27. $3c - 3 = 2c \quad or \quad 3c - 3 = -2c$
$$c - 3 = \quad 0 \quad or \quad 5c - 3 = 0$$
$$c = \quad 3 \quad or \quad 5c = 3$$
$$c = \frac{3}{5}$$
The solution set is $\left\{\frac{3}{5}, 3\right\}$.

28. $2t - 1 = 3t \quad or \quad 2t - 1 = -3t$
$$-t - 1 = 0 \quad or \quad 5t - 1 = 0$$
$$-t = 1 \quad or \quad 5t = 1$$
$$t = -1 \quad or \quad t = \frac{1}{5}$$
-1 is not a solution
since $|2(-1) - 1| \neq 3(-1)$.
The solution set is $\left\{\frac{1}{5}\right\}$.

29. $2a = 4 - a \quad or \quad -2a = 4 - a$
$$3a = 4 \quad or \quad -a = 4$$
$$a = \frac{4}{3} \quad or \quad a = -4$$
-4 is not a solution
since $|4 - (-4)| \neq 2(-4)$.
The solution set is $\left\{\frac{4}{3}\right\}$.

30. $s = 2 - 3s \quad or \quad -s = 2 - 3s$
$$4s = 2 \quad or \quad 2s = 2$$
$$s = \frac{1}{2} \quad or \quad s = 1$$
The solution set is $\left\{\frac{1}{2}, 1\right\}$.

31. $|4 - 2n + 2| = 3, |6 - 2n| = 3;$
$$6 - 2n = \quad 3 \quad or \quad 6 - 2n = -3$$
$$-2n = -3 \quad or \quad -2n = -9$$
$$n = \quad \frac{3}{2} \quad or \quad n = \frac{9}{2}$$
The solution set is $\left\{\frac{3}{2}, \frac{9}{2}\right\}$.

32. $|7x - 9x - 1| = 5, |-2x + 1| = 5;$
$$-2x - 1 = \quad 5 \quad or \quad -2x - 1 = -5$$
$$-2x = \quad 6 \quad or \quad -2x = -4$$
$$x = -3 \quad or \quad x = 2$$
The solution set is $\{-3, 2\}$.

33. Let x represent the number.
$$|x| - 5 = 27, |x| = 32;$$
$$x = 32 \text{ or } x = -32$$
$$\{-32, 32\}$$
The number is -32 or 32.

34. Let x represent the number.

$|2x - 5| = 33$

$2x - 5 = 33$ $\quad or \quad$ $2x - 5 = -33$

$2x = 38$ $\quad or \quad$ $2x = -28$

$x = 19$ $\quad or \quad$ $x = -14$

$\{-14, 19\}$ The number is -14 or 19.

35. Let x represent the number.

$5|4x - 4| = 20, |4x - 4| = 4$

$4x - 4 = 4$ $\quad or \quad$ $4x - 4 = -4$

$4x = 8$ $\quad or \quad$ $4x = 0$

$x = 2$ $\quad or \quad$ $x = 0$

$\{0, 2\}$ The number is 0 or 2.

36. Let x represent the number.

$2|2x - 2| - 4 = 24, |2x - 2| = 14$

$2x - 2 = 14$ $\quad or \quad$ $2x - 2 = -14$

$x = 8$ $\quad or \quad$ $x = -6$

$\{-6, 8\}$ The number is -6 or 8.

37. $\dfrac{c - 1}{2} = 5$ $\quad or \quad$ $\dfrac{c - 1}{2} = -5$

$c - 1 = 10$ $\quad or \quad$ $c - 1 = -10$

$c = 11$ $\quad or \quad$ $c = -9$

The solution set is $\{-9, 11\}$.

38. $\dfrac{2 - d}{3} = 21$ $\quad or \quad$ $\dfrac{2 - d}{3} = -21$

$2 - d = 63$ $\quad or \quad$ $2 - d = -63$

$-d = 61$ $\quad or \quad$ $-d = -65$

$d = -61$ $\quad or \quad$ $d = 65$

The solution set is $\{-61, 65\}$.

39. $\left|\dfrac{4 - m}{3}\right| = 5;$

$\dfrac{4 - m}{3} = 5$ $\quad or \quad$ $\dfrac{4 - m}{3} = -5$

$4 - m = 15$ $\quad or \quad$ $4 - m = -15$

$-m = 11$ $\quad or \quad$ $-m = -19$

$m = -11$ $\quad or \quad$ $m = 19$

The solution set is $\{-11, 19\}$.

40. $\left|\dfrac{p + 1}{5}\right| = 5;$

$\dfrac{p + 1}{5} = 5$ $\quad or \quad$ $\dfrac{p + 1}{5} = -5$

$p + 1 = 25$ $\quad or \quad$ $p + 1 = -25$

$p = 24$ $\quad or \quad$ $p = -26$

The solution set is $\{-26, 24\}$.

41. $3y - 2 = 2y - 1$ $\ or \ 3y - 2 = -(2y - 1)$

$y - 2 = -1$ $\quad or \ 3y - 2 = -2y + 1$

$y = 1$ $\quad or \ 5y - 2 = 1$

$5y = 3$

$y = \dfrac{3}{5}$

The solution set is $\left\{\dfrac{3}{5}, 1\right\}$.

42. $2z - 1 = 3z + 4$ $or \ 2z - 1 = -(3z + 4)$

$-1 = z + 4$ $\quad or \ 2z - 1 = -3z - 4$

$-5 = z$ $\quad or \ 5z - 1 = -4$

$z = -\dfrac{3}{5}$

-5 is not a solution since

$|2(-5) - 1| \neq 3(-5) + 4.$

The solution set is $\left\{-\dfrac{3}{5}\right\}$.

43. Let x represent the acceptable amount of juice per case.

$|x - 12(64)| = 2.5, |x - 768| = 2.5$

$x - 768 = 2.5$ $\quad or \quad$ $x - 768 = -2.5$

$x = 770.5$ $\quad or \quad$ $x = 765.5$

Maximum acceptable amount: 770.5 oz

Minimum acceptable amount: 765.5 oz

44. Let x represent the acceptable weight of a nickel roll.
$|x - 0.05 - 40(0.18)| = 0.015, |x - 7.25| = 0.015$
$x - 7.25 = 0.015 \quad or \quad x - 7.25 = -0.015$
$\qquad x = 7.265 \quad or \qquad\quad x = 7.235$
Maximum acceptable weight: 7.265 oz; minimum: 7.235 oz

45. $\left\{-2, -1, -\dfrac{1}{2}, 0, 1\right\}$ **46.** $\left\{-\dfrac{1}{2}, \dfrac{1}{2}\right\}$ **47.** Ø

48. $y = -\dfrac{5x + 16z}{3}$ **49.** $l = \dfrac{41k}{6}$

50. $x + x + 6 = 60$; 27, 29, 31, 33 **51.** $-2, -1, 0, 1, 2$

page 210 Capsule Review

1. $0 < x < 5$ **2.** $y \geq \dfrac{1}{4}$ **3.** $-3 < n < 0$

page 212 Class Exercises

1. b **2.** d **3.** a **4.** c

5. c **6.** a **7.** d **8.** b

pages 212–213 Practice Exercises

A 1. $y - 4 > 6 \quad or \quad y - 4 < -6$
$\quad y > 10 \quad or \qquad y < -2$
The solution set is $\{y < -2 \text{ or } y > 10\}$.

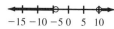
$-15 \ -10 \ -5 \ 0 \quad 5 \quad 10$

2. $x - 3 > 7 \quad or \quad x - 3 < -7$
$\quad x > 10 \quad or \qquad x < -4$
The solution set is $\{x < -4 \text{ or } x > 10\}$.

$-15 \ -10 \ -5 \ 0 \quad 5 \quad 10$

3. $z + 6 > 8 \quad or \quad z + 6 < -8$
$\quad z > 2 \quad or \qquad z < -14$
The solution set is
$\{z < -14 \text{ or } z > 2\}$.

$-16 \ -12 \ -8 \ -4 \quad 0 \quad 4$

4. $a + 1 > 5 \quad or \quad a + 1 < -5$
$\quad a > 4 \quad or \qquad a < -6$
The solution set is
$\{a < -6 \text{ or } a > 4\}$.

$-12 \ -8 \ -4 \ 0 \quad 4 \quad 8$

5. $c + 3 < 9 \quad and \quad c + 3 > -9$
$\quad c < 6 \quad and \qquad c > -12$
The solution set is $\{c: -12 < c < 6\}$.

$-12 \ -8 \ -4 \ 0 \quad 4 \quad 8$

6. $b + 5 < 10 \quad and \quad b + 5 > -10$
$\quad b < 5 \quad and \qquad b > -15$
The solution set is $\{b: -15 < b < 5\}$.

$-15 \ -10 \ -5 \ 0 \quad 5 \quad 10$

7. $t - 2 < 5 \quad and \quad t - 2 > -5$
$\quad t < 7 \quad and \qquad t > -3$
The solution set is $\{t: -3 < t < 7\}$.

$-2 \ 0 \quad 2 \quad 4 \quad 6 \quad 8$

8. $w - 1 < 8 \quad and \quad w - 1 > -8$
$\quad w < 9 \quad and \qquad w > -7$
The solution set is $\{w: -7 < w < 9\}$.

$-6 \ -3 \ 0 \quad 3 \quad 6 \quad 9$

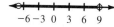

9. $x - 3 < 4$ and $x - 3 < -4$
$x < 7$ and $x < -1$
The solution set is $\{x: -1 < x < 7\}$.

$-2\ 0\ 2\ 4\ 6\ 8$

10. $a - 2 < 5$ and $a - 2 > -5$
$a < 8$ and $a > -3$
The solution set is $\{a: -3 < a < 8\}$.

$-4\ -2\ 0\ 2\ 4\ 6\ 8$

11. $2t - 3 < 0$ and $2t - 3 > 0$
$2t < 3$ and $2t > 3$
$t < \dfrac{3}{2}$ and $t > \dfrac{3}{2}$
The solution set is Ø.

12. $3t - 1 < 5$ and $3t - 1 > -5$
$3t < 6$ and $3t > -4$
$t < 2$ and $t > -\dfrac{4}{3}$

$\left\{t: -\dfrac{4}{3} < t < 2\right\}$
$-3\ -2\ -1\ 0\ 1\ 2$

13. $2m - 9 > 1$ or $2m - 9 < -1$
$2m > 10$ or $2m < 8$
$m > 5$ or $m < 4$
The solution set is
$\{m < 4 \text{ or } m > 5\}$.

$0\ 1\ 2\ 3\ 4\ 5$

14. $4q - 1 > 0$ or $4q - 1 < 0$
$4q > 1$ or $4q < 1$
$q > \dfrac{1}{4}$ or $q < \dfrac{1}{4}$

$\left\{q < \dfrac{1}{4} \text{ or } q > \dfrac{1}{4}\right\}$
$-2\ -1\ 0\ 1\ 2\ 3$

15. $2t + 3 \le 6$ and $2t + 3 \ge -6$
$2t \le 3$ and $2t \ge -9$
$t \le \dfrac{3}{2}$ and $t \ge -\dfrac{9}{2}$

$\left\{t: -\dfrac{9}{2} \le t \le \dfrac{3}{2}\right\}$
$-4\ -3\ -2\ -1\ 0\ 1$

16. $3x + 1 \le 8$ and $3x + 1 \ge -8$
$3x \le 7$ and $3x \ge -9$
$x \le \dfrac{7}{3}$ and $x \ge -3$

$\left\{x: -3 \le x \le \dfrac{7}{3}\right\}$
$-3\ -2\ -1\ 0\ 1\ 2$

17. $5a - 2 \le 9$ and $5a - 2 \ge -9$
$a \le \dfrac{11}{5}$ and $a \ge -\dfrac{7}{5}$

$\left\{a: -\dfrac{7}{5} \le a \le \dfrac{11}{5}\right\}$
$-2\ -1\ 0\ 1\ 2\ 3$

18. $6b - 3 \le 12$ and $6b - 3 \ge -12$
$b \le \dfrac{5}{2}$ and $b \ge -\dfrac{3}{2}$

$\left\{b: -\dfrac{3}{2} \le b \le \dfrac{5}{2}\right\}$
$-2\ -1\ 0\ 1\ 2\ 3$

19. $a + 3 \ge 7$ or $a + 3 \le -7$
$a \le 4$ or $a \le -10$
$\{a \le -10 \text{ or } a \ge 4\}$
$-12\ -8\ -4\ 0\ 4\ 8$

20. $b + 5 \ge 2$ or $b + 5 \le -2$
$b \ge -3$ or $b \le -7$
$\{b \le -7 \text{ or } b \ge -3\}$
$-8\ -6\ -4\ -2\ 0\ 2$

21. $4x - 8 \le 2$ and $4x - 8 \ge -2$
$x \le \dfrac{5}{2}$ and $x \ge \dfrac{3}{2}$

$\left\{x: \dfrac{3}{2} \le x \le \dfrac{5}{2}\right\}$
$-2\ -1\ 0\ 1\ 2\ 3$

22. $2q - 1 \le 5$ and $2q - 1 \ge -5$
$q \le 3$ and $q \ge -2$
$\{q: -2 \le q \le 3\}$
$-4\ -2\ 0\ 2\ 4\ 6$

23. $2t \geq 0$ *or* $2t \leq 0$
$\quad\ t \geq 0$ *or* $\quad t \leq 0$
{all real numbers}

$-2\ -1\ \ 0\ \ 1\ \ 2\ \ 3$

24. $3x \geq 0$ *or* $3x \leq 0$
$\quad\ x \geq 0$ *or* $\quad x \leq 0$
{all real numbers}

$-2\ -1\ \ 0\ \ 1\ \ 2\ \ 3$

B 25. $-|-z + 3z - 3| \geq -2,\ |2z - 3| \leq 2;$
$\quad 2z - 3 \leq 2 \quad and \quad 2z - 3 \geq -2$
$\qquad 2z \leq 5 \quad and \qquad\ 2z \geq 1$
$\qquad\ z \leq \dfrac{5}{2} \quad and \qquad\ z \geq \dfrac{1}{2}$
$\left\{ z: \dfrac{1}{2} \leq z \leq \dfrac{5}{2} \right\}$

$-2\ -1\ \ 0\ \ 1\ \ 2\ \ 3$

26. $14 - |-t + 2t - 6| \geq 9,$
$\quad |t - 6| \leq 5;$
$\quad t - 6 \leq 5 \quad and \quad t - 6 \geq -5$
$\qquad\ t \leq 11 \quad and \qquad\ t \geq 1$
The solution set is $\{ t: 1 \leq t \leq 11 \}$.

$-3\ \ 0\ \ 3\ \ 6\ \ 9\ \ 12$

27. $-|-x + 2x - 4| \geq -6,\ |x - 4| \leq 6;$
$\quad x - 4 \leq 6 \quad and \quad x - 4 \geq -6$
$\qquad\ x \leq 10 \quad and \qquad\ x \geq -2$
$\{ x: -2 \leq x \leq 10 \}$

$-10\ -5\ \ 0\ \ 5\ \ 10\ \ 15$

28. $-|-m + 3m - 12| \geq 2,\ |2m - 12| \leq 2;$
$\quad 2m - 12 \leq 2 \quad and \quad 2m - 12 \geq -2$
$\qquad\ m \leq 7 \quad and \qquad\ m \geq 5$
$\{ m: 5 \leq m \leq 7 \}$

$-2\ \ 0\ \ 2\ \ 4\ \ 6\ \ 8$

29. $|2 - w| < -1;$
$\quad 2 - w < -1 \quad and \quad 2 - w > 1$
$\qquad\ w > 3 \quad and \qquad\ w < 1$
The solution set is $\emptyset$.

30. $|m - 4| > -2,$
$\quad m - 4 > -2 \quad or \quad m - 4 < 2$
$\qquad\ m > 2 \quad or \qquad\ m < 6$
{all real numbers}

$-2\ -1\ \ 0\ \ 1\ \ 2\ \ 3$

31. $\dfrac{2q - 4}{3} \leq 1 \quad and \quad \dfrac{2q - 4}{3} \geq -1$
$\quad 2q - 4 \leq 3 \quad and \quad 2q - 4 \geq -3$
$\qquad\ 2q \leq 7 \quad and \qquad\ 2q \geq 1$
$\qquad\ q \leq \dfrac{7}{2} \quad and \qquad\ q \geq \dfrac{1}{2}$
$\left\{ q: \dfrac{1}{2} \leq q \leq \dfrac{7}{2} \right\}$

$-1\ \ 0\ \ 1\ \ 2\ \ 3\ \ 4$

32. $\dfrac{2 - 3b}{2} \leq 1 \quad and \quad \dfrac{2 - 3b}{2} \geq -1$
$\quad 2 - 3b \leq 2 \quad and \quad 2 - 3b \geq -2$
$\qquad\ -3b \leq 0 \quad and \qquad\ -3b \geq -4$
$\qquad\ b \geq 0 \quad and \qquad\ b \leq \dfrac{4}{3}$
$\left\{ 0 \leq b \leq \dfrac{4}{3} \right\}$

$-1\ \ 0\ \ 1\ \ 2\ \ 3\ \ 4$

33. $\dfrac{2 - 2c}{3} \geq 0 \quad and \quad \dfrac{2 - 2c}{3} \leq 0$
$\quad 2 - 2c \geq 0 \quad and \quad 2 - 2c \leq 0$
$\qquad -2c \geq -2 \quad and \qquad -2c \leq -2$
$\qquad\ c \leq 1 \quad and \qquad\ c \geq 1$
{all real numbers}

$-2\ -1\ \ 0\ \ 1\ \ 2\ \ 3$

34. $\dfrac{3 - 4m}{4} \geq 1 \quad or \quad \dfrac{3 - 4m}{4} \leq 1$
$\quad 3 - 4m \geq 4 \quad or \quad 3 - 4m \leq -4$
$\qquad -4m \geq 1 \quad or \qquad -4m \leq -7$
$\qquad\ m \leq -\dfrac{1}{4} \quad or \qquad\ m \geq \dfrac{7}{4}$
$\left\{ m \leq -\dfrac{1}{4} \text{ or } m \geq \dfrac{7}{4} \right\}$

$-2\ -1\ \ 0\ \ 1\ \ 2\ \ 3$

35. $|5 - 3d + 4| > 6, |9 - 3d| > 6;$

$\quad 9 - 3d > 6 \quad$ or $\quad 9 - 3d < -6$

$\quad\quad -3d > -3 \quad$ or $\quad\quad -3d < -15$

$\quad\quad\quad d < 1 \quad$ or $\quad\quad\quad d > 5$

The solution set is

$\{d: d < 1 \text{ or } d > 5\}.$

36. $|2 - 4 + 2f| > 4, |-2 + 2f| > 4;$

$\quad -2 + 2f > 4 \quad$ or $\quad -2 + 2f < -4$

$\quad\quad\quad 2f > 6 \quad$ or $\quad\quad\quad 2f < -2$

$\quad\quad\quad f > 3 \quad$ or $\quad\quad\quad f < -1$

The solution set is

$\{f: f < -1 \text{ or } f > 3\}.$

37. $|2t + 5 - 5t| \le 4, |-3t + 5| \le 4;$

$\quad -3t + 5 \le 4 \quad$ and $\quad -3t + 5 \ge -4$

$\quad\quad -3t \le -1 \quad$ and $\quad\quad -3t \ge -9$

$\quad\quad\quad t \ge \dfrac{1}{3} \quad$ and $\quad\quad\quad t \le 3$

The solution set is $\left\{t: \dfrac{1}{3} \le t \le 3\right\}.$

38. $|3 - 3v - 6| \le 2, |-3v - 3| \le 2;$

$\quad -3v - 3 \le 2 \quad$ and $\quad -3v - 3 \ge -2$

$\quad\quad -3v \le 5 \quad$ and $\quad\quad -3v \ge 1$

$\quad\quad\quad v \ge -\dfrac{5}{3} \quad$ and $\quad\quad\quad v \le -\dfrac{1}{3}$

The solution set is

$\left\{v: -\dfrac{5}{3} \le v \le -\dfrac{1}{3}\right\}.$

39. $\left|\dfrac{p + 1}{2}\right| \ge 0;$

$\quad \dfrac{p + 1}{2} \ge 0 \quad$ or $\quad \dfrac{p + 1}{2} \le 0$

$\quad\quad p + 1 \ge 0 \quad$ or $\quad p + 1 \le 0$

$\quad\quad\quad p \ge -1 \quad$ or $\quad\quad\quad p \le -1$

The solution set is

$\{p: p \text{ is a real number}\}.$

40. $\left|\dfrac{1 - 2x}{3}\right| \ge 0;$

$\quad \dfrac{1 - 2x}{3} \ge 0 \quad$ or $\quad \dfrac{1 - 2x}{3} \le 0$

$\quad 1 - 2x \ge 0 \quad$ or $\quad 1 - 2x \le 0$

$\quad\quad -2x \ge -1 \quad$ or $\quad\quad -2x \le -1$

$\quad\quad\quad x \le \dfrac{1}{2} \quad$ or $\quad\quad\quad x \ge \dfrac{1}{2}$

The solution set is

$\{x: x \text{ is a real number}\}.$

C 41. $|3z - 5| > -1;$

$\quad 3z - 5 > -1 \quad$ or $\quad 3z - 5 < 1$

$\quad\quad 3z > 4 \quad$ or $\quad\quad 3z < 6$

$\quad\quad\quad z > \dfrac{4}{3} \quad$ or $\quad\quad\quad z < 2$

The solution set is

$\{z: z \text{ is a real number}\}.$

42. $|2 - 5m| \ge -2;$

$\quad 2 - 5m \ge -2 \quad$ or $\quad 2 - 5m \le 2$

$\quad\quad -5m \ge -4 \quad$ or $\quad\quad -5m \le 0$

$\quad\quad\quad m \le \dfrac{4}{5} \quad$ or $\quad\quad\quad m \ge 0$

The solution set is

$\{m: m \text{ is a real number}\}.$

43. $g - 1 < 3g \quad$ and $\quad g - 1 > -3g$

$\quad -1 < 2g \quad$ and $\quad 4g - 1 > 0$

$\quad -\dfrac{1}{2} < g \quad$ and $\quad\quad 4g > 1$

$\quad\quad\quad\quad\quad\quad g > \dfrac{1}{4}$

The solution set is $\left\{g : g > \dfrac{1}{4}\right\}.$

44. $4 - m \ge 2m \quad$ or $\quad 4 - m \le -2m$

$\quad\quad 4 \ge 3m \quad$ or $\quad\quad 4 \le -m$

$\quad\quad \dfrac{4}{3} \ge m \quad$ or $\quad\quad -4 \ge m$

The solution set is

$\left\{m: m \le \dfrac{4}{3}\right\}.$

45. $2 - y > 2 - y$ *or* $2 - y < -(2 - y)$
$\qquad 2 > 2 \qquad$ *or* $\;\; 2 - y < -2 + y$
$\qquad\qquad\qquad\qquad\qquad 4 < 2y$
$\qquad\qquad\qquad\qquad\qquad 2 < y$

$2 > 2$ is a false inequality.
The solution set is $\{y: y > 2\}$.

46. $p + 6 \geq p + 6 \quad$ *or* $\quad -(p + 6) \leq p + 6$
$\qquad 6 \geq 6 \qquad\;$ *or* $\quad -p - 6 \leq p + 6$
$\qquad\qquad\qquad\qquad\qquad\qquad -12 \leq 2p$
$\qquad\qquad\qquad\qquad\qquad\qquad\; -6 \leq p$

The solution set is $\{p: p \geq -6\}$.

47. $3k - 2 \leq 2k - 3$ *and* $3k - 2 \geq -(2k - 3)$
$\qquad k \leq -1 \qquad$ *and* $3k - 2 \geq -2k + 3$
$\qquad\qquad\qquad\qquad\qquad\qquad k \geq 1$

The solution set is Ø.

48. $2t - 1 > 3t + 5$ *or* $2t - 1 < -(3t + 5)$
$\qquad -1 > t + 5 \;$ *or* $2t - 1 < -3t - 5$
$\qquad -6 > t \qquad$ *or* $5t - 1 < -5$
$\qquad\qquad\qquad\qquad\qquad\;\; 5t < -4$
$\qquad\qquad\qquad\qquad\qquad\qquad t < -\dfrac{4}{5}$

The solution set is $\left\{t: t < -\dfrac{4}{5}\right\}$.

49. Let t represent the temperature, $|t - 0| < 5$.

50. Let x represent the percentage of sportscasters who watched the Superbowl, $|x - 85| \leq 3$.

page 213 Algebra in Mechanics

1. $|x - 6| \leq 0.001$;
$\quad x - 6 \leq 0.001 \quad$ *and* $\;\; x - 6 \geq -0.001$
$\qquad x \leq 6.001 \quad$ *and* $\qquad x \geq 5.999$
The solution set is $\{x: 5.999 \leq x \leq 6.001\}$.

2. $|x - 14| \leq 0.5$;
$\quad x - 14 \leq 0.5 \quad$ *and* $\;\; x - 14 \geq -0.5$
$\qquad x \leq 14.5 \quad$ *and* $\qquad x \geq 13.5$
The solution set is $\{x: 13.5 \leq x \leq 14.5\}$.

3. $|x - 2.75| \leq 0.001$;
$\quad x - 2.75 \leq 0.001 \quad$ *and* $\;\; x - 2.75 \geq -0.001$
$\qquad x \leq 2.751 \quad$ *and* $\qquad x \geq 2.749$
The solution set is $\{x: 2.749 \leq x \leq 2.751\}$.

page 215 Class Exercises

1. Let x represent the number of boxes. $60x \leq 2000$, $x \leq 33\frac{1}{3}$, The elevator can hold 33 boxes.

2. Let x represent the amount Sang-Ho receives on his third sale. $150 + 225 + x > 500$, $x + 375 > 500$, $x > 125$, The commission must be greater than $125.

pages 215–216   Practice Exercises

A 1. Let x represent each salary, $8x + 110{,}000 \le 350{,}000$, then $8x \le 240{,}000$, $x \le 30{,}000$. Each employee will receive at most \$30,000.

2. Let x represent the amount of Sally's sales in one week, then $0.03x + 250 \ge 460$, $0.03x \ge 210$, $x \ge 7000$. Sally must sell \$7000 or more in one week.

3. Let x represent the score on Lisa's fifth exam, then $\dfrac{80 + 92 + 86 + 78 + x}{5} > 85$, $\dfrac{336 + x}{5} > 85$, $336 + x > 425$, $x > 89$. Lisa must score at least a 90.

4. Let x represent the amount of rainfall in the fourth year, then $\dfrac{65 + 72 + 59 + x}{4} \ge 68$, $\dfrac{196 + x}{4} \ge 68$, $196 + x \ge 272$, $x \ge 76$. The amount of rainfall must be at least 76 in.

5. Let x represent the number of hours Freddy worked, then $6x + 100 \ge 244$, $6x \ge 144$, $x \ge 24$. Freddy must work at least 24 h in one week at the job that pays \$6 per hour.

6. Let x represent the width, then $x + 2$ represents the length. $2x + 2(x + 2) > 16$, $4x + 4 > 16$, $4x > 12$, $x > 3$. The width is at least 4 ft, and the length is at least 6 ft.

7. Let x represent the width, then $2x + 4$ represents the length. $2x + 2(2x + 4) \le 38$, $6x + 8 \le 38$, $6x \le 30$, $x \le 5$. The greatest possible value for the width is 5 m.

B 8. Let x represent the number of 25¢ stamps, then $x - 10$ represents the number of 30¢ stamps. $0.25x + 0.30(x - 10) \ge 19$, $0.55x - 3 \ge 19$, $0.55x \ge 22$, $x \ge 40$. Manuel can buy no more than 40 stamps at 25¢ and no more than 30 stamps at 30¢.

9. Let x represent each quarter mile traveled by Jim, then $0.75 + 0.35x < 7.40$, $0.35x < 6.65$, $x < 19$. Jim traveled less than 19 quarter miles or less than $4\frac{3}{4}$ miles. To the nearest mile Jim rode less than 5 mi.

10. Let x represent the number of one-half hours worked, then $27 + 18x \le 100$, $18x \le 73$, $x \le 4\frac{1}{18}$. The technician worked less than 5 one-half hours, or no more than $2\frac{1}{2}$ h.

C **11.** Let x represent the amount invested in bonds, then $10,000 - x$ represents the amount invested in stocks. $0.045x + 0.025(10,000 - x) \geq 300$, $0.02x + 250 \geq 300$, $0.02x \geq 50$, $x \geq 2500$. Mrs. Charis invested at least $2500 in bonds, and $7500 in stocks.

12. Let x represent the amount invested at $4\frac{1}{2}\%$, then $450 - x$ represents the amount invested at 5%. $0.045 + 0.05(450 - x) \geq 21.50$, $-0.005x + 22.50 \geq 21.50$, $-0.005x \geq -1$, $x \leq 200$. At most $200 must be invested at $4\frac{1}{2}\%$ interest.

page 216   Mixed Problem Solving Review

1. Let x represent the number of tapes John bought, then $5.69x + 2.50 = 70.78$, $5.69x = 68.28$, $x = 12$. John bought 12 tapes.

2. Let r represent the rate of one bus, then $r + 10$ represents the rate of the other bus. $3r + 3(r + 10) = 270$, $6r + 30 = 270$, $6r = 240$, $r = 40$. One bus travels at 40 mi/h and the other bus at 50 mi/h.

3. Let x represent the first odd integer, then $x + 2$, $x + 4$, $x + 6$ represent the next three odd integers. $x + x + 2 + x + 4 + x + 6 > 48$, $4x + 12 > 48$, $4x > 36$, $x > 9$. The four consecutive odd integers are 11, 13, 15, 17.

page 217   Project

Answers may vary. Check students' work.

page 217   Test Yourself

1. $3p > -3$ or $p < -3$, $p > -1$ or $p < -3$, $\{p: p > -1 \text{ or } p < -3\}$,

2. $-4 < 2m < 8$, $-2 < m < 4$, $\{m: -2 < m < 4\}$,

3. $-2y > -10$ or $3y > 12$, $y < 5$ or $y > 4$, $\{y: y \text{ is any real number}\}$,

4. $-7k > 7$ and $\frac{3}{4}k < -9$, $k < -1$ and $k < -12$, $\{k: k < -12\}$,

5. $x - 3 = 4 \quad or \quad x - 3 = -4$
$x = 7 \quad or \quad x = -1$
The solution set is $\{-1, 7\}$.

6. $3t + 2 = 7 \quad or \quad 3t + 2 = -7$
$3t = 5 \quad or \quad 3t = -9$
$t = \frac{5}{3} \quad or \quad t = -3$
The solution set is $\left\{-3, \frac{5}{3}\right\}$.

7. $m + \dfrac{2}{3} = 1 \quad$ or $\quad m + \dfrac{2}{3} = -1$

$\qquad m = \dfrac{1}{3} \quad$ or $\qquad m = -1\dfrac{2}{3}$

The solution set is $\left\{-1\dfrac{2}{3}, \dfrac{1}{3}\right\}$.

8. $-5|6 - y| = -10, |6 - y| = 2;$

$\qquad 6 - y = 2 \quad$ or $\quad 6 - y = -2$

$\qquad -y = -4 \quad$ or $\qquad -y = -8$

$\qquad y = 4 \quad$ or $\qquad y = 8$

The solution set is $\{4, 8\}$.

9. $y > \dfrac{1}{2}$ or $y < -\dfrac{1}{2}, \left\{y: y > \dfrac{1}{2} \text{ or } y < -\dfrac{1}{2}\right\}$,

$\qquad -2\ -1\ \ 0\ \ 1\ \ 2\ \ 3$

10. $x - 4 < 2 \quad$ and $\quad x - 4 > -2$

$\qquad x < 6 \quad$ and $\qquad x > 2$

The solution set is $\{x: 2 < x < 6\}$.

$\qquad -4\ -2\ \ 0\ \ 2\ \ 4\ \ 6$

11. $7 + f > 13 \quad$ or $\quad 7 + f < -13$

$\qquad f > 6 \quad$ or $\qquad f < -20$

The solution set is $\{f < -20 \text{ or } f > 6\}$.

$\qquad -24 -18 -12 -6\ 0\ \ 6$

12. $|10 - 3g| > -1;$

$\qquad 10 - 3g > -1 \quad$ or $\quad 10 - 3g < 1$

$\qquad -3g > -11 \quad$ or $\qquad -3g < -9$

$\qquad g < \dfrac{11}{3} \quad$ or $\qquad g > 3$

The solution set is $\{g: g \text{ is any real number}\}$.

$\qquad -3\ -2\ -1\ \ 0\ \ 1\ \ 2$

13. Let n represent the number, then $11 - \dfrac{1}{3}n > 5$, $-\dfrac{1}{3}n > -6$, $n < 18$, The number is less than 18.

14. Let s represent Bill's weekly sales, then $0.15s \geq 240$, $s \geq 1600$. Bill must sell at least $1600 worth of items.

15. Let x represent Margot's time in her next race, then
$$\dfrac{13.1 + 12.8 + 13.0 + 13.3 + 12.7 + x}{6} < 12.8, \ 64.9 + x < 76.8, \ x < 11.9.$$
Margot must run her next race in less than 11.9s.

pages 218–219   Integrating Algebra: Marathon Running

1. Let m represent the number of miles Carlos will run each day, then $6m$ represents the number of miles Carlos runs for 6 days. $6m + 3 \geq 27$, $6m \geq 24$, $m \geq 4$. Carlos must run at least 4 mi.

2. Let m represent the number of miles, then $7m \geq 35$, $m \geq 5$, Carlos must run at least 5 mi a day. The previous week $7m \geq 27$, $m \geq 3\dfrac{6}{7}$, Carlos ran at least $3\dfrac{6}{7}$ mi. The difference is $5 - 3\dfrac{6}{7}$, or $1\dfrac{1}{7}$ mi more each day.

3. Let m represent the number of miles Christine will run, then $10 + 8.5 + 9 + 3m \geq 50$, $27.5 + 3m \geq 50$, $3m \geq 22.5$, $m \geq 7.5$. Christine runs 7.5 mi or more.

4. Let m represent the number of miles Katrina runs, then $12 \leq 2 + 3 + 4m \leq 18$, $12 \leq 5 + 4m \leq 18$, $7 \leq 4m \leq 13$, $\frac{7}{4} \leq m \leq \frac{13}{4}$. Katrina must run between $1\frac{3}{4}$ mi and $3\frac{1}{4}$ mi.

5. Let m represent the number of miles Adrienne runs, then $10 \leq 2\frac{1}{2} + 2\frac{1}{4} + 2m \leq 12$, $10 \leq 4\frac{3}{4} + 2m \leq 12$, $5\frac{1}{4} \leq 2m \leq 7\frac{1}{4}$, $2\frac{5}{8} \leq m \leq 3\frac{5}{8}$. Adrienne will run between $2\frac{5}{8}$ mi and $3\frac{5}{8}$ mi.

6. Let r = rate of a runner, then the rate of Gelinda Bordin is found using the equation $(2h\ 10\ \text{min}\ 32\ s)r = 26\ \text{mi}\ 385\ \text{yd}$, converting to common units of h and mi; $2.1756r = 26.21875$, $r \approx 12.05$; 12 mi/h. Rosa Mota: $2.4278r = 26.21875$, $r \approx 10.8$; 11 mi/h. Carlos Lopes: $2.1653r = 26.21875$, $r \approx 12.1$; 12 mi/h. Joan Benoit: $2.1444r = 26.21875$, $r \approx 10.9$; 11 mi/h.

page 220   Summary and Review

1. $-3\ -2\ -1\ 0\ \ 1\ \ 2$ 2. $-2\ -1\ 0\ \ 1\ \ 2\ \ 3$ 3. $-3\ -2\ -1\ 0\ \ 1\ \ 2$ 4. $-6\ -4\ -2\ \ 0\ \ 2$

5. $0 - 5 > x + 5 - 5$, $-5 > x$, $\{x < -5\}$, $-10\ -5\ 0\ \ 5\ \ 10\ \ 15$

6. $y - \frac{1}{2} + \frac{1}{2} < 1 + \frac{1}{2}$, $y < 1\frac{1}{2}$, $\left\{y < \frac{3}{2}\right\}$, $-2\ -1\ 0\ \ 1\ \ 2\ \ 3$

7. $-1.5a \div -1.5 < 3 \div -1.5$, $a < -2$, $\{a < -2\}$, $-8\ -6\ -4\ -2\ 0\ \ 2$

8. $4 - 2x + 3x \geq 0$, $4 + x \geq 0$, $4 + x - 4 \geq 0 - 4$, $x \geq -4$, $\{x \geq -4\}$, $-4\ -2\ 0\ \ 2\ \ 4\ \ 6$

9. $g - 6g - 2 > 20 + 5g + 8$, $-5g - 2 > 28 + 5g$, $-5g - 2 - 28 > 28 - 28 + 5g$, $-5g - 30 > 5g$, $-5g + 5g - 30 > 5g + 5g$, $-30 > 10g$, $-3 > g$, $\{g < -3\}$, $-4\ -3\ -2\ -1\ 0\ \ 1$

10. $3y < -15$ or $6 - 2y < -4$
    $y < -5$ or $-2y < -10$
                              $y > 5$
    The solution set is $\{y : y < -5 \text{ or } y > 5\}$.

     $-10\ -5\ 0\ \ 5\ \ 10\ \ 15$

11. $-5 + 2x + 3 \leq 0$ and $14 - 2x \leq x + 17$
    $2x - 2 \leq 0$ and $-3x \leq 3$
    $2x \leq 2$ and $x \geq -1$
    $x \leq 1$ and
    The solution set is $\{x : -1 \leq x \leq 1\}$.

     $-2\ -1\ 0\ \ 1\ \ 2\ \ 3$

12. $-3 < 2 - 3v < 6$, $-5 < -3v < 4$, $\dfrac{5}{3} > v > -\dfrac{4}{3}$,

$\left\{v: -\dfrac{4}{3} < v < \dfrac{5}{3}\right\}$,

13. $m - 2 = 7$ or $m - 2 = -7$
$m = 9$ or $m = -5$
The solution set is $\{-5, 9\}$.

14. $m + 3 = \dfrac{5}{3}$ or $m + 3 = -\dfrac{5}{3}$
$m = -\dfrac{4}{3}$ or $m = -\dfrac{14}{3}$
The solution set is $\left\{-\dfrac{14}{3}, -\dfrac{4}{3}\right\}$.

15. $|6x - 8x - 1| = 9$, $|-2x - 1| = 9$;
$-2x - 1 = 9$ or $-2x - 1 = -9$
$-2x = 10$ or $-2x = -8$
$x = -5$ or $x = 4$
The solution set is $\{-5, 4\}$.

16. $g - 3 > 5$ or $g - 3 < -5$
$g > 8$ or $g < -2$

The solution set is
$\{g: g < -2 \text{ or } g > 8\}$.

17. $-|4 - y| \le -2$, $|4 - y| \ge 2$
$4 - y \ge 2$ or $4 - y \le -2$
$-y \ge -2$ or $-y \le -6$
$y \le 2$ or $y \ge 6$
The solution set is $\{y: y \le 2 \text{ or } y \ge 6\}$.

18. $\dfrac{y - 8}{4} \le 4$ and $\dfrac{y - 8}{4} \ge -4$
$y - 8 \le 16$ and $y - 8 \ge -16$
$y \le 24$ and $y \ge -8$
The solution set is $\{y: -8 \le y \le 24\}$.

19. Let x represent the number, $\dfrac{1}{2}x + 3 > 2$, $\dfrac{1}{2}x > -1$, $x > -2$.

20. Let x represent Wanda's commission for the sixth week,
$\dfrac{17 + 21 + 19 + 25 + 12 + x}{6} \ge 20$, $\dfrac{94 + x}{6} \ge 20$, $94 + x \ge 120$, $x \ge 26$;
Wanda must earn at least \$26.

21. Let w represent the width of the rectangle, then $w + 3$ represents the
length, $2w + 2(w + 3) < 52$, $4w + 6 < 52$, $4w < 46$, $w < 11\dfrac{1}{2}$. The width
is 11 ft and the length is 14 ft.

1. 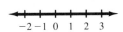   **2.**   **3.**

4. $x - 5 + 5 > 8 + 5$, $x > 13$, $\{x: x > 13\}$,

5. $1 - 1 - 3y > 7 - 1$, $-3y > 6$, $y < -2$, $\{y: y < -2\}$,

6. $18 - 3x > 0$ *or* $-16 > -8x$
     $-3x > -18$ *or*
       $x < 6$ $2 < x$
The solution set is $\{x:$ all real numbers$\}$.

7. $3y - 4 < 5$ *and* $2y > 1$
     $3y < 9$
     $y < 3$ *and* $y > \dfrac{1}{2}$

The solution set is $\left\{y: \dfrac{1}{2} < y < 3\right\}$.

8. $b = 3$ or $b = -3$, The solution set is $\{3, -3\}$.

9. $y - 3 = 7$ *or* $y - 3 = -7$
     $y = 10$ *or* $y = -4$
The solution set is $\{-4, 10\}$.

10. $|x - 3x - 2| = 3$, $|-2x - 2| = 3$
     $-2x - 2 = 3$ *or* $-2x - 2 = -3$
     $-2x = 5$ *or* $-2x = -1$
     $x = -\dfrac{5}{2}$ *or* $x = \dfrac{1}{2}$

The solution set is $\left\{-\dfrac{5}{2}, \dfrac{1}{2}\right\}$.

11. $a > 4$ or $a < -4$, The solution set is $\{a: a > 4$ or $a < -4\}$.

12. $3 - 2x \geq 5$ *or* $3 - 2x \leq -5$
     $-2x \geq 2$ *or* $-2x \leq -8$
     $x \leq -1$ *or* $x \geq 4$
The solution set is $\{x: x \leq -1$ or $x \geq 4\}$.

13. $8 + a < 2$ *and* $8 + a > -2$
     $a < -6$ *and* $a > -10$
The solution set is $\{a: -10 < a < -6\}$.

14. Let x represent the amount of money Amy can spend on the other items. $75 + 45 + x \leq 200$, $120 + x \leq 200$, $x \leq 80$, Amy can spend no more than \$80 on the other items.

15. Let x represent the first odd integer, then $x + 2$ represents the second odd integer. $x + x + 2 > -15$, $2x + 2 > -15$, $2x > -17$, $x > -\dfrac{17}{2}$. The two consecutive odd integers whose sum is greater than -15 are $-7, -5$.

16. Let t represent the amount of time Seth will spend cycling. $12(t + 2) > 3(12t)$, $12t + 24 > 36t$, $24 > 24t$, $1 > t$. Seth should spend less than 1 h cycling.

17. Let x represent the amount of money in Oscar's account. $x - 125 < 500$, $x < 625$. Oscar had less than \$625 in his account.

18. Let r represent Lisa's rate. $(r + 15)2 > 3r$, $2r + 30 > 3r$, $30 > r$. Lisa must travel at a rate less than 30 mi/h.

Challenge

1. $(x - 1 < 0.5$ and $x - 1 > -0.5)$ and $(x - 2 < 0.5$ and $x - 2 > -0.5)$

$(x < 1.5$ and $x > 0.5)$ and $(x < 2.5$ and $x > 1.5)$

$0.5 < x < 1.5$ and $1.5 < x < 2.5$

no solution

2. $(x + 1 < 1$ and $x + 1 > -1)$ or $(x + 2 > 1$ or $x + 2 < -1)$

$(x < 0$ and $x > -2)$ or $(x > -1$ or $x < -3)$

$-2 < x < 0$ or $x > -1$ or $x < -3$

$\{x: x < -3$ or $x > -2\}$,

$-4\ -3\ -2\ -1\ \ 0\ \ 1$

page 223   Preparing for Standardized Tests

1. D; $x + 3x + 4 = 2x - 6$, $4x + 4 = 2x - 6$, $2x + 4 = -6$, $2x = -10$;
$x = -5$

2. E; $8.03 + 11.02 + 7.863 + 0.007 = 27.100$

3. A; $3x + 5 < x + 17$, $2x + 5 < 17$, $2x < 12$; $x < 6$

4. C; $39.80 - 38.42 = 1.38$

5. D; $|3a + 2b^2| = |3(-2) + 2(-4)^2| = |-6 + 32| = 26$

6. A; since all points to the right of and including -2 are part of the graph, the sentence $x \geq -2$ is the correct choice.

7. E; $m^2n - n^3p^2 = 3^2(-2) - (-2)^3(-1)^2 = 9(-2) - (-8)(1) = -18 + 8 = -10$

8. A; $3(x - 2) - 5(x + 4) = 4(2x + 11)$, $3x - 6 - 5x - 20 = 8x + 44$,
$-2x - 26 = 8x + 44$, $-10x - 26 = 44$, $-10x = 70$; $x = -7$

9. C; \$16.95 × 24 = \$406.80, therefore, \$406.80 was received for the 24 clippers. Then, when the \$299.95 is subtracted, the profit is \$106.85.

10. E; $s = \frac{1}{2}(a + b + c)$, so $2s = a + b + c$ and $b = 2s - a - c$.

11. B; The sum of the four amounts is \$36.72, and the $5\frac{1}{2}\%$ sales tax on this amount is $0.055(\$36.72) = \$2.01960$ or \$2.02.

page 224 Mixed Review

1. $10 \cdot 10 \cdot 10 \cdot 10 \cdot 10 = 10{,}000$

2. $2 \cdot 2 \cdot 2 = 8$

3. $-1 \cdot (-1) \cdot (-1) \cdot (-1) = 1$

4. $5 \cdot 5 = 25$

5. $-4 \cdot (-4) \cdot (-4) = -64$

6. $-[-3 \cdot (-3) \cdot (-3) \cdot (-3)] = -[81] = -81$

7. $3^2 \cdot 3^2 \cdot 3^2 = 9 \cdot 9 \cdot 9 = 729$

8. $2^3 \cdot 2^3 \cdot 2^3 = 8 \cdot 8 \cdot 8 = 512$

9. $[(-1)^3] \cdot [(-1)^3] = [-1 \cdot (-1) \cdot (-1)] \cdot [-1 \cdot (-1) \cdot (-1)] = [-1] \cdot [-1] = 1$

10. $[(-2)^2] \cdot [(-2)^2] = [-2 \cdot (-2)] \cdot [-2 \cdot (-2)] = [4] \cdot [4] = 16$

11. $(4)^2 \cdot (4)^2 = (4 \cdot 4) \cdot (4 \cdot 4) = 16 \cdot 16 = 256$

12. $(-3)^2 \cdot (-3)^2 \cdot (-3)^2 = 9 \cdot 9 \cdot 9 = 729$

13. $\frac{7 \cdot 1}{7 \cdot 8} = \frac{1}{8}$

14. $\frac{3 \cdot 3}{3 \cdot 7} = \frac{3}{7}$

15. $\frac{5 \cdot 3}{5 \cdot 5} = \frac{3}{5}$

16. $\frac{12 \cdot 1}{12 \cdot 3} = \frac{1}{3}$

17. $\frac{4 \cdot 5}{4 \cdot 6} = \frac{5}{6}$

18. $\frac{7 \cdot 2}{7 \cdot 3} = \frac{2}{3}$

19. $(-2x^2 - 3x^2) + (3xy + 2xy) = -5x^2 + 5xy$

20. $(5m^2 - 3m^2) + (mn - 2mn) = 2m^2 - mn$

21. $2c^2 - 5cd - 3c^2 + c^2 = (2c^2 + c^2 - 3c^2) - 5cd = -5cd$

22. $7s^2 - 3s^2 + 3st + st = (7s^2 - 3s^2) + (3st + st) = 4s^2 + 4st$

23. $2p^2 + 4q^2 - 4pq + 2q^2 = 2p^2 - 4pq + (4q^2 + 2q^2) = 2p^2 - 4pq + 6q^2$

24. $-6ab + 3b^2 + 2ab = (-6ab + 2ab) + 3b^2 = -4ab + 3b^2 = 3b^2 - 4ab$

25. Let i represent interest, then $i = (0.12)(3500)(1) = 420$, the total cost is $3500 + 420 = 3920$. The loan costs \$3920.

26. Let s represent the amount of sales tax, then $s = 0.08\,(24.98 + 25.89)$, $s = 0.08\,(50.87)$, $s = 4.07$. The total cost is $50.87 + 4.07 = 54.94$. The amount spent was \$54.94.

27. Let i represent the amount of simple interest, then $i = (0.07)(1800)(3)$, $i = 378$. The amount of interest is \$378.

Chapter 6   Monomials and Multiplying Monomials

page 226   Capsule Review

1. $4 \cdot 4 = 16$

2. $5 \cdot 5 \cdot 5 = 125$

3. $-(2 \cdot 2 \cdot 2 \cdot 2 \cdot 2 \cdot 2) = -64$

4. $(-8)(-8) = 64$

5. $-[(-10)(-10)(-10)] = 1000$

page 228   Class Exercises

1. $7 \cdot y \cdot y \cdot y$

2. $3 \cdot 3 \cdot s \cdot t \cdot t \cdot t \cdot t \cdot t$

3. $-(4 \cdot 4 \cdot x \cdot x \cdot z)$

4. yes

5. no

6. yes

7. yes

8. a^7

9. 3^9 or $19{,}683$

10. $(2 \cdot 2^3)(x^3 \cdot x)(y \cdot y) = 2^4x^4y^2$ or $16x^4y^2$

11. Associative property for multiplication is used to regroup the factors. Commutative property for multiplication is used to change their order.
$a^2(b \cdot a)b^4 = a^2(a \cdot b)b^4 = (a^2 \cdot a) \cdot (b \cdot b^4) = a^3b^5$

12. Not possible; monomials must have the same base; $x^6 \cdot y^5$ or x^6y^5.

pages 228–229   Practice Exercises

A **1.** $3 \cdot 3 \cdot 3 \cdot z \cdot z$

2. $2 \cdot 2 \cdot 2 \cdot y \cdot y$

3. $-7 \cdot a \cdot a \cdot a \cdot b \cdot b \cdot b \cdot b$

4. $-5 \cdot c \cdot c \cdot c \cdot d \cdot d \cdot d \cdot d$

5. yes

6. yes

7. no

8. no

9. yes

10. yes

11. yes

12. yes

13. $(3 \cdot 5) \cdot (x^2 \cdot x^5) \cdot (y \cdot y^2) = 15x^7y^3$

14. $(7 \cdot 2) \cdot (m^2 \cdot m^5) \cdot (n \cdot n^2) = 14m^7n^3$

15. $(6 \cdot 3) \cdot (r^3 \cdot r^4) \cdot (s \cdot s^5) = 18r^7s^6$

16. $(8 \cdot 2) \cdot (a^3 \cdot a^4) \cdot (b \cdot b^5) = 16a^7b^6$

17. $(4 \cdot 5) \cdot (g^3 \cdot g^5) = 20g^8$

18. $(7 \cdot 3) \cdot (z^3 \cdot z^5) = 21z^8$

19. $(3 \times 5) \times (10^2 \times 10^7) = 15 \times 10^9$

20. $(4 \times 8) \times (10^5 \times 10^8) = 32 \times 10^{13}$

B **21.** $(2 \times 1.5) \times (10^5 \times 10^4) = 3 \times 10^9$

22. $(1.2 \times 2) \times (10^4 \times 10^6) = 2.4 \times 10^{10}$

23. $(4.2 \times 0.5) \times (10^6 \times 10^2) = 2.1 \times 10^8$

24. $(2.1 \times 0.2) \times (10^3 \times 10^3) = 0.42 \times 10^6$

25. k^{m+4}

26. h^{1+b}

27. $b^{x+1+2} = b^{x+3}$

28. $x^{a+4+3} = x^{a+7}$

29. $2^{1+m-4+3} = 2^m$

30. $3^{1+x-5+2} = 3^{x-2}$

C 31. $(-19)^5 = -2,476,099$ **32.** $(171)^2 = 29,241$ **33.** $(-19)^4 = 130,321$

34. $(171)^3 = 5,000,211$ **35.** $2 \cdot (r^{3x} \cdot r^{2x}) \cdot d^x = 2r^{5x}d^x$

36. $3 \cdot (e^{3y} \cdot e^y) \cdot (f^x \cdot f^x) = 3e^{4y}f^{2x}$

37. $[9(-3)] \cdot (z^{3a} \cdot z^{2a}) \cdot (y^{2m} \cdot y^{4m}) = -27z^{5a}y^{6m}$

38. $\left[6\left(\frac{1}{2}\right)^3\right] \cdot 2^{(-3)(-1)} = 6 \cdot \frac{1}{8} \cdot 2^3 = 6$ **39.** $[5(-3)^2] \cdot [3^2\left(\frac{1}{3}\right)] = 5 \cdot 9 \cdot 3 = 135$

40. $(-5)^{(10)\left(\frac{1}{2}\right)+(-3)-(-1)} = (-5)^{5-3+1} = -125$ **41.** $(-3)^{(-3)+6\left(\frac{1}{2}\right)-3(-1)} = (-3)^{-3+3+3} = -27$

42. $(-5)^{6\left(\frac{1}{2}\right)} \cdot (-1)^{-(-3)} = (-5)^3 \cdot (-1)^3 = -125 \cdot (-1) = 125$

43. $(-3)(-3)^{-(-1)} \div \left(\frac{1}{2}\right)^5 = (-3)(-3) \div \frac{1}{32} = 9 \cdot 32 = 288$

44. $(2a) \cdot (4a) = 8a^2$ **45.** $(x) \cdot \left(\frac{1}{2}x\right) = \frac{1}{2}x^2$ **46.** $2[3 - 14] = -22$

47. $-3[29 + 9] = -114$ **48.** $\frac{4(-2)}{-8} = 1$ **49.** $\frac{4 \div 24}{-6 + -4} = -\frac{1}{60}$

50. $\frac{5 + 5}{25 + 80 + 64} = \frac{10}{169}$ **51.** $2(-6)^2 + (-3 \cdot 3)^3 = -657$

52. $\frac{(-6 + 3)^2 - 3^2}{2 \cdot 9} = 0$ **53.** $\frac{(9 + -6 - 3)^2}{5(3 \cdot -6 - 9)} = 0$

54. $a^{\frac{1}{2}} \cdot a^{\frac{1}{2}} = a^{\frac{1}{2}+\frac{1}{2}} = a^1 = a$. Therefore, for example, $4^{\frac{1}{2}} \cdot 4^{\frac{1}{2}} = 4$, implying $4^{\frac{1}{2}} = \sqrt{4}$ (or 2). Conclusion: $a^{\frac{1}{2}} = \sqrt{a}$.

page 230   Capsule Review

1. $\frac{3}{5}$ **2.** $\frac{4}{3}$ **3.** $\frac{2}{5}$ **4.** $\frac{4}{3}$ **5.** $\frac{3}{4}$ **6.** $\frac{3}{2}$

page 232   Class Exercises

1. $x^{6-2} = x^4$ **2.** $\frac{1}{a^{7-3}} = \frac{1}{a^4}$ **3.** $m^{8-8} = m^0 = 1$

4. $3^{3-2} = 3$ **5.** $\frac{1}{9^{3-1}} = \frac{1}{9^2} = \frac{1}{81}$ **6.** $10^{2-1} = 10$

7. $\frac{-2}{3} \cdot \frac{x^2}{x^2} \cdot \frac{y}{y} = -\frac{2}{3}$ **8.** $\frac{3^{3-1} \cdot a^{2-1} b^{3-2}}{c^4} = \frac{3^2 \cdot a \cdot b}{c^4} = \frac{9ab}{c^4}$

9. If any factor equals 0, then the denominator of the fraction equals 0. Division by zero is undefined.

10. $a^0 = \frac{a^n}{a^n} = 1$. If $a = 0$, then $0^0 = \frac{0^n}{0^n}$, which is undefined.

A 1. $x^{7-4} = x^3$

2. $a^{9-2} = a^7$

3. $\dfrac{1}{y^{9-1}} = \dfrac{1}{y^8}$

4. $\dfrac{1}{i^{8-1}} = \dfrac{1}{i^7}$

5. $b^{2-2} = b^0 = 1$

6. $n^{3-3} = n^0 = 1$

7. $8^{2-1} = 8$

8. $2^{4-1} = 2^3 = 8$

9. $\dfrac{1}{10^{7-3}} = \dfrac{1}{10^4}$

10. $\dfrac{1}{10^{5-2}} = \dfrac{1}{10^3}$

11. $\dfrac{6}{2} \cdot \dfrac{a^2}{a^6} \cdot \dfrac{b^5}{b^2} = \dfrac{3 \cdot b^{5-2}}{a^{6-2}} = \dfrac{3b^3}{a^4}$

12. $\dfrac{9}{3} \cdot \dfrac{x^2}{x^6} \cdot \dfrac{y^5}{y^2} = \dfrac{3 \cdot y^{5-2}}{x^{6-2}} = \dfrac{3y^3}{x^4}$

13. $\dfrac{3}{7} \cdot \dfrac{a}{a^3} \cdot c^5 = \dfrac{3c^5}{7a^{3-1}} = \dfrac{3c^5}{7a^2}$

14. $\dfrac{5}{9} \cdot \dfrac{m}{m^4} \cdot n^7 = \dfrac{5n^7}{9m^{4-1}} = \dfrac{5n^7}{9m^3}$

15. $4 \cdot \dfrac{t}{t^2} \cdot h = \dfrac{4h}{t^{2-1}} = \dfrac{4h}{t}$

16. $8 \cdot \dfrac{s}{s^2} \cdot t = \dfrac{8t}{s^{2-1}} = \dfrac{8t}{s}$

17. $\dfrac{20}{5} \times \dfrac{10^7}{10^4} = 4 \times 10^{7-4} = 4 \times 10^3$

18. $\dfrac{18}{9} \times \dfrac{10^6}{10^5} = 2 \times 10^{6-5} = 2 \times 10$

19. $\dfrac{6}{2} \times \dfrac{10^4}{10^3} = 3 \times 10^{4-3} = 3 \times 10$

20. $\dfrac{15}{3} \times \dfrac{10^7}{10^2} = 5 \times 10^{7-2} = 5 \times 10^5$

21. 1

22. 1

23. -1

24. -1

B 25. $\dfrac{1}{2}$

26. $-4 \cdot 1 = -4$

27. $3 \cdot 1 = 3$

28. $-(1) = -1$

29. $\dfrac{-3}{6} \cdot \dfrac{a^5}{a^4} \cdot \dfrac{b^7}{b^8} = \dfrac{-a^{5-4}}{2b^{8-7}} = -\dfrac{a}{2b}$

30. $-\dfrac{2}{4} \cdot \dfrac{c^6}{c^2} \cdot \dfrac{d^4}{d^5} = \dfrac{-c^{6-2}}{2d^{5-4}} = -\dfrac{c^4}{2d}$

31. $\dfrac{5}{-3} \cdot \dfrac{x^2}{x^3} \cdot \dfrac{y^4}{y^2} = \dfrac{-5y^{4-2}}{3x^{3-2}} = -\dfrac{5y^2}{3x}$

32. $\dfrac{-4}{7} \cdot \dfrac{x^3}{x^4} \cdot \dfrac{y^5}{y} = \dfrac{-4y^{5-1}}{7x^{4-3}} = -\dfrac{4y^4}{7x}$

33. $\dfrac{7}{3} \cdot \dfrac{x^2}{x^4} \cdot y^3 \cdot \dfrac{z^7}{z^4} = \dfrac{7y^3z^{7-4}}{3x^{4-2}} = \dfrac{7y^3z^3}{3x^2}$

34. $\dfrac{9}{-3} \cdot \dfrac{k^4}{k^3} \cdot \dfrac{m^5}{m} \cdot n = -3k^{4-3}m^{5-1}n = -3km^4n$

35. $-\dfrac{4}{2} \cdot \dfrac{p^2}{p^2} \cdot \dfrac{q^3}{q} \cdot \dfrac{r^4}{r} = -2 \cdot p^{2-2} \cdot q^{3-1} \cdot r^{4-1} = -2p^0q^2r^3 = -2q^2r^3$

36. $\dfrac{-6}{-4} \cdot \dfrac{d}{d^2} \cdot \dfrac{e}{e} \cdot \dfrac{f}{f^5} = \dfrac{3e^{1-1}}{2d^{2-1}f^{5-1}} = \dfrac{3e^0}{2df^4} = \dfrac{3}{2df^4}$

37. $\dfrac{4.2}{3} \times \dfrac{10^{14}}{10^5} = 1.4 \times 10^9$

38. $\dfrac{1.6}{0.8} \times \dfrac{10^{11}}{10^3} = 2 \times 10^8$

39. $\dfrac{-1.2}{0.2} \times \dfrac{10^4}{10^5} = -6 \times \dfrac{1}{10} = -\dfrac{6}{10} = -\dfrac{3}{5}$ **40.** $\dfrac{-4.6}{2.3} \times \dfrac{10^6}{10^9} = \dfrac{-2}{10^3} = -\dfrac{1}{500}$

C 41. x^{a-2} **42.** $\dfrac{1}{z^{3-c}}$ **43.** $\dfrac{1}{y^{2k-1}}$ **44.** $r^{m-1-1} = r^{m-2}$

45. $\dfrac{1}{3^{9-x}} = \dfrac{1}{3^3}$, $9 - x = 3$, $-x = -6$, $x = 6$ **46.** $\dfrac{1}{r^{x-1}} = \dfrac{1}{r^5}$, $x - 1 = 5$, $x = 6$

47. $t^{5-x-x} = t^3$, $5 - 2x = 3$, $-2x = -2$, $x = 1$

48. $2^{3+x-1} = 2^7$, $2 + x = 7$, $x = 5$

49. $x^6 = x^3 \cdot x \cdot h$, $x^6 = x^4 \cdot h$, $\dfrac{x^6}{x^4} = h$, $x^{6-4} = h$, $x^2 = h$

50. $36x^8 = 9x^4 \cdot w \cdot 2x^2$, $w = \dfrac{36x^8}{9x^4 \cdot 2x^2}$, $w = \dfrac{36}{18} \cdot x^{8-4-2}$, $w = 2x^2$

page 233   Algebra in Astronomy

$V = \dfrac{4 \cdot (5)^3}{3}\pi$ cubic miles, $V = \dfrac{4 \cdot 125}{3}\pi \cdot 5280^3 \cdot 12^3$ cubic inches; weight $=$

$\dfrac{4 \cdot 125}{3}\pi \cdot 5280^3 \cdot 12^3 \cdot 10^9$ tons, weight $= 1.3318157 \times 10^{26}$ tons

page 234   Capsule Review

1. $c^{5+5} = c^{10}$

2. $4 \cdot 2 \cdot p^3 \cdot p^2 = 8p^{3+2} = 8p^5$

3. $a \cdot a^2 \cdot b \cdot b^3 = a^{1+2} \cdot b^{1+3} = a^3b^4$

4. $(-3)(-3)a^{1+1} = 9a^2$

page 235   Class Exercises

1. $-d^{2\cdot3} = -d^6$

2. $d^{2\cdot3} = d^6$

3. $-3^{2\cdot3} = -3^6 = -729$

4. $-3^{3\cdot2} = -3^6 = -729$

5. $3^2 \cdot x^{2\cdot2} \cdot y^2 = 9x^4y^2$

6. $(-4)^3y^{2\cdot3}z^{5\cdot3} = -64y^6z^{15}$

7. $\dfrac{(-a)^5}{b^{2\cdot5}} = \dfrac{-a^5}{b^{10}} = -\dfrac{a^5}{b^{10}}$

8. $\dfrac{(-m)^8}{b^{3\cdot8}} = \dfrac{m^8}{b^{24}}$

page 236   Practice Exercises

A 1. a. $x^{5\cdot6} = x^{30}$

   b. $z^{3\cdot9} = z^{27}$

2. a. $3^{2\cdot3} = 3^6 = 729$

   b. $2^{3\cdot4} = 2^{12} = 4096$

3. $-(10^{2\cdot5}) = -10^{10}$

4. $-(10^{3\cdot7}) = -10^{21}$

5. $(-3)^{2\cdot2} = (-3)^4 = 81$

6. $(-2)^{3\cdot3} = (-2)^9 = -512$

7. $\dfrac{(-a)^5}{b^{2\cdot5}} = \dfrac{-a^5}{b^{10}} = -\dfrac{a^5}{b^{10}}$

8. $\dfrac{(-m)^8}{b^{3\cdot8}} = \dfrac{m^8}{b^{24}}$

9. $(-2)^4 \times 10^{3\cdot4} = 16 \times 10^{12}$

10. $(-5)^3 \times 10^{2\cdot3} = -125 \times 10^6$

11. $2^3 \cdot x^{2\cdot3} \cdot 3x^4 = 8 \cdot 3 \cdot x^6 \cdot x^4 = 8 \cdot 3 \cdot x^{6+4} = 24x^{10}$

12. $3^2 \cdot z^{3\cdot2} \cdot 2z^5 = 9 \cdot 2 \cdot z^6 \cdot z^5 = 9 \cdot 2 \cdot z^{6+5} = 18z^{11}$

13. $\dfrac{(-3)^3a^3}{b^{2\cdot3}} = -\dfrac{27a^3}{b^6}$

14. $\dfrac{(-2)^2c^2}{7^2 \cdot d^2} = \dfrac{4c^2}{49d^2}$

B 15. $\dfrac{4^2 \cdot y^2}{5^2 \cdot x^2} = \dfrac{16y^2}{25x^2}$

16. $\dfrac{7^2 \cdot w^2}{12^2 \cdot y^2} = \dfrac{49w^2}{144y^2}$

17. $\dfrac{(-2)^3k^{4\cdot3}}{3^3j^{3\cdot3}} = -\dfrac{8k^{12}}{27j^9}$

18. $\dfrac{5^2x^{3\cdot2}}{(-2)^2y^{4\cdot2}} = \dfrac{25x^6}{4y^8}$

19. $\left(\dfrac{1}{9b^{3-1}}\right)^2 = \left(\dfrac{1}{9b^2}\right)^2 = \dfrac{1^2}{9^2b^{2\cdot2}} = \dfrac{1}{9^2b^4} = \dfrac{1}{81b^4}$

20. $(5d^{5-2})^2 = (5d^3)^2 = 5^2d^{3\cdot2} = 25d^6$

21. $2^2a^2b^{3\cdot2} = 4a^2b^6$

22. $3^3a^{4\cdot3}b^3 = 27a^{12}b^3$

23. $(-2)(5)^3m^{3+3}n^{1+2\cdot3} = -250m^6n^7$

24. $(-3)(9)^2y^{1+2}z^{4+2\cdot3} = -243y^3y^{10}$

25. $\dfrac{3^3a^{2\cdot3}a^2}{-9} = \dfrac{3^3a^{2\cdot3+2}}{-9} = -3a^8$

26. $\dfrac{4^2a^{2\cdot2}b^2}{(-3)^2a^{5\cdot2}b^{2\cdot2}} = \dfrac{16a^4b^2}{9a^{10}b^4} = \dfrac{16}{9a^{10-4}b^{4-2}} = \dfrac{16}{9a^6b^2}$

27. $\dfrac{-(x^5y^5)}{2^5x^{2\cdot5}y^{4\cdot5}} = \dfrac{-x^5y^5}{2^5x^{10}y^{20}} = -\dfrac{1}{32x^{10-5}y^{20-5}} = -\dfrac{1}{32x^5y^{15}}$

28. $\dfrac{(-2)^5h^{3\cdot5}}{2h^4} = -\dfrac{32h^{15}}{2h^4} = -16h^{15-4} = -16h^{11}$

29. z^{4k}

30. $y^{3\cdot2j} = y^{6j}$

31. $(-2)^5k^{5\cdot m} = -32k^{5m}$

C 32. $(-3^tj^t)(-3^{t+1}j^{2(t+1)}) = (-3)^{t+t+1}j^{t+2(t+1)} = -3^{2t+1}j^{3t+2}$

33. $\dfrac{a^{2(2m+3)}}{a^{2(m+1)}} = a^{4m+6-(2m+2)} = a^{2m+4}$

34. $\dfrac{b^{4(2a+1)}}{b^{4a}} = b^{8a+4-4a} = b^{4a+4}$

35. $[(-2)^3 \cdot 3^2 \cdot a^{2\cdot3} \cdot b^2]^2 = (-2)^{3\cdot2} \cdot 3^{2\cdot2} \cdot a^{2\cdot3\cdot2} \cdot b^{2\cdot2} = 64 \cdot 81a^{12}b^4 = 5184a^{12}b^4$

36. $[5^2c^{3\cdot2}(-2)^3d^{2\cdot3}]^2 = 5^4(-2)^6c^{6\cdot2}d^{6\cdot2} = 40{,}000c^{12}d^{12}$

37. The repeating pattern is formed by the digits 1, 4, 9, 6, 5, 6, 9, 4, 1, 0, . . .

38. The difference between each term and the previous term is 4 greater than the previous difference (for example: $25 - 13 = 12$, which is 4 greater than $13 - 5$, which is 8). To predict the value for $n = 101$, use the values generated by the table: $20201 + (20201 - 19801) + 4 = 20605$.

39. Examine the pattern of the last digits of the powers of 2: 2, 4, 8, 6, 2, 4, 8, 6 . . .. If n is a whole number, then the last digit of 2^{1+4n} is 2, the last digit of 2^{2+4n} is 4, the last digit of 2^{3+4n} is 8, and the last digit of 2^{4+4n} is 6. Since $27 = 3 + 4 \cdot 6$, the last digit of 2^{27} is 8.

40. Examine the pattern of the last digits of the powers of 4: 4, 6, 4, 6, 4, 6, ... If p is a positive integer, then the last digit of 4^p is 4 when p is odd, and 6 when p is even. If the last digit of 4^n is 6, then n is even.

page 237   Capsule Review

1. 10^8 **2.** 39×10^9 **3.** 64×10^6 **4.** 40×10^4

page 239   Class Exercises

1. $3^{-4} = \dfrac{1}{3^4} = \dfrac{1}{81}$

2. $x^{-2}y = \dfrac{1}{x^2} \cdot y = \dfrac{y}{x^2}$

3. $(m^{-3})^3 = m^{-9} = \dfrac{1}{m^9}$

4. $\dfrac{p^{-4}}{p^{-5}} = p^{[-4-(-5)]} = p^{-4+5} = p$

5. 1×10^5 **6.** 1.05×10^{-3} **7.** 7.28345×10^4 **8.** 6.2×10^0

page 239   Practice Exercises

A 1. $\dfrac{1}{5^2} \cdot \dfrac{1}{x} \cdot y = \dfrac{y}{25x}$

2. $\dfrac{1}{4^3} \cdot j \cdot \dfrac{1}{k^5} = \dfrac{j}{64k^5}$

3. $m^{-2}n^2 = \dfrac{1}{m^2} \cdot n^2 = \dfrac{n^2}{m^2}$

4. $(64x^{-6})(xy^{-4}) = 64x^{-5}y^{-4} = \dfrac{64}{x^5y^4}$

5. No, $27 > 10$ **6.** yes   **7.** No, $0.009 < 1$ **8.** No, $0.7 < 1$

9. 2×10^6 **10.** 8×10^6 **11.** 7.65×10^{-3} **12.** 4.38×10^{-3}

13. 3.98×10^9 **14.** 8.68×10^3

B 15. $0.9 \times 10^7 = 9 \times 10^6$ **16.** $0.9 \times 10^7 = 9 \times 10^6$

17. $0.6 \times 10^{10} = 6 \times 10^9$ **18.** $0.7 \times 10^6 = 7 \times 10^5$

19. $(-27u^3v^{-6})(5u^{-4}v) = -135u^{-1}v^{-5} = -135 \cdot \dfrac{1}{u} \cdot \dfrac{1}{v^5} = -\dfrac{135}{uv^5}$

20. $\dfrac{10x^6y^{-8}}{16x^{-4}y^6} = \dfrac{5}{8}x^{10}y^{-14} = \dfrac{5}{8}x^{10} \cdot \dfrac{1}{y^{14}} = \dfrac{5x^{10}}{8y^{14}}$

21. $\dfrac{8a^{-15}b^{-12}}{36a^6b^{-12}} = \dfrac{2}{9}a^{-21}b^0 = \dfrac{2}{9} \cdot \dfrac{1}{a^{21}} \cdot 1 = \dfrac{2}{9a^{21}}$

22. $\dfrac{(2)(4)}{6} \times 10^{-3+8-2} = \dfrac{4}{3} \times 10^3 = 1.\overline{3} \times 10^3$ **23.** $\dfrac{(1.5)(2)}{3} \times 10^{-11+15-9} = 1 \times 10^{-5}$

C 24. $(-1.5)^2 \times (1.6)^3 \times 10^{2 \cdot 3 + 2 \cdot 3} = 9.216 \times 10^{12}$

25. $(1.2)^3 \times (-2.4)^2 \times 10^{2 \cdot 3 + 3 \cdot 2} = 9.95328 \times 10^{12}$

26. $8 \times 10^8 \times 60 = 480 \times 10^8 = 4.8 \times 10^{10}$ per min; $4.8 \times 10^{10} \times 60 \times 24 =$ $6912 \times 10^{10} = 6.912 \times 10^{13}$ per day; $6.912 \times 10^{13} \times 365 =$ $2522.88 \times 10^{13} = 2.52288 \times 10^{16}$ per yr

27. $27y - 51 = 300, 27y = 351, y = 13$　　　**28.** $316 = 312 - 4n, 4 = -4n, n = -1$

page 240　Integrating Algebra: Fluid Motion

1. 1.428×10^{33}　　　　**2.** 2.142×10^{34}　　　　**3.** 4.284×10^{34}

4. 8.568×10^{34}　　　　**5.** 1.1064×10^{-15} min

page 241　Capsule Review

1. $5a$　　　　　　**2.** $-\dfrac{13}{3}x^3$　　　　　　**3.** not possible　　　　**4.** 0

page 243　Class Exercises

1. 1　　　　　**2.** 9　　　　　**3.** 0　　　　　**4.** 4　　　　　**5.** 1

6. 1　　　　　**7.** 2　　　　　**8.** 4　　　　　**9.** 4　　　　　**10.** 5

11. $-a^2 + 10a$　　　　　　　　　**12.** $-3ab^2$

13. $\dfrac{1}{3}a^3c - \dfrac{9}{3}a^3c + a = -\dfrac{8}{3}a^3c + a$

14. $\dfrac{1}{2}a^4 - 2a^5 + 4a^3 - 2 + a^2 = -2a^5 + \dfrac{1}{2}a^4 + 4a^3 + a^2 - 2$

pages 243–244　Practice Exercises

A 1. 1　　　　　**2.** 1　　　　　**3.** 7　　　　　**4.** 9　　　　　**5.** 0

6. 0　　　　　**7.** 4　　　　　**8.** 4　　　　　**9.** 5　　　　　**10.** 5

11. $-4x^3y + 6x^2y^2 + 2xy^3$　　　　　　**12.** $-5x^3y + 3x^2y^2 + 9xy^3$

13. $5x^2y^2 - 2xy^3$　　　　　　**14.** $4x^3y - 2xy^3$

15. $3x^3 - 15 - 20x^3 - 25x^2 - 5 + 30 = -17x^3 - 25x^2 + 10$

16. $5x^3 - 30 - 21x^3 - 21x^2 - 7 + 15 = -16x^3 - 21x^2 - 22$

B 17. $4x^2z - \dfrac{15}{8}xz^2$　　　　　　　　　**18.** $0.9xy^2$

19. $-\dfrac{23}{5}x^2z + 2xz^2$

20. $-0.6x^2y$

21. $\dfrac{1}{2}cde - \dfrac{1}{2}cd^2 - \dfrac{1}{2}cde = -\dfrac{1}{2}cd^2$

22. $fg + \dfrac{1}{3}g^2 - fg + \dfrac{1}{3}g^2 = \dfrac{2}{3}g^2$

23. $4x^2 + 3x^2 - 9 - 8 + 14x^2 = 21x^2 - 17$

24. $ab^2 - 4a - 4b + 2ab^2 - 2b = 3ab^2 - 6b - 4a$

25. $8xy^2 + 12x^2y - 4yx^2 - 10y^2x = -2xy^2 + 8x^2y$

26. $6xz^3 + 6x^2z - 7z^2x + 14x^3z = 6xz^3 - 7xz^2 + 14x^3z + 6x^2z$

27. $-x^{3a} + x^{2a} + 2x^a$

28. $-x^{4a} + x^a$

C 29. $x^{2a} - x^a$

30. $x^{2a} + 2x^a$

31. $-x^a y + 3xy^a + y^a; \ -(-2)^2\left(\dfrac{1}{2}\right) + 3(-2)\left(\dfrac{1}{2}\right)^2 + \left(\dfrac{1}{2}\right)^2 = -2 - \dfrac{3}{2} + \dfrac{1}{4} = -\dfrac{13}{4}$

32. $2x^a y^a - 2y^a - 6y; \ 2(-2)^2\left(\dfrac{1}{2}\right)^2 - 2\left(\dfrac{1}{2}\right)^2 - 6\left(\dfrac{1}{2}\right) = 2 - \dfrac{1}{2} - 3 = -\dfrac{3}{2}$

33. $2(3w + 4 + w) = 8w + 8$

34. $4(a + 1) = 4a + 4$

35. $(x + 3) + (3x - 5) + (2x - 1) = 6x - 3$

page 244   Test Yourself

1. No, the expression has 2 terms.

2. yes

3. yes

4. No, a variable is in the denominator.

5. $2^4 \cdot 2a^{3+2}b^{2+1} = 2^5a^5b^3 = 32a^5b^3$

6. -81

7. $a^{-11-7} = a^{-18} = \dfrac{1}{a^{18}}$

8. $\dfrac{3}{-1.5}x^{[1-(-4)]}y^{4-5} = -2x^5y^{-1} = -2x^5 \cdot \dfrac{1}{y} = -\dfrac{2x^5}{y}$

9. 3.12×10^{-4}

10. 1.9385×10^6

11. $23 \times 10^{-5} = 2.3 \times 10^{-4}$

12. $7x^3 - 21 - 24x^3 - 18x^2 - 12 + 9 = -17x^3 - 18x^2 - 24$

13. $-6x^3y - 3x^2y^2 + 8xy^3$

page 245   Capsule Review

1. $4a^2 + 4a$

2. $-16m$

3. $-2x^3 - 7x^2$

4. $m^3n - 2m^2n^2 - 3m$

5. $-\dfrac{4}{3}s^2 + \dfrac{3}{2}s$

page 246   Class Exercises

1. $12r + 5s + t$

2. $2a + 5b$

3. $4d - 2e + 3f$

4. $5x + 3$

5. $5c^2 + 9c$

6. $10h + k + 7$

7. $11m^2 + 19m - 4$ **8.** $c^2 + 2c$ **9.** $k + 2d$

10. $-11a^2 + 2a - 1 - 7a^2 - 4a + 1 = -18a^2 - 2a$

11. $6r^3 - 3r + 7 - 6r^3 + 3r - 7 = 0$

pages 246–247 Practice Exercises

A 1. $5x^3 + 2x^2 - 9x + 2$ **2.** $8x^3 + 2x^2 - 13x + 2$

3. $7r + 6s - 5t$ **4.** $6j - 3k + 9m$

5. $2x^3 + 9x^2 - x + 2$ **6.** $5x^3 + 7x^2 + 9x$

7. $(5x^3 + x^2 - 3x - 7) - (x^2 - 6x + 9) = 5x^3 + x^2 - 3x - 7 - x^2 + 6x - 9 = 5x^3 + 3x - 16$

8. $(8x^3 + x^2 - 5x - 9) - (x^2 - 8x + 11) = 8x^3 + x^2 - 5x - 9 - x^2 + 8x - 11 = 8x^3 + 3x - 20$

9. $(7r + 2s - 2t) - (4s - 3t) = 7r + 2s - 2t - 4s + 3t = 7r - 2s + t$

10. $(8j - 3k + 6m) - (-2j + 3m) = 8j - 3k + 6m + 2j - 3m = 10j - 3k + 3m$

11. $(2x^3 + 6x^2 + x) - (3x^2 - 2x + 2) = 2x^3 + 6x^2 + x - 3x^2 + 2x - 2 = 2x^3 + 3x^2 + 3x - 2$

12. $(3x^3 + 7x^2 + 6x) - (2x^3 + 3x) = 3x^3 + 7x^2 + 6x - 2x^3 - 3x = x^3 + 7x^2 + 3x$

13. $5x^3 + x^2 - 3x - 8 - x^2 + 7x - 9 = 5x^3 + 4x - 17$

14. $6x^3 + 3x^2 - x - 6 - x^2 + 8x - 9 = 6x^3 + 2x^2 + 7x - 15$

15. $3m^3 - m^7 - 6 - 2m^3 + m^7 + 9 = m^3 + 3$

16. $7n^3 - n^2 - 8 - 6n^3 + n^2 + 10 = n^3 + 2$

B 17. $5c^3 - 4c$ **18.** $15x^2 + 13$ **19.** $13a + b + 2c$

20. $3x^2 - x + 5$ **21.** $9r + 6s + 10$ **22.** $5x + 12y + 7$

23. $(3j + 2k + 4m) - (2j + k - 2m) = 3j + 2k + 4m - 2j - k + 2m = j + k + 6m$

24. $(6a + 5b - 3c) - (4a - 3b + 2c) = 6a + 5b - 3c - 4a + 3b - 2c = 2a + 8b - 5c$

25. $(3x^2 + 3x + 5) - (2x^2 - 2x + 5) = 3x^2 + 3x + 5 - 2x^2 + 2x - 5 = x^2 + 5x$

26. $(7a^2 - 11a + 7) - (5a^2 - 11a - 3) = 7a^2 - 11a + 7 - 5a^2 + 11a + 3 = 2a^2 + 10$

27. $(3x^4 - 7x^3 - 3x^2) - (6x^3 + 4x^2 + 3x) = 3x^4 - 7x^3 - 3x^2 - 6x^3 - 4x^2 - 3x = 3x^4 - 13x^3 - 7x^2 - 3x$

28. $(-7z^2 + 11z^2 + 12z) - (4z^3 + 3z^2 + 4) = -7z^3 + 11z^2 + 12z - 4z^3 - 3z^2 - 4 = -11z^3 + 8z^2 + 12z - 4$

C 29. $11a^2 + 3a + 8 + 6a^2 - 2a + 4 - 3a^2 - 5 = 14a^2 + a + 7$

30. $3y^2 - 7y + 9 + 2y^2 + 8y - 4 - 2y + 3 = 5y^2 - y + 8$

31. $17x^3 + 3x^2 + 4x - 7x^3 + 2x - 4 - 4x^2 + 3 = 10x^3 - x^2 + 6x - 1$

32. $4z^3 + 7z^2 + 8z - 2z^3 + 11z - 3 - 2z^2 - 4 = 2z^3 + 5z^2 + 19z - 7$

33. $(n) + (n + 1) + (n + 2) + (n + 3) = 4n + 6$

34. $w + w + (w + 4) + (w + 4) = 4w + 8$

page 247 Algebra in Health

The bacteria will start to grow at 0 h (immediately) and again at 22 h. It will start to decay at 6 h.

page 248 Capsule Review

1. $3m^2 - 21m + 9$

2. $-12n^2 - 15p^2$

3. $\frac{1}{2}a + b$

4. $0.2x^2 + 0.8x - 1.2$

5. $-12x^3 + 2x^2 - 0.6x + 0.2$

6. $-8y^2 + 6y - 12$

page 249 Class Exercises

1. $12xy - 8y^2$

2. $18r^2st + 6rs^2t + 4rst^2$

3. $-6a^2b + 12ab^2$

4. $-6d^3f + 12d^2ef - 18d^2f^2$

5. $6x^2 - 4x$

6. $-6c^2 - 18c$

7. $-15c^2 + 25c$

8. $6k^2 + 9k$

9. $-77a^4b^3 - 33a^2b^4 - 11a^2b^3c$

10. $48r^4s^4 + 16rs^6 - 8r^2s^5$

pages 249–250 Practice Exercises

A 1. $-5x^5 + 30x^4 - 40x^3 + 25x^2$

2. $-2x^5 + 14x^4 - 18x^3 + 16x^2$

3. $-6p^7 + 4p^5 - 10p^3$

4. $-6t^6 - 42t^4 - 24t^3$

5. $12p^3q + 6p^2q^2 - 3p^2qr$

6. $6a^3b + 8a^2b^2 - 10a^2bc$

7. $2x^5 - 2x^3 + 6x^2$

8. $-10y^7 + 10y^5 - 20y^3$

9. $15m^4n^2 - 10mn^4 - 5m^2n^3 + 5mn^2p$

10. $14xy^5 - 21xy^4 - 7x^2y^3 + 7xy^2z$

11. $49h^2k - 7hk^2 + 56hk$

12. $10m^2n + 15mn^2 - 25mn$

13. $-6a^2b^3 + 4a^3b^2 + 5a^2b^3 - 15a^3b^2 + 40a^2b = -a^2b^3 - 11a^3b^2 + 40a^2b$

14. $-20x^2y^3 + 12x^3y^2 + 3x^2y^3 - 15x^3y^2 + 27x^2y = -17x^2y^3 - 3x^3y^2 + 27x^2y$

B 15. $-2y^2 + 2y + 5y - 10 = -2y^2 + 7y - 10$

16. $6x - 9x^2 + 12x - 4 = -9x^2 + 18x - 4$

17. $2m - 6m^3 + 12m^3 - 6m^2 = 6m^3 - 6m^2 + 2m$

18. $15g^2 - 10g^3 + g^2 + 3g^3 = -7g^3 + 16g^2$

19. $6a^2b^3 - 4a^3b^2$

20. $6d^4e^4 - 7d^3e^3$

21. $15r^4s^4 - 10r^2s^6 + 5rs^5$

22. $42a^4b + 18a^3b^2 - 6a^2b^2$

23. $-12a^3bc^4 + 6ab^3c^6 - 3a^2b^2c^3$

24. $24x^4y^2z + 56x^3yz^3 - 8x^2y^2z^2$

C 25. $-7m^3 + 7m^2 + 2m^2n + 2m^3 = -5m^3 + 7m^2 + 2m^2n$

26. $-9r^3t^2 + 9st^2 + 3r^3t^2 + 15st^2 = -6r^3t^2 + 24st^2$

27. $24a^{2x} - 8a^{x+2} + 16a^x$

28. $14y^{2b} + 6y^{b+3} - 16y^b$

29. $j(j + 2) = j^2 + 2j$

30. $e(e - 2) = e^2 - 2e$

31. $w(w + 2) = w^2 + 2w$

32. $w(2w + 5) = 2w^2 + 5w$

33. $l(3l - 10) = 3l^2 - 10l$

34. $-2x < 6, x > -3$

35. $x \le 5$

36. $2y + 2 \ge -12 + 3y, -y \ge -14, y \le 14$

37. $-6 < 4x < 5, -\dfrac{3}{2} < x < \dfrac{5}{4}$

38. $3 - 2n \ge 1$ or $3 - 2n \le -1, -2n \ge -2$ or $-2n \le -4, n \le 1$ or $n \ge 2$

39. $\dfrac{5 - 2n}{5} \le 2$ and $\dfrac{5 - 2n}{5} \ge -2, 5 - 2n \le 10$ and $5 - 2n \ge -10, -2n \le 5$

and $-2n \ge -15, n \ge -\dfrac{5}{2}$ and $n \le \dfrac{15}{2}, -\dfrac{5}{2} \le n \le \dfrac{15}{2}$

40. Let t = the time for the second bus, then $t + 1$ = the time for the first bus. $54t = 48(t + 1), 54t = 48t + 48, 6t = 48, t = 8$; The second bus will overtake the first bus in 8 h.

41. Let r = the rate of the northbound jogger, then $r + 3$ = the rate of the southbound jogger. $r(1.75) + (r + 3)(1.75) = 24.5, 1.75r + 1.75r + 5.25 = 24.5, 3.5r + 5.25 = 24.5, 3.5r = 19.25, r = 5.5$; The rate of the northbound jogger: 5.5 km/h; the southbound jogger: 8.5 km/h.

42. Let x = the number of ounces of pure acid to be added. $(0.012)(64) + x = (0.03)(64 + x), 0.768 + x = 1.92 + 0.03x, 0.97x = 1.152, x \approx 1.2$; Add approximately 1.2 oz of acid to increase the concentration.

43. Let x = the number of mL of water to be added. $(0.15)(80) + 0 = (0.06)(80 + x), 12 = 4.8 + 0.06x, 7.2 = 0.06x, x = 120$; Add 120 mL of water to decrease the concentration of salt.

1. $5z^2 + 33z - 14$ 2. $6y^3 - 2y^2 + 4y$

3. $2a^2b - ab^2 + ab$ 4. $8x^2 + 6x - 27$

page 253   Class Exercises

1. $x^2 - 5x + 1x - 5 = x^2 - 4x - 5$ 2. $y^2 - 4y + 2y - 8 = y^2 - 2y - 8$

3. $ce + cf + de + df$ 4. $ac - ad - bc + bd$

5. $6c^2 - 15c + 8c - 20 = 6c^2 - 7c - 20$

6. $3g^2 + 7g - 15g - 35 = 3g^2 - 8g - 35$

7. $-2x^2 + x^3 - 3x - 6x + 3x^2 - 9 = x^3 + x^2 - 9x - 9$

8. $-7z^2 + z^3 - 6z - 14z + 2z^2 - 12 = z^3 - 5z^2 - 20z - 12$

9. $4h^2 - 6h - 6h + 9 = 4h^2 - 12h + 9$

10. $3x^2 + 9x - 2x - 6 = 3x^2 + 7x - 6$ 11. $4x^2 + 8x - 8x - 16 = 4x^2 - 16$

12. $xa + ya + xb + yb$; Yes, with both methods each term of one binomial is multiplied by each term of the other binomial.

13. $3y + 4 - 6y^2 - 8y = -6y^2 - 5y + 4$; The product would have descending order of exponents, which makes it easier to combine like terms.

14. The product will be a trinomial when like terms can be combined and when the binomials are not in the form $(a + b)(a - b)$, which yields $a^2 - b^2$.

pages 253–255   Practice Exercises

A   1. $a^2 - 8a + 5a - 40 = a^2 - 3a - 40$ 2. $x^2 - 6x + 5x - 30 = x^2 - x - 30$

3. $h^2 - 7h + h - 7 = h^2 - 6h - 7$ 4. $m^2 - 2m + 4m - 8 = m^2 + 2m - 8$ ✔

5. $3a^2 - 4a + 15a - 20 = 3a^2 + 11a - 20$

6. $4p^2 - 3p + 24p - 18 = 4p^2 + 21p - 18$ ✔

7. $10m^2 + 12m - 45m - 54 = 10m^2 - 33m - 54$

8. $18n^2 + 15n - 12n - 10 = 18n^2 + 3n - 10$ ✔

9. $7r^2 - 35r - 3r + 15 = 7r^2 - 38r + 15$

10. $3j^2 - 6j - 7j + 14 = 3j^2 - 13j + 14$ ✔

11. $10x^2 - 15x - 16x + 24 = 10x^2 - 31x + 24$

12. $14y^2 - 6y - 35y + 15 = 14y^2 - 41y + 15$ ✔

13. $4y^2 - 8yz + yz - 2z^2 = 4y^2 - 7yz - 2z^2$

14. $5b^2 - 10bd + bd - 2d^2 = 5b^2 - 9bd - 2d^2$ ✔

15. $8j^2 + 2jk - 12jk - 3k^2 = 8j^2 - 10jk - 3k^2$

16. $2p^2 + 3pq - 2pq - 3q^2 = 2p^2 + pq - 3q^2$ ✓

17. $(3b + 7)(-2b + 3) = -6b^2 + 9b - 14b + 21 = -6b^2 - 5b + 21$

18. $(4a + 5)(-4a + 2) = -16a^2 + 8a - 20a + 10 = -16a^2 - 12a + 10$ ✓

19. $(-4y^3 + y^2 - 2y)(6y - 3) = -24y^4 + 6y^3 - 12y^2 + 12y^3 - 3y^2 + 6y =$
$-24y^4 + 18y^3 - 15y^2 + 6y$

20. $(-3x^3 + x^2 - 5x)(4x - 6) = -12x^4 + 4x^3 - 20x^2 + 18x^3 - 6x^2 + 30x =$ ✓
$-12x^4 + 22x^3 - 26x^2 + 30x$

21. $(-5z^3 + z^2 + 4z)(2z + 3) = -10z^4 + 2z^3 + 8z^2 - 15z^3 + 3z^2 + 12z =$
$-10z^4 - 13z^3 + 11z^2 + 12z$

22. $(-2m^3 + 3m^2 + 8m)(m + 5) = -2m^4 + 3m^3 + 8m^2 - 10m^3 + 15m^2 + 40m =$ ✓
$-2m^4 - 7m^3 + 23m^2 + 40m$

23. $x^3 + 3x^2 - 5x + 2x^2 + 6x - 10 = x^3 + 5x^2 + x - 10$

24. $y^3 + 11y^2 - 12y + 5y^2 + 55y - 60 = y^3 + 16y^2 + 43y - 60$ ✓

25. $z^3 - 5z^2 - 12z - 3z^2 + 15z + 36 = z^3 - 8z^2 + 3z + 36$

26. $w^3 - 5w^2 + w - 4w^2 + 20w - 4 = w^3 - 9w^2 + 21w - 4$ ✓

27. $c^3 + 2c^2 + c - 5c^2 - 10c - 5 = c^3 - 3c^2 - 9c - 5$

28. $d^3 - 2d^2 + 3d + 2d^2 - 4d + 6 = d^3 - d + 6$ ✓

B 29. $-k^3 + 7k^2 + 8k - 3k^2 + 21k + 24 = -k^3 + 4k^2 + 29k + 24$

30. $-n^3 + 3n^2 + 2n - 5n^2 + 15n + 10 = -n^3 - 2n^2 + 17n + 10$ ✓

31. $(y - 3)(y^2 - 2y + 1) = y^3 - 2y^2 + y - 3y^2 + 6y - 3 = y^3 - 5y^2 + 7y - 3$

32. $(x + 2)(x^2 + 3x - 1) = x^3 + 3x^2 - x + 2x^2 + 6x - 2 = x^3 + 5x^2 + 5x - 2$ ✓

33. $(3x - 4)(3x - 4) = 9x^2 - 12x - 12x + 16 = 9x^2 - 24x + 16$

34. $(2y - 7)(2y - 7) = 4y^2 - 14y - 14y + 49 = 4y^2 - 28y + 49$ ✓

35. $(2a - 3b)(2a - 3b) = 4a^2 - 6ab - 6ab + 9b^2 = 4a^2 - 12ab + 9b^2$

36. $(3p - 4q)(3p - 4q) = 9p^2 - 12pq - 12pq + 16q^2 = 9p^2 - 24pq + 16q^2$ ✓

37. $6a^3 + 4a^2 - 3a^2 - 2a + 15a + 10 = 6a^3 + a^2 + 13a + 10$

38. $20b^3 + 12b^2 + 5b^2 + 3b - 10b - 6 = 20b^3 + 17b^2 - 7b - 6$ ✓

39. $-2c^3 + 10c^2 - 4c - c^2 + 5c - 2 = -2c^3 + 9c^2 + c - 2$

40. $-3d^3 - 2d^2 - 9d^2 - 6d + 3d + 2 = -3d^3 - 11d^2 - 3d + 2$ ✓

41. $10g^3 - 5g^2 + 6g - 3 - 4g^2 + 2g = 10g^3 - 9g^2 + 8g - 3$

42. $6h^3 - 15h^2 + 12h - 2h^2 + 5h - 4 = 6h^3 - 17h^2 + 17h - 4$

43. $x^4 - x^3 + 3x^2 - 5x^3 + 5x^2 - 15x = x^4 - 6x^3 + 8x^2 - 15x$

44. $(y - 3)(-y^3 + 2y^2 + y) = -y^4 + 2y^3 + y^2 + 3y^3 - 6y^2 - 3y =$
$-y^4 + 5y^3 - 5y^2 - 3y$

45. $(2y^3 - y^2 + 2y)(2y - 3) = 4y^4 - 6y^3 - 2y^3 + 3y^2 + 4y^2 - 6y =$
$4y^4 - 8y^3 + 7y^2 - 6y$

46. $(3x^3 - 2x^2 - x)(3x - 4) = 9x^4 - 12x^3 - 6x^3 + 8x^2 - 3x^2 + 4x =$
$9x^4 - 18x^3 + 5x^2 + 4x$

47. $5t^4 + 10t^3 - 6t^3 - 12t^2 - 2t^2 - 4t = 5t^4 + 4t^3 - 14t^2 - 4t$

48. $6v^4 + 3v^3 - 4v^3 - 2v^2 - 6v^2 - 3v = 6v^4 - v^3 - 8v^2 - 3v$

C 49. $a^3 + a^2b - a^2b - ab^2 + ab^2 + b^3 = a^3 + b^3$

50. $c^3 - c^2d + c^2d - cd^2 + cd^2 - d^3 = c^3 - d^3$

51. $y^4 + 3y^3 - 2y^3 - 6y^2 + 3y + 9 = y^4 + y^3 - 6y^2 + 3y + 9$

52. $-4(x^2 + 2x - 3x - 6) = -4(x^2 - x - 6) = -4x^2 + 4x + 24$

53. $(x^2 + 3x + 3x + 9)(x + 3) = (x^2 + 6x + 9)(x + 3) = x^3 + 3x^2 + 6x^2 +$
$18x + 9x + 27 = x^3 + 9x^2 + 27x + 27$

54. $(x^2 - 2x - 2x + 4)(x - 2) = (x^2 - 4x + 4)(x - 2) = x^3 - 2x^2 - 4x^2 +$
$8x + 4x - 8 = x^3 - 6x^2 + 12x - 8$

55. $(4b - 1)(4b - 1)(4b - 1) = (16b^2 - 4b - 4b + 1)(4b - 1) = (16b^2 - 8b + 1)$
$(4b - 1) = 64b^3 - 16b^2 - 32b^2 + 8b + 4b - 1 = 64b^3 - 48b^2 + 12b - 1$

56. $(3d - 1)(3d - 1)(3d - 1) = (9d^2 - 3d - 3d + 1)(3d - 1) = (9d^2 - 6d + 1)$
$(3d - 1) = 27d^3 - 9d^2 - 18d^2 + 6d + 3d - 1 = 27d^3 - 27d^2 + 9d - 1$

57. $3x^5 - x^3 - 3x^3 + x + 6x^2 - 2 = 3x^5 - 4x^3 + 6x^2 + x - 2$

58. $5y^7 + 5y^5 - 5y^3 - y^5 - y^3 + y = 5y^7 + 4y^5 - 6y^3 + y$

59. $10a^5 - 4a^4 + 2a^3 - 5a^4 + 2a^3 - a^2 + 15a^2 - 6a + 3 = 10a^5 - 9a^4 +$
$4a^3 + 14a^2 - 6a + 3$

60. $12b^5 - 6b^4 - 18b^2 + 4b^4 - 2b^3 - 6b - 8b^3 + 4b^2 + 12 = 12b^5 - 2b^4 -$
$10b^3 - 14b^2 - 6b + 12$

61. $(n + 1)(n + 2); n^2 + 2n + 1n + 2 = n^2 + 3n + 2$

62. $(x + 2)(x + 2 + 3) = (x + 2)(x + 5); x^2 + 5x + 2x + 10 = x^2 + 7x + 10$

63. $(x + 1)[5 + 2(x + 1)] = (x + 1)(5 + 2x + 2) = (x + 1)(2x + 7);$
$2x^2 + 7x + 2x + 7 = 2x^2 + 9x + 7$

64. $(2x + 1)(x + 4) = 2x^2 + 8x + x + 4 = 2x^2 + 9x + 4; (2x - 9)(x + 9) =$
$2x^2 + 18x - 9x - 81 = 2x^2 + 9x - 81;$ The first rectangle has the greater
area; $(2x^2 + 9x + 4) - (2x^2 + 9x - 81) = 4 + 81 = 85;$ The first
rectangle is 85 square units greater.

65. $(e + 2)(e + 4) = e^2 + 4e + 2e + 8 = e^2 + 6e + 8$

   a. Mercury: $(4.34 \times 10^7) \div (1.86 \times 10^5) = 2.33 \times 10^2 = 233$ s;
      Venus: $(6.77 \times 10^7) \div (1.86 \times 10^5) = 3.64 \times 10^2 = 364$ s;
      Earth: $(9.46 \times 10^7) \div (1.86 \times 10^5) = 5.09 \times 10^2 = 509$ s;
      Mars: $(1.55 \times 10^8) \div (1.86 \times 10^5) = 0.833 \times 10^3 = 833$ s;
      Jupiter: $(5.07 \times 10^8) \div (1.86 \times 10^5) = 2.726 \times 10^3 = 2726$ s;
      Saturn: $(9.37 \times 10^8) \div (1.86 \times 10^5) = 5.04 \times 10^3 = 5{,}040$ s;
      Uranus: $(1.86 \times 10^9) \div (1.86 \times 10^5) = 1 \times 10^4 = 10{,}000$ s;
      Neptune: $(2.82 \times 10^9) \div (1.86 \times 10^5) = 1.5161 \times 10^4 = 15{,}161$ s;
      Pluto: $(4.55 \times 10^9) \div (1.86 \times 10^5) = 2.4462 \times 10^4 = 24{,}462$ s.

   b. Mercury: $(2.86 \times 10^7) \div (1.86 \times 10^5) = 1.54 \times 10^2 = 154$ s;
      Venus: $(6.68 \times 10^7) \div (1.86 \times 10^5) = 3.59 \times 10^2 = 359$ s;
      Earth: $(9.14 \times 10^7) \div (1.86 \times 10^5) = 4.91 \times 10^2 = 491$ s;
      Mars: $(1.29 \times 10^8) \div (1.86 \times 10^5) = 0.694 \times 10^3 = 694$ s;
      Jupiter: $(4.61 \times 10^8) \div (1.86 \times 10^5) = 2.48 \times 10^3 = 2480$ s;
      Saturn: $(8.38 \times 10^8) \div (1.86 \times 10^5) = 4.51 \times 10^3 = 4510$ s;
      Uranus: $(1.67 \times 10^9) \div (1.86 \times 10^5) = 0.8978 \times 10^4 = 8978$ s;
      Neptune: $(2.76 \times 10^9) \div (1.86 \times 10^5) = 1.4839 \times 10^4 = 14{,}839$ s;
      Pluto: $(2.76 \times 10^9) \div (1.86 \times 10^5) = 1.4839 \times 10^4 = 14{,}839$ s.

  1. $(-12)(-12) = 144$ **2.** $(3)^2(r)^2 = 9r^2$

  3. $\dfrac{(a)^2}{(2)^2} = \dfrac{a^2}{4}$ **4.** $\dfrac{(2)^2(x^4)^2}{(3)^2} = \dfrac{4x^8}{9}$

  5. $\left(\dfrac{1}{3}\right)^2(s^4)^2(t)^2 = \dfrac{1}{9}s^8t^2$ **6.** $(6)^2(s)^2(t^2)^2(u)^2 = 36s^2t^4u^2$

  7. $(0.2)^2(x)^2(y)^2 = 0.04x^2y^2$ **8.** $(-1.2)^2(g)^2(h^5)^2 = 1.44g^2h^{10}$

  9. $\dfrac{(2)^2}{(5)^2}(d^2)^2(e^2)^2 = \dfrac{4}{25}d^4e^4$ **10.** $(0.1)^2(a^5)^2(b^3)^2(c^2)^2 = 0.01a^{10}b^6c^4$

  1. $c^2 + 2(c \cdot d) + d^2 = c^2 + 2cd + d^2$ **2.** $r^2 + 2(r \cdot 3s) + 9s^2 = r^2 + 6rs + 9s^2$

  3. $m^2 + 2[m \cdot (-n)] + n^2 = m^2 - 2mn + n^2$

  4. $4r^2 + 2[2r \cdot (-s)] + s^2 = 4r^2 - 4rs + s^2$

  5. $u^2 + uv - uv - v^2 = u^2 - v^2$ **6.** $49m^2 + 7m - 7m - 1 = 49m^2 - 1$

7. $(60 + 2)^2 = 60^2 + 2 \cdot 60 \cdot 2 + 2^2 = 3600 + 240 + 4 = 3844$

8. $(90 - 1)^2 = 90^2 + 2 \cdot 90(-1) + 1^2 = 8100 - 180 + 1 = 7921$

9. $30^2 + 2 \cdot 30 \cdot 3 + 3^2 = 900 + 180 + 9 = 1089$

10. $4900 - 1 = 4899$ **11.** $(3x^3)^2 - (4y^2)^2 = 9x^6 - 16y^4$

12. $(5x^3)^2 - (2y^2)^2 = 25x^6 - 4y^4$ **13.** $(3m^2)^2 - (16m)^2 = 9m^4 - 256m^2$

14. $(10x^4)^2 - (15x^2)^2 = 100x^8 - 225x^4$ **15.** $(4y^5)^2 - (3y^4)^2 = 16y^{10} - 9y^8$

16. $(16a^2b^2)^2 - (4ab)^2 = 256a^4b^4 - 16a^2b^2$

17. Write 102^2 as $(100 + 2)^2$, then square the binomial.

18. No, it might be simpler to use $(50 - 5)^2$.   **19.** true

pages 258–259   Practice Exercises

A   **1.** $(3x)^2 + 2 \cdot 3x \cdot 2y + (2y)^2 = 9x^2 + 12xy + 4y^2$

     2. $(5a)^2 + 2 \cdot 5a \cdot 3b + (3b)^2 = 25a^2 + 30ab + 9b^2$

     3. $(8m)^2 + 2 \cdot 8m \cdot 2n + (2n)^2 = 64m^2 + 32mn + 4n^2$

     4. $(4j)^2 + 2 \cdot 4j \cdot 6k + (6k)^2 = 16j^2 + 48jk + 36k^2$

     5. $(4m^2)^2 + 2(4m^2)(-6) + (-6)^2 = 16m^4 - 48m^2 + 36$

     6. $(6r^2)^2 + 2(6r^2)(-4) + (-4)^2 = 36r^4 - 48r^2 + 16$

     7. $(7s^2)^2 + 2(7s^2)(-3) + (-3)^2 = 49s^4 - 42s^2 + 9$

     8. $(3j^2)^2 + 2(3j^2)(-7) + (-7)^2 = 9j^4 - 42j^2 + 49$

     9. $(20 + 4)^2 = 20^2 + 2 \cdot 20 \cdot 4 + 4^2 = 400 + 160 + 16 = 576$

     10. $(40 + 5)^2 = 40^2 + 2 \cdot 40 \cdot 5 + 5^2 = 1600 + 400 + 25 = 2025$

     11. $(10 + 2)^2 = 10^2 + 2 \cdot 10 \cdot 2 + 2^2 = 100 + 40 + 4 = 144$

     12. $(50 + 2)^2 = 50^2 + 2 \cdot 50 \cdot 2 + 2^2 = 2500 + 200 + 4 = 2704$

     13. $x^2 - 4^2 = x^2 - 16$ **14.** $h^2 - 9^2 = h^2 - 81$

     15. $y^2 - 5^2 = y^2 - 25$ **16.** $m^2 - 2^2 = m^2 - 4$

     17. $(4x^3)^2 - 3^2 = 16x^6 - 9$ **18.** $(5b^3)^2 - 8^2 = 25b^6 - 64$

B   **19.** $(2t)^2 + 2(2t)(-u) + u^2 = 4t^2 - 4tu + u^2$

     20. $(3j)^2 + 2(3j)(-h) + h^2 = 9j^2 - 6jh + h^2$

     21. $(d^2)^2 + 2 \cdot d^2 \cdot e^2 + (e^2)^2 = d^4 + 2d^2e^2 + e^4$

     22. $(f^2)^2 + 2 \cdot f^2 \cdot g^2 + (g^2)^2 = f^4 + 2f^2g^2 + g^4$

23. $(h^2)^2 - (j^2)^2 = h^4 - j^4$

24. $(k^2)^2 - (m^2)^2 = k^4 - m^4$

25. $1^2 + 2(1)(-12g^3h^2) + (-12g^3h^2)^2 = 1 - 24g^3h^2 + 144g^6h^4$

26. $2^2 + 2(2)(-6x^3y^2) + (-6x^3y^2)^2 = 4 - 24x^3y^2 + 36x^6y^4$

27. $(5ab^3)^2 + 2(5ab^3)(6c^2d^4) + (6c^2d^4)^2 = 25a^2b^6 + 60ab^3c^2d^4 + 36c^4d^8$

28. $(3ef^3)^2 + 2(3ef^3)(4g^2h^4) + (4g^2h^4)^2 = 9e^2f^6 + 24ef^3g^2h^4 + 16g^4h^8$

29. $(60x^3y^2z + x^3y^2z)^2 = (60x^3y^2z)^2 + 2(60x^3y^2z)(x^3y^2z) + (x^3y^2z)^2 =$
$3600x^6y^4z^2 + 120x^6y^4z^2 + x^6y^4z^2 = 3721x^6y^4z^2$

30. $(80a^4b^0c^5 + 2a^4b^0c^5)^2 = (80a^4b^0c^5)^2 + 2(80a^4b^0c^5)(2a^4b^0c^5) + (2a^4b^0c^5)^2 =$
$6400a^8c^{10} + 320a^8c^{10} + 4a^8c^{10} = 6724a^8c^{10}$

C 31. $(4)(4) + 4a^x - 4a^x - a^{2x} = 16 - a^{2x}$

32. $y^{2b} - 2y^b + 2y^b - 4 = y^{2b} - 4$

33. $(x^{a+1} + x)(x^{a+1} + x) = x^{2(a+1)} + x^{a+2} + x^{a+2} + x^2 = x^{2a+2} + 2x^{a+2} + x^2$

34. $(a^{x+2} + 2a)(a^{x+2} + 2a) = a^{2(x+2)} + 2a^{x+3} + 2a^{x+3} + 4a^2 = a^{2x+4} + 4a^{x+3} + 4a^2$

35. $(3^{2y+1} - 2)(3^{2y+1} - 2) = 3^{2(2y+1)} - (2)(3^{2y+1}) - (2)(3^{2y+1}) + 4 =$
$3^{4y+2} - 4(3^{2y+1}) + 4$

36. $(5^{1-3x} - 4)(5^{1-3x} - 4) = 5^{2(1-3x)} - 4(5^{1-3x}) - 4(5^{1-3x}) + 16 =$
$5^{2-6x} - 8(5^{1-3x}) + 16$

37. $(a + 1)(a + 1) = a^2 + a + a + 1 = a^2 + 2a + 1$

38. $(b + 3)(b + 3) = b^2 + 3b + 3b + 9 = b^2 + 6b + 9$

39. $(2x - 3)(2x + 3) = 4x^2 + 6x - 6x - 9 = 4x^2 - 9$

40. $(4m - 3)(4m + 3) = 16m^2 + 12m - 12m - 9 = 16m^2 - 9$

page 259   Test Yourself

1. $3x^3 - 4x$

2. $-2x^2y + x^3y^3 + 9x = x^3y^3 - 2x^2y + 9x$

3. $(5y^2 - 3y^2) + (6 - 10) = 2y^2 - 4$

4. $a^3 + 5a^2 - 6 + 4a^2 + a - 9 = a^3 + 9a^2 + a - 15$

5. $6y^3 - 2y^2$

6. $(-3x^4)(4x^3) + (-3x^4)(2x^2) + (-3x^4)(-1) = -12x^7 - 6x^6 + 3x^4$

7. $(11a^2b)(5a^3b^2) + (11a^2b)(-3a^2b^3) + (11a^2b)(7ab^4) = 55a^5b^3 - 33a^4b^4 + 77a^3b^5$

8. $-3t + 6t^3 + 40t - 32 = 6t^3 + 37t - 32$

9. $3r^2 + 33r + 7r + 77 = 3r^2 + 40r + 77$

10. $3 + 4j - 15j - 20j^2 = 3 - 11j - 20j^2$

11. $24n^2 + 40n - 45n - 75 = 24n^2 - 5n - 75$

12. $-2z^2 + 3z^3 + 9z + 12z - 18z^2 - 54 = 3z^3 - 20z^2 + 21z - 54$

13. $(9y + 3)(9y + 3) = 81y^2 + 27y + 27y + 9 = 81y^2 + 54y + 9$

14. $4x^2 - 6x + 6x - 9 = 4x^2 - 9$

15. $(5a - 9b)(5a - 9b) = 25a^2 - 45ab - 45ab + 81b^2 = 25a^2 - 90ab + 81b^2$

16. $b^8 + 11ab^4 - 11ab^4 - 121a^2 = b^8 - 121a^2$

page 262   Class Exercises

1. The pattern is the addition of the two previous terms. $13 + 21 = 34$; $21 + 34 = 55$; $34 + 55 = 89$; 34, 55, 89

2. The pattern is the previous term plus two. $14 + 2 = 16$; $16 + 2 = 18$; $18 + 2 = 20$; 16, 18, 20

3. The pattern is the addition of the two previous terms. $44 + 71 = 115$; $71 + 115 = 186$; $115 + 186 = 301$; 115, 186, 301

4. The pattern is the addition of the two previous terms. $21 + 34 = 55$; $34 + 55 = 89$; $55 + 89 = 144$; 55, 89, 144

pages 262–263   Practice Exercises

A   1. The pattern is the previous term times 10.
$10{,}000 \times 10 = 100{,}000$; $100{,}000 \times 10 = 1{,}000{,}000$;
$1{,}000{,}000 \times 10 = 10{,}000{,}000$; 100,000, 1,000,000, 10,000,000

2. The pattern is the previous term times 3.
$162 \times 3 = 486$; $486 \times 3 = 1458$; $1458 \times 3 = 4374$; 486, 1458, 4374

3. The pattern is adding successfully larger odd integers to the previous term, starting with 3 and then $+5$, $+7$, $+9$, $+11$, ...; $24 + 11 = 35$; $35 + 13 = 48$; $48 + 15 = 63$; 35, 48, 63

4. $5 + 9 + 14 + 23 + 37 + 60 + 97 + 157 + 254 + 411 + 665 + 1076 + 1741 = 4549$

5. $1076 + 1741 = 2817$; $1741 + 2817 = 4558$

6. $4558 - 9 = 4549$; yes

7. 6, 12, 18, 30, 48, 78, 126, ...

8. 6, 12, 18, 30, 48, 78, 126, 204, 330, 534, 864, 1398, 2262, 3660, 5922; the difference of the 15th term and the 2nd term is $5922 - 12 = 5910$; the sum of the first 13 terms is $6 + 12 + 18 + 30 + 48 + 78 + 126 + 204 + 330 + 534 + 864 + 1398 + 2262 = 5910$; yes

B **9.** $1 + 2 = 3 = \dfrac{(2)(2 + 1)}{2}$

$1 + 2 + 3 = 6 = \dfrac{(3)(3 + 1)}{2}$

$1 + 2 + 3 + 4 = 10 = \dfrac{(4)(4 + 1)}{2}$

$1 + 2 + 3 + 4 + 5 = 15 = \dfrac{(5)(5 + 1)}{2}$

$1 + 2 + 3 + \ldots + n = \dfrac{n(n + 1)}{2}$

10. 2, 2, 4, 6, 10, 16, 26, 42, . . .   **11.** It is twice as large.

12. $a + b + (a + b) + (a + 2b) + (2a + 3b) + (3a + 5b) + (5a + 8b) +$
$(8a + 13b) + (13a + 21b) + 21a + 34b = 55a + 88b = 11(5a + 8b)$

C **13.** $233 + 377 + 610 + 987 + 1597 + 2584 + 4181 + 6765 + 10{,}946 +$
$17{,}711 = 45{,}991$

14. $1; 1 + 6 = 7; 6 + 15 = 21; 15 + 20 = 35; 20 + 15 = 35; 15 + 6 = 21;$
$6 + 1 = 7; 1;$ The numbers in the eighth row are 1, 7, 21, 35, 35, 21, 7, 1.

15. $1 + 1 = 2; 1 + 2 = 3; 1 + 3 + 1 = 5$

page 263   Mixed Problem Solving Review

1. Let t = the time it takes for the two jets to be 4000 km apart;

$850(t) + 750(t) = 4000, 1600t = 4000, t = 2\frac{1}{2}.$ The two jets will be 4000 km

apart after $2\frac{1}{2}$ h or at 4:30 pm.

2. Let x = the grade Lauren must get on her last test. $\dfrac{83 + 74 + x}{3} \geq 80,$

$\dfrac{157 + x}{3} \geq 80, 157 + x \geq 240, x \geq 83.$ Lauren must get at least 83 on her

last test in order to have an average for the term of no less than 80.

page 263   Project

All even numbers can be expressed by writing an addition sentence using only prime numbers. Yes. This conjecture is known as Goldbach's conjecture.

page 264   Summary and Review

1. $a^{3+4} = a^7$ **2.** $3^1 \cdot 3^2 = 3^{1+2} = 3^3 = 27$

3. $-15a^{1+3}c^{2+3} = -15a^4c^5$ **4.** $-7.5 \times 10^{2+5} = -7.5 \times 10^7$

5. $x^{5-3} = x^2$

6. $y^{2-8} = y^{-6} = \dfrac{1}{y^6}$

7. $x^{12-12} = x^0 = 1$

8. $\dfrac{4x^{3-2}}{-3y^{9-4}} = \dfrac{4x}{-3y^5} = -\dfrac{4x}{3y^5}$

9. 1

10. 1

11. $\dfrac{25a^2}{c^8}$

12. $\dfrac{-z^{7-4}}{3x^{3-2}} = \dfrac{-z^3}{3x} = \dfrac{z^3}{3x}$

13. $\dfrac{1}{7^2} \cdot a \cdot \dfrac{1}{b} = \dfrac{a}{49b}$

14. $x^{-3+5}y^{1+(-4)} = x^2y^{-3} = \dfrac{x^2}{y^3}$

15. $(64c^{10}d^{-4})(2c^{-4}d) = 128c^6d^{-3} = \dfrac{128c^6}{d^3}$

16. $\dfrac{20m^{-6}n^4}{32m^{-5}n^{-10}} = \dfrac{5}{8}m^{-6-(-5)}n^{4-(-10)} = \dfrac{5}{8}m^{-1}n^{14} = \dfrac{5n^{14}}{8m}$

17. 2.89×10^5

18. 4.1×10^{-2}

19. $16.0 \times 10^5 = 1.6 \times 10^6$

20. $12.6 \times 10^6 = 1.26 \times 10^7$

21. $0.9 \times 10^4 = 9.0 \times 10^3$

22. $0.3 \times 10^3 = 3.0 \times 10^2$

23. 1; 1

24. 8; 3; 0; 8

25. $5 - x^2 + 3x^3 = 3x^3 - x^2 + 5$

26. $-2x^3y^2 - x^3y^2 + 3x^4y - xy + y^3 - 4y^3 = -3x^3y^2 + 3x^4y - xy - 3y^3 = 3x^4y - 3x^3y^2 - xy - 3y^3$

27. $9x + 2y$

28. $9r - 5s + 3t$

29. $2y + 3 - 3y + 2 = -y + 5$

30. $-9a^3 + 6a^2 - 3 - 5a^2 - a + 3 = -9a^3 + a^2 - a$

31. $3x(x^2) + 3x(2y) + 3x(-1) = 3x^3 + 6xy - 3x$

32. $-2a^7 + 6a^6 - 10a^5$

33. $-35x^3y^2 + 5x^2y^3$

34. $2a(a) + 2a(7) + 4(a) + 4(7) = 2a^2 + 14a + 4a + 28 = 2a^2 + 18a + 28$

35. $b(3b) + b(-8) - 3(3b) - 3(-8) = 3b^2 - 8b - 9b + 24 = 3b^2 - 17b + 24$

36. $c(5c) + c(-2d) + 4d(5c) + 4d(-2d) = 5c^2 - 2cd + 20cd - 8d^2 = 5c^2 + 18cd - 8d^2$

37. $(d - 9)(d - 9) = d(d) + 2(d)(-9) + (-9)(-9) = d^2 - 18d + 81$

38. $(3z^2 + 7)(3z^2 + 7) = 3z^2(3z^2) + 2(3z^2)(7) + 7(7) = 9z^4 + 42z^2 + 49$

39. $5x(5x) + 5x(4) - 4(5x) - 4(4) = 25x^2 + 20x - 20x - 16 = 25x^2 - 16$

40. The pattern is the previous term multiplied by 4; 512, 2048, 8192.

41. The pattern is adding successively larger integers beginning with 4, then 5, 6, 7, ...; $22 + 8 = 30$; $30 + 9 = 39$; $39 + 10 = 49$; 30, 39, 49.

page 266   Chapter Test

1. $3^{3+2} = 3^5$ or 243

2. $\dfrac{1}{c^{7-1}} = \dfrac{1}{c^6}$

3. 1

4. $\dfrac{1}{2^4} \cdot p \cdot \dfrac{1}{q^2} = \dfrac{p}{16q^2}$

5. $2^{2\cdot3} = 2^6 = 64$

6. $24x^{1+3}y^{2+5} = 24x^4y^7$

7. $2c^{4-(-5)}d^{-3-1}e^{2-2} = 2c^9d^{-4}e^0 = \dfrac{2c^9}{d^4}$

8. $\dfrac{-27a^3}{2} = -\dfrac{27a^3}{2}$

9. $\dfrac{625b^{2\cdot4}c^{3\cdot4}}{10,000b^{4\cdot4}} = \dfrac{625b^8c^{12}}{10,000b^{16}} = \dfrac{1c^{12}}{16b^{16-8}} = \dfrac{c^{12}}{16b^8}$

10. scientific notation

11. 5.1×10^3

12. $5 \times 10^{5-(-2)} = 5 \times 10^7$

13. 1.034×10^0

14. $12.0 \times 10^{1+3} = 12.0 \times 10^4 = 1.2 \times 10^5$

15. $(0.2 \times 10^2)(0.2 \times 10^2)(0.2 \times 10^2) = 0.008 \times 10^{2+2+2} = 0.008 \times 10^6 = 8.0 \times 10^3$

16. 0; 0

17. 3, 4, 4, 5; 5

18. $-3y^2 + y^2 + 2y - 9 - y^3 = -2y^2 + 2y - 9 - y^3 = -y^3 - 2y^2 + 2y - 9$

19. $5x^2y - 3xy^2 - 3y^3 + 6x^2y + x^3 = 11x^2y - 3xy^2 - 3y^3 + x^3 = -3y^3 - 3xy^2 + 11x^2y + x^3$

20. $4x - 3y$

21. $3a + 4b - 3c$

22. $2x^2 - x + 7 + x^2 + 2x - 3 = 3x^2 + x + 4$

23. $2c^4 + c^2 - c - c^2 - 2c = 2c^4 - 3c$

24. $2x(x) + 2x(-3y) + 2x(z) = 2x^2 - 6xy + 2xz$

25. $-3b^3(2b^2) - 3b^3(3b) - 3b^3(-1) = -6b^5 - 9b^4 + 3b^3$

26. $k(2k) + k(5) + 3(2k) + 3(5) = 2k^2 + 5k + 6k + 15 = 2k^2 + 11k + 15$

27. $(4y - 3)(4y - 3) = 4y(4y) + 4y(-3) - 3(4y) - 3(-3) = 16y^2 - 12y - 12y + 9 = 16y^2 - 24y + 9$

28. $5r^2(5r^2) + 5r^2(1) - 1(5r^2) - 1(1) = 25r^4 + 5r^2 - 5r^2 - 1 = 25r^4 - 1$

29. $c^2(7c^2) + c^2(-2c) + c^2(3) + 6(7c^2) + 6(-2c) + 6(3) = 7c^4 - 2c^3 + 3c^2 + 42c^2 - 12c + 18 = 7c^4 - 2c^3 + 45c^2 - 12c + 18$

Challenge $(3x^3 - 2y^2)(3x^3 - 2y^2)(3x^3 - 2y^2) = [(3x^3)(3x^3) + (3x^3)(-2y^2) - (2y^2)(3x^3) - (2y^2)(-2y^2)](3x^3 - 2y^2) = [9x^6 - 6x^3y^2 - 6x^3y^2 + 4y^4](3x^3 - 2y^2) = (9x^6 - 12x^3y^2 + 4y^4)(3x^3 - 2y^2) = (3x^3)(9x^6) + (3x^3)(-12x^3y^2) + (3x^3)(4y^4) - (2y^2)(9x^6) - (2y^2)(-12x^3y^2) - (2y^2)(4y^4) = 27x^9 - 36x^6y^2 + 12x^3y^4 - 18x^6y^2 + 24x^3y^4 - 8y^6 = 27x^9 - 54x^6y^2 + 36x^3y^4 - 8y^6$

page 267   Preparing for Standardized Tests

1. B; $\dfrac{\dfrac{1}{6} + \dfrac{3}{2}}{2} = \dfrac{1 + 9}{6} \cdot \dfrac{1}{2} = \dfrac{5}{3} \cdot \dfrac{1}{2} = \dfrac{5}{6}$

2. D; $13s^2 = 13$, $5ns = 5 \cdot 3 = 15$, $n^2 + s^2 = 9 + 1 = 10$, $2n^2 = 18$, and $-4(n + s) = -4(-4) = 16$.

3. A; $\frac{2}{3}(0.60 \times 150) = \frac{2}{3}(90) = 60$

4. C; $5xy + 3yz - (3xy - 2xz + 5yz) = 5xy + 3yz - 3xy + 2xz - 5yz = 2xy + 2xz - 2yz$

5. A; $7{,}130{,}000 = 7.13 \times 10^6$

6. B; $20.00 - (4.50 + 7.95 + 1.98 + 0.79 + 0.05)$ becomes $20.00 - 15.27 = 4.73$, so the change is \$4.73.

7. D; $2(x + 3) > x + 4$, $2x + 6 > x + 4$, $2x > x - 2$, $x > -2$; So the solution set is $\{x{:}x > -2\}$.

8. D; x of $480 = 2.4$, $x \cdot 480 = 2.4$, $x = \dfrac{2.4}{480} = 0.005$, or $\frac{1}{2}\%$

9. C; $2(x - 1) = 3(x + 2) - 8$, $2x - 2 = 3x + 6 - 8$, $2x - 2 = 3x - 2$, $-x = 0$; $x = 0$

10. E; $(2x + 5y)(x - 3y) = 2x(x) + 2x(-3y) + 5y(x) + 5y(-3y) = 2x^2 - 6xy + 5xy - 15y^2 = 2x^2 - xy - 15y^2$

11. B; $(-2a^2b^3c)(7ab^2c^4) = -14a^{2+1}b^{3+2}c^{1+4} = -14a^3b^5c^5$

12. A; From 7:47 to 8:39 is 13 minutes + 39 minutes = 52 minutes.

page 268   Cumulative Review (Chapters 1–6)

1. -0.5 **2.** 8.0 **3.** $-21 \div -3 = 7$ **4.** -2.1

5. $2x - 3 - 7x + 4 = -5x + 1$ **6.** $8h^2 - 6 + 7 - 5h^2 = 3h^2 + 1$

7. $9m^2 + 8 - m - 6 = 9m^2 - m + 2$

8. $3y(y^2) + 3y(-2y) + 3y(4) = 3y^3 - 6y^2 + 12y$

9. $-2b^2(-2b) - 2b^2(3a) - 2b^2(-1) = 4b^3 - 6ab^2 + 2b^2$

10. $g(g) + g(-3) + 2(g) + 2(-3) = g^2 - 3g + 2g - 6 = g^2 - g - 6$

11. $2n(3n) + 2n(-4) - 1(3n) - 1(-4) = 6n^2 - 8n - 3n + 4 = 6n^2 - 11n + 4$

12. $p(p^2) + p(3p) + p(-2) + 1(p^2) + 1(3p) + 1(-2) = p^3 + 3p^2 - 2p + p^2 + 3p - 2 = p^3 + 4p^2 + p - 2$

13. $3 + 12 = 15$ **14.** $[2 - (-8)]^2 = [2 + 8]^2 = [10]^2 = 100$

15. $|-3| + 5 = 3 + 5 = 8$ **16.** $-4a^2 + 5a$

17. $-3xy - xy = -4xy$ **18.** $d^{5+3} = d^8$

19. $6n^{2+3} = 6n^5$

20. $s^{2+1}t^2 = s^3t^2$

21. $8p^{2\cdot3} = 8p^6$

22. $b^{2-3} = b^{-1} = \dfrac{1}{b}$

23. $\dfrac{1}{4}m^{4-2} = \dfrac{1}{4}m^2 = \dfrac{m^2}{4}$

24. $\dfrac{81k^{10}}{18k^3} = \dfrac{9k^{10-3}}{2} = \dfrac{9k^7}{2}$

25. $-5 - 9 = r, r = -14$

26. $c > 1 + 8, c > 9$; {all real numbers greater than 9}

27. $\dfrac{-3m}{-3} = \dfrac{24}{-3} = m = -8$

28. $\dfrac{-2p}{-2} < \dfrac{12}{-2}, p > -6$; {all real numbers greater than -6}

29. $\dfrac{x}{2} = 11 - 1, \dfrac{x}{2} = 10, 2 \cdot \dfrac{x}{2} = 10 \cdot 2, x = 20$

30. $3a - 5 = 7$ or $3a - 5 = -7$

$\qquad\quad 3a = 12 \qquad\quad 3a = -2$

$\qquad\qquad a = 4 \qquad\qquad a = -\dfrac{2}{3}$

The solution set is $\left\{-\dfrac{2}{3}, 4\right\}$.

31. $3t - 6 = 9, 3t = 15, t = 5$

32. $5d \geq -5, d \geq -1$; {all real numbers greater than or equal to -1}

33. $5n - 10 = 45, 5n = 55, n = 11$

34. $8 = 4y - 4, 4y = 12, y = 3$

35. $b - 2b + 3 + 5b - 2 = 17, 4b + 1 = 17, 4b = 16, b = 4$

36. $1 \leq f < 3$; {1 and all real numbers between 1 and 3}

37. {all real numbers less than -2 or greater than or equal to 0}

38. Let x = the first integer, then $x + 1$ = the second consecutive integer, and $x + 2$ = the third consecutive integer. The equation is $(x + 1) + (x + 2) = 23, 2x + 3 = 23, 2x = 20, x = 10$; then $x + 1 = 11$, $x + 2 = 12$. The three consecutive integers are 10, 11, 12.

39. Let r = the rate of first cyclist, and then $r + 5$ = the rate of second cyclist. $2(r) + 2(r + 5) = 90, 2r + 2r + 10 = 90, 4r + 10 = 90, 4r = 80$, $r = 20; r + 5 = 25$; The rate of the two cyclists are 20 km/h, and 25 km/h.

40. Let w = the width of the rectangle, then $3w + 2$ is the length of the rectangle. $2(3w + 2) + 2w \geq 24, 6w + 4 + 2w \geq 24, 8w + 4 \geq 24, 8w \geq 20, w \leq 2.5$. The least possible value for the width is 2.5 m.

Chapter 7 Factoring Polynomials

page 270 **Capsule Review**

1. 2^4 **2.** 5^3 **3.** $2 \cdot 3 \cdot 5^2$ **4.** $3 \cdot 5 \cdot 7^3$ **5.** 1^5

6. 3^5 **7.** $3 \cdot 5^2$ **8.** $2^2 \cdot 7^2$ **9.** $10^3 \cdot 11^2$

page 217 **Class Exercises**

1. 1, 3, 5, 15   **2.** 1, 2, 3, 6   **3.** 1, 2, 4, 8, 16

4. 1, 2, 3, 4, 6, 8, 12, 24   **5.** prime   **6.** composite

7. neither   **8.** composite   **9.** $14 = 2 \cdot 7$

10. $18 = 2 \cdot 9$
$\quad = 2 \cdot 3 \cdot 3$
$\quad = 2 \cdot 3^2$

11. prime

12. $60 = 2 \cdot 30$
$\quad = 2 \cdot 2 \cdot 15$
$\quad = 2 \cdot 2 \cdot 3 \cdot 5$
$\quad = 2^2 \cdot 3 \cdot 5$

13. No, prime factorization is unique.

14. 1 does not have exactly two positive integral factors, so it is not prime. 1 does not have more than two positive integral factors so it is not composite.

page 272 **Practice Exercises**

A 1. $26 = 2 \cdot 13$ **2.** $33 = 3 \cdot 11$ **3.** $52 = 2 \cdot 26$
$\quad\quad\quad\quad\quad\quad\quad\quad\quad\quad\quad\quad\quad\quad\quad\quad\quad\quad = 2 \cdot 2 \cdot 13$
$\quad\quad\quad\quad\quad\quad\quad\quad\quad\quad\quad\quad\quad\quad\quad\quad\quad\quad = 2^2 \cdot 13$

4. $70 = 2 \cdot 35$ **5.** $154 = 2 \cdot 77$ **6.** $135 = 3 \cdot 45$
$\quad = 2 \cdot 5 \cdot 7$ $= 2 \cdot 7 \cdot 11$ $= 3 \cdot 3 \cdot 15$
$\quad\quad\quad\quad\quad\quad\quad\quad\quad\quad\quad\quad\quad\quad\quad\quad\quad\quad\quad = 3 \cdot 3 \cdot 3 \cdot 5$
$\quad\quad\quad\quad\quad\quad\quad\quad\quad\quad\quad\quad\quad\quad\quad\quad\quad\quad\quad = 3^3 \cdot 5$

7. $195 = 3 \cdot 65$ **8.** $315 = 3 \cdot 105$ **9.** $105 = 3 \cdot 35$
$\quad = 3 \cdot 5 \cdot 13$ $= 3 \cdot 3 \cdot 35$ $= 3 \cdot 5 \cdot 7$
$\quad\quad\quad\quad\quad\quad\quad\quad\quad\quad\quad = 3^2 \cdot 5 \cdot 7$

10. $165 = 3 \cdot 55$ **11.** $143 = 11 \cdot 13$ **12.** $273 = 3 \cdot 91$
$\quad = 3 \cdot 5 \cdot 11$ $= 3 \cdot 7 \cdot 13$

13. $3575 = 5 \cdot 715$ **14.** $1925 = 5 \cdot 385$
$\quad\quad = 5 \cdot 5 \cdot 143$ $= 5 \cdot 5 \cdot 77$
$\quad\quad = 5 \cdot 5 \cdot 11 \cdot 13$ $= 5 \cdot 5 \cdot 7 \cdot 11$
$\quad\quad = 5^2 \cdot 11 \cdot 13$ $= 5^2 \cdot 7 \cdot 11$

15. $9625 = 5 \cdot 1925$
$= 5 \cdot 5 \cdot 385$
$= 5 \cdot 5 \cdot 5 \cdot 77$
$= 5 \cdot 5 \cdot 5 \cdot 7 \cdot 11$
$= 5^3 \cdot 7 \cdot 11$

16. $7425 = 5 \cdot 1485$
$= 5 \cdot 5 \cdot 297$
$= 5 \cdot 5 \cdot 9 \cdot 33$
$= 5 \cdot 5 \cdot 3 \cdot 3 \cdot 3 \cdot 11$
$= 3^3 \cdot 5^2 \cdot 11$

B 17. $2205 = 3 \cdot 735$
$= 3 \cdot 5 \cdot 147$
$= 3 \cdot 5 \cdot 3 \cdot 49$
$= 3 \cdot 5 \cdot 3 \cdot 7 \cdot 7$
$= 3^2 \cdot 5 \cdot 7^2$

18. $1155 = 5 \cdot 231$
$= 3 \cdot 5 \cdot 77$
$= 3 \cdot 5 \cdot 7 \cdot 11$

19. $1760 = 20 \cdot 88$
$= 4 \cdot 5 \cdot 8 \cdot 11$
$= 2 \cdot 2 \cdot 5 \cdot 2 \cdot 4 \cdot 11$
$= 2 \cdot 2 \cdot 5 \cdot 2 \cdot 2 \cdot 2 \cdot 11$
$= 2^5 \cdot 5 \cdot 11$

20. $1716 = 13 \cdot 132$
$= 13 \cdot 11 \cdot 12$
$= 13 \cdot 11 \cdot 3 \cdot 4$
$= 13 \cdot 11 \cdot 3 \cdot 2 \cdot 2$
$= 2^2 \cdot 3 \cdot 11 \cdot 13$

21. $242 = 2 \cdot 121$
$= 2 \cdot 11 \cdot 11$
$= 2 \cdot 11^2$

22. $338 = 2 \cdot 169$
$= 2 \cdot 13 \cdot 13$
$= 2 \cdot 13^2$

23. $288 = 2 \cdot 144$
$= 2 \cdot 2 \cdot 72$
$= 2 \cdot 2 \cdot 8 \cdot 9$
$= 2 \cdot 2 \cdot 2 \cdot 4 \cdot 3 \cdot 3$
$= 2 \cdot 2 \cdot 2 \cdot 2 \cdot 2 \cdot 3 \cdot 3$
$= 2^5 \cdot 3^2$

24. $455 = 5 \cdot 91$
$= 5 \cdot 7 \cdot 13$

25. prime

26. prime

27. $1000 = 10 \cdot 100$
$= 2 \cdot 5 \cdot 10 \cdot 10$
$= 2 \cdot 5 \cdot 2 \cdot 5 \cdot 2 \cdot 5$
$= 2^3 \cdot 5^3$

28. $1875 = 25 \cdot 75$
$= 5 \cdot 5 \cdot 3 \cdot 25$
$= 5 \cdot 5 \cdot 3 \cdot 5 \cdot 5$
$= 3 \cdot 5^4$

29. 59   **30.** 73   **31.** 109   **32.** 131

C 33. Yes, $3x$ is the product of two primes.

34. Yes, $x + x = 2x$, $2x$ is the product of two primes.

35. Yes, $x \cdot x = x^2$ is a product of two primes.

36. No, if $x = 3$, $x + 4 = 7$, and 7 is a prime number.

37–40. Spreadsheets should be generated to display all pairs of integers whose sum is one-half the given number. The product of these pairs should be displayed as the area.

41. Yes; the closest dimensions produce the greatest amount of area.

page 273 Biography

1. Answers may vary. Possible answers are: 3, 4, 5; 9, 12, 15; 18, 24, 30

2. (8, 8), (11, 3), multiply x by 5, multiply y by 3, and then add to get 64.

3. Answers may vary.

page 274 Capsule Review

1. x^3 **2.** $4y^3$ **3.** $-3m^2$ **4.** a^4b^2 **5.** $-5x^4y$

6. $8s$ **7.** $8m^2n$ **8.** $-5x^2$ **9.** $52n^3$

page 276 Class Exercises

1. $14 = 2 \cdot 7$
$24 = 2 \cdot 2 \cdot 2 \cdot 3 = 2^3 \cdot 3$
GCF is 2

2. $15 = 3 \cdot 5$
$19 = 1 \cdot 19$
GCF is 1

3. $24 = 2 \cdot 2 \cdot 2 \cdot 3 = 2^3 \cdot 3$
$36 = 2 \cdot 2 \cdot 3 \cdot 3 = 2^2 \cdot 3^2$
$42 = 2 \cdot 3 \cdot 7$
GCF is $2 \cdot 3 = 6$

4. $7 = 1 \cdot 7$
$35 = 5 \cdot 7$
$105 = 3 \cdot 5 \cdot 7$
GCF is 7

5. $3(p) + 3(2) = 3(p + 2)$

6. $5(2z) + 5(3) = 5(2z + 3)$

7. $4(2k) - 4(1) = 4(2k - 1)$

8. $9(3r) + 9(t) = 9(3r + t)$

9. $x(2) + x(7x) = x(2 + 7x)$

10. $y(14) - y(y^3) = y(14 - y^3)$

11. $4a(2a) + 4a(3) = 4a(2a + 3)$

12. $12(1) - 12(a) = 12(1 - a)$

13. GCF is $3pq$; $3pq(q) + 3pq(6p) = 3pq(q + 6p)$

14. GCF is j^3k^2; $j^3k^2(11jk^3) - j^3k^2(13) = j^3k^2(11jk^3 - 13)$

pages 276–277 Practice Exercises

A   **1.** $35 = 3 \cdot 7$
$49 = 7 \cdot 7$
GCF is 7

2. $39 = 3 \cdot 13$
$52 = 2^2 \cdot 13$
GCF is 13

3. $144 = 2^4 \cdot 3^2$
$126 = 2 \cdot 7 \cdot 3^2$
GCF is $2 \cdot 3^2$ or 18

4. $154 = 2 \cdot 7 \cdot 11$
$198 = 2 \cdot 3^2 \cdot 11$
GCF is $2 \cdot 11$ or 22

5. $25 = 5^2$
$75 = 3 \cdot 5^2$
$100 = 2^2 \cdot 5^2$
GCF is 5^2 or 25

6. $32 = 2^5$
$40 = 2^3 \cdot 5$
$56 = 2^3 \cdot 7$
GCF is 2^3 or 8

7. $112 = 2^4 \cdot 7$
$224 = 2^5 \cdot 7$
$104 = 2^3 \cdot 13$
GCF is 2^3 or 8

8. $174 = 2 \cdot 3 \cdot 29$
$216 = 2^3 \cdot 3^3$
$162 = 2 \cdot 3^4$
GCF is $2 \cdot 3$ or 6

9. $2x^2y(x) - 2x^2y(6y^3) = 2x^2y(x - 6y^3)$

10. $5a^2b(1) - 5a^2b(3ab^2) = 5a^2b(1 - 3ab^2)$

11. $7cd^3(1) + 7cd^3(2c^2d^2) = 7cd^3(1 + 2c^2d^2)$

12. $6mn^2(n^2) + 6mn^2(3m^4) = 6mn^2(n^2 + 3m^4)$

13. $13x^2y^2(y) + 13x^2y^2(2) = 13x^2y^2(y + 2)$

14. $11w^2y^2(3w) + 11w^2y^2(1) = 11w^2y^2(3w + 1)$

15. $6mn^3(2m^3n^2) - 6mn^3(3) = 6mn^3(2m^3n^2 - 3)$

16. $a^2b^3(7) - a^2b^3(9a^2) = a^2b^3(7 - 9a^2)$

17. $2jk(2j^2) - 2jk(3k) + 2jk(4) = 2jk(2j^2 - 3k + 4)$

18. $5mn(m) - 5mn(3n) + 5mn(2) = 5mn(m - 3n + 2)$

19. $4ab(a) + 4ab(2ab) + 4ab(3) = 4ab(a + 2ab + 3)$

20. $3cd(3d) + 3cd(2c) + 3cd(1) = 3cd(3d + 2c + 1)$

B **21.** $12x^3y^4(1) + 12x^3y^4(3x) - 12x^3y^4(5y) = 12x^3y^4(1 + 3x - 5y)$

22. $5x^4y^4(2) + 5x^4y^4(3y^3) - 5x^4y^4(5xy^2) = 5x^4y^4(2 + 3y^3 - 5xy^2)$

23. $3m^2n^2(8m) + 3m^2n^2(7n) - 3m^2n^2(13n^2) = 3m^2n^2(8m + 7n - 13n^2)$

24. $7jk^2(7) - 7jk^2(3jk) + 7jk^2(12j^2k^2) = 7jk^2(7 - 3jk + 12j^2k^2)$

25. $x^3y^3(13x^2y) - x^3y^3(11y) + x^3y^3(17x) = x^3y^3(13x^2y - 11y + 17x)$

26. $s^2t^3(23st) - s^2t^3(19t^2) + s^2t^3(29s^2) = s^2t^3(23st - 19t^2 + 29s^2)$

27. $6l^2m(7lm) - 6l^2m(6m^4) - 6l^2m(9l^2) = 6l^2m(7lm - 6m^4 - 9l^2)$

28. $4c^2d(18d^4) - 4c^2d(16c) - 4c^2d(7d^3) = 4c^2d(18d^4 - 16c - 7d^3)$

29. $33x^5y^7(1) - 33x^5y^7(3xy^2) - 33x^5y^7(2y) = 33x^5y^7(1 - 3xy^2 - 2y)$

30. $38ab(2a^3b^4) - 38ab(a^4b^4) - 38ab(3) = 38ab(2a^3b^4 - a^4b^4 - 3)$

31. $r^2s^3(89s^6) + r^2s^3(113rs^4) + r^2s^3(73r^2) = r^2s^3(89s^6 + 113rs^4 + 73r^2)$

32. $x^4y^6(71x^4y^3) + x^4y^6(119x^3) + x^4y^6(97y^2) = x^4y^6(71x^4y^3 + 119x^3 + 97y^2)$

C **33.** $-5a^2b^4(3a) - 5a^2b^4(7a^2b) - 5a^2b^4(11) = -5a^2b^4(3a + 7a^2b + 11)$

34. $-2wz^4(11w^2) - 2wz^4(14wz) - 2wz^4(17) = -2wz^4(11w^2 + 14wz + 17)$

35. $-5x^4y^4(23) - 5x^4y^4(45xy) - 5x^4y^4(57x^2y^2) = -5x^4y^4(23 + 45xy + 57x^2y^2)$

36. $-5m^4n^6(29n) - 5m^4n^6(56m) - 5m^4n^6(62m^2n^3) = -5m^4n^6(29n + 56m + 62m^2n^3)$

37. $7c^4d^2e^3(11d^4) + 7c^4d^2e^3(4ce) = 7c^4d^2e^3(11d^4 + 4ce)$

38. $8j^2kl^3(7jkl) + 8j^2kl^3(9) = 8j^2kl^3(7jkl + 9)$

39. $18x^2y^2z^2(3y^3z) - 18x^2y^2z^2(2x) = 18x^2y^2z^2(3y^3z - 2x)$

40. $7s^2t^2u^3(6tu) + 7s^2t^2u^3(5s) = 7s^2t^2u^3(6tu + 5s)$

41. $\text{Area}_{\text{shaded}} = \text{Area}_{\text{square}} + \dfrac{1}{2}\,\text{Area}_{\text{circle}} = (2r)^2 + \dfrac{1}{2}(\pi)r^2 = 4r^2 + \dfrac{1}{2}\pi r^2 =$
$r^2\left(4 + \dfrac{1}{2}\pi\right)$; The area of the shaded region is $r^2\left(4 + \dfrac{1}{2}\pi\right)$.

42. $\text{Area}_{\text{shaded}} = \text{Area}_{\text{square}} + \dfrac{1}{2}\,\text{Area}_{\text{4 circles}} = (2r)^2 + \dfrac{1}{2}(4\pi r^2) = 4r^2 + 2\pi r^2 =$
$2r^2(2 + \pi)$; The area of the shaded region is $2r^2(2 + \pi)$.

43. $\text{Area}_{\text{shaded}} = \text{Area}_{\text{square}} - \text{Area}_{\text{2 circles}} = (4r)^2 - \pi r^2 - \pi r^2 = 16r^2 - 2\pi r^2 =$
$2r^2(8 - \pi)$; The area of the shaded region is $2r^2(8 - \pi)$.

44. $9 - x^2 + 4 = -x^2 + 13$

45. $5n - 3 - n - 6 = 4n - 9$

46. $3c - 7 - 2c + 5 = c - 2$

47. 2.037×10^6

48. 9.17×10^{-4}

49. 1.53×10^{-2}

50.
$$
\begin{array}{r}
4\ \text{r}\ 17 \\
187\overline{)765} \\
-748 \\
\hline
17
\end{array}
\qquad
\begin{array}{r}
11\ \text{r}\ 0 \\
17\overline{)187} \\
-187 \\
\hline
0
\end{array}
\qquad
\text{GCF} = 17;\ 17(45x + 11)
$$

51.
$$
\begin{array}{r}
1\ \text{r}\ 119 \\
136\overline{)255} \\
-136 \\
\hline
119
\end{array}
\qquad
\begin{array}{r}
1\ \text{r}\ 17 \\
119\overline{)136} \\
-119 \\
\hline
17
\end{array}
\qquad
\begin{array}{r}
7\ \text{r}\ 0 \\
17\overline{)119} \\
-119 \\
\hline
0
\end{array}
\qquad
\text{GCF} = 17;\ 17(8x + 15)
$$

52.
$$
\begin{array}{r}
2\ \text{r}\ 138 \\
161\overline{)460} \\
-322 \\
\hline
138
\end{array}
\qquad
\begin{array}{r}
1\ \text{r}\ 23 \\
138\overline{)161} \\
-138 \\
\hline
23
\end{array}
\qquad
\begin{array}{r}
6\ \text{r}\ 0 \\
23\overline{)138} \\
-138 \\
\hline
0
\end{array}
\qquad
\text{GCF} = 23;\ 23(20x + 7)
$$

53.
$$
\begin{array}{r}
2\ \text{r}\ 406 \\
435\overline{)1276} \\
-870 \\
\hline
406
\end{array}
\qquad
\begin{array}{r}
1\ \text{r}\ 29 \\
406\overline{)435} \\
-406 \\
\hline
29
\end{array}
\qquad
\begin{array}{r}
14\ \text{r}\ 0 \\
29\overline{)406} \\
-406 \\
\hline
0
\end{array}
\qquad
\text{GCF} = 29;\ 29(44x + 15)
$$

page 278   Capsule Review

1. $x^2 + 2x + x + 2 = x^2 + 3x + 2$

2. $t^2 - 5t - 2t + 10 = t^2 - 7t + 10$

3. $y^2 + 4y + 3y + 12 = y^2 + 7y + 12$

4. $m^2 - 50m - 10m + 500 = m^2 - 60m + 500$

5. $800 + 100z + 8z + z^2 = 800 + 108z + z^2$

6. $110 - 11a - 10a + a^2 = 110 - 21a + a^2$

page 280   Class Exercises

1. 3   **2.** 12   **3.** 1; 7   **4.** 1; 17

5. $(x + 1)(x + 3)$ **6.** $(m + 2)(m + 1)$ **7.** $(y - 5)(y - 1)$

8. $(p - 1)(p - 1)$ **9.** $(7 + m)(1 + m)$ **10.** $(11 - d)(1 - d)$

pages 280–281   Practice Exercises

A   1. $(x + 1)(x + 9)$ **2.** $(y + 1)(y + 13)$ **3.** $(a + 1)(a + 11)$

4. $(r + 2)(r + 4)$ **5.** $(z + 1)(z + 29)$ **6.** $(t + 3)(t + 13)$

7. $(y + 3)(y + 11)$ **8.** $(x + 7)(x + 13)$ **9.** $(s - 1)(s - 6)$

10. $(k - 1)(k - 4)$ **11.** $(m - 1)(m - 41)$ **12.** $(b - 1)(b - 19)$

13. $(y - 2)(y - 11)$ **14.** $(n - 2)(n - 7)$ **15.** $(z - 3)(z - 17)$

16. $(g - 5)(g - 7)$ **17.** $(13 - b)(5 - b)$ **18.** $(13 - n)(2 - n)$

19. $(12 - z)(3 - z)$ **20.** $(18 - a)(3 - a)$ **21.** $(3 + y)(16 + y)$

22. $(3 + d)(4 + d)$ **23.** $(8 - f)(9 - f)$ **24.** $(6 - r)(9 - r)$

25. $(a + 3b)(a + 9b)$ **26.** $(n + 5p)(n + 7p)$ **27.** $(x - y)(x - 7y)$

28. $(s - t)(s - 13t)$ **29.** $(r + 2t)(r + 3t)$ **30.** $(z + 6x)(z + 2x)$

B 31. $(m - n)(m - 25n)$ **32.** $(r + s)(r + 32s)$ **33.** $(s + 2t)(s + 9t)$

34. $(j - 3k)(j - 4k)$ **35.** $(y - 2z)(y - 8z)$ **36.** $(m + 4n)(m + 5n)$

37. $(x - 2y)(x - 25y)$ **38.** $(s - 4t)(s - 12t)$ **39.** $(r + 4t)(r + 16t)$

40. $(a + 2b)(a + 30b)$ **41.** $(m - 3n)(m - 24n)$ **42.** $(n - 4q)(n - 25q)$

43. $(m - 2n)(m - 4n)$ **44.** $(s - 3t)(s - 6t)$ **45.** $(c^2 + 2)(c^2 + 6)$

46. $(u^2 - 2)(u^2 - 14)$ **47.** $(y^2 - 3)(y^2 - 8)$ **48.** $(n^2 - 2)(n^2 - 16)$

C 49. $(a^2 + 2a + 1) + 8a + 8 + 7 = a^2 + 10a + 16 = (a + 8)(a + 2)$

50. $x^2 + 4x + 4 + 4x + 8 + 3 = x^2 + 8x + 15 = (x + 5)(x + 3)$

51. $x^2 - 2x + 1 + 2x - 2 + 1 = x^2$

52. $a^2 + 4a + 4 + 2a + 4 + 1 = a^2 + 6a + 9 = (a + 3)(a + 3) = (a + 3)^2$

53. $36 - 15z - 15 + z^2 + 2z + 1 = 22 - 13z + z^2 = (2 - z)(11 - z)$

54. $44 - 15x - 30 + x^2 + 4x + 4 = 18 - 11x + x^2 = (9 - x)(2 - x)$

55. $(a^x + 1)(a^x + 2)$ **56.** $(b^y + 2)(b^y + 3)$

57. $(x^{2n} - 3)(x^{2n} - 4)$ **58.** $(y^{2m} - 1)(y^{2m} - 12)$

59. $(x + 4)(x + 1)$, length $= x + 4$; width $= x + 1$

60. $(y + 3)(y + 2)$, length $= y + 3$; width $= y + 2$

61. $(m + 3)(m + 1)$, length $= m + 3$; width $= m + 1$

62. $(n + 8)(n + 2)$, length $= n + 8$; width $= n + 2$

63. $(x + 2a)(x + a)$, length $= x + 2a$; width $= x + a$

64. $(b + 3c)(b + c)$, length $= b + 3c$; width $= b + c$

65. When the first term of a trinomial has a coefficient of 1, and the third term is greater than 0, use the factors of the last term that add up to the coefficient of the middle term.

66.

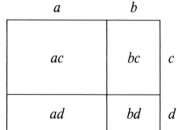

$(a + b)(c + d) = ac + ad + bc + bd$

page 282   Capsule Review

1. $x^2 - 2x + 4x - 8 = x^2 + 2x - 8$

2. $x^2 + 7x - 5x - 35 = x^2 + 2x - 35$

3. $x^2 + 5x - 4x - 20 = x^2 + x - 20$

4. $x^2 + 6x - 7x - 42 = x^2 - x - 42$

5. $y^2 + 11y - 9y - 99 = y^2 + 2y - 99$

6. $s^2 - 12s + 3s - 36 = s^2 - 9s - 36$

page 283   Class Exercises

1. $+; -$ **2.** $+; -$ **3.** $+; -$ **4.** $-; +$

5. 3; 2   **6.** 1; 16   **7.** 9; 1   **8.** 3; 6

9. $(x - 1)(x + 5)$ **10.** $(m + 2)(m - 5)$ **11.** $(y + 2)(y - 8)$

12. If the middle term is positive, then the second term in the binomial factors that is larger in absolute value will be positive. If the middle term is negative, then the second term in the binomial factors that is larger in absolute value will be negative.

page 284   Practice Exercises

A   **1.** $(x + 5)(x - 1)$ **2.** $(a + 7)(a - 1)$ **3.** $(m - 10)(m + 1)$

     4. $(z - 24)(z + 1)$ **5.** $(x + 4)(x - 2)$ **6.** $(a + 9)(a - 2)$

     7. $(m - 6)(m + 2)$ **8.** $(z - 10)(z + 2)$ **9.** $(x + 3)(x - 1)$

    10. $(y + 13)(y - 1)$ **11.** $(a + 6)(a - 1)$ **12.** $(z + 9)(z - 1)$

    13. $(m - 4)(m + 3)$ **14.** $(b - 8)(b + 7)$ **15.** $(k - 15)(k + 2)$

    16. $(s - 6)(s + 4)$ **17.** $(x - 3y)(x + 2y)$ **18.** $(m - 5n)(m + 3n)$

    19. $(r + 3s)(r - 7s)$ **20.** $(n - 14p)(n + 3p)$ **21.** $(x - 10y)(x + 4y)$

B **22.** $(a - 9b)(a + 3b)$ **23.** $(k - 9j)(k + 4j)$ **24.** $(w - 12z)(w + 4z)$

    25. $(x - 2y)(x + y)$ **26.** $(s - 5t)(s + t)$ **27.** $(r + 7t)(r - 2t)$

    28. $(a - 9b)(a + 3b)$ **29.** $(y - 16z)(y + 2z)$ **30.** $(n - 8p)(n + 6p)$

    31. $(x - 10y)(x + 2y)$ **32.** $(s - 7t)(s + 6t)$ **33.** $(r + 9t)(r - 6t)$

    34. $(a - 13b)(a + 4b)$ **35.** $(y - 20z)(y + 5z)$ **36.** $(n - 16p)(n + 5p)$

C **37.** $(y^2 + 25)(y^2 - 2)$ **38.** $(z^3 - 15)(z^3 + 5)$

    39. $a^2 - 2a + 1 - 4a + 4 - 32 = a^2 - 6a - 27 = (a - 9)(a + 3)$

    40. $b^2 + 4b + 4 + 15b + 30 - 54 = b^2 + 19b - 20 = (b + 20)(b - 1)$

    41. $(a + 35b)(a - 10b)$; $[200 + 35(10)][200 - 10(10)] = (200 + 350)(200 - 100)$; 550 ft × 100 ft

    42. $32a^3b^4$ **43.** $-12xy^3$

    44. $3x^{-2}b^3 = \dfrac{3b^3}{x^2}$ **45.** $-7(1) = -7$

    46. $-18x^{-2}y^{-3} = -\dfrac{18}{x^2y^3}$ **47.** $\dfrac{12a^7b^{-6}}{16a^4b^{-6}} = \dfrac{3a^3}{4}$

    48. $\dfrac{27x^{-15}b^{-9}}{36x^8b^{-10}} = \dfrac{3}{4}x^{-23}b = \dfrac{3b}{4x^{23}}$

    49. Commutative property for mult.   **50.** Associative property for mult.

1. $2m^2 + 3m + 1$ | 2. $3t^2 + 10t + 3$ | 3. $3y^2 - 4y - 4$

4. $5y^2 + 18y - 8$ | 5. $3h^2 - 22h + 7$ | 6. $6x^2 + 11x + 3$

7. $3a^2 - 2ab - b^2$ | 8. $6s^2 + st - 2t^2$ | 9. $12u^2 - 25uv + 12v^2$

page 287   Class Exercises

1. $+; -$ | 2. $5; 1$ | 3. $+; -$ | 4. $1; 7$

5. $(3x - 1)(x + 7)$ | 6. $(2m - 7)(m + 2)$

7. $(6y + 7)(y + 3)$ | 8. $(2x - 3)(2x + 1)$

page 287–288   Practice Exercises

A 1. $(3x + 2)(x - 8)$ | 2. $(2m + 3)(m - 7)$ | 3. $(5z + 2)(z - 3)$

4. $(7r + 5)(r - 4)$ | 5. $(2m + 5)(m + 1)$ | 6. $(3s + 5)(s + 4)$

7. $(5r + 3)(3r + 7)$ | 8. $(7k + 4)(2k + 3)$ | 9. $(9l + 3)(9l + 5)$

10. $(3q + 5)(3q + 4)$ | 11. $(z - 1)(6z + 5)$ | 12. $(4c - 5)(3c + 2)$

13. $(2m - 9)(2m + 1)$ | 14. $(4x + 3)(4x - 5)$ | 15. $(2x + 11)(x - 1)$

16. $(5a + 7)(a - 1)$ | 17. $(2d + 3)(7d - 5)$ | 18. $(2k - 1)(5k + 4)$

19. $(8x - 5)(3x - 4)$ | 20. $(17d + 5)(2d - 3)$ | 21. $(5z + 6)(4z + 5)$

22. $(22n + 3)(n + 2)$ | 23. $(13c - 5)(2c + 3)$ | 24. $(9n - 8)(2n + 3)$

B 25. $(3x + 2y)(x - y)$ | 26. $(7s + t)(s - 3t)$ | 27. $(2r + t)(r + 7t)$

28. $(2y + 3z)(2y + z)$ | 29. $(3a - b)(a - 5b)$ | 30. $(7s - 5t)(s - 2t)$

31. $(3n - 2p)(2n + p)$ | 32. $(8n + 7p)(n - 2p)$ | 33. $(5x - 2y)(x + 4y)$

34. prime | 35. prime | 36. $(10c - d)(c + 10d)$

37. $(4x + 9y)(3x + 2y)$ | 38. $(12g + 7h)(2g + 3h)$ | 39. $(7k - 3m)(2k - 11m)$

40. $(20r + 7t)(r - 3t)$ | 41. $(9a + 8b)(3a - 4b)$ | 42. $(6m + 7n)(6m - 5n)$

C 43. $4(x^2 + 4x + 4) + 11x + 22 + 6 = 4x^2 + 16x + 16 + 11x + 22 + 6 =$
$4x^2 + 27x + 44 = (4x + 11)(x + 4)$

44. $6(y^2 + 10y + 25) + 11y + 55 - 10 = 6y^2 + 60y + 150 + 11y + 55 - 10 =$
$6y^2 + 71y + 195 = (3y + 13)(2y + 15)$

45. $12(a^2 - 6a + 9) - 19a + 57 - 21 = 12a^2 - 72a + 108 - 19a + 57 - 21 =$
$12a^2 - 91a + 144 = (4a - 9)(3a - 16)$

46. $15(b^2 - 4b + 4) - 37b + 74 + 18 = 15b^2 - 60b + 60 - 37b + 74 + 18 =$
$15b^2 - 97b + 152 = (5b - 19)(3b - 8)$

47. $(3x^k + 2)(2x^k + 7)$

48. $(5c^{2r} + 8)(3c^{2r} - 2)$

49. $(5x^{2k+3} + 4)(2x^{2k+3} - 3)$

50. $(5d^{r+8} - 2)(4d^{r+8} - 3)$

51. $(y - 12)(6y + 1)$; length = $6y + 1$; width = $y - 12$

52. $(b - 18)(16b - 1)$; length = $16b - 1$; width = $b - 18$

53. $(5c + 17)(5c + 17)$; length = $5c + 17$; width = $5c + 17$

54. $(7 - 19x)(7 - 19x)$; length = $7 - 19x$; width = $7 - 19x$

page 288   Test Yourself

1. $110 = 2 \cdot 55$
$= 2 \cdot 5 \cdot 11$

2. $105 = 3 \cdot 35$
$= 3 \cdot 5 \cdot 7$

3. $180 = 2 \cdot 90$
$= 2 \cdot 2 \cdot 45$
$= 2 \cdot 2 \cdot 3 \cdot 15$
$= 2 \cdot 2 \cdot 3 \cdot 3 \cdot 5$
$= 2^2 \cdot 3^2 \cdot 5$

4. $250 = 2 \cdot 125$
$= 2 \cdot 5 \cdot 25$
$= 2 \cdot 5 \cdot 5 \cdot 5$
$= 2 \cdot 5^3$

5. $147 = 3 \cdot 49$
$= 3 \cdot 7 \cdot 7$
$= 3 \cdot 7^2$

6. $9(a) + 9(2) = 9(a + 2)$

7. $14x(3) - 14x(x^2) = 14x(3 - x^2)$

8. $3pq(17p) + 3pq(1) = 3pq(17p + 1)$

9. $(a + 7)(a + 5)$

10. $(t + 18)(t + 2)$

11. $(m - 10)(m - 6)$

12. $(x - 5)(x + 10)$

13. $(y - 15)(y + 4)$

14. $(m - 3n)(m + 2n)$

15. $(2x + 1)(x + 12)$

16. $(2y + 1)(4y - 3)$

17. $(3x - 2)(x + 2)$

page 289   Capsule Review

1. $x^2 + 2(2)(x) + 2^2 = x^2 + 4x + 4$

2. $m^2 - 2(5)(m) + (-5)^2 = m^2 - 10m + 25$

3. $(2 - 3x)(2 - 3x) = 2^2 - 2(6x) + (-3x)^2 = 4 - 12x + 9x^2$

page 290   Class Exercises

1. $(2x)^2 - (3y)^2 = (2x - 3y)(2x + 3y)$

2. $(10a)^2 - (4b^2z^4)^2 = (10a - 4b^2z^4)(10a + 4b^2z^4)$

3. $(4rs)^2 - (9t^5)^2 = (4rs - 9t^5)(4rs + 9t^5)$

4. $x^2 + 2(x \cdot y) + y^2 = (x + y)(x + y) = (x + y)^2$

5. $(5m)^2 + 2(5m \cdot 2) + 2^2 = (5m + 2)(5m + 2) = (5m + 2)^2$

6. $(9m^2)^2 + 2(9m^2 \cdot 4n) + (4n)^2 = (9m^2 + 4n)(9m^2 + 4n) = (9m^2 + 4n)^2$

pages 290–291 Practice Exercises

A 1. perfect square trinomial; $x^2 + 2x + 1 = (x)^2 + 2(x \cdot 1) + (1)^2$

2. perfect square trinomial; $y^2 + 8y + 16 = (y)^2 + 2(y \cdot 4) + (4)^2$

3. difference of two squares; $x^2 - 9 = (x)^2 - (3)^2$

4. difference of two squares; $a^2 - 121 = (a)^2 - (11)^2$

5. neither   **6.** neither

7. $x^2 - 2(x \cdot 6) + 6^2 = (x - 6)(x - 6) = (x - 6)^2$

8. $x^2 + 2(x \cdot 7) + 7^2 = (x + 7)(x + 7) = (x + 7)^2$

9. $(3p)^2 - 4^2 = (3p + 4)(3p - 4)$

10. $(5y)^2 - 2^2 = (5y + 2)(5y - 2)$

11. $(2d)^2 + 2(2d \cdot 9) + 9^2 = (2d + 9)(2d + 9) = (2d + 9)^2$

12. $(3h)^2 + 2(3h \cdot 4) + 4^2 = (3h + 4)(3h + 4) = (3h + 4)^2$

13. $(2c)^2 - (7d)^2 = (2c + 7d)(2c - 7d)$

14. $(11m)^2 - n^2 = (11m + n)(11m - n)$

15. $k^2 + 2(k \cdot 10) + 10^2 = (k + 10)(k + 10) = (k + 10)^2$

16. $s^2 + 2(s \cdot 13) + 13^2 = (s + 13)(s + 13) = (s + 13)^2$

17. $(9y)^2 - 2(9y \cdot 2) + 2^2 = (9y - 2)(9y - 2) = (9y - 2)^2$

18. $(4n)^2 - 2(4n \cdot 7) + 7^2 = (4n - 7)(4n - 7) = (4n - 7)^2$

B 19. $(a^5b^2)^2 - 4^2 = (a^5b^2 - 4)(a^5b^2 + 4)$

20. $(m^8m^4)^2 - 5^2 = (m^8n^4 - 5)(m^8n^4 + 5)$

21. $(5t)^2 + 2(5t \cdot 1) + 1^2 = (5t + 1)(5t + 1) = (5t + 1)^2$

22. $(7r)^2 + 2(7r \cdot 1) + 1^2 = (7r + 1)(7r + 1) = (7r + 1)^2$

23. $(10k^5)^2 + 2(10k^5 \cdot 1) + 1^2 = (10k^5 + 1)(10k^5 + 1) = (10k^5 + 1)^2$

24. $(6y^5)^2 + 2(6y^5 \cdot 1) + 1^2 = (6y^5 + 1)(6y^5 + 1) = (6y^5 + 1)^2$

25. $(x^9y^5)^2 - 6^2 = (x^9y^5 - 6)(x^9y^5 + 6)$

26. $(c^6d^3)^2 - 8^2 = (c^6d^3 - 8)(c^6d^3 + 8)$

27. $(5y^4)^2 + 2(5y^4 \cdot 1) + 1^2 = (5y^4 + 1)(5y^4 + 1) = (5y^4 + 1)^2$

28. $(3x^6)^2 + 2(3x^6 \cdot 1) + 1^2 = (3x^6 + 1)(3x^6 + 1) = (3x^6 + 1)^2$

29. $1^2 - (x^5y^6)^2 = (1 - x^5y^6)(1 + x^5y^6)$

30. $1^2 - (5a^3b^6)^2 = (1 - 5a^3b^6)(1 + 5a^3b^6)$

31. $(12x^2y)^2 - 25^2 = (12x^2y - 25)(12x^2y + 25)$

32. $(11m^3n)^2 - 9^2 = (11m^3n - 9)(11m^3n + 9)$

33. $(7rs)^2 + 2(7rs \cdot 1) + 1^2 = (7rs + 1)(7rs + 1) = (7rs + 1)^2$

34. $(5ab)^2 + 2(5ab \cdot 2) + 2^2 = (5ab + 2)(5ab + 2) = (5ab + 2)^2$

35. $(a^3bc^2)^2 - d^2 = (a^3bc^2 - d)(a^3bc^2 + d)$

36. $(e^3f^2g^3)^2 - h^2 = (e^3f^2g^3 - h)(e^3f^2g^3 + h)$

37. $9(4x^6y^4 - 4x^3y^2 + 1) = 9[(2x^3y)^2 - 2(2x^3y^2 \cdot 1) + 1^2] =$
$9[(2x^3y^2 - 1)(2x^3y^2 - 1)] = 9(2x^3y^2 - 1)^2$

38. $4(16c^6d^8 - 8c^3d^4 + 1) = 4[(4c^3d^4)^2 - 2(4c^3d^4 \cdot 1) + 1^2] =$
$4[(4c^3d^4 - 1)(4c^3d^4 - 1)] = 4(4c^3d^4 - 1)^2$

39. $(2p^2q^2r^4)^2 - 9^2 = (2p^2q^2r^4 - 9)(2p^2q^2r^4 + 9)$

40. $(3x^4y^2z)^2 - 8^2 = (3x^4y^2z - 8)(3x^4y^2z + 8)$

41. $(e^{31}f^{50})^2 - (g^{72}h^{18})^2 = (e^{31}f^{50} - g^{72}h^{18})(e^{31}f^{50} + g^{72}h^{18})$

42. $(5x^{25})^2 + (7y^{100})^2 = (5x^{25} - 7y^{100})(5x^{25} + 7y^{100})$

C 43. $(a^2 + 2a + 1) - a^2 = 2a + 1$

44. $(x^2 - 4x + 4) - x^2 = -4x + 4 = -4(x - 1)$

45. $(x^2 - 2x + 1) + 2x - 2 + 1 = x^2$

46. $(a^2 + 4a + 4) + 2a + 4 + 1 = a^2 + 6a + 9 = (a + 3)(a + 3) = (a + 3)^2$

47. $(a + b)^2 - c^2 = (a + b - c)(a + b + c)$

48. $m^2 - (n + p)^2 = [m + (n + p)][m - (n + p)] = (m + n + p)(m - n - p)$

49. $x^2 + 2x + 1 + 2(2x^2 + 5x + 3) + (4x^2 + 12x + 9) =$
$x^2 + 2x + 1 + 4x^2 + 10x + 6 + 4x^2 + 12x + 9 = 9x^2 + 24x + 16 =$
$(3x + 4)(3x + 4) = (3x + 4)^2$

50. $\text{Area}_{\text{shaded}} = \text{Area}_{\text{big square}} - \text{Area}_{\text{little square}} = m^2 - n^2 = (m - n)(m + n)$; The
area of the shaded region is $(m - n)(m + n)$.

51. $\text{Area}_{\text{shaded}} = \text{Area}_{\text{square}} - \text{Area}_{\text{rectangle}} = t^2 - rs$; The area of the shaded
region is $t^2 - rs$.

52. $x^2 - 2(94 \cdot x) + 94^2 = (x - 94)^2$

53. $x^2 - (546.5)^2 = (x + 546.5)(x - 546.5)$ **54.** $5x^3 + x^2 - 9x + 2$

55. $2x^3 + 3x^2 + x + 2$ **56.** $-2n^2 + 9n - 10$

57. $64x^3 + 48x^2y + 36xy^2 - 48x^2y - 36xy^2 - 27y^3 = 64x^3 - 27y^3$

page 292   Capsule Review

1. $-1(1 - m) = -1 + m$ **2.** $-1(3t - 4) = -3t + 4$
$= m - 1;$ Yes $= 4 - 3t;$ Yes

3. $-1(3y - 2) = -3y + 2$
   $-2 - 3y \neq -3y + 2;$ No

4. $-1(5y - 2) = -5y + 2$
   $5y + 2 \neq -5y + 2;$ No

5. $-1(4h^2 + 5) = -4h^2 - 5$
   $5 - 4h^2 \neq -4h^2 - 5;$ No

6. $-1(-2x - 3) = 2x + 3$
   $2x + 3 = 2x + 3;$ Yes

page 294 Class Exercises

1. $2x(x) + 2x(2);$ yes; $2x$

2. $3x(5x) + 3x(4) + 3x(2x^2);$ yes; $3x$

3. yes; $j + k$ **4.** no

5. no   **6.** yes; $2(m - n)$

7. $(x + y)(8 - 3) = 5(x + y)$

8. $(x - 3)(7 + 2) = 9(x - 3)$

9. $11(j - 2) + 17(-1)(j - 2) = (j - 2)(11 - 17) = -6(j - 2)$

10. $12(-1)(m - 4) - 3(m - 4) = (m - 4)(-12 - 3) = -15(m - 4)$

11. $b(a + b) + c(a + b) = (a + b)(b + c)$

12. $y(y - x) + (y - x) = (y - x)(y + 1)$

13. $r(1 - t) + s(t - 1) = r(1 - t) + s(-1)(1 - t) = (1 - t)(r - s)$

14. $9(1 + c) - d^2(1 + c) = (9 - d^2)(1 + c) = (3 - d)(3 + d)(1 + c)$

15. $4^2 5^2 3^2 - 4^2 5^2 5^2 = 4^2 5^2(3^2 - 5^2) = (4)^2(5)^2(3 - 5)(3 + 5) = 4^2 \cdot 5^2(-2)(8) =$
   $-4^4 \cdot 5^2$

pages 294–295 Practice Exercises

A   **1.** $(7 - 5 + 3)(a + 3) = 5(a + 3)$

2. $(13 - 2 + 8)(g + 2) = 19(g + 2)$

3. $(5 + 2 - 3)(x - 3) = 4(x - 3)$

4. $(7 + 5 - 4)(r - 4) = 8(r - 4)$

5. $(3x + 2x - 4)(x - 2) = (5x - 4)(x - 2)$

6. $2m(n - 2p) + 4m(n - 2p) - 2(n - 2p) = (2m + 4m - 2)(n - 2p) =$
   $(6m - 2)(n - 2p) = 2(3m - 1)(n - 2p)$

7. $11(j - 2) - 6(-1)(j - 2) + 2(j - 2) = (11 + 6 + 2)(j - 2) = 19(j - 2)$

8. $9(x - 5) - 4(-1)(x - 5) + 5(x - 5) = (9 + 4 + 5)(x - 5) = 18(x - 5)$

9. $9a(b - c) + 2(-1)(b - c) + 3a(b - c) = (9a - 2 + 3a)(b - c) =$
   $(12a - 2)(b - c) = 2(6a - 1)(b - c)$

10. $4x(y - z) + 5(-1)(y - z) + 7x(y - z) = (4x - 5 + 7x)(y - z) =$
   $(11x - 5)(y - z)$

11. $4(-1)(j - 3) + 2j(j - 3) + 8(j - 3) = (-4 + 2j + 8)(j - 3) =$
   $(2j + 4)(j - 3) = 2(j + 2)(j - 3)$

12. $5(-1)(m - 2) + 3m(m - 2) + 7(m - 2) = (-5 + 3m + 7)(m - 2) =$
   $(3m + 2)(m - 2)$

13. $10(a - b) - 12(a - b) = (10 - 12)(a - b) = -2(a - b)$

14. $21(x - y) + 15(x - y) = (21 + 15)(x - y) = 36(x - y)$

15. $13(r - s) + 8(-s + r) = 13(r - s) + 8(r - s) = (13 + 8)(r - s) = 21(r - s)$

16. $11(m - n) - 10(-n + m) = 11(m - n) - 10(m - n) = $
$(11 - 10)(m - n) = m - n$

17. $6(h - k) - h^3(h - k) = (6 - h^3)(h - k)$

18. $5(m - n) - m^4(m - n) = (5 - m^4)(m - n)$

19. $r(s + t) - 3(s + t) = (r - 3)(s + t)$

20. $n(m + p) + 2(m + p) = (n + 2)(m + p)$

21. $t(3r + s) + w(3r + s) = (t + w)(3r + s)$

22. $b(2a - 1) + 7(2a - 1) = (2a - 1)(b + 7)$

23. $a(c - 2) + 3b(c - 2) = (c - 2)(a + 3b)$

24. $2w(2x - 3y) + 3z(2x - 3y) = (2w + 3z)(2x - 3y)$

25. $36y^2 - (9x^2 + 24x + 16) = (6y)^2 - (3x + 4)^2 = $
$[6y - (3x + 4)][6y + (3x + 4)] = (6y - 3x - 4)(6y + 3x + 4)$

26. $49w^2 - (16x^2 + 24x + 9) = (7m)^2 - (4x + 3)^2 = $
$[7w - (4x + 3)][7w + (4x + 3)] = (7w - 4x - 3)(7w + 4x + 3)$

B 27. $(x^z + zx) + (zy + xy) = x(x + z) + y(z + x) = (x + y)(z + x)$

28. $3ba + 4bc + 9a + 12c = b(3a + 4c) + 3(3a + 4c) = (b + 3)(3a + 4c)$

29. $3b(5a - 3c) + 4c(5a - 3c) = (3b + 4c)(5a - 3c)$

30. $6mn - 5np + 12mp - 10p^2 = n(6m - 5p) + 2p(6m - 5p) = (n + 2p)$
$(6m - 5p)$

31. $6rt - 15rs + 2t - 5s = 3r(2t - 5s) + 1(2t - 5s) = (3r + 1)(2t - 5s)$

32. $q(2p - 5r) - 2(2p - 5r) = (q - 2)(2p - 5r)$

33. $[r - (4s + t)][r + (4s + t)] = (r - 4s - t)(r + 4s + t)$

34. $[m - (2n - 3p)][m + (2n - 3p)] = (m - 2n + 3p)(m + 2n - 3p)$

35. $(a + 3)^2 - c^2 = [(a + 3) - c][(a + 3) + c] = (a + 3 - c)(a + 3 + c)$

36. $(x - 2)^2 - y^2 = [(x - 2) - y][(x - 2) + y] = (x - 2 - y)(x - 2 + y)$

37. $25y^2 - (4x^2 - 12x + 9) = (5y)^2 - (2x - 3)^2 = $
$[5y - (2x - 3)][5y + (2x - 3)] = (5y - 2x + 3)(5y + 2x - 3)$

38. $121m^2 - (4n^2 + 52n + 169) = (11m)^2 - (2n + 13)^2 = [11m - (2n + 13)]$
$[11m + (2n + 13)] = (11m - 2n - 13)(11m + 2n + 13)$

C **39.** $x^{2r+1} + 3x^{2r}z^{3r} + xy + 3z^{3r}y = x^{2r}(x + 3z^{3r}) + y(x + 3z^{3r}) = (x^{2r} + y)$
$(x + 3z^{3r})$

40. $2a^{k+3} + a^{k+1}c^{2k+3} + 2a^2b^k + b^kc^{2k+3} = a^{k+1}(2a^2 + c^{2k+3}) + b^k(2a^2 + c^{2k+3}) =$
$(a^{k+1} + b^k)(2a^2 + c^{2k+3})$

41. $m^rp + m^rn^{2r} + np + n^{2r+1} = m^r(p + n^{2r}) + n(p + n^{2r}) = (m^r + n)(p + n^{2r})$

42. $2a^{n+2} + a^nc^{2n} + 2a^2b^2 + b^2c^{2n} = a^n(2a^2 + c^{2n}) + b^2(2a^2 + c^{2n}) = (a^n + b^2)$
$(2a^2 + c^{2n})$

43. $(x^{2a} + 2x^ay^b + y^{2b}) - 1 = (x^a + y^b)^2 - 1 = [(x^a + y^b) - 1][(x^a + y^b) + 1] =$
$(x^a + y^b - 1)(x^a + y^b + 1)$

44. $(m^{2x} - 2m^xn^y + n^{2y}) - 4 = (m^x - n^y)^2 - 2^2 = [(m^x - n^y) - 2][(m^x - n^y) + 2] =$
$(m^x - n^y - 2)(m^x - n^y + 2)$

45. $49y^2 - (9x^2 - 6xy + y^2) = (7y)^2 - (3x - y)^2 = [7y - (3x - y)][7y +$
$(3x - y)] = (7y - 3x + y)(7y + 3x - y) = (8y - 3x)(6y + 3x) =$
$3(x + 2y)(8y - 3x)$

46. $484r^2 - (4r^2 - 32r + 64) = 4[121r^2 - (r^2 - 8r + 16)] =$
$4[(11r)^2 - (r - 4)^2] = 4[11r - (r - 4)][11r + (r - 4)] =$
$4[11r - r + 4][11r + r - 4] = 4[10r + 4][12r - 4] = 4(2)[5r + 2](4)[3r - 1] =$
$4(2)(4)(5r + 2)(3r - 1) = 32(5r + 2)(3r - 1)$

page 295   Historical Note

Mersenne primes are expressed in the form $2^n - 1$. The largest prime yet found has
65,050 digits when $n = 216,091$.

page 296   Capsule Review

1. $5(x) + 5(2y) = 5(x + 2y)$

2. $2h^3(h) + 2h^3(2) = 2h^3(h + 2)$

3. $3c(3c) - 3c(1) = 3c(3c - 1)$

4. $2ab^2(3) + 2ab^2(4a^2) = 2ab^2(3 + 4a^2)$

5. $2m^3n(1) - 2m^3n(4m^2n) = 2m^3n(1 - 4m^2n)$

6. $5st(s) - 5st(3t^3) = 5st(s - 3t^3)$

7. $xy^2(9xy) - xy^2(12y^2) + xy^2(8) = xy^2(9xy - 12y^2 + 8)$

8. $5b^2(3a^4b^2) - 5b^2(2a^4) - 5b^2(b^2) = 5b^2(3a^4b^2 - 2a^4 - b^2)$

pages 297–298   Class Exercises

1. $2(x^2) + 2(2x) + 2(3) = 2(x^2 + 2x + 3);\ 2$

2. $2r^2(r) + 2r^2(2) = 2r^2(r + 2);\ 2r^2$

3. $3xy(y) + 3xy(3x^2) = 3xy(y + 3x^2);\ 3xy$

4. $2ab^3(6a) + 2ab^3(7b) = 2ab^3(6a + 7b);\ 2ab^3$

5. $rs^2(9rs) + rs^2(12s^2) + rs^2(8r) = rs^2(9rs + 12s^2 + 8r); rs^2$

6. $16xyz(8x^2z^2) + 16xyz(y^4z^3) + 16xyz(3x^3y^2) = 16xyz(8x^2z^2 + y^4z^3 + 3x^3y^2);$
$16xyz$

7. $3x^2(1) + 3x^2(7x^2) = 3x^2(1 + 7x^2)$

8. $3(b^2 + 2b + 1) = 3(b + 1)(b + 1) = 3(b + 1)^2$

9. $4(z^2 - 1) = 4(z - 1)(z + 1)$ **10.** $2(w^2 + 4w - 12) = 2(w + 6)(w - 2)$

11. $t(t^2 + 7t + 12) = t(t + 4)(t + 3)$ **12.** $4h(h^2 + 3h + 2) = 4h(h + 2)(h + 1)$

pages 298–299   Practice Exercises

A   **1.** $5(x^2 - 4) = 5(x + 2)(x - 2)$ **2.** $6(a^2 - 1) = 6(a + 1)(a - 1)$

3. $3(k^2 - 49) = 3(k + 7)(k - 7)$ **4.** $5(m^2 - 25) = 5(m + 5)(m - 5)$

5. $4y(y^2 - 9z^2) = 4y(y + 3z)(y - 3z)$ **6.** $27p(p^2 - 4q^2) = 27p(p - 2q)(p + 2q)$

7. $6x(x^2 - 4y^2) = 6x(x + 2y)(x - 2y)$ **8.** $3r(r^2 - 16s^2) = 3r(r - 4s)(r + 4s)$

9. $2(x^2 + 3x - 10) = 2(x + 5)(x - 2)$ **10.** $3(x^2 - 2x - 8) = 3(x + 2)(x - 4)$

11. $6(k^2 + 2k + 1) = 6(k + 1)(k + 1) = 6(k + 1)^2$

12. $8(c^2 - 3c + 2) = 8(c - 2)(c - 1)$

13. $5(2k^2 + 7k + 3) = 5(2k + 1)(k + 3)$

14. $3(4r^2 - 15r - 4) = 3(4r + 1)(r - 4)$

15. $-4(x^2 + x - 6) = -4(x + 3)(x - 2)$

16. $-3(d^2 + 2d - 8) = -3(d - 2)(d + 4)$

17. $-10(m^2 - 4m - 21) = -10(m + 3)(m - 7)$

18. $-6(x^2 - 6x - 16) = -6(x + 2)(x - 8)$

19. $3x(25x^2 - 10x + 1) = 3x(5x - 1)(5x - 1) = 3x(5x - 1)^2$

20. $3y(16y^2 - 8y + 1) = 3y(4y - 1)(4y - 1) = 3y(4y - 1)^2$

21. $12x(x^2 + 2x + 1) = 12x(x + 1)(x + 1) = 12x(x + 1)^2$

22. $3m(9m^2 + 12m + 4) = 3m(3m + 2)(3m + 2) = 3m(3m + 2)^2$

23. $4x^2y(2x^2 + x - 3) = 4x^2y(2x + 3)(x - 1)$

24. $12m^2n(2m^2 - m - 1) = 12m^2n(2m + 1)(m - 1)$

25. $6r^2s^2(2r^2 + r - 1) = 6r^2s^2(r + 1)(2r - 1)$

26. $5j^2k^2(9j^2 + 9j - 4) = 5j^2k^2(3j + 4)(3j - 1)$

27. $3x^3y^3(6x^2 - 5xy - 6y^2) = 3x^3y^3(3x + 2y)(2x - 3y)$

28. $16x^3y^2(2x^2 - 3xy - 2y^2) = 16x^3y^2(2x + y)(x - 2y)$

29. $4x^4y^4(4x^2 - 12xy + 9y^2) = 4x^4y^4(2x - 3y)(2x - 3y) = 4x^4y^4(2x - 3y)^2$

30. $2a^3b^3(9a^2 - 30ab + 25b^2) = 2a^3b^3(3a - 5b)(3a - 5b) = 2a^3b^3(3a - 5b)^2$

B 31. $-3m^5p^2(4m^2 + 20mp + 25p^2) = -3m^5p^2(2m + 5p)(2m + 5p) =$
$-3m^5p^2(2m + 5p)^2$

32. $-4d^3f^5(9d^2 + 24df + 16f^2) = -4d^3f^5(3d + 4f)(3d + 4f) = -4d^3f^5(3d + 4f)^2$

33. $2x(x^8 - 25) = 2x(x^4 - 5)(x^4 + 5)$ **34.** $3r(r^8 - 9) = 3r(r^4 - 3)(r^4 + 3)$

35. $2x(x^6 - 16) = 2x(x^3 - 4)(x^3 + 4)$ **36.** $3x(x^6 - 25) = 3x(x^3 - 5)(x^3 + 5)$

37. $3m(m^4 - 20m^2 + 64) = 3m(m^2 - 16)(m^2 - 4) =$
$3m(m + 4)(m - 4)(m + 2)(m - 2)$

38. $2m(m^4 - 34m^2 + 225) = 2m(m^2 - 9)(m^2 - 25) =$
$2m(m - 3)(m + 3)(m + 5)(m - 5)$

39. $2r^2(9r^4 - 85r^2 + 36) = 2r^2(9r^2 - 4)(r^2 - 9) =$
$2r^2(3r - 2)(3r + 2)(r + 3)(r - 3)$

40. $2r^2(4r^4 - 25r^2 + 36) = 2r^2(4r^2 - 9)(r^2 - 4) =$
$2r^2(2r - 3)(2r + 3)(r + 2)(r - 2)$

41. $3x^4y^5(64x^2 - 48x + 9) = 3x^4y^5(8x - 3)(8x - 3) = 3x^4y^5(8x - 3)^2$

42. $3x^4y^5(25x^2 - 20x + 4) = 3x^4y^5(5x - 2)(5x - 2) = 3x^4y^5(5x - 2)^2$

43. $6(a^2 - 2a + 1) - 15a + 15 - 9 = 6a^2 - 12a + 6 - 15a + 15 - 9 =$
$6a^2 - 27a + 12 = 3(2a^2 - 9a + 4) = 3(2a - 1)(a - 4)$

44. $4(b^2 + 4b + 4) + 20b + 40 - 24 = 4b^2 + 16b + 16 + 20b + 40 - 24 =$
$4b^2 + 36b + 32 = 4(b^2 + 9b + 8) = 4(b + 8)(b + 1)$

C 45. $5(a^2 + 2a + 1) - 5a + 5 - 20 = 5a^2 + 10a + 5 - 5a + 5 - 20 =$
$5a^2 + 5a - 10 = 5(a^2 + a - 2) = 5(a + 2)(a - 1)$

46. $3(a^2 + 4a + 4) - 3a + 3 - 27 = 3a^2 + 12a + 12 - 3a + 3 - 27 =$
$3a^2 + 9a - 12 = 3(a^2 + 3a - 4) = 3(a + 4)(a - 1)$

47. $(a^2 - 9)(4a + a^2 + 4) = (a^2 - 9)(a^2 + 4a + 4) = (a - 3)(a + 3)(a + 2)(a + 2) =$
$(a - 3)(a + 3)(a + 2)^2$

48. $(a^2 - 16)(a^2 + 4a + 4) = (a + 4)(a - 4)(a + 2)(a + 2) =$
$(a + 4)(a - 4)(a + 2)^2$

49. $3a(a^{16} + 10a^8 + 25) = 3a(a^8 + 5)(a^8 + 5) = 3a(a^8 + 5)^2$

50. $4a(a^{18} + 6a^9 + 9) = 4a(a^9 + 3)(a^9 + 3) = 4a(a^9 + 3)^2$

51. $x^{k+7}(x^{2k+14} + 2x^{k+7} + 1) = x^{k+7}(x^{k+7} + 1)(x^{k+7} + 1) = x^{k+7}(x^{k+7} + 1)^2$

52. $a^{3z-9}(3a^{6z-18} - 13a^{3z-9} - 10) = a^{3z-9}(3a^{3z-9} + 2)(a^{3z-9} - 5)$

53. $x(x^2 + 3x + 2) = x(x + 2)(x + 1)$ **54.** $y(y^2 + 8y + 15) = y(y + 5)(y + 3)$

55. $2a(a^2 + 4a + 3) = 2a(a + 3)(a + 1)$ **56.** $2b(3b^2 + 7b + 2) = 2b(3b + 1)(b + 2)$

page 299   Algebra in Geometry

1. $\text{Area}_{shaded} = \text{Area}_{circle\ x} - \text{Area}_{circle\ y} = \pi x^2 - \pi y^2 = \pi(x^2 - y^2) = \pi(x - y)(x + y)$; The area of the shaded region is $\pi(x - y)(x + y)$.

2. $\text{Area}_{shaded} = \text{Area}_{square\ x} - 4\ \text{Area}_{square\ y} = x^2 - 4y^2 = (x - 2y)(x + 2y)$; The area of the shaded region is $(x - 2y)(x + 2y)$.

3. $\text{Area}_{shaded} = \text{Area}_{circle\ x} - 9\ \text{Area}_{circle\ y} = \pi x^2 - 9\pi y^2 = \pi(x^2 - 9y^2) = \pi(x - 3y)(x + 3y)$; The area of the shaded region is $\pi(x - 3y)(x + 3y)$.

page 300   Capsule Review

1. $3(4)^2 - 12 \stackrel{?}{=} 0$, $3(16) - 12 \stackrel{?}{=} 0$, $48 - 12 \stackrel{?}{=} 0$, $36 \neq 0$; 4 is not a solution
$3(-2)^2 - 12 \stackrel{?}{=} 0$, $3(4) - 12 \stackrel{?}{=} 0$, $12 - 12 \stackrel{?}{=} 0$, $0 = 0$; -2 is a solution

2. $3^2 - 7(3) \stackrel{?}{=} -12$, $9 - 21 \stackrel{?}{=} -12$, $-12 = -12$; 3 is a solution
$4^2 - 7(4) \stackrel{?}{=} -12$, $16 - 28 \stackrel{?}{=} -12$, $-12 = -12$; 4 is a solution

3. $9(0)^2 \stackrel{?}{=} 3(0)$, $0 = 0$; 0 is a solution
$9\left(-\dfrac{1}{3}\right)^2 \stackrel{?}{=} 3\left(-\dfrac{1}{3}\right)$, $9\left(\dfrac{1}{9}\right) \stackrel{?}{=} 3\left(-\dfrac{1}{3}\right)$, $1 \neq -1$; $-\dfrac{1}{3}$ is not a solution

4. $\left(\dfrac{1}{2}\right)^2 \stackrel{?}{=} \dfrac{1}{4}$, $\dfrac{1}{4} = \dfrac{1}{4}$; $\dfrac{1}{2}$ is a solution
$\left(-\dfrac{1}{2}\right)^2 \stackrel{?}{=} \dfrac{1}{4}$, $\dfrac{1}{4} = \dfrac{1}{4}$; $-\dfrac{1}{2}$ is a solution

page 302   Class Exercises

1. $2r(r - 3) = 0$
$2r = 0 \mid r - 3 = 0$
$r = 0 \mid\quad\ r = 3$

2. $(x - 4)(3x + 2) = 0$
$x - 4 = 0 \mid 3x + 2 = 0$
$x = 4 \mid\quad 3x = -2$
$\mid\qquad\quad x = -\dfrac{2}{3}$

3. $m(3m - 1) = 0$
$m = 0 \mid 3m - 1 = 0$
$\mid\qquad 3m = 1$
$\mid\qquad\quad m = \dfrac{1}{3}$

4. $z(2z + 3) = 0$
$z = 0 \mid 2z + 3 = 0$
$\mid\qquad 2z = -3$
$\mid\qquad\quad z = -\dfrac{3}{2}$

5. $(y + 6)(y + 1) = 0$

$y + 6 = 0 \mid y + 1 = 0$

$y = -6 \mid \quad y = -1$

6. $(2x - 1)(x + 7) = 0$

$2x - 1 = 0 \mid x + 7 = 0$

$2x = 1 \mid \quad x = -7$

$x = \dfrac{1}{2} \mid$

7. $(2m - 1)(m + 1) = 0$

$2m - 1 = 0 \mid m + 1 = 0$

$2m = 1 \mid \quad m = -1$

$m = \dfrac{1}{2} \mid$

8. $(c + 4)(c + 4) = 0$

$c + 4 = 0 \mid c + 4 = 0$

$c = -4 \mid \quad c = -4$

9. $x^2 - 4x - 12 = 0$

$(x - 6)(x + 2) = 0$

$x - 6 = 0 \mid x + 2 = 0$

$x = 6 \mid \quad x = -2$

10. $3w^2 + 5w - 12 = 0$

$(3w - 4)(w + 3) = 0$

$3w - 4 = 0 \mid w + 3 = 0$

$3w = 4 \mid \quad w = -3$

$w = \dfrac{4}{3} \mid$

pages 302–303   Practice Exercises

A **1.** $(3x - 2)(x + 4) = 0$

$3x - 2 = 0 \mid x + 4 = 0$

$3x = 2 \mid \quad x = -4$

$x = \dfrac{2}{3} \mid$

2. $(4z - 3)(z + 5) = 0$

$4z - 3 = 0 \mid z + 5 = 0$

$4z = 3 \mid \quad z = -5$

$z = \dfrac{3}{4} \mid$

3. $(2x - 6)(x + 6) = 0$

$2x - 6 = 0 \mid x + 6 = 0$

$2x = 6 \mid \quad x = -6$

$x = 3 \mid$

4. $(5x - 8)(x + 9) = 0$

$5x - 8 = 0 \mid x + 9 = 0$

$5x = 8 \mid \quad x = -9$

$x = \dfrac{8}{5} \mid$

5. $(5x + 3)(x - 2) = 0$

$5x + 3 = 0 \mid x - 2 = 0$

$5x = -3 \mid \quad x = 2$

$x = -\dfrac{3}{5} \mid$

6. $(7x + 5)(x - 3) = 0$

$7x + 5 = 0 \mid x - 3 = 0$

$7x = -5 \mid \quad x = 3$

$x = -\dfrac{5}{7} \mid$

7. $m(m + 4) = 0$

$m = 0 \mid m + 4 = 0$

$\quad \mid \quad m = -4$

8. $n(n + 5) = 0$

$n = 0 \mid n + 5 = 0$

$\quad \mid \quad n = -5$

9. $2w(w - 4) = 0$

$2w = 0 \mid w - 4 = 0$

$w = 0 \mid \quad w = 4$

10. $3z(z - 2) = 0$

$3z = 0 \mid z - 2 = 0$

$z = 0 \mid \quad z = 2$

11. $5x(x + 3) = 0$
$5x = 0 \mid x + 3 = 0$
$x = 0 \mid \quad x = -3$

12. $4r(r + 4) = 0$
$4r = 0 \mid r + 4 = 0$
$r = 0 \mid \quad r = -4$

13. $x^2 + 2x - 8 = 0$
$(x + 4)(x - 2) = 0$
$x + 4 = 0 \mid x - 2 = 0$
$x = -4 \mid \quad x = 2$

14. $z^2 - 4z - 21 = 0$
$(z - 7)(z + 3) = 0$
$z - 7 = 0 \mid z + 3 = 0$
$z = 7 \mid \quad z = -3$

15. $3m^2 + 4m + 1 = 0$
$(3m + 1)(m + 1) = 0$
$3m + 1 = 0 \mid m + 1 = 0$
$3m = -1 \mid \quad m = -1$
$m = -\dfrac{1}{3}$

16. $2z^2 + 4z - 16 = 0$
$2(z^2 + 2z - 8) = 0$
$2(z + 4)(z - 2) = 0$
$z + 4 = 0 \mid z - 2 = 0$
$z = -4 \mid \quad z = 2$

17. $4m^2 - 11m - 3 = 0$
$(4m + 1)(m - 3) = 0$
$4m + 1 = 0 \mid m - 3 = 0$
$4m = -1 \mid \quad m = 3$
$m = -\dfrac{1}{4}$

18. $5m^2 - 7m - 6 = 0$
$(5m + 3)(m - 2) = 0$
$5m + 3 = 0 \mid m - 2 = 0$
$5m = -3 \mid \quad m = 2$
$m = -\dfrac{3}{5}$

19. $r^2 - r - 90 = 0$
$(r - 10)(r + 9) = 0$
$r - 10 = 0 \mid r + 9 = 0$
$r = 10 \mid \quad r = -9$

20. $t^2 - 3t - 18 = 0$
$(t - 6)(t + 3) = 0$
$t - 6 = 0 \mid t + 3 = 0$
$t = 6 \mid \quad t = -3$

21. $x^2 - 8x + 15 = 0$
$(x - 5)(x - 3) = 0$
$x - 5 = 0 \mid x - 3 = 0$
$x = 5 \mid \quad x = 3$

22. $3k^2 + 17k + 10 = 0$
$(3k + 2)(k + 5) = 0$
$3k + 2 = 0 \mid k + 5 = 0$
$3k = -2 \mid \quad k = -5$
$k = -\dfrac{2}{3}$

23. $3m^2 - 7m - 20 = 0$
$(3m + 5)(m - 4) = 0$
$3m + 5 = 0 \mid m - 4 = 0$
$3m = -5 \mid \quad m = 4$
$m = -\dfrac{5}{3}$

24. $2n^2 + 13n - 24 = 0$
$(2n - 3)(n + 8) = 0$
$2n - 3 = 0 \mid n + 8 = 0$
$2n = 3 \mid \quad n = -8$
$n = \dfrac{3}{2}$

B 25. $4x^2 + 4x + 1 = 0$
$(2x + 1)(2x + 1) = 0$
$2x + 1 = 0 \; | \; 2x + 1 = 0$
$2x = -1 \; | \quad 2x = -1$
$x = -\dfrac{1}{2} \; | \quad x = -\dfrac{1}{2}$

26. $9x^2 - 12x + 4 = 0$
$(3x - 2)(3x - 2) = 0$
$3x - 2 = 0 \; | \; 3x - 2 = 0$
$3x = 2 \; | \quad 3x = 2$
$x = \dfrac{2}{3} \; | \quad x = \dfrac{2}{3}$

27. $3r^2 + 4r - 15 = 0$
$(3r - 5)(r + 3) = 0$
$3r - 5 = 0 \; | \; r + 3 = 0$
$3r = 5 \; | \quad r = -3$
$r = \dfrac{5}{3} \;$

28. $2x^2 - 11x - 21 = 0$
$(2x + 3)(x - 7) = 0$
$2x + 3 = 0 \; | \; x - 7 = 0$
$2x = -3 \; | \quad x = 7$
$x = -\dfrac{3}{2} \;$

29. $6y^2 + 11y - 10 = 0$
$(2y + 5)(3y - 2) = 0$
$2y + 5 = 0 \; | \; 3y - 2 = 0$
$2y = -5 \; | \quad 3y = 2$
$y = -\dfrac{5}{2} \; | \quad y = \dfrac{2}{3}$

30. $15s^2 - s - 28 = 0$
$(5s - 7)(3s + 4) = 0$
$5s - 7 = 0 \; | \; 3s + 4 = 0$
$5s = 7 \; | \quad 3s = -4$
$s = \dfrac{7}{5} \; | \quad s = -\dfrac{4}{3}$

31. $14a^2 - 29a - 15 = 0$
$(7a + 3)(2a - 5) = 0$
$7a + 3 = 0 \; | \; 2a - 5 = 0$
$7a = -3 \; | \quad 2a = 5$
$a = -\dfrac{3}{7} \; | \quad a = \dfrac{5}{2}$

32. $6b^2 + b - 35 = 0$
$(2b + 5)(3b - 7) = 0$
$2b + 5 = 0 \; | \; 3b - 7 = 0$
$2b = -5 \; | \quad 3b = 7$
$b = -\dfrac{5}{2} \; | \quad b = \dfrac{7}{3}$

33. $24d^2 + 18d + 4d - 2 = 0$
$24d^2 + 22d - 2 = 0$
$2(12d^2 + 11d - 1) = 0$
$2(12d - 1)(d + 1) = 0$
$12d - 1 = 0 \; | \; d + 1 = 0$
$d = \dfrac{1}{12} \; | \quad d = -1$

34. $2t^2 + 12t - 18t - 108 = 0$
$2t^2 - 6t - 108 = 0$
$2(t^2 - 3t - 54) = 0$
$2(t - 9)(t + 6) = 0$
$t - 9 = 0 \; | \; t + 6 = 0$
$t = 9 \; | \quad t = -6$

35. $16x^2 + 30x - 6x + 9 = 0$
$16x^2 + 24x + 9 = 0$
$(4x + 3)(4x + 3) = 0$
$4x + 3 = 0 \; | \; 4x + 3 = 0$
$x = -\dfrac{3}{4} \; | \quad x = -\dfrac{3}{4}$

36. $25x^2 - 12x - 8x + 4 = 0$
$25x^2 - 20x + 4 = 0$
$(5x - 2)(5x - 2) = 0$
$5x - 2 = 0 \; | \; 5x - 2 = 0$
$x = \dfrac{2}{5} \; | \quad x = \dfrac{2}{5}$

37. $n(n^2 + 4n - 21) = 0$
$n(n + 7)(n - 3) = 0$
$n = 0 \mathrel{\vert} n + 7 = 0 \mathrel{\vert} n - 3 = 0$
$\phantom{n = 0 \mathrel{\vert}} n = -7 \mathrel{\vert} n = 3$

38. $m(m^2 - m - 20) = 0$
$m(m - 5)(m + 4) = 0$
$m = 0 \mathrel{\vert} m - 5 = 0 \mathrel{\vert} m + 4 = 0$
$\phantom{m = 0 \mathrel{\vert}} m = 5 \mathrel{\vert} m = -4$

39. $x(6x^2 - 7x - 20) = 0$
$x(3x + 4)(2x - 5) = 0$
$x = 0 \mathrel{\vert} 3x + 4 = 0 \mathrel{\vert} 2x - 5 = 0$
$\phantom{x = 0 \mathrel{\vert}} x = -\dfrac{4}{3} \mathrel{\vert} x = \dfrac{5}{2}$

40. $n(10n^2 - 29n - 21) = 0$
$n(5n + 3)(2n - 7) = 0$
$n = 0 \mathrel{\vert} 5n + 3 = 0 \mathrel{\vert} 2n - 7 = 0$
$\phantom{n = 0 \mathrel{\vert}} n = -\dfrac{3}{5} \mathrel{\vert} n = \dfrac{7}{2}$

41. $16j^3 + 44j^2 - 126j = 0$
$2j(8j^2 + 22j - 63) = 0$
$2j(2j + 9)(4j - 7) = 0$
$2j = 0 \mathrel{\vert} 2j + 9 = 0 \mathrel{\vert} 4j - 7 = 0$
$j = 0 \mathrel{\vert} j = -\dfrac{9}{2} \mathrel{\vert} j = \dfrac{7}{4}$

42. $15k^3 - 114k^2 - 189k = 0$
$3k(5k^2 - 38k - 63) = 0$
$3k(5k + 7)(k - 9) = 0$
$3k = 0 \mathrel{\vert} 5k + 7 = 0 \mathrel{\vert} k - 9 = 0$
$k = 0 \mathrel{\vert} k = -\dfrac{7}{5} \mathrel{\vert} k = 9$

43. $30x^3 - 21x^2 - 135x = 0$
$3x(10x^2 - 7x - 45) = 0$
$3x(5x + 9)(2x - 5) = 0$
$3x = 0 \mathrel{\vert} 5x + 9 = 0 \mathrel{\vert} 2x - 5 = 0$
$x = 0 \mathrel{\vert} x = -\dfrac{9}{5} \mathrel{\vert} x = \dfrac{5}{2}$

44. $12b^3 - 86b^2 - 80b = 0$
$2b(6b^2 - 43b - 40) = 0$
$2b(6b + 5)(b - 8) = 0$
$2b = 0 \mathrel{\vert} 6b + 5 = 0 \mathrel{\vert} b - 8 = 0$
$b = 0 \mathrel{\vert} b = -\dfrac{5}{6} \mathrel{\vert} b = 8$

45. $72a^3 - 132a^2 + 108a^2 - 198a = 0$
$72a^3 - 24a^2 - 198a = 0$
$6a(12a^2 - 4a - 33) = 0$
$6a(2a + 3)(6a - 11) = 0$
$6a = 0 \mathrel{\vert} 2a + 3 = 0 \mathrel{\vert} 6a - 11 = 0$
$a = 0 \mathrel{\vert} a = -\dfrac{3}{2} \mathrel{\vert} a = \dfrac{11}{6}$

46. $120y^3 + 72y^2 - 140y^2 - 84y = 0$
$120y^3 - 68y^2 - 84y = 0$
$4y(30y^2 - 17y - 21) = 0$
$4y(5y + 3)(6y - 7) = 0$
$4y = 0 \mathrel{\vert} 5y + 3 = 0 \mathrel{\vert} 6y - 7 = 0$
$y = 0 \mathrel{\vert} y = -\dfrac{3}{5} \mathrel{\vert} y = \dfrac{7}{6}$

C 47. $t^2 - 4t + 4 + 7t - 14 + 12 = 0$
$t^2 + 3t + 2 = 0$
$(t + 2)(t + 1) = 0$
$t = -2 \mathrel{\vert} t = -1$

48. $y^2 + 8y + 16 + 3y + 12 - 10 = 0$
$y^2 + 11y + 18 = 0$
$(y + 9)(y + 2) = 0$
$y = -9 \mathrel{\vert} y = -2$

49. $a(a^4 - 10a^2 + 9) = 0$
$a(a^2 - 9)(a^2 - 1) = 0$
$a(a + 3)(a - 3)(a + 1)(a - 1) = 0$
$a = 0 \mathrel{\vert} a + 3 = 0 \mathrel{\vert} a - 3 = 0 \mathrel{\vert} a + 1 = 0 \mathrel{\vert} a - 1 = 0$
$\phantom{a = 0 \mathrel{\vert}} a = -3 \mathrel{\vert} a = 3 \mathrel{\vert} a = -1 \mathrel{\vert} a = 1$

50. $2b(b^4 - 50b^2 + 49) = 0$
$2b(b^2 - 49)(b^2 - 1) = 0$
$2b(b - 7)(b + 7)(b - 1)(b + 1) = 0$
$b = 0 \mid b = 7 \mid b = -7 \mid b = 1 \mid b = -1$

51. $(x + 3)[(x + 3)^2 + 2(x + 3) - 8] = 0$
$(x + 3)[x^2 + 6x + 9 + 2x + 6 - 8] = 0$
$(x + 3)[x^2 + 8x + 7] = 0$
$(x + 3)(x + 7)(x + 1) = 0$
$x = -3 \mid x = -7 \mid x = -1$

52. $(z - 3)[(z - 3)^2 + 9(z - 3) + 14] = 0$
$(z - 3)[z^2 - 6z + 9 + 9z - 27 + 14] = 0$
$(z - 3)[z^2 + 3z - 4] = 0$
$(z - 3)(z + 4)(z - 1) = 0$
$z = 3 \mid z = -4 \mid z = 1$

53. $n^2 + 5n = 24$ **54.** $n^2 - 3n = 18$ **55.** $2n^2 = n - 10$

page 303   Test Yourself

1. $2(y - 7)(y + 7)$ **2.** $5y(1 + 5y^2)$ **3.** $4a(a - 2)(a - 1)$

4. prime   **5.** $x(1 - y) + 2x(1 - y) = 3x(1 - y)$

6. $-(4x^2 - 12x + 9) = (3y)^2 - (2x - 3)^2 = [3y - (2x - 3)][3y + (2x - 3)] =$
$(3y - 2x + 3)(3y + 2x - 3)$

7. $16(x^2 + 4)$ **8.** $m(m - 4) = 0$ **9.** $(a - 10)(a + 1) = 0$
                                  $m = 0 \mid m = 4$ $a = 10 \mid a = -1$

10. $24b^2 - 32b + 12 = 9b$ **11.** $(2x + 3)(x - 4) = 0$
     $24b^2 - 41b + 12 = 0$ $2x + 3 = 0 \mid x - 4 = 0$
     $(8b - 3)(3b - 4) = 0$ $2x = -3 \mid$ $x = 4$
     $8b - 3 = 0 \mid 3b - 4 = 0$
             $b = \dfrac{3}{8} \mid$ $b = \dfrac{4}{3}$ $x = -\dfrac{3}{2} \mid$

page 305   Class Exercises

1. Given: length of the rectangular cloth is 3 in. more than the width and the area is 130 in.²; Find the width.

2. Area of rectangle = length × width

3.

w

$w + 3$

4. $w(w + 3) = 130$

5. $w^2 + 3w = 130$
$w^2 + 3w - 130 = 0$
$(w - 10)(w + 13) = 0$
$w = 10$ in.; $w + 3 =$
$10 + 3 = 13$ in.

6. If the solution set includes a negative number then it means that the width is negative, which cannot be the case.

page 306   Practice Exercises

A   **1.** Let $w =$ the width, then $w + 4 =$ the length of the rectangular garden.
$A = (w + 4)w$, $140 = (w + 4)w$, $140 = w^2 + 4w$, $0 = w^2 + 4w - 140$,
$(w + 14)(w - 10) = 0$
$w + 14 = 0 \quad | \quad w - 10 = 0$
$\qquad w = -14 \; | \qquad w = 10$
$w + 4 = 10 + 4 = 14$
The dimensions of the rectangular garden are 14 ft $\times$ 10 ft.

2. Let $w =$ the width, then $w + 3 =$ the length of the rectangular rug.
$A = (w + 3)w$, $180 = (w + 3)w$, $180 = w^2 + 3w$, $0 = w^2 + 3w - 180$,
$(w - 12)(w + 15) = 0$
$w - 12 = 0 \; | \; w + 15 = 0$
$\qquad w = 12 \; | \qquad w = -15$
$w + 3 = 12 + 3 = 15$
The dimensions of the rectangular rug are 15 ft $\times$ 12 ft.

3. Let $p =$ the perimeter, $A =$ the area, $l =$ the length and $w =$ the width of the rectangle.
$p = 2l + 2w$, $A = lw$
$40 = 2l + 2w$, $84 = lw$, $l = \dfrac{84}{w}$; $40 = 2\left(\dfrac{84}{w}\right) + 2w$,
$40w = 168 + 2w^2$, $2w^2 - 40w + 168 = 0$, $w^2 - 20w + 84 = 0$,
$(w - 14)(w - 6) = 0$
$w - 14 = 0 \; | \; w - 6 = 0$
$\qquad w = 14 \; | \qquad w = 6$
$l = \dfrac{84}{w}$; $l = \dfrac{84}{14} = 6$ or $l = \dfrac{84}{6} = 14$
The dimensions of a rectangular piece of tin are 14 cm $\times$ 6 cm.

4. Let $p =$ the perimeter, $A =$ the area, $l =$ the length, and $w =$ the width of the patio.

$p = 2l + 2w, A = lw$

$44 = 2l + 2w, 96 = lw, l = \dfrac{96}{w}; 44 = 2\left(\dfrac{96}{w}\right) + 2w,$

$44w = 192 + 2w^2, 2w^2 - 44w + 192 = 0, w^2 - 22w + 96 = 0,$

$(w - 16)(w - 6) = 0$

$w - 16 = 0 \quad w - 6 = 0$

$\qquad w = 16 \qquad w = 6$

$l = \dfrac{96}{w}; l = \dfrac{96}{16} = 6 \text{ or } l = \dfrac{96}{6} = 16$

The dimensions of the rectangular patio are 16m × 6m.

5. Let x = the first even integer, then $x + 2$ = the second consecutive even integer. The equation is: $x (x + 2) = 48$; solve $x^2 + 2x = 48$,

$x^2 + 2x - 48 = 0$

$(x + 8)(x - 6) = 0$

$x + 8 = 0 \quad x - 6 = 0$

$\qquad x = -8 \qquad x = 6$

Then $x + 2 = -6$ or $x + 2 = 8$.

The two consecutive even integers are $-8, -6$ or $6, 8$.

6. Let x = the first even integer, then $x + 2$ = the second consecutive even integer. The equation is: $x (x + 2) = 120$; solve $x^2 + 2x = 120$,

$x^2 + 2x - 120 = 0$

$(x + 12)(x - 10) = 0$

$x + 12 = 0 \quad x - 10 = 0$

$\qquad x = -12 \qquad x = 10$

Then $x + 2 = -10$ or $x + 2 = 12$.

The two consecutive even integers are $-12, -10$ or $10, 12$.

7. Let x = the first odd integer, then $x + 2$ = the second consecutive odd integer. The equation is: $x (x + 2) = 63$; solve $x^2 + 2x = 63$,

$x^2 + 2x - 63 = 0$

$(x + 9)(x - 7) = 0$

$x + 9 = 0 \quad x - 7 = 0$

$\qquad x = -9 \qquad x = 7$

Then $x + 2 = -7$ or $x + 2 = 9$.

The two consecutive odd integers are $-9, -7$ or $7, 9$.

8. Let x = the first odd integer, then $x + 2$ = the second consecutive odd integer. The equation is: $x (x + 2) = 143$; solve $x^2 + 2x = 143$,

$x^2 + 2x - 143 = 0$

$(x + 13)(x - 11) = 0$

$x + 13 = 0 \quad x - 11 = 0$

$\qquad x = -13 \qquad x = 11$

Then $x + 2 = -11$ or $x + 2 = 13$.
The two consecutive odd integers are $-13, -11$, or $11, 13$.

9. Let $x =$ the first even integer, then $x + 2 =$ the second consecutive even integer, and $x + 4 =$ the third consecutive even integer. The equation is:
$x(x + 4) = 7(x + 2) + 4$; solve $x^2 + 4x = 7x + 14 + 4$,
$x^2 + 4x = 7x + 18$, $x^2 - 3x - 18 = 0$
$(x - 6)(x + 3) = 0$
$x - 6 = 0 \mathrel{\vdots} x + 3 = 0$
$\quad x = 6 \mathrel{\vdots} \quad x = -3$
Since -3 is an odd integer, the least of the three consecutive even integers is 6.

B 10. Let $x =$ the first odd integer, then $x + 2 =$ the second consecutive odd integer, and $x + 4 =$ the third consecutive odd integer.
The equation is: $(x + 2)(x + 4) = 13(x) + 8$; solve $x^2 + 4x + 2x + 8 = 13x + 8$, $x^2 + 6x + 8 = 13x + 8$, $x^2 - 7x = 0$
$x(x - 7) = 0$
$x = 0 \mathrel{\vdots} x - 7 = 0$
$\quad \mathrel{\vdots} \quad x = 7$
Then $x + 2 = 9$ and $x + 4 = 11$.
The greatest consecutive odd integer is 11.

11. Let $x =$ the amount of increase/decrease. $(10 + x)(25 - x) = 216$,
$250 - 10x + 25x - x^2 = 216$, $x^2 - 15x - 34 = 0$
$(x - 17)(x + 2) = 0$
$x - 17 = 0 \mathrel{\vdots} x + 2 = 0$
$\quad x = 17 \mathrel{\vdots} \quad x = -2$
Then $10 + x = 27$ and $25 - x = 8$.
The dimensions of the garden are 27m $\times$ 8m.

12. Let $s =$ the side of the garden. $(s + 5)(s + 2) = 130$, $s^2 + 2s + 5s + 10 = 130$, $s^2 + 7s + 10 = 130$, $s^2 + 7s - 120 = 0$
$(s + 15)(s - 8) = 0$
$s + 15 = 0 \mathrel{\vdots} s - 8 = 0$
$\quad s = -15 \mathrel{\vdots} \quad s = 8$
Then $s + 5 = 8 + 5 = 13$ and $s + 2 = 8 + 2 = 10$.
The dimensions of the garden are 13 ft $\times$ 10 ft.

13. Let $s =$ the side of the table. $(s + 10)(s + 12) = 1680$, $s^2 + 22s + 120 = 1680$, $s^2 + 22s - 1560 = 0$
$(s + 52)(s - 30) = 0$
$s + 52 = 0 \mathrel{\vdots} s - 30 = 0$
$\quad s = -52 \mathrel{\vdots} \quad s = 30$
The dimensions of the table are 30 in. $\times$ 30 in.

14. Let s = the side of the table. $(s + 10)(s + 6) = 1085$, $s^2 + 16s + 60 = 1085$, $s^2 + 16s - 1025 = 0$

$(s - 25)(s + 41) = 0$

$s - 25 = 0 \mid s + 41 = 0$

$\quad\quad s = 25 \mid \quad\quad s = -41$

The dimensions of the table are 25 in. $\times$ 25 in.

C 15. Let x = width of the photograph, then $2x$ = length of the photograph.

$(2x + 2)(x + 2) = 60$, $2x^2 + 6x + 4 = 60$, $2x^2 + 6x - 56 = 0$,

$x^2 + 3x - 28 = 0$

$(x + 7)(x - 4) = 0$

$x + 7 = 0 \mid x - 4 = 0$

$\quad x = -7 \mid \quad x = 4$

Then $2x = 8$.

The dimensions of the photograph are 8 in. $\times$ 4 in.

16. Let w = width of the deck. $(25 + w + w)(10 + w + w) = 594$, $(25 + 2w)(10 + 2w) = 594$, $250 + 70w + 4w^2 = 594$, $4w^2 + 70w - 344 = 0$,

$2w^2 + 35w - 172 = 0$

$(2w + 43)(w - 4) = 0$

$2w + 43 = 0 \quad \mid w - 4 = 0$

$\quad\quad 2w = -43 \mid \quad\quad w = 4$

$\quad\quad w = -\dfrac{43}{2} \mid$

Then w = 4. The deck is 4 ft wide.

page 307   Mixed Problem Solving Review

1. Let w = the width, then $6w$ = the length of the rectangle.

$p = 2l + 2w$; then $70 = 2(6w) + 2(w)$, $70 = 12w + 2w$, $70 = 14w$,

$w = 5$; $l = 6w$, $l = 6(5)$, $l = 30$; The dimensions of the rectangle are 30 cm $\times$ 5 cm.

2. Let x = the amount of milliliters to be added. $(8\%)x + 10(20\%) = (x + 10)(12\%)$, $x(0.08) + 10(0.2) = (x + 10)(0.12)$, $0.08x + 2 = 0.12x + 1.2$, $0.04x = 0.8$, $x = 20$; Therefore, 20mL must be added to get a 12% solution.

3. Let t = the time it takes for them to meet. $64\left(t - \dfrac{45}{60}\right) + 56(t) = 228$,

$64t - 48 + 56t = 228$, $120t = 276$, $t = 2.3h$

Lauren and Erik will meet in 2.3h.

4. Let j = the weight of Joanne, then $j + 10$ = the weight of David. $j + j + 10 < 100$, $2j + 10 < 100$, $2j < 90$, $j < 45$; Joanne weighs less than 45 kg; $j + 10 = 55$, David weighs less than 55 kg.

page 307   Project

1. $h = vt - 5t^2$
$h = 45(6) - 5(6)^2$
$h = 270 - 180$
$h = 90$
The height will be 90 m.

2. $h = vt - 5t^2$
$50 = v(10) - 5(10)^2$
$50 = 10v - 500$
$10v = 550$
$v = 55$
The initial velocity should be 55 m/s.

3. $h = vt - 5t^2$
$0 = 50(t) - 5(t^2)$
$0 = 50t - 5t^2$
$5t^2 - 50t = 0$
$5t(t - 10) = 0$
$5t = 0 \mid t - 10 = 0$
$t = 0 \mid \quad t = 10$
The time will be 10 s.

4. $h = vt - 5t^2$
$0 < v(32) - 5(32)^2$
$0 < 32v - 5120$
$5120 < 32v$
$v > 160$
The velocity is greater than 160 m/s.

page 309   Integrating Algebra: Agriculture

1. $(5040)(0.8) = 4032$; 4032 bushels

2. number of bushels $= d^2(h)(0.625) = 16^2 12(0.625) = 1920$ bushels

3. Let $w =$ the width , then $2w =$ the length of the rectangular bin.
$450 = (2w)(w)$, $2w^2 = 450$, $w^2 = 225$, $w = 15$; $l = 2(15)$, $l = 30$;
$(l)(w)(h)(0.08) =$ number of bushels, $(30)(15)(14)(0.8) = 5040$; $l = 30$ ft;
$w = 15$ ft; 5040 bushels

4. Estimate: 2mi is about 10,000 ft.
$2l + 2w = 10,000$
$l = 5000 - w$
150 acres is about 6,000,000 ft², $l \cdot w = 6,000,000$; $(5000 - w)(w) =$
6,000,000; $w^2 - 5000w + 6,000,000 = 0$,
$(w - 3000)(w - 2000) = 0$
$w - 3000 = 0 \quad \mid w - 2000 = 0$
$w = 3000 \mid \quad w = 2000$
The dimensions of the field are 3000 ft × 2000ft.

5. Let $w =$ the width, then $4w =$ the length of the cornfield.   3 mi $= 15,840$
ft; $15,840 = 2(4w) + 2w$, $15,840 = 10w$, $w = 1584$; $l = 4(1584)$, $l =$
6336; Area $= (6336)(1584) = 10,036,224$ ft²; 1 acre $= 43,530$ ft²;
$10,036,224 \div 43,530 = 230$ acres; $(28,000)(230) = 6,400,000$; Approximately
6,400,000 plants can be grown in the cornfield.

6. Let w = the width of the cornfield. 2 mi = 10,560 ft, 10,560 = $2w$ + 1 mi, 1 mi = 5280, 10,560 = $2w$ + 5280, 5280 = $2w$, w = 2640; Area = (5280)(2640) = 13,939,200 ft^2; 1 acre = 43,530 ft^2; 13,939,200 ÷ 43,530 = 320.3; A farmer can enclose approximately 320 acres with 2 mi of fencing.

page 310–311   Summary and Review

1. prime   **2.** composite   **3.** composite   **4.** prime

5. 48 = 2 · 24
     = 2 · 2 · 12
     = 2 · 2 · 3 · 4
     = 2 · 2 · 3 · 2 · 2
     = 2^4 · 3

6. 300 = 3 · 100
     = 3 · 2 · 50
     = 3 · 2 · 2 · 25
     = 3 · 2 · 2 · 5 · 5
     = 2^2 · 3 · 5^2

7. 120 = 2 · 60
     = 2 · 2 · 30
     = 2 · 2 · 2 · 15
     = 2^3 · 3 · 5

8. 64 = 2 · 32
     = 2 · 2 · 16
     = 2 · 2 · 2 · 2 · 2 · 2
     = 2^6

9.   35 = 5 · 7
   200 = 2^3 · 5^2
   GCF is 5

10. $12x^2 = 2^2 · 3 · x^2$
   $16x^3y = 2^4 · x^3 · y$
   GCF is 2^2x^2 or $4x^2$

11. $3ab^2(1) - 3ab^2(3a) =$
$3ab^2(1 - 3a)$

12. $6m^2n(mn) + 6m^2n(2n) - 6m^2n(3) =$
$6m^2n(mn + 2n - 3)$

13. $(x + 3)(x + 2)$ **14.** $(y - 11)(y - 3)$ **15.** $(a + 11b)(a + b)$

16. $(c + 8)(c - 7)$ **17.** $(y - 12)(y + 2)$ **18.** $(x - 3y)(x + 5y)$

19. $(2m + 1)(m + 7)$ **20.** $(3x - 4)(x + 3)$ **21.** $(3x - 5)(2x + 3)$

22. $(w + 7)(w - 7)$ **23.** $4(25m^2 - 4n) =$
$4(5m - 2n)(5m + 2n)$ **24.** $(y + 1)(y + 1) =$
$(y + 1)^2$

25. $7(m^2 - 4) =$
$7(m + 2)(m - 2)$ **26.** $xy(x^2 - 9) =$
$xy(x + 3)(x - 3)$ **27.** $3a(2a^2 + 4a + 3)$

28. $(3w - 2)(z + 3)$ **29.** $9(x - y) - 4(x - y) =$
$(9 - 4)(x - y) =$
$5(x - y)$ **30.** $c(x - d) - y(x - d) =$
$(c - y)(x - d)$

31. $4x(9x^2 - 6x + 1) =$
$4x(3x - 1)^2$ **32.** $3(25x^2 - 1) =$
$3(5x - 1)(5x + 1)$ **33.** $9(2x^2 - 5)$

34. $m(m + 6) = 0$
$m = 0 \mathrel{\vert} m + 6 = 0$
          $m = -6$

35. $(r - 10)(r + 8) = 0$
$r - 10 = 0 \mathrel{\vert} r + 8 = 0$
$r = 10 \mathrel{\vert}$ $r = -8$

36. $t^2 - 3t - 18 = 0$
$(t - 6)(t + 3) = 0$
$t - 6 = 0 \mathrel{\vert} t + 3 = 0$
$t = 6 \mathrel{\vert}$ $t = -3$

37. Let x = the first even integer, then $x + 2$ = the second consecutive even integer. The equation is: $x(x + 2) = 728$; solve $x^2 + 2x - 728 = 0$,

$(x + 28)(x - 26) = 0$

$x + 28 = 0 \quad | \quad x - 26 = 0$

$\qquad x = -28 \; | \qquad x = 26$

Then $x + 2 = -26$ or $x + 2 = 28$.

The two consecutive even integers are $-28, -26$, or $26, 28$.

38. Let w = the width , then $w + 3$ = the length of the rug. $270 = w(w + 3)$,

$w^2 + 3w - 270 = 0$

$(w + 18)(w - 15) = 0$

$w + 18 = 0 \quad | \quad w - 15 = 0$

$\qquad w = -18 \; | \qquad w = 15$

Then $w + 3 = 18$;

The dimensions of the rug are 18 ft × 15 ft.

page 312　Chapter Test

1. composite

2. prime

3. prime

4. composite

5. $24 = 2 \cdot 12$
$\quad = 2 \cdot 2 \cdot 6$
$\quad = 2 \cdot 2 \cdot 2 \cdot 3$
$\quad = 2^3 \cdot 3$

6. $200 = 2 \cdot 100$
$\quad = 2 \cdot 2 \cdot 50$
$\quad = 2 \cdot 2 \cdot 2 \cdot 25$
$\quad = 2 \cdot 2 \cdot 2 \cdot 5 \cdot 5$
$\quad = 2^3 \cdot 5^2$

7. $34 = 2 \cdot 17$

8. $75 = 3 \cdot 25$
$\quad = 3 \cdot 5 \cdot 5$
$\quad = 3 \cdot 5^2$

9. $3(3a) + 3(2b) =$
$3(3a + 2b)$

10. $5(5) - 5(y^2) =$
$5(5 - y^2)$

11. $2ab^2(4ab) - 2ab^2(3) =$
$2ab^2(4ab - 3)$

12. $(a - 4)(a + 4)$

13. $(x - 5y)(x + 5y)$

14. $(y + 2)(y + 2) =$
$(y + 2)^2$

15. $(x + 5)(x + 1)$

16. $(n - 5)(n - 11)$

17. prime

18. $(x + 7y)(x - 3y)$

19. $(2m + 1)(m + 3)$

20. $2x(3x^2 - x - 10) =$
$2x(3x + 5)(x - 2)$

21. $(2d - 1)(f + 2)$

22. $5(a - b) + 2(a - b) =$
$(5 + 2)(a - b) =$
$7(a - b)$

23. $d(y - e) + z(e - y) =$
$d(y - e) - z(y - e) =$
$(y - e)(d - z)$

24. $(16x^2 - 8xy + y^2) - 9 =$
$(4x - y)^2 - 3^2 =$
$[(4x - y) - 3][(4x - y) + 3] =$
$(4x - y - 3)(4x - y + 3)$

25. $x - 2 = 0 \quad 3x + 1 = 0$
$\qquad x = 2 \quad\quad 3x = -1$
$$\qquad\qquad\qquad x = -\frac{1}{3}$$

26. $b(b + 5) = 0$
$\qquad b = 0 \mid b + 5 = 0$
$\qquad\qquad\qquad b = -5$

27. $(s - 4)(s - 2) = 0$
$\quad s - 4 = 0 \mid s - 2 = 0$
$\quad\quad s = 4 \mid \quad\quad s = 2$

28. $z^2 - 14z + 49 = 0$
$\quad (z - 7)(z - 7) = 0$
$\quad z - 7 = 0 \mid z - 7 = 0$
$\quad\quad z = 7 \mid \quad\quad z = 7$

29. Let x = the negative number, then $x + x^2 = 72$, $x^2 + x - 72 = 0$,
$(x + 9)(x - 8) = 0$
$x + 9 = 0 \mid x - 8 = 0$
$\quad x = -9 \mid \quad x = 8$
The negative number is -9.

30. Let x = the first odd integer, then $x + 2$ = the second consecutive odd
integer. The equation is: $x(x + 2) = 255$; solve $x^2 + 2x - 255 = 0$,
$(x + 17)(x - 15) = 0$
$x + 17 = 0 \mid x - 15 = 0$
$\quad x = -17 \mid \quad x = 15$
Then $x + 2 = -15$ or $x + 2 = 17$.
The two consecutive odd integers are $-17, -15$ or $15, 17$.

31. Let w = the width, then $3w$ = the length of the garden.
$(w + 1)(3w + 3) = 75$, $3w^2 + 6w - 72 = 0$,
$(3w - 12)(w + 6) = 0$

$3w - 12 = 0 \mid w + 6 = 0$
$\quad 3w = 12 \mid \quad w = -6$
$\quad\quad w = 4$
Then $w + 1 = 5$ and $3w + 3 = 15$.
The new dimensions of the rectangle are 15 ft $\times$ 5 ft.

Challenge   **a.** A: x^2; B: $5x$; C: $2x$; D: 10   **b.** $x^2 + 5x + 2x + 10$

         c. $x^2 + 7x + 10$ **d.** $(x + 5)(x + 2)$

page 313   **Preparing for Standardized Tests**

1. A; $\dfrac{3}{2} + \dfrac{4}{5} = \dfrac{15 + 8}{10} = \dfrac{23}{10}$ and $\dfrac{2}{3} + \dfrac{5}{4} = \dfrac{8 + 15}{12} = \dfrac{23}{12}; \dfrac{23}{10} > \dfrac{23}{12}$

2. A; $\dfrac{1}{3}$ of $276 = 92$ and $\dfrac{3}{5}$ of $150 = 90$; $92 > 90$

3. D; Since n can be any value between 12 and 42, it follows that the
average must satisfy $22 <$ average < 32.

4. B; $a^2b - ab = (-2)^2 \cdot 3 - (-2) \cdot 3 = 4 \cdot 3 - (-6) = 12 + 6 = 18$
$1 - ab^2 = 1 - (-2) \cdot 3^2 = 1 - (-2) \cdot 9 = 1 + 18 = 19$

5. A; $0.0064 = \dfrac{64}{10,000} = \dfrac{4}{625}; \dfrac{4}{625} > \dfrac{2}{625}$

6. B; If $n > 5$ and $t > n$, then $t > 5$.

7. A; 30% of \$650 equals \$195.00, and 25% of 765 equals \$191.25; \$195 > \$191.25

8. C; $3.4 \times 10^3 = 3400$

9. B; $|3(2 - 6) + 4| = |3(-4) + 4| = |-12 + 4| = |-8| = 8$
$|6 - 4(-2)| = |6 + 8| = 14$

10. D; If $x = -1$, then $x = \dfrac{1}{x}$, but otherwise they are unequal.

11. C; $(a + b)(a - b) = a^2 - b^2$

12. A; The unknown percentage is added to 48% leaving 52% for C grades. Thus, $x = 52$. $52 > 48$.

13. B; 8% of $35 = 0.08 \times 35 = 2.80$ which rounds to 3; $4 > 3$.

14. A; Since 8% A's plus 20% B's plus 52% C's gives a total of 80% having grades of C or better, a total of $0.80 \times 35 = 28$ of the 35 are in this group. $28 > 25$.

page 314   Mixed Review

1. $\dfrac{\overset{1}{\cancel{3}}}{\underset{2}{\cancel{4}}} \times \dfrac{\overset{1}{\cancel{2}}}{\underset{1}{\cancel{3}}} = \dfrac{1 \times 1}{2 \times 1} = \dfrac{1}{2}$

2. $\dfrac{\overset{1}{\cancel{2}}}{7} \times \dfrac{\overset{1}{\cancel{5}}}{\underset{1}{\cancel{10}}} = \dfrac{1 \times 1}{7 \times 1} = \dfrac{1}{7}$

3. $\dfrac{\overset{1}{\cancel{5}}}{\underset{2}{\cancel{8}}} \times \dfrac{\overset{1}{\cancel{4}}}{\underset{5}{\cancel{25}}} = \dfrac{1 \times 1}{2 \times 5} = \dfrac{1}{10}$

4. $\dfrac{3 \times 3}{4 \times 7} = \dfrac{9}{28}$

5. $\dfrac{\overset{2}{\cancel{4}}}{\underset{1}{\cancel{5}}} \times \dfrac{\overset{3}{\cancel{15}}}{\underset{7}{\cancel{14}}} = \dfrac{2 \times 3}{1 \times 7} = \dfrac{6}{7}$

6. $\dfrac{\overset{1}{\cancel{5}}}{\underset{4}{\cancel{8}}} \cdot \dfrac{\overset{5}{\cancel{10}}}{\underset{3}{\cancel{9}}} = \dfrac{1 \times 5}{4 \times 3} = \dfrac{5}{12}$

7. $\dfrac{5 \times 5}{7 \times 2} = \dfrac{25}{14}$

8. $\dfrac{\overset{1}{\cancel{5}}}{\underset{3}{\cancel{6}}} \times \dfrac{\overset{4}{\cancel{8}}}{\underset{1}{\cancel{5}}} = \dfrac{1 \times 4}{3 \times 1} = \dfrac{4}{3}$

9. $\dfrac{2}{3} = \dfrac{2}{3} \cdot \dfrac{3}{3} = \dfrac{6}{9}; \dfrac{5}{9}, \dfrac{6}{9}$

10. $\dfrac{6}{11} = \dfrac{6}{11} \cdot \dfrac{3}{3} = \dfrac{18}{33}$ and $\dfrac{1}{3} = \dfrac{1}{3} \cdot \dfrac{11}{11} = \dfrac{11}{33}; \dfrac{18}{33}, \dfrac{11}{33}$

11. $\dfrac{5}{18} = \dfrac{5}{18} \cdot \dfrac{2}{2} = \dfrac{10}{36}$ and $\dfrac{7}{12} = \dfrac{7}{12} \cdot \dfrac{3}{3} = \dfrac{21}{36}$; $\dfrac{10}{36}, \dfrac{21}{36}$

12. $\dfrac{3}{8} = \dfrac{3}{8} \cdot \dfrac{3}{3} = \dfrac{9}{24}$ and $\dfrac{5}{6} = \dfrac{5}{6} \cdot \dfrac{4}{4} = \dfrac{20}{24}$; $\dfrac{9}{24}, \dfrac{20}{24}$

13. $\dfrac{2}{3} \cdot \dfrac{8}{8} + \dfrac{3}{8} \cdot \dfrac{3}{3} = \dfrac{16}{24} + \dfrac{9}{24} = \dfrac{25}{24}$

14. $\dfrac{7}{12} - \dfrac{1}{6} = \dfrac{7}{12} - \dfrac{1}{6} \cdot \dfrac{2}{2} = \dfrac{7}{12} - \dfrac{2}{12} = \dfrac{5}{12}$

15. $\dfrac{3}{10} \cdot \dfrac{6}{6} + \dfrac{5}{12} \cdot \dfrac{5}{5} = \dfrac{18}{60} + \dfrac{25}{60} = \dfrac{43}{60}$

16. $\dfrac{3}{4} \cdot \dfrac{3}{3} - \dfrac{7}{12} = \dfrac{9}{12} - \dfrac{7}{12} = \dfrac{2}{12} = \dfrac{1}{6}$

17. $A = lw = (10)(4) = 40$ in.^{2}

18. $A = \dfrac{1}{2}bh = \dfrac{1}{2}(9)(5) = \dfrac{1}{2}(45) = 22\dfrac{1}{2}$ in.^{2}

19. $A = s^2 = (3.5)^2 = 12.25$ cm^2

20. $A = \dfrac{1}{2} h (b_1 + b_2) = \dfrac{1}{2}(7)(8 + 11) =$
$\dfrac{1}{2}(7)(19) = 66.5$ cm^2

Chapter 8 Rational Expressions

1. $a(a + 3) = 0$; $a = 0$; $a = -3$

2. $b(2b - 3) = 0$; $b = 0$; $b = \dfrac{3}{2}$

3. $c(5c - 1) = 0$; $c = 0$; $c = \dfrac{1}{5}$

4. $(x - 6)(x - 1)$; $x = 6$; $x = 1$

5. $(2y - 3)(y + 2) = 0$; $y = \dfrac{3}{2}$; $y = -2$

6. $(3z - 5)(z + 4) = 0$; $z = \dfrac{5}{3}$; $z = -4$

page 318 Class Exercises

1. $d - 2 = 0$, $d = 2$

2. $2x + 1 = 0$, $x = -\dfrac{1}{2}$

3. $(m - 5)(3m - 2) = 0$; $m = 5$; $m = \dfrac{2}{3}$

4. $(y - 4)(y + 3) = 0$; $y = 4$; $y = -3$

5. $\dfrac{15b^3}{25b^2} = \dfrac{3 \cdot 5 \cdot b^2 \cdot b}{5 \cdot 5 \cdot b^2} = \dfrac{3b}{5}$; $b \neq 0$

6. $\dfrac{18m^3n^2}{6m^2n^3} = \dfrac{3 \cdot 6 \cdot m^2 \cdot m \cdot n^2}{6 \cdot m^2 \cdot n^2 \cdot n} = \dfrac{3m}{n}$; $m \neq 0$, $n \neq 0$

7. $\dfrac{4(d - 5)}{12} = \dfrac{d - 5}{3}$

8. $\dfrac{3(x + 4)}{4 + x} = 3$; $x \neq -4$

9. $\dfrac{6(2c + 3)}{3(3c + 2)} = \dfrac{2(2c + 3)}{3c + 2} = \dfrac{4c + 6}{3c + 2}$; $c \neq -\dfrac{2}{3}$

10. $\dfrac{-1(z - 3)}{(z + 5)(z - 3)} = -\dfrac{1}{z + 5}$; $z \neq -5, 3$

11. $\dfrac{3(a - 3)}{(a - 3)(a + 2)} = \dfrac{3}{a + 2}$; $a \neq -2, 3$

12. $\dfrac{(4 + y)(4 - y)}{(y - 4)(y - 3)} = \dfrac{-1(4 + y)(y - 4)}{(y - 4)(y - 3)} = -\dfrac{y + 4}{y - 3}$; $y \neq 3, 4$

pages 318–319 Practice Exercises

A 1. $m + 4 = 0$; $m = -4$

2. $a + 7 = 0$; $a = -7$

3. $10x + 4 = 0$; $x = -\dfrac{2}{5}$

4. $8n - 6 = 0$; $n = \dfrac{3}{4}$

5. $(3x + 1)(3x - 1) = 0$; $x = -\dfrac{1}{3}$, $x = \dfrac{1}{3}$

6. $(4m + 1)(4m - 1) = 0$; $m = -\dfrac{1}{4}$, $m = \dfrac{1}{4}$

7. $(b + 3)(b - 3) = 0$; $b = -3$, $b = 3$

8. $(y + 4)(y - 4) = 0$; $y = -4$, $y = 4$

9. $\dfrac{3c}{12c^2} = \dfrac{3c}{3 \cdot 4 \cdot c \cdot c} = \dfrac{1}{4c}$; $c \neq 0$

10. $\dfrac{4x}{28x^2} = \dfrac{4x}{4 \cdot 7 \cdot x \cdot x} = \dfrac{1}{7x}$; $x \neq 0$

11. $\dfrac{3(2a + 3)}{12} = \dfrac{2a + 3}{4}$

12. $\dfrac{6(4z + 3)}{36} = \dfrac{4z + 3}{6}$

13. $\dfrac{7(a - 2)}{a - 2} = 7$; $a \neq 2$

14. $\dfrac{5(r - 3)}{r - 3} = 5$; $r \neq 3$

15. $\dfrac{-1(2m - 5)}{3(2m - 5)} = -\dfrac{1}{3}$; $m \neq \dfrac{5}{2}$

16. $-\dfrac{2(12 - p)}{4(12 - p)} = -\dfrac{1}{2}$; $p \neq 12$

17. $\dfrac{3 + x}{(2x - 1)(x + 3)} = \dfrac{1}{2x - 1}$; $x \neq \dfrac{1}{2}, -3$

18. $\dfrac{2 + x}{(3x - 4)(x + 2)} = \dfrac{1}{3x - 4}$; $x \neq \dfrac{4}{3}, -2$

19. $\dfrac{x + 2}{(5x - 3)(x + 2)} = \dfrac{1}{5x - 3}$; $x \neq \dfrac{3}{5}, -2$

20. $\dfrac{(k + 6)(k - 2)}{(2 + k)(2 - k)} = \dfrac{(k + 6)(k - 2)}{(-1)(k + 2)(k - 2)} = -\dfrac{k + 6}{k + 2}$; $k \neq -2, 2$

21. $\dfrac{(c + 6)(c - 3)}{(3 + c)(3 - c)} = \dfrac{(c + 6)(c - 3)}{(-1)(c + 3)(c - 3)} = -\dfrac{c + 6}{c + 3}$; $c \neq -3, 3$

22. $\dfrac{(a + 4)(a - 2)}{(4 - a)(4 + a)} = \dfrac{a - 2}{4 - a}$; $a \neq -4, 4$

23. $\dfrac{(y + 2)(y + 3)}{(y + 2)(y - 2)} = \dfrac{y + 3}{y - 2}$; $y \neq -2, 2$

24. $\dfrac{(h + 4)(h - 3)}{(h + 3)(h - 3)} = \dfrac{h + 4}{h + 3}$; $h \neq -3, 3$

25. $\dfrac{(m + 5)(m - 4)}{(m + 5)(m - 5)} = \dfrac{m - 4}{m - 5}$; $m \neq 5, -5$

B 26. $\dfrac{18xy^3}{24x^3y^2} = \dfrac{6 \cdot 3 \cdot x \cdot y^2 \cdot y}{6 \cdot 4 \cdot x^2 \cdot x \cdot y^2} = \dfrac{3y}{4x^2}$; $24x^3y^2 = 0$; $x = 0$; $y = 0$

27. $\dfrac{25a^2c^5}{15a^4c^3} = \dfrac{5 \cdot 5 \cdot a^2 \cdot c^3 \cdot c^2}{5 \cdot 3 \cdot a^2 \cdot a^2 \cdot c^2 \cdot c} = \dfrac{5c^2}{3a^2}$; $15a^4c^3 = 0$; $a = 0$; $c = 0$

28. $\dfrac{32a^3}{8a(2a - 1)} = \dfrac{4a^2}{2a - 1}$; $8a(2a - 1) = 0$; $a = 0$; $a = \dfrac{1}{2}$

29. $\dfrac{4m^2(3 + 2n)}{28m^5n^3} = \dfrac{3 + 2n}{7m^3n^3}$; $28m^5n^3 = 0$; $m = 0$; $n = 0$

30. $\dfrac{(2r - 5)(2r + 3)}{(2r - 5)(r - 4)} = \dfrac{2r + 3}{r - 4}$; $(2r - 5)(r - 4) = 0$; $r = 2\dfrac{1}{2}$; $r = 4$

31. $\dfrac{(7z + 2)(z + 3)}{(z + 3)(z - 1)} = \dfrac{7z + 2}{z - 1}$; $(z + 3)(z - 1) = 0$; $z = 1$; $z = -3$

32. $\dfrac{(2r - 1)(r + 5)}{(r + 5)(r + 5)} = \dfrac{2r - 1}{r + 5}; r + 5 = 0; r = -5$

33. $\dfrac{(2c + 3)(2c + 3)}{(2c + 3)(c - 7)} = \dfrac{2c + 3}{c - 7}; (2c + 3)(c - 7) = 0; c = -1\frac{1}{2}; c = 7$

34. $\dfrac{-1(3s^2 + s - 4)}{6s^2 - s - 2} = \dfrac{-1(3s + 4)(s - 1)}{(2s + 1)(3s - 2)}; (2s + 1)(3s - 2) = 0; s = -\frac{1}{2}; s = \frac{2}{3}$

35. $\dfrac{(2a - 1)(2a + 5)}{(5 - 2a)(3 + a)}; (5 - 2a)(3 + a) = 0; a = 2\frac{1}{2}; a = -3$

36. $\dfrac{(4 + 3g)(4 + g)}{(g - 7)(g + 4)} = \dfrac{4 + 3g}{g - 7}; (g - 7)(g + 4) = 0; g = 7; g = -4$

37. $\dfrac{(9 + 2x)(1 + x)}{(x - 11)(x + 1)} = \dfrac{9 + 2x}{x - 11}; (x - 11)(x + 1) = 0; x = 11; x = -1$

38. $\dfrac{(5z - 4)(z + 2)}{(3z - 1)(z + 2)} = \dfrac{5z - 4}{3z - 1}; (3z - 1)(z + 2) = 0; z = \frac{1}{3}; z = -2$

39. $\dfrac{(3r - 5)(2r - 1)}{(3r + 1)(2r - 1)} = \dfrac{3r - 5}{3r + 1}; (3r + 1)(2r - 1) = 0; r = -\frac{1}{3}; r = \frac{1}{2}$

40. $\dfrac{(5 + 3c)(2 - c)}{(5c + 4)(c - 2)} = \dfrac{-1(5 + 3c)}{5c + 4} = \dfrac{-5 - 3c}{5c + 4}; (5c + 4)(c - 2) = 0; c = 2; c = -\frac{4}{5}$

C 41. $\dfrac{(a - 3b)(a - 2b)}{(a + 4b)(a - 2b)} = \dfrac{a - 3b}{a + 4b}; (a + 4b)(a - 2b) = 0; a = -4b; a = 2b$

42. $\dfrac{(c - 4d)(c - 2d)}{(c + 6d)(c - 2d)} = \dfrac{c - 4d}{c + 6d}; (c + 6d)(c - 2d) = 0; c = -6d; c = 2d$

43. $\dfrac{(x - y)(x + y)}{(x + 3y)(x + y)} = \dfrac{x - y}{x + 3y}; (x + 3y)(x + y) = 0; x = -3y; x = -y$

44. $\dfrac{(m + n)(m - n)}{(m + n)(m + 10n)} = \dfrac{m - n}{m + 10n}; (m + n)(m + 10n) = 0; m = -n; m = -10n$

45. $\dfrac{(3r - 4s)(3r + 4s)}{(3r - 4s)(2r - s)} = \dfrac{3r + 4s}{2r - s}; (3r - 4s)(2r - s) = 0; r = \frac{4s}{3}; r = \frac{s}{2}$

46. $\dfrac{(6g - 7h)(6g + 7h)}{(3g - 2h)(6g + 7h)} = \dfrac{6g - 7h}{3g - 2h}; (3g - 2h)(6g + 7h) = 0; g = \frac{2h}{3}; g = -\frac{7h}{6}$

47. $\dfrac{6s^2}{s^3} = \dfrac{6}{s}$

48. $\dfrac{2\pi r^2}{\frac{2}{3}\pi r^3} = \dfrac{2\pi r^2 \cdot 3}{2\pi r^3} = \dfrac{3}{r}$

49. sometimes true

50. sometimes true

51. never true

page 320 Capsule Review

1. $\dfrac{5}{m}$

2. $\dfrac{3y}{x}$

3. $\dfrac{3(m+2)}{2(m+2)} = \dfrac{3}{2}$

4. $\dfrac{2x-3}{2-x}$

page 322 Class Exercises

1. $\dfrac{5}{18}$

2. $\dfrac{\overset{2}{\cancel{4}}}{7} \cdot \dfrac{\overset{3}{\cancel{21}}}{\underset{3}{\cancel{6}}} = 2$
$\quad\underset{1}{}$

3. $\dfrac{m^2}{12}$

4. $\dfrac{\overset{1}{\cancel{a}}}{\underset{1}{\cancel{b}}} \cdot \dfrac{\overset{1}{\cancel{b}}}{\underset{-1}{-\cancel{a}}} = -1$

5. $\dfrac{\overset{1}{\cancel{x\!\!\not{v}}}}{\underset{\underset{1}{\cancel{a}}}{\cancel{x^2}}} \cdot \dfrac{\overset{1}{\cancel{x\!v^2}}}{\cancel{\not{v}}} = v^2$

6. $\dfrac{3\overset{b}{\cancel{b^2}}}{\underset{1}{\cancel{a}}} \cdot \dfrac{2\overset{a}{\cancel{a^2}}}{\underset{1}{\cancel{b}}} = 6ab$

7. $\dfrac{8a\overset{}{\cancel{b^2}}}{\underset{1}{\cancel{c^2}}} \cdot \dfrac{3a\overset{c}{\cancel{c^3}}}{\underset{2\;\; b}{\cancel{10b^3}}} = \dfrac{3a^2c}{2b}$

8. $\dfrac{\cancel{y}(x+1)}{\cancel{xy}} \cdot \dfrac{\cancel{x}}{y} = \dfrac{x+1}{y}$

9. $\dfrac{\overset{1}{s(s+3)}}{\underset{1}{6(s+2)}} \cdot \dfrac{\overset{1}{2(s+2)}}{\underset{1}{(s+3)}} = \dfrac{s}{3}$

10. $\dfrac{\overset{1}{(t+3)(t+2)}}{t-3} \cdot \dfrac{\overset{1}{(t-3)}\overset{1}{(t+1)}}{\underset{1}{(t+2)}\underset{1}{(t+1)}} = t+3$

pages 322–323 Practice Exercises

A 1. $\dfrac{6}{5x^2y}$

2. $\dfrac{12}{7a^2b}$

3. $\dfrac{12}{5x^2y}$

4. $\dfrac{21}{4ab^2}$

5. $\dfrac{30x^3y}{6xy^3} = \dfrac{5x^2}{y^2}$

6. $\dfrac{40ab^3}{20a^3b} = \dfrac{2b^2}{a^2}$

7. $\dfrac{12r^2st}{10s^2t^2} = \dfrac{6r^2}{5st}$

8. $\dfrac{90xy^2z}{15x^2z^2} = \dfrac{6y^2}{xz}$

9. $\dfrac{6q}{r^2} \cdot \dfrac{q}{p} = \dfrac{6q^2}{pr^2}$

10. $\dfrac{8b}{3c^2} \cdot \dfrac{b}{a} = \dfrac{8b^2}{3ac^2}$

11. $\dfrac{4m^2n}{1} \cdot \dfrac{6n^2p}{1} = 24m^2n^3p$

12. $\dfrac{(m-2)(2)(m+3)}{(3)(m+3)(2)(m-2)} = \dfrac{1}{3}$

13. $\dfrac{(x-5)(3)(2x+3)}{2(2x+3)(3)(x-5)} = \dfrac{1}{2}$

14. $\dfrac{(n+3)(6)(n-4)}{2(n-4)(2n+1)} = \dfrac{3(n+3)}{2n+1} = \dfrac{3n+9}{2n+1}$

15. $\dfrac{2(c+2)(c-5)}{2(3c-4)(c+2)} = \dfrac{c-5}{3c-4}$

16. $\dfrac{(r+4)(3)(r+9)}{2(r+9)(r-7)} = \dfrac{3(r+4)}{2(r-7)} = \dfrac{3r+12}{2r-14}$

17. $\dfrac{(b+2)(b+12)}{3(b+12)(2b-3)} = \dfrac{b+2}{3(2b-3)} = \dfrac{b+2}{6b-9}$

18. $\dfrac{2(x^2 - x - 2)}{8x} \cdot \dfrac{-8(x + 2)}{(x + 2)(x - 2)} = \dfrac{2(x - 2)(x + 1)(-8)(x + 2)}{8x(x + 2)(x - 2)} = -\dfrac{2(x + 1)}{x} = -\dfrac{2x + 2}{x}$

19. $\dfrac{(3r + 2)(r - 4)}{2r} \cdot \dfrac{-2(r + 4)}{(r + 4)(r - 4)} = \dfrac{-2(3r + 2)(r - 4)(r + 4)}{2r(r + 4)(r - 4)} = -\dfrac{3r + 2}{r}$

20. $\dfrac{(x - 5)(x + 1)}{(5 - x)(5 + x)} \cdot \dfrac{(x + 5)(x - 3)}{(x - 3)(x + 1)} = \dfrac{(x - 5)(x + 1)(x + 5)(x - 3)}{(-1)(x - 5)(x + 5)(x - 3)(x + 1)} = -1$

21. $\dfrac{(2z - 3)(z + 2)}{(2 - z)(z + 2)} \cdot \dfrac{(z + 5)(z - 2)}{(2z - 3)(z + 5)} = \dfrac{(2z - 3)(z + 2)(z + 5)(z - 2)}{(-1)(z - 2)(z + 2)(2z - 3)(z + 5)} = -1$

B 22. $\dfrac{32 \cdot 28 \cdot a^5 b^3 c^3 d^2}{7 \cdot 24 \cdot a^2 b^5 c^2 d^3} = \dfrac{4 \cdot 4 \cdot a^3 c}{1 \cdot 3 \cdot b^2 d} = \dfrac{16a^3 c}{3b^2 d}$ **23.** $\dfrac{45 \cdot 24 \cdot q^2 r^3 s^5 t^2}{8 \cdot 18 \cdot q^4 r s^3 t^5} = \dfrac{5 \cdot 3 \cdot r^2 s^2}{1 \cdot 2 \cdot q^2 t^3} = \dfrac{15r^2 s^2}{2q^2 t^3}$

24. $\dfrac{7t(t - 4)}{(2t + 3)(t - 4)} \cdot \dfrac{(2t + 3)(3t - 5)}{49t^3} = \dfrac{3t - 5}{7t^2}$

25. $\dfrac{6i(3i - 1)}{3i^4} \cdot \dfrac{(5i + 4)(2i - 3)}{(3i - 1)(5i + 4)} = \dfrac{2(2i - 3)}{i^3} = \dfrac{4i - 6}{i^3}$

26. $\dfrac{(2\theta + 1)(\theta + 3)}{(3\theta + 5)(\theta + 3)} \cdot \dfrac{(2\theta + 3)(3\theta + 5)}{(2\theta + 1)(2\theta + 3)} = 1$ **27.** $\dfrac{(5n + 2)(3n + 2)}{(5n + 2)(4n + 7)} \cdot \dfrac{(4n + 7)(2n + 9)}{(3n + 2)(2n + 9)} = 1$

28. $\dfrac{(a + 5)(9a - 2)}{(3a + 2)(9a - 2)} \cdot \dfrac{(3a + 2)(2a - 5)}{(5a + 4)(a + 5)} = \dfrac{2a - 5}{5a + 4}$

29. $\dfrac{(10l + 3)(l - 7)}{(2l - 9)(10l + 3)} \cdot \dfrac{(3l - 7)(l + 5)}{(3l - 7)(l - 7)} = \dfrac{l + 5}{2l - 9}$

30. $\dfrac{(5a + 4)(2a - 3)}{(3 - 2a)(3 + 2a)} \cdot \dfrac{(2a + 3)(a - 2)}{(7a + 4)(a - 2)} = \dfrac{-1(5a + 4)}{7a + 4} = -\dfrac{5a + 4}{7a + 4}$

31. $\dfrac{(7l - 2)(l + 5)}{(3 - 4l)(5 + l)} \cdot \dfrac{(7l + 3)(4l - 3)}{(7l - 2)(7l + 3)} = \dfrac{(4l - 3)}{-1(4l - 3)} = -1$

32. $\dfrac{(3w + 1)(2w - 7)}{(2w - 7)(w + 5)} \cdot \dfrac{(3 - 8w)(5 + w)}{(8w - 3)(3w + 1)} = \dfrac{-1(8w - 3)}{8w - 3} = -1$

33. $\dfrac{(2a + 1)(a + 2)}{(5 - 3a)(2 + 7a)} \cdot \dfrac{(3a - 5)(a + 4)}{(a + 4)(a + 2)} = \dfrac{-1(2a + 1)}{(2 + 7a)} = -\dfrac{2a + 1}{2 + 7a}$

34. $\dfrac{2(2y^2 + 7y + 3)}{3(6y^2 + 23y + 7)} \cdot \dfrac{6(2y^2 + 17y + 35)}{y^2 + 4y + 3} = \dfrac{12(2y + 1)(y + 3)(2y + 7)(y + 5)}{3(2y + 7)(3y + 1)(y + 3)(y + 1)} =$

$\dfrac{4(2y + 1)(y + 5)}{(3y + 1)(y + 1)} = \dfrac{8y^2 + 44y + 20}{3y^2 + 4y + 1}$

35. $\dfrac{2(s - 1)(2s + 1)}{(3s - 5)(s + 2)} \cdot \dfrac{(3s - 5)(s - 4)}{4(2s + 1)(s + 3)} = \dfrac{(s - 1)(s - 4)}{2(s + 2)(s + 3)} = \dfrac{s^2 - 5s + 4}{2s^2 + 10s + 12}$

C 36. $\dfrac{(f-r)(f+r)}{(f+2r)(f-r)} \cdot \dfrac{(f+2r)(f+r)}{(f+r)(f+r)} = 1$

37. $\dfrac{(a+2e)(a+3e)}{(2a+e)(a+3e)} \cdot \dfrac{(2a+e)(a+5e)}{(a+5e)(a+2e)} = 1$

38. $\dfrac{(2c+3d)(c-d)}{(2c+5d)(c-7d)} \cdot \dfrac{(2c+5d)(c+d)}{(2c+3d)(c+d)} = \dfrac{c-d}{c-7d}$

39. $\dfrac{(3t+u)(2t+u)}{(2t+u)(t+u)} \cdot \dfrac{(7t-3u)(t+u)}{(3t+u)(2t+9u)} = \dfrac{7t-3u}{2t+9u}$

40. $\dfrac{(2\theta+7c)(\theta+c)}{(3\theta-c)(\theta+c)} \cdot \dfrac{(3\theta+c)(\theta+4c)}{(2\theta+7c)(\theta+3c)} = \dfrac{(\theta+4c)(3\theta+c)}{(3\theta-c)(\theta+3c)} = \dfrac{30\theta^2+13\theta c+4c^2}{3\theta^2+8\theta c-3c^2}$

41. $\dfrac{(3r+2e)(r-2e)}{(2r-3e)(2r+e)} \cdot \dfrac{(2r-3e)(r+2e)}{(3r+2e)(2r-e)} = \dfrac{(r-2e)(r+2e)}{(2r+e)(2r-e)} = \dfrac{r^2-4e^2}{4r^2-e^2}$

42. $\dfrac{x-2}{(x+7)(x-5)} \cdot \dfrac{3x+2}{4} \cdot \dfrac{x-5}{3x+2} = \dfrac{x-2}{4(x+7)} = \dfrac{x-2}{4x+28}$

43. $\dfrac{(3x-1)(x+3)}{(2x-1)(x+3)} \cdot \dfrac{2(2x-3)}{3x-1} \cdot \dfrac{2x-1}{(2x-3)(x+5)} = \dfrac{2}{x+5}$

44. $7y = 21,\ y = 3$

45. $-9 = x - 4,\ -5 = x,\ x = -5$

46. $15 - 3n = 27,\ -3n = 12,\ n = -4$

47. $-6x - 10 = 0,\ -6x = 10,\ x = -\dfrac{5}{3}$

48. $12 - 9x = -7x,\ 12 = 2x,\ 6 = x,\ x = 6$

49. $8n + 4 = 4,\ 8n = 0,\ n = 0$

50. $2a - 10 + 3a = -3a + 35 - a,\ 5a - 10 = -4a + 35,\ 9a - 10 = 35,\ 9a = 45,\ a = 5$

51. $t - t - 13 = 9t + 15 + 5t,\ -13 = 14t + 15,\ -28 = 14t,\ -2 = t,\ t = -2$

52. Let x, $x + 1$ and $x + 2$ represent the consecutive integers. $x + x + 1 + x + 2 = 60$, $3x + 3 = 60$, $3x = 57$, $x = 19$; The integers are 19, 20, and 21.

53. Let x represent the first even integer, then $x + 2$ and $x + 4$ represent the next two consecutive even integers. $x + x + 4 = -20$, $2x + 4 = -20$, $2x = -24$, $x = -12$; The integers are -12, -10, and -8.

54. An example: To multiply *rational expressions*, first *factor* each of the *polynomials*. Next, *reduce* by any factors common to a numerator and a denominator. *Multiply* the remaining factors, and *simplify*.

page 324   Capsule Review

1. $\dfrac{2}{3} \cdot \dfrac{5}{4} = \dfrac{5}{6}$

2. $\dfrac{6}{1} \cdot \dfrac{8}{3} = 16$

3. $\dfrac{3}{4} \cdot \dfrac{1}{2} = \dfrac{3}{8}$

4. $\dfrac{7}{2} \cdot \dfrac{1}{4} = \dfrac{7}{8}$

page 325   Class Exercises

1. $\dfrac{3}{4} \cdot \dfrac{8}{9} = \dfrac{2}{3}$

2. $\dfrac{\overset{1}{3mh}}{\underset{1}{7}} \cdot \dfrac{\overset{3}{21}mh^{\overset{1}{3}}}{2m^{\underset{m}{\underset{mh^2}{3}}}} = \dfrac{9}{2m}$

3. $\dfrac{d}{e} \cdot \dfrac{d^3}{e^2} = \dfrac{d^4}{e^3}$

4. $\dfrac{\overset{1}{\cancel{a^2}}\overset{1}{b}}{2\cancel{a}} \cdot \dfrac{1}{\cancel{ab^2}} = \dfrac{1}{2b}$
   $\quad{}_{1}\quad{}_{1\;b}$

5. $\dfrac{\overset{1}{\cancel{a-2}}}{ab} \cdot \dfrac{\overset{1}{\cancel{a}}}{\cancel{a-2}} = \dfrac{1}{b}$
   $\quad{}_{1}\qquad{}_{1}$

6. $\dfrac{\overset{1}{x-3}}{\cancel{6}} \cdot \dfrac{\overset{1}{\cancel{2}}}{-1(\cancel{x-3})} = -\dfrac{1}{3}$
   $\quad{}_{3}\qquad\quad{}_{1}$

7. $\dfrac{y+3}{y+2} \cdot \dfrac{1}{y+2} = \dfrac{y+3}{(y+2)(y+2)}$

8. $\dfrac{(\overset{1}{\cancel{x+2}})(x+1)}{(\cancel{x-3})(x-1)} \cdot \dfrac{\overset{1}{x-3}}{\cancel{x+2}} = \dfrac{x+1}{x-1}$
   $\quad{}_{1}\qquad\qquad{}_{1}$

9. $\dfrac{p}{\cancel{q}} \cdot \dfrac{\overset{1}{\cancel{q}}}{r} \cdot \dfrac{p}{q} = \dfrac{p^2}{rq}$
   $\;{}_{1}$

pages 326–327   Practice Exercises

A 1. $\dfrac{15}{a} \cdot \dfrac{2a}{3b} = \dfrac{5}{1} \cdot \dfrac{2}{b} = \dfrac{10}{b}$

2. $\dfrac{9}{x} \cdot \dfrac{4x}{3y} = \dfrac{3}{1} \cdot \dfrac{4}{y} = \dfrac{12}{y}$

3. $\dfrac{6}{2c} \cdot \dfrac{4c}{3d} = \dfrac{2}{1} \cdot \dfrac{2}{d} = \dfrac{4}{d}$

4. $\dfrac{r^2}{7s^2} \cdot \dfrac{28s}{3r} = \dfrac{r}{s} \cdot \dfrac{4}{3} = \dfrac{4r}{3s}$

5. $\dfrac{d^2}{3e^2} \cdot \dfrac{2e}{4d} = \dfrac{d}{3e} \cdot \dfrac{1}{2} = \dfrac{d}{6e}$

6. $\dfrac{15a^2}{4b^2} \cdot \dfrac{2b}{5a} = \dfrac{3a}{2b} \cdot \dfrac{1}{1} = \dfrac{3a}{2b}$

7. $\dfrac{3(z-17)}{2z+5} \cdot \dfrac{1}{z-17} = \dfrac{3}{2z+5}$

8. $\dfrac{11(k+11)}{7k-15} \cdot \dfrac{1}{k+11} = \dfrac{11}{7k-15}$

9. $\dfrac{(x+11)(x-1)}{(x+11)(x+1)} \cdot \dfrac{1}{x-1} = \dfrac{1}{x+1}$

10. $\dfrac{(z+5)(z-3)}{(z+5)(z+4)} \cdot \dfrac{1}{z-3} = \dfrac{1}{z+4}$

11. $\dfrac{3(r-7)}{5(r+3)} \cdot \dfrac{7(r+3)}{3(r+2)} = \dfrac{7(r-7)}{5(r+2)}$

12. $\dfrac{5(a+2)}{2(a-10)} \cdot \dfrac{2(7a-10)}{7(a+2)} = \dfrac{5(7a-10)}{7(a-10)}$

13. $\dfrac{(x+5)(x+2)}{(x+6)(x-6)} \cdot \dfrac{x-6}{x+5} = \dfrac{x+2}{x+6}$

14. $\dfrac{(a+3)(a-3)}{(a-6)(a+4)} \cdot \dfrac{a-6}{a-3} = \dfrac{a+3}{a+4}$

15. $\dfrac{2a^2}{3b} \cdot \dfrac{b}{a} \cdot \dfrac{2a}{5b^2} = \dfrac{4a^3b}{15ab^3} = \dfrac{4a^2}{15b^2}$

16. $\dfrac{r^4s}{3} \cdot \dfrac{s^2}{r^3} \cdot \dfrac{4}{3s} = \dfrac{4r^4s^3}{9r^3s} = \dfrac{4rs^2}{9}$

17. $\dfrac{b-a}{2b+a} \cdot \dfrac{a+2b}{b+a} \cdot \dfrac{b+a}{b-a} = 1$

18. $\dfrac{2r+s}{3s-1} \cdot \dfrac{-1(3s-1)}{s-3} \cdot \dfrac{s-3}{s+2r} = -1$

B 19. $\dfrac{18a^3c^2}{25b^2} \cdot \dfrac{5b}{12a^2c} = \dfrac{18 \cdot 5a^3bc^2}{25 \cdot 12a^2b^2c} = \dfrac{3ac}{10b}$

20. $\dfrac{24x^5y^3}{18z^2} \cdot \dfrac{12z}{15x^2y} = \dfrac{24 \cdot 12x^5y^3z}{18 \cdot 15x^2yz^2} = \dfrac{16x^3y^2}{15z}$

21. $\dfrac{5x^2}{(y+6)(y-6)} \cdot \dfrac{(y-6)(y-1)}{25x(y-1)} = \dfrac{x}{5(y+6)}$

22. $\dfrac{4a^3}{(b-2)(b+2)} \cdot \dfrac{(b-3)(b+2)}{6a(b-3)} = \dfrac{2a^2}{3(b-2)}$

23. $\dfrac{-1(c-4)(c+4)}{1} \cdot \dfrac{6(2c-9)}{(2c-9)(c+4)} = -6(c-4) = 6(4-c)$

24. $\dfrac{-3(r+3)(r-3)}{1} \cdot \dfrac{9(5r+6)}{(5r+6)(r-3)} = -27(r+3)$

25. $\dfrac{(5h-2)(2h+5)}{(4h+3)(3h-4)} \cdot \dfrac{(4h+3)(h+2)}{(2h+5)(h+2)} = \dfrac{5h-2}{3h-4}$

26. $\dfrac{(2n-3)(3n+8)}{(2n-3)(n+3)} \cdot \dfrac{(8n-3)(n+3)}{(8-3n)(8+3n)} = \dfrac{8n-3}{8-3n} = -\dfrac{8n-3}{3n-8}$

27. $\dfrac{(4t-5)(2t+3)}{2(3t+2)(t-1)} \cdot \dfrac{(2t-3)(3t+2)}{(4t-5)(2t-3)} = \dfrac{2t+3}{2(t-1)} = \dfrac{2t+3}{2t-2}$

28. $\dfrac{(3a-2)(a+3)}{(2a+5)(2a-1)} \cdot \dfrac{(2a-1)(a+1)}{-1(3a-2)(a+3)} = \dfrac{a+1}{-1(2a+5)} = -\dfrac{a+1}{2a+5}$

29. $\dfrac{5(x+3)(x-1)}{(x-5)(x-1)} \cdot \dfrac{(2x+1)(2x-5)}{(2x+1)(x+3)} = \dfrac{5(2x-5)}{x-5}$

30. $\dfrac{(4x+5)(3x+1)}{(2x-1)(x-3)} \cdot \dfrac{3x-2}{5(2x+1)} \cdot \dfrac{(2x+1)(x-3)}{(4x-3)(3x+1)} = \dfrac{(4x+5)(3x-2)}{5(2x-1)(4x-3)}$

31. $\dfrac{(7y+3)(2y+1)}{3(5y-7)(2y+1)} \cdot \dfrac{(5y-7)(5y-3)}{(3y-2)(2y+5)} \cdot \dfrac{2y+5}{5y-3} = \dfrac{7y+3}{3(3y-2)}$

32. $\dfrac{(x+4)(x+2)}{(x+2)(x-1)} \cdot \dfrac{2(x+2)}{x+4} \cdot \dfrac{x-1}{x+3} = \dfrac{2(x+2)}{x+3}$

33. $\dfrac{(2y+1)(y-3)}{(2y-7)(2y+1)} \cdot \dfrac{2y-7}{4y+5} \cdot \dfrac{3y-1}{y-3} = \dfrac{3y-1}{4y+5}$

C 34. $\dfrac{3a^2}{(b-4)(b+4)} \cdot \dfrac{(b+4)(b+2)}{3a(b+2)} = \dfrac{a}{b-4}$

35. $\dfrac{5x^2}{(y+5)(y-5)} \cdot \dfrac{(y-5)(y-5)}{5x(y-5)} = \dfrac{x}{y+5}$

36. $\dfrac{(2a+3b)(a-2b)}{(2b-a)(b+5a)} \cdot \dfrac{(a+2b)(a-2b)}{(2a+3b)(a+2b)} =$

$\dfrac{(2a+3b)(a-2b)(a+2b)(a-2b)}{-1(a-2b)(b+5a)(2a+3b)(a+2b)} = \dfrac{a-2b}{-1(b+5a)} = -\dfrac{a-2b}{b+5a}$

37. $\dfrac{4(g-3h)(g+h)}{2(4h+g)(h+g)} \cdot \dfrac{(g-4h)(g+4h)}{6(g+h)(g+2h)} = \dfrac{(g-3h)(g-4h)}{3(g+h)(g+2h)}$

38. $\dfrac{(m+6n)(m-n)}{m(6m-n)} \cdot \dfrac{(2m+n)(m-3n)}{(m-n)(m+n)} \cdot \dfrac{(3m-2n)(m+n)}{(m+6n)(m-3n)} =$

$\dfrac{(2m+n)(3m-2n)}{m(6m-n)}$

39. $\dfrac{(3c + 2d)(c - 3d)}{(c + 3d)(c + 3d)} \cdot \dfrac{(c - 3d)(c + 3d)}{(3c - 8d)(c - 3d)} \cdot \dfrac{(3c - 8d)(c + 3d)}{(2c + 3d)(3c - d)} =$

$\dfrac{(3c + 2d)(c - 3d)}{(3c - d)(2c + 3d)}$

40. Stress $= \dfrac{4x - 6}{2x^2 + 5x - 12} = \dfrac{2(2x - 3)}{(2x - 3)(x + 4)} = \dfrac{2}{x + 4}$ lb/ft^2

41. Efficiency $= \dfrac{3x^2 + 5x - 2}{x + 2} = \dfrac{(3x - 1)(x + 2)}{x + 2} = 3x - 1$

42. Rate $= \dfrac{x^2 - 9}{x + 3} = \dfrac{(x + 3)(x - 3)}{x + 3} = x - 3$ mi/h

43. 1. Take the reciprocal of the divisor.   **2.** Factor if possible.   **3.** Divide by common factors.   **4.** Multiply.

44. Yes, step 1 must be done first.   **45.** steps 2–4

page 328   Capsule Review

1. $36 = 6 \cdot 6 = 2 \cdot 2 \cdot 3 \cdot 3$

2. $12g^2h = 2 \cdot 6g^2h = 2 \cdot 2 \cdot 3 \cdot g \cdot g \cdot h$

3. $140x^3 = 2 \cdot 70x^3 = 2 \cdot 2 \cdot 35x^3 = 2 \cdot 2 \cdot 5 \cdot 7 \cdot x \cdot x \cdot x$

4. $825abc = 25 \cdot 33abc = 3 \cdot 5 \cdot 5 \cdot 11 \cdot a \cdot b \cdot c$

5. $23m^3np^2 = 23 \cdot m \cdot m \cdot m \cdot n \cdot p \cdot p$

6. $1875e^2f = 25 \cdot 75e^2f = 5 \cdot 5 \cdot 25 \cdot 3e^2f = 3 \cdot 5 \cdot 5 \cdot 5 \cdot 5 \cdot e \cdot e \cdot f$

7. $1000x^2y^3z = 10 \cdot 100 \cdot x^2y^3z = 2 \cdot 2 \cdot 2 \cdot 5 \cdot 5 \cdot 5 \cdot x \cdot x \cdot y \cdot y \cdot y \cdot z$

8. $100{,}000n^5 = 10 \cdot 10 \cdot 10 \cdot 10 \cdot 10n^5 = 2 \cdot 2 \cdot 2 \cdot 2 \cdot 2 \cdot 5 \cdot 5 \cdot 5 \cdot 5 \cdot 5 \cdot n \cdot n \cdot n \cdot n \cdot n$

page 330   Class Exercises

1. $8 = 2 \cdot 2 \cdot 2$; $24 = 2 \cdot 2 \cdot 2 \cdot 3$; LCD is $2 \cdot 2 \cdot 2 \cdot 3 = 24$

2. $12 = 2 \cdot 2 \cdot 3$; $18 = 2 \cdot 3 \cdot 3$; LCD is $2 \cdot 2 \cdot 3 \cdot 3 = 36$

3. $5x = 5 \cdot x$; $20x^2 = 2 \cdot 2 \cdot 5 \cdot x \cdot x$; LCD is $2^2 \cdot 5 \cdot x^2 = 20x^2$

4. $3a^2 = 3 \cdot a \cdot a$; $5b^2 = 5 \cdot b \cdot b$; LCD is $3 \cdot 5 \cdot a^2 \cdot b^2 = 15a^2b^2$

5. $18m^2n = 2 \cdot 3 \cdot 3 \cdot m \cdot m \cdot n$; $24n^2 = 2 \cdot 2 \cdot 2 \cdot 3 \cdot n \cdot n$; LCD is $2^3 \cdot 3^2 \cdot m^2 \cdot n^2 = 72m^2n^2$

6. $2a = 2 \cdot a$; $2a + 4 = 2 \cdot (a + 2)$; LCD is $2 \cdot a \cdot (a + 2) = 2a(a + 2)$

7. $x - 3 = 1 \cdot (x - 3)$; $x + 5 = 1 \cdot (x + 5)$; LCD is $(x - 3)(x + 5)$

8. $3m + 9 = 3 \cdot (m + 3)$; $m^2 + 4m + 3 = (m + 3)(m + 1)$;
LCD is $3 \cdot (m + 3)(m + 1) = 3(m + 3)(m + 1)$

9. LCD is $12x^2y$; $\dfrac{4}{3xy} \cdot \dfrac{4x}{4x} = \dfrac{16x}{12x^2y}$; $\dfrac{7}{12x^2} \cdot \dfrac{y}{y} = \dfrac{7y}{12x^2y}$

10. $a^2 - 16 = (a + 4)(a - 4)$; $12 + a - a^2 = -1(a - 4)(a + 3)$;
LCD is $-1(a + 4)(a - 4)(a + 3)$; $\dfrac{3}{(a + 4)(a - 4)} \cdot \dfrac{-1(a + 3)}{-1(a + 3)} =$

$\dfrac{3(a + 3)}{(a + 4)(a - 4)(a + 3)}$; $\dfrac{5}{(-1)(a - 4)(a + 3)} \cdot \dfrac{a + 4}{a + 4} =$

$\dfrac{-5(a + 4)}{(a + 4)(a - 4)(a + 3)}$

pages 331–332   Practice Exercises

A 1. $5ab = 5 \cdot a \cdot b$; $2a = 2 \cdot a$; LCD is $2 \cdot 5 \cdot a \cdot b = 10ab$

2. $10rs = 2 \cdot 5 \cdot r \cdot s$; $12r = 2 \cdot 2 \cdot 3 \cdot r$; LCD is $2 \cdot 2 \cdot 3 \cdot 5 \cdot r \cdot s = 60rs$

3. $6x = 2 \cdot 3 \cdot x$; $14y = 2 \cdot 7 \cdot y$; LCD is $2 \cdot 3 \cdot 7 \cdot x \cdot y = 42xy$

4. $6c = 2 \cdot 3 \cdot c$; $9d = 3 \cdot 3 \cdot d$; LCD is $2 \cdot 3 \cdot 3 \cdot c \cdot d = 18cd$

5. $14a^3b = 2 \cdot 7 \cdot a \cdot a \cdot a \cdot b$; $21ab^2 = 3 \cdot 7 \cdot a \cdot b \cdot b$;
LCD is $2 \cdot 3 \cdot 7 \cdot a \cdot a \cdot a \cdot b \cdot b = 42a^3b^2$

6. $24s^3t = 2^3 \cdot 3 \cdot s^3t$; $6st^2 = 2 \cdot 3 \cdot s \cdot t^2$; LCD is $2^3 \cdot 3 \cdot s^3 \cdot t^2 = 24s^3t^2$

7. $18xy^3 = 2 \cdot 3^2 \cdot x \cdot y^3$; $6x^2y = 2 \cdot 3 \cdot x^2 \cdot y$; LCD is $2 \cdot 3^2 \cdot x^2 \cdot y^3 = 18x^2y^3$

8. $13cd^3 = 1 \cdot 13 \cdot c \cdot d^3$; $3c^2d = 1 \cdot 3 \cdot c^2 \cdot d$; LCD is $3 \cdot 13 \cdot c^2 \cdot d^3 = 39c^2d^3$

9. $3n + 3 = 3(n + 1)$; $n^2 + 6n + 5 = (n + 5)(n + 1)$; LCD is $3(n + 1)(n + 5)$

10. $6x + 6 = 6(x + 1)$; $x^2 + 9x + 8 = (x + 8)(x + 1)$; LCD is $6(x + 1)(x + 8)$

11. $3r - 27 = 3(r - 9)$; $4r + 10 = 2(2r + 5)$; LCD is $2 \cdot 3 \cdot (r - 9)(2r + 5) =$
$6(r - 9)(2r + 5)$

12. $8y - 36 = 4(2y - 9)$; $5y + 15 = 5(y + 3)$; LCD is $4 \cdot 5 \cdot (2y - 9)(y + 3) =$
$20(2y - 9)(y + 3)$

13. $t^2 - 36 = (t + 6)(t - 6)$; $t - 6 = 1(t - 6)$; LCD is $(t + 6)(t - 6)$

14. $x^2 - 4 = (x - 2)(x + 2)$; $x - 2 = 1(x - 2)$; LCD is $(x + 2)(x - 2)$

15. $12mn^2 = 2^2 \cdot 3mn^2$; $4n = 2^2 \cdot n$; LCD is $2^2 \cdot 3 \cdot mn^2 = 12mn^2$;
$\dfrac{5}{12mn^2} \cdot \dfrac{1}{1} = \dfrac{5}{12mn^2}$; $\dfrac{3m}{4n} \cdot \dfrac{3mn}{3mn} = \dfrac{9m^2n}{12mn^2}$

16. $45ab^2 = 3^2 \cdot 5ab^2$; $18b = 3^2 \cdot 2b$; LCD is $2 \cdot 3^2 \cdot 5ab^2 = 90ab^2$;

$\dfrac{1}{45ab^2} \cdot \dfrac{2}{2} = \dfrac{2}{90ab^2}, \dfrac{5a}{18b} \cdot \dfrac{5ab}{5ab} = \dfrac{25a^2b}{90ab^2}$

17. $7y + 7 = 7(y + 1)$; $y^2 + 4y + 3 = (y + 1)(y + 3)$; LCD is $7(y + 1)(y + 3)$;

$\dfrac{5}{7(y + 1)} \cdot \dfrac{y + 3}{y + 3} = \dfrac{5(y + 3)}{7(y + 1)(y + 3)}, \dfrac{3}{(y + 1)(y + 3)} \cdot \dfrac{7}{7} = \dfrac{21}{7(y + 1)(y + 3)}$

18. $2h + 2 = 2(h + 1)$; $h^2 + 4h + 3 = (h + 1)(h + 3)$; LCD is $2(h + 1)(h + 3)$;

$\dfrac{5}{2(h + 1)} \cdot \dfrac{h + 3}{h + 3} = \dfrac{5(h + 3)}{2(h + 1)(h + 3)}, \dfrac{2}{(h + 1)(h + 3)} \cdot \dfrac{2}{2} = \dfrac{4}{2(h + 1)(h + 3)}$

19. $y^2 - 9 = (y + 3)(y - 3)$; $y - 3 = 1(y - 3)$; LCD is $(y + 3)(y - 3)$;

$\dfrac{3y}{(y + 3)(y - 3)} \cdot \dfrac{1}{1} = \dfrac{3y}{(y + 3)(y - 3)}, \dfrac{2}{y - 3} \cdot \dfrac{y + 3}{y + 3} = \dfrac{2(y + 3)}{(y - 3)(y + 3)}$

20. $a^2 - 25 = (a + 5)(a - 5)$; $5 - a = -1(a - 5)$; LCD is $(a + 5)(a - 5)$;

$\dfrac{3a}{(a + 5)(a - 5)} \cdot \dfrac{1}{1} = \dfrac{3a}{(a + 5)(a - 5)}, \dfrac{-1}{a - 5} \cdot \dfrac{a + 5}{a + 5} = \dfrac{-1(a + 5)}{(a - 5)(a + 5)}$

B 21. $x^2 + 5x + 6 = (x + 2)(x + 3)$; $x^2 + 7x + 10 = (x + 5)(x + 2)$;
LCD is $(x + 2)(x + 3)(x + 5)$

22. $r^2 - r - 12 = (r - 4)(r + 3)$; $r^2 - 4r - 21 = (r - 7)(r + 3)$;
LCD is $(r - 7)(r + 3)(r - 4)$

23. $12c^2 + 13c - 35 = (3c + 7)(4c - 5)$; $3c^2 - 11c - 42 = (3c + 7)(c - 6)$;
LCD is $(3c + 7)(4c - 5)(c - 6)$

24. $6h^2 - 17h + 12 = (3h - 4)(2h - 3)$; $8h^2 + 2h - 21 = (2h - 3)(4h + 7)$;
LCD is $(3h - 4)(2h - 3)(4h + 7)$

25. $16x^2 - 43x - 15 = (x - 3)(16x + 5)$; $2x^2 + 7x + 3 = (2x + 1)(x + 3)$;
LCD is $(x - 3)(16x + 5)(2x + 1)(x + 3)$

26. $12x^2 - 7x - 12 = (4x + 3)(3x - 4)$; $6x^2 - 5x - 6 = (3x + 2)(2x - 3)$;
LCD is $(4x + 3)(3x - 4)(3x + 2)(2x - 3)$

27. $2a^2$; $3b$; $2 \cdot 3ab^2$; LCD is $6a^2b^2$; $\dfrac{3}{2a^2} \cdot \dfrac{3b^2}{3b^2} = \dfrac{9b^2}{6a^2b^2}, \dfrac{9}{3b} \cdot \dfrac{2a^2b}{2a^2b} = \dfrac{18a^2b}{6a^2b^2}$;

$\dfrac{a + b}{6ab^2} \cdot \dfrac{a}{a} = \dfrac{a(a + b)}{6a^2b^2}$

28. $3n^2$; $5 \cdot 3m^2n$; $2 \cdot 2mn$; LCD is $2^2 \cdot 3 \cdot 5m^2n^2 = 60m^2n^2$; $\dfrac{m}{3n^2} \cdot \dfrac{20m^2}{20m^2} =$

$\dfrac{20m^3}{60m^2n^2}, \dfrac{n + 1}{15m^2n} \cdot \dfrac{4n}{4n} = \dfrac{4n(n + 1)}{60m^2n^2}, \dfrac{3}{4mn} \cdot \dfrac{15mn}{15mn} = \dfrac{45mn}{60m^2n^2}$

29. $5 - 3n = -1(3n - 5)$; $9n^2 - 25 = (3n + 5)(3n - 5)$; LCD is $(3n + 5)(3n - 5)$;

$$\dfrac{8n}{-1(3n - 5)} \cdot \dfrac{3n + 5}{3n + 5} = \dfrac{-8n(3n + 5)}{(3n - 5)(3n + 5)}; \dfrac{5n^2}{(3n - 5)(3n + 5)} \cdot \dfrac{-1}{-1} = \dfrac{5n^2}{(3n - 5)(3n + 5)}$$

30. $3 - 2r = -1(2r - 3)$; $4r^2 - 9 = (2r - 3)(2r + 3)$; LCD is $-1(2r - 3)(2r + 3)$;

$$\dfrac{5r}{-1(2r - 3)} \cdot \dfrac{2r + 3}{2r + 3} = \dfrac{-5r(2r + 3)}{(2r + 3)(2r - 3)}; \dfrac{3r^2}{(2r - 3)(2r + 3)} \cdot \dfrac{-1}{-1} = \dfrac{3r^2}{(2r - 3)(2r + 3)}$$

31. $2d^2 - 4d = 2d(d - 2)$; $d^2 - d - 2 = (d - 2)(d + 1)$; LCD is $2d(d - 2)(d + 1)$;

$$\dfrac{3d}{2d(d - 2)} \cdot \dfrac{d + 1}{d + 1} = \dfrac{3d(d + 1)}{2d(d - 2)(d + 1)}; \dfrac{d + 3}{(d - 2)(d + 1)} \cdot \dfrac{2d}{2d} = \dfrac{2d(d + 3)}{2d(d - 2)(d + 1)}$$

32. $x^2 + 3x - 4 = (x + 4)(x - 1)$; $3x^2 + 12x = 3x(x + 4)$; LCD is $3x(x + 4)$

$(x - 1)$; $\dfrac{5x + 1}{(x + 4)(x - 1)} \cdot \dfrac{3x}{3x} = \dfrac{3x(5x + 1)}{3x(x + 4)(x - 1)}; \dfrac{x}{3x(x + 4)} \cdot \dfrac{x - 1}{x - 1} = \dfrac{x(x - 1)}{3x(x + 4)(x - 1)}$

33. $3c + 3 = 3(c + 1)$; $c^2 + 3c + 2 = (c + 2)(c + 1)$; $c + 2 = 1(c + 2)$;

LCD is $3(c + 1)(c + 2)$; $\dfrac{c}{3(c + 1)} \cdot \dfrac{c + 2}{c + 2} = \dfrac{c(c + 2)}{3(c + 1)(c + 2)}; \dfrac{c^2}{(c + 2)(c + 1)} \cdot \dfrac{3}{3} =$

$\dfrac{3c^2}{3(c + 2)(c + 1)}; \dfrac{3c}{c + 2} \cdot \dfrac{3(c + 1)}{3(c + 1)} = \dfrac{9c(c + 1)}{3(c + 2)(c + 1)}$

34. $5y + 15 = 5(y + 3)$; $y^2 + 5y + 6 = (y + 3)(y + 2)$; $y + 2 = 1(y + 2)$;

LCD is $5(y + 3)(y + 2)$; $\dfrac{y + 1}{5(y + 3)} \cdot \dfrac{y + 2}{y + 2} = \dfrac{(y + 1)(y + 2)}{5(y + 3)(y + 2)}; \dfrac{4}{(y + 3)(y + 3)} \cdot \dfrac{5}{5} =$

$\dfrac{20}{5(y + 3)(y + 2)}; \dfrac{2y}{y + 2} \cdot \dfrac{5(y + 3)}{5(y + 3)} = \dfrac{10y(y + 3)}{5(y + 2)(y + 3)}$

C 35. $15x^2 - xy - 28y^2 = (5x - 7y)(3x + 4y)$; $12x^2 + 7xy - 12y^2 = (3x + 4y)(4x - 3y)$; LCD is $(3x + 4y)(5x - 7y)(4x - 3y)$

36. $8x^2 - 14xy - 15y^2 = (2x - 5y)(4x + 3y)$; $4x^2 - 25y^2 = (2x + 5y)(2x - 5y)$; LCD is $(2x - 5y)(4x + 3y)(2x + 5y) = 6(a^2 + b^2)(a^2 - b^2)$

37. $8a^4 - 8b^4 = 8(a^2 + b^2)(a - b)(a + b)$; $(3a - 3b)^2 = 3 \cdot 3(a - b)(a - b)$; LCD is $72(a^2 + b^2)(a - b)^2(a + b)$

38. $3a^4 + 6a^2b^2 + 3b^4 = 3(a^2 + b^2)(a^2 + b^2)$; $2a^4 - 2b^4 = 2(a^2 + b^2)(a - b)(a + b)$; LCD is $6(a^2 + b^2)^2(a - b)(a + b) = 6(a^2 + b^2)^2(a^2 - b^2)$

39. $15m^2 - 2m - 8 = (5m - 4)(3m + 2)$; $3m^2 - 10m - 8 = (3m + 2)(m - 4)$; $6m - 24 = 6(m - 4)$; LCD is $6(5m - 4)(3m + 2)(m - 4)$

40. $2t^2 - 2t - 12 = 2(t - 3)(t + 2)$; $4t^2 + 10t + 4 = 2(2t + 1)(t + 2)$; $6t + 3 = 3(2t + 1)$; LCD is $6(t - 3)(t + 2)(2t + 1)$

41. $x^2 = x \cdot x$; $2x = 2 \cdot x$; $3x^3 = 3 \cdot x \cdot x \cdot x$; LCD is $2 \cdot 3 \cdot x^3 = 6x^3$

42. $\pi l = \pi \cdot l$; $3l^3 = 3 \cdot l \cdot l \cdot l$; LCD is $3\pi l^3$;

$$1 \cdot \frac{3\pi l^3}{3\pi l^3} = \frac{3\pi l^3}{3\pi l^3}; \frac{4c}{\pi l} \cdot \frac{3l^2}{3l^2} = \frac{12cl^2}{3\pi l^3}; \frac{c^3}{3l^3} \cdot \frac{\pi}{\pi} = \frac{c^3\pi}{3\pi l^3}; \frac{3\pi l^3}{3\pi l^3} - \frac{12cl^2}{3\pi l^3} + \frac{\pi c^3}{3\pi l^3}$$

43. Area of shaded region = Area of triangle − Area of circle = $\frac{1}{2}bh - \pi r^2 =$

$\frac{1}{2}(6)(6) - \pi (2)^2 = 18 - 4\pi \approx 5.4$ in.²

44. Area of shaded region = Area of circle − Area of rectangle = $\pi r^2 - (l \cdot w)$
   $= \pi(5)^2 - (8)(6) = 25\pi - 48 \approx 30.5$ cm²

45. Area of shaded region = Area of rectangle − Area of triangle = $(l \cdot w)$ −

$\frac{1}{2}bh = (12)(10) - \frac{1}{2}(12)(5) = 120 - 30 = 90$ cm²

46. $-3 \le a < 2$ $\quad$ **47.** all real numbers

48. $5 < 3x - 7 \quad$ and $\quad 3x - 7 < 8$ $\qquad$ **49.** $y + 5 < 6 \quad$ or $\quad -y \ge 3$

$4 < x \qquad$ and $\qquad x < 5$ $\qquad\qquad\qquad$ $y < 1 \quad$ or $\quad y \le -3$

$4 < x < 5$ $\qquad\qquad$ $y < 1$

50. $a(16a - 25)$ $\qquad$ **51.** $(d + 11)(d - 11)$ $\qquad$ **52.** $(n - 15)(n - 15)$ or $(n - 15)^2$

53. $(y + 8)(y - 7)$ $\qquad$ **54.** $x(x - 6)(x + 5)$ $\qquad$ **55.** $(a - b)(m - 1)$

56. If the denominators are differences with the terms reversed, then they are
opposites, and both will be factors of -1 times either denominator.

page 333 Capsule Review

1. $6x$ $\qquad$ **2.** a^2b^3 $\qquad$ **3.** $2xy(x + 3)$ $\qquad$ **4.** $(d - 3)(d + 3)$

page 335 Class Exercises

1. $\frac{6}{11}$ $\qquad$ **2.** $\frac{4x}{4} = x$ $\qquad$ **3.** $\frac{-2}{2ab} = -\frac{1}{ab}$ $\qquad$ **4.** $\frac{m + 2}{m - 3}$

5. LCD is $6cd^2$; $\frac{2c}{3d^2} \cdot \frac{2c}{2c} + \frac{3}{2cd} \cdot \frac{3d}{3d} = \frac{4c^2}{6cd^2} + \frac{9d}{6cd^2} = \frac{4c^2 + 9d}{6cd^2}$

6. LCD is $(r + 3)(r - 3)$; $\frac{3r}{(r + 3)(r - 3)} - \frac{5r}{r + 3} \cdot \frac{r - 3}{r - 3} =$

$\frac{3r}{(r + 3)(r - 3)} - \frac{5r^2 - 15r}{(r + 3)(r - 3)} = \frac{3r - 5r^2 + 15r}{(r + 3)(r - 3)} = \frac{18r - 5r^2}{(r + 3)(r - 3)}$

pages 335–336 Practice Exercises

A **1.** $\frac{8}{2m} = \frac{4}{m}$ $\qquad$ **2.** $\frac{12}{6n} = \frac{2}{n}$ $\qquad$ **3.** $\frac{2}{4x} = \frac{1}{2x}$ $\qquad$ **4.** $\frac{5}{5z} = \frac{1}{z}$

5. $\dfrac{2z + 4}{z + 2} = \dfrac{2(z + 2)}{z + 2} = 2$

6. $\dfrac{4c + 28}{c + 7} = \dfrac{4(c + 7)}{c + 7} = 4$

7. $\dfrac{2 - a}{a - 2} = \dfrac{-1(a - 2)}{a - 2} = -1$

8. $\dfrac{3 - b}{b - 3} = \dfrac{-1(b - 3)}{b - 3} = -1$

9. $\dfrac{3}{2a} \cdot \dfrac{2a}{2a} + \dfrac{2}{4a^2} - \dfrac{1}{a} \cdot \dfrac{4a}{4a} = \dfrac{6a}{4a^2} + \dfrac{2}{4a^2} - \dfrac{4a}{4a^2} = \dfrac{2a + 2}{4a^2} = \dfrac{2(a + 1)}{4a^2} = \dfrac{a + 1}{2a^2}$

10. $\dfrac{3}{3x} \cdot \dfrac{3x}{3x} + \dfrac{6}{9x^2} - \dfrac{1}{x} \cdot \dfrac{9x}{9x} = \dfrac{9x}{9x^2} + \dfrac{6}{9x^2} - \dfrac{9x}{9x^2} = \dfrac{6}{9x^2} = \dfrac{2}{3x^2}$

11. $\dfrac{1}{3m^2} \cdot \dfrac{2}{2} - \dfrac{4}{6m} \cdot \dfrac{m}{m} - \dfrac{1}{m} \cdot \dfrac{6m}{6m} = \dfrac{2}{6m^2} - \dfrac{4m}{6m^2} - \dfrac{6m}{6m^2} = \dfrac{2 - 10m}{6m^2} =$

$\dfrac{2(1 - 5m)}{6m^2} = \dfrac{1 - 5m}{3m^2}$

12. $\dfrac{8}{16r^2} - \dfrac{3}{8r} \cdot \dfrac{2r}{2r} - \dfrac{1}{r} \cdot \dfrac{16r}{16r} = \dfrac{8}{16r^2} - \dfrac{6r}{16r^2} - \dfrac{16r}{16r^2} = \dfrac{8 - 22r}{16r^2} = \dfrac{4 - 11r}{8r^2}$

13. $\dfrac{4}{2(a + 4)} - \dfrac{a}{5(a + 4)} = \dfrac{4}{2(a + 4)} \cdot \dfrac{5}{5} - \dfrac{a}{5(a + 4)} \cdot \dfrac{2}{2} =$

$\dfrac{20}{10(a + 4)} - \dfrac{2a}{10(a + 4)} = \dfrac{20 - 2a}{10(a + 4)} = \dfrac{2(10 - a)}{10(a + 4)} = \dfrac{10 - a}{5(a + 4)}$

14. $\dfrac{a}{a + 3} \cdot \dfrac{a + 5}{a + 5} - \dfrac{3}{a + 5} \cdot \dfrac{a + 3}{a + 3} = \dfrac{a^2 + 5a}{(a + 3)(a + 5)} - \dfrac{3a + 9}{(a + 3)(a + 5)} =$

$\dfrac{a^2 + 5a - 3a - 9}{(a + 3)(a + 5)} = \dfrac{a^2 + 2a - 9}{(a + 3)(a + 5)}$

15. $\dfrac{5}{l + 4} \cdot \dfrac{l - 1}{l - 1} + \dfrac{3}{l - 1} \cdot \dfrac{l + 4}{l + 4} = \dfrac{5l - 5}{(l + 4)(l - 1)} + \dfrac{3l + 12}{(l + 4)(l - 1)} = \dfrac{8l + 7}{(l + 4)(l - 1)}$

16. $\dfrac{i}{7(i + 2)} + \dfrac{6}{3(i + 2)} = \dfrac{i}{7(i + 2)} \cdot \dfrac{3}{3} + \dfrac{6}{3(i + 2)} \cdot \dfrac{7}{7} = \dfrac{3i + 42}{21(i + 2)} = \dfrac{3(i + 14)}{21(i + 2)} = \dfrac{i + 14}{7(i + 2)}$

17. $\dfrac{5x + 1}{-1(x^2 - 25)} + \dfrac{5}{x - 5} = \dfrac{-1(5x + 1)}{(x - 5)(x + 5)} + \dfrac{5}{x - 5} = \dfrac{-1(5x + 1)}{(x - 5)(x + 5)} + \dfrac{5}{x - 5} \cdot \dfrac{(x + 5)}{(x + 5)} =$

$\dfrac{-1(5x + 1)}{(x - 5)(x + 5)} + \dfrac{5(x + 5)}{(x - 5)(x + 5)} = \dfrac{-5x - 1 + 5x + 25}{(x - 5)(x + 5)} = \dfrac{24}{(x - 5)(x + 5)}$

18. $\dfrac{3z + 2}{-1(z^2 - 16)} + \dfrac{3}{z - 4} = \dfrac{-1(3z + 2)}{(z + 4)(z - 4)} + \dfrac{3}{(z - 4)} \cdot \dfrac{z + 4}{z + 4} = \dfrac{-3z - 2}{(z + 4)(z - 4)} +$

$\dfrac{3z + 12}{(z + 4)(z - 4)} = \dfrac{10}{(z + 4)(z - 4)}$

19. $\dfrac{4r + 1}{-1(r^2 - 9)} + \dfrac{4}{r - 3} = \dfrac{-1(4r + 1)}{(r + 3)(r - 3)} + \dfrac{4}{r - 3} \cdot \dfrac{r + 3}{r + 3} = \dfrac{-4r - 1}{(r + 3)(r - 3)} +$

$\dfrac{4r + 12}{(r + 3)(r - 3)} = \dfrac{11}{(r + 3)(r - 3)}$

20. $\dfrac{y + 3}{-1(y^2 - 4)} + \dfrac{1}{y - 2} = \dfrac{-1(y + 3)}{(y + 2)(y - 2)} + \dfrac{1}{y - 2} \cdot \dfrac{y + 2}{y + 2} = \dfrac{-y - 3}{(y + 2)(y - 2)} +$

$\dfrac{y + 2}{(y + 2)(y - 2)} = \dfrac{-1}{(y + 2)(y - 2)}$

21. $\dfrac{2}{y - 1} + \dfrac{6y - 2}{(y + 3)(y - 1)} = \dfrac{2}{y - 1} \cdot \dfrac{y + 3}{y + 3} + \dfrac{6y - 2}{(y + 3)(y - 1)} = \dfrac{2y + 6}{(y + 3)(y - 1)} +$

$\dfrac{6y - 2}{(y + 3)(y - 1)} = \dfrac{8y + 4}{(y + 3)(y - 1)}$

22. $\dfrac{5}{d - 3} - \dfrac{d - 4}{(d - 3)(d + 2)} = \dfrac{5}{d - 3} \cdot \dfrac{d + 2}{d + 2} - \dfrac{d - 4}{(d - 3)(d + 2)} =$

$\dfrac{5d + 10}{(d - 3)(d + 2)} - \dfrac{d - 4}{(d - 3)(d + 2)} = \dfrac{4d + 14}{(d - 3)(d + 2)}$

B 23. $\dfrac{2a^2 - a}{(2a + 3)(a - 1)} - \dfrac{6}{(2a + 3)(a - 1)} = \dfrac{2a^2 - a - 6}{(2a + 3)(a - 1)} = \dfrac{(2a + 3)(a - 2)}{(2a + 3)(a - 1)} = \dfrac{a - 2}{a - 1}$

24. $\dfrac{l^2}{(2l + 3)(l - 2)} + \dfrac{l - 6}{(2l + 3)(l - 2)} = \dfrac{l^2 + l - 6}{(2l + 3)(l - 2)} = \dfrac{(l + 3)(l - 2)}{(2l + 3)(l - 2)} = \dfrac{l + 3}{2l + 3}$

25. $\dfrac{g^3 + 3g}{g(g + 2)} + \dfrac{g^2 - g}{g(g + 2)} - \dfrac{4}{g(g + 2)} = \dfrac{g^3 + g^2 + 2g - 4}{g(g + 2)}$

26. $\dfrac{e^3 + e^2}{2e(e + 3)} + \dfrac{e^3 + 2e^2}{2e(e + 3)} - \dfrac{e^2 + 12}{2e(e + 3)} = \dfrac{2e^3 + 2e^2 - 12}{2e(e + 3)} = \dfrac{2(e^3 + e^2 - 6)}{2e(e + 3)} = \dfrac{e^3 + e^2 - 6}{e(e + 3)}$

27. $\dfrac{b - 2}{2b} \cdot \dfrac{3b}{3b} + \dfrac{b + 3}{3b} \cdot \dfrac{2b}{2b} - \dfrac{b - 2}{6b^2} \cdot \dfrac{1}{1} = \dfrac{3b^2 - 6b}{6b^2} + \dfrac{2b^2 + 6b}{6b^2} - \dfrac{b - 2}{6b^2} =$

$\dfrac{5b^2 - b + 2}{6b^2}$

28. $\dfrac{r+3}{4r} \cdot \dfrac{3r}{3r} - \dfrac{r+2}{3r^2} \cdot \dfrac{4}{4} + \dfrac{r-4}{12r^2} \cdot \dfrac{1}{1} = \dfrac{3r^2+9r}{12r^2} - \dfrac{4r+8}{12r^2} + \dfrac{r-4}{12r^2} =$

$\dfrac{3r^2+6r-12}{12r^2} = \dfrac{3(r^2+2r-4)}{12r^2} = \dfrac{r^2+2r-4}{4r^2}$

29. $\dfrac{u-5}{2(u+2)} \cdot \dfrac{3(u-2)}{3(u-2)} + \dfrac{u+3}{3(u-2)} \cdot \dfrac{2(u+2)}{2(u+2)} =$

$\dfrac{3(u-5)(u-2)}{6(u+2)(u-2)} + \dfrac{2(u+3)(u+2)}{6(u-2)(u+2)} = \dfrac{3u^2-21u+30}{6(u-2)(u+2)} + \dfrac{2u^2+10u+12}{6(u-2)(u+2)} =$

$\dfrac{5u^2-11u+42}{6(u-2)(u+2)}$

30. $\dfrac{n+5}{4(n+4)} \cdot \dfrac{3(n-3)}{3(n-3)} + \dfrac{n-1}{3(n-3)} \cdot \dfrac{4(n+4)}{4(n+4)} =$

$\dfrac{3(n+5)(n-3)}{12(n+4)(n-3)} + \dfrac{4(n-1)(n+4)}{12(n+4)(n-3)} = \dfrac{3n^2+6n-45}{12(n+4)(n-3)} + \dfrac{4n^2+12n-16}{12(n+4)(n-3)} =$

$\dfrac{7n^2+18n-61}{12(n+4)(n-3)}$

31. $\dfrac{3a-2}{(a-4)(a+3)} \cdot \dfrac{1}{1} + \dfrac{a+3}{a-4} \cdot \dfrac{a+3}{a+3} = \dfrac{3a-2}{(a-4)(a+3)} + \dfrac{a^2+6a+9}{(a-4)(a+3)} =$

$\dfrac{a^2+9a+7}{(a-4)(a+3)}$

32. $\dfrac{n^2+1}{(n-5)(n+3)} \cdot \dfrac{1}{1} + \dfrac{n-1}{n+3} \cdot \dfrac{n-5}{n-5} = \dfrac{n^2+1}{(n-5)(n+3)} + \dfrac{n^2-6n+5}{(n-5)(n+3)} =$

$\dfrac{2n^2-6n+6}{(n-5)(n+3)}$

C 33. $\dfrac{7d-2}{(d+4)(d-2)} \cdot \dfrac{1}{1} - \dfrac{4}{d+4} \cdot \dfrac{d-2}{d-2} - \dfrac{d}{d-2} \cdot \dfrac{d+4}{d+4} = \dfrac{7d-2}{(d+4)(d-2)} -$

$\dfrac{4(d-2)}{(d+4)(d-2)} - \dfrac{d(d+4)}{(d+4)(d-2)} = \dfrac{7d-2}{(d+4)(d-2)} - \dfrac{4d-8}{(d+4)(d-2)} -$

$\dfrac{d^2+4d}{(d+4)(d-2)} = \dfrac{-d^2-d+6}{(d+4)(d-2)} = \dfrac{-1(d+3)(d-2)}{(d+4)(d-2)} = \dfrac{-1(d+3)}{d+4} = -\dfrac{d+3}{d+4}$

34. $\dfrac{i-24}{(i-6)(i+3)} \cdot \dfrac{1}{1} - \dfrac{3}{i+3} \cdot \dfrac{i-6}{i-6} + \dfrac{i}{i-6} \cdot \dfrac{i+3}{i+3} = \dfrac{i-24}{(i-6)(i+3)} -$

$\dfrac{3i-18}{(i-6)(i+3)} + \dfrac{i^2+3i}{(i-6)(i+3)} = \dfrac{i^2+i-6}{(i-6)(i+3)} = \dfrac{(i+3)(i-2)}{(i-6)(i+3)} = \dfrac{i-2}{i-6}$

35. $\dfrac{t}{(4t-1)(2t+3)} \cdot \dfrac{t+5}{t+5} - \dfrac{3}{(4t-1)(t+5)} \cdot \dfrac{2t+3}{2t+3} =$

$\dfrac{t^2+5t}{(4t-1)(t+5)(2t+3)} - \dfrac{6t+9}{(4t-1)(t+5)(2t+3)} = \dfrac{t^2-t-9}{(4t-1)(t+5)(2t+3)}$

36. $\dfrac{a}{(2a-5)(3a+4)} \cdot \dfrac{a-2}{a-2} - \dfrac{2}{(3a+4)(a-2)} \cdot \dfrac{2a-5}{2a-5} =$

$\dfrac{a^2-2a}{(2a-5)(3a+4)(a-2)} - \dfrac{4a-10}{(2a-5)(3a+4)(a-2)} = \dfrac{a^2-6a+10}{(2a-5)(3a+4)(a-2)}$

37. $\dfrac{n+2}{(n+3)(n-2)} \cdot \dfrac{3n+4}{3n+4} + \dfrac{n-5}{(3n+4)(n+3)} \cdot \dfrac{n-2}{n-2} =$

$\dfrac{3n^2+10n+8}{(n+3)(n-2)(3n+4)} + \dfrac{n^2-7n+10}{(n+3)(n-2)(3n+4)} = \dfrac{4n^2+3n+18}{(n+3)(n-2)(3n+4)}$

38. $\dfrac{2t+1}{(3t+2)(t-11)} \cdot \dfrac{4t-9}{4t-9} + \dfrac{t-7}{(4t-9)(3t+2)} \cdot \dfrac{t-11}{t-11} =$

$\dfrac{8t^2-14t-9}{(3t+2)(t-11)(4t-9)} + \dfrac{t^2-18t+77}{(3t+2)(t-11)(4t-9)} = \dfrac{9t^2-32t+68}{(3t+2)(t-11)(4t-9)}$

39. $\dfrac{6}{(k+5)(k-3)} \cdot \dfrac{k(k+3)}{k(k+3)} + \dfrac{3}{k(k-3)} \cdot \dfrac{(k+5)(k+3)}{(k+5)(k+3)} + \dfrac{5}{(k+3)(k-3)} \cdot \dfrac{k(k+5)}{k(k+5)} =$

$\dfrac{6k^2+18k}{k(k+5)(k+3)(k-3)} + \dfrac{3k^2+24k+45}{k(k+5)(k+3)(k-3)} + \dfrac{5k^2+25k}{k(k+5)(k+3)(k-3)} =$

$\dfrac{14k^2+67k+45}{k(k+5)(k+3)(k-3)}$

40. $\dfrac{b}{(2b-3)(b+2)} \cdot \dfrac{b(b-2)}{b(b-2)} - \dfrac{2}{b(2b-3)} \cdot \dfrac{(b+2)(b-2)}{(b+2)(b-2)} + \dfrac{b+1}{(b+2)(b-2)} \cdot \dfrac{b(2b-3)}{b(2b-3)} =$

$\dfrac{b^3-2b^2}{b(2b-3)(b+2)(b-2)} - \dfrac{2b^2-8}{b(2b-3)(b+2)(b-2)} + \dfrac{2b^3-b^2-3b}{b(2b-3)(b+2)(b-2)} =$

$\dfrac{3b^3-5b^2-3b+8}{b(2b-3)(b+2)(b-2)}$

41. a. $w = \dfrac{2}{a+4} \cdot \dfrac{a}{a} - \dfrac{2}{a} \cdot \dfrac{a+4}{a+4} = \dfrac{2a}{a(a+4)} - \dfrac{2a+8}{a(a+4)} = \dfrac{-8}{a(a+4)}$

b. $w = \dfrac{3}{h-1} \cdot \dfrac{h+1}{h+1} - \dfrac{3}{h+1} \cdot \dfrac{h-1}{h-1} = \dfrac{3h+3}{(h+1)(h-1)} - \dfrac{3h-3}{(h+1)(h-1)} =$

$\dfrac{6}{(h+1)(h-1)}$

c. $w = \dfrac{2d}{(d-3)(d+3)} \cdot \dfrac{1}{1} - \dfrac{5}{d-3} \cdot \dfrac{d+3}{d+3} = \dfrac{2d}{(d-3)(d+3)} - \dfrac{5d+15}{(d-3)(d+3)} =$

$\dfrac{-3d-15}{(d-3)(d+3)}$

page 336 Algebra in Physics

1. Series circuit: $R = 0.5 + 3.0 = 3.5$ ohms; Parallel circuit: $R = \dfrac{(0.5)(3.0)}{0.5 + 3.0} =$

$\dfrac{1.5}{3.5} = 0.429$ ohms; The parallel circuit provides less resistance (3.071 ohms).

2. No; $R_1 + R_2 > \dfrac{R_1 \times R_2}{R_1 + R_2}$ for all positive values of R_1 and R_2.

page 337 Capsule Review

1. $2 + \dfrac{2}{3} = \dfrac{2}{1} \cdot \dfrac{3}{3} + \dfrac{2}{3} = \dfrac{6}{3} + \dfrac{2}{3} = \dfrac{8}{3}$

2. $3 + \dfrac{1}{4} = \dfrac{3}{1} \cdot \dfrac{4}{4} + \dfrac{1}{4} = \dfrac{12}{4} + \dfrac{1}{4} = \dfrac{13}{4}$

3. $1 + \dfrac{3}{8} = \dfrac{1}{1} \cdot \dfrac{8}{8} + \dfrac{3}{8} = \dfrac{8}{8} + \dfrac{3}{8} = \dfrac{11}{8}$

4. $11 + \dfrac{14}{15} = \dfrac{11}{1} \cdot \dfrac{15}{15} + \dfrac{14}{15} = \dfrac{165}{15} + \dfrac{14}{15} = \dfrac{179}{15}$

5. $2 + \dfrac{1}{100} = \dfrac{2}{1} \cdot \dfrac{100}{100} + \dfrac{1}{100} = \dfrac{200}{100} + \dfrac{1}{100} = \dfrac{201}{100}$

6. $99 + \dfrac{99}{100} = \dfrac{99}{1} \cdot \dfrac{100}{100} + \dfrac{99}{100} = \dfrac{9900}{100} + \dfrac{99}{100} = \dfrac{9999}{100}$

pages 338–339 Class Exercises

1. $\dfrac{4}{1} \cdot \dfrac{3}{3} + \dfrac{1}{3} = \dfrac{12}{3} + \dfrac{1}{3} = \dfrac{13}{3}$

2. $\dfrac{a}{1} \cdot \dfrac{c}{c} + \dfrac{b}{c} = \dfrac{ac + b}{c}$

3. $\dfrac{x + 3}{1} \cdot \dfrac{x - 2}{x - 2} - \dfrac{4}{x - 2} = \dfrac{x^2 + x - 6}{x - 2} - \dfrac{4}{x - 2} = \dfrac{x^2 + x - 10}{x - 2}$

4. $\dfrac{t}{t + 1} + \dfrac{5t}{1} \cdot \dfrac{t + 1}{t + 1} = \dfrac{t}{t + 1} + \dfrac{5t^2 + 5t}{t + 1} = \dfrac{5t^2 + 6t}{t + 1}$

5. $\dfrac{2}{3} \div \dfrac{5}{6} = \dfrac{2}{3} \cdot \dfrac{6}{5} = \dfrac{4}{5}$

6. $\dfrac{3s}{8} \div \dfrac{s}{4} = \dfrac{3s}{8} \cdot \dfrac{4}{s} = \dfrac{3}{2}$

7. $\dfrac{\left(\dfrac{2n}{m^2} - \dfrac{1}{m}\right) \cdot m^2}{\left(1 + \dfrac{2n}{m^2}\right) \cdot m^2} = \dfrac{2n - m}{m^2 + 2n}$

8. $\dfrac{\left(h + \dfrac{3}{2h + 5}\right) \cdot (2h + 5)}{\left(h - \dfrac{4h + 3}{2h + 5}\right) \cdot (2h + 5)} = \dfrac{h(2h + 5) + 3}{h(2h + 5) - (4h + 3)} = \dfrac{2h^2 + 5h + 3}{2h^2 + 5h - 4h - 3} =$

$\dfrac{2h^2 + 5h + 3}{2h^2 + h - 3} = \dfrac{(2h + 3)(h + 1)}{(2h + 3)(h - 1)} = \dfrac{h + 1}{h - 1}$

A 1. $\dfrac{2}{1} \cdot \dfrac{x}{x} + \dfrac{3}{x} = \dfrac{2x + 3}{x}$

2. $\dfrac{4}{1} \cdot \dfrac{a}{a} - \dfrac{8}{a} = \dfrac{4a - 8}{a}$

3. $\dfrac{3}{1} \cdot \dfrac{x}{x} - \dfrac{10}{x} = \dfrac{3x - 10}{x}$

4. $\dfrac{4}{1} \cdot \dfrac{b}{b} - \dfrac{5}{b} = \dfrac{4b - 5}{b}$

5. $\dfrac{2z}{1} \cdot \dfrac{z}{z} - \dfrac{z + 1}{z} = \dfrac{2z^2 - z - 1}{z}$

6. $\dfrac{3c}{1} \cdot \dfrac{c}{c} - \dfrac{c + 1}{c} = \dfrac{3c^2 - c - 1}{c}$

7. $\dfrac{2y + 3}{4y} + \dfrac{y - 3}{1} \cdot \dfrac{4y}{4y} = \dfrac{2y + 3 + 4y^2 - 12y}{4y} = \dfrac{4y^2 - 10y + 3}{4y}$

8. $\dfrac{3r + 5}{3r} + \dfrac{2 - r}{1} \cdot \dfrac{3r}{3r} = \dfrac{3r + 5 + 6r - 3r^2}{3r} = \dfrac{-3r^2 + 9r + 5}{3r}$

9. $\dfrac{d}{1} \cdot \dfrac{2d + 1}{2d + 1} + \dfrac{d - 3}{2d + 1} = \dfrac{2d^2 + d + d - 3}{2d + 1} = \dfrac{2d^2 + 2d - 3}{2d + 1}$

10. $\dfrac{c}{1} \cdot \dfrac{3c - 1}{3c - 1} + \dfrac{c - 2}{3c - 1} = \dfrac{3c^2 - c + c - 2}{3c - 1} = \dfrac{3c^2 - 2}{3c - 1}$

11. $\dfrac{2u - 1}{u + 2} + \dfrac{u}{1} \cdot \dfrac{u + 2}{u + 2} = \dfrac{2u - 1 + u^2 + 2u}{u + 2} = \dfrac{u^2 + 4u - 1}{u + 2}$

12. $\dfrac{3v - 1}{v + 3} - \dfrac{v}{1} \cdot \dfrac{v + 3}{v + 3} = \dfrac{3v - 1 - v^2 - 3v}{v + 3} = \dfrac{-v^2 - 1}{v + 3}$

13. $\dfrac{\left(\dfrac{1}{x} + \dfrac{1}{z}\right) \cdot 2xz}{\left(\dfrac{1}{2x} + \dfrac{1}{2z}\right) \cdot 2xz} = \dfrac{\dfrac{1}{x} \cdot 2xz + \dfrac{1}{z} \cdot 2xz}{\dfrac{1}{2x} \cdot 2xz + \dfrac{1}{2z} \cdot 2xz} = \dfrac{2z + 2x}{z + x} = \dfrac{2(z + x)}{z + x} = 2$

14. $\dfrac{\left(\dfrac{1}{r} + \dfrac{1}{s}\right) \cdot 5rs}{\left(\dfrac{3}{5r} + \dfrac{3}{5s}\right) \cdot 5rs} = \dfrac{\dfrac{1}{r} \cdot 5rs + \dfrac{1}{s} \cdot 5rs}{\dfrac{3}{5r} \cdot 5rs + \dfrac{3}{5s} \cdot 5rs} = \dfrac{5s + 5r}{3s + 3r} = \dfrac{5(s + r)}{3(s + r)} = \dfrac{5}{3}$

15. $\dfrac{\left(\dfrac{1}{u} - \dfrac{1}{v}\right) \cdot 2uv}{\left(\dfrac{5}{2u} - \dfrac{5}{2v}\right) \cdot 2uv} = \dfrac{\dfrac{1}{u} \cdot 2uv - \dfrac{1}{v} \cdot 2uv}{\dfrac{5}{2u} \cdot 2uv - \dfrac{5}{2v} \cdot 2uv} = \dfrac{2v - 2u}{5v - 5u} = \dfrac{2(v - u)}{5(v - u)} = \dfrac{2}{5}$

16. $\dfrac{\left(\dfrac{1}{a} - \dfrac{1}{b}\right) \cdot 4ab}{\left(\dfrac{3}{4a} - \dfrac{3}{4b}\right) \cdot 4ab} = \dfrac{\dfrac{1}{a} \cdot 4ab - \dfrac{1}{b} \cdot 4ab}{\dfrac{3}{4a} \cdot 4ab - \dfrac{3}{4b} \cdot 4ab} = \dfrac{4b - 4a}{3b - 3a} = \dfrac{4(b - a)}{3(b - a)} = \dfrac{4}{3}$

17. $\dfrac{\left(\dfrac{m}{n} + \dfrac{2}{n^2}\right) \cdot n^2}{\left(2 - \dfrac{m}{n^2}\right) \cdot n^2} = \dfrac{\dfrac{m}{n} \cdot n^2 + \dfrac{2}{n^2} \cdot n^2}{2 \cdot n^2 - \dfrac{m}{n^2} \cdot n^2} = \dfrac{mn + 2}{2n^2 - m}$

18. $\dfrac{\left(\dfrac{5}{v^2} + \dfrac{u}{v}\right) \cdot v^2}{\left(\dfrac{u}{v} - 3\right) \cdot v^2} = \dfrac{\dfrac{5}{v^2} \cdot v^2 + \dfrac{u}{v} \cdot v^2}{\dfrac{u}{v} \cdot v^2 - 3 \cdot v^2} = \dfrac{5 + uv}{uv - 3v^2}$

19. $\dfrac{\left(1 + \dfrac{4}{a}\right) \cdot a^2}{\left(1 - \dfrac{16}{a^2}\right) \cdot a^2} = \dfrac{1 \cdot a^2 + \dfrac{4}{a} \cdot a^2}{1 \cdot a^2 - \dfrac{16}{a^2} \cdot a^2} = \dfrac{a^2 + 4a}{a^2 - 16} = \dfrac{a(a + 4)}{(a + 4)(a - 4)} = \dfrac{a}{a - 4}$

20. $\dfrac{\left(1 + \dfrac{5}{b}\right) \cdot b^2}{\left(1 - \dfrac{25}{b^2}\right) \cdot b^2} = \dfrac{1 \cdot b^2 + \dfrac{5}{b} \cdot b^2}{1 \cdot b^2 - \dfrac{25}{b^2} \cdot b^2} = \dfrac{b^2 + 5b}{b^2 - 25} = \dfrac{b(b + 5)}{(b + 5)(b - 5)} = \dfrac{b}{b - 5}$

21. $\dfrac{\left(3 - \dfrac{12}{d + 4}\right) \cdot (d + 4)}{\left(2 - \dfrac{8}{d + 4}\right) \cdot (d + 4)} = \dfrac{3(d + 4) - \dfrac{12}{d + 4}(d + 4)}{2(d + 4) - \dfrac{8}{d + 4}(d + 4)} = \dfrac{3d + 12 - 12}{2d + 8 - 8} = \dfrac{3d}{2d} = \dfrac{3}{2}$

22. $\dfrac{\left(5 - \dfrac{25}{c + 5}\right) \cdot (c + 5)}{\left(3 - \dfrac{15}{c + 5}\right) \cdot (c + 5)} = \dfrac{5(c + 5) - \dfrac{25}{c + 5}(c + 5)}{3(c + 5) - \dfrac{15}{c + 5}(c + 5)} = \dfrac{5c + 25 - 25}{3c + 15 - 15} = \dfrac{5c}{3c} = \dfrac{5}{3}$

23. $\dfrac{\left(u - \dfrac{4}{u + 3}\right) \cdot (u + 3)}{\left(1 + \dfrac{1}{u + 3}\right) \cdot (u + 3)} = \dfrac{u(u + 3) - \dfrac{4}{u + 3}(u + 3)}{1(u + 3) + \dfrac{1}{u + 3}(u + 3)} = \dfrac{u^2 + 3u - 4}{u + 3 + 1} =$

$\dfrac{(u + 4)(u - 1)}{u + 4} = u - 1$

24. $\dfrac{\left(v - \dfrac{3}{v + 2}\right) \cdot (v + 2)}{\left(1 + \dfrac{1}{v + 2}\right) \cdot (v + 2)} = \dfrac{v(v + 2) - \dfrac{3}{v + 2}(v + 2)}{1(v + 2) + \dfrac{1}{v + 2}(v + 2)} = \dfrac{v^2 + 2v - 3}{v + 2 + 1} =$

$\dfrac{(v + 3)(v - 1)}{v + 3} = v - 1$

B 25. $\dfrac{\left(x - \dfrac{3}{3x - 4}\right) \cdot (3x - 4)}{\left(3x - 1 - \dfrac{9}{3x - 4}\right) \cdot (3x - 4)} = \dfrac{x(3x - 4) - \dfrac{3}{3x - 4}(3x - 4)}{(3x - 1)(3x - 4) - \dfrac{9}{3x - 4}(3x - 4)} =$

$\dfrac{3x^2 - 4x - 3}{9x^2 - 15x + 4 - 9} = \dfrac{3x^2 - 4x - 3}{9x^2 - 15x - 5}$

26. $\dfrac{\left(b - \dfrac{6}{2b - 4}\right) \cdot (2b - 4)}{\left(2b - 1 - \dfrac{10}{2b - 4}\right) \cdot (2b - 4)} = \dfrac{b(2b - 4) - \dfrac{6}{2b - 4}(2b - 4)}{(2b - 1)(2b - 4) - \dfrac{10}{2b - 4}(2b - 4)} =$

$\dfrac{2b^2 - 4b - 6}{4b^2 - 10b + 4 - 10} = \dfrac{2b^2 - 4b - 6}{4b^2 - 10b - 6} = \dfrac{2(b - 3)(b + 1)}{2(2b + 1)(b - 3)} = \dfrac{b + 1}{2b + 1}$

27. $\dfrac{\left(z - \dfrac{20z + 10}{3z + 7}\right) \cdot (3z + 7)}{\left(z - \dfrac{7z + 5}{3z + 7}\right) \cdot (3z + 7)} = \dfrac{z(3z + 7) - \dfrac{20z + 10}{3z + 7}(3z + 7)}{z(3z + 7) - \dfrac{7z + 5}{3z + 7}(3z + 7)} =$

$\dfrac{3z^2 + 7z - 20z - 10}{3z^2 + 7z - 7z - 5} = \dfrac{3z^2 - 13z - 10}{3z^2 - 5}$

28. $\dfrac{\left(a - \dfrac{36a + 6}{5a + 11}\right) \cdot (5a + 11)}{\left(a - \dfrac{3a - 3}{5a + 11}\right) \cdot (5a + 11)} = \dfrac{a(5a + 11) - \dfrac{36a + 6}{5a + 11}(5a + 11)}{a(5a + 11) - \dfrac{3a - 3}{5a + 11}(5a + 11)} =$

$\dfrac{5a^2 + 11a - 36a - 6}{5a^2 + 11a - 3a + 3} = \dfrac{5a^2 - 25a - 6}{5a^2 + 8a + 3}$

29. $\dfrac{x + 1}{1} \cdot \dfrac{x + 1}{x + 1} - \dfrac{4}{x + 1} = \dfrac{x^2 + 2x + 1 - 4}{x + 1} = \dfrac{x^2 + 2x - 3}{x + 1}$

30. $\dfrac{y + 3}{1} \cdot \dfrac{y + 3}{y + 3} - \dfrac{1}{y + 3} = \dfrac{y^2 + 6y + 9 - 1}{y + 3} = \dfrac{y^2 + 6y + 8}{y + 3}$

31. $\dfrac{a - 2}{1} \cdot \dfrac{a + 1}{a + 1} + \dfrac{3}{a + 1} = \dfrac{a^2 - a - 2 + 3}{a + 1} = \dfrac{a^2 - a + 1}{a + 1}$

32. $\dfrac{b - 4}{1} \cdot \dfrac{b + 3}{b + 3} - \dfrac{2}{b + 3} = \dfrac{b^2 - b - 12 - 2}{b + 3} = \dfrac{b^2 - b - 14}{b + 3}$

33. $\dfrac{m - 3}{2m + 5} + \dfrac{m - 4}{1} \cdot \dfrac{2m + 5}{2m + 5} = \dfrac{m - 3 + 2m^2 - 3m - 20}{2m + 5} = \dfrac{2m^2 - 2m - 23}{2m + 5}$

34. $\dfrac{n + 2}{5n - 3} + \dfrac{n - 6}{1} \cdot \dfrac{5n - 3}{5n - 3} = \dfrac{n + 2 + 5n^2 - 33n + 18}{5n - 3} = \dfrac{5n^2 - 32n + 20}{5n - 3}$

35.
$$\frac{\left(1 + \dfrac{1}{c} - \dfrac{6}{c^2}\right) \cdot c^2}{\left(\dfrac{1}{c} - \dfrac{2}{c^2}\right) \cdot c^2} = \frac{1 \cdot c^2 + \dfrac{1}{c} \cdot c^2 - \dfrac{6}{c^2} \cdot c^2}{\dfrac{1}{c} \cdot c^2 - \dfrac{2}{c^2} \cdot c^2} = \frac{c^2 + c - 6}{c - 2} = \frac{(c+3)(c-2)}{(c-2)} = c + 3$$

36.
$$\frac{\left(1 - \dfrac{2}{z} - \dfrac{8}{z^2}\right) \cdot z^2}{\left(\dfrac{1}{z} + \dfrac{2}{z^2}\right) \cdot z^2} = \frac{1 \cdot z^2 - \dfrac{2}{z} \cdot z^2 - \dfrac{8}{z^2} \cdot z^2}{\dfrac{1}{z} \cdot z^2 + \dfrac{2}{z^2} \cdot z^2} = \frac{z^2 - 2z - 8}{z + 2} = \frac{(z-4)(z+2)}{z+2} = z - 4$$

37.
$$\frac{\left(a - 1 - \dfrac{3}{a+1}\right) \cdot (a+1)}{\left(a + 5 + \dfrac{3}{a+1}\right) \cdot (a+1)} = \frac{(a-1)(a+1) - \dfrac{3}{a+1}(a+1)}{(a+5)(a+1) + \dfrac{3}{a+1}(a+1)} =$$

$$\frac{a^2 - 1 - 3}{a^2 + 6a + 5 + 3} = \frac{a^2 - 4}{a^2 + 6a + 8} = \frac{(a+2)(a-2)}{(a+2)(a+4)} = \frac{a-2}{a+4}$$

38.
$$\frac{\left(b - 2 - \dfrac{25}{b-2}\right) \cdot (b-2)}{\left(b + 2 - \dfrac{5}{b-2}\right) \cdot (b-2)} = \frac{(b-2)(b-2) - \dfrac{25}{b-2}(b-2)}{(b+2)(b-2) - \dfrac{5}{b-2}(b-2)} =$$

$$\frac{b^2 - 4b + 4 - 25}{b^2 - 4 - 5} = \frac{b^2 - 4b - 21}{b^2 - 9} = \frac{(b-7)(b+3)}{(b-3)(b+3)} = \frac{b-7}{b-3}$$

39.
$$\frac{\left(\dfrac{2n+6}{n+3} - \dfrac{n-2}{n+2}\right) \cdot (n+2)(n+3)}{\left(\dfrac{n^2-9}{(n+2)(n+3)}\right) \cdot (n+2)(n+3)} = \frac{(2n+6)(n+2) - (n-2)(n+3)}{n^2 - 9} =$$

$$\frac{2n^2 + 10n + 12 - n^2 - n + 6}{n^2 - 9} = \frac{n^2 + 9n + 18}{n^2 - 9} = \frac{(n+3)(n+6)}{(n+3)(n-3)} = \frac{n+6}{n-3}$$

40.
$$\frac{\left(\dfrac{2t+2}{t} - \dfrac{t-5}{t-3}\right) \cdot t(t-3)}{\dfrac{t^2-4}{t(t-3)} \cdot t(t-3)} = \frac{(2t+2)(t-3) - t(t-5)}{t^2 - 4} =$$

$$\frac{2t^2 - 4t - 6 - t^2 + 5t}{t^2 - 4} = \frac{t^2 + t - 6}{t^2 - 4} = \frac{(t+3)(t-2)}{(t+2)(t-2)} = \frac{t+3}{t+2}$$

41.
$$\frac{\left(\dfrac{u-1}{u+1} - \dfrac{u+1}{u-1}\right) \cdot (u+1)(u-1)}{\left(\dfrac{u-1}{u+1} + \dfrac{u+1}{u-1}\right) \cdot (u+1)(u-1)} = \frac{(u-1)(u-1) - (u+1)(u+1)}{(u-1)(u-1) + (u+1)(u+1)} =$$

$$\frac{u^2 - 2u + 1 - u^2 - 2u - 1}{u^2 - 2u + 1 + u^2 + 2u + 1} = \frac{-4u}{2u^2 + 2} = \frac{-4u}{2(u^2 + 1)} = \frac{-2u}{u^2 + 1}$$

42. $\dfrac{\left(\dfrac{v+2}{v-2}+\dfrac{v-2}{v+2}\right)\cdot(v-2)(v+2)}{\left(\dfrac{v+2}{v-2}-\dfrac{v-2}{v+2}\right)\cdot(v-2)(v+2)}=\dfrac{(v+2)(v+2)+(v-2)(v-2)}{(v+2)(v+2)-(v-2)(v-2)}=$

$\dfrac{v^2+4v+4+v^2-4v+4}{v^2+4v+4-v^2+4v-4}=\dfrac{2v^2+8}{8v}=\dfrac{2(v^2+4)}{8v}=\dfrac{v^2+4}{4v}$

43. $\dfrac{\left(\dfrac{x+2}{x}-\dfrac{3}{x+1}\right)\cdot x(x+1)}{\left(\dfrac{x-1}{x(x+1)}+\dfrac{3}{x+1}\right)\cdot x(x+1)}=\dfrac{(x+2)(x+1)-3x}{x-1+3x}=$

$\dfrac{x^2+3x+2-3x}{4x-1}=\dfrac{x^2+2}{4x-1}$

44. $\dfrac{\left(\dfrac{r+3}{r(r+4)}+\dfrac{r-1}{r+4}\right)\cdot r(r+4)}{\left(\dfrac{r-2}{r+4}-\dfrac{3}{r}\right)\cdot r(r+4)}=\dfrac{r+3+r(r-1)}{r(r-2)-3(r+4)}=\dfrac{r+3+r^2-r}{r^2-2r-3r-12}=$

$\dfrac{r^2+3}{r^2-5r-12}$

45. $\dfrac{\left[\dfrac{2}{(a+2)(a-2)}+\dfrac{2}{(a+2)(a-1)}\right]\cdot\dfrac{(a+2)(a-2)(a-1)(a+1)}{(a+2)(a-2)(a-1)(a+1)}}{\left[\dfrac{6}{(a+1)(a-1)}-\dfrac{3}{(a-2)(a+1)}\right]\cdot\dfrac{(a+2)(a-2)(a-1)(a+1)}{(a+2)(a-2)(a-1)(a+1)}}=$

$\dfrac{2(a-1)(a+1)+2(a-2)(a+1)}{6(a+2)(a-2)-3(a+2)(a-1)}=\dfrac{2a^2-2+2a^2-2a-4}{6a^2-24-3a^2-3a+6}=\dfrac{4a^2-2a-6}{3a^2-3a-18}$

46. $\dfrac{\dfrac{2}{3x}+\dfrac{1}{x^2}}{2}=\dfrac{\left(\dfrac{2}{3x}+\dfrac{1}{x^2}\right)\cdot6x^2}{2\cdot6x^2}=\dfrac{4x+6}{12x^2}=\dfrac{2(2x+3)}{12x^2}=\dfrac{2x+3}{6x^2}$

47. $\dfrac{\left(\dfrac{3}{a}\right)\left(\dfrac{a+3}{a^3}\right)}{\dfrac{3}{a}+\dfrac{a+3}{a^3}}=\dfrac{\left(\dfrac{3a+9}{a^4}\right)\cdot a^4}{\left(\dfrac{3}{a}+\dfrac{a+3}{a^3}\right)\cdot a^4}=\dfrac{3a+9}{3a^3+a(a+3)}=\dfrac{3a+9}{3a^3+a^2+3a}=$

$\dfrac{3(a+3)}{a(3a^2+a+3)}$

page 340 Test Yourself

1. $\dfrac{(m+6)(m-4)}{6m}\cdot\dfrac{2(m+3)}{(m-4)(m+3)}=\dfrac{m+6}{3m}$

2. $\dfrac{-1(x-y)}{2y+3x} \cdot \dfrac{3x+2y}{x+y} \cdot \dfrac{x+y}{x-y} = -1$

3. LCD is $10ab^2$; $\dfrac{3}{10ab^2}$; $\dfrac{2a}{5b} \cdot \dfrac{2ab}{2ab} = \dfrac{4a^2b}{10ab^2}$

4. LCD is $(p-2)(p-3)$; $\dfrac{4}{p-2} \cdot \dfrac{p-3}{p-3} = \dfrac{4p-12}{(p-2)(p-3)}$; $\dfrac{p}{(p-2)(p-3)}$

5. $\dfrac{m-1}{3m^2} \cdot \dfrac{2}{2} + \dfrac{m+3}{2m} \cdot \dfrac{3m}{3m} - \dfrac{m-2}{6m^2} = \dfrac{2m-2+3m^2+9m-m+2}{6m^2} =$

$\dfrac{3m^2+10m}{6m^2} = \dfrac{m(3m+10)}{6m^2} = \dfrac{3m+10}{6m}$

6. $\dfrac{5x-x+3}{x+4} = \dfrac{4x+3}{x+4}$

7. $y \cdot \dfrac{3}{3} - \dfrac{1}{3} = \dfrac{3y}{3} - \dfrac{1}{3} = \dfrac{3y-1}{3}$

8. $\dfrac{\dfrac{r+1}{(r+2)(r-2)} \cdot (r+2)(r-2)}{\dfrac{(r+1)(r-1)}{r+2} \cdot (r+2)(r-2)} = \dfrac{r+1}{(r+1)(r-1)(r-2)} = \dfrac{1}{(r-1)(r-2)}$

page 341 Integrating Algebra: Interest

1. $m = \dfrac{(1500)\left(\dfrac{0.12}{12}\right)\left(1+\dfrac{0.12}{12}\right)^{18}}{\left(1+\dfrac{0.12}{12}\right)^{18}-1} = 91.47307;\ \$91.47$

2. $m = \dfrac{(3000)\left(\dfrac{0.095}{12}\right)\left(1+\dfrac{0.095}{12}\right)^{24}}{\left(1+\dfrac{0.095}{12}\right)^{24}-1} = 137.7435;\ \$137.74$

3. Car A $= \dfrac{(3200)\left(\dfrac{0.08}{12}\right)\left(1+\dfrac{0.08}{12}\right)^{24}}{\left(1+\dfrac{0.08}{12}\right)^{24}-1} = 144.73$

Car B $= \dfrac{(2800)\left(\dfrac{0.095}{12}\right)\left(1+\dfrac{0.095}{12}\right)^{24}}{\left(1+\dfrac{0.095}{12}\right)^{24}-1} = 128.56$

Car C $= \dfrac{(4200)\left(\dfrac{0.06}{12}\right)\left(1+\dfrac{0.06}{12}\right)^{24}}{\left(1+\dfrac{0.06}{12}\right)^{24}-1} = 186.14$

Car A or Car B should be considered.

4. $m = \dfrac{(40{,}000)\left(\dfrac{0.18}{12}\right)\left(1 + \dfrac{0.18}{12}\right)^{36}}{\left(1 + \dfrac{0.18}{12}\right)^{36} - 1} = 1446.0958;\ \$1446.10;$

$(36)(1446.10) - 40{,}000 = 52{,}059.60 - 40{,}000 = 12{,}059.60;$
Harvey will pay \$12,059.60.

5. Down payment $= (145{,}000)(0.20) = 29{,}000;$

$m = \dfrac{(116{,}000)\left(\dfrac{0.114}{12}\right)\left(1 + \dfrac{0.114}{12}\right)^{240}}{\left(1 + \dfrac{0.114}{12}\right)^{240} - 1} = 1229.0756;$

The monthly payment will be \$1229.08.

page 342 Capsule Review

1. 32 R: 79   2. 28 R: 156   3. 90 R: 150

4. 777 R: 77   5. 101 R: 5   6. 108 R: 125

page 343 Class Exercises

1. $\dfrac{8q^2}{2q} - \dfrac{32q^3}{2q} = 4q - 16q^2$

2. $\dfrac{14t^4}{7t^2} - \dfrac{28t^3}{7t^2} + \dfrac{35t^2}{7t^2} - \dfrac{7t}{7t^2} = 2t^2 - 4t + 5 - \dfrac{1}{t}$

3. $\dfrac{(n-4)(n-1)}{(n-4)} = n - 1$

4.
$$
\begin{array}{r}
3s + 1 + \dfrac{8}{2s-3} \\[2pt]
2s - 3\overline{\smash{)}6s^2 - 7s + 5} \\
-(6s^2 - 9s\ ) \\ \hline
2s + 5 \\
-(2s - 3\ ) \\ \hline
8
\end{array}
$$

5.
$$
\begin{array}{r}
2x^2 + 3x - 1 \\
x - 3\overline{\smash{)}2x^3 - 3x^2 - 10x + 3} \\
-(2x^3 - 6x^2) \\ \hline
3x^2 - 10x \\
-(3x^2 - 9x) \\ \hline
-x + 3 \\
-(-x + 3) \\ \hline
0
\end{array}
$$

6.
$$
\begin{array}{r}
3a^2 + 6a + 12 + \dfrac{8}{a-2} \\
a - 2\overline{\smash{)}3a^3 + 0a^2 + 0a - 16} \\
-(3a^3 - 6a^2) \\ \hline
6a^2 + 0a \\
-(6a^2 - 12a) \\ \hline
12a - 16 \\
-(12a - 24) \\ \hline
8
\end{array}
$$

1. $\dfrac{6x^4}{3x^3} + \dfrac{3x^3}{3x^3} - \dfrac{x^2}{3x^3} = 2x + 1 - \dfrac{1}{3x}$

2. $\dfrac{10z^4}{5z^3} + \dfrac{5z^3}{5z^3} - \dfrac{z^2}{5z^3} = 2z + 1 - \dfrac{1}{5z}$

3. $\dfrac{x^3}{x} - \dfrac{18x^2}{x} + \dfrac{3x}{x} - \dfrac{7}{x} = x^2 - 18x + 3 - \dfrac{7}{x}$

4. $\dfrac{c^3}{c} + \dfrac{11c^2}{c} - \dfrac{15c}{c} + \dfrac{8}{c} = c^2 + 11c - 15 + \dfrac{8}{c}$

5. $\dfrac{3y^4}{3y} - \dfrac{15y^3}{3y} + \dfrac{21y^2}{3y} + \dfrac{6y}{3y} - \dfrac{9}{3y} = y^3 - 5y^2 + 7y + 2 - \dfrac{3}{y}$

6. $\dfrac{4m^4}{2m} + \dfrac{6m^3}{2m} - \dfrac{20m^2}{2m} + \dfrac{8m}{2m} - \dfrac{12}{2m} = 2m^3 + 3m^2 - 10m + 4 - \dfrac{6}{m}$

7. $\dfrac{(n + 6)(n - 5)}{n + 6} = n - 5$

8. $\dfrac{(a - 6)(a + 4)}{a + 4} = a - 6$

9.
$$
\begin{array}{r}
k + 5 + \dfrac{-12}{k + 6} \\[2pt]
\hline
k + 6\,)\,k^2 + 11k + 18 \\
-(k^2 + 6k) \\
\hline
5k + 18 \\
-(5k + 30) \\
\hline
-12
\end{array}
$$

10.
$$
\begin{array}{r}
r + 1 + \dfrac{3}{r + 2} \\[2pt]
\hline
r + 2\,)\,r^2 + 3r + 5 \\
-(r^2 + 2r) \\
\hline
r + 5 \\
-(r + 2) \\
\hline
3
\end{array}
$$

11.
$$
\begin{array}{r}
3j + 16 + \dfrac{43}{j - 3} \\[2pt]
\hline
j - 3\,)\,3j^2 + 7j - 5 \\
-(3j^2 - 9j) \\
\hline
16j - 5 \\
-(16j - 48) \\
\hline
43
\end{array}
$$

12.
$$
\begin{array}{r}
s + 10 + \dfrac{56}{s - 8} \\[2pt]
\hline
s - 8\,)\,s^2 + 2s - 24 \\
-(s^2 - 8s) \\
\hline
10s - 24 \\
-(10s - 80) \\
\hline
56
\end{array}
$$

13.
$$
\begin{array}{r}
5m + 12 \\[2pt]
\hline
2m - 3\,)\,10m^2 + 9m - 36 \\
-(10m^2 - 15m) \\
\hline
24m - 36 \\
-(24m - 36) \\
\hline
0
\end{array}
$$

14.
$$
\begin{array}{r}
3h - 10 \\[2pt]
\hline
5h - 4\,)\,15h^2 - 62h + 40 \\
-(15h^2 - 12h) \\
\hline
-50h + 40 \\
-(-50h + 40) \\
\hline
0
\end{array}
$$

15.
$$
\begin{array}{r}
d - 4 + \dfrac{-8}{d - 3} \\[2pt]
\hline
d - 3\,)\,d^2 - 7d + 4 \\
-(d^2 - 3d) \\
\hline
-4d + 4 \\
-(-4d + 12) \\
\hline
-8
\end{array}
$$

16.
$$
\begin{array}{r}
2m + 1 + \dfrac{7}{m - 2} \\[2pt]
\hline
m - 2\,)\,2m^2 - 3m + 5 \\
-(2m^2 - 4m) \\
\hline
m + 5 \\
-(m - 2) \\
\hline
7
\end{array}
$$

17.
$$
b - 1 \overline{)2b^3 + 0b^2 + 8b + 0} \quad \frac{2b^2 + 2b + 10 + \dfrac{10}{b-1}}{}
$$

$$
\begin{array}{r}
2b^2 + 2b + 10 + \dfrac{10}{b - 1} \\
b - 1 \overline{)2b^3 + 0b^2 + 8b + 0} \\
\underline{-(2b^3 - 2b^2)} \\
2b^2 + 8b \\
\underline{-(2b^2 - 2b)} \\
10b + 0 \\
\underline{-(10b - 10)} \\
10
\end{array}
$$

18.
$$
\begin{array}{r}
3a^2 + 3a + 12 + \dfrac{12}{a - 1} \\
a - 1 \overline{)3a^3 + 0a^2 + 9a + 0} \\
\underline{-(3a^3 - 3a^2)} \\
3a^2 + 9a \\
\underline{-(3a^2 - 3a)} \\
12a + 0 \\
\underline{-(12a - 12)} \\
12
\end{array}
$$

19.
$$
\begin{array}{r}
3x^2 - 3x + 3 - \dfrac{6}{x + 1} \\
x + 1 \overline{)3x^3 + 0x^2 + 0x - 3} \\
\underline{-(3x^3 + 3x^2)} \\
-3x^2 + 0x \\
\underline{-(-3x^2 - 3x)} \\
3x - 3 \\
\underline{-(3x + 3)} \\
-6
\end{array}
$$

20.
$$
\begin{array}{r}
3q^2 + 6q + 12 + \dfrac{30}{q - 2} \\
q - 2 \overline{)3q^3 + 0q^2 + 0q + 6} \\
\underline{-(3q^3 - 6q^2)} \\
6q^2 + 0q \\
\underline{-(6q^2 - 12q)} \\
12q + 6 \\
\underline{-(12q - 24)} \\
30
\end{array}
$$

B 21. $\dfrac{18s^2}{3s} - \dfrac{51s}{3s} = 6s - 17$

22. $\dfrac{108a^3}{9a^3} - \dfrac{72a^4}{9a^3} = 12 - 8a$

23. $\dfrac{18p^5q^4}{9pq^2} - \dfrac{54p^3q^3}{9pq^2} + \dfrac{27p^2q^5}{9pq^2} - \dfrac{9pq^2}{9pq^2} = 2p^4q^2 - 6p^2q + 3pq^3 - 1$

24. $\dfrac{21a^3b^2}{7a^2b} + \dfrac{56a^5b}{7a^2b} - \dfrac{28a^2b}{7a^2b} + \dfrac{63a^5b^2}{7a^2b} = 3ab + 8a^3 - 4 + 9a^3b$

25.
$$
\begin{array}{r}
p^2 + p - 7 + \dfrac{10}{p + 2} \\
p + 2 \overline{)p^3 + 3p^2 - 5p - 4} \\
\underline{-(p^3 + 2p^2)} \\
p^2 - 5p \\
\underline{-(p^2 + 2p)} \\
-7p - 4 \\
\underline{-(-7p - 14)} \\
10
\end{array}
$$

26.
$$
\begin{array}{r}
a^2 - a + 7 + \dfrac{-26}{a + 3} \\
a + 3 \overline{)a^3 + 2a^2 + 4a - 5} \\
\underline{-(a^3 + 3a^2)} \\
-a^2 + 4a \\
\underline{-(-a^2 - 3a)} \\
7a - 5 \\
\underline{-(7a + 21)} \\
-26
\end{array}
$$

27.
$$
\begin{array}{r}
3h^2 + h - 5 + \dfrac{-2}{2h + 1} \\
2h + 1 \overline{)6h^3 + 5h^2 - 9h - 7} \\
\underline{-(6h^3 + 3h^2)} \\
2h^2 - 9h \\
\underline{-(2h^2 + h)} \\
-10h - 7 \\
\underline{-(-10h - 5)} \\
-2
\end{array}
$$

28.
$$
\begin{array}{r}
3t^2 - 3t - 1 + \dfrac{7}{3t + 2} \\
3t + 2 \overline{)9t^3 - 3t^2 - 9t + 5} \\
\underline{-(9t^3 + 6t^2)} \\
-9t^2 - 9t \\
\underline{-(-9t^2 - 6t)} \\
-3t + 5 \\
\underline{-(-3t - 2)} \\
7
\end{array}
$$

29.
$$x - 4 \overline{)\, x^3 + 0x^2 - 3x + 5}$$

quotient: $x^2 + 4x + 13 + \dfrac{57}{x - 4}$

$$\begin{aligned}
&\underline{-(x^3 - 4x^2)}\\
&\quad 4x^2 - 3x\\
&\quad \underline{-(4x^2 - 16x)}\\
&\qquad\quad 13x + 5\\
&\qquad\quad \underline{-(13x - 52)}\\
&\qquad\qquad\quad 57
\end{aligned}$$

30.
$$y - 3 \overline{)\, y^3 + 0y^2 + 4y - 7}$$

quotient: $y^2 + 3y + 13 + \dfrac{32}{y - 3}$

$$\begin{aligned}
&\underline{-(y^3 - 3y^2)}\\
&\quad 3y^2 + 4y\\
&\quad \underline{-(3y^2 - 9y)}\\
&\qquad\quad 13y - 7\\
&\qquad\quad \underline{-(13y - 39)}\\
&\qquad\qquad\quad 32
\end{aligned}$$

31.
$$3h - 1 \overline{)\, 9h^3 + 0h^2 - 4h + 2}$$

quotient: $3h^2 + h - 1 + \dfrac{1}{3h - 1}$

$$\begin{aligned}
&\underline{-(9h^3 - 3h^2)}\\
&\quad 3h^2 - 4h\\
&\quad \underline{-(3h^2 - h)}\\
&\qquad\quad -3h + 2\\
&\qquad\quad \underline{-(-3h + 1)}\\
&\qquad\qquad\quad 1
\end{aligned}$$

32.
$$2j + 3 \overline{)\, 4j^3 + 0j^2 - 5j - 3}$$

quotient: $2j^2 - 3j + 2 + \dfrac{-9}{2j + 3}$

$$\begin{aligned}
&\underline{-(4j^3 + 6j^2)}\\
&\quad -6j^2 - 5j\\
&\quad \underline{-(-6j^2 - 9j)}\\
&\qquad\quad 4j - 3\\
&\qquad\quad \underline{-(4j + 6)}\\
&\qquad\qquad\quad -9
\end{aligned}$$

C 33.
$$3t - 7 \overline{)\, 27t^3 + 0t^2 + 0t - 343}$$

quotient: $9t^2 + 21t + 49$

$$\begin{aligned}
&\underline{-(27t^3 - 63t^2)}\\
&\quad 63t^2 + 0t\\
&\quad \underline{-(63t^2 - 147t)}\\
&\qquad\quad 147t - 343\\
&\qquad\quad \underline{-(147t - 343)}\\
&\qquad\qquad\quad 0
\end{aligned}$$

34.
$$4h - 5 \overline{)\, 64h^3 + 0h^2 + 0h - 125}$$

quotient: $16h^2 + 20h + 25$

$$\begin{aligned}
&\underline{-(64h^3 - 80h^2)}\\
&\quad 80h^2 + 0h\\
&\quad \underline{-(80h^2 - 100h)}\\
&\qquad\quad 100h - 125\\
&\qquad\quad \underline{-(100h - 125)}\\
&\qquad\qquad\quad 0
\end{aligned}$$

35.
$$5x - y \overline{)\, 15x^2 + 7xy - 2y^2}$$

quotient: $3x + 2y$

$$\begin{aligned}
&\underline{-(15x^2 - 3xy)}\\
&\quad 10xy - 2y^2\\
&\quad \underline{-(10xy - 2y^2)}\\
&\qquad\quad 0
\end{aligned}$$

36.
$$2m - n \overline{)\, 10m^2 - 11mn + 3n^2}$$

quotient: $5m - 3n$

$$\begin{aligned}
&\underline{-(10m^2 - 5mn)}\\
&\quad -6mn + 3n^2\\
&\quad \underline{-(-6mn + 3n^2)}\\
&\qquad\quad 0
\end{aligned}$$

37.
$$3p + 2q \overline{)\, 6p^2 + 10pq + 6q^2}$$

quotient: $2p + 2q + \dfrac{2q^2}{3p + 2q}$

$$\begin{aligned}
&\underline{-(6p^2 + 4pq)}\\
&\quad 6pq + 6q^2\\
&\quad \underline{-(6pq + 4q^2)}\\
&\qquad\quad 2q^2
\end{aligned}$$

38.
$$3c + 4d \overline{)\, 12c^2 + 25cd + 12d^2}$$

quotient: $4c + 3d$

$$\begin{aligned}
&\underline{-(12c^2 + 16cd)}\\
&\quad 9cd + 12d^2\\
&\quad \underline{-(9cd + 12d^2)}\\
&\qquad\quad 0
\end{aligned}$$

39. $x^2 - 1 = p + p\left(\dfrac{1}{x}\right)(x^2),\ x^2 - 1 = p + px,\ x^2 - 1 = p(1 + x)\ ,\ p = \dfrac{x^2 - 1}{1 + x},$

$p = \dfrac{(x + 1)(x - 1)}{(x + 1)},\ p = x - 1;\ p = 101 - 1 = \$100$

page 344   Algebra in Health

Answers may vary.

page 345   Capsule Review

1. $\dfrac{9}{16}$

2. $\dfrac{3}{126} = \dfrac{1}{42}$

3. $\dfrac{5}{4} \cdot \dfrac{1}{2} = \dfrac{5}{8}$

4. $\dfrac{7}{2} \cdot \dfrac{1}{7} = \dfrac{1}{2}$

5. $\dfrac{6d}{6} = \dfrac{-20}{6},\ d = -\dfrac{20}{6};\ d = -\dfrac{10}{3}$

6. $\dfrac{12y}{12} = \dfrac{72}{12};\ y = 6$

7. $\dfrac{4}{3} \cdot \dfrac{3}{4}a = 3 \cdot \dfrac{4}{3};\ a = 4$

8. $-3 \cdot \dfrac{-x}{3} = \dfrac{1}{4} \cdot -3;\ x = -\dfrac{3}{4}$

page 347   Class Exercises

1. $\dfrac{0.2}{3} = \dfrac{2}{30} = \dfrac{1}{15}$

2. $\dfrac{2.5}{3} = \dfrac{25}{30} = \dfrac{5}{6}$

3. $\dfrac{4a^2}{2ab} = \dfrac{2a}{b}$

4. $\dfrac{1.5}{45} = \dfrac{15}{45} = \dfrac{1}{3}$

5. $(2)(15) = (5)(6)$; true

6. $(2)(31) \neq (7)(9)$; false

7. $(2)\left(\dfrac{3}{2}\right) = (1)(3)$; true

8. $(2)(18) \neq (9)(3)$; false

9. $2x = (8)(5),\ 2x = 40,\ \dfrac{1}{2}(2x) = \dfrac{1}{2}(40),\ x = 20$

10. $(10)(3n) = (2)(-9),\ 30n = -18,\ \dfrac{1}{30}(30n) = \left(\dfrac{1}{30}\right)(-18),\ n = \dfrac{-18}{30},\ n = -\dfrac{3}{5}$

11. $3(a + 1) = (9)(2),\ 3a + 3 = 18,\ 3a = 15,\ \dfrac{1}{3}(3a) = \dfrac{1}{3}(15),\ a = 5$

12. $8(m - 3) = 40,\ 8m - 24 = 40,\ 8m - 24 + 24 = 40 + 24,\ 8m = 64,$

$\dfrac{1}{8}(8m) = \dfrac{1}{8}(64),\ m = \dfrac{64}{8} = 8$

13. $\dfrac{\frac{1}{2}}{20} = \dfrac{4}{x},\ \dfrac{1}{2}x = 4(20),\ \dfrac{1}{2}x = 80,\ x = 160$

A **1.** $\dfrac{6}{8} = \dfrac{3}{4}$

2. $\dfrac{48}{32} = \dfrac{3}{2}$

3. $\dfrac{\frac{1}{2}}{3} = \dfrac{1}{2} \cdot \dfrac{1}{3} = \dfrac{1}{6}$

4. $\dfrac{4}{\frac{8}{3}} = 4 \cdot \dfrac{3}{8} = \dfrac{3}{2}$

5. $\dfrac{0.5}{2.5} = \dfrac{5}{25} = \dfrac{1}{5}$

6. $\dfrac{7.5}{6} \cdot \dfrac{75}{60} = \dfrac{5}{4}$

7. $\dfrac{2}{4x} = \dfrac{1}{2x}$

8. $\dfrac{3y}{y^2} = \dfrac{3}{y}$

9. $\dfrac{5\text{ft}}{2\text{yd}} = \dfrac{5\text{ft}}{6\text{ft}} = \dfrac{5}{6}$

10. $\dfrac{2.4 \text{ cm}}{36 \text{ mm}} = \dfrac{24 \text{ mm}}{36 \text{ mm}} = \dfrac{2 \text{ mm}}{3 \text{ mm}} = \dfrac{2}{3}$

11. $\dfrac{3.20}{0.80} = \dfrac{32}{8} = \dfrac{4}{1}$

12. $\dfrac{0.72}{0.6} = \dfrac{72}{60} = \dfrac{12}{10} = \dfrac{6}{5}$

13. $(5)(7) \neq (12)(3)$; false

14. $(4)(16) \neq (9)(7)$; false

15. $\dfrac{6}{15} = \dfrac{16}{40}$, $(6)(40) = (15)(16)$, $240 = 240$, true

16. $(9)(55) \neq (30)(16)$; false

17. $\dfrac{3}{0.93} = \dfrac{5}{1.55}$, $(3)(1.55) = (5)(0.93)$, $4.65 = 4.65$; true

18. $\dfrac{5}{0.95} = \dfrac{8}{1.52}$, $(5)(1.52) = (8)(0.95)$, $7.60 = 7.60$; true

19. $5s = (6)(30)$, $5s = 180$, $s = 36$

20. $(7)(28) = 4r$, $196 = 4r$, $49 = r$

21. $4(s + 4) = (12)(7)$, $4s + 16 = 84$, $4s = 68$, $s = 17$

22. $9(t + 2) = (5)(18)$, $9t + 18 = 90$, $9t = 72$, $t = 8$

B **23.** $(8)(3) = 16a$, $24 = 16a$, $\dfrac{24}{16} = a$, $a = \dfrac{6}{4} = \dfrac{3}{2}$

24. $(5)(12) = 24x$, $60 = 24x$, $\dfrac{60}{24} = x$, $x = \dfrac{16}{4} = \dfrac{5}{2}$

25. $(9)(18p) = (12)(54)$, $162p = 648$, $p = 4$

26. $(7)(12c) = (28)(15)$, $84c = 420$, $c = 5$

27. $5(5n - 2) = 5(3n + 5)$, $25n - 10 = 15n + 25$, $25n - 10 + 10 = 15n + 25 + 10$, $25n = 15n + 35$, $25n - 15n = 15n - 15n + 35$, $10n = 35$, $n = \dfrac{35}{10} = \dfrac{7}{2}$

28. $5(7b + 1) = 3(3 + b)$, $35b + 5 = 9 + 3b$, $35b + 5 - 5 = 9 - 5 + 3b$,

$35b = 4 + 3b$, $35b - 3b = 4 + 3b - 3b$, $32b = 4$, $b = \dfrac{4}{32} = \dfrac{1}{8}$

29. $4(2x + 1) = 2(x + 3)$, $8x + 4 = 2x + 6$, $8x + 4 - 4 = 2x + 6 - 4$,

$8x = 2x + 2$, $8x - 2x = 2x - 2x + 2$, $6x = 2$, $x = \dfrac{2}{6} = \dfrac{1}{3}$

30. $4(3y + 2) = 2(y + 4)$, $12y + 8 = 2y + 8$, $12y + 8 - 8 = 2y + 8 - 8$,
$12y = 2y$, $12y - 2y = 2y - 2y$, $10y = 0$, $y = 0$

31. $3(s - 2) = 5(2s + 3)$, $3s - 6 = 10s + 15$, $3s - 6 + 6 = 10s + 15 + 6$,
$3s = 10s + 21$, $3s - 10s = 10s - 10s + 21$, $-7s = 21$, $s = -3$

32. $6(t + 3) = 2(2t - 5)$, $6t + 18 = 4t - 10$, $6t + 18 - 18 = 4t - 10 - 18$,
$6t = 4t - 28$, $6t - 4t = 4t - 4t - 28$, $2t = -28$, $t = -14$

C 33. $(2)(4) = (2m + 3)(3m - 2)$, $8 = 6m^2 - 4m + 9m - 6$, $8 = 6m^2 + 5m - 6$,
$0 = 6m^2 + 5m - 14$, $(6m - 7)(m + 2) = 0$, $6m - 7 = 0$ or $m + 2 = 0$,

$6m = 7$, $m = \dfrac{7}{6}$ or $m = -2$

34. $5a^2 = 4(a + 3)$, $5a^2 = 4a + 12$, $5a^2 - 4a - 12 = 0$,
$(5a + 6)(a - 2) = 0$

$5a + 6 = 0 \quad \mid a - 2 = 0$

$5a = -6 \quad \mid$

$a = -\dfrac{6}{5} \quad \mid \quad a = 2$

35. $(q)(q + 4) = 12$, $q^2 + 4q = 12$, $q^2 + 4q - 12 = 0$,
$(q + 6)(q - 2) = 0$

$q + 6 = 0 \quad \mid q - 2 = 0$

$q = -6 \quad \mid \quad q = 2$

36. $(7)(5) = (p)(p + 2)$, $35 = p^2 + 2p$, $p^2 + 2p - 35 = 0$,
$(p + 7)(p - 5) = 0$

$p + 7 = 0 \quad \mid p - 5 = 0$

$p = -7 \quad \mid \quad p = 5$

37. $(w - 3)(w + 5) = (3)(3)$, $w^2 - 3w + 5w - 15 = 9$, $w^2 + 2w - 15 = 9$,
$w^2 + 2w - 24 = 0$, $(w + 6)(w - 4) = 0$,
$w + 6 = 0 \quad \mid w - 4 = 0$

$w = -6 \quad \mid \quad w = 4$

38. $(11)(2) = (u + 7)(u - 2)$, $22 = u^2 + 7u - 2u - 14$, $u^2 + 5u - 14 = 22$,
$u^2 + 5u - 36 = 0$, $(u + 9)(u - 4) = 0$,

$$u + 9 = 0 \mid u - 4 = 0$$
$$u = -9 \mid \quad u = 4$$

39. a. 2 qt pineapple juice = 8c of pineapple juice; $\dfrac{8}{1\frac{3}{4}} = \dfrac{8}{\frac{7}{4}}$, $\dfrac{8}{1} \cdot \dfrac{4}{7} = \dfrac{32}{7}$

b. 3 pt of orange juice = 6c of orange juice; $\dfrac{2\frac{1}{2}}{6} = \dfrac{\frac{5}{2}}{6}$, $\dfrac{5}{2} \cdot \dfrac{1}{6} = \dfrac{5}{12}$

c. 2 qt of pineapple juice = 4 pt of pineapple juice; $\dfrac{3}{4}$

40. $\dfrac{1.5 \text{ cm}}{6 \text{ km}} = \dfrac{3.25 \text{ cm}}{x}$, $(1.5 \text{ cm})(x) = (6 \text{ km})(3.25 \text{ cm})$, $x = \dfrac{(6 \text{ km})(3.25 \text{ cm})}{1.5 \text{ cm}}$;
$x = 13 \text{ km}$

41. $\dfrac{2.5 \text{ cm}}{10 \text{ km}} = \dfrac{4.25 \text{ cm}}{x}$, $(2.5 \text{ cm})(x) = (10 \text{ km})(4.25 \text{ cm})$, $x = \dfrac{(10 \text{ km})(4.25 \text{ cm})}{2.5 \text{ cm}}$;
$x = 17 \text{ km}$

42. $\dfrac{66 \text{ mi}}{1\frac{1}{2} \text{ h}} = \dfrac{x}{2 \text{ h}}$; $x = \dfrac{(66 \text{ mi})(2 \text{ h})}{\frac{3}{2} \text{ h}}$; $x = 88 \text{ mi}$

43. $\dfrac{45 \text{ mi}}{2\frac{1}{2} \text{ h}} = \dfrac{153 \text{ mi}}{x}$, $(x)(45 \text{ mi}) = (153 \text{ mi})\left(\dfrac{5}{2} \text{ h}\right)$, $x = \dfrac{(153 \text{ mi})\left(\frac{5}{2} \text{ h}\right)}{45 \text{ mi}}$; $x = 8.5 \text{ h}$

page 349   Biography: Jean-Victor Poncelet

Answers may vary.

page 350   Capsule Review

1. LCD is $5 \cdot x = 5x$

2. LCD is $a^2 \cdot b^3 = a^2 b^3$

3. LCD is $4 \cdot 11 \cdot m \cdot m = 44m^2$

4. LCD is $2y(y + 2)$

5. LCD is $4(r + s)$

6. LCD is $(d + 3)(d - 3)$

page 352   Class Exercises

1. $(3)\left(\dfrac{y}{3}\right) + (3)\left(\dfrac{2}{3}\right) = (3)(1)$, $y + 2 = 3$; $y = 1$

2. $(6)\left(\dfrac{1}{3}\right) + (6)\left(\dfrac{5z}{6}\right) = (6)(2)$, $2 + 5z = 12$, $5z = 10$; $z = 2$

3. $(a)\left(\dfrac{3}{a}\right) - (a)\left(\dfrac{5}{a}\right) = (2)(a)$, $3 - 5 = 2a$, $-2 = 2a$; $a = -1$

4. $(3m)\left(\dfrac{2}{3}\right) - (3m)\left(\dfrac{5}{m}\right) = (3m)\left(\dfrac{1}{3m}\right)$, $2m - 15 = 1$, $2m = 16$; $m = 8$

5. $(6)(x) + (6)\left(\dfrac{2}{3}\right) = (6)\left(\dfrac{5x}{6}\right)$, $6x + 4 = 5x$, $4 = -x$; $x = -4$

6. $(p)(5) + (p)\left(\dfrac{2}{p}\right) = (p)\left(\dfrac{17}{p}\right)$, $5p + 2 = 17$, $5p = 15$; $p = 3$

7. $(b + 3)\left(\dfrac{b}{b + 3}\right) = (b + 3)(2) - (b + 3)\left(\dfrac{3}{b + 3}\right)$, $b = 2b + 6 - 3$,

$b = 2b + 3$, $-b = 3$; $b = -3$; Check: $\dfrac{-3}{-3 + 3} = 2 - \dfrac{3}{-3 + 3}$;

The denominator is 0; no solution.

8. $(f)(f + 1)\left(\dfrac{5}{f}\right) + (f)(f + 1)\left(\dfrac{3}{f + 1}\right) = (f)(f + 1)\left(\dfrac{7}{f}\right)$, $5(f + 1) + 3f =$

$7(f + 1)$, $5f + 5 + 3f = 7f + 7$, $8f + 5 = 7f + 7$, $8f = 7f + 2$, $f = 2$

9. $(g)(g - 3)\left(\dfrac{1}{g - 3}\right) + (g)(g - 3)(1) = (g)(g - 3)\left[\dfrac{3}{g(g - 3)}\right]$, $g + g^2 - 3g =$

3, $g^2 - 2g - 3 = 0$, $(g - 3)(g + 1) = 0$, $g = 3$ or $g = -1$; Check: $\dfrac{1}{3 - 3} + 1 =$

$\dfrac{3}{3^2 - (3)(3)}$; The denominator is 0, 3 is not a solution; the solution is -1.

page 352–353 Practice Exercises

A **1.** $(12)(3m) - (12)\left(\dfrac{3}{4}\right) = (12)\left(\dfrac{2m}{3}\right)$, $36m - 9 = 8m$, $28m = 9$; $m = \dfrac{9}{28}$

2. $(8)(2y) - 8\left(\dfrac{3}{4}\right) = (8)\left(\dfrac{3y}{8}\right)$, $16y - 6 = 3y$, $13y = 6$; $y = \dfrac{6}{13}$

3. $(6)(5x) - (6)\left(\dfrac{2}{3}\right) = (6)\left(\dfrac{-5x}{6}\right)$, $30x - 4 = -5x$, $35x = 4$; $x = \dfrac{4}{35}$

4. $(15)(4t) - (15)\left(\dfrac{3}{5}\right) = (15)\left(\dfrac{-2t}{3}\right)$, $60t - 9 = -10t$, $70t = 9$; $t = \dfrac{9}{70}$

5. $(6)\left(\dfrac{5e}{3}\right) - (6)\left(\dfrac{7e}{6}\right) = (6)(2)$, $10e - 7e = 12$, $3e = 12$; $e = 4$

6. $(15)\left(\dfrac{7f}{3}\right) - (15)\left(\dfrac{8f}{15}\right) = (15)(9)$, $35f - 8f = 135$, $27f = 135$; $f = 5$

7. $(b + 3)\left(\dfrac{b}{b + 3}\right) = (b + 3)(5) - (b + 3)\left(\dfrac{3}{b + 3}\right)$, $b = 5b + 15 - 3$,

$-4b = 12$; $b = -3$; Check: $\dfrac{-3}{-3 + 3} = 5 - \dfrac{3}{-3 + 3}$; The denominator is 0;

no solution.

8. $(x + 5)\left(\dfrac{x}{x + 5}\right) = (x + 5)(3) - (x + 5)\left(\dfrac{5}{x + 5}\right)$, $x = 3x + 15 - 5$,

$-2x = 10$; $x = -5$; Check: $\dfrac{-5}{-5 + 5} = 3 - \dfrac{5}{-5 + 5}$; The denominator is 0;

no solution.

9. $(t - 4)\left(\dfrac{2t}{t - 4}\right) = (t - 4)(5) - (t - 4)\left(\dfrac{1}{t - 4}\right)$, $2t = 5t - 20 - 1$,

$2t = 5t - 21$, $-3t = -21$; $t = 7$

10. $(s - 5)\left(\dfrac{3s}{s - 5}\right) = (s - 5)(7) - (s - 5)\left(\dfrac{1}{s - 5}\right)$, $3s = 7s - 35 - 1$,

$-4s = -36$; $s = 9$

11. $i\left(\dfrac{2}{i}\right) - i\left(\dfrac{8}{i}\right) = (i)(-15)$, $2 - 8 = -15i$, $-6 = -15i$, $i = \dfrac{6}{15} = \dfrac{2}{5}$

12. $(j)\left(\dfrac{7}{j}\right) - (j)\left(\dfrac{9}{j}\right) = (j)(-14)$; $7 - 9 = -14j$, $-2 = -14j$; $j = \dfrac{1}{7}$

13. $m \cdot 2 - m \cdot \dfrac{8}{m} = m \cdot 6$; $2m - 8 = 6m$, $4m = -8$; $m = -2$

14. $k \cdot 1 - k \cdot \dfrac{9}{k} = k \cdot 4$, $k - 9 = 4k$, $3k = -9$; $k = -3$

15. $\left(\dfrac{5}{2s}\right)(4s) + \left(\dfrac{3}{4}\right)(4s) = \left(\dfrac{9}{4s}\right)(4s)$, $10 + 3s = 9$, $3s = -1$; $s = -\dfrac{1}{3}$

16. $\left(\dfrac{2}{3t}\right)(12t) + \left(\dfrac{1}{2}\right)(12t) = \dfrac{3}{4t}(12t)$, $8 + 6t = 9$, $6t = 1$; $t = \dfrac{1}{6}$

17. $\left(\dfrac{u + 1}{u}\right)(2u) + \left(\dfrac{1}{2u}\right)(2u) = (4)(2u)$, $2u + 2 + 1 = 8u$, $6u = 3$; $u = \dfrac{1}{2}$

18. $\left(\dfrac{v + 2}{v}\right)(3v) + \left(\dfrac{4}{3v}\right)(3v) = (11)(3v)$, $3v + 6 + 4 = 33v$, $30v = 10$; $v = \dfrac{1}{3}$

19. $\left(\dfrac{4w + 5}{w - 4}\right)(w - 4) = \left(\dfrac{5w}{w - 4}\right)(w - 4)$, $4w + 5 = 5w$; $w = 5$

20. $\left(\dfrac{2x+4}{x-3}\right)(x-3) = \left(\dfrac{3x}{x-3}\right)(x-3)$, $2x+4 = 3x$; $x = 4$

B 21. $\left(\dfrac{y}{y-3}\right)(y-3) = \left(\dfrac{3}{y-3}\right)(y-3) - 1(y-3)$, $y = 3 - y + 3$; $y = 3$;

Check: $y = 3$, $\dfrac{3}{3-3} = \dfrac{3}{3-3} - 1$; The denominator is 0; no solution.

22. $\left(\dfrac{z}{z+2}\right)(z+2) = 3(z+2) - \left(\dfrac{2}{z+2}\right)(z+2)$, $z = 3z + 6 - 2$; $z = -2$;

Check: $z = -2$, $\dfrac{-2}{-2+2} = 3 - \dfrac{2}{-2+2}$; The denominator is 0; no solution.

23. $\left(\dfrac{c}{c+6}\right)(c+6)(c+2) = \left(\dfrac{1}{c+2}\right)(c+6)(c+2)$, $c(c+2) = 1(c+6)$,

$c^2 + 2c = c + 6$, $c^2 + c - 6 = 0$, $(c-2)(c+3) = 0$, $c = 2$ or $c = -3$

24. $\left(\dfrac{h}{h+5}\right)(h+5) = \left(\dfrac{2}{h+5}\right)(h+5)$, $h(h+5) = 2(h+5)$, $h^2 + 5h = 2h + 10$,

$h^2 + 3h - 10 = 0$, $(h+5)(h-2) = 0$; $h = -5$ or $h = 2$

25. $\left(\dfrac{3}{e-1}\right)(e-1)(e+4) = \left(\dfrac{2e}{e+4}\right)(e-1)(e+4)$, $3(e+4) = 2e(e-1)$,

$3e + 12 = 2e^2 - 2e$, $2e^2 - 5e - 12 = 0$, $(2e+3)(e-4) = 0$; $e = -\dfrac{3}{2}$ or $e = 4$

26. $\left(\dfrac{4}{c-4}\right)(c-4)(c+3) = \left(\dfrac{3c}{c+3}\right)(c-4)(c+3)$, $4(c+3) = 3c(c-4)$,

$4c + 12 = 3c^2 - 12c$, $3c^2 - 16c - 12 = 0$, $(3c+2)(c-6) = 0$; $c = -\dfrac{2}{3}$ or $c = 6$

27. $\left(\dfrac{4}{k}\right)(3k) + \left(\dfrac{2}{3}\right)(3k) = (10k)(3k)$, $12 + 2k = 30k^2$, $30k^2 - 2k - 12 = 0$,

$15k^2 - k - 6 = 0$, $(5k+3)(3k-2) = 0$; $k = -\dfrac{3}{5}$ or $k = \dfrac{2}{3}$

28. $(5f)(2f) = \left(\dfrac{7}{2}\right)(2f) + \left(\dfrac{6}{f}\right)(2f)$, $10f^2 = 7f + 12$, $10f^2 - 7f - 12 = 0$,

$(5f+4)(2f-3) = 0$; $f = -\dfrac{4}{5}$ or $f = \dfrac{3}{2}$

29. $\left(\dfrac{2}{e-2}\right)(e)(e-2) = 2(e)(e-2) - \left(\dfrac{4}{e}\right)(e)(e-2)$, $2e = 2e^2 - 4e - 4e + 8$,

$2e^2 - 10e + 8 = 0$, $e^2 - 5e + 4 = 0$, $(e-4)(e-1) = 0$; $e = 4$ or $e = 1$

30. $(6)(r)(r-3) - \left(\dfrac{2}{r}\right)(r)(r-3) = \left(\dfrac{-5}{r-3}\right)(r)(r-3)$, $6r^2 - 18r - 2r + 6 = -5r$,

$6r^2 - 15r + 6 = 0$, $2r^2 - 5r + 2 = 0$, $(2r-1)(r-2) = 0$; $r = \dfrac{1}{2}$ or $r = 2$

31. $\left(\dfrac{2e}{e-4}\right)(e-4)(e+5) - (2)(e-4)(e+5) = \left(\dfrac{4}{e+5}\right)(e-4)(e+5),$

$2e(e+5) - 2(e^2 + e - 20) = 4(e-4),\ 2e^2 + 10e - 2e^2 - 2e + 40 =$

$4e - 16,\ 4e = -56;\ e = -14$

C 32. $\left(\dfrac{r+1}{r-1}\right)(3)(r-1) = \left(\dfrac{r}{3}\right)(3)(r-1) + \left(\dfrac{2}{r-1}\right)(3)(r-1),\ 3(r+1) =$

$(r)(r-1) + 6,\ 3r+3 = r^2 - r + 6,\ r^2 - 4r + 3 = 0,\ (r-3)(r-1) = 0;\ r = 3$

or $r = 1$; Check: $r = 1;\ \dfrac{1+1}{1-1} = \dfrac{1}{3} + \dfrac{2}{1-1};$ The denominator is 0; the

solution is $r = 3$.

33. $\left(\dfrac{2}{a+2}\right)(a+2)(a-2) = \left(\dfrac{a}{a-2}\right)(a+2)(a-2) + \left[\dfrac{-13}{(a+2)(a-2)}\right](a+2)(a-2),$

$2(a-2) = a(a+2) - 13,\ 2a - 4 = a^2 + 2a - 13,\ a^2 - 9 = 0,\ (a-3)(a+3) = 0;$

$a = 3$ or $a = -3$

34. $\left(\dfrac{e+1}{e+2}\right)(e+2)(e-3) = \left(\dfrac{-1}{e-3}\right)(e+2)(e-3) + \left[\dfrac{e-1}{(e-3)(e+2)}\right](e+2)(e-3),$

$(e+1)(e-3) = -1(e+2) + e - 1,\ e^2 - 2e - 3 = -e - 2 + e - 1,$

$e^2 - 2e = 0,\ e(e-2) = 0;\ e = 0$ or $e = 2$

35. $\left(\dfrac{f-2}{f-4}\right)(f-4)(f+2) = \left(\dfrac{1}{f+2}\right)(f-4)(f+2) + \left[\dfrac{f+3}{(f-4)(f+2)}\right](f-4)(f+2),$

$(f-2)(f+2) = f - 4 + f + 3,\ f^2 - 4 = 2f - 1,\ f^2 - 2f - 3 = 0;\ (f-3)(f+1) = 0;$

$f = 3$ or $f = -1$

36. $\dfrac{u}{2(u-1)} \cdot 2(2u+1)(u-1) + \dfrac{u+1}{2u+1} \cdot 2(2u+1)(u-1) =$

$\dfrac{-3u}{(2u+1)(u-1)} \cdot 2(2u+1)(u-1),\ u(2u+1) + 2(u+1)(u-1) =$

$2(-3u),\ 4u^2 + 7u - 2 = 0;\ (4u-1)(u+2) = 0;\ u = \dfrac{1}{4}$ or $u = -2$

37. $\left(\dfrac{s}{3s+2}\right)(2)(3s+2)(s-2) + \left(\dfrac{s+3}{2s-4}\right)(2)(3s+2)(s-2) =$

$\left[\dfrac{-2s}{(3s+2)(s-2)}\right](2)(3s+2)(s-2),\ 2s(s-2) + (s+3)(3s+2) = 2(-2s),$

$5s^2 + 11s + 6 = 0,\ (5s+6)(s+1) = 0;\ s = -\dfrac{6}{5}$ or $s = -1$

38. Let a, b, and c represent the lengths of the sides of the triangle. $a + b +$

$c = 24,\ c = 2 + a,\ b = \dfrac{3}{5}c,\ a + \dfrac{3}{5}c + 2 + a = 24,\ a + \dfrac{3}{5}(2 + a) + 2 + a =$

$24,\ a + \dfrac{6}{5} + \dfrac{3}{5}a + 2 + a = 24,\ (5)\left(2a + 2 + \dfrac{3}{5}a + \dfrac{6}{5}\right) = (5)(24),\ 10a +$

$10 + 3a + 6 = 120,\ 13a + 16 = 120,\ 13a = 104;\ a = 8;\ b = 6;\ c = 10$

39. Let T represent the total book allowance. $\frac{1}{3}T + \frac{1}{4}T + \frac{1}{5}T = 940$,

$$60\left(\frac{T}{3}\right) + 60\left(\frac{T}{4}\right) + 60\left(\frac{T}{5}\right) = (60)(940),\ 20T + 15T + 12T = 56{,}400,$$

$47T = 56{,}400;\ T = 1200;$ Amount left $= 1200 - 940;\ \$260.00$.

40. $2^2 \cdot 3 \cdot 5$ **41.** $7 \cdot 13$ **42.** $3 \cdot 5 \cdot 7$ **43.** $2^2 \cdot 3^3 \cdot 5^2$

44. $(x + 3)(2x - 3) = 0$ **45.** $a(25a - 16) = 0$ **46.** $n^2 - 14n + 49 = 0$

$x + 3 = 0 \mid 2x - 3 = 0$ $a = 0 \mid 25a - 16 = 0$ $(n - 7)(n - 7) = 0$

$\qquad x = -3 \mid \quad x = \dfrac{3}{2}$ $a = \dfrac{16}{25}$ $n = 7 \mid n = 7$

47. $81x^2 - 25 = 0$ **48.** $y^2 - y - 42 = 0$ **49.** $(2c + 1)(c + 3) = 0$

$(9x + 5)(9x - 5) = 0$ $(y + 6)(y - 7) = 0$ $2c + 1 = 0 \mid c + 3 = 0$

$9x + 5 = 0 \mid 9x - 5 = 0$ $y + 6 = 0 \mid y - 7 = 0$ $2c = -1 \mid \quad c = -3$

$\quad x = -\dfrac{5}{9} \mid \quad x = \dfrac{5}{9}$ $y = -6 \mid \quad y = 7$ $\quad c = -\dfrac{1}{2} \mid$

50. Let x represent the number. $x + x^2 = 90$,   **51.** Let x and $x + 2$ represent the integers.

$x^2 + x - 90 = 0$, $x(x + 2) = 168,\ x^2 + 2x = 168$,

$(x + 10)(x - 9) = 0$ $x^2 + 2x - 168 = 0$

$x + 10 = 0 \mid x - 9 = 0$ $(x + 14)(x - 12) = 0$

$\quad x = -10 \mid \quad x = 9$ $x + 14 = 0 \mid x - 12 = 0$

The integer is 9. $\quad x = -14 \mid \quad x = 12$

The integers are -14 and -12.

page 356   Class Exercises

1. Given: Each person's time working alone, and that Barry and Carrie work for 2 h. Find the time needed by Harry to finish the job.

2. Barry: $\dfrac{1}{12}$; Carrie: $\dfrac{1}{10}$; Harry: $\dfrac{1}{9}$

3. Barry: $\dfrac{2}{12} = \dfrac{1}{6}$; Carrie: $\dfrac{2}{10} = \dfrac{1}{5}$, together: $\dfrac{1}{6} + \dfrac{1}{5} = \dfrac{5}{30} + \dfrac{6}{30} = \dfrac{11}{30}$

4. $\dfrac{11}{30} + \dfrac{x}{9} = 1,\ 90\left(\dfrac{x}{9}\right) + 90\left(\dfrac{11}{30}\right) = 90(1),\ 10x + 33 = 90;\ x = \dfrac{57}{10}$, 5 h 42 min

pages 356–357   Practice Exercises

A **1.**

	work rate	· time worked	= part of job done
Marian	$\dfrac{1}{3}$	x	$\dfrac{x}{3}$
Robin	$\dfrac{1}{4}$	x	$\dfrac{x}{4}$

$\frac{x}{3} + \frac{x}{4} = 1$, $4x + 3x = 12$, $7x = 12$, $x = \frac{12}{7}$; It will take them $\frac{12}{7}$ h to weed a garden if they work together.

2.

work rate	· time worked	= part of job done
David	$\frac{1}{20}$	x	$\frac{x}{20}$
Allie	$\frac{1}{35}$	x	$\frac{x}{35}$

$\frac{x}{20} + \frac{x}{35} = 1$, $140\left(\frac{x}{20} + \frac{x}{35}\right) = 140 \cdot 1$, $7x + 4x = 140$, $11x = 140$, $x = \frac{140}{11} = 12\frac{8}{11}$; It will take $12\frac{8}{11}$ min to unload a delivery truck if they work together.

3.

work rate	· time worked	= part of job done
Art	$\frac{1}{5}$	x	$\frac{x}{5}$
Mother	$\frac{1}{4}$	x	$\frac{x}{4}$

$\frac{x}{5} + \frac{x}{4} = 1$, $20\left(\frac{x}{5} + \frac{x}{4}\right) = 20 \cdot 1$, $4x + 5x = 20$, $9x = 20$, $x = \frac{20}{9} = 2\frac{2}{9}$;

It will take $2\frac{2}{9}$ h to paint a set of kitchen cabinets if they work together.

4.

work rate	· time worked	= part of job done
Peggy	$\frac{1}{45}$	x	$\frac{x}{45}$
Peter	$\frac{1}{75}$	x	$\frac{x}{75}$

$\frac{x}{45} + \frac{x}{75} = 1$, $225\left(\frac{x}{45} + \frac{x}{75}\right) = 225 \cdot 1$, $5x + 3x = 225$, $8x = 225$, $x = \frac{225}{8} = 28\frac{1}{8}$; It will take $28\frac{1}{8}$ min to gather a bushel of apples if they work together.

5.

	Rate	× Time	= Distance
Rhoda	$r + 7$	$\frac{275}{r + 7}$	275
Van	r	$\frac{240}{r}$	240

$\frac{275}{r + 7} = \frac{240}{r}$, $275r = 240r + 1680$, $35r = 1680$, $r = 48$; Van drove at 48 mi/h; Rhoda drove at 55 mi/h.

6.

	Rate	× Time	= Distance
Earl	$\frac{135}{t + 3}$	$t + 3$	135
Alice	$\frac{90}{t}$	t	90

$\frac{135}{t + 3} = \frac{90}{t}$, $135t = 90t + 270$, $45t = 270$, $t = 6$; Earl rode for 9 h.

Rate	×	Time	=	Distance
Car	r	$\dfrac{75}{r}$		75
Plane	$12r$	$\dfrac{2100}{12r}$		2100

$\dfrac{75}{r} + \dfrac{2100}{12r} = 4$, $(12)(75) + 2100 =$
$(4)(12r)$, $900 + 2100 = 48r$, $3000 = 48r$,
$r = \dfrac{3000}{48} = 62\dfrac{1}{2}$; Maggie drove $62\dfrac{1}{2}$ mi/h.

8.

Rate	×	Time	=	Distance
Frwy	$3r$	$\dfrac{135}{3r}$		135
Road	r	$\dfrac{45}{r}$		45

$\dfrac{135}{3r} + \dfrac{45}{r} = 5$, $3r\left(\dfrac{135}{3r}\right) + 3r\left(\dfrac{45}{r}\right) =$
$3r(5)$, $135 + 135 = 15r$, $270 = 15r$,
$r = 18$; Fred drove 54 mi/h on the freeway.

9.

	work rate	·	time worked	=	part of job done
Large pipe	$\dfrac{1}{r}$		9		$\dfrac{9}{r}$
Small pipe	$\dfrac{1}{3r}$		9		$\dfrac{9}{3r}$

$\dfrac{9}{3r} + \dfrac{9}{r} = 1$, $9 + 27 = 3r$, $36 = 3r$, $r = 12$; It would take the large pipe
12 h to fill the tank.

10.

	work rate	·	time worked	=	part of job done
hot	$\dfrac{1}{2r}$		20		$\dfrac{20}{2r}$
cold	$\dfrac{1}{r}$		20		$\dfrac{20}{r}$

$\dfrac{20}{r} + \dfrac{20}{2r} = 1$, $40 + 20 = 2r$, $60 = 2r$, $r = 30$; It would take the cold water
faucet 30 min to fill the bath tub.

C 11.

	work rate	·	time worked	=	part of job done
Apprentice	$\dfrac{1}{r}$		2 h 16 min		$\dfrac{136}{r}$
Sumi	$\dfrac{1}{\frac{3}{4}r}$		2 h 16 min		$\dfrac{136}{\frac{3}{4}r}$
Sumi alone	$\dfrac{1}{\frac{3}{4}r}$		4 h 32 min		$\dfrac{272}{\frac{3}{4}r}$

$\dfrac{136}{r} + \dfrac{136}{\frac{3}{4}r} + \dfrac{272}{\frac{3}{4}r} = 1$, $\dfrac{136}{r} + (136)\left(\dfrac{4}{3r}\right) + 272\left(\dfrac{4}{3r}\right) = 1$, $\dfrac{136}{r} + \dfrac{544}{3r} + \dfrac{1088}{3r} = 1$,

$408 + 544 + 1088 = 3r$, $3r = 2040$, $r = 680$; It would take the apprentice 680 min or 11 h 20 min to wash all the windows.

12.

	work rate ·	time worked =	part of job done
Tom	$\dfrac{1}{r}$	1 h 11 min	$\dfrac{71}{r}$
Tim	$\dfrac{1}{\frac{2}{3}r}$	1 h 11 min	$\dfrac{71}{\frac{2}{3}r}$
Tom alone	$\dfrac{1}{r}$	35 min 30 s	$\dfrac{35.5}{r}$

$\dfrac{71}{r} + \dfrac{71}{\frac{2}{3}r} + \dfrac{35.5}{r} = 1$, $\dfrac{71}{r} + 71\left(\dfrac{3}{2r}\right) + \dfrac{35.5}{r} = 1$, $\dfrac{71}{r} + \dfrac{213}{2r} + \dfrac{35.5}{r} = 1$,

$71 + 106.5 + 35.5 = r$, $r = 213$; $\dfrac{2}{3}r = 142$; It would take Tim 142 min

or 2 h 22 min working alone to trim the ten trees.

13. $j = \left(\dfrac{3}{2}\right)(-6k)$, $j = -9k$

14. $17 - a = 5b$, $\dfrac{17 - a}{5} = b$, $b = \dfrac{17 - a}{5}$

15. $a = 3 - 9c$, $a - 3 = -9c$, $\dfrac{a - 3}{-9} = c$, $c = \dfrac{-a + 3}{9}$

16. $-30m + 6n = p$, $6n = p + 30m$, $n = \dfrac{p + 30m}{6}$

17. Let x represent the number of students taking the course now. $x = 80 + (0.30)(80)$, $x = 80 + 24$, $x = 104$; One hundred four students take the course now.

18. Let x represent the amount of the investment. $(0.052)(x) = 66.30$, $x = 1275$; The amount of the investment is \$1275.

page 359–360   Class Exercises

1. $p = 18n - 18u - \dfrac{tu}{n}$

2. $p = 15n - 15(0) - \dfrac{0 \cdot t}{n}$, $p = 15n$

pages 360–361   Practice Exercises

A 1. $t = 12{,}225$, $n = 650$, $u = 14$; $p = 15(650) - 15(14) - \dfrac{(14)(12{,}225)}{650}$, $p =$

$9750 - 210 - 263.31$, $p = 9276.69$; A profit of \$9276.69 will be made.

2. $t = 11{,}500$, $n = 545$, $u = 23$; $p = 15(545) - 15(23) - \dfrac{(23)(11{,}500)}{545}$, $p =$

$8175 - 345 - 485.32$, $p = 7344.68$; A profit of \$7344.68 will be made.

3. $n = 500$, $p = (15)(500) - 150u - \dfrac{ut}{500}$, $p = \dfrac{(15)(500)^2 - (15)(500)(u) - ut}{500}$,

$p = \dfrac{(15)(500)^2 - 7500u - ut}{500}$, $p = \dfrac{3{,}750{,}000 - (7500)(45) - (45)(11{,}000)}{500}$,

$p = 5835$; A profit of \$5835.00 will be made.

4. $p = 15n - 15u - \dfrac{10{,}000u}{n}$, $p = \dfrac{15n^2 - 15un - 10{,}000u}{n}$,

$p = \dfrac{(15)475^2 - (15)(45)(475) - (10{,}000)(45)}{475}$, $p = 5502.6316$; A profit of

\$5502.63 will be made.

5. Total income = income from the deli + income from the service station.
Total income = $r(5.50) + s(4.50)$.

6. Amount saved = amount at self service − amount at full service,
$s = 12.5c - 12.5z$; $s = 12.5(c - z)$.

B 7. Commuting time = time for 10 mi + time for next 7 mi, $t = \dfrac{10}{u} + \dfrac{7}{v}$;

$t = \dfrac{10v + 7u}{uv}$.

8.

	work rate	· time worked	= part of job done
Juana	$\dfrac{1}{j}$	t	$\dfrac{t}{j}$
Kate	$\dfrac{1}{k}$	t	$\dfrac{t}{k}$

$\dfrac{t}{j} + \dfrac{t}{k} = 1$, $jk\left(\dfrac{t}{j} + \dfrac{t}{k}\right) = jk \cdot 1$, $\dfrac{kt + jt}{jk} = 1$, $t(j + k) = jk$, $t = \dfrac{jk}{j + k}$

9. Amount Sue received = amount Sue sold × price for each pattern sold.
Amount Sue sold = number bought − 3; $n - 3$.

Price for each pattern sold = cost of each pattern bought + 2; $\dfrac{200}{n} + 2$.

$r = (n - 3)\left(\dfrac{200}{n} + 2\right)$

10. Amount of his sales = amount sold × price of each sold.
$a = (h - 11)\left(\dfrac{125}{h} + 0.25\right)$

C 11. $c = \dfrac{t}{n}$; $p = 15n - 15u - uc$

12. profit = total sales − total cost.

$p = (n - u)\left(\dfrac{t}{n} + 15\right) + u\left(\dfrac{t}{n} + 5\right) - t$, $p = t + 15n - \dfrac{ut}{n} - 15u + \dfrac{ut}{n} +$ $5u - 1$, $p = 15n - 15u + 5u$; $p = 15n - 10u$

page 361 Mixed Problem Solving Review

1.

	Rate	× Time	= Distance
Express	0.75	t	$0.75t$
Local	0.50	$t + 5$	$0.5(t + 5)$

$0.75t = 0.5(t + 5)$, $0.75t = 0.5t + 2.5$, $0.25t = 2.5$, $t = 10$
The express will catch the local in 10 min.

2.

	Substance	Total no. oz	No. of oz pure alcohol
Start with	15%	25	$(0.15)(25)$
Add	100%	x	x
Finish with	30%	$25 + x$	$0.30(25 + x)$

$0.30(25 + x) = (0.15)(25) + x$, $7.5 + 0.3x = 3.75 + x$, $0.7x = 3.75$, $x =$
5.357; The solution must have 5.357 oz of pure alcohol added.

3. Let x represent the width of the uniform border, then $36 - 2x$ represents
the length and $24 - 2x$ represents the width of the rug.
$(36 - 2x)(24 - 2x) = 540$, $864 - 120x + 4x^2 = 540$, $x^2 - 30x + 81 = 0$,
$(x - 27)(x - 3) = 0$, $x = 27$; $x = 3$. The border of 27 ft would be to great,
so the border is 3 ft.

page 361 Project

Answers may vary.

page 361 Test Yourself

1. $\dfrac{18y^3}{3y^3} - \dfrac{3y^2}{3y^3} + \dfrac{15y}{3y^3} = 6 - \dfrac{1}{y} + \dfrac{5}{y^2}$

2.
$$\begin{array}{r} 3m^2 - 14m + 42 + \dfrac{-135}{m + 3} \\ m + 3\overline{)3m^3 - 5m^2 + 0m - 9} \\ \underline{-(3m^3 + 9m^2)} \\ -14m^2 + 0m \\ \underline{-(-14m^2 - 42m)} \\ 42m - 9 \\ \underline{- (42m + 126)} \\ -135 \end{array}$$

3. $\dfrac{7.2 \text{ mm}}{18 \text{ cm}} = \dfrac{72 \text{ mm}}{1800 \text{ mm}} = \dfrac{8 \text{ mm}}{200 \text{ mm}} = \dfrac{2}{50} = \dfrac{1}{25}$

4. $\dfrac{24 \text{ pt}}{4 \text{ pt}} = \dfrac{6}{1}$; 6:1

5. $\dfrac{2.5 \text{ cm}}{100 \text{ m}} = \dfrac{32.5 \text{ cm}}{x}$, $x = \dfrac{(32.5 \text{ cm})(100 \text{ m})}{2.5 \text{ cm}}$; $x = 1300$ m

6. $\left(\dfrac{5}{2y}\right)(2y) - \left(\dfrac{12}{y}\right)(2y) = (-19)(2y)$, $5 - 24 = -38y$, $-38y = -19$; $y = \dfrac{1}{2}$

7. $\left(\dfrac{3z}{z + 4}\right)(z + 4)(z - 5) - 2(z + 4)(z - 5) = \left(\dfrac{3}{z - 5}\right)(z + 4)(z - 5)$, $3z(z - 5) -$

$2(z + 4)(z - 5) = 3(z + 4)$, $3z^2 - 15z - 2z^2 + 2z + 40 = 3z + 12$,

$z^2 - 16z + 28 = 0$, $(z - 14)(z - 2) = 0$; $z = 14$; $z = 2$

8.

	work rate	· time worked	= part of job done
Said	$\dfrac{1}{2}$	t	$\dfrac{t}{2}$
Sister	$\dfrac{1}{3}$	t	$\dfrac{t}{3}$

$\dfrac{t}{2} + \dfrac{t}{3} = 1$, $6\left(\dfrac{t}{2} + \dfrac{t}{3}\right) = 6 \cdot 1$, $3t + 2t = 6$, $5t = 6$, $t = \dfrac{6}{5}$; $t = 1.2$ h; It

should take them 1 h 12 min to clean a room working together.

9. Let n represent the number sold, and b represent the number bought.

profit = total sales − total cost;

$$p = n\left(\dfrac{25}{b} + 1.50\right) - 25$$

pages 362–363   Summary and Review

1. $\dfrac{b}{3}$; $b \neq 0$

2. $\dfrac{2 \cdot 2}{2(x + 3)} = \dfrac{2}{x + 3}$; $x \neq -3$

3. $\dfrac{1}{m + 2}$; $m \neq 1, -2$

4. $\dfrac{-1(y - 4)}{(y - 4)(y + 4)} = \dfrac{-1}{y + 4}$; $y \neq -4, 4$

5. $\dfrac{4x}{5z} \cdot \dfrac{4}{1} = \dfrac{16x}{5z}$

6. $\dfrac{2w}{3} \cdot \dfrac{9}{4w} = \dfrac{3}{2}$

7. $\dfrac{a + 2}{a + 1} \cdot \dfrac{(a + 1)(a - 1)}{(a + 2)(a - 2)} = \dfrac{a - 1}{a - 2}$

8. $\dfrac{(2r - 1)(r + 5)}{r - 3} \cdot \dfrac{r}{(r + 5)(r - 3)} = \dfrac{r(2r - 1)}{(r - 3)^2}$

9. LCD is $2 \cdot 2 \cdot 2 \cdot 3 \cdot m^3n^2 = 24m^3n^2$; $\left(\dfrac{1}{8m^3n}\right)\left(\dfrac{3n}{3n}\right) = \dfrac{3n}{24m^3n^2}$;

$\left(\dfrac{5}{12mn^2}\right)\left(\dfrac{2m^2}{2m^2}\right) = \dfrac{10m^2}{24m^3n^2}$

10. LCD is $(c + 1)(c - 2)$; $\left(\dfrac{5}{c + 1}\right)\left(\dfrac{c - 2}{c - 2}\right) = \dfrac{5(c - 2)}{(c + 1)(c - 2)}$; $\left(\dfrac{3}{c - 2}\right)\left(\dfrac{c + 1}{c + 1}\right) =$

$\dfrac{3(c + 1)}{(c + 1)(c - 2)}$

11. LCD is $(x + 3)(x - 3)$; $\dfrac{2x}{(x + 3)(x - 3)}$; $\left(\dfrac{3}{x + 3}\right)\left(\dfrac{x - 3}{x - 3}\right) = \dfrac{3(x - 3)}{(x + 3)(x - 3)}$

12. $\left(\dfrac{1}{8m^3n}\right)\left(\dfrac{3n^2}{3n^2}\right) + \left(\dfrac{5}{12mn^3}\right)\left(\dfrac{2m^2}{2m^2}\right) = \dfrac{3n^2}{24m^3n^3} + \dfrac{10m^2}{24m^3n^3} = \dfrac{3n^2 + 10m^2}{24m^3n^3}$

13. $\left(\dfrac{5}{c + 1}\right)\left(\dfrac{c - 2}{c - 2}\right) - \left(\dfrac{3}{c - 2}\right)\left(\dfrac{c + 1}{c + 1}\right) = \dfrac{5c - 10}{(c + 1)(c - 2)} - \dfrac{3c + 3}{(c + 1)(c - 2)} =$

$\dfrac{2c - 13}{(c + 1)(c - 2)}$

14. $\left(\dfrac{3}{x + 3}\right)\left(\dfrac{x - 3}{x - 3}\right) - \dfrac{2x}{(x + 3)(x - 3)} = \dfrac{3x - 9}{(x + 3)(x - 3)} - \dfrac{2x}{(x + 3)(x - 3)} =$

$\dfrac{x - 9}{(x + 3)(x - 3)}$

15. $(u + 2)\left(\dfrac{u + 2}{u + 2}\right) - \dfrac{1}{u + 2} = \dfrac{u^2 + 4u + 4 - 1}{u + 2} = \dfrac{u^2 + 4u + 3}{u + 2}$

16. $\dfrac{\left(1 - \dfrac{1}{r}\right) \cdot r^2}{\left(\dfrac{1}{r^2}\right) \cdot r^2} = \dfrac{r^2 - r}{1} = r^2 - r$

17. $\dfrac{\left(\dfrac{y}{x^2} - \dfrac{1}{y}\right) \cdot x^2y}{\left(\dfrac{1}{xy} + \dfrac{1}{x^2}\right) \cdot x^2y} = \dfrac{y^2 - x^2}{x + y} = \dfrac{(y + x)(y - x)}{y + x} = y - x$

18. $\dfrac{12z^2}{6z} + \dfrac{42z}{6z} = 2z + 7$

19.

$$
\begin{array}{r}
2r^2 + 3r - 15 + \dfrac{50}{r + 5} \\[4pt]
r + 5\overline{)2r^3 + 13r^2 + 0r - 25} \\
\underline{-(2r^3 + 10r^2)} \\
3r^2 + 0r \\
\underline{-(3r^2 + 15r)} \\
-15r - 25 \\
\underline{-(-15r - 75)} \\
50
\end{array}
$$

20. $30 = 15y$, $y = 2$

21. $30 = 12n$, $n = \dfrac{30}{12} = \dfrac{5}{2}$

22. $7(x + 3) = (28)(9)$, $7x + 21 = 252$, $7x = 231$, $x = 33$

23. $\left(\dfrac{d}{5}\right)(10) + \left(\dfrac{3d}{2}\right)(10) = (17)(10)$, $2d + 15d = 170$, $17d = 170$, $d = 10$

24. $\left(\dfrac{2x - 5}{x - 5}\right)(2)(x - 5) + \left(\dfrac{x}{2}\right)(2)(x - 5) = \left(\dfrac{5}{x - 5}\right)(2)(x - 5)$, $2(2x - 5) +$

$x(x - 5) = 10$, $4x - 10 + x^2 - 5x = 10$, $x^2 - x - 10 = 10$, $x^2 - x - 20$

$= 0$, $(x - 5)(x + 4) = 0$; $x = 5$ or $x = -4$; Check: $x = 5$; $\dfrac{2(5) - 5}{5 - 5} + \dfrac{5}{2} =$

$\dfrac{5}{5 - 5}$; The denominator is 0; the solution is $x = -4$.

25.

work rate	· time worked	= part of job done
James $\dfrac{1}{45}$	t	$\dfrac{t}{45}$
Bertha $\dfrac{1}{30}$	t	$\dfrac{t}{30}$

$\dfrac{t}{45} + \dfrac{t}{30} = 1$, $2t + 3t = 90$, $5t = 90$, $t = 18$. It will take 18 min to wash a

car if they work together.

26.

	Rate	× Time	= Distance
plane	$r + 400$	$\dfrac{1125}{r + 400}$	1125
car	r	$\dfrac{125}{r}$	125

$\dfrac{1125}{r + 400} = \dfrac{125}{r}$, $1125r = 125(r + 400)$, $1125r = 125r + 50{,}000$,

$1000r = 50{,}000$, $r = 50$. The car was traveling at a rate of 50 mi/h.

page 364   Chapter Test

1. $\dfrac{9m}{3} - \dfrac{15}{3} = 3m - 5$

2. $\dfrac{(3x - 1)(x + 3)}{(x + 3)(x - 4)} = \dfrac{3x - 1}{x - 4}$

3. $\dfrac{5b}{c} \cdot \dfrac{4}{1} = \dfrac{20b}{c}$

4. $\dfrac{2d}{5} \cdot \dfrac{15}{4d} = \dfrac{3}{2}$

5. $\dfrac{5}{x + 1} \cdot \dfrac{2(x + 1)}{6} = \dfrac{5}{3}$

6. $\dfrac{(a + 2)(a - 1)}{(a + 3)(a - 3)} \cdot \dfrac{a + 3}{a + 2} = \dfrac{a - 1}{a - 3}$

7. $\dfrac{4 - y}{y - 4} = \dfrac{-1(y - 4)}{y - 4} = -1$

8. $\left(\dfrac{5}{6k}\right)\left(\dfrac{2k}{2k}\right) + \left(\dfrac{3}{4k^2}\right)\left(\dfrac{3}{3}\right) - \left(\dfrac{2}{3k}\right)\left(\dfrac{4k}{4k}\right) = \dfrac{10k + 9 - 8k}{12k^2} = \dfrac{2k + 9}{12k^2}$

9. $\left(\dfrac{1}{a + 1}\right)\left(\dfrac{a - 3}{a - 3}\right) - \left(\dfrac{2}{a - 3}\right)\left(\dfrac{a + 1}{a + 1}\right) = \dfrac{a - 3 - 2(a + 1)}{(a + 1)(a - 3)} =$

$\dfrac{a - 3 - 2a - 2}{(a + 1)(a - 3)} = \dfrac{-a - 5}{(a + 1)(a - 3)}$

10. $\dfrac{3}{u} - (6)\left(\dfrac{u}{u}\right) = \dfrac{3 - 6u}{u}$

11. $(g + 5)\left(\dfrac{g + 5}{g + 5}\right) - \dfrac{1}{g + 5} = \dfrac{g^2 + 10g + 25 - 1}{g + 5} = \dfrac{g^2 + 10g + 24}{g + 5}$

12. $\dfrac{\left(\dfrac{1}{x^2} - 4\right) \cdot x^2}{\dfrac{1}{x} \cdot x^2} = \dfrac{1 - 4x^2}{x}$

13. $\dfrac{14p^2}{7p} + \dfrac{21p}{7p} = 2p + 3$

14.
$$
\begin{array}{r}
c + 1 - \dfrac{2}{c - 2} \\[4pt]
c - 2\overline{)c^2 - c - 4} \\
\underline{-(c^2 - 2c)} \\
c - 4 \\
\underline{-(c - 2)} \\
-2
\end{array}
$$

15.
$$
\begin{array}{r}
3x^2 - 4x + 2 \\[4pt]
x + 2\overline{)3x^3 + 2x^2 - 6x + 4} \\
\underline{-(3x^3 + 6x^2)} \\
-4x^2 - 6x \\
\underline{-(-4x^2 - 8x)} \\
2x + 4 \\
\underline{-(2x + 4)} \\
0
\end{array}
$$

16. $\dfrac{8}{20} = \dfrac{2}{5}$

17. $\dfrac{3 \text{ pt}}{4 \text{ pt}} = \dfrac{3}{4}$

18. $\dfrac{15 \text{ mm}}{30 \text{ mm}} = \dfrac{1}{2}$

19. $\left(\dfrac{b}{5}\right)(15) + \left(\dfrac{5b}{3}\right)(15) = (2)(15),\ 3b + 25b = 30,\ 28b = 30,\ b = \dfrac{30}{28} = \dfrac{15}{14}$

20. $\left(\dfrac{x + 1}{x}\right)(2x) - (7)(2x) = \left(\dfrac{9}{2x}\right)(2x),\ 2x + 2 - 14x = 9,\ -12x + 2 = 9,$

$-12x = 7,\ x = -\dfrac{7}{12}$

21. $\left(\dfrac{1}{r}\right)(r)(r - 1) + \left(\dfrac{2}{r - 1}\right)(r)(r - 1) = (-2)(r)(r - 1),\ r - 1 + 2r =$

$-2r^2 + 2r,\ 3r - 1 = -2r^2 + 2r;\ 2r^2 + r - 1 = 0,\ (2r - 1)(r + 1) = 0,$

$r = \dfrac{1}{2}$ or $r = -1$

22.

	work rate	· time worked	= part of job done
Sadie	$\dfrac{1}{40}$	t	$\dfrac{t}{40}$
Alvin	$\dfrac{1}{30}$	t	$\dfrac{t}{30}$

$\dfrac{t}{40} + \dfrac{t}{30} = 1,\ 3t + 4t = 120,\ 7t = 120,\ t = \dfrac{120}{7}$

It will take $17\dfrac{1}{7}$ min to make a pizza if they work together.

23. Sales = amount sold × price of each sold. Price of each sold = cost of each pennant + 0.75 = $\dfrac{150}{p}$ + 0.75; $s = (p - 5)\left(\dfrac{150}{p} + 0.75\right)$

Challenge

	work rate	· time worked	= part of job done
hot	$\dfrac{1}{15}$	t	$\dfrac{t}{15}$
cold	$\dfrac{1}{12}$	t	$\dfrac{t}{12}$
drain	$-\dfrac{1}{20}$	t	$-\dfrac{t}{20}$

$\dfrac{t}{15} + \dfrac{t}{12} - \dfrac{t}{20} = 1$, $4t +$ $5t - 3t = 60$, $6t = 60$; $t = 10$
The tub will be filled in 10 min if both faucets and the drain are open.

page 365   Preparing for Standardized Tests

1. $\dfrac{5}{6} + \dfrac{2}{5} - \dfrac{7}{10} = \dfrac{25 + 12 - 21}{30} = \dfrac{16}{30} = \dfrac{8}{15}$ **2.** $\dfrac{8 \times 10^8}{4 \times 10^6} = 2 \times 10^2 = 200$

3. $\dfrac{16}{7} = \dfrac{x}{28}$, $(16)(28) = 7x$, $448 = 7x$, $x = 64$

4. $33\dfrac{1}{3}\%$ of $450 = \dfrac{1}{3}$ of $450 = \$150$ discount, leaving $300 for the sale price of the stereo. The sales tax is 6% of $300 or $18.00 and the total cost of the stereo is $318.

5. $\dfrac{x}{y} = \dfrac{7}{3}$, $2\left(\dfrac{x}{y}\right) = 2\left(\dfrac{7}{3}\right)$, $\dfrac{2x}{y} = \dfrac{14}{3}$; The ratio of $2x$ to y is $\dfrac{14}{3}$.

6. $6x - 3 + 4x + 4 = 6x + 8 - 9$, $10x + 1 = 6x - 1$, $4x = -2$, $x = -\dfrac{1}{2}$

7. 15

8. 25 sold 80 tickets + 20 sold 90 tickets + 5 sold 100 tickets = 50 sold 80 or more

9. $10(60) + 15(70) + 25(80) + 20(90) + 5(100) = 600 + 1050 + 2000 + 1800 + 500 = 5950$ tickets sold

10. $\dfrac{80 + 92 + 86 + 78 + 82 + x}{6} > 85$, $\dfrac{418 + x}{6} > 85$, $418 + x > 510$, $x > 92$; The lowest grade she can receive is 93.

11. Perimeter = 2 × length + 2 × width, $84 = 2(w + 4) + 2w$, $84 = 2w + 8 + 2w$, $84 = 4w + 8$, $76 = 4w$, $w = 19$; length = $w + 4 = 19 + 4 = 23$

12. If m is an even integer and n is the next consecutive even integer, then $n = m + 2$. Since $mn = 728$, $m(m + 2) = 728$, $m^2 + 2m = 728$, $m^2 + 2m - 728 = 0$, $(m + 28)(m - 26) = 0$. Since m is negative, $m = -28$. $n = m + 2 = -28 + 2 = -26$. The integers are -28 and -26.

13. The remainder when 103 is divided by 6 is 1. One glass will be left for the next packing.

14. 4 (27, 36, 45, and 54)

15. The five consecutive integers can be represented by x, $x + 1$, $x + 2$, $x + 3$, and $x + 4$. $x + x + 1 = 19$, $2x + 1 = 19$, $2x = 18$, $x = 9$. Therefore, the sum of the last two integers is $x + 3 + x + 4 = 9 + 3 + 9 + 4 = 25$.

pages 366–368 Cumulative Review (Chapters 1–8)

1. -1

2. 0

3. -2

4. $\dfrac{1}{2}$

5. $2\dfrac{1}{2}$

6. 4

7. $-3\dfrac{1}{2}$

8. $-2\dfrac{1}{2}$

9. $x \geq -1$

10. $x < 0$

11. $x \neq -2$

12. $-2 < x < 1$

13. $x \leq -1$ or $x > 0$

14. $-1 \leq x \leq 1$

15. $5x = -15$, $x = -3$

16. $\dfrac{a}{3} = 8$, $a = 24$

17. $t + 3 \geq 9$, $t \geq 6$

18. $m > -4$

19. $|y| = -4$, no solution

20. $3b = b + 5$, $2b = 5$ $b = \dfrac{5}{2}$

21. $\dfrac{2n}{3} - \dfrac{15}{3} = \dfrac{1}{3}$, $\dfrac{2n}{3} = \dfrac{16}{3}$, $2n = 16$, $n = 8$

22. $2c > 23$, $c > \dfrac{23}{2}$

23. $-\dfrac{4}{r} = -2$, $-2r = -4$, $r = 2$

24. $z(z + 2) = 0$
$z = 0 \mathrel{\vdots} z + 2 = 0$
$\phantom{z = 0 \mathrel{\vdots}} z = -2$

25. $-2 - 4 \leq 3p + 4 - 4 \leq 13 - 4$
$-6 \leq 3p \leq 9$
$-2 \leq p \leq 3$

26. $|4d - 7| = 29$
$4d - 7 = 29$ or $4d - 7 = -29$
$d = 9$ or $\quad d = -\dfrac{11}{2}$

27. $8n - 6n + 2 = 26$
$2n + 2 = 26$
$n = 12$

28. $(g + 6)(g - 4) = 0$,
$g + 6 = 0 \mathrel{\vdots} g - 4 = 0$
$g = -6 \mathrel{\vdots} \quad g = 4$

29. $2.7 - 0.2k + 0.2k = 0.7k + 0.2k$
$2.7 = 0.9k$
$3 = k$

30. $2y > 0$ or $3y \le -3$

$\quad\quad\; y > 0$ or $\;\; y \le -1$

31. $5(m - 1) = 3(m + 3)$

$\quad\quad 5m - 5 = 3m + 9$

$\quad\quad\; 2m - 5 = 9$

$\quad\quad\quad\;\; 2m = 14$

$\quad\quad\quad\quad m = 7$

32. $\left(\dfrac{1}{p + 1}\right)(p - 2)(p + 1) - \left(\dfrac{1}{p - 2}\right)(p - 2)(p + 1) = (0)(p - 2)(p + 1),$

$\quad\; p - 2 - (p + 1) = 0,\, p - 2 - p - 1 = 0,\, -3 \ne 0;$ no solution

33. 6×10^6 $\quad\quad\quad$ **34.** 4.5×10^2 $\quad\quad\quad$ **35.** 1.28×10^4 $\quad\quad\quad$ **36.** 7.9×10^8

37. $\left(\dfrac{2}{3}\right)(-6) - 1 = -4 - 1 = -5$

38. $\dfrac{2}{\frac{1}{2}} - (-6) = 2 \div \left(\dfrac{1}{2}\right) + 6 = 2 \cdot 2 + 6 = 4 + 6 = 10$

39. $2\left(\dfrac{1}{2} - 1\right) + (-6)^2 = 2\left(-\dfrac{1}{2}\right) + 36 = -1 + 36 = 35$

40. $\dfrac{2}{3}\left(\dfrac{\frac{1}{2} + 3}{-6 - 1}\right) = \dfrac{2}{3}\left(\dfrac{\frac{7}{2}}{-7}\right) = \dfrac{2}{3}\left[\dfrac{7}{2} \div (-7)\right] = \left(\dfrac{2}{3}\right)\left(\dfrac{7}{2} \cdot -\dfrac{1}{7}\right) = \left(\dfrac{2}{3}\right)\left(-\dfrac{1}{2}\right) = -\dfrac{1}{3}$

41. $\left(\dfrac{1}{2}\right)^2\left(\dfrac{2}{3}\right)(-6) = \left(\dfrac{1}{4}\right)\left(\dfrac{2}{3}\right)\left(-\dfrac{6}{1}\right) = -\dfrac{12}{12} = -1$

42. $\dfrac{\frac{2}{3} - (-6)}{\frac{2}{3} + 1} - \left(\dfrac{1}{2}\right)^2 = \dfrac{\frac{2}{3} + 6}{\frac{2}{3} + 1} - \dfrac{1}{4} = \dfrac{\frac{20}{3}}{\frac{5}{3}} - \dfrac{1}{4} = \dfrac{20}{3} \div \dfrac{5}{3} - \dfrac{1}{4} = \dfrac{20}{5} - \dfrac{1}{4} =$

$\quad\; \dfrac{80}{20} - \dfrac{5}{20} = \dfrac{75}{20} = \dfrac{15}{4}$

43. $\left[(-6)\left(\dfrac{1}{2} + \dfrac{2}{3}\right) + 5\right]^3 = \left[(-6)\left(\dfrac{3}{6} + \dfrac{4}{6}\right) + 5\right]^3 = \left[(-6)\left(\dfrac{7}{6}\right) + 5\right]^3 =$

$\quad\; [-7 + 5]^3 = [-2]^3 = -8$

44. $\left(\dfrac{\frac{1}{2}}{\frac{2}{3}} + \dfrac{\frac{2}{3}}{\frac{1}{2}}\right)\dfrac{-6}{5} = \left[\left(\dfrac{1}{2}\right)\left(\dfrac{3}{2}\right) + \left(\dfrac{2}{3}\right)\left(\dfrac{2}{1}\right)\right]\left(-\dfrac{6}{5}\right) = \left[\dfrac{3}{4} + \dfrac{4}{3}\right]\left(-\dfrac{6}{5}\right) =$

$\quad\; \left[\dfrac{9}{12} + \dfrac{16}{12}\right]\left(-\dfrac{6}{5}\right) = \left(\dfrac{25}{12}\right)\left(-\dfrac{6}{5}\right) = -\dfrac{5}{2}$

45. $|[2 + (-6)]^2 - 20| = |(-4)^2 - 20| = |16 - 20| = |-4| = 4$

46. $\left(\dfrac{1}{2}\right)^2 - \left(\dfrac{2}{3}\right)^2 = \dfrac{1}{4} - \dfrac{4}{9} = \dfrac{9}{36} - \dfrac{16}{36} = -\dfrac{7}{36}$

47. $(-6)\left(\dfrac{1}{2}\right)^2\left(\dfrac{2}{3}\right)^2 = (-6)\left(\dfrac{1}{4}\right)\left(\dfrac{4}{9}\right) = -\dfrac{6}{9} = -\dfrac{2}{3}$

48. $\dfrac{\left|\dfrac{1}{2} - \dfrac{2}{3}\right| \cdot \left|\dfrac{1}{2} + \dfrac{2}{3}\right|}{-6} = \dfrac{\left|\dfrac{3}{6} - \dfrac{4}{6}\right| \cdot \left|\dfrac{3}{6} + \dfrac{4}{6}\right|}{-6} = \dfrac{\left|-\dfrac{1}{6}\right| \cdot \left|\dfrac{7}{6}\right|}{-6} = \dfrac{\dfrac{1}{6} \cdot \dfrac{7}{6}}{-6} = \dfrac{\dfrac{7}{36}}{-6} =$

$\dfrac{7}{36} \cdot -\dfrac{1}{6} = -\dfrac{7}{216}$

49. $4 + 12 \div (-2) = 4 + -6 = -2$

50. $[-8 + 9]^5 = 1$

51. $|-5| - 9 = 5 - 9 = -4$

52. $6m^5n^3$

53. $4b^6$

54. $(2p + 3)(2p + 3) = 4p^2 + 12p + 9$

55. $\dfrac{x}{3y}$

56. $\dfrac{3(c + 3)}{15} = \dfrac{c + 3}{5}$

57. $\dfrac{a - b}{-1(a - b)} = \dfrac{1}{-1} = -1$

58. $\dfrac{5(g^2 - 1)}{g + 1} = \dfrac{5(g + 1)(g - 1)}{g + 1} = 5(g - 1)$

59. $\dfrac{z + 2}{(z + 2)(z - 2)} = \dfrac{1}{z - 2}$

60. $\dfrac{(d + 3)(d - 3)}{-1(d - 3)} = -1(d + 3) = -(d + 3)$

61. $5a^2 + 3a - 2a^2 + 2a = 3a^2 + 5a$

62. $6t^2 - 8t - (t^2 - 2t + 1) = 6t^2 - 8t - t^2 + 2t - 1 = 5t^2 - 6t - 1$

63. $\dfrac{(r + 5)(r - 5)}{r + 5} + 2r - 10 = r - 5 + 2r - 10 = 3r - 15$

64. $\dfrac{f - 2}{(f + 2)(f - 2)} + \dfrac{f + 1}{f + 2} = \dfrac{1}{f + 2} + \dfrac{f + 1}{f + 2} = \dfrac{f + 2}{f + 2} = 1$

65. $\dfrac{(h + 3)(h - 3)}{h + 1} \cdot \dfrac{(h + 1)(h + 1)}{h - 3} = \dfrac{(h + 3)(h + 1)}{1} = (h + 3)(h + 1)$

66. $\dfrac{(h + 5)(h - 3)}{(h + 4)(h - 3)} \cdot \dfrac{(h + 4)(h - 2)}{(h + 5)(h - 2)} = 1$

67. $\dfrac{(k + 5)(k - 5)}{(k + 2)(k + 2)} \cdot \dfrac{(k + 2)(k - 2)}{(k + 5)(k + 5)} = \dfrac{(k - 5)(k - 2)}{(k + 5)(k + 2)}$

68. $\dfrac{(m - 5)(m + 2)}{(m + 5)(m - 5)} \cdot \dfrac{(m + 5)(m - 3)}{(m + 2)(m - 3)} = 1$

69.

Statement	Reason
1. $x = y$	**1.** Given
2. $x + z = x + z$	**2.** Reflexive property
3. $x + z = y + z$	**3.** Substitution property
4. $x + z = z + y$	**4.** Commutative property for addition

70.

	Substance	Total no. mL	No. of mL pure acid
Start with	35%	70 mL	(0.35)(70)
Add	100% water	x mL	0
Finish with	25%	$70 + x$	$0.25(70 + x)$

$0.25(70 + x) = (0.35)(70)$, $17.5 + 0.25x = 24.5$, $0.25x = 7$, $x = 28$
Add 28 mL of water to the solution.

71.

	Substance	Total no. oz	No. of oz pure salt
Start with	20%	40 oz	(0.20)(40)
Add	100% salt	x	x
Finish with	50%	$40 + x$	$0.50(40 + x)$

$0.50(40 + x) = (0.20)(40) + x$, $20 + 0.5x = 8 + x$, $12 = 0.5x$, $x = 24$
Add 24 oz of salt to the solution.

72.

	Substance	Total no. mL	No. of mL pure sugar
Start with	20%	70 mL	(0.20)(70)
Add	100% water	30 mL	0 mL
Finish with	x	100 mL	$100x$

$100x = (0.20)(70)$, $100x = 14$, $x = \dfrac{14}{100}$, $x = 0.14$, $x = 14\%$

The solution will have a 14% concentration of sugar.

73. $4(a - 2b)$ **74.** $3(4x + 5y)$ **75.** $3c(2c - 3)$

76. $5m(2m + 1)$ **77.** $7p(3p^2 + 2p - 4)$ **78.** $(k + 3)(k - 3)$

79. $(d + 4)^2$ **80.** $5(t^2 - 4) = 5(t + 2)(t - 2)$

81. $(g + 5)(g - 3)$ **82.** $(3a + 5)(3a - 5)$ **83.** $(r - 6)(r - 4)$

84. $(3y - 2)(2y + 3)$ **85.** LCD is $2 \cdot 3 \cdot a^3 = 6a^3$ **86.** LCD is a^2b^2

87. LCD is $3x(x + 1)$ **88.** $\left(-\dfrac{3}{2}\right)\left(\dfrac{2}{2}\right) + \dfrac{3}{4} - \left(-\dfrac{1}{2}\right)\left(\dfrac{2}{2}\right) = -\dfrac{6}{4} + \dfrac{3}{4} - \left(-\dfrac{2}{4}\right) = -\dfrac{1}{4}$

89. $\dfrac{2}{3}\left[\dfrac{1}{2} - \dfrac{5}{6}\right] = \left(\dfrac{2}{3}\right)\left(\dfrac{1}{2}\right) - \left(\dfrac{2}{3}\right)\left(\dfrac{5}{6}\right) = \dfrac{1}{3} - \dfrac{5}{9} = \dfrac{3}{9} - \dfrac{5}{9} = -\dfrac{2}{9}$

90. $\left(\dfrac{3}{5}\right)\left(-\dfrac{1}{3}\right)\left(\dfrac{6}{1}\right) = -\dfrac{6}{5}$ **91.** $\left(\dfrac{5}{4b}\right)\left(\dfrac{b}{b}\right) + \left(\dfrac{3}{2b^2}\right)\left(\dfrac{2}{2}\right) = \dfrac{5b}{4b^2} + \dfrac{6}{4b^2} = \dfrac{5b + 6}{4b^2}$

92. $\dfrac{t}{(t+3)(t-3)} - \dfrac{3}{-1(t-3)} = \dfrac{t}{(t+3)(t-3)} + \left(\dfrac{3}{t-3}\right)\left(\dfrac{t+3}{t+3}\right) =$

$\dfrac{t+3t-9}{(t+3)(t-3)} = \dfrac{4t-9}{(t+3)(t-3)} = \dfrac{4t+9}{t^2+9}$

93. $(5h)\left(\dfrac{h+2}{h+2}\right) + \dfrac{h+3}{h+2} = \dfrac{5h^2+10h+h+3}{h+2} = \dfrac{5h^2+11h+3}{h+2}$

94. $\dfrac{m-2}{m+2} - \dfrac{2(m^2+3)}{(m+2)(m-2)} - \dfrac{m+2}{-1(m-2)} = \left(\dfrac{m-2}{m+2}\right)\left(\dfrac{m-2}{m-2}\right) -$

$\dfrac{2(m^2+3)}{(m-2)(m-2)} + \left(\dfrac{m+2}{m-2}\right)\left(\dfrac{m+2}{m+2}\right) =$

$\dfrac{m^2-4m+4-2m^2-6+m^2+4m+4}{(m+2)(m-2)} = \dfrac{2}{(m+2)(m-2)}$

95. $\dfrac{(a-2)^2-a+2}{(a-2)(a-3)} = \dfrac{a^2-4a+4-a+2}{(a-2)(a-3)} = \dfrac{a-5a+6}{(a-2)(a-3)} =$

$\dfrac{(a-2)(a-3)}{(a-2)(a-3)} = 1$

96. $10m^2 - 4m + 25m - 10 = 10m^2 + 21m - 10$

97. $4x^3y - 8x^2y^3 + 4x^2y$

98. $3d(d^2 + 4d + 4) = 3d^3 + 12d^2 + 12d$

99. $3a^2 - 2a^2 + 5a + 2a - 3 - 1 = a^2 + 7a - 4$

100. $\dfrac{4p^5}{2p^2} + \dfrac{6p^3}{2p^2} - \dfrac{2p^2}{2p^2} = 2p^3 + 3p - 1$

101. Let x = the amount of discount. $(65)(0.15) = x$, $x = 9.75$; $65 - 9.75 =$ 55.25; The jacket's new price is $55.25.

102. Let m = the number of miles Frank can drive. $200 \geq 128 + 0.12m$, $72 \geq 0.12m$, $600 \geq m$; Frank will need to drive 600 mi or less.

103. Let x = first odd integer, then $x + 2$ is the second odd integer and $x + 4$ is the third odd integer. The equation is: $x(x + 4) = 4(x + 2) + 1$; Solve $x^2 + 4x = 4x + 9$, $x^2 - 9 = 0$, $(x + 3)(x - 3) = 0$, $x = -3$ or $x = 3$.
The three consecutive integers are 3, 5, 7 or $-3, -1, 1$.

104. Let x = number of miles, then $\dfrac{\frac{1}{4}}{15} = \dfrac{2\frac{1}{2}}{x}$, $\dfrac{1}{4}x = (15)\left(\dfrac{5}{2}\right)$, $\dfrac{x}{4} = \dfrac{75}{2}$,

$2x = 300$, $x = 150$. The scale of $2\frac{1}{2}$ in. is equal to 150 mi.

105.

	work rate ·	time worked	= part of job done
Allison	$\dfrac{1}{2}$	t	$\dfrac{t}{2}$
Jason	$\dfrac{1}{2.5}$	t	$\dfrac{t}{2.5}$

$\dfrac{t}{2} + \dfrac{t}{2.5} = 1, \dfrac{t}{2} + \dfrac{10t}{25} = 1, 25t + 20t = 50, 45t = 50, t = \dfrac{50}{45} = \dfrac{10}{9}$.

It will take $\dfrac{10}{9}$ h or $66\dfrac{2}{3}$ min to cut the lawn if they work together.

106.

	Rate ×	Time	= Distance
Train 1	75	$x + 1.5$	$75(x + 1.5)$
Train 2	85	x	$85x$

$75(x + 1.5) = 85x, 75x + 112.5 = 85x, 112.5 = 10x, 11.25 = x$;
The second train will catch the first train after 11.25 h or 11 h 15 min.

107.

	Rate ×	Time	= Distance
Going	50	$3\dfrac{1}{2}$	175
Return	x	$4\dfrac{1}{5}$	$\dfrac{21}{5}x$

$\dfrac{21}{5}x = 175, x = 175 \cdot \dfrac{5}{21}, x = 41\dfrac{2}{3}$. The Ward family returned at an

average rate of $41\dfrac{2}{3}$ mi/h.

Chapter 9 Linear Equations

page 370 Capsule Review

1.
```
←————+————————•————→
      0          15
                 A
```

2.
```
    B
←——•————————+————→
   -2        0
```

3.
```
        C
←————————•————————→
         0
```

4.
```
        D
←——+———•+———+————→
  -2   -1   0
```

page 372 Class Exercises

1-4.

```
        U •      ↑y
      (-2, 5)    ┤
                 ┤       T (4, 1)
                 ┤      •      x
      ┼┼┼┼┼┼┼┼┼┼┼┼┼┼┼┼┼→
                 ┤
         R       ┤
      (-3, -6)┤    • V (2, -5)
          •
```

5. (3, 5), I **6.** (0, −3), y-axis **7.** (6, −5), IV

8. (−1, 4), II **9.** (−3, 0), x-axis **10.** (−5, −2), III

11. (0, 6), y-axis **12.** (5, 0), x-axis

13. B **14.** E **15.** O **16.** G

17. P **18.** L **19.** M **20.** N

21. $x + 2y = 2$, $0 + 2(1) \stackrel{?}{=} 2$, $0 + 2 \stackrel{?}{=} 2$, $2 = 2$; (0, 1) is a solution.

22. $3x - y = -1$, $3(0) - 1 \stackrel{?}{=} -1$, $0 - 1 \stackrel{?}{=} -1$, $-1 = -1$; (0, 1) is a solution.

23. $x = y + 1$, $0 \stackrel{?}{=} 1 + 1$, $0 \neq 2$; (0, 1) is not a solution.

24. $-x + y = -1$, $0 + 1 \stackrel{?}{=} -1$, $1 \neq -1$; (0, 1) is not a solution.

25. $4x + y = 3$, $4(0) + 1 \stackrel{?}{=} 3$, $0 + 1 \stackrel{?}{=} 3$, $1 \neq 3$; (0,1) is not a solution.

26. $xy = 0$, $0(1) \stackrel{?}{=} 0$, $0 = 0$; (0, 1) is a solution.

pages 372–373 Practice Exercises

A 1-4.

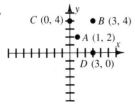

5-8.

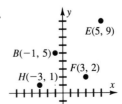

9–12.

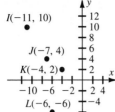

13–16.

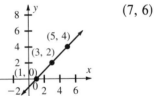

17. $3x - y = -2$, $3(-3) - (-7) \overset{?}{=} -2$, $-9 + 7 \overset{?}{=} -2$, $-2 = -2$; $(-3, -7)$ is a solution.

$3x - y = -2$, $3(3) - 5 \overset{?}{=} -2$, $9 - 5 \overset{?}{=} -2$, $4 \neq 2$; $(3, 5)$ is not a solution.

18. $5x - y = -7$, $5(2) - 5 \overset{?}{=} -7$, $10 - 5 \overset{?}{=} -7$, $5 \neq -7$; $(2, 5)$ is not a solution.

$5x - y = -7$, $5(0) - 7 \overset{?}{=} -7$, $0 - 7 \overset{?}{=} -7$, $-7 = -7$; $(0, 7)$ is a solution.

19. $x + 3y = 6$, $3 + 3(1) \overset{?}{=} 6$, $3 + 3 \overset{?}{=} 6$, $6 = 6$; $(3, 1)$ is a solution.

$x + 3y = 6$, $4 + 3(2) \overset{?}{=} 6$, $4 + 6 \overset{?}{=} 6$, $10 \neq 6$; $(4, 2)$ is not a solution.

20. $x + 5y = 1$, $0 + 5\left(\dfrac{1}{5}\right) \overset{?}{=} 1$, $0 + 1 \overset{?}{=} 1$, $1 = 1$; $\left(0, \dfrac{1}{5}\right)$ is a solution.

$x + 5y = 1$, $3 + 5(2) \overset{?}{=} 1$, $3 + 10 \overset{?}{=} 1$, $13 \neq 1$; $(3, 2)$ is not a solution.

21.

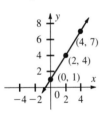

(6, 10)

22.

(7, 6)

23.

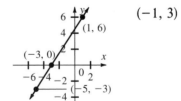

(5, 8)

24.

(−1, 3)

B 25.

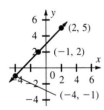

(−4, −1)

26.

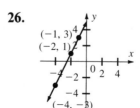

(−3, −1)

27. 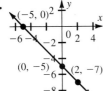 $(-2, -3)$

28. 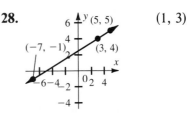 $(1, 3)$

29. y-axis

30. x-axis

For Exercises 31–34, answers may vary. Possible answers are:

31. $y = \frac{1}{2}x - \frac{7}{2}; y = 3x - 6$

32. $y = 4x - 4; y = 8x - 6$

33. $y = \frac{1}{8}x - 1; y = \frac{1}{4}x + 1$

34. $y = 2x + 35; y = 6x + 7$

C 35.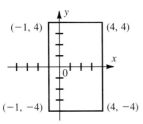

36.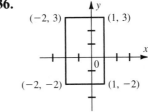

37. parallel, perpendicular

38. perpendicular, parallel

39. one, $(-2, 4)$

40.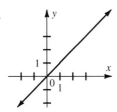

x and y coordinates are equal; $y = x$.

41.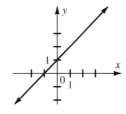

y-coordinate is one more than the x-coordinate; $y = x + 1$.

42. $d = 24t, 60 \stackrel{?}{=} 24(2.5), 60 = 60; (2.5, 60)$ is a solution.
$d = 24t, 42 \stackrel{?}{=} 24(1.75), 42 = 42; (1.75, 42)$ is a solution.
$d = 24t, 14 \stackrel{?}{=} 24(0.5), 14 \neq 12; (0.5, 14)$ is not a solution.
$d = 24t, 84 \stackrel{?}{=} 24(3.5), 84 = 84; (3.5, 84)$ is a solution.
$d = 24t, 0 \stackrel{?}{=} 24(0), 0 = 0; (0, 0)$ is a solution.

43. $s = 50h + 100$, $125 \overset{?}{=} 50(0.5) + 100$,
$125 \overset{?}{=} 25 + 100$, $125 = 125$; $(0.5, 125)$ is a solution.
$s = 50h + 100$, $162 \overset{?}{=} 50(1.25) + 100$,
$162 \overset{?}{=} 62.5 + 100$, $162 \neq 162.5$; $(1.25, 162)$ is not a solution.
$s = 50h + 100$, $225 \overset{?}{=} 60(2.5) + 100$, $225 \overset{?}{=} 125 + 100$,
$225 = 225$; $(2.5, 225)$ is a solution.
$s = 50h + 100$, $300 \overset{?}{=} 50(3.0) + 100$, $300 \overset{?}{=} 150 + 100$,
$300 \neq 250$; $(3.0, 300)$ is not a solution.
$s = 50h + 100$, $375 \overset{?}{=} 50(5.5) + 100$, $375 \overset{?}{=} 275 + 100$,
$375 = 375$; $(5.5, 375)$ is a solution.

44. $480 = (8)(12)(h)$, $480 = 96 \, h$, $h = 5$ **45.** $30 = \dfrac{5}{9}(F - 32)$, $54 = F - 32$, $F = 86$

46. $65 = \dfrac{1}{2}(a)(13)$, $65 = \dfrac{13}{2}a$, $a = 10$

47. $54 = 2(17) + 2w$, $54 = 34 + 2w$, $20 = 2w$, $w = 10$

48. $226.08 = (\pi)(6^2)(h)$, $226.08 = 36\pi h$, $h \approx 2$

49. $2l + 2(19) = 68$, $2l + 38 = 68$, $2l = 30$, $l = 15$; 15 cm

50. $3n + 4 = 73$, $3n = 69$, $n = 23$ **51.** $60h = 210$, $h = 3.5$; 3.5 h

52. $11n = -176$, $n = -16$ **53.** $\dfrac{18xz^2}{12xyz} = \dfrac{3z}{2y}$

54. $\dfrac{15c^2}{5ab} \cdot \dfrac{10a^2b}{3c} = \dfrac{15a^2bc^2}{15abc} = 10ac$ **55.** $\dfrac{3(m + 2)}{m^2n} \cdot \dfrac{m^3}{m(m + 2)} = \dfrac{3}{n}$

56. $\dfrac{(x + 3)(x - 1)}{(x + 3)(x + 2)} \cdot \dfrac{4(x + 2)}{x - 1} = 4$ **57.** $\dfrac{(a + 7)(a - 7)}{a + 7} \cdot \dfrac{8}{a - 7} = 8$

58. $\dfrac{2(y + 2)(y - 2)}{y(2y + 1)(2y - 1)} \cdot \dfrac{y^3(2y - 1)}{6(y - 2)} = \dfrac{y^2(y + 2)}{3(2y + 1)}$ or $\dfrac{y^3 + 2y^2}{6y + 3}$

page 375 Capsule Review

1. 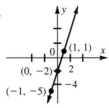 **2.** 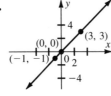 **3.** 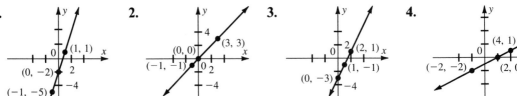 **4.**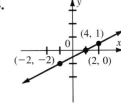

page 378 Class Exercises

1. $y - 3x = 4$, $0 - 3x = 4$, $x = -\dfrac{4}{3}$; x-intercept is $-\dfrac{4}{3}$.

$y - 3(0) = 4$, $y = 4$; y-intercept is 4.

2. $x + y = 8$, $x + 0 = 8$, $x = 8$; x-intercept is 8. $0 + y = 8$, $y = 8$; y-intercept is 8.

3. $y = -7$, no x-intercept; y-intercept is -7.   **4.** $x = 3$, x-intercept is 3; no y-intercept.

5. Two points determine a line. The third point acts as a check when you graph an equation.

6. If $A = 0$, the graph is a horizontal line. If $B = 0$, the graph is a vertical line.

7. Choosing multiples of 4 for x produced integral values for y (since x was multiplied by $-\frac{3}{4}$).

pages 378–379   Practice Exercises

A 1.

x	y
0	4
3	2
-3	6

2.

x	y
0	4
5	0
-5	8

3.

x	y
0	-2
4	0
-4	-4

4.

x	y
0	-5
3	-4
6	-3

5. $x = \frac{1}{2}y + 3$, $2x = y + 6$, $2x - y = 6$;

$$2x - 0 = 6 \qquad\qquad 2(0) - y = 6$$
$$2x = 6 \qquad\qquad -y = 6$$
$$x = 3 \qquad\qquad y = -6$$

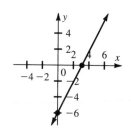

6. $x = \frac{2}{3}y + 5$, $3x = 2y + 15$, $3x - 2y = 15$;

$$3x - 2(0) = 15 \qquad\qquad 3(0) - 2y = 15$$
$$3x - 0 = 15 \qquad\qquad 0 - 2y = 15$$
$$3x = 15 \qquad\qquad -2y = 15$$
$$x = 5 \qquad\qquad y = -7.5$$

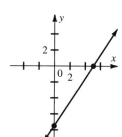

7. $2x + y = 8$;

$$2x + 0 = 8 \qquad 2(0) + y = 8$$
$$2x = 8 \qquad 0 + y = 8$$
$$x = 4 \qquad y = 8$$

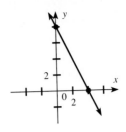

8. $4x + y = 16$;

$$4x + 0 = 16 \qquad 4(0) + y = 16$$
$$4x = 16 \qquad 0 + y = 16$$
$$x = 4 \qquad y = 16$$

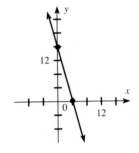

9. $y = \frac{2}{5}x + 2$, $5y = 2x + 10$, $-2x + 5y = 10$;

$$-2x + 5(0) = 10 \qquad -2(0) + 5y = 10$$
$$-2x = 10 \qquad 0 + 5y = 10$$
$$x = -5 \qquad 5y = 10$$
$$\qquad\qquad\qquad y = 2$$

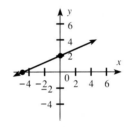

10. $y = \frac{1}{2}x + 2$, $2y = x + 4$, $-x + 2y = 4$;

$$-x + 2(0) = 4 \qquad 0 + 2y = 4$$
$$-x = 4 \qquad 2y = 4$$
$$x = -4 \qquad y = 2$$

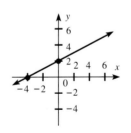

11. $y = \frac{1}{4}x - 1$, $4y = x - 4$, $-x + 4y = -4$;

$$-x + 4(0) = -4 \qquad 0 + 4y = -4$$
$$-x = -4 \qquad 4y = -4$$
$$x = 4 \qquad y = -1$$

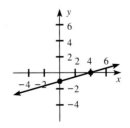

12. $y = \frac{3}{4}x - 3$, $4y = 3x - 12$, $-3x + 4y = -12$;

$$-3x + 4(0) = -12 \qquad\qquad -3(0) + 4y = -12$$
$$-3x + 0 = -12 \qquad\qquad 0 + 4y = -12$$
$$-3x = -12 \qquad\qquad 4y = -12$$
$$x = 4 \qquad\qquad y = -3$$

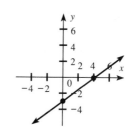

For Exercises 13–20, solutions may vary. Possible solutions are:

13.

x	y
0	2
−1	2
1	2

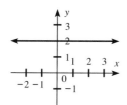

14.

x	y
−2	0
0	0
2	0

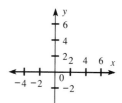

15.

x	y
1	−1
1	0
1	1

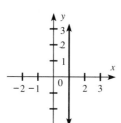

16.

x	y
5	−2
5	0
5	3

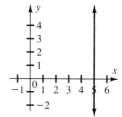

17.

x	y
2	0
0	4
1	2

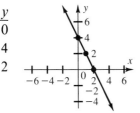

18.

x	y
3	0
0	9
1	6

19.

x	y
0	0
3	−1
−3	1

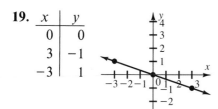

20.

x	y
0	0
4	−1
−4	1

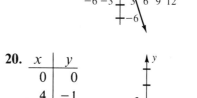

21.

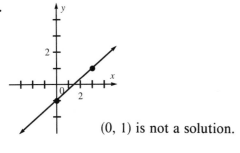

(0, 1) is not a solution.

22. (3, 1) is a solution.

23. $(-3, -3)$ is a solution.

24. $(1, 0)$ is not a solution.

B 25.

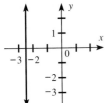

26.

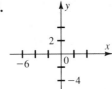

27.

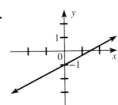

28.

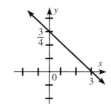

29.

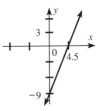

30.

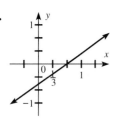

31.

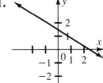

32.

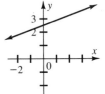

C 33. $(0, y) = (0, 3)$

34. $(-2, y) = (-2, 1)$

35. $(x, 0) = (-3, 0)$

36. $(x, -1) = (-4, -1)$

37. $y = 3x,\ y = 3(2a^2),\ y = 6a^2;\ (2a^2, 6a^2)$

38. $y = 2x + 1,\ y = 2(b^2) + 1,\ y = 2b^2 + 1;\ (b^2, 2b^2 + 1)$

39. $y = \frac{1}{2}x - 3,\ 4d^4 = \frac{1}{2}x - 3,\ 8d^4 = x - 6,\ 8d^4 + 6 = x;\ (8d^4 + 6, 4d^4)$

40. $y = \frac{1}{4}x + 1,\ 8b^3 = \frac{1}{4}x + 1,\ 32b^3 = x + 4,\ 32b^3 - 4 = x;\ (32b^3 - 4, 8b^3)$

41.

s	w
0	75
1	75.25
2	75.5

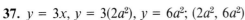

42. The w-intercept is 75. The w-intercept represents the weekly salary in a week in which there were no sales.

page 379   Algebra in Aviation

1. Check students' work.   2. New York to Miami

page 381   Class Exercises

1. $22.50   2. 200 mi   3. $40

4. Jill charged $17.50 for $2\frac{1}{2}$ h

   and $35.00 for 6 h.

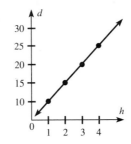

5. The given or independent variable is represented along the horizontal axis. The variable that is to be solved or the dependent variable is represented along the vertical axis.

6. You should examine the data so that it can be graphed within the scales chosen. The two axes may have different scales or they may be the same, depending on the data.

pages 381–382   Practice Exercises

A 1. $F = 25 + 15d$, $F = 25 + 15(2)$,
       $F = 25 + 30$; $F = \$55$
       $F = 25 + 15d$, $F = 25 + 15(3)$,
       $F = 25 + 45$; $F = \$70$

2

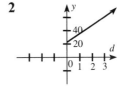

3. $115 for 6 days   4. $200 for 11 days

5. $C = x + 0.02x$, $C = 100 + 0.02(100)$,
   $C = 100 + 2$; $C = \$102$
   $C = x + 0.02x$, $C = 200 + 0.02(200)$,
   $C = 200 + 4$; $C = \$204$

6.

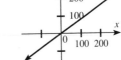

7. $306   8. $459

B 9. $C = 3p$, $C = 3(10)$; $C = 30$
       $C = 3p$, $C = 3(12)$; $C = 36$

10–11.

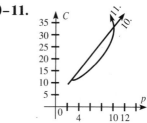

12. Yes, the claim is justified.

C 13.

i	p
0	0
0.5	37.5
1	50
1.5	37.5
2	0

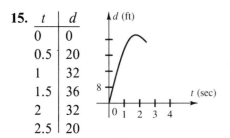

14. 21.9 watts; 21.9 watts

15.

t	d
0	0
0.5	20
1	32
1.5	36
2	32
2.5	20

16. The stone will hit the ground after 3 s.

page 383   Mixed Problem Solving Review

1. Let x represent the width of a rectangle.

$$\frac{4}{w^2 - 4} = \frac{2}{w + 2} \cdot x, \; \frac{w + 2}{2} \cdot \frac{4}{w^2 + 4} = \frac{2}{w + 2} \cdot \frac{w + 2}{2} \cdot x, \; x = \frac{4(w + 2)}{2(w^2 - 4)},$$

$$x = \frac{4(w + 2)}{2(w + 2)(w - 2)}, \; x = \frac{2}{w - 2}$$

2.

	rate	× time =	distance
jet	960 km/h	t	4560
prop plane	560 km/h	t	4560

$960t + 560t = 4560$, $1520t = 4560$, $t = 3$; 3 h

3. Let x represent the first even integer, then $x + 2$ represents the second consecutive even integer.

$p = x + x + 2 + x + x + 2$, $p = 4x + 4$, $108 = 4x + 4$, $104 = 4x$, $26 = x$; then $x + 2 = 28$; The dimensions of the parallelogram are 26 m and 28 m.

4. Let w represent the width of the rectangle, then $w + 5$ represents the length of the rectangle.

$A = lw$, $176 = (w + 5)w$, $176 = w^2 + 5w$, $0 = w^2 + 5w - 176$,
$0 = (w + 16)(w - 11)$, $w + 16 = 0$, $w = -16$ is not a solution;
$w - 11 = 0$, $w = 11$; width $= 11$ cm, length $= w + 5 = 11 + 5 = 16$ cm

5. $\dfrac{1120}{2} = \dfrac{x}{13}$, $2x = 14{,}560$, $x = 7280$; 7280 km

page 383 Project

1. Check students' work.   **2.** Check students' graphs.   **3.** Answers may vary.

page 384 Capsule Review

1. $\dfrac{24}{32} = \dfrac{24 \div 8}{32 \div 8} = \dfrac{3}{4}$

2. $\dfrac{51}{17} = \dfrac{51 \div 17}{17 \div 17} = \dfrac{3}{1}$

3. $\dfrac{-18}{54} = \dfrac{-18 \div 18}{54 \div 18} = -\dfrac{1}{3}$

4. $\dfrac{16}{-6} = \dfrac{16 \div 2}{-6 \div 2} = -\dfrac{8}{3}$

5. $\dfrac{-11}{-11} = \dfrac{-11 \div -11}{-11 \div -11} = \dfrac{1}{1}$

pages 386–387 Class Exercises

1. positive slope

2. no slope

3. negative slope

4. zero

5. $m = \dfrac{y_2 - y_1}{x_2 - x_1} = \dfrac{3 - 7}{1 - 6} = \dfrac{-4}{-5} = \dfrac{4}{5}$

6. $m = \dfrac{y_2 - y_1}{x_2 - x_1} = \dfrac{4 - 3}{2 - (-1)} = \dfrac{1}{3}$

7. $m = \dfrac{y_2 - y_1}{x_2 - x_1} = \dfrac{1 - 3}{2 - (-2)} = \dfrac{-2}{4} = -\dfrac{1}{2}$

8. $m = \dfrac{y_2 - y_1}{x_2 - x_1} = \dfrac{-3 - 3}{5 - 1} = \dfrac{-6}{4} = -\dfrac{3}{2}$

9. $m = \dfrac{y_2 - y_1}{x_2 - x_1} = \dfrac{2 - 2}{3 - (-5)} = \dfrac{0}{8} = 0$

10. $m = \dfrac{y_2 - y_1}{x_2 - x_1} = \dfrac{4 - 2}{-1 - (-1)} = \dfrac{2}{0}$
undefined; no slope

pages 387–388 Practice Exercises

A **1.** $m = \dfrac{y_2 - y_1}{x_2 - x_1} = \dfrac{6 - 2}{5 - 3} = \dfrac{4}{2} = 2$

2. $m = \dfrac{y_2 - y_1}{x_2 - x_1} = \dfrac{4 - 8}{1 - 3} = \dfrac{-4}{-2} = 2$

3. $m = \dfrac{y_2 - y_1}{x_2 - x_1} -= \dfrac{14 - 9}{5 - 2} = \dfrac{5}{3}$

4. $m = \dfrac{y_2 - y_1}{x_2 - x_1} = \dfrac{11 - 7}{7 - 4} = \dfrac{4}{3}$

5. $m = \dfrac{y_2 - y_1}{x_2 - x_1} = \dfrac{-5 - 4}{2 - (-4)} = \dfrac{-9}{6} = -\dfrac{3}{2}$

6. $m = \dfrac{y_2 - y_1}{x_2 - x_1} = \dfrac{-4 - 1}{3 - (-3)} = \dfrac{-5}{6} = -\dfrac{5}{6}$

7. $m = \dfrac{y_2 - y_1}{x_2 - x_1} = \dfrac{-5 - 1}{3 - (-2)} = \dfrac{-6}{5} = -\dfrac{6}{5}$

8. $m = \dfrac{y_2 - y_1}{x_2 - x_1} = \dfrac{-4 - 2}{2 - (-5)} = \dfrac{-6}{7} = -\dfrac{6}{7}$

9. $m = \dfrac{y_2 - y_1}{x_2 - x_1} = \dfrac{4 - (-2)}{3 - 9} = \dfrac{6}{-6} = -1$

10. $m = \dfrac{y_2 - y_1}{x_2 - x_1} = \dfrac{2 - (-3)}{1 - 6} = \dfrac{5}{-5} = -1$

11. $m = \dfrac{y_2 - y_1}{x_2 - x_1} = \dfrac{3 - (-1)}{2 - 7} = \dfrac{4}{-5} = -\dfrac{4}{5}$

12. $m = \dfrac{y_2 - y_1}{x_2 - x_1} = \dfrac{3 - (-2)}{4 - 5} = \dfrac{5}{-1} = -5$

13. $m = \dfrac{y_2 - y_1}{x_2 - x_1} = \dfrac{3 - 3}{2 - 1} = \dfrac{0}{1} = 0$; horizontal line

14. $m = \dfrac{y_2 - y_1}{x_2 - x_1} = \dfrac{5 - 5}{3 - 0} = \dfrac{0}{3} = 0$; horizontal line

15. $m = \dfrac{y_2 - y_1}{x_2 - x_1} = \dfrac{1 - 2}{-5 - (-5)} = \dfrac{-1}{0}$, undefined, no slope; vertical line

16. $m = \dfrac{y_2 - y_1}{x_2 - x_1} = \dfrac{2 - 0}{-3 - (-3)} = \dfrac{2}{0}$, undefined, no slope; vertical line

17. $m = \dfrac{y_2 - y_1}{x_2 - x_1} = \dfrac{7 - 7}{3 - 4} = \dfrac{0}{-1} = 0$; horizontal line

18. $m = \dfrac{y_2 - y_1}{x_2 - x_1} = \dfrac{1 - 1}{0 - 2} = \dfrac{0}{-2} = 0$; horizontal line

19. 1. Plot $R(3, 4)$.
    2. From R, go down 1 unit (for the rise, -1) since $-\dfrac{1}{2} = \dfrac{-1}{2}$. Go to the right 2 units (for the run, 2). Mark the point S.
    3. Draw a line through R and S.

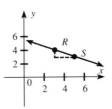

20. 1. Plot $P(2, 5)$.
    2. From P, go down 4 units (for the rise, -4) since $-\dfrac{4}{3} = \dfrac{-4}{3}$. Go to the right 3 units (for the run, 3). Mark the point Q.
    3. Draw a line through P and Q.

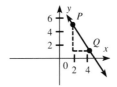

21. 1. Plot $T(-1, 5)$.
    2. From T, go down 2 units (for the rise, -2) since $-\dfrac{2}{3} = \dfrac{-2}{3}$. Go to the right 3 units (for the run, 3). Mark the point S.
    3. Draw a line through T and S.

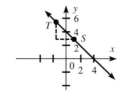

22. 1. Plot $U(-2, 3)$.

    2. From U, go down 1 unit (for the rise, -1) since $-\dfrac{1}{3} = \dfrac{-1}{3}$. Go to the right 3 units (for the run, 3). Mark the point V.

    3. Draw a line through U and V.

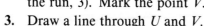

23. Slope of $\overleftrightarrow{AB} = m = \dfrac{-2 - (-5)}{1 - (-1)} = \dfrac{3}{2}$

    Slope of $\overleftrightarrow{BC} = m = \dfrac{4 - (-2)}{5 - 1} = \dfrac{6}{4} = \dfrac{3}{2}$

    Since the slope of $\overleftrightarrow{AB}$ equals the slope of $\overleftrightarrow{BC}$, the points are collinear.

24. Slope of $\overleftrightarrow{XY} = m = \dfrac{-2 - (-6)}{0 - (-2)} = \dfrac{4}{2} = 2$

    Slope of $\overleftrightarrow{YZ} = m = \dfrac{10 - (-2)}{6 - 0} = \dfrac{12}{6} = 2$

    Since the slope of $\overleftrightarrow{XY}$ equals the slope of $\overleftrightarrow{YZ}$, the points are collinear.

B 25. Slope of $\overleftrightarrow{DE} = m = \dfrac{2 - 4}{0 - (-3)} = -\dfrac{2}{3}$

    Slope of $\overleftrightarrow{EF} = m = \dfrac{0 - 2}{-3 - 0} = \dfrac{-2}{-3} = \dfrac{2}{3}$

    Since the slope of $\overleftrightarrow{DE}$ does not equal the slope of $\overleftrightarrow{EF}$, the points are not collinear.

26. Slope of $\overleftrightarrow{GH} = m = \dfrac{-1 - 3}{1 - 3} = \dfrac{-4}{-2} = 2$

    Slope of $\overleftrightarrow{HI} = m = \dfrac{0 - (-1)}{0 - 1} = \dfrac{1}{-1} = -1$

    Since the slope of $\overleftrightarrow{GH}$ does not equal the slope of $\overleftrightarrow{HI}$, the points are not collinear.

27. Slope of $\overleftrightarrow{JK} = m = \dfrac{4 - 1}{-2 - (-2)} = \dfrac{3}{0}$, undefined

    Slope of $\overleftrightarrow{KL} = m = \dfrac{4 - 4}{2 - (-2)} = \dfrac{0}{4} = 0$

    Since the slope of $\overleftrightarrow{JK}$ does not equal the slope of $\overleftrightarrow{KL}$, the points are not collinear.

28. Slope of $\overleftrightarrow{MN} = m = \dfrac{4 - 1}{0 - (-2)} = \dfrac{3}{2}$

    Slope of $\overleftrightarrow{NP} = m = \dfrac{7 - 4}{2 - 0} = \dfrac{3}{2}$

    Since the slope of $\overleftrightarrow{MN}$ equals the slope of $\overleftrightarrow{NP}$, the points are collinear.

29. Slope of $\overleftrightarrow{RS} = m = \dfrac{-5 - (-2)}{-1 - 1} = \dfrac{-3}{-2} = \dfrac{3}{2}$

Slope of $\overleftrightarrow{ST} = m = \dfrac{4 - (-5)}{5 - (-1)} = \dfrac{9}{6} = \dfrac{3}{2}$

Since the slope of $\overleftrightarrow{RS}$ equals the slope of $\overleftrightarrow{ST}$, the points are collinear.

30. Slope of $\overleftrightarrow{SM} = m = \dfrac{2 - 4}{0 - (-3)} = -\dfrac{2}{3}$

Slope of $\overleftrightarrow{MC} = m = \dfrac{0 - 2}{-3 - 0} = \dfrac{-2}{-3} = \dfrac{2}{3}$

Since the slope of $\overleftrightarrow{SM}$ does not equal the slope of $\overleftrightarrow{MC}$, the points are not collinear.

31. $m_{AC} = \dfrac{y_2 - y_1}{x_2 - x_1} = \dfrac{-1 - 3}{5 - 0} = -\dfrac{4}{5}$;

$m_{BC} = \dfrac{y_2 - y_1}{x_2 - x_1} = \dfrac{-1 - 3}{5 - 2} = -\dfrac{4}{3}$;

$m_{AB} = \dfrac{y_2 - y_1}{x_2 - x_1} = \dfrac{3 - 3}{2 - 0} = \dfrac{0}{2} = 0$

32. $m_{AD} = \dfrac{y_2 - y_1}{x_2 - x_1} = \dfrac{-2 - 4}{-1 - (-2)} = \dfrac{-6}{1} = -6$;

$m_{DC} = \dfrac{y_2 - y_1}{x_2 - x_1} = \dfrac{-3 - (-2)}{0 - (-1)} = \dfrac{-1}{1} = -1$;

$m_{CB} = \dfrac{y_2 - y_1}{x_2 - x_1} = \dfrac{3 - (-3)}{1 - 0} = \dfrac{6}{1} = 6$;

$m_{AB} = \dfrac{y_2 - y_1}{x_2 - x_1} = \dfrac{3 - 4}{1 - (-2)} = -\dfrac{1}{3}$

33. 1. Plot $P(6, -1)$.

    2. From P, go down 1 unit (for the rise, -1) since $-1 = \dfrac{-1}{1}$. Go to the right 1 unit (for the run, 1). Mark the point Q.

    3. Draw a line through P and Q.

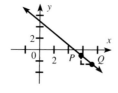

34. 1. Plot $R\left(0, \dfrac{1}{2}\right)$.

    2. From R, go up 6 units (for the rise, 6) since $6 = \dfrac{6}{1}$. Go to the right 1 unit (for the run, 1). Mark the point S.

    3. Draw a line through R and S.

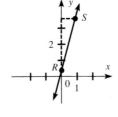

35. 1. Plot $S\left(\dfrac{1}{2}, -\dfrac{3}{2}\right)$,

    2. From S, do not move (for the rise, 0) since $0 = \dfrac{0}{1}$. Go to the right 1 unit (for the run, 1).

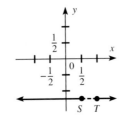

Mark the point T.

3. Draw a line through S and T.

36. 1. Plot $T\left(1\frac{1}{3}, -2\frac{2}{3}\right)$.

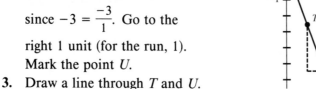

2. From T go down 3 units. (for the rise -3), since $-3 = \dfrac{-3}{1}$. Go to the right 1 unit (for the run, 1). Mark the point U.

3. Draw a line through T and U.

C 37. $m = \dfrac{y_2 - y_1}{x_2 - x_1}, -\dfrac{1}{2} = \dfrac{12 - y}{8 - 4}, -\dfrac{1}{2} = \dfrac{12 - y}{4}, -4 = 2(12 - y), -4 = 24 - 2y,$

$-28 = -2y;\ 14 = y$

38. $m = \dfrac{y_2 - y_1}{x_2 - y_1}, \dfrac{1}{4} = \dfrac{1 - (-3)}{2 - x}, -\dfrac{1}{4} = \dfrac{4}{2 - x}, 2 - x = 16, -x = 14;$

$x = -14$

39. $m = \dfrac{y_2 - y_1}{x_2 - x_1}, 2 = \dfrac{4 - 0}{3 - x}, \dfrac{2}{1} = \dfrac{4}{3 - x}, 2(3 - x) = 4, 6 - 2x = 4,$

$-2x = -2;\ x = 1$

40. $m = \dfrac{y_2 - y_1}{x_2 - x_1}, 3 = \dfrac{-1 - y}{1 - (-1)}, \dfrac{3}{1} = \dfrac{-1 - y}{2}, -1 - y = 6, -y = 7; y = -7$

41. $m_{AB} = \dfrac{y_2 - y_1}{x_2 - x_1} = \dfrac{3b - 3a}{b - a} = \dfrac{3(b - a)}{b - a} = 3$

42. $m_{CD} = \dfrac{y_2 - y_1}{x_2 - x_1} = \dfrac{d + 4 - (c + 4)}{d - c} = \dfrac{d + 4 - c - 4}{d - c} = \dfrac{d - c}{d - c} = 1$

43. Since $RSTU$ is a parallelogram, $m_{RS} = m_{TU}$ and $m_{ST} = m_{RU}$.

$m_{RS} = \dfrac{y_2 - y_1}{x_2 - x_1} = \dfrac{6 - 4}{2 - (-2)} = \dfrac{2}{4} = \dfrac{1}{2};$

$m_{TU} = \dfrac{y_2 - y_1}{x_2 - x_1} = \dfrac{0 - 2}{x - 7} = \dfrac{-2}{x - 7} = \dfrac{2}{7 - x};$ Since $m_{RS} = m_{TU}, \dfrac{1}{2} = \dfrac{2}{7 - x}, x = 3$

$m_{ST} = \dfrac{y_2 - y_1}{x_2 - x_1} = \dfrac{2 - 6}{7 - 2} = \dfrac{-4}{5} = -\dfrac{4}{5};$

$m_{RU} = \dfrac{y_2 - y_1}{x_2 - x_1} = \dfrac{0 - 4}{x - (-2)} = \dfrac{-4}{x + 2} = -\dfrac{4}{x + 2} = -\dfrac{4}{5}$

44. 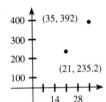 Plot ordered pairs (35, 392) and (21, 235.2).

$$m = \frac{y_2 - y_1}{x_2 - x_1} = \frac{235.2 - 392}{21 - 35} = \frac{-156.8}{-14} \approx 11.2;$$

The slope represents g/cm³.

page 388 Test Yourself

1. $2x + 3y = -2$, $2(5) + 3(-4) \stackrel{?}{=} -2$, $10 + (-12) \stackrel{?}{=} -2$, $-2 = -2$; $(5, -4)$ is a solution.

2. $y - x = 2$, $\frac{3}{2} - \frac{1}{2} \stackrel{?}{=} 2$, $\frac{2}{2} \stackrel{?}{=} 2$, $1 \neq 2$; $\left(\frac{1}{2}, \frac{3}{2}\right)$ is not a solution.

3. 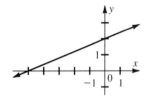 **4.** 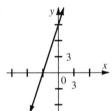 **5.**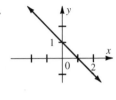

6. 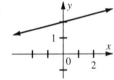 **7.** 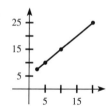 **8.** 50°C

9. 95 min

10. Slope of $\overleftrightarrow{AB} = m = \frac{2 - 3}{3 - 5} = \frac{-1}{-2} = \frac{1}{2}$

11. Slope of $\overleftrightarrow{CD} = m = \frac{-3 - (-7)}{4 - (-2)} = \frac{4}{6} = \frac{2}{3}$

page 389 Integrating Algebra: Radio Waves

1. Let $r = 186,000$ and $t = \frac{1}{10,000}$. $d = 186,000 \times \frac{1}{10,000} = 18.6$ mi;

The signal traveled 18.6 mi altogether. So the distance between the ships is one-half of the distance or 9.3 mi.

2. A signal must travel twice the distance of 7 mi to verify that the ships are 3.5 mi apart.

3. Let r = 186,000 and t = 0.0002. d = 186,000(0.0002) = 37.2 mi; The signal traveled 37.2 mi altogether. So the distance between the ships is one-half of the distance or 18.6 mi.

4. Let r = 186,000 and d = 500.
500 = 186,000(t), t = 500 ÷ 186,000 = 0.0027
It takes 0.0027 s for the signal to reach the station.

5. Let d = 240,000 and r = 186,000.
240,000 = 186,000(t), t = 240,000 ÷ 186,000, t = 1.29;
It will take the signal 1.29 s to reach the moon and 1.29 s to return. The signal will travel 2.58 s altogether.

6. Let r = 186,000 and t = 0.0003. d = 186,000(0.003), d = 558;
The signal traveled 558 mi altogether. So the distance between the satellite and the earth is one-half of 558 mi or 279 mi.

7. Let r = 186,000 and d = 150.
150 = 186,000(0.0014 − t), 150 = 260.4 − 186,000t, 186,000t = 110.4,
t = 0.000594
Let t = 0.000594 and r = 186,000. d = 186,000(0.000594), d = 110.4 mi;
The satellite is 110.4 mi from the earth.

8. The spreadsheet begins with the values displayed in the text, and ends with an optical horizon distance of 100, a radar distance of 107, and a geometric distance of 93.04348, all in row 88.

9. Check students' work.

page 391   Capsule Review

1. $y − 2x = 8$, $y = 2x + 8$

2. $y + 2x = −5$, $y = −2x − 5$

3. $3y − x = 9$, $3y = x + 9$, $y = \frac{1}{3}x + 3$

4. $2y + 3x = 4$, $2y = −3x + 4$, $y = −\frac{3}{2}x + 2$

5. $2y + x + 5 = 0$, $2y = −x − 5$, $y = −\frac{1}{2}x − \frac{5}{2}$

6. $3y + 2x − 7 = 0$, $3y = −2x + 7$, $y = −\frac{2}{3}x + \frac{7}{3}$

pages 393–394   Class Exercises

1. $y = −\frac{2}{3}x$, $m = −\frac{2}{3}$; $b = 0$

2. $x + 5y = 10$, $5y = -x + 10$, $y = -\dfrac{1}{5}x + 2$, $m = -\dfrac{1}{5}$; $b = 2$

3. $8x - y = 2$, $-y = -8x + 2$, $y = 8x - 2$, $m = 8$; $b = -2$

4. $x = 3y + 7$, $-3y = -x + 7$, $y = \dfrac{1}{3}x - \dfrac{7}{3}$, $m = \dfrac{1}{3}$; $b = -\dfrac{7}{3}$

5. $y = -\dfrac{4}{3}x + 2$, $m = -\dfrac{4}{3}$; $b = 2$

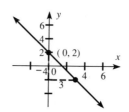

6. $y - 2x = 0$, $y = 2x$, $m = 2$, $b = 0$

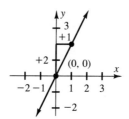

7. $2x + 5y = 10$, $5y = -2x + 10$,

$y = -\dfrac{2}{5}x + 2$, $m = -\dfrac{2}{5}$, $b = 2$

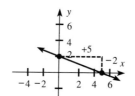

8. $y - 4 = 0$, $y = 4$, $m = 0$, $b = 4$

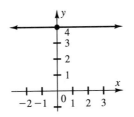

9. It is the same except that when you use the y-intercept you are locating a particular point—the point where the line intersects the y-axis.

10. All three lines intersect the y-axis at the same point.

11. $y = 2x + 5$, $m = 2$; $-2x + y = 9$, $y = 2x + 9$, $m = 2$; Lines are parallel since the slopes are equal; $m = 2$.

12. $x + 3y = 7$, $3y = -x + 7$, $y = -\dfrac{1}{3}x + \dfrac{7}{3}$, $m = -\dfrac{1}{3}$;

$y = \dfrac{1}{3}x - 4$, $m = \dfrac{1}{3}$; Lines are not parallel since the slopes are not equal.

13. $x - y = 3$, $-y = -x + 3$, $y = x - 3$, $m = 1$;
$y - x = 2$, $y = x + 2$, $m = 1$; Lines are parallel since the slopes are equal; $m = 1$.

14. $y = 8$, $m = 0$; $y + 2 = 3$, $y = 1$, $m = 0$; Lines are parallel since the slopes are equal; $m = 0$.

pages 394–395 Practice Exercises

A 1. $3x + 2y = 6$, $2y = -3x + 6$, $y = -\frac{3}{2}x + 3$, $m = -\frac{3}{2}$; $b = 3$

2. $5x + 3y = 15$, $3y = -5x + 15$, $y = \frac{-5}{3}x + 5$, $m = -\frac{5}{3}$; $b = 5$

3. $x - 3y = 9$, $-3y = -x + 9$, $y = \frac{1}{3}x - 3$, $m = \frac{1}{3}$; $b = -3$

4. $x - 4y = 20$, $-4y = -x + 20$, $y = \frac{1}{4}x - 5$; $m = \frac{1}{4}$, $b = -5$

5. $y = -2x$, $m = -2$; $b = 0$ **6.** $y = -5x$, $m = -5$; $b = 0$

7. $y = 3x + 6$, $m = 3$; $b = 6$ **8.** $y = 4x + 8$, $m = 4$; $b = 8$

9. $m = -2$, $b = 0$, $y = mx + b$, $y = -2x + 0$; $y = -2x$

10. $m = -5$, $b = 2$, $y = mx + b$; $y = -5x + 2$

11. $m = 1$, $b = 3$, $y = mx + b$; $y = x + 3$

12. $m = 4$, $b = 2$, $y = mx + b$; $y = 4x + 2$

13. $m = \frac{2}{3}$, $b = 4$, $y = mx + b$; $y = \frac{2}{3}x + 4$

14. $m = \frac{3}{4}$, $b = 3$, $y = mx + b$; $y = \frac{3}{4}x + 3$

15. $m = 0$, $b = -1$, $y = mx + b$, $y = 0 \cdot x + (-1)$, $y = 0 + (-1)$; $y = -1$

16. $m = 0$, $b = -3$, $y = mx + b$; $y = 0 \cdot x + (-3)$, $y = 0 + (-3)$; $y = -3$

17.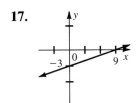

18.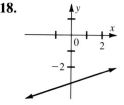

19.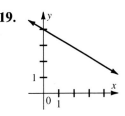

20.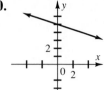

21.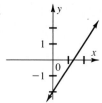

22.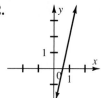

23.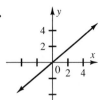

24.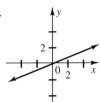

25. $y = -3x + 2$, $m = -3$; $3x + 2y = 7$, $2y = -3x + 7$, $y = -\frac{3}{2}x + \frac{7}{2}$,

$m = -\frac{3}{2}$; Lines are not parallel since the slopes are not equal.

26. $x - y = 2$, $-y = -x + 2$, $y = x - 2$, $m = 1$; $4x = 4y + 6$, $4x - 6 = 4y$,

$x - \frac{6}{4} = y$, $y = x - \frac{3}{2}$, $m = 1$; The lines are parallel since the slopes are

equal; $m = 1$.

27. $2x - 3y = 9$, $-3y = -2x + 9$, $y = \frac{2}{3}x - 3$, $m = \frac{2}{3}$;

$-4x + 6y = 12$, $6y = 4x + 12$, $y = \frac{4}{6}x + 2$, $y = \frac{2}{3}x + 2$, $m = \frac{2}{3}$;

The lines are parallel since the slopes are equal; $m = \frac{2}{3}$.

28. $2x = 4y - 7$, $2x + 7 = 4y$, $\frac{2}{4}x + \frac{7}{4} = y$, $y = \frac{1}{2}x + \frac{7}{4}$, $m = \frac{1}{2}$;

$2y = 6x + 1$, $y = \frac{6}{2}x + \frac{1}{2}$, $y = 3x + \frac{1}{2}$, $m = 3$;

The lines are not parallel since the slopes are not equal.

B 29. $3y = 6x$, $y = 2x$, $m = 2$; $b = 0$

30. $2y = -x + 7$, $y = -\frac{1}{2}x + \frac{7}{2}$, $m = -\frac{1}{2}$; $b = \frac{7}{2}$

31. $3x - 5y = 15$, $-5y = -3x + 15$, $y = \frac{3}{5}x - 3$; $m = \frac{3}{5}$; $b = -3$

32. $2x - 3y = 7$, $-3y = 2x + 7$, $y = \frac{2}{3}x - \frac{7}{3}$; $m = \frac{2}{3}$; $b = -\frac{7}{3}$

33. $\frac{x}{2} = \frac{y}{5}$, $2y = 5x$, $y = \frac{5}{2}x$; $m = \frac{5}{2}$; $b = 0$

34. $\frac{y}{2} - \frac{x}{3} = \frac{1}{4}$, $6y - 4x = 3$, $6y = 4x + 3$, $y = \frac{4}{6}x + \frac{3}{6}$, $y = \frac{2}{3}x + \frac{1}{2}$; $m = \frac{2}{3}$; $b = \frac{1}{2}$

35. $2x + 3y = 4y$, $2x = y$, $y = 2x$; $m = 2$; $b = 0$

36. $x + 4 = y - 1$, $x + 5 = y$, $m = 1$; $b = 5$

37. $5x + 2y + 10 = 0$
$2y = -5x - 10$
$y = -\frac{5}{2}x - 5$;
$m = -\frac{5}{2}$; $b = -5$

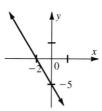

38. $2y - 3x - 2 = 0$
$2y = 3x + 2$
$y = \frac{3}{2}x + 1$;
$m = \frac{3}{2}$; $b = 1$

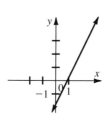

39. $4x = 2y + 3$
$4x - 3 = 2y$
$2x - \frac{3}{2} = y$;
$m = 2$, $b = -\frac{3}{2}$

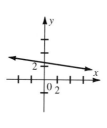

40. $3x = 4y + 1$
$3x - 1 = 4y$
$\frac{3}{4}x - \frac{1}{4} = y$;
$m = \frac{3}{4}$; $b = -\frac{1}{4}$

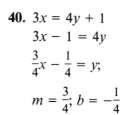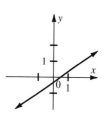

41. $\frac{x}{2} + 6y = 15$
$x + 12y = 30$
$12y = -x + 30$
$y = -\frac{1}{12}x + \frac{30}{12}$
$y = -\frac{1}{12}x + \frac{5}{2}$;
$m = -\frac{1}{12}$; $b = \frac{5}{2}$

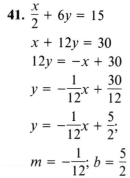

42. $2x - 3y = 2$
$-3y = -2x + 2$
$y = \frac{2}{3}x - \frac{2}{3}$;
$m = \frac{2}{3}$; $b = -\frac{2}{3}$

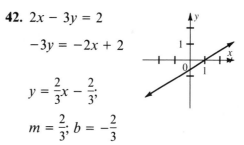

C 43. $y = 2ax + 4$, $m = -1$, $2a = -1$; $a = -\frac{1}{2}$

44. $3y = 2ax - 2$, $m = -2$, $y = \frac{2ax}{3} - \frac{2}{3}$, $\frac{2a}{3} = -2$, $2a = -6$; $a = -3$

45. $ax + 2y = 3$, $m = 4$, $2y = -ax + 3$, $y = -\frac{a}{2}x + \frac{3}{2}$, $-\frac{a}{2} = 4$, $-a = 8$; $a = -8$

46. $y - 3ax = 5$, $m = \frac{1}{3}$, $y = 3ax + 5$, $3a = \frac{1}{3}$, $9a = 1$; $a = \frac{1}{9}$

47. $ax - by = c,\ -by = -ax + c,\ y = \dfrac{a}{b}x - \dfrac{c}{b},\ m_1 = \dfrac{a}{b};$

$-ax + by = d,\ by = ax + d,\ y = \dfrac{a}{b}x + \dfrac{d}{b},\ m_2 = \dfrac{a}{b};$ The lines are parallel since the slopes are equal.

48. $ax + by = c,\ by = -ax + c,\ y = -\dfrac{a}{b}x + \dfrac{c}{b},\ m_1 = -\dfrac{a}{b};$

$bx - ay = d,\ -ay = -bx + d,\ y = \dfrac{b}{a}x - \dfrac{d}{a},\ m_2 = \dfrac{b}{a};$

The lines are not parallel since the slopes are different.

49.

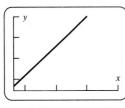

x scl: 50   y scl: 50

50. a. 24° **b.** 38° **c.** 97°

51. a. 167° **b.** 212° **c.** 405°

52. $32 - 22 + 1 = 10 + 1 = 11$

53. $29 - 3 - 2(9) = 29 - 3 - 18 = 8$

54. $(24)(3)(3) = 216$

55. $52 - 0 + 52 = 104$

56. $9(m + 2) = 90,\ 9m + 18 = 90,\ 9m = 72,\ m = 8$

57. $6(2n + 5) = 6(5n - 2),\ 12n + 30 = 30n - 12,\ 42 = 18n,\ n = \dfrac{7}{3}$

58. a.

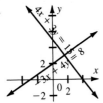

$m_1 = \dfrac{3}{4},\ m_2 = -\dfrac{4}{3}$

$m_1 \cdot m_2 = \left(\dfrac{3}{4}\right)\left(-\dfrac{4}{3}\right) = -1$

$m_1 = -\dfrac{1}{2},\ m_2 = 2$

$m_1 \cdot m_2 = \left(-\dfrac{1}{2}\right)(2) = -1$

b. The graphs are perpendicular to each other.

c. Answers may vary; slopes should be negative reciprocals of each other. An example: $3x + 2y = 6$ and $2x - 3y = 6$.

59. Two parallel lines cannot have the same y-intercept because parallel lines do not intersect.

60. Two perpendicular lines can have the same y-intercept because perpendicular lines can intersect on the y-axis.

page 396   Capsule Review

1. $m = \dfrac{y_2 - y_1}{x_2 - x_1} = \dfrac{3 - 5}{6 - 9} = \dfrac{-2}{-3} = \dfrac{2}{3}$

2. $m = \dfrac{y_2 - y_1}{x_2 - x_1} = \dfrac{1 - 3}{3 - 6} = \dfrac{-2}{-3} = \dfrac{2}{3}$

3. y-intercept is -1

4. x-intercept is $\dfrac{3}{2}$

page 398 Class Exercises

1. $m = \dfrac{1}{3}$, $b = 1$ $y = mx + b$; $y = \dfrac{1}{3}x + 1$

2. $m = -2$, $b = -2$ $y = mx + b$; $y = -2x - 2$

3. $y = 3x - 1$; $-3x + y = -1$

4. $y = \dfrac{1}{4}x + \dfrac{3}{8}$, $8y = 2x + 3$; $-2x + 8y = 3$

5. $x = \dfrac{2}{3}y$, $3x = 2y$; $3x - 2y = 0$

6. $\dfrac{x}{4} + \dfrac{y}{3} = 8$; $3x + 4y = 96$

7. Write the equation in slope-intercept form. $y = 3x + b$, $-4 = 3(-2) + b$, (Substitute -4 for y and -2 for x), $-4 = -6 + b$ (Solve for b), $2 = b$, $y = 3x + 2$, $-3x + y = 2$; $3x - y = -2$.

8. Write the equation in slope-intercept form. $y = x + b$, $-1 = -4 + b$ (Substitute -1 for y and -4 for x and solve for b), $3 = b$, $y = x + 3$, $-x + y = 3$; $x - y = -3$.

9. Write the equation in slope-intercept form. $y = \dfrac{3}{4}x + b$, $1 = \dfrac{3}{4}(1) + b$

(Substitute 1 for y and 1 for x), $4 = 3 + 4b$, $1 = 4b$, $\dfrac{1}{4} = b$ (Solve for b),

$y = \dfrac{3}{4}x + \dfrac{1}{4}$, $4y = 3x + 1$; $-3x + 4y = 1$

10. Write the equation in slope-intercept form. $y = mx + b$, $y = -\dfrac{1}{2}x + b$,

$3 = -\dfrac{1}{2}(0) + b$ (Substitute 3 for y and 0 for x), $3 = 0 + b$ (Solve for b),

$3 = b$, $y = -\dfrac{1}{2}x + 3$, $2y = -x + 6$; $x + 2y = 6$

11. Find the slope of the line through the two points, $m = \dfrac{y_2 - y_1}{x_2 - x_1} =$

$\dfrac{12 - 8}{-1 - (-3)} = \dfrac{4}{2}$; $m = 2$; Write in slope-intercept form. $y = mx + b$,

$y = 2x + b$, $8 = 2(-3) + b$ (Substitute the coordinates of either point for x and y), $8 = -6 + b$ (Solve for b), $14 = b$; $y = 2x + 14$; $-2x + y = 14$

12. $y - 2x = 4$, $y = 2x + 4$; $m = 2$; (Lines are parallel; slopes are equal.) $y = mx + b$, $y = 2x + b$, $-2 = 2(2) + b$, $-2 = 4 + b$, $-6 = b$, $y = 2x - 6$; $-2x + y = -6$

A   **1.** $m = \dfrac{y_2 - y_1}{x_2 - x_1}$, $m = \dfrac{-6 - (-3)}{-2 - 0} = \dfrac{-3}{-2} = \dfrac{3}{2}$;

$m = \dfrac{3}{2}$; $b = -3$, $y = mx + b$; $y = \dfrac{3}{2}x - 3$

2. $m = \dfrac{y_2 - y_1}{x_2 - x_1}$, $m = \dfrac{5 - 4}{2 - 0} = \dfrac{1}{2}$; $m = \dfrac{1}{2}$; $b = 4$, $y = mx + b$; $y = \dfrac{1}{2}x + 4$

3. Write the equation in slope-intercept form.

$y = mx + b$, $y = \dfrac{1}{2}x + b$, $0 = \dfrac{1}{2}(0) + b$ (Substitute 0 for y and 0 for x),

$0 = b$ (Solve for b), $y = \dfrac{1}{2}x$, $2y = x$; $-x + 2y = 0$

4. Write the equation in slope-intercept form. $y = mx + b$, $y = \dfrac{3}{4}x + b$,

$6 = \dfrac{3}{2}(2) + b$ (Substitute 6 for y and 2 for x), $6 = 3 + b$, $3 = b$

(Solve for b), $y = \dfrac{3}{2}x + 3$, $2y = 3x + 6$; $-3x + 2y = 6$

5. Write the equation in slope-intercept form. $y = mx + b$, $y = -2x + b$,
$1 = -2(-1) + b$ (Substitute 1 for y and -1 for x), $1 = 2 + b$, $-1 = b$
(Solve for b), $y = -2x - 1$; $2x + y = -1$

6. Write the equation in slope-intercept form. $y = mx + b$, $y = -3x + b$,
$5 = -3(-1) + b$ (Substitute 5 for y and -1 for x), $5 = 3 + b$, $2 = b$
(Solve for b), $y = -3x + 2$; $3x + y = 2$

7. Write the equations in slope-intercept form. $y = mx + b$, $y = -\dfrac{5}{4}x + b$,

$-3 = -\dfrac{5}{4}(4) + b$ (Substitute -3 for y and 4 for x), $-3 = -5 + b$, $2 = b$

(Solve for b), $y = -\dfrac{5}{4}x + 2$, $4y = -5x + 8$; $5x + 4y = 8$

8. Write the equation in slope-intercept form. $y = mx + b$, $y = -\dfrac{1}{5}x + b$,

$1 = -\dfrac{1}{5}(6) + b$ (Substitute 1 for y and 6 for x), $1 = -\dfrac{6}{5} + b$, $\dfrac{11}{5} = b$ (Solve

for b), $y = -\dfrac{1}{5}x + \dfrac{11}{5}$, $5y = -x + 11$; $x + 5y = 11$

9. Write the equation in slope-intercept form. $y = mx + b$, $y = \frac{7}{3}x + b$,

$-\frac{1}{6} = \frac{7}{3}(0) + b$ (Substitute $-\frac{1}{6}$ for y and 0 for x), $-\frac{1}{6} = b$ (Solve for b),

$y = \frac{7}{3}x - \frac{1}{6}$, $6y = 14x - 1$; $-14x + 6y = -1$

10. Write the equation in slope-intercept form. $y = mx + b$, $y = \frac{5}{2}x + b$,

$-\frac{1}{4} = \frac{5}{2}(3) + b$ (Substitute $-\frac{1}{4}$ for y and 3 for x), $-\frac{1}{4} = \frac{15}{2} + b$, $-\frac{1}{4} - \frac{15}{2} = b$,

$\frac{-1 - 30}{4} = b$, $-\frac{31}{4} = b$ (Solve for b), $y = \frac{5}{2}x - \frac{31}{4}$, $4y = 10x - 31$; $-10x + 4y = -31$

11. Find the slope of the line through the two points.
$m = \frac{y_2 - y_1}{x_2 - x_1} = \frac{8 - (-7)}{2 - (-1)} = \frac{15}{3} = 5$; $m = 5$;
Write in slope-intercept form. $y = mx + b$, $y = 5x + b$,
$8 = 5(2) + b$, $8 = 10 + 6$; $-2 = b$; $y = 5x - 2$

12. Find the slope of the line through the two points.
$m = \frac{y_2 - y_1}{x_2 - x_1} = \frac{8 - (-9)}{1 - (-2)} = \frac{17}{3}$; $m = \frac{17}{3}$;
Write in slope-intercept form. $y = mx + b$, $y = \frac{17}{3}x + b$,
$8 = \frac{17}{3}(1) + b$, $8 = \frac{17}{3} + b$, $\frac{24}{3} - \frac{17}{3} = b$, $\frac{7}{3} = b$; $y = \frac{17}{3}x + \frac{7}{3}$

13. Find the slope of the line through the two points.
$m = \frac{y_2 - y_1}{x_2 - x_1} = \frac{9 - 6}{6 - 5} = \frac{3}{1} = 3$; $m = 3$;
Write in slope-intercept form. $y = mx + b$, $y = 3x + b$, $6 = 3(5) + b$,
$6 = 15 + b$, $-9 = b$; $y = 3x - 9$

14. Find the slope of the line through the two points.
$m = \frac{y_2 - y_1}{x_2 - x_1} = \frac{8 - 1}{4 - 3} = \frac{7}{1} = 7$; $m = 7$;
Write in slope-intercept form. $y = mx + b$, $y = 7x + b$, $1 = 7(3) + b$,
$1 = 21 + b$, $-20 = b$; $y = 7x - 20$

15. Find the slope of the line through the two points.
$m = \frac{y_2 - y_1}{x_2 - x_1} = \frac{11 - 5}{-1 - 1} = \frac{6}{-2} = -3$; $m = -3$;
Write in slope-intercept form. $y = mx + b$, $y = -3x + b$, $5 = -3(1) + b$,
$5 = -3 + b$, $8 = b$; $y = -3x + 8$

16. Find the slope of the line through the two points.

$m = \dfrac{y_2 - y_1}{x_2 - x_1} = \dfrac{8 - 0}{-2 - 2} = \dfrac{8}{-4} = -2; \; m = -2;$

Write in slope-intercept form. $y = mx + b, \; y = -2x + b, \; 0 = -2(2) + b,$
$0 = -4 + b, \; 4 = b; \; y = -2x + 4$

17. Find the slope of the line through the two points.

$m = \dfrac{y_2 - y_1}{x_2 - x_1} = \dfrac{9 - 3}{-6 - 3} = \dfrac{6}{-9} = -\dfrac{2}{3}; \; m = -\dfrac{2}{3};$

Write in slope-intercept form. $y = mx + b, \; y = -\dfrac{2}{3}x + b, \; 3 = -\dfrac{2}{3}(3) + b,$

$3 = -2 + b, \; 5 = b; \; y = -\dfrac{2}{3}x + 5$

18. Find the slope of the line through the two points.

$m = \dfrac{y_2 - y_1}{x_2 - x_1} = \dfrac{2 - 2}{-2 - 7} = \dfrac{0}{-9} = 0; \; m = 0;$

Write in slope-intercept form. $y = mx + b, \; y = 0x + b, \; 2 = 0(7) + b,$
$2 = b; \; y = 2$

19. $y = 3x - 2, \; m = 3; \; y = mx + b, \; y = 3x + b,$ (Parallel lines have equal slopes.) $-3 = 3(2) + b, \; -3 = 6 + b, \; -9 = b, \; y = 3x - 9; \; -3x + y = -9$

20. $y = 3x - 5; \; m = 3; \; y = mx + b, \; y = 3x + b,$ (Parallel lines have equal slopes.) $-4 = 3(1) + b, \; -4 = 3 + b, \; -7 = b, \; y = 3x - 7; \; -3x + y = -7$

B 21. $2x + 5y = 3, \; 5y = -2x + 3, \; y = -\dfrac{2}{5}x + \dfrac{3}{5}; \; m = -\dfrac{2}{5}; \; y = mx + b, \; 0 = -\dfrac{2}{5}(-2) + b,$

$0 = \dfrac{4}{5} + b, \; -\dfrac{4}{5} = b, \; y = -\dfrac{2}{5}x - \dfrac{4}{5}, \; 5y = -2x - 4; \; 2x + 5y = -4$

22. $-3x + y - 4 = 0, \; y = 3x + 4; \; m = 3; \; y = mx + b, \; m = 3; \; b = -\dfrac{1}{2},$

$y = 3x - \dfrac{1}{2}, \; 2y = 6x - 1; \; -6x + 2y = -1$

23. Find the slope of the line through the two points.

$m = \dfrac{y_2 - y_1}{x_2 - x_1} = \dfrac{2 - (-12)}{5 - (-2)} = \dfrac{14}{7} = 2; \; m = 2;$

Write in slope-intercept form. $y = mx + b, \; y = 2x + b, \; 2 = 2(5) + b,$
$2 = 10 + b, \; -8 = b, \; y = 2x - 8; \; -2x + y = -8$

24. Find the slope of the line through the two points.

$m = \dfrac{y_2 - y_1}{x_2 - x_1} = \dfrac{16 - 8}{-9 - (-6)} = \dfrac{8}{-3} = -\dfrac{8}{3}$

Write in slope-intercept form. $y = mx + b$, $y = -\dfrac{8}{3}x + b$, $y = -\dfrac{8}{3}(-6) + b$,

$8 = 16 + b$, $-8 = b$, $y = -\dfrac{8}{3}x - 8$, $3y = -8x - 24$; $8x + 3y = -24$

25. Find the slope of the line through the two points.

$m = \dfrac{y_2 - y_1}{x_2 - x_1} = \dfrac{-10 - (-7)}{-1 - 2} = \dfrac{-3}{-3} = 1$; $m = 1$;

Write in slope-intercept form. $y = mx + b$, $y = 1x + b$, $-7 = 2 + b$, $-9 = b$,

$y = x - 9$; $-x + y = -9$

26. Find the slope of the line through the two points.

$m = \dfrac{y_2 - y_1}{x_2 - x_1} = \dfrac{8 - (-1)}{-1 - 5} = \dfrac{9}{-6} = -\dfrac{3}{2}$;

Write in slope-intercept form. $y = mx + b$, $y = -\dfrac{3}{2}x + b$, $8 = -\dfrac{3}{2}(-1) + b$,

$8 = \dfrac{3}{2} + b$, $\dfrac{16}{2} - \dfrac{3}{2} = b$, $\dfrac{13}{2} = b$, $y = -\dfrac{3}{2}x + \dfrac{13}{2}$, $2y = -3x + 13$; $3x + 2y = 13$

27. Find the slope of the line through the two points.

$m = \dfrac{y_2 - y_1}{x_2 - x_1} = \dfrac{2\frac{1}{2} - 0}{3 - \frac{1}{2}} = \dfrac{\frac{5}{2}}{\frac{5}{2}} = 1$; $m = 1$;

Write in slope-intercept form. $y = mx + b$, $y = 1x + b$, $0 = \dfrac{1}{2} + b$, $-\dfrac{1}{2} = b$,

$y = x - \dfrac{1}{2}$, $2y = 2x - 1$; $-2x + 2y = -1$

28. Find the slope of the line through the two points.

$m = \dfrac{y_2 - y_1}{x_2 - x_1} = \dfrac{5 - 1}{2\frac{2}{3} - \frac{1}{3}} = \dfrac{4}{2\frac{1}{3}} = \dfrac{4}{\frac{7}{3}} = 4 \cdot \dfrac{3}{7} = \dfrac{12}{7}$; $m = \dfrac{12}{7}$;

Write in slope-intercept form. $y = mx + b$, $y = \dfrac{12}{7}x + b$, $1 = \dfrac{12}{7}\left(\dfrac{1}{3}\right) + b$,

$1 = \dfrac{4}{7} + b$, $\dfrac{7}{7} - \dfrac{4}{7} = b$, $\dfrac{3}{7} = b$, $y = \dfrac{12}{7}x + \dfrac{3}{7}$, $7y = 12x + 3$; $-12x + 7y = 3$

29. Find the slope of the line through the two points.

$m = \dfrac{y_2 - y_1}{x_2 - x_1} = \dfrac{4 - 2}{1\frac{1}{2} - \left(-1\frac{1}{2}\right)} = \dfrac{2}{3}$, $m = \dfrac{2}{3}$;

Write in slope-intercept form. $y = mx + b$, $y = \dfrac{2}{3}x + b$, $4 = \dfrac{2}{3}\left(\dfrac{3}{2}\right) + b$,

$4 = 1 + b$, $3 = b$, $y = \dfrac{2}{3}x + 3$, $3y = 2x + 9$; $-2x + 3y = 9$

30. Find the slope of the line through two points.

$$m = \frac{y_2 - y_1}{x_2 - x_1} = \frac{3 - 1\frac{1}{2}}{0 - (-1)} = \frac{\frac{3}{2}}{1} = \frac{3}{2}; \ m = \frac{3}{2};$$

Write the slope-intercept form. $y = mx + b$, $y = \frac{3}{2}x + b$, $3 = \frac{3}{2}(0) + b$, $3 = b$,

$y = \frac{3}{2}x + 3$, $2y = 3x + 6$; $-3x + 2y = 6$

31. x-intercept $(-5, 0)$, y-intercept $(0, 3)$, $m = \frac{y_2 - y_1}{x_2 - x_1} = \frac{3 - 0}{0 - (-5)} = \frac{3}{5}$; $y =$

$mx + b$, $y = \frac{3}{5}x + b$, $0 = \frac{3}{5}(-5) + b$, $0 = -3 + b$, $3 = b$, $y = \frac{3}{5}x + 3$,

$5y = 3x + 15$; $-3x + 5y = 15$

32. $m = \frac{y_2 - y_1}{x_2 - x_1} = \frac{6 - 4}{2 - 1} = \frac{2}{1} = 2$, $m = 2$; $y = mx + b$, $y = 2x + b$,

$7 = 2(-3) + b$, $7 = -6 + b$, $13 = b$, $y = 2x + 13$; $-2x + y = 13$

C 33. $y + 3x - 8 = 0$, $y = -3x + 8$, The lines are perpendicular, so $m_1 m_2 = -1$.

$m_1 = -3$, $-3m_2 = -1$, $m_2 = \frac{1}{3}$; $y = \frac{1}{3}x + b$, $-5 = \frac{1}{3}(2) + b$, $-5 = \frac{2}{3} + b$,

$-\frac{15}{3} - \frac{2}{3} = b$, $-\frac{17}{3} = b$, $y = \frac{1}{3}x - \frac{17}{3}$, $3y = x - 17$; $-x + 3y = -17$

34. $3y + 4x = 6$, $3y = -4x + 6$, $y = -\frac{4}{3}x + 2$, The lines are perpendicular,

so $m_1 m_2 = -1$. $m_1 = -\frac{4}{3}$, $-\frac{4}{3}m_2 = -1$, $-4m_2 = -3$, $m_2 = \frac{3}{4}$; $y = m_2 x + b$,

$4c = \frac{3}{4}c + b$, $16c = 3c + 4b$, $13c = 4b$, $\frac{13c}{4} = b$; $y = \frac{3}{4}x + \frac{13c}{4}$, $4y = 3x + 13c$;

$-3x + 4y = 13c$

35. Write the equation in slope-intercept form.
$y = mx + b$, $y = 2gx + b$, $7 = 2g(2) + b$, $7 = 4g + b$, $7 - 4g = b$;
Write in slope-intercept form. $y = mx + b$; $y = 2gx + 7 - 4g$

36. Write the equation in slope-intercept form.
$y = mx + b$, $y = -5x + b$, $-3k = -5(0) + b$, $-3k = b$;
Write in slope-intercept form. $y = mx + b$; $y = -5x - 3k$

37. Write the equation in slope-intercept form.
$y = mx + b$, $y = \frac{1}{n}x + b$, $1 = \frac{1}{n}(-2) + b$, $\frac{2}{n} + 1 = b$;

Write in slope-intercept form. $y = mx + b$; $y = \frac{1}{n}x + \frac{2}{n} + 1$

38. $(t_1, s_1) = (20,100); (t_2, s_2) = (70,170),$

$$m = \frac{s_2 - s_1}{t_2 - t_1} = \frac{170 - 100}{70 - 20} = \frac{70}{50} = \frac{7}{5};$$

$$s = mt + b, s = \frac{7}{5}t + b, 100 = \frac{7}{5}(20) + b, 100 = 28 + b, 72 = b; s = \frac{7}{5}t + 72$$

39. $\{x : x \geq -13\}$ **40.** $\{a : a < 2\}$ **41.** $\{b : b < 11\}$ **42.** $\{p : p > 5\}$

43. $\{t : t \geq -18\}$ **44.** 4   **45.** $\dfrac{1}{x - 4}$

46. $x = \dfrac{5}{4}$ or $x = 2$ **47.** $z = -3$ or $z = 8$ **48.** $b = 9$

49. $s = -3$ **50.** $a = -1$ **51.** $c = -12$

52. As the slope of a line gets closer to zero, the line gets flatter. As the slope gets further from zero, the line gets steeper. The greater the absolute value of the slope, the steeper the line.

page 401 Capsule Review

1. $b = 0, m = 1, y = mx + b; y = x$ or $-x + y = 0$ **2.** $b = 2, m = 0, y = mx + b;$ $y = 2$ **3.** $b = -2, m = -2, y = mx + b; y = -2x -2$ or $2x + y = -2$

page 403 Class Exercises

1. $-1 \overset{?}{>} 1 + 1, -1 \not> 2; (1, -1)$ does not belong to the graph of $y > x + 1$.

2. $2 \overset{?}{\leq} 2(3) + 4, 2 \overset{?}{\leq} 6 + 4, 2 \leq 10; (3, 2)$ does belong to the graph of $y \leq 2x + 4$.

3. $-6 \overset{?}{<} 2(-3), -6 \not< -6; (-3, -6)$ does not belong to the graph of $y < 2x$.

4. $-1 \not> 5; (6, -1)$ does not belong to the graph of $y > 5$.

5. $-3 \not\geq -1; (-3, 0)$ does not belong to the graph of $x \geq -1$.

6. $-1 \leq 0; (-1, 4)$ does belong to the graph of $x \leq 0$.

7. $y \leq x$; closed half-plane   **8.** $y > -x + 2$; open half-plane

9. $x \geq 2$; closed half-plane

pages 403–404 Practice Exercises

A 1. $y > x + 5, 1 \overset{?}{>} 3 + 5, 1 \not> 8; y > x + 5, 0 \overset{?}{>} -6 + 5, 0 > -1;$ $(-6, 0)$ belongs to the graph of $y > x + 5$.

2. $y > x + 1, 1 \overset{?}{>} 3 + 1, 1 \not> 4; y > x + 1, 2 \overset{?}{>} 0 + 1, 2 > 1;$ $(0, 2)$ belongs to the graph of $y > x + 1$.

3. $y > x + 1, 0 + 2 \overset{?}{>} 0 + 1, 2 > 1; y > x + 1, 5 + 2 \overset{?}{>} 3 + 1, 7 > 4;$ $(0, 0)$ and $(3, 5)$ belong to the graph of $y + 2 > x + 1$.

4. $y + 3 \geq x + 4$, $2 + 3 \overset{?}{\geq} 1 + 4$, $5 \geq 5$; (1, 2) belongs to the graph of $y + 3 \geq x + 4$.

5. $m = -1$, $b = 2$; $y < -x + 2$; open half-plane

6. $m = -1$, $b = 3$; $y > -x + 3$; open half-plane

7. $m = 1$, $b = 0$; $y \leq x$; closed half-plane

8. $m = \dfrac{1}{2}$, $b = 0$; $y \leq \dfrac{1}{2}x$; closed half-plane

9. m is undefined, no y-intercept; $x \leq 2$; closed half-plane.

10. $m = 0$, $b = 4$; $y \leq 4$; closed half-plane

11. Draw the graph of $y = 3x + 2$ as a dashed line.
Test a point in each half-plane.
Try (0, 0).   Try (−1, 2).
$0 \overset{?}{<} 3(0) + 2$ $2 \overset{?}{<} 3(-1) + 2$
$0 < 2$ $2 \not< -1$
Shade the half-plane containing the point (0, 0).

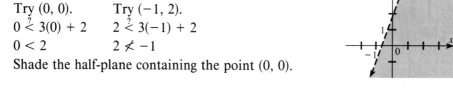

12. Draw the graph of $y = 4x + 1$ as a dashed line.
Test a point in each half-plane.
Try (0, 0).   Try (−1, 3).
$0 \overset{?}{<} 4(0) + 1$ $3 \overset{?}{<} 4(-1) + 1$
$0 < 1$ $3 \not< -3$
Shade the half-plane containing the point (0, 0).

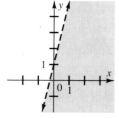

13. Draw the graph of $y = x + 3$ as a dashed line.
Test a point in each half-plane.
Try (0, 0).   Try (0, 4).
$0 \overset{?}{>} 0 + 3$ $4 \overset{?}{>} 0 + 3$
$0 \not> 3$ $4 > 3$
Shade the half-plane containing the point (0, 4).

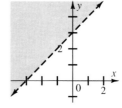

14. Draw the graph of $y = 2x + 1$ as a dashed line.
Test a point in each half-plane.
Try (0, 0).   Try (0, 3).
$0 \overset{?}{>} 2(0) + 1$ $3 \overset{?}{>} 2(0) + 1$
$0 \not> 1$ $3 > 1$
Shade the half-plane containing the point (0, 3).

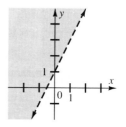

15. Draw the graph of $x = 3$ as a solid line.
Test a point in each half-plane.
Try (0, 0). Try (4, 0).
$0 \not\geq 3$ $4 \geq 3$
Shade the half-plane containing the point (4, 0).

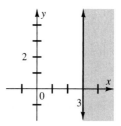

16. Draw the graph of $x = 5$ as a solid line.
Test a point in each half-plane.
Try (0, 0). Try (6, 0).
$0 \not\geq 5$ $6 \geq 5$
Shade the half-plane containing the point (6, 0).

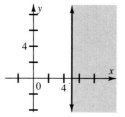

17. Draw the graph of $y = 3x + 4$ as a dashed line.
Test a point in each half-plane.
Try (0, 0). Try (0, 5).
$0 \overset{?}{>} 3(0) + 4$ $5 \overset{?}{>} 3(0) + 4$
$0 > 4$ $5 \not> 4$
Shade the half-plane containing the point (0, 5).

18. Draw the graph of $y = 2x + 5$ as a dashed line.
Test a point in each half-plane.
Try (0, 0). Try (−3, 6).
$0 \overset{?}{>} 2(0) + 5$ $6 \overset{?}{>} 2(-3) + 5$
$0 \not> 5$ $6 > -1$
Shade the half-plane containing the point (−3, 6).

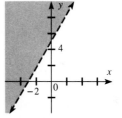

19. Draw the graph of $y = 1$ as a solid line.
Test a point in each half-plane.
Try (0, 0). Try (0, 2).
$0 \not\geq 1$ $2 \geq 1$
Shade the half-plane containing the point (0, 2).

20. Draw the graph of $y = -4$ as a solid line.
Test a point in each half-plane.
Try (0, 0). Try (0, −5).
$0 \geq -4$ $-5 \not\geq -4$
Shade the half-plane containing the point (0, 0).

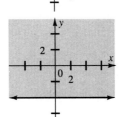

21. Draw the graph of $y = 3x - 1$ as a dashed line.
Test a point in each half-plane.

Try $(0, 0)$. Try $(2, -2)$.

$0 \overset{?}{<} 3(0) - 1$ $-2 \overset{?}{<} 3(2) - 1$

$0 \not< -1$ $-2 < 5$

Shade the half-plane containing the point $(2, -2)$.

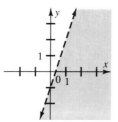

22. Draw the graph of $y = 2x - 3$ as a dashed line.
Test a point in each half-plane.

Try $(0, 0)$. Try $(3, -2)$.

$0 \overset{?}{<} 2(0) - 3$ $-2 \overset{?}{<} 2(3) - 3$

$0 \not< -3$ $-2 < 3$

Shade the half-plane containing the point $(3, -2)$.

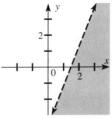

B 23. $y - 1 \le x + 3, y \le x + 4$
Draw the graph of $y = x + 4$ as a solid line.
Test a point in each half-plane.

Try $(0, 0)$. Try $(0, 5)$.

$0 \overset{?}{\le} 0 + 4$ $5 \overset{?}{\le} 0 + 4$

$0 \le 4$ $5 \not\le 4$

Shade the half-plane containing the point $(0, 0)$.

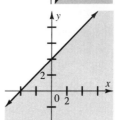

24. $y - 4 \le x + 5, y \le x + 9$
Draw the graph of $y = x + 9$ as a solid line.
Test a point in each half-plane.

Try $(0, 0)$. Try $(0, 10)$.

$0 - 4 \overset{?}{\le} 0 + 5$ $10 - 4 \overset{?}{\le} 0 + 5$

$-4 \le 5$ $6 \not\le 5$

Shade the half-plane containing the point $(0, 0)$.

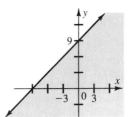

25. $y + 3 < x - 3, y < x - 6$
Draw the graph of $y = x - 6$ as a dashed line.
Test a point in each half-plane.

Try $(0, 0)$. Try $(0, -7)$.

$0 + 3 \overset{?}{<} 0 - 3$ $-7 + 3 \overset{?}{<} 0 - 3$

$0 \not< -3$ $-4 < -3$

Shade the half-plane containing the point $(0, -7)$.

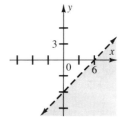

26. $y + 2 < x - 1, y < x - 3$
Draw the graph of $y = x - 3$ as a dashed line.
Test a point in each half-plane.

Try $(0, 0)$. Try $(0, -4)$.

$0 + 2 \overset{?}{<} 0 - 1$ $-4 + 2 \overset{?}{<} 0 - 1$

$2 \not< -1$ $-2 < -1$

Shade the half-plane containing the point $(0, -4)$.

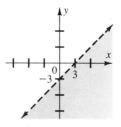

27. $y + 6 > 2x - 1,\ y > 2x - 7$

Draw the graph of $y = 2x - 7$ as a dashed line.
Test a point in each half-plane.

Try $(0, 0)$.   Try $(5, 0)$.

$0 + 6 \overset{?}{>} 2(0) - 1$ $0 + 6 \overset{?}{>} 2(5) - 1$

$6 > -1$ $6 \not> 9$

Shade the half-plane containing the point $(0, 0)$.

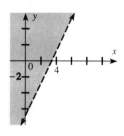

28. $y + 3 > x - 4,\ y > 3x - 7$

Draw the graph of $y = 3x - 7$ as a dashed line.
Test a point in each half-plane.

Try $(0, 0)$.   Try $(3, 0)$.

$0 + 3 \overset{?}{>} 3(0) - 4$ $0 + 3 \overset{?}{>} 3(3) - 4$

$3 > -4$ $3 \not> 5$

Shade the half-plane containing the point $(0, 0)$.

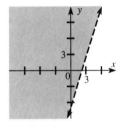

29. $y - 5 > 5x - 10,\ y > 5x - 5$

Draw the graph of $y = 5x - 5$ as a dashed line.
Test a point in each half-plane.

Try $(0, 0)$.   Try $(2, 0)$.

$0 - 5 \overset{?}{>} 5(0) - 10$ $0 - 5 \overset{?}{>} 5(2) - 10$

$-5 > -10$ $-5 \not> 0$

Shade the half-plane containing the point $(0, 0)$.

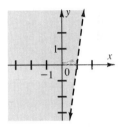

30. $y - 3 > 4x - 9,\ y > 4x - 6$

Draw the graph of $y = 4x - 6$ as a dashed line.
Test a point in each half-plane.

Try $(0, 0)$.   Try $(3, 0)$.

$0 - 3 \overset{?}{>} 4(0) - 9$ $0 \overset{?}{>} 4(3) - 6$

$-3 > -9$ $0 \not> 6$

Shade the half-plane containing the point $(0, 0)$.

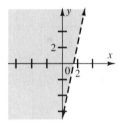

31. $y + 8 > x + 9,\ y > x + 1$

Draw the graph of $y = x + 1$ as a dashed line.
Test a point in each half-plane.

Try $(0, 0)$.   Try $(0, 2)$.

$0 + 8 \overset{?}{>} 0 + 9$ $2 + 8 \overset{?}{>} 0 + 9$

$8 \not> 9$ $10 > 9$

Shade the half-plane containing the point $(0, 2)$.

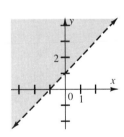

32. $m = \dfrac{1}{4},\ b = 0;\ y < \dfrac{1}{4}x$

33. $m = \dfrac{5}{2},\ b = -3;\ y < \dfrac{5}{2}x - 3$

34. $m = -1,\ b = 3;\ y \geq -x + 3$

C 35. $m = \frac{1}{4}$, $y = \frac{1}{4}x + b$, $1 = \frac{1}{4}(1) + b$, $(1, 1)$ is a point on the graph.

$1 = \frac{1}{4} + b$, $\frac{3}{4} = b$, $y = \frac{1}{4}x + \frac{3}{4}$; $y \geq \frac{1}{4}x + \frac{3}{4}$

36. $m = \frac{3}{4}$, $1 = \frac{3}{4}x + b$, $1 = \frac{3}{4}(1) + b$, $(1, 1)$ is a point on the graph.

$\frac{1}{4} = b$, $y = \frac{3}{4}x + \frac{1}{4}$; $y \geq \frac{3}{4}x + \frac{1}{4}$

37. $m = 0$, $b = 0$; $y \geq 0$

38. $2y + 3 < \frac{1}{2}x + 5$, $2y < \frac{1}{2}x + 2$; $y < \frac{1}{4}x + 1$

Draw the graph of $y = \frac{1}{4}x + 1$ as a dashed line.

Test a point in each half-plane.
Try $(0, 0)$.　　　　　Try $(0, 2)$.

$2(0) + 3 \overset{?}{<} \frac{1}{2}(0) + 5$　　$2(2) + 3 \overset{?}{<} \frac{1}{2}(0) + 5$

$3 < 5$　　　　　　$7 \not< 5$

Shade the half-plane containing the point $(0, 0)$.

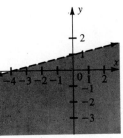

open half-plane

39. $3y + 4 > \frac{1}{3}x + 7$, $3y < \frac{1}{3}x + 3$; $y < \frac{1}{9}x + 1$

Draw the graph of $y = \frac{1}{9}x + 1$ as a dashed line.

Test a point in each half-plane.
Try $(0, 0)$.　　　　　Try $(0, 3)$.

$3(0) + 4 \overset{?}{<} \frac{1}{3}(0) + 7$　　$3(3) + 4 \overset{?}{<} \frac{1}{3}(0) + 7$

$4 < 7$　　　　　　$13 \not< 7$

Shade the half-plane containing the point $(0, 0)$.

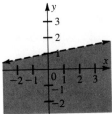

open half-plane

40. $2y - 3 \geq \frac{1}{4}x + 4$, $2y \geq \frac{1}{4}x + 7$; $y \geq \frac{1}{8}x + \frac{7}{2}$

Draw the graph of $y = \frac{1}{8}x + \frac{7}{2}$ as a solid line.

Test a point in each half-plane.
Try $(0, 0)$.　　　　　Try $(0, 4)$.

$2(0) - 3 \overset{?}{\geq} \frac{1}{4}(0) + 4$　　$2(4) - 3 \overset{?}{\geq} \frac{1}{4}(0) + 4$

$-3 \not\geq 4$　　　　　$5 \geq 4$

Shade the half-plane containing the point $(0, 4)$.

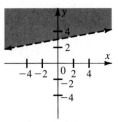

closed half-plane

41. $4y - 1 \geq \frac{1}{8}x + 6$, $4y \geq \frac{1}{8}x + 7$; $y \geq \frac{1}{32}x + \frac{7}{4}$

Draw the graph of $y = \frac{1}{32}x + \frac{7}{4}$ as a solid line.

Test a point in each half-plane.

Try (0, 0).   Try (0, 4).

$4(0) - 1 \overset{?}{\geq} \frac{1}{8}(0) + 6$ $4(4) - 1 \overset{?}{\geq} \frac{1}{8}(0) + 6$

$-1 \not\geq 6$ $15 \geq 6$

Shade the half-plane containing the point (0, 4).

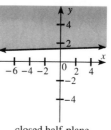

closed half-plane

42. $\frac{3}{5}y + 1 < \frac{1}{3}x + 2$, $9y + 15 < 5x + 30$, $9y < 5x + 15$, $y < \frac{5}{9}x + \frac{15}{9}$; $y < \frac{5}{9}x + \frac{5}{3}$

Draw the graph of $y = \frac{5}{9}x + \frac{5}{3}$ as a dashed line.

Test a point in each half-plane.

Try (0, 0).   Try (0, 5).

$\frac{3}{5}(0) + 1 \overset{?}{<} \frac{1}{3}(0) + 2$ $\frac{3}{5}(5) + 1 \overset{?}{<} \frac{1}{3}(0) + 2$

$1 < 2$ $4 \not< 2$

Shade the half-plane containing the point (0, 0).

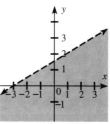

open half-plane

43. $\frac{2}{3}y + 2 < \frac{2}{5}x + 1$, $10y + 30 < 6x + 15$, $10y - 6x < -15$, $10y \leq 6x - 15$,

$y \leq \frac{6}{10}x - \frac{15}{10}$; $y \leq \frac{3}{5}x - \frac{3}{2}$

Draw the graph of $y = \frac{3}{5}x - \frac{3}{2}$ as a dashed line.

Test a point in each half-plane.

Try (0, 0).   Try (0, -3).

$\frac{2}{3}(0) + 2 \overset{?}{<} \frac{2}{5}(0) + 1$ $\frac{2}{3}(-3) + 2 \overset{?}{<} \frac{2}{5}(0) + 1$

$2 \not< 1$ $0 < 1$

Shade the half-plane containing the point (0, -3).

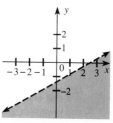

open half-plane

44. $4x + 8y \leq 24$, $8y \leq -4x + 24$, $y \leq -\frac{1}{2}x + 3$

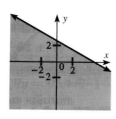

Answers may vary. Possible purchases: 2 lb peanuts, 1 lb cashews; 1 lb peanuts, 2 lb cashews; 4 lb peanuts, 1 lb cashews

45. $3x + 10y > 250$, $10y > -3x + 250$, $y > -\dfrac{3}{10}x + 25$

Answers may vary. Possible combinations: 11 nylon, 22 canvas; 20 nylon, 20 canvas; 30 nylon, 25 canvas

46. Paragraph should include these ideas: The graph of a linear inequality is a half-plane consisting of all points above or below (or, in the case of a vertical line, to the left of or to the right of) the graph of the equation of the boundary line. The boundary line itself may or may not be included in the graph, depending upon the relation symbol used in the inequality.

47. $y < \dfrac{1}{2}x - 5$, $y + 5 < \dfrac{1}{2}x$, $2y + 10 < x$, $x > 2y + 10$; Its x-coordinate is greater than 10 more than twice its y-coordinate.

page 405   Test Yourself

1. $4x + y = 2$
$y = -4x + 2$
$m = -4$, $b = 2$

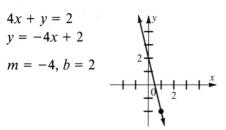

2. $3x - 2y = 6$
$-2y = -3x + 6$
$y = \dfrac{3}{2}x - 3$
$m = \dfrac{3}{2}$; $b = -3$

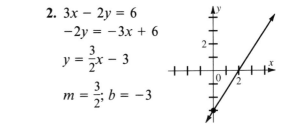

3. Write the equation in slope-intercept form. $y = mx + b$, $y = -2x + b$, $2 = -2(3) + b$ (Substitute 2 for y and 3 for x), $2 = -6 + b$, $8 = b$ (Solve for b), $y = -2x + 8$; $2x + y = 8$

4. Find the slope of the line through the two points.
$$m = \frac{y_2 - y_1}{x_2 - x_1} = \frac{-4 - (-1)}{3 - 2} = \frac{-3}{1} = -3; \; m = -3;$$
Write in slope-intercept form. $y = -3x + b$, $-1 = -3(2) + b$ (Substitute the coordinates of either point for x and y and solve for b), $-1 = -6 + b$, $5 = b$, $y = -3x + 5$; $3x + y = 5$

5. $y > 2x + 3$ $y > 2x + 3$
$3 \overset{?}{>} 2(0) + 3$ $7 \overset{?}{>} 2(1) + 3$
$3 \not> 3$ $7 > 5$
$(1, 7)$ belongs to the graph of $y > 2x + 3$.

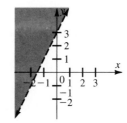

6. $y < 3x - 2$ $\qquad$ $y < 3x - 2$

$-7 \overset{?}{<} 3(-1) - 2$ $\qquad$ $0 \overset{?}{<} 3(-2) - 2$

$-7 < -5$ $\qquad$ $0 \not< -8$

$(-1, -7)$ belongs to the graph of $y < 3x - 2$.

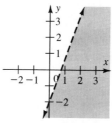

7. $y \le x + 5$ $\qquad$ $y \le x + 5$

$6 \overset{?}{\le} 0 + 5$ $\qquad$ $3 \overset{?}{\le} 1 + 5$

$6 \not\le 5$ $\qquad$ $3 \le 6$

$(1, 3)$ belongs to the graph of $y \le x + 5$.

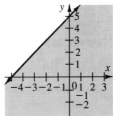

8. $y \ge \frac{1}{2}x - 4$ $\qquad$ $y \ge \frac{1}{2}x - 4$

$4 \overset{?}{\ge} \frac{1}{2}(0) - 4$ $\qquad$ $-5 \overset{?}{\ge} \frac{1}{2}(2) - 4$

$4 \ge -4$ $\qquad$ $-5 \not\ge -3$

$(0, 4)$ belongs to the graph of $y \ge \frac{1}{2}x - 4$.

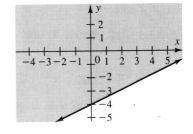

pages 406–407 Summary and Review

1–4.

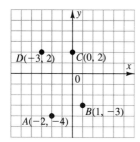

D(-3, 2) C(0, 2) B(1, -3) A(-2, -4)

5. $2x + y = 7$ $\qquad$ $2x + y = 7$

$2(2) + 3 \overset{?}{=} 7$ $\qquad$ $2(4) + 2 \overset{?}{=} 7$

$7 = 7$ $\qquad$ $10 \ne 7$

$(2, 3)$ is a solution of $2x + y = 7$.

6. $-x - y = -3$ $\qquad$ $-x - y = -3$

$-(-3) - 0 \overset{?}{=} -3$ $\qquad$ $-3 - 0 \overset{?}{=} -3$

$3 \ne -3$ $\qquad$ $-3 = -3$

$(3, 0)$ is a solution of $-x - y = -3$.

7.

x	y
0	1
2	0
-2	2

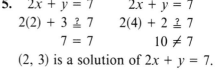

8. $2x - y = 4$, $-y = -2x + 4$, $y = 2x - 4$

x	y
0	-4
2	0
3	1

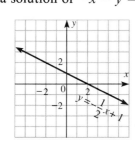

9. $3x + 2y = 6$, $2y = -3x + 6$; $y = -\dfrac{3}{2}x + 3$ **10.**

x	y
0	-3
2	0
-2	-6

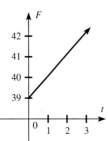

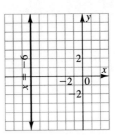

11. $F = t + 39$

t	F
0	39
1	40
2	41

12. If the outdoor temperature is 95°F, you would expect to hear 56 chirps in 15 s.

13. $m = \dfrac{y_2 - y_1}{x_2 - x_1} = \dfrac{6 - 5}{1 - (-1)} = \dfrac{1}{2}$

14. $m = \dfrac{y_2 - y_1}{x_2 - x_1} = \dfrac{-2 - (-4)}{1 - (-2)} = \dfrac{2}{3}$

15.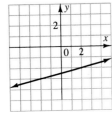

16.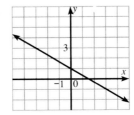

17. $m = \dfrac{1}{2}$; $b = -4$

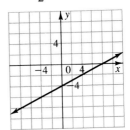

18. $m = -\dfrac{3}{4}$; $b = 3$

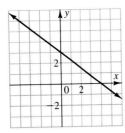

19. $m = -\dfrac{1}{2}$; $b = 3$

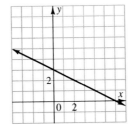

20. $m = \dfrac{3}{5}$; $b = -3$

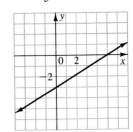

21. $5x - y = 1$ $5x + y = 7$

$\quad\;\; -y = -5x + 1$ $y = -5x + 7$

$\quad\;\;\;\; y = 5x - 1$

$\quad\;\;\; m = 5$ $\qquad\; m = -5$

Since the slopes are not equal, the lines are not parallel.

22. $3y = -2x + 2$ $6y = -4x - 4$

$\quad\;\; y = \dfrac{-2}{3}x + \dfrac{2}{3}$ $y = \dfrac{-4}{6}x - \dfrac{4}{6}$

$\qquad\qquad\qquad\qquad\;\; y = \dfrac{-2}{3}x - \dfrac{2}{3}$

$\quad\; m = -\dfrac{2}{3}$ $\qquad m = -\dfrac{2}{3}$

Since the slopes are equal, the lines are parallel.

23. $m = \dfrac{2}{3}, \, b = -4, \, y = mx + b, \, y = \dfrac{2}{3}x - 4, \, 3y = 2x - 12; \, -2x + 3y = -12$

24. $m = \dfrac{y_2 - y_1}{x_2 - x_1} = \dfrac{-2 - 4}{-4 - (-1)} = \dfrac{-6}{-3} = 2;$

$y = mx + b, \, y = 2x + b, \, 4 = 2(-1) + b, \, 4 = -2 + b, \, 6 = b,$

$y = 2x + 6; \, -2x + y = 6$

25.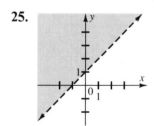

26.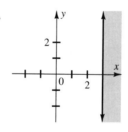

27.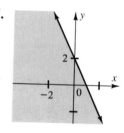

page 408 Chapter Test

1-4.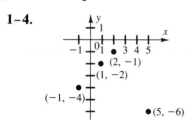

5-8.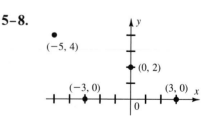

9.
$$x - 3y = 6 \qquad\qquad x - 3y = 6$$
$$3 - 3(-1) \stackrel{?}{=} 6 \qquad -1 - 3(6) \stackrel{?}{=} 6$$
$$6 = 6 \qquad\qquad -19 \neq 6$$

$(3, -1)$ is a solution of $x - 3y = 6$.

10.
$$y = \frac{1}{2}x - 4 \qquad\qquad y = \frac{1}{2}x - 4$$
$$-5 \stackrel{?}{=} \frac{1}{2}(-2) - 4 \qquad -3 \stackrel{?}{=} \frac{1}{2}(1) - 4$$
$$-5 = -5 \qquad\qquad -3 \neq -3\frac{1}{2}$$

$(-2, -5)$ is a solution of $y = \frac{1}{2}x - 4$.

11.

x	y
0	−1
1	1
−1	−3

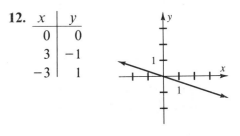

12.

x	y
0	0
3	−1
−3	1

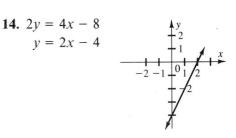

13.
$$3x - y = 9$$
$$-y = -3x + 9$$
$$y = 3x - 9$$

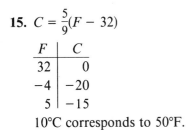

14.
$$2y = 4x - 8$$
$$y = 2x - 4$$

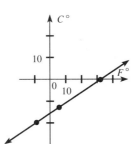

15. $C = \dfrac{5}{9}(F - 32)$

F	C
32	0
−4	−20
5	−15

10°C corresponds to 50°F.

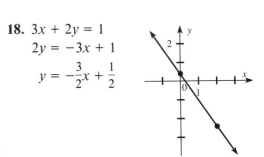

16. $m = \dfrac{y_2 - y_1}{x_2 - x_1} = \dfrac{1 - (-3)}{4 - 8} = \dfrac{4}{-4} = -1$

17.

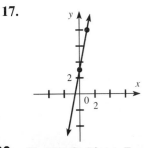

18.
$$3x + 2y = 1$$
$$2y = -3x + 1$$
$$y = -\frac{3}{2}x + \frac{1}{2}$$

19. $y = 4x - 1$, $m = 4$; $16x - 4y = 12$, $-4y = -16x + 12$, $y = 4x - 3$, $m = 4$; Since the slopes are equal, the lines are parallel.

20. $y = mx + b$, $y = \frac{1}{4}x + b$, $-2 = \frac{1}{4}(-1) + b$, $-2 = -\frac{1}{4} + b$, $-\frac{7}{4} = b$;

$y = \frac{1}{4}x - \frac{7}{4}$, $4y = x - 7$; $-x + 4y = -7$

21. $m = \frac{y_2 - y_1}{x_2 - x_1} = \frac{2 - 6}{1 - (-5)} = \frac{-4}{6} = -\frac{2}{3}$; $y = mx + b$, $y = -\frac{2}{3}x + b$, $2 = -\frac{2}{3}(1) + b$,

$2 = -\frac{2}{3} + b$, $\frac{8}{3} = b$; $y = mx + b$, $y = -\frac{2}{3}x + \frac{8}{3}$, $3y = -2x + 8$; $2x + 3y = 8$

22.

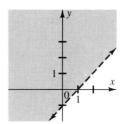

23.

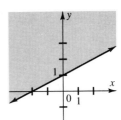

Challenge

$4y + 3x = 6$, $4y = -3x + 6$, $y = -\frac{3}{4}x + \frac{6}{4}$, $y = -\frac{3}{4}x + \frac{3}{2}$, $m_1 = -\frac{3}{4}$; $m_1 m_2 = -1$, $-\frac{3}{4}m_2 = $

-1, $-3m_2 = -4$, $m_2 = \frac{4}{3}$; $y = mx + b$, $y = \frac{4}{3}x + b$, $3a = \frac{4}{3}(a) + b$, $9a = 4a + 3b$, $5a = $

$3b$, $\frac{5a}{3} = b$; $y = mx + b$, $y = \frac{4}{3}x + \frac{5a}{3}$, $3y = 4x + 5a$; $-4x + 3y = 5a$

page 409   Preparing for Standardized Tests

1. A; Using $(3, 1)$ and $m = 2$, $y - 1 = 2(x - 3)$, $y - 1 = 2x - 6$, $y = 2x - 5$

2. B; $\frac{518}{28} = \frac{37}{2} = 18.5$ gal

3. D; $|3x - 2y + 4| = |3(-5) - 2(3) + 4| = |-15 - 6 + 4| = |-17| = 17$

4. C; $\frac{1}{x} + \frac{3}{2x} - \frac{4}{3x} = \frac{6 + 9 - 8}{6x} = \frac{7}{6x}$

5. B; $m = \frac{5 - 2}{1 - (-3)} = \frac{3}{4}$

6. E; $(2x - 3)(3x + 2) = 6x^2 - 5x - 6$

7. D; 4 tubes at $11.95 = $47.80
1 brush at   8.50 =   8.50
2 brushes at   6.75 =   13.50
total spent = $69.80

8. D; $\dfrac{3}{x} + \dfrac{2}{x} = \dfrac{5}{2}$

$\dfrac{5}{x} = \dfrac{5}{2}$

$x = 2$

9. A; $\left(\dfrac{3x^2y}{2ab}\right) \cdot \left(\dfrac{4ab^2}{6xy^2}\right) = \dfrac{xb}{y} = \dfrac{3(-2)}{2} = -3$

10. E; All of the points check out in the equation except $(3, -1)$ since
$2(3) - 3(-1) \neq 8; \ 9 \neq 8$.

11. E; \$500 interest on the loan of \$4000 gives $\dfrac{500}{4000} \times 100 = 12.5\%$ for the
annual rate of interest.

12. B; Since $x = 0.30y$ and $y = 0.20z$, it follows that $x = 0.30(0.20z)$,
$x = 0.06z$, so x is 6% of z.

page 410   Mixed Review

1. $(2, 1)$ **2.** $(0, 2)$ **3.** $(1, -3)$

4. $(2, -1)$ **5.** $(0, -2)$ **6.** $(-2, -2)$

7–12.
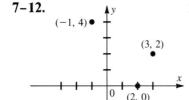

13.
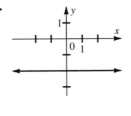

14.
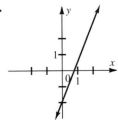

15.
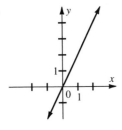

16.
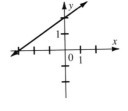

17. $3x - 2 = 3(-1) - 2 = -3 - 2 = -5$ **18.** $2x + 5 = 2(2) + 5 = 4 + 5 = 9$

19. $-3x + 3 = -3(-3) + 3 = 9 + 3 = 12$ **20.** $x^2 + 3 = (4)^2 + 3 = 16 + 3 = 19$

21. $-2x^2 - 3 = -2(2)^2 - 3 = -2(4) - 3 = -8 - 3 = -11$

22. $3x^2 - 2 = 3(-3)^2 - 2 = 3(9) - 2 = 27 - 2 = 25$

23. $\frac{1}{2}x^2 - 2 = \frac{1}{2}(3)^2 - 2 = \frac{1}{2}(9) - 2 = \frac{9}{2} - 2 = \frac{9}{2} - \frac{4}{2} = \frac{5}{2}$

24. $|-2| = 2$ **25.** $|1| - 1 = 0$ **26.** $\dfrac{75}{125} = \dfrac{75 \div 25}{125 \div 25} = \dfrac{3}{5}$

27. Let x represent how many students voted in the election.

$\frac{7}{10}x = 42, \ \frac{10}{7} \cdot \frac{7}{10}x = 42 \cdot \frac{10}{7}, \ x = 60$ students

28. Let x represent the total amount of boys.

$\frac{6}{7} = \frac{x}{350}, \ \frac{7x}{7} = \frac{6 \cdot 350}{7}; \ x = 300$ boys.

Chapter 10   Relations, Functions, and Variation

page 412   Capsule Review

   1. $1, -1, 2, 0, -2$
                                     2. $\{(-3, 1), (-1, -1), (0, 2), (2, 0), (5, -2)\}$

page 414   Class Exercises

   1. No, it is not a set of one or more ordered pairs of numbers.

   2. $\{(5, 1.80)\}$
                         3. $D = \{4\}$; $R = \{1, 2\}$

   4. $D = \{-2, -1, 1, 2, 3, 5\}$; $R = \{-2, 0, 1, 2, 3\}$

   5. $D = \{-2, 0, 1, 3\}$; $R = \{0, 1, 6\}$

   6. $D = \{\text{Sept., Oct., Nov., Dec.}\}$; $R = \{140, 180, 235, 255\}$

pages 414–416   Practice Exercises

A **1.** $\{(-3, 5), (-1, 4), (1, 4), (3, -3)\}$; $D = \{-3, -1, 1, 3\}$; $R = \{-3, 4, 5\}$

   2. $\{(2, -1), (5, 3), (7, -1), (9, -7)\}$; $D = \{2, 5, 7, 9\}$; $R = \{-7, -1, 3\}$

   3. $\left\{\left(\frac{1}{3}, -1\right), (0, 0), \left(\frac{1}{3}, 2\right), \left(\frac{1}{2}, 2\frac{1}{2}\right)\right\}$; $D = \left\{0, \frac{1}{3}, \frac{1}{2}\right\}$; $R = \left\{-1, 0, 2, 2\frac{1}{2}\right\}$

   4. $\left\{(2, -1), \left(2\frac{1}{2}, 0\right), (3, -1), \left(3\frac{1}{2}, -3\frac{1}{2}\right)\right\}$; $D = \left\{2, 2\frac{1}{2}, 3, 3\frac{1}{2}\right\}$; $R = \left\{-3\frac{1}{2}, -1, 0\right\}$

   5. $\{(-3, -4), (1, 6), (4, 30), (5, 24)\}$; $D = \{-3, 1, 4, 5\}$; $R = \{-4, 6, 24, 30\}$

   6. $\{(1, 1), (-1, 1), (2, 4), (-2, 4)\}$; $D = \{-2, -1, 1, 2\}$; $R = \{1, 4\}$

   7. $\{(-2, 0), (-1, 1), (0, 2), (1, 1), (2, 0), (3, -1), (4, -2)\}$; $D = \{-2, -1, 0, 1, 2, 3\}$;
      $R = \{-2, -1, 0, 1, 2\}$

   8. $\{(1, 2), (2, 1), (2, 3), (3, 0), (3, 4), (4, 1), (4, 3), (5, 2)\}$, $D = \{1, 2, 3, 4, 5\}$;
      $R = \{0, 1, 2, 3, 4\}$

   9. $y = 3(-1) + 5, y = -3 + 5, y = 2$
      $y = 3(0) + 5, y = 0 + 5, y = 5$
      $y = 3(2) + 5, y = 6 + 5, y = 11$
      $R = \{2, 5, 11\}$

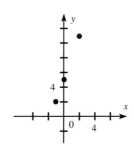

10. $y = 2(-1) - 3, y = -2 - 3, y = -5$
$y = 2(0) - 3, y = 0 - 3, y = -3$
$y = 2(2) - 3, y = 4 - 3, y = 1$
$R = \{-5, -3, 1\}$

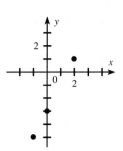

11. $y = 9 - 2(-1), y = 9 + 2, y = 11$
$y = 9 - 2(0), y = 9 - 0, y = 9$
$y = 9 - 2(2), y = 9 - 4, y = 5$
$R = \{5, 9, 11\}$

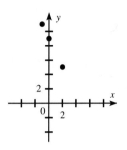

B 12. $x - 14 = 2y, y = \dfrac{x - 14}{2};$

$y = \dfrac{-1 - 14}{2}, y = -\dfrac{15}{2} = -7\dfrac{1}{2}$

$y = \dfrac{0 - 14}{2}, y = -\dfrac{14}{2} = -7$

$y = \dfrac{2 - 14}{2}, y = -\dfrac{12}{2} = -6$

$R = \left\{ -\dfrac{15}{2}, -7, -6 \right\}$

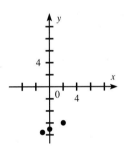

13. $y = \dfrac{-1}{2} + 1, y = -\dfrac{1}{2} + \dfrac{2}{2}, y = \dfrac{1}{2}$

$y = \dfrac{0}{2} + 1, y = 0 + 1, y = 1$

$y = \dfrac{2}{2} + 1, y = 1 + 1, y = 2$

$R = \left\{ \dfrac{1}{2}, 1, 2 \right\}$

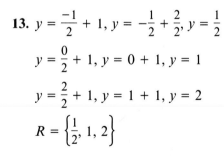

14. $y = \dfrac{-1}{2} - 2, y = -\dfrac{1}{2} - \dfrac{4}{2}, y = -\dfrac{5}{2}$

$y = \dfrac{0}{2} - 2, y = 0 - 2, y = -2$

$y = \dfrac{2}{2} - 2, y = 1 - 2, y = -1$

$R = \left\{ -\dfrac{5}{2}, -2, -1 \right\}$

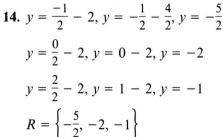

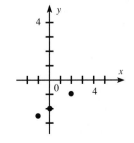

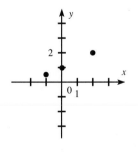

15. {(1890, 0.07), (1910, 0.08), (1930, 0.14), (1950, 0.21), (1970, 0.33)};
D = {1890, 1910, 1930, 1950, 1970};
R = {0.07, 0.08, 0.14, 0.21, 0.33}

16. $\left\{\left(10, 22\frac{1}{2}\right), (11, 23), \left(12, 23\frac{1}{4}\right), \left(1, 22\frac{5}{8}\right), (2, 23)\right\}$;
D = {10 A.M., 11 A.M., Noon, 1 P.M., 2 P.M.};
$R = \left\{22\frac{1}{2}, 22\frac{5}{8}, 23, 23\frac{1}{4}\right\}$

17. {(−1, 2), (−1, 8), (0, 4), (1, 4), (2, 6), (3, 4)};
D = {−1, 0, 1, 2, 3}; R = {2, 4, 6, 8}

18. {(−3, 6), (4, 8), (2, 4), (−5, 10), (0, 0)};
D = {−5, −3, 0, 2, 4}; R = {0, 4, 6, 8, 10}

C 19. $\frac{1}{2}y = 8 - 8x$, $y = 16 - 16x$;

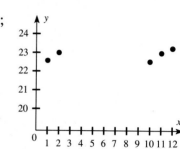

$y = 16 - 16\left(-\frac{5}{4}\right)$, $y = 16 + 20$, $y = 36$

$y = 16 - 16\left(\frac{1}{2}\right)$, $y = 16 - 8$, $y = 8$

$y = 16 - 16(2)$, $y = 16 - 32$, $y = -16$
R = {−16, 8, 36}

20. $\frac{1}{2}y = 4x + 5,\ y = 8x + 10;$

$y = 8\left(-\frac{5}{4}\right) + 10,\ y = -10 + 10,\ y = 0$

$y = 8\left(\frac{1}{2}\right) + 10,\ y = 4 + 10,\ y = 14$

$y = 8(2) + 10,\ y = 16 + 10,\ y = 26$

$R = \{0, 14, 26\}$

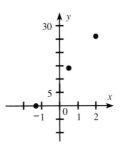

21. $y = 4x - 2;$

$y = 4\left(-\frac{5}{4}\right) - 2,\ y = -5 - 2,\ y = -7$

$y = 4\left(\frac{1}{2}\right) - 2,\ y = 2 - 2,\ y = 0$

$y = 4(2) - 2,\ y = 8 - 2,\ y = 6$

$R = \{-7, 0, 6\}$

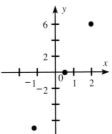

22. $2y = 16x + 4,\ y = 8x + 2;$

$y = 8\left(-\frac{5}{4}\right) + 2,\ y = -10 + 2,\ y = -8$

$y = 8\left(\frac{1}{2}\right) + 2,\ y = 4 + 2,\ y = 6$

$y = 8(2) + 2,\ y = 16 + 2,\ y = 18$

$R = \{-8, 6, 18\}$

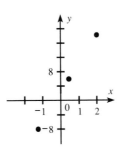

23. $D = \{1, 2, 3, 4, 5\};\ R = \{5000, 5200, 5400, 5600, 5800\};$
$\{(1, 5000), (2, 5200), (3, 5400), (4, 5600), (5, 5800)\}$

24. $D = \{1, 2, 3, 4, 5, 6\};\ R = \{6, 12, 18, 24, 30, 36\};\ \{(1, 6), (2, 12), (3, 18),$
$(4, 24), (5, 30), (6, 36)\}$

25. $D = \{1, 2, 3, 4, 5\};\ R = \{4, 5, 6, 7, 8\};\ \{(1, 4), (2, 5), (3, 6), (4, 7), (5, 8)\}$

page 416   Algebra in Taxation

1. $(11,000)(0.025) = 275;\ \$275$

2. $(35,000)(0.28) = 9800;\ \$9800$

3. $\dfrac{0.0751\,(48,000)}{50,000} = 0.0721;$ Marcia paid 7.51% and Alex paid 7.21%.

page 417   Capsule Review

1. $\{(1, 1), (2, 2), (3, 3), (4, 4)\}$

2. $\{(1, 1), (2, 1), (3, 1), (4, 1)\}$

3. $\{(1, 1), (1, 2), (2, 3), (3, 4), (4, 4), (4, 5)\}$

page 419   Class Exercises

1. No, each person may have more than one car.

2. Yes, each person has only one birthday.

3. Yes, each person has just one social security number.

pages 419–421   Practice Exercises

A 1. function   **2.** not a function   **3.** function   **4.** not a function

5. function   **6.** function   **7.** function   **8.** function

9. not a function   **10.** not a function   **11.** function   **12.** $f(2) = 2 + 2 = 4$

13. $f(3) = 3 + 2 = 5$ **14.** $f(-1) = -1 + 2 = 1$ **15.** $f(-4) = -4 + 2 = -2$

16.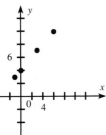
$R = \{3, 4, 7, 10\}$; function

17.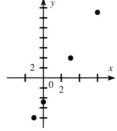
$R = \{-8, -5, 4, 13\}$; function

18.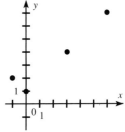
$R = \{2, 1, 4, 7\}$; function

B 19.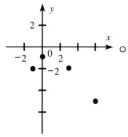
$R = \{-5, -2, -1\}$; function

20.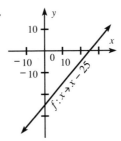
$R = \{\text{all real numbers}\}$; function

21.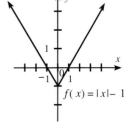
$R = \{\text{all real numbers} \geq -1\}$; function

22. $f(-2) = (-2)^2 + 1 = 4 + 1 = 5$ **23.** $g(1) = 5(1) - 3 = 5 - 3 = 2$

24. $f(-2) = 5, g(1) = 2; f(-2) + g(1) = 5 + 2 = 7$

25. $f(2) = (2)^2 + 1 = 5, g(0) = 5(0) - 3 = -3; f(2) \times f(0) = 5 \times (-3) = -15$

26. $f(4) = (4)^2 + 1 = 17, g(1) = 5(1) - 3 = 2; f(4) \times g(1) = 17 \times 2 = 34$

27. $f(-2) = (-2)^2 + 1 = 5, g(1) = 5(1) - 3 = 2; f(-2) - g(1) = 5 - 2 = 3$

28.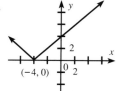

29.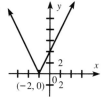

30.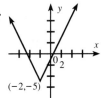

C 31.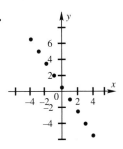

function

32.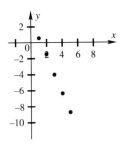

function

33.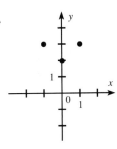

function

34.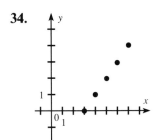

function

35.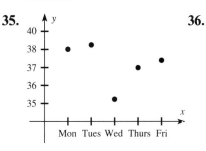

yes

36.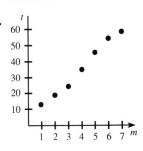

yes

37. $14b + 12$

38. $7x^2 - 14x - 12x - 20 = 7x^2 - 26x - 20$

39. $3(5a + 2a + 2c) - 8c = 3(7a + 2c) - 8c = 21a + 6c - 8c = 21a - 2c$

40. $-16m^5y - 8m^3 + 24m^5y = 8m^5y - 8m^3$ **41.** $8d^2 - 4d + 10d - 5 = 8d^2 + 6d - 5$

42. $(y^2 - 49)(y^2 + 2) = y^4 + 2y^2 - 49y^2 - 98 = y^4 - 47y^2 - 98$

43. $\dfrac{4 - 2}{5 - (-3)} = \dfrac{2}{8} = \dfrac{1}{4}$

44. $\dfrac{-12 - (-6)}{4 - 1} = \dfrac{-6}{3} = -2$

45. $\dfrac{-5 - 4}{-1 - (-10)} = \dfrac{-9}{9} = -1$

46. $\dfrac{10 - (-2)}{9 - (-6)} = \dfrac{12}{15} = \dfrac{4}{5}$

47. $\dfrac{-2 - (-5)}{-5 - (-2)} = \dfrac{3}{-3} = -1$

48. $\dfrac{-5 - 0}{-7 - 10} = \dfrac{-5}{-17} = \dfrac{5}{17}$

49. $m = \dfrac{3}{4}$

$b = 6$

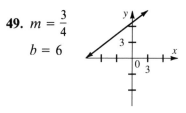

50. $m = \dfrac{1}{2}$

$b = -4$

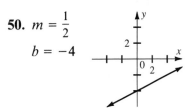

51. $m = \dfrac{2}{3}$

$b = -3$

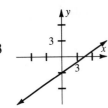

52. $m = -4$

$b = 2$

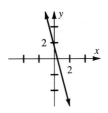

page 422   Capsule Review

1. $y = -\dfrac{1}{3}x + 3$

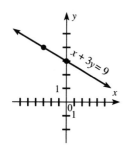

2. $y = -\dfrac{1}{5}x - 2$

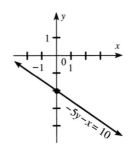

3. $y = \dfrac{2}{3}x - \dfrac{8}{3}$

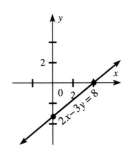

page 423   Class Exercises

1.

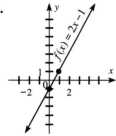

yes; yes; no

2.

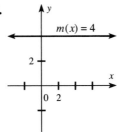

yes; yes; yes

3.

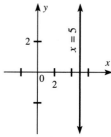

not a function

4. $g(2) = 6(2) + 5 = 12 + 5 = 17$
$f(17) = 2(17) + 2 = 34 + 2 = 36$
$f(g(7)) = 36$

5. $g(7) = 6(7) + 5 = 42 + 5 = 47$
$f(47) = 2(47) + 2 = 94 + 2 = 96$
$f(g(7)) = 96$

6. $f(-1) = 2(-1) + 2 = -2 + 2 = 0$
$g(0) = 6(0) + 5 = 0 + 5 = 5$
$g(f(-1)) = 5$

7. $f(-4) = 2(-4) + 2 = -8 + 2 = -6$
$g(-6) = 6(-6) + 5 = -36 + 5 = -31$
$g(f(-4)) = -31$

A 1.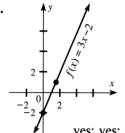

yes; yes; no

2.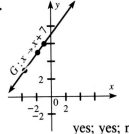

yes; yes; no

3.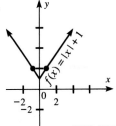

yes; no; no

4.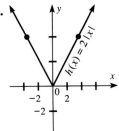

yes; no; no

5.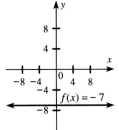

yes; yes; yes

6.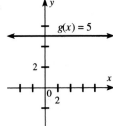

yes; yes; yes

7.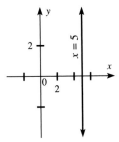

not a function

8.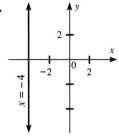

not a function

9.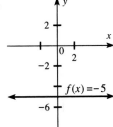

yes; yes; yes

10.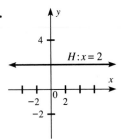

yes; yes; yes

11.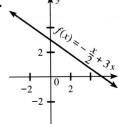

yes; yes; no

12.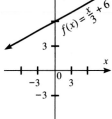

yes; yes; no

13. $f(-1) = 2(-1) + 1 = -2 + 1 = -1$

14. $g(-1) = 4(-1) - 5 = -4 - 5 = -9$

15. $f(-2) = 2(-2) + 1 = -4 + 1 = -3$

16. $g(-2) = 4(-2) - 5 = -8 - 5 = -13$

17. $g(1) = 4(1) - 5 = 4 - 5 = -1$
$f(-1) = 2(-1) + 1 = -2 + 1 = -1$
$f(g(1)) = -1$

18. $g(2) = 4(2) - 5 = 8 - 5 = 3$
$f(3) = 2(3) + 1 = 6 + 1 = 7$
$f(g(2)) = 7$

19. $f(3) = 2(3) + 1 = 6 + 1 = 7$
$g(7) = 4(7) - 5 = 28 - 5 = 23$
$g(f(3)) = 23$

20. $f(4) = 2(4) + 1 = 8 + 1 = 9$
$g(9) = 4(9) - 5 = 36 - 5 = 31$
$g(f(4)) = 31$

B 21.

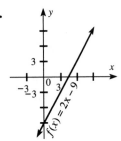

yes; yes; no

22.

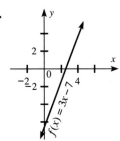

yes; yes; no

23.

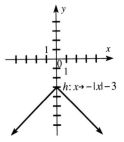

yes; no; no

24.

yes; no; no

25.

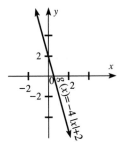

yes; no; no

26.

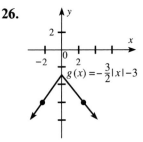

yes; no; no

C 27. $g(1) = 1^2 - 3(1) + 1 = 1 - 3 + 1 = -1$
$f(-1) = 5(-1) - 3 = -5 - 3 = -8$
$f(g(1)) = -8$

28. $g(2) = 2^2 - 3(2) + 1 = 4 - 6 + 1 = -1$
$f(-1) = 5(-1) - 3 = -5 - 3 = -8$
$f(g(2)) = -8$

29. $f(0) = 5(0) - 3 = 0 - 3 = -3$
$g(-3) = (-3)^2 - 3(-3) + 1 = 19$
$g(f(0)) = 19$

30. $f(3) = 5(3) - 3 = 15 - 3 = 12$
$g(12) = 12^2 - 3(12) + 1 = 109$
$g(f(3)) = 109$

31. $g(3) = 3^2 - 3(3) + 1 = 9 - 9 + 1 = 1$
$f(1) = 5(1) - 3 = 5 - 3 = 2$
$f(g(3)) = 2$

32. $g(-1) = (-1)^2 - 3(-1) + 1 = 5$
$f(5) = 5(5) - 3 = 25 - 3 = 22$
$f(g(-1)) = 22$

33. $f(1) = 5(1) - 3 = 5 - 3 = 2$
$g(2) = 2^2 - 3(2) + 1 = 4 - 6 + 1 = -1$
$g(f(1)) = -1$

34. $f(2) = 5(2) - 3 = 10 - 3 = 7$
$g(7) = 7^2 - 3(7) + 1 = 29$
$g(f(2)) = 29$

35. $t(15) = \dfrac{1}{15}$

$f\left(\dfrac{1}{15}\right) = \left|-\dfrac{1}{15}\right| - 2 = \dfrac{1}{15} - \dfrac{30}{15} = -\dfrac{29}{15}$

$f(t(15)) = -\dfrac{29}{15}$

36. $r(-2) = -1.1$
$g(-1.1) = |-1.1| + -1.1 = 1.1 - 1.1 = 0$
$g(r(-2)) = 0$

37. $f(15) = 0.75(15) + 1.45 = 11.25 + 1.45 = \$12.70$

page 424   Test Yourself

1. $y = 2(-3) + 3, \; y = -6 + 3, \; y = -3$
$y = 2(0) + 3, \; y = 0 + 3, \; y = 3$
$y = 2(4) + 3, \; y = 8 + 3, \; y = 11$
$R = \{-3, \, 3, \, 11\}$

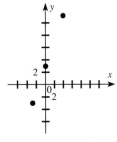

2. $y = 3(-3) - 4, \; y = -9 - 4, \; y = -13$
$y = 3(0) - 4, \; y = 0 - 4, \; y = -4$
$y = 3(4) - 4, \; y = 12 - 4, \; y = 8$
$R = \{-13, \, -4, \, 8\}$

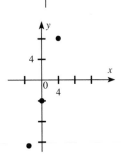

3. $y = \dfrac{1}{4}(-3) - 1, \; y = -\dfrac{3}{4} - 1, \; y = -\dfrac{7}{4}$

$y = \dfrac{1}{4}(0) - 1, \; y = 0 - 1, \; y = -1$

$y = \dfrac{1}{4}(4) - 1, \; y = 1 - 1, \; y = 0$

$R = \left\{-\dfrac{7}{4}, \, -1, \, 0\right\}$

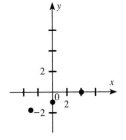

4. $y = \frac{1}{3}(-3) + 1$, $y = -1 + 1$, $y = 0$

$y = \frac{1}{3}(0) + 1$, $y = 0 + 1$, $y = 1$

$y = \frac{1}{3}(4) + 1$, $y = \frac{4}{3} + \frac{3}{3}$, $y = \frac{7}{3}$

$R = \left\{0, 1, \frac{7}{3}\right\}$

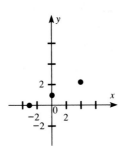

5.

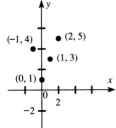

yes; no; no

6.

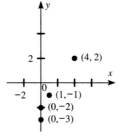

no; no; no

7.

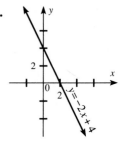

yes; yes; no

8.

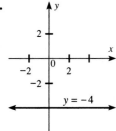

yes; yes; yes

9.

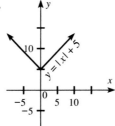

yes; no; no

10.

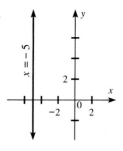

no; no; no

11. $f(-3) = 3(-3) + 2 = -9 + 2 = -7$

12. $g(4) = 4(4) - 1 = 16 - 1 = 15$

13. $g(2) = 4(2) - 1 = 8 - 1 = 7$
$f(7) = 3(7) + 2 = 21 + 2 = 23$
$f(g(2)) = 23$

14. $f(3) = 3(3) + 2 = 9 + 2 = 11$
$g(11) = 4(11) - 1 = 44 - 1 = 43$
$g(f(3)) = 43$

page 425   Integrating Algebra: Cost Analysis

1. Let $c(x)$ = cost function, and x = amount of guitar strings purchased.
$c(x) = 100 + 0.45(x - 180)$, $c(x) = 100 + 0.45x - 81$,
$c(x) = 0.45x + 19$; $c(600) = 0.45(600) + 19 = 270 + 19 = 289$;
The cost of 600 guitar strings would be \$289.

2. Let $c(x)$ = cost function, and x = number of pounds of finishing nails purchased.

$c(x) = 8 + 0.75(x - 10)$, $c(x) = 8 + 0.75x - 7.5$,

$c(x) = 0.75x + 0.5$; $c(25) = 0.75(25) + 0.5 = 18.75 + 0.5 = 19.25$;

The cost of 25 lb of finishing nails is $19.25.

page 426   Capsule Review

1. $y = 4 + x$ **2.** $\frac{1}{2}y = \frac{1}{2}x - 1$, $y = x - 2$ **3.** $4y = 3x + 3$, $y = \dfrac{3x + 3}{4}$

page 428   Class Exercises

1. $5.1 = k(3)$, $k = 1.7$ **2.** $3.2 = k(0.8)$, $k = 4$

3. yes; $(2, 1)$, $(4, 2)$ **4.** No, not in the form $y = kx$.

5. No, not in the form $y = kx$.

pages 428–429   Practice Exercises

A 1. yes   **2.** no   **3.** no   **4.** yes

5. $4 = k(1)$, $k = 4$

$12 = k(3)$, $k = 4$

$20 = k(5)$, $k = 4$

$36 = k(9)$, $k = 4$

yes; $k = 4$, $y = 4x$

6. $4 = k(8)$, $k = \frac{1}{2}$

$9 = k(18)$, $k = \frac{1}{2}$

$10 = k(20)$, $k = \frac{1}{2}$

$11 = k(21)$, $k = \frac{11}{24}$

y does not vary directly as x.

7. $0.2 = k(0.6)$, $k = \frac{1}{3}$

$1.5 = k(4.5)$, $k = \frac{1}{3}$

$2.1 = k(6.3)$, $k = \frac{1}{3}$

$5.2 = k(15.6)$, $k = \frac{1}{3}$

yes; $k = \frac{1}{3}$; $y = \frac{1}{3}x$

8. $1.2 = k(1.6)$, $k = 0.75$; $1.75 = k(2.2)$, $k = 0.795$; y does not vary directly as x.

9. $\dfrac{5}{25} = \dfrac{y}{55}$, $25y = 275$, $y = 11$ **10.** $\dfrac{3}{2} = \dfrac{y}{34}$, $2y = 102$, $y = 51$

11. $\dfrac{4}{150} = \dfrac{B}{200}$, $150B = 800$, $B = 5\frac{1}{3}$ **12.** $\dfrac{84}{12} = \dfrac{N}{14}$, $12N = 1176$, $N = 98$

13. $\dfrac{5}{6} = \dfrac{y}{42}$, $6y = 210$, $y = 35$ **14.** $\dfrac{35}{56} = \dfrac{y}{8}$, $56y = 280$, $y = 5$

B 15. $\dfrac{110}{17.6} = \dfrac{80}{M} = 110M = 1408$, $M = 12.8$ **16.** $\dfrac{3.78}{3} = \dfrac{C}{4.5}$, $3C = 17.01$, $C = 5.67$

17. $\dfrac{9}{12.5} = \dfrac{W}{16.5}$, $12.5W = 148.5$, $W = 11.88$ **18.** $\dfrac{2.25}{12} = \dfrac{F}{50}$, $12F = 112.5$, $F = 9.375$

19. $\dfrac{3.9}{3} = \dfrac{l}{40}$, $3l = 156$, $l = 52$

20. $\dfrac{36}{0.9} = \dfrac{d}{11}$, $0.9d = 396$, $d = 440$

21. $\dfrac{5.4}{1.8} = \dfrac{A}{3}$, $1.8A = 16.2$, $A = 9$

22. $\dfrac{1.9}{0.213} = \dfrac{C}{10}$, $0.213C = 19$, $C = 89.2$

23. $\dfrac{2.56}{3.2} = \dfrac{y}{25}$, $3.2y = 64$, $y = 20$

24. $\dfrac{7.29}{0.9} = \dfrac{y}{1.2}$, $0.9y = 8.748$, $y = 9.72$

C 25. $\dfrac{1.2}{2^2} = \dfrac{y}{5^2}$, $4y = 30$, $y = 7.5$

26. $\dfrac{40.5}{9^2} = \dfrac{y}{12^2}$, $81y = 5832$, $y = 72$

27. $A = \pi r^2$, no

28. $P = 4s$, yes

29. $\dfrac{5000}{350} = \dfrac{8000}{I}$, $5000I = 2{,}800{,}000$, $= \$560$ **30.** $\dfrac{3\frac{1}{2}}{208} = \dfrac{1\frac{3}{4}}{d}$, $\left(3\frac{1}{2}\right)\!\left(d\right) = 364$, $d = 104$ mi

31. $5y + 3 = -2y - 18$, $7y = -21$, $y = -3$

32. $2x - 16 + 12 = 3x - 15$, $2x - 4 = 3x - 15$, $-x = -11$, $x = 11$

33. $8 - 6f - 3f - 6 = 7f$, $2 - 9f = 7f$, $-16f = -2$, $f = \dfrac{1}{8}$

34. $-2a + 3 + 13 = 8 + 2a$, $-2a + 16 = 8 + 2a$, $-4a = -8$, $a = 2$

35. $7(a^2 - 4) = 7(a + 2)(a - 2)$ **36.** $3(x^2 + 2x + 1) = 3(x + 1)(x + 1)$

37. $b^2(4b^2 + 19b + 12) = b^2(4b + 3)(b + 4)$

38. $2xy(6x^2 + 13x - 5) = 2xy(3x - 1)(2x + 5)$

39. $m = \dfrac{2 - 8}{-1 - 2} = \dfrac{-6}{-3} = 2$; $y = mx + b$,   **40.** $m = -2$; $y = mx + b$, $3 = -2(-4) + b$,
$8 = 2(2) + b$, $8 = 4 + b$, $b = 4$; $\qquad$ $3 = 8 + b$, $b = -5$; $y = -2x - 5$,
$y = 2x + 4$, $2x - y = -4$ $\qquad\qquad$ $2x + y = -5$

page 430 Capsule Review

1. $8 = k(2)$, $k = 4$; $y = 4x$ **2.** $-3 = k\left(\dfrac{2}{3}\right)$, $k = -\dfrac{9}{2}$; $y = -\dfrac{9}{2}x$ **3.** $28 = k(7)$, $k = 4$; $y = 4x$

pages 431–432 Class Exercises

1. $6(4) = k$, $k = 24$
$\quad$ $8(3) = k$, $k = 24$
$\quad$ $1.2(20) = k$, $k = 24$
$\quad$ $48(0.5) = k$, $k = 24$
$\quad$ $k = 24$; $xy = 24$

2. $3(15) = k$, $k = 45$
$\quad$ $4.5(10) = k$, $k = 45$
$\quad$ $5(9) = k$, $k = 45$
$\quad$ $1.2(38) = k$, $k = 45.6$
$\quad$ No, y does not vary inversely as x.

3. $4(8) = 32$

4. $3(3.3) = 9.9$

5. $0.7(8.1) = 5.67$

6. $0.8(6) = 3(y)$, $4.8 = 3y$, $y = 1.6$

7. $32(9) = 24(w)$, $288 = 24w$, $w = 12$

pages 432–433   **Practice Exercises**

A 1. $2(6) = 12$
$(-3)(-4) = 12$
$12(1) = 12$
yes; $xy = 12$

2. $9(4) = 36$
$(-2)(-18) = 36$
$3(12) = 36$
yes; $xy = 36$

3. $(-0.8)(1.8) = -1.44$
$1.2(-1.2) = -1.44$
$(-3.6)(0.4) = -1.44$
yes; $xy = -1.44$

4. $(-1.6)0.3 = -0.48$
$0.4(7.2) = 2.88$
No, y does not vary inversely as x.

5. $6(12) = 9y$, $72 = 9y$, $y = 8$

6. $18(8) = 12(y)$, $144 = 12y$, $y = 12$

7. $30(1.6) = 12(l)$, $48 = 12l$, $l = 4$

8. $8(25) = 0.5d$, $200 = 0.5d$, $d = 400$

9. $6(3.2) = 0.8(w)$, $19.2 = 0.8w$, $w = 24$

10. $1.6(24) = 0.12(C)$, $38.4 = 0.12C$, $C = 320$

11. $3.9(4.2) = 2.6(y)$, $16.38 = 2.6y$, $y = 6.3$

12. $3.6(2.25) = 0.3(y)$, $8.1 = 0.3y$, $y = 27$

13. $2.5(9) = 4.25(j)$, $22.5 = 4.25j$, $j = 5.29$

14. $\frac{1}{6}\left(\frac{1}{4}\right) = \frac{2}{3}(h)$, $\frac{1}{24} = \frac{2}{3}h$, $h = \frac{1}{16}$

B 15. $15(1.5) = 2.25(v)$, $22.5 = 2.25v$, $v = 10$ in.^{3}

16. $12(1.75) = 4(p)$, $21 = 4p$, $p = 5.25$ atm

17. $m(2.5) = 35(3)$, $2.5m = 105$, $m = 42$ kg

18. $60(3) = 45(x)$, $180 = 45x$, $x = 4$ m

19. $6(\sqrt{144}) = 132(\sqrt{t})$, $6(12) = 132(\sqrt{t})$, $72 = 132(\sqrt{t})$, $\frac{72}{132} = \sqrt{t}$,

$\sqrt{t} = \frac{6}{11}$, $t = \frac{36}{121}$

20. $2^2(9) = 90z^2$, $36 = 90z^2$, $\frac{36}{90} = z^2$, $\frac{6}{15} = z^2$, $z = 0.63$

21. $9.3(4.2)^3 = (3.2)^3 n$, $689.0184 = 32.768n$, $n = 21.03$

22. $4(1.6)^2 = (6.2)^2 h$, $10.24 = 38.44h$, $h \approx 0.27$

C 23. yes; $r_1 t_1 = r_2 t_2$

24. yes; $\$20 \times 25$ h $= \$10 \times 50$ h

25. $42(5) = r(6)$, $210 = 6r$, $r = 35$ mi/h

26. $160\left(3\frac{1}{2}\right) = 200(t)$, $560 = 200t$, $t = 2.8$ h

27. $A = \frac{1}{2}bh$, yes

28. $C = 2\pi r$, no

29. 18 in. = 1.5 ft, 15 in. = 1.25 ft, (15)(1.5) = (x)(1.25), 22.5 = 1.25x,
x = 18 ft

30. $\left(48\right)\left(2\frac{1}{2}\right)$ = 40t, 120 = 40t, t = 3 h

31. Answers may vary.

page 435   Class Exercises

1. $A = lw$

2. $A = \frac{1}{2}bh$

3. $y = mx + b$

4. $t = 18(0.86)$, $t = \$15.48$

5. $I = prt$, $I = (5000)(0.0825)(4)$, $I = \$1650$

6. $V = C\left(1 - \frac{n}{N}\right)$, $V = 5000\left(1 - \frac{3}{10}\right)$, $V = 5000\left(\frac{7}{10}\right)$, $V = (500)(7)$, $V = \$3500$

pages 435–436   Practice Exercises

A   **1.** $A = s^2$

2. $C = 2\pi r$

3. $V = C\left(1 - \frac{n}{N}\right)$

4. $I = prt$

5. $V = \frac{4}{3}\pi r^3$

6. $V = s^3$

7. $t = 36(1.12)$, $t = \$40.32$

8. $F = \frac{9}{5}(20) + 32$, $F = 36 + 32 = 68$; $68°F$

9. $b = \frac{169}{130}$, $b = \$1.30$

10. $m = \frac{9.65}{5}$, $m = \$1.93$

11. $9 = \frac{1}{2}g(1)^2$, $9 = \frac{1}{2}g$, $g = 18$

12. $36 = \frac{1}{2}g(2)^2$, $36 = \frac{1}{2}g(4)$, $36 = 2g$, $g = 18$

13. $64 = \frac{1}{2}g(4)^2$, $64 = \frac{1}{2}g(16)$, $64 = 8g$, $g = 8$

14. $81 = \frac{1}{2}g(3)^2$, $81 = \frac{1}{2}g(9)$, $81 = \frac{9}{2}g$, $g = 18$

15. $V = 5000\left(1 - \frac{1}{5}\right)$, $V = 5000\left(\frac{4}{5}\right)$, $V = 1000(4)$, $V = \$4000$

16. $V = 20{,}000\left(1 - \frac{4}{4}\right)$, $V = 20{,}000(1 - 1)$, $V = 0$

B 17. $V = 3{,}000\left(1 - \frac{3}{3}\right)$, $V = 3000(1 - 1)$, $V = 0$

18. $V = 15{,}000\left(1 - \frac{8}{10}\right)$, $V = 15{,}000\left(\frac{2}{10}\right)$, $V = (1500)(2)$, $V = \$3000$

19. $V = 50{,}000\left(1 - \frac{15}{20}\right)$, $V = 50{,}000\left(\frac{5}{20}\right)$, $V = (2500)(5)$, $V = \$12{,}500$

20. $V = 60{,}000\left(1 - \frac{17}{25}\right)$, $V = 60{,}000\left(\frac{8}{25}\right)$, $V = (2400)(8)$, $V = \$19{,}200$

21. $64 = \frac{1}{3}\pi(3)^2 h$, $64 = \frac{1}{3}\pi 9(h)$, $64 = 3\pi h$, $64 = 3(3.14)h$, $9.42h = 64$, $h = 6.79$

22. $64 = \frac{1}{3}\pi(r)^2(6)$, $64 = 2\pi r^2$, $64 = 2(3.14)r^2$, $64 = 6.28(r^2)$, $r^2 = 10.19$, $r = 3.19$

23. $V = \frac{1}{3}\pi\left(\frac{1}{4}\right)^2\left(\frac{1}{3}\right)$, $V = \frac{1}{9}(\pi)\frac{1}{16}$, $V = \frac{1}{9}(3.14)\left(\frac{1}{16}\right)$, $V = 0.0218$

24. $0.07 = \frac{1}{3}\pi(r^2)(0.03)$, $0.07 = (0.01)(3.14)(r^2)$, $r^2 = 2.23$, $r = 1.49$

C 25. $\dfrac{1800 \text{ mi/h}}{740 \text{ mi/h}} = 2.4$ **26.** $M = \dfrac{A}{s}$

page 436   Mixed Problem Solving Review

1. Let x represent gals per week William uses. $x = \dfrac{648}{18} = 36$

Let y represent the gals over a 7-week period William uses. $y = 36(7) = 252$
Let p represent how much William spends on gasoline over a 7-week period.
$p = 252(0.96)$, $p = 241.92$
John spends $241.92 on gasoline over a 7-week period.

2. Let x represent the number of hours working together.

$\dfrac{x}{6} + \dfrac{x}{7} = 1$, $7x + 6x = 42$, $13x = 42$, $x = 3.23$

Working together it would take 3.23h to paint a room.

3. Let w = the width, then $6w + 3$ = the length of the rectangle.
$48 = 2(6w + 3) + 2w$, $48 = 12w + 6 + 2w$, $48 = 14w + 6$,
$14w = 42$, $w = 3$ in. Then $6w + 3 = 21$ in.
The dimensions of the rectangle are 21 in. × 3 in.

Answers may vary.

1. Yes, $y = -3x$; The equation is in the form $y = kx$.

2. No, $x = y + 2$ is not in the form $y = kx$.

3. Yes, $y = -\frac{1}{2}x$; The equation is in the form $y = kx$.

4. No, $y = \frac{4}{7}x + 3$ is not in the form of $y = kx$.

5. $\frac{3}{2} = \frac{y}{4}$, $2y = 12$, $y = 6$

6. $\frac{5}{4} = \frac{y}{8}$, $4y = 40$, $y = 10$

7. $4(3.5) = k$, $k = 14$
$5.6(2.5) = k$, $k = 14$
$20(0.7) = k$, $k = 14$
$2(7) = k$, $k = 14$
yes; $k = 14$, $xy = 14$

8. $6(2.5) = k$, $k = 15$
$1.2(12.5) = k$, $k = 15$
$30(0.5) = k$, $k = 15$
$3(5) = k$, $k = 15$
yes; $k = 15$, $xy = 15$

9. $-2.3(0.3) = k$, $k = -0.69$
$0.5(4.3) = k$, $k = 2.15$
No, y does not vary inversely as x.

10. $3.2(0.5) = k$, $k = 1.6$
$1.6(2) = k$, $k = 3.2$
No, y does not vary inversely as x.

11. $4(5) = 2(x)$, $20 = 2x$, $x = 10$

12. $10(8) = 5(x)$, $80 = 5x$, $x = 16$

13. $8.98(5) = \$44.90$

14. $\frac{6.45}{5} = \$1.29$

15. $V = 4000\left(1 - \frac{1}{4}\right)$, $V = 4000\left(\frac{3}{4}\right)$, $V = 1000(3)$, $V = \$3000$

16. $V = 12{,}000\left(1 - \frac{8}{10}\right)$, $V = 12{,}000\left(\frac{2}{10}\right)$, $V = 1200(2)$, $V = \$2400$

1. $\{(2, -2), (5, -5), (2, 0), (-2, 1)\}$; $D = \{-2, 2, 5\}$; $R = \{-5, -2, 0, 1\}$; not a function

2. $\{(3, -1), (-2, 0), (-2, 1)\}$; $D = \{-2, 3\}$; $R = \{-1, 0, 1\}$; not a function

3.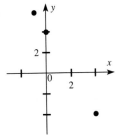

function; no; no

4.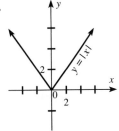

function; no; no

5.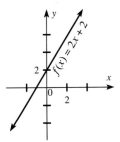

function; yes; no

6.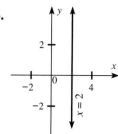

not a function

7.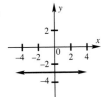

function; yes; yes

8.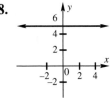

function; yes; yes

9.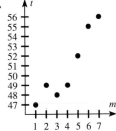

function; no

10. $f(-2) = (-2)^2 = 4$,
$g(0) = 2(0) - 1 = -1$
$f(-2) + g(0) = 4 + (-1) = 3$

11. $f(g(1)) = f(4(1) - 7) = f(4 - 7) = f(-3) = 3(-3)^2 + 2 = 3(9) + 2 = 27 + 2 = 29$

12. direct variation; $y = 4x$; $k = 4$

13. inverse variation; $y = \dfrac{18}{x}$; $k = 18$

14. $\dfrac{5}{25} = \dfrac{y}{55}$, $25y = 275$, $y = 11$

15. $46(1.5) = r(2)$, $2r = 69$, $r = 34.5$

16. $\dfrac{65,000}{650} = \dfrac{48,000}{x}$, $65,000x = (48,000)(650)$, $65,000x = 31,200,000$, $x = \$480$

17. $\dfrac{20}{87} = \dfrac{x}{240}$, $87x = 20(240)$, $87x = 4800$, $x \approx 55.2$ L

18. $6(4) = l(3)$, $24 = 3l$, $l = 8$ m

19. $C = \frac{5}{9}(77 - 32) = \frac{5}{9}(45) = 25°C$

page 440 Chapter Test

1. $\{(2, -5), (5, -14), (0, 1), (-2, 7)\}$; $D = \{-2, 0, 2, 5\}$; $R = \{-14, -5, 1, 7\}$; function

2. $\left\{\left(-\frac{1}{4}, 0\right), (5, 1), \left(6\frac{2}{3}, 0\right)\right\}$; $D = \left\{-\frac{1}{4}, 5, 6\frac{2}{3}\right\}$; $R = \{0, 1\}$; function

3.

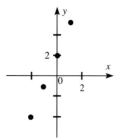

function; no; no

4.

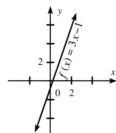

function; yes; no

5.

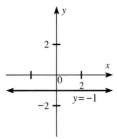

function; yes; yes

6.

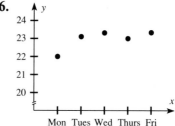

function; no

7. $f(0) = (0)^2 + 1 = 1$, $g(-2) = 3(-2) = -6$
$f(0) + g(-2) = 1 + (-6) = -5$

8. $g(h(3)) = g(-2(3)^2 + 7) = g(-2(9) + 7) = g(-18 + 7) = g(-11) =$
$5(-11) + 2 = -55 + 2 = -53$

9. direct variation; $y = -2x$; $k = -2$

10. $18(0.5) = 9$
$3(3) = 9$
$4(2.25) = 9$
$45(0.2) = 9$
inverse variation; $xy = 9$; $k = 9$

11. $9(6) = 2.7(y)$, $2.7y = 54$, $y = 20$

12. $\frac{3.78}{3} = \frac{c}{4.5}$, $3c = 17.01$, $c = 5.67$

13. $\frac{6}{2} = \frac{c}{5}$, $2c = 30$, $c = 15$ cups

14. $(40)(6) = 30d$, $240 = 30d$, $d = 8$ m

Challenge

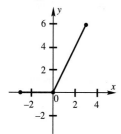

page 441 Preparing for Standardized Tests

1. D; $m = \dfrac{7 - (-1)}{0 - 4} = \dfrac{8}{-4} = -2$; $y - 7 = -2(x - 0)$, $y - 7 = -2x$, $y = -2x + 7$

2. A; $\dfrac{2}{3}$ of 690 = 460 and $\dfrac{3}{5}$ of 865 = 519. Then 519 − 460 = 59.

3. D; \$29.95 + 14.50 + 8.90 + 2.75 + 0.79 = \$56.89; \$100 − \$56.89 = \$43.11 in change.

4. E; $\dfrac{12}{33} = \dfrac{x}{198}$, $x = \dfrac{2376}{33}$; $x = 72$

5. A; $x = -1$

6. C; $f(-2) = 2(-2)^2 + 4(-2) + 7 = 2(4) - 8 + 7 = 7$

7. D; The total time was 3 hours and the total number of miles she traveled was 30, so the average rate was $\dfrac{30}{3}$ or 10 mi/h.

8. B; Since $x \geq 1$, it follows that $5x \geq 5$, $5x - 2 \geq 5 - 2$, $5x - 2 \geq 3$; so $y \geq 3$.

9. D; $1\dfrac{1}{2}$ oz of type A contains $1\dfrac{1}{2}(160) = 240$ mg of sodium; 2 oz of type B contains $2(110) = 220$ mg of sodium; Thus, the serving of type A has 20 mg more than the serving of type B.

page 442 Cumulative Review (Chapters 1–10)

1. $x + 3 = 0$, $x = -3$

2. $5y + 3 = 0$, $5y = -3$, $y = -\dfrac{3}{5}$

3. $(m + 2)(m + 6) = 0$,
   $m + 2 = 0 \ \vdots \ m + 6 = 0$
   $m = -2 \ \vdots \ \ \ m = -6$

4. $(s + 4)(s - 4) = 0$
   $s + 4 = 0 \ \vdots \ s - 4 = 0$
   $s = -4 \ \vdots \ \ \ s = 4$

5. $3xy^2$ **6.** $8rs^3$ **7.** $\dfrac{2m(7m+n)}{14m^2n} = \dfrac{7m+n}{7mn}$

8. $\dfrac{4x(4x^2+y)}{8x^3y} = \dfrac{4x^2+y}{2x^2y}$ **9.** $\dfrac{4}{8} = \dfrac{1}{2}$; 1 to 2

10. $\dfrac{0.3}{2.1} = \dfrac{3}{21} = \dfrac{1}{7}$; 1:7   **11.** 3 ft = 1 yd, $\dfrac{1}{5}$; 1 to 5

12. $\dfrac{\$4.40}{\$0.22} = \$20$; 20 to 1   **13.** $5 \cdot \dfrac{x}{x} + 7x \cdot \dfrac{x}{x} - \dfrac{3}{x} = \dfrac{5x + 7x^2 - 3}{x}$

14. $\dfrac{\dfrac{3n}{n^2} - \dfrac{1}{n} \cdot \dfrac{n}{n}}{\dfrac{n^2}{n^2} + \dfrac{3m}{n^2}} = \dfrac{\dfrac{3n-n}{n^2}}{\dfrac{n^2+3m}{n^2}} = \dfrac{\dfrac{2n}{n^2}}{\dfrac{n^2+3m}{n^2}} = \dfrac{2n}{n^2+3m}$

15. $\dfrac{\dfrac{4x}{x^2} - \dfrac{1}{x} \cdot \dfrac{x}{x}}{\dfrac{x^2}{x^2} + \dfrac{4y}{x^2}} = \dfrac{\dfrac{4x-x}{x^2}}{\dfrac{x^2+4y}{x^2}} = \dfrac{\dfrac{3x}{x^2}}{\dfrac{x^2+4y}{x^2}} = \dfrac{3x}{x^2+4y}$

16. $4x^2(x^2-4) = 4x^2(x+2)(x-2)$ **17.** $(2x+3)(2x-5)$

18. $8x^2y^2(2x-1+3y)$

19. $y-(-3) = 3(x-0)$, $y+3 = 3x$, $-3x+y = -3$

20. $y-0 = -\dfrac{1}{2}(x-6)$, $y = -\dfrac{1}{2}x+3$, $y+\dfrac{1}{2}x = 3$, $2y+x = 6$

21. Let m = the slope of the line. $m = \dfrac{-10-(-9)}{4-6}$, $m = \dfrac{-1}{-2}$, $m = \dfrac{1}{2}$

    $y-(-9) = \dfrac{1}{2}(x-6)$, $y+9 = \dfrac{1}{2}x-3$, $y = \dfrac{1}{2}x-12$, $\dfrac{1}{2}x-y = 12$, $x-2y = 24$

22. Let m = the slope of the line. $m = \dfrac{7-(-13)}{-9-11}$, $m = \dfrac{20}{-20}$, $m = -1$

    $y-(-13) = -1(x-11)$, $y+13 = -x+11$ $y = -x-2$, $x+y = -2$

23. 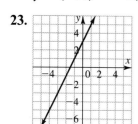   **24.** 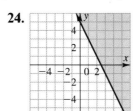   **25.**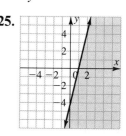

26. $f(-1) = \dfrac{1}{1 + 3(-1)} = \dfrac{1}{1 - 3} = -\dfrac{1}{2}$

$f(0) = \dfrac{1}{1 + 3(0)} = \dfrac{1}{1} = 1$

$f(1) = \dfrac{1}{1 + 3(1)} = \dfrac{1}{1 + 3} = \dfrac{1}{4}$

$R = \left\{ -\dfrac{1}{2}, \dfrac{1}{4}, 1 \right\}$

27. $g: -4 \rightarrow (-4)^2 - 16$, $g: -4 \rightarrow 16 - 16$, $g: -4 \rightarrow 0$

$g: \dfrac{1}{2} \rightarrow \left(\dfrac{1}{2}\right)^2 - 16$, $g: \dfrac{1}{2} \rightarrow \dfrac{1}{4} - 16$, $g: \dfrac{1}{2} \rightarrow \dfrac{1}{4} - \dfrac{64}{4}$, $g: \dfrac{1}{2} \rightarrow -\dfrac{63}{4}$

$g: 3 \rightarrow (3)^2 - 16$, $g: 3 \rightarrow 9 - 16$, $g: 3 \rightarrow -7$

$R = \left\{ -\dfrac{63}{4}, -7, 0 \right\}$

28. $2m - 3 = 17$ or $2m - 3 = -17$

$\qquad 2m = 20 \quad \vdots \quad 2m = -14$

$\qquad m = 10 \quad \vdots \quad m = -7$

The solution set is $\{-7, 10\}$.

29. $8 - 3t < 6$ and $8 - 3t > -6$

$\qquad -3t < -2 \quad \vdots \quad -3t > -14$

$\qquad t > \dfrac{2}{3} \quad \vdots \quad t < \dfrac{14}{3}$

The solution set is $\left\{ \dfrac{2}{3} < t < \dfrac{14}{3} \right\}$.

30. $4a - 7 \le 1$ and $4a - 7 \ge -1$

$\qquad 4a \le 8 \quad \vdots \quad 4a \ge 6$

$\qquad a \le 2 \quad \vdots \quad a \ge \dfrac{3}{2}$

The solution set is $\left\{ \dfrac{3}{2} \le a \le 2 \right\}$.

31. Let $t =$ the time for train 2, then $t + 1 =$ the time for train 1.

$110(t + 1) = 120t$, $110t + 110 = 120t$, $10t = 110$, $t = 11$h

Train 2 will overtake train 1 in 11 h.

32. Let $x =$ the first consecutive even integer, then $x + 2 =$ the second consecutive even integer, $x + 4 =$ the third consecutive even integer, and $x + 6 =$ the fourth consecutive even integer.

The equation is $x + (x + 2) + (x + 4) + (x + 6) = 52$; solve

$4x + 12 = 52$, $4x = 40$, $x = 10$; Then $x + 2 = 12$, $x + 4 = 14$, $x + 6 = 16$

The four consecutive even integers are 10, 12, 14, 16.

33. Let $w =$ the width, then $w + 5 =$ the length of the rectangle.

$110 = 2w + 2(w + 5)$, $110 = 2w + 2w + 10$, $110 = 4w + 10$, $4w = 100$,

$w = 25$; $w + 5 = 30$; The dimensions of the rectangle are 30 cm $\times$ 25 cm.

Chapter 11   Systems of Linear Equations

page 444   Capsule Review

1. $b = 4$

$m = -\dfrac{1}{2}$

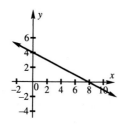

2. $b = 1$

$m = -2$

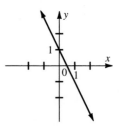

3. $y = x + 3$

$b = 3$

$m = 1$

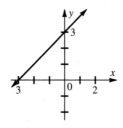

4. $y = -\dfrac{5}{2}x$

$b = 0$

$m = -\dfrac{5}{2}$

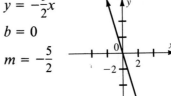

page 446   Class Exercises

1. $(2, 4)$

2. $(-1, -5)$

3.

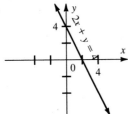

$\{(x, y): 2x + y = 4\}$

4.

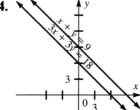

no solution

5. same slope (3); different y-intercept (6 and 0); parallel lines; no solution

6. same slope $\left(-\dfrac{1}{2}\right)$; same y-intercept (-1); same line; infinitely many solutions

7. different slopes (1 and -2); different y-intercepts (-4 and 1); intersecting lines; one solution

8. same slope $\left(\dfrac{1}{2}\right)$; different y-intercepts (4 and -3); parallel lines; no solution

9. different slopes $\left(-\dfrac{3}{4} \text{ and } -\dfrac{1}{2}\right)$; different y-intercepts $\left(\dfrac{5}{4} \text{ and } -2\right)$; intersecting lines; one solution

10. same slope (3); same y-intercept (2); same line; infinitely many solutions

11. Systems with non-integer solutions or nonlinear systems would be easier to solve by using a graphing utility.

pages 446–449 Practice Exercises

A **1.**

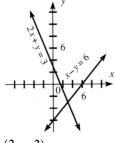

$(3, -3)$

2.

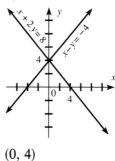

$(0, 4)$

3.

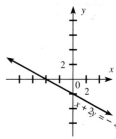

$\{(x, y): x + 2y = -4\}$

4.

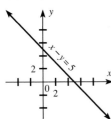

$\{(x, y): x - y = 5\}$

5.

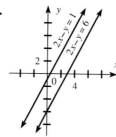

no solution

6.

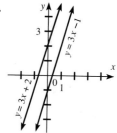

no solution

7.

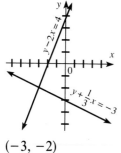

$(-3, -2)$

8.

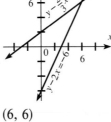

$(6, 6)$

9.

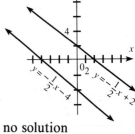

no solution

10.

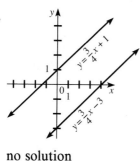

no solution

11.

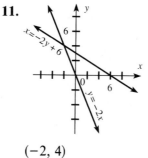

$(-2, 4)$

12.

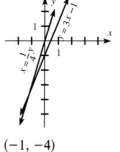

$(-1, -4)$

13. independent and consistent

14. independent and consistent

15. independent and inconsistent

16. independent and inconsistent

17. dependent and consistent

18. dependent and consistent

19. independent and consistent

20. independent and consistent

21. dependent and consistent

22. dependent and consistent

23. independent and inconsistent

24. independent and inconsistent

25. intersecting lines; one solution

26. intersecting lines; one solution

27. parallel lines; no solution

28. same line; infinitely many solutions

29. parallel lines; no solution

30. parallel lines; no solution

B 31. parallel lines; no solution

32. same line; infinitely many solutions

33. parallel lines; no solution

34. intersecting lines; one solution

35. intersecting lines; one solution

36. parallel lines; no solution

37.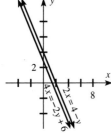

$(0, -5)$

38.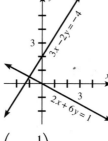

$(0, 6)$

39.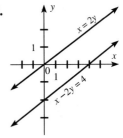

no solution

40.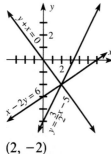

no solution

41.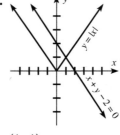

$\left(-1, \frac{1}{2}\right)$

42.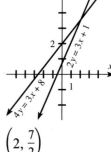

$\left(2, \frac{7}{2}\right)$

C 43.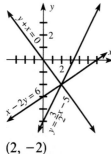

$(2, -2)$

44.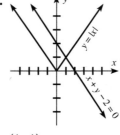

$(1, 1)$

45.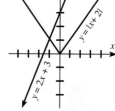

$(-1, 1)$

46. Dependent systems have the same slope, $m_1 = m_2$, and the same y-intercept, $b_1 = b_2$.

47. Inconsistent systems have the same slope, $m_1 = m_2$, and different y-intercepts $b_1 \neq b_2$.

48. Independent systems have different slopes, $m_1 \neq m_2$, and may or may not have different y-intercepts.

49. Consistent systems have the same slope, $m_1 = m_2$, and the same y-intercept, $b_1 = b_2$. Or, consistent systems have different slopes, $m_1 \neq m_2$ and may or may not have the same y-intercept, $b_1 = b_2$ or $b_1 \neq b_2$.

50. a. $(-3.2, -3.2)$ **b.** no solution   **c.** $(-1.1, 0.9)$

51.

2 min

52.

g: \$6/h; s: \$4/h

53.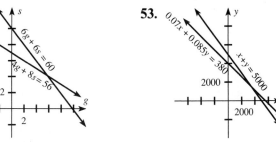

CD = \$3000; mutual fund = \$2000

54.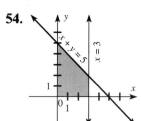

Area rectangle = 3(2) = 6; Area triangle = $\frac{1}{2}$ (3)(3) = $4\frac{1}{2}$;

Area trapezoid = Area rectangle + Area triangle = $6 + 4\frac{1}{2} = 10\frac{1}{2}$

page 449   Career: Air Traffic Controller

1. Let r_1 represent the rate of the plane leaving Atlanta, then $r_1 = \dfrac{800}{3\frac{1}{4}}$, $r_1 \approx 246.15$ mi/h.

Let r_2 represent the rate of the plane leaving Dallas, then $r_2 = \dfrac{800}{2}$, $r_2 = 400$ mi/h.

Let t represent the time after takeoff that the two planes pass each other, then $246.15t + 400t = 800$, $646.15t = 800$; $t = 1.24$. The two planes pass each other after 1.24 h.

Let d represent the distance from Atlanta, then $(246.15)(1.24) = d$, $d = 305.23$; The two planes meet at a distance of 305 mi from Atlanta.

2. Let r_1 represent the rate of the plane from Boston, then $r_1 = \dfrac{1200}{3}$, $r_1 = 400$.

Let r_2 represent the rate of the plane from Miami, then $r_2 = \dfrac{1200}{4\frac{1}{2}}$, $r_2 = 266\frac{2}{3}$.

Let t represent the time after take off that the two planes will pass each other, then $400t + 266\frac{2}{3}t = 1200$, $666\frac{2}{3}t = 1200$, $t = 1.8$. The two planes will meet in 1.8 h.

Let d represent the distance from Boston that the two planes will pass each other, then $d = (1.8)(400) = 720$; The distance from Boston at that time will be 720 mi.

3. Let r_1 represent the speed from Denver to Chicago, then $0.75r_1$ represents the speed from Chicago to Denver.

Let t represent the time for the first flight, then $4 - t$ represents the time for the return trip.

$$r_1 t = 920, \quad t = \dfrac{920}{r_1}, \quad (0.75r_1)(4 - t) = 920,$$

$$(0.75\,r_1)\left(4 - \dfrac{920}{r_1}\right) = 920, \quad 3r_1 - 690 = 920, \quad 3r_1 = 1610, \quad r_1 = 536.\overline{666};$$

The speed from Denver to Chicago was 536.7 mi/h, and the speed from Chicago to Denver was $(0.75)(536.7) = 402.5$ mi/h.

page 450 Integrating Algebra: Break-Even Point

1. The loss is $200 during the first month if only 500 copies are sold.

2. The profit is $200 during the second month if 1500 copies are sold.

3.

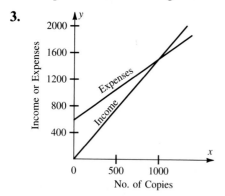

4. 1000 copies must be sold before there is a profit.

5. The loss is $420 if only 300 copies are sold. **6.** The average profit is $600 per month.

page 451 Capsule Review

1. $y = x - 3$

2. $y = \dfrac{x}{2} - 2$

3. $2x = 3y + 5$, $x = \dfrac{3}{2}y + \dfrac{5}{2}$

4. $x = 3y - 9$ **5.** $4\left(\dfrac{x}{2}\right) = 4\left(\dfrac{y}{4}\right), y = 2x$ **6.** $\left(\dfrac{1}{2}x\right)2 = (4y)2, x = 8y$

page 452 Class Exercises

1. Solve first equation for y: $y = 2x$; substitute $7x - 2x = 35$, $5x = 35$, $x = 7$; $y = 2(7) = 14$; $(7, 14)$

2. Solve first equation for a: $a = -3b + 5$; substitute $2(-3b + 5) - 4b = -5$, $-6b + 10 - 4b = -5$, $-10b + 10 = -5$, $-10b = -15$, $b = \dfrac{3}{2}$; $a = -3\left(\dfrac{3}{2}\right) + 5$, $a = -\dfrac{9}{2} + \dfrac{10}{2} = \dfrac{1}{2}$; $\left(\dfrac{1}{2}, -\dfrac{3}{2}\right)$

3. Solve first equation for m: $3m = -6n - 15$, $m = -2n - 5$; substitute $2(-2n - 5) - 3n = 4$, $-4n - 10 - 3n = 4$, $-7n - 10 = 4$, $-7n = 14$, $n = -2$; $m = -2(-2) - 5$, $m = 4 - 5 = -1$; $(-1, -2)$

4. Solve first equation for f: $-2f = -5g - 8$, $f = \dfrac{5}{2}g + 4$; substitute $3\left(\dfrac{5}{2}g + 4\right) + 5g = -3$, $\dfrac{15}{2}g + 12 + 5g = -3$, $15g + 24 + 10g = -6$, $25g + 24 = -6$, $25g = -30$, $g = -\dfrac{6}{5}$; $f = \dfrac{5}{2}\left(-\dfrac{6}{5}\right) + 4$, $f = -3 + 4$, $f = 1$; $\left(1, -\dfrac{6}{5}\right)$

5. $y = 3x + 2$; substitute $3x - (3x + 2) = -2$, $0 - 2 = -2$, $-2 = -2$; $\{(x, y): -3x + y = 2\}$; the lines coincide

6. $-v = -3u$, $v = 3u$; substitute $6u - 2(3u) = 5$, $6u - 6u = 5$, $0 \neq 5$; no solution; the lines are parallel

pages 452–454 Practice Exercises

A **1.** $x = y$; $x + x = 2$, $2x = 2$, $x = 1$; $1 - y = 0$, $y = 1$; $(1, 1)$

2. $x = -3y - 4$; $y + (-3y - 4) = 0$, $y - 3y - 4 = 0$, $-2y - 4 = 0$, $-2y = 4$, $y = -2$; $x = -3(-2) - 4$, $x = 6 - 4 = 2$; $(2, -2)$

3. $-v = -3u + 17$, $v = 3u - 17$; $(3u - 17) + 2u = 8$, $5u - 17 = 8$, $5u = 25$, $u = 5$; $v = 3(5) - 17$, $v = 15 - 17 = -2$; $(5, -2)$

4. $p = -5q + 5$; $(-5q + 5) + q = -3$, $-4q + 5 = -3$, $-4q = -8$, $q = 2$; $p = -5(2) + 5$, $p = -10 + 5 = -5$; $(-5, 2)$

5. $x = -y + 3$; $3(-y + 3) + 2y = 9$, $-3y + 9 + 2y = 9$, $-y + 9 = 9$, $-y = 0$, $y = 0$; $x = -0 + 3 = 3$; $(3, 0)$

6. $x = -3y - 4$; $2y + 3(-3y - 4) = 3$, $2y - 9y - 12 = 3$, $-7y - 12 = 3$, $-7y = 15$, $y = -\dfrac{15}{7}$, $x = -3\left(-\dfrac{15}{7}\right) - 4$, $x = \dfrac{45}{7} - 4$, $x = \dfrac{45}{7} - \dfrac{28}{7} = \dfrac{17}{7}$; $\left(\dfrac{17}{7}, -\dfrac{15}{7}\right)$

7. $x = -2$; $2y - 3(-2) = 4$, $2y + 6 = 4$, $2y = -2$, $y = -1$; $(-2; -1)$

8. $y = 3$; $2x = 5(3) + 7$, $2x = 15 + 7$, $2x = 22$, $x = 11$; $(11, 3)$

9. $e = 3d - 13$; $2d = 5(3d - 13)$, $2d = 15d - 65$, $-13d = -65$, $d = 5$;
   $e = 3(5) - 13$, $e = 15 - 13 = 2$; $(5, 2)$

10. $g = 2f + 46$; $5(2f + 46) = 3f + 6$, $10f + 230 = 3f + 6$, $7f + 230 = 6$,
    $7f = -224$, $f = -32$, $g = 2(-32) + 46$, $g = -64 + 46 = -18$; $(-32, -18)$

11. $k = 10 - 2j$; $10 - 2j = 4j + 36$, $-6j = 26$, $j = -\dfrac{26}{6} = -\dfrac{13}{3}$;
    $k = 10 - 2\left(-\dfrac{13}{3}\right)$, $k = 10 + \dfrac{26}{3}$, $k = \dfrac{30}{3} + \dfrac{26}{3} = \dfrac{56}{3}$; $\left(-\dfrac{13}{3}, \dfrac{56}{3}\right)$

12. $n = 39 - 3m$; $39 - 3m = 2m - 61$, $100 = 5m$, $20 = m$; $n = 39 - 3(20)$,
    $n = 39 - 60 = -21$; $(20, -21)$

13. $5q = -15r + 60$, $q = -3r + 12$; $25r - 10(-3r + 12) = 100$, $25r + 30r - 120 = 100$, $55r = 220$, $r = 4$; $q = -3(4) + 12$, $q = -12 + 12 = 0$; $(0, 4)$

14. $10n = 40m + 80$, $n = 4m + 8$; $10m - 5(4m + 8) = -20$, $10m - 20m - 40 = -20$, $-10m = 20$, $m = -2$; $n = 4(-2) + 8$, $n = -8 + 8 = 0$; $(-2, 0)$

15. $d = 6c + 6$; $3c + 4(6c + 6) = 6$, $3c + 24c + 24 = 6$, $27c + 24 = 6$,
    $27c = -18$, $c = -\dfrac{18}{27} = -\dfrac{2}{3}$; $d = 6\left(-\dfrac{2}{3}\right) + 6$, $d = -4 + 6 = 2$; $\left(-\dfrac{2}{3}, 2\right)$

16. $f = -18e + 14$; $12e + 5(-18e + 14) = 5$, $12e - 90e + 70 = 5$, $-78e + 70 = 5$,
    $-78e = -65$, $e = \dfrac{65}{78} = \dfrac{5}{6}$; $f = -18\left(\dfrac{5}{6}\right) + 14$, $f = -15 + 14 = -1$; $\left(\dfrac{5}{6}, -1\right)$

17. $-x = -4y + 2$, $x = 4y - 2$; $-2(4y - 2) + 12y = 17$, $-8y + 4 + 12y = 17$,
    $4y + 4 = 17$, $4y = 13$, $y = \dfrac{13}{4}$; $x = 4\left(\dfrac{13}{4}\right) - 2$, $x = 13 - 2 = 11$; $\left(11, \dfrac{13}{4}\right)$

18. $x = \dfrac{y}{18} - \dfrac{10}{18}$, $x = \dfrac{y}{18} - \dfrac{5}{9}$; $18\left(\dfrac{y}{18} - \dfrac{5}{9}\right) = 30y + 19$, $y - 10 = 30y + 19$,
    $-29 = 29y$, $-1 = y$; $x = \dfrac{-1}{18} - \dfrac{5}{9}$, $x = -\dfrac{1}{18} - \dfrac{10}{18} = -\dfrac{11}{18}$; $\left(-\dfrac{11}{18}, -1\right)$

19. $x = -3y$; $3y + (-3y) = -1$, $0 \neq 1$; no solution

20. $y = x + 4$; $x + 3 = x + 4$, $3 \neq 4$; no solution

21. $\dfrac{5}{2}b - 1 = a$; $3\left(\dfrac{5}{2}b - 1\right) + 6 = 25b$, $\dfrac{15}{2}b - 3 + 6 = 25b$, $\dfrac{15}{2}b + 3 = 25b$,
    $15b + 6 = 50b$, $6 = 35b$, $\dfrac{6}{35} = b$; $a = \dfrac{5}{2}\left(\dfrac{6}{35}\right) - 1$, $a = \dfrac{3}{7} - 1$, $a = \dfrac{3}{7} - \dfrac{7}{7} = -\dfrac{4}{7}$;
    $\left(-\dfrac{4}{7}, \dfrac{6}{35}\right)$

22. $c = 4d + 3$; $2(4d + 3) - 6 = 8d$, $8d + 6 - 6 = 8d$, $8d = 8d$, $0 = 0$, $\{(c,d): c - 4d = 3\}$

23. $t = -2s - 1$; $4s - 3(-2s - 1) = 8$, $4s + 6s + 3 = 8$, $10s + 3 = 8$, $10s = 5$, $s = \frac{1}{2}$; $t = -2\left(\frac{1}{2}\right) - 1$, $t = -1 - 1 = -2$; $\left(\frac{1}{2}, -2\right)$

24. $j = -6k$; $4(-6k) - 3k = 9$, $-24k - 3k = 9$, $-27k = 9$, $k = -\frac{9}{27} = -\frac{1}{3}$, $j = -6\left(-\frac{1}{3}\right) = 2$; $\left(2, -\frac{1}{3}\right)$

B 25. $-2f = -4e + 16f$, $f = 2e - 8$; $5e - 7(2e - 8) = 1$, $5e - 14e + 56 = 1$, $-9e + 56 = 1$, $-9e = -55$, $e = \frac{55}{9}$; $f = 2\left(\frac{55}{9}\right) - 8$, $f = \frac{110}{9} - \frac{72}{9} = \frac{38}{9}$; $\left(\frac{55}{9}, \frac{38}{9}\right)$

26. $-3h = -6g - 3$, $h = 2g + 1$; $13(2g + 1) - 10g = 45$, $26g + 13 - 10g = 45$, $16g + 13 = 45$, $16g = 32$, $g = 2$, $h = 2(2) + 1$, $h = 4 + 1 = 5$; $(2, 5)$

27. $x = y + 1$; $20(y + 1) - 15y = 17$, $20y + 20 - 15y = 17$, $5y + 20 = 17$, $5y = -3$, $y = -\frac{3}{5}$, $x = -\frac{3}{5} + 1 = \frac{2}{5}$; $\left(\frac{2}{5}, -\frac{3}{5}\right)$

28. $y = 3x - 1$; $8x - 1 = 4(3x - 1)$, $8x - 1 = 12x - 4$, $8x + 3 = 12x$, $3 = 4x$, $\frac{3}{4} = x$, $y = 3\left(\frac{3}{4}\right) - 1$, $y = \frac{9}{4} - 1 = \frac{5}{4}$; $\left(\frac{3}{4}, \frac{5}{4}\right)$

29. $2x = 3y + 19$, $x = \frac{3}{2}y + \frac{19}{2}$; $5y - 2\left(\frac{3}{2}y + \frac{19}{2}\right) = -37$, $5y - 3y - 19 = -37$, $2y - 19 = -37$, $2y = -18$, $y = -9$, $x = \frac{3}{2}(-9) + \frac{19}{2}$, $x = -\frac{27}{2} + \frac{19}{2}$, $x = -\frac{8}{2} = -4$; $(-4, -9)$

30. $3y = -5x - 3$, $y = -\frac{5}{3}x - 1$; $2x - 3\left(-\frac{5}{3}x - 1\right) = -30$, $2x + 5x + 3 = -30$, $7x + 3 = -30$, $7x = -33$, $x = -\frac{33}{7}$, $y = -\frac{5}{3}\left(-\frac{33}{7}\right) - 1$, $y = \frac{55}{7} - 1 = \frac{48}{7}$; $\left(-\frac{33}{7}, \frac{48}{7}\right)$

31. $n = -18m$; $15m + 2(-18m) = -7$, $15m - 36m = -7$, $-21m = -7$, $m = \frac{1}{3}$, $n = -18\left(\frac{1}{3}\right) = -6$; $\left(\frac{1}{3}, -6\right)$

32. $r = 8t$; $-5(8t) + 14t = 13$, $-40t + 14t = 13$, $-26t = 13$, $t = -\dfrac{1}{2}$,

$r = 8\left(-\dfrac{1}{2}\right) = -4$; $\left(-4, -\dfrac{1}{2}\right)$

33. $y = 2x - 1$; $2x - 5(2x - 1) = -1$, $2x - 10x + 5 = -1$, $-8x + 5 = -1$,

$-8x = -6$, $x = \dfrac{3}{4}$, $y = 2\left(\dfrac{3}{4}\right) - 1$, $y = \dfrac{3}{2} - 1 = \dfrac{1}{2}$; $\left(\dfrac{3}{4}, \dfrac{1}{2}\right)$

34. $x = 2y - 2$; $12y - 3(2y - 2) = 11$, $12y - 6y + 6 = 11$, $6y + 6 = 11$,

$6y = 5$, $y = \dfrac{5}{6}$, $x = 2\left(\dfrac{5}{6}\right) - 2$, $x = \dfrac{5}{3} - 2 = -\dfrac{1}{3}$; $\left(-\dfrac{1}{3}, \dfrac{5}{6}\right)$

35. $3u = -4v - 6$, $u = -\dfrac{4}{3}v - 2$; $5v - 6\left(-\dfrac{4}{3}v - 2\right) = -27$, $5v + 8v + 12 = -27$,

$13v = -39$, $v = -3$, $u = -\dfrac{4}{3}(-3) - 2$, $u = 4 - 2 = 2$; $(2, -3)$

36. $2s = 5t + 6$, $s = \dfrac{5}{2}t + 3$; $4\left(\dfrac{5}{2}t + 3\right) + 3t = -1$, $10t + 12 + 3t = -1$,

$13t + 12 = -1$, $13t = -13$, $t = -1$, $s = \dfrac{5}{2}(-1) + 3$, $s = -\dfrac{5}{2} + \dfrac{6}{2} = \dfrac{1}{2}$; $\left(\dfrac{1}{2}, -1\right)$

37. $-7y = -5x - 21$, $y = \dfrac{5}{7}x + 3$; $14\left(\dfrac{5}{7}x + 3\right) - 5x = 22$, $10x + 42 - 5x = 22$,

$5x + 42 = 22$, $5x = -20$, $x = -4$, $y = \dfrac{5}{7}(-4) + 3$, $y = -\dfrac{20}{7} + \dfrac{21}{7} = \dfrac{1}{7}$; $\left(-4, \dfrac{1}{7}\right)$

38. $-2y = -5x + 4$, $y = \dfrac{5}{2}x - 2$; $8\left(\dfrac{5}{2}x - 2\right) + 15x = \dfrac{8}{3}$, $20x - 16 + 15x = \dfrac{8}{3}$,

$35x - 16 = \dfrac{8}{3}$, $105x - 48 = 8$, $105x = 56$, $x = \dfrac{56}{105}$, $x = \dfrac{8}{15}$, $y = \dfrac{5}{2}\left(\dfrac{8}{15}\right) - 2$,

$y = \dfrac{4}{3} - 2 = -\dfrac{2}{3}$; $\left(\dfrac{8}{15}, -\dfrac{2}{3}\right)$

39. $-3a + 3b = -6$, $-3a = -3b - 6$, $a = b + 2$; $4(b + 2) - 2b - 8 = 2b$,

$4b + 8 - 2b - 8 = 2b$, $2b = 2b$, $\{(a, b): a - b = 2\}$

40. $2d + 4c + 7 = -1$, $2d + 4c = -8$, $2d = -4c - 8$, $d = -2c - 4$;

$-7c + 2(-2c - 4) - 12 = 6 - 8c$, $-7c - 4c - 8 - 12 = 6 - 8c$,

$-11c - 20 = 6 - 8c$, $-3c - 20 = 6$, $-3c = 26$, $c = -\dfrac{26}{3}$, $d = -2\left(-\dfrac{26}{3}\right) - 4$,

$d = \dfrac{52}{3} - 4 = \dfrac{40}{3}$; $\left(-\dfrac{26}{3}, \dfrac{40}{3}\right)$

41. $3b - 5 = 6a - 5$, $3b = 6a$, $b = 2a$; $-8(2a) + 2a + 3 = 10 - 6(2a)$,
$-16a + 2a + 3 = 10 - 12a$, $-14a + 3 = 10 - 12a$, $-2a + 3 = 10$,
$-2a = 7$, $a = -\dfrac{7}{2}$, $b = 2\left(-\dfrac{7}{2}\right) = -7$; $\left(-\dfrac{7}{2}, -7\right)$

C 42. $\dfrac{3}{4}x - 1 = -y + \dfrac{1}{2}x$, $\dfrac{1}{4}x - 1 = -y$, $-\dfrac{1}{4}x + 1 = y$; $-\left[\left(-\dfrac{1}{4}x + 1\right) - 3x\right] =$
$2x - \left(-\dfrac{1}{4}x + 1\right)$, $-\left[-3\dfrac{1}{4}x + 1\right] = 2x + \dfrac{1}{4}x - 1$, $x - 1 = -1$, $x = 0$;
$y = -\dfrac{1}{4}(0) + 1$; $(0, 1)$

43. $3y - 6x = 5y - 10x - 8$, $-2y - 6x = -10x - 8$, $-2y = -4x - 8$,
$y = 2x + 4$; $-\dfrac{1}{3}x + \dfrac{5}{3}(2x + 4) = 2x + 4 - x - 4$, $-\dfrac{1}{3}x + \dfrac{5}{3}(2x + 4) = x$,
$3\left[-\dfrac{1}{3}x + \dfrac{5}{3}(2x + 4)\right] = 3(x)$, $-x + 5(2x + 4) = 3x$, $-x + 10x + 20 = 3x$,
$9x + 20 = 3x$, $6x = -20$, $x = -\dfrac{20}{6} = -\dfrac{10}{3}$, $y = 2\left(-\dfrac{10}{3}\right) + \dfrac{12}{3}$,
$y = -\dfrac{20}{3} + \dfrac{12}{3} = -\dfrac{8}{3}$; $\left(-\dfrac{10}{3}, -\dfrac{8}{3}\right)$

44. $r - s = 4$, $r = s + 4$; $\dfrac{1}{3}(s + 4) - \dfrac{1}{3} = \dfrac{3}{2}(s + 4) + \dfrac{s}{2}$, $6\left[\dfrac{1}{3}(s + 4) - \dfrac{1}{3}\right] =$
$6\left[\dfrac{3}{2}(s + 4) + \dfrac{s}{2}\right]$, $2(s + 4) - 2 = 9(s + 4) + 3s$, $2s + 8 - 2 = 9s + 36 + 3s$,
$2s + 6 = 12s + 36$, $-10s = 30$, $s = -3$, $r = -3 + 4 = 1$; $(1, -3)$

45. $y - 3x = -2$, $y = 3x - 2$; $\dfrac{4}{3}x + \dfrac{2}{3} - \dfrac{9}{2}(3x - 2) + \dfrac{3}{2} = -1$,
$6\left(\dfrac{4}{3}x + \dfrac{2}{3} - \dfrac{9}{2}(3x - 2) + \dfrac{3}{2}\right) = 6(-1)$, $8x + 4 - 27(3x - 2) + 9 = -6$,
$8x - 27(3x - 2) + 13 = -6$, $8x - 27(3x - 2) = -19$, $8x - 81x + 54 = -19$,
$-73x + 54 = -19$, $-73x = -73$, $x = 1$, $y = 3(1) - 2 = 1$; $(1, 1)$

46. $j + 4k = -6$, $j = -4k - 6$; $-\dfrac{1}{4}[(-4k - 6) + 2] = \dfrac{1}{2}[(-4k - 6) + 2k]$,
$-\dfrac{1}{4}(-4k - 4) = \dfrac{1}{2}(-2k - 6)$, $4\left[-\dfrac{1}{4}(-4k - 4)\right] = 4\left[\dfrac{1}{2}(-2k - 6)\right]$,
$-(-4k - 4) = 2(-2k - 6)$, $4k + 4 = -4k - 12$, $8k + 4 = -12$,
$8k = -16$, $k = -2$; $j = -4(-2) - 6$, $j = 8 - 6 = 2$; $(2, -2)$

47. $-2y - x = 3$, $x = -2y - 3$; $\dfrac{3}{4}[1 - 5(-2y - 3)] - \dfrac{1}{3}(2y - 1) = 4$,
$\dfrac{3}{4}[10y + 16] - \dfrac{1}{3}(2y - 1) = 4$, $12\left[\dfrac{3}{4}(10y + 16) - \dfrac{1}{3}(2y - 1)\right] = 12(4)$,

$9(10y + 16) - 4(2y - 1) = 48, 90y + 144 - 8y + 4 = 48, 82y + 148 = 48, 82y =$

$-100, y = -\dfrac{100}{82} = -\dfrac{50}{41}; x = -2\left(-\dfrac{50}{41}\right) - 3, x = \dfrac{100}{41} - \dfrac{123}{41} = -\dfrac{23}{41}; \left(-\dfrac{23}{41}, -\dfrac{50}{41}\right)$

48. $\begin{cases} c + d = \dfrac{1}{3} \\ 2c - 3d = \dfrac{1}{4} \end{cases}$ $c = -d + \dfrac{1}{3}; 2\left(-d + \dfrac{1}{3}\right) - 3d = \dfrac{1}{4}, -2d + \dfrac{2}{3} - 3d = \dfrac{1}{4}, -5d + \dfrac{2}{3} = \dfrac{1}{4},$

$\qquad\qquad 12\left(-5d + \dfrac{2}{3}\right) = 12\left(\dfrac{1}{4}\right), -60d + 8 = 3, -60d = -5, d = \dfrac{1}{12};$

$\qquad\qquad c = -\left(\dfrac{1}{12}\right) + \dfrac{1}{3} = \dfrac{1}{4}, c = \dfrac{1}{4} = \dfrac{1}{x}, x = 4, d = \dfrac{1}{12} = \dfrac{1}{y}, y = 12; (4, 12)$

49. $\begin{cases} c - d = -\dfrac{5}{6} \\ 3c + 2d = -\dfrac{5}{6} \end{cases}$ $c = d - \dfrac{5}{6}; 3\left(d - \dfrac{5}{6}\right) + 2d = -\dfrac{5}{6}, 3d - \dfrac{15}{6} + 2d =$

$\qquad\qquad -\dfrac{5}{6}, 5d = \dfrac{10}{6}, 5d = \dfrac{5}{3}, d = \dfrac{1}{3}; c = \dfrac{1}{3} - \dfrac{5}{6} = -\dfrac{1}{2},$

$\qquad\qquad c = -\dfrac{1}{2} = \dfrac{1}{x}, x = -2, d = \dfrac{1}{3} = \dfrac{1}{y}, y = 3; (-2, 3)$

50. $\begin{cases} y = 1.45x + 1972 \\ y = 9.95x \end{cases}$ $9.95x = 1.45x + 1972, 8.5x = 1972, x = 232;$
$\qquad\qquad\qquad\qquad$ They must sell 232 tapes to break even.

51. $2y + 3 = -11$ or $2y + 3 = 11$
$\qquad y = -7$ or $\qquad y = 4$
$\qquad \{-7, 4\}$

52. $a - 4 \geq -2$ and $a - 4 \leq 2$
$\qquad a \geq 2$ and $a \leq 6$
$\qquad \{a: 2 \leq a \leq 6\}$

53. $4b - 2 < -10$ or $4b - 2 > 10$
$\qquad b < -2$ or $b > 3$
$\qquad \{b: b < -2$ or $b > 3\}$

54. $|3 - c| \leq 1$
$\qquad 3 - c \geq -1$ and $3 - c \leq 1$
$\qquad c \leq 4$ and $c \geq 2$
$\qquad \{c: 2 \leq c \leq 4\}$

55. $\dfrac{6\,(a - 3)}{(a + 3)(a - 3)} = \dfrac{6}{a + 3}$

56. $\dfrac{(2x - 5)(x + 3)}{2(x + 3)} = \dfrac{2x - 5}{2}$

57. $\dfrac{m^2\left(\dfrac{4}{m}\right) + m^2(3)}{m^2\left(\dfrac{1}{m^2}\right)} = 4m + 3m^2$

58.

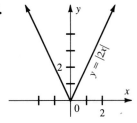

function; not linear

59.

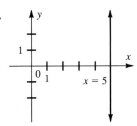

not a function

60.

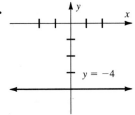

function; linear

61. Answers may vary. Graphically, (3, 4) represents the point at which the graphs of the two equations intersect. Algebraically, these are the only values of x and y ($x = 3$, $y = 4$) that satisfy both equations. To check the solution, substitute these values into both equations.

page 456   Class Exercises

1. Let x = a number and y = a second number.
$$\begin{cases} x = 3y \\ x - y = 10 \end{cases}$$

2. Let g = the number of girls and b = the number of boys.
$$\begin{cases} g + b = 26 \\ g = 2b - 4 \end{cases}$$

3. Let s = the length of the equal sides and b = the length of the base.
$$\begin{cases} 2s + b = 48 \\ b = \frac{1}{2}s - 1 \end{cases}$$

pages 457–458   Practice Exercises

1. Let l = the length and w = the width of the rectangle.
$$\begin{cases} 2l + 2w = 68 \\ l = 3w - 2 \end{cases}$$
$2(3w - 2) + 2w = 68$, $6w - 4 + 2w = 68$, $8w - 4 = 68$, $8w = 72$, $w = 9$
$l = 3(9) - 2 = 25$; $l = 25$ ft and $w = 9$ ft

2. Let l = the length and w = the width of the rectangle.
$$\begin{cases} 2l + 2w = 10 \\ 2w = \frac{1}{2}l \end{cases}$$
$w = \frac{1}{4}l$; $2l + 2\left(\frac{1}{4}l\right) = 10$, $2l + \frac{1}{2}l = 10$,
$\frac{5}{2}l = 10$, $l = 4$, $w = \frac{1}{4}(4) = 1$;
$l = 4$ m and $w = 1$ m

3. Let x = the smaller number and y = the larger number.
$$\begin{cases} x + y = 48 \\ y - x = 12 \end{cases}$$
$y = 12 + x$; $x + 12 + x = 48$,
$2x + 12 = 48$, $2x = 36$, $x = 18$;
$y = 12 + 18 = 30$; The numbers are 18 and 30.

4. Let x = one number and y = the second number.
$$\begin{cases} x + y = -64 \\ 2x = 1 + y \end{cases}$$
$2x - 1 = y$; $x + 2x - 1 = -64$,
$3x - 1 = -64$, $3x = -63$,
$x = -21$; $y = 2(-21) - 1 = -43$;
The numbers are -21 and -43.

5. Let s = the number of coins Said has and f = the number of coins his father has.

$$\begin{cases} f = 13 + 5s \\ s + f = 247 \end{cases}$$

$s + 13 + 5s = 247$, $6s + 13 = 247$, $6s = 234$, $s = 39$, $f = 13 + 5(39) = 208$; Said has 39 coins and his father has 208 coins.

6. Let b = the no. of tickets Bella sold and i = the no. of tickets Irina sold.

$$\begin{cases} i + b = 137 \\ i = 2b - 10 \end{cases}$$

$2b - 10 + b = 137$, $3b - 10 = 137$, $3b = 147$, $b = 49$, $i = 2(49) - 10 = 88$; Bella sold 49 tickets and Irina sold 88 tickets.

7. Let n = the number of students in one group and m = number of students in the second group.

$$\begin{cases} n + m = 27 \\ n = 2m + 3 \end{cases}$$

$2m + 3 + m = 27$, $3m + 3 = 27$, $3m = 24$, $m = 8$; $n = 2(8) + 3$, $n = 16 + 3 = 19$; One group had 19 students and the second group had 8 students.

8. Let x = the length of one piece of rope and y = the length of the second piece of rope.

$$\begin{cases} x + y = 50 \\ x = 9y \end{cases}$$

$9y + y = 50$, $10y = 50$, $y = 5$, $x = 9(5) = 45$; The length of the longer piece of rope is 45 ft.

9. Let x = a number and y = the other number.

$$\begin{cases} y - x = 5 \\ x = \dfrac{2}{3} \end{cases}$$

$y - \dfrac{2}{3}y = 5$, $\dfrac{1}{3}y = 5$, $y = 15$; $15 - x = 5$, $x = 10$; The two numbers are 10 and 15.

10. Let x = a number and y = the other number.

$$\begin{cases} y - x = 2 \\ y = 4 + \dfrac{1}{2}x \end{cases}$$

$4 + \dfrac{1}{2}x - x = 2$, $4 - \dfrac{1}{2}x = 2$, $-\dfrac{1}{2}x = -2$, $x = 4$; $y - 4 = 2$; $y = 6$

or

$$\begin{cases} y - x = 2 \\ x = 4 + \dfrac{1}{2}y \end{cases}$$

$y - \left(4 + \dfrac{1}{2}y\right) = 2$, $y - 4 - \dfrac{1}{2}y = 2$, $\dfrac{1}{2}y = 6$, $y = 12$; $12 - x = 2$; $x = 10$

The two numbers of 6 and 4 or 12 and 10.

B 11. Let x = the amount invested at 7.5% and y = the amount invested at 8.0%.

$$\begin{cases} 0.075x + 0.08y = 469.75 \\ x + y = 6000 \end{cases}$$

$x = 6000 - y$; $0.075(6000 - y) + 0.08y = 469.75$, $450 - 0.075y + 0.08y =$ 469.75, $0.005y = 19.75$, $y = 3950$; $x = 6000 - 3950 = 2025$;

The amount invested at 7.5% was $2050 and the amount invested at 8.0% was $3950.

12. Let t = the federal tax amount and s = the state tax amount.

$$\begin{cases} t + s = 2700 \\ t = 8s \end{cases}$$

$8s + s = 2700$, $9s = 2700$, $s = 300$; The amount of state tax was $300.

13. Let l = Leila's time in hours and j = Josh's time in hours.

$$\begin{cases} l = j + 1 \\ 36j = 24l \end{cases}$$

$36j = 24(j + 1)$, $36j = 24j + 24$, $12j = 24$, $j = 2$; It took Josh 2 h to overtake Leila.

14. Let g = the rate of Henri going and r = the rate of Henri returning.

$$\begin{cases} r = 1 + g \\ 45r = 50g \end{cases}$$

$45(1 + g) = 50g$, $45 + 45g = 50g$, $45 = 5g$, $9 = g$; $r = 1 + 9 = 10$;

The rate of his return trip was 10 mi/h.

C 15. Let l = length of a side and w = length of the other side of a rectangle.

$$\begin{cases} 2l = w - 14 \\ 4(2l) = 84 \end{cases}$$

$8l = 84$, $l = 10.5$; $2(10.5) = w - 14$, $21 = w - 14$, $35 = w$;

The dimensions of the rectangle are 35 m and 10.5 m.

16. Let s = the length of the side of a square.

$2(3s) + 2(s + 7) = 96$, $6s + 2s + 14 = 96$, $8s + 14 = 96$, $8s = 82$, $s = 10.25$;

The dimensions of the original square are 10.25 cm × 10.25 cm.

17. Let a = number of blocks Alice is from the sports center and j = number of blocks Jeanine is from the sports center.

$$\begin{cases} a + j = 18 \\ a = 3 + \dfrac{1}{4}j \end{cases}$$

$3 + \dfrac{1}{4}j + j = 18$, $3 + \dfrac{5}{4}j = 18$, $\dfrac{5}{4}j = 15$, $j = 12$; $a = 3 + \dfrac{1}{4}(12) = 6$;

Alice lives 6 blocks and Jeanine lives 12 blocks from the sports center.

1. Let x = the new percent of salt solution.

	Substance	Total No. L	Total No. L pure salt
Start with	10% salt solution	10	(0.10)(10)
Add	water	2.5	0
Finish with	x% salt solution	12.5	x(12.5)

$x(12.5) = 0.10(10)$, $x(12.5) = 1$, $x = \dfrac{1}{12.5}$, $x = 0.08$; The new solution will

have an 8% concentration of salt.

2.

rate $\times$ time = distance

	rate	time	distance
going	r	$2\frac{1}{2}$	$2\frac{1}{2}(r)$
returning	$7 + r$	2	$2(7 + r)$

$\dfrac{5}{2}r = 2(7 + r)$, $\dfrac{5}{2}r = 14 + 2r$, $\dfrac{1}{2}r = 14$, $r = 28$

The rate going was 28 mi/h, and returning was 35 mi/h.

3. Let x = the number of hours Ian works at his second job.
   $20(4.50) + (x)(6.25) \geq 160$, $90 + 6.25x \geq 160$, $6.25x \geq 70$, $x \geq 11.2$;
   Ian must work at least 11.2 h.

1. Single variable: Let n represent the numerator, then $\dfrac{n}{n + 12}$ is the fraction.

   $\dfrac{n - 5}{n + 12 - 5} = \dfrac{2}{3}, \dfrac{n - 5}{n + 7} = \dfrac{2}{3}$, $3(n - 5) = 2(n + 7)$, $3n - 15 = 2n + 14$,

   $n = 29$; The fraction is $\dfrac{29}{41}$.

   Two variables: Let n represent the numerator and d represent the denominator.

   $d = n + 12, \dfrac{n - 5}{d - 5} = \dfrac{2}{3}, \dfrac{n - 5}{n + 12 - 5} = \dfrac{2}{3}, \dfrac{n - 5}{n + 7} = \dfrac{2}{3}$, $3n - 15 = 2n + 14$, $n = 29$;

   $d = 29 + 12 = 41$; The fraction is $\dfrac{29}{41}$.

2. Single variable: Let x = a number, then $x + 0.3x$ = the second number.
   $x + (x + 0.3x) = 161$, $2x + 0.3x = 161$, $2.3x = 161$, $x = 70$; The
   numbers are 70 and 91.
   Two variables: Let x = a number and y = the second number.
   $\begin{cases} x + 0.3x = y \\ x + y = 161 \end{cases}$
   $y = 161 - x$, $x + 0.3x = 161 - x$, $x + x + 0.3x = 161$, $2.3x = 161$,
   $x = 70$; $y = 161 - 70 = 91$; The numbers are 70 and 91.

3. Single variable: this problem is not possible to solve using a single variable, because the problem does not provide enough information. Two variables: Let x = number of days worked by one girl and y = the number of days worked by the other girl.

$$\begin{cases} \dfrac{4}{x} + \dfrac{4}{y} = 1 \\[2mm] \dfrac{8}{x} + \dfrac{2}{y} = 1 \end{cases}$$

$\dfrac{4}{x} + \dfrac{4}{y} = \dfrac{8}{x} + \dfrac{2}{y}, \dfrac{4}{y} - \dfrac{2}{y} = \dfrac{8}{x} - \dfrac{4}{x}, \dfrac{2}{y} = \dfrac{4}{x}, \dfrac{8}{x} + \dfrac{4}{x} = 1, 8 + 4 = x, x = 12;$

$\dfrac{4}{12} + \dfrac{4}{y} = 1, \dfrac{4}{y} = 1 - \dfrac{1}{3}, \dfrac{4}{y} = \dfrac{2}{3}, 4 = \dfrac{2}{3}y, y = 6$

It would take one girl 12 days and the other 6 days working alone.

page 459 Capsule Review

1. $5x + 9y + 4 = 0,\ 5x + 9y = -4$ **2.** $y + 3x = -5,\ 3x + y = -5$

3. $14 - x + 0y = 0,\ -x + 0y = -14$ or $x - 0y = 14$

pages 460–461 Class Exercises

1. addition: $x + y = 4$
$$\underline{x - y = -10}$$
$$2x = -6$$
$$x = -3$$
$-3 + y = 4, y = 7; (-3, 7)$

2. subtraction: $x + y = 4$
$$\underline{2x + y = 5}$$
$$-x = -1$$
$$x = 1$$
$1 + y = 4, y = 3; (1, 3)$

3. subtraction: $-7a + b = 1$
$$\underline{-7a - 3b = -3}$$
$$4b = 4$$
$$b = 1$$
$-7a + 1 = 1, -7a = 0, a = 0; (0, 1)$

4. subtraction: $2p - q = 1$
$$\underline{2p - q = -3}$$
$$0 \neq 4$$
no solution

5. subtraction: $3m - 4n = 1$
$$\underline{3m - 2n = -1}$$
$$-2n = 2$$
$$n = -1$$
$3m - 4(-1) = 1, 3m + 4 = 1,$
$3m = -3, m = -1; (-1, -1)$

6. addition: $9s + 4t = -13$
$$\underline{-9s - \dfrac{1}{2}t = -1}$$
$$\dfrac{7}{2}t = -14$$
$$t = -4$$

$9s + 4(-4) = -13, 9s - 16 = -13,$
$9s = 3, s = \dfrac{1}{3}; \left(\dfrac{1}{3}, -4\right)$

7. $x = 4 - y$; $4 - y - y = -10$, $4 - 2y = -10$, $-2y = -14$, $y = 7$; $x = 4 - 7$, $x = -3$; $(-3, 7)$

8. The addition method can be used to solve for y, and the subtraction method can be used to solve for x.

9. There is no ordered pair which satisfies the system.

pages 461–462   Practice Exercises

A 1. $x - y = -8$
$\underline{x + y = 12}$
$2x = 4$
$x = 2$
$2 + y = 12$
$y = 10$; $(2, 10)$

2. $x + y = 10$
$\underline{x - y = -2}$
$2x = 8$
$x = 4$
$4 + y = 10$
$y = 6$; $(4, 6)$

3. $a + b = 0$
$\underline{a - b = -6}$
$2a = -6$
$a = -3$
$-3 + b = 0$
$b = 3$; $(-3, 3)$

4. $a - b = -14$
$\underline{a + b = -4}$
$2a = -18$
$a = -9$
$-9 + b = -4$
$b = 5$; $(-9, 5)$

5. $p + q = -2$
$\underline{-p + q = 10}$
$2q = 8$
$q = 4$
$p + 4 = -2$
$p = -6$; $(-6, 4)$

6. $-p + q = 16$
$\underline{-p - q = 0}$
$-2p = 16$
$p = -8$
$-(-8) + q = 16$
$q = 8$; $(-8, 8)$

7. $3a - b = 21$
$\underline{2a + b = 4}$
$5a = 25$
$a = 5$
$2(5) + b = 4$
$10 + b = 4$
$b = -6$; $(5, -6)$

8. $7s + t = 22$
$\underline{5s - t = 14}$
$12s = 36$
$s = 3$
$7(3) + t = 22$
$21 + t = 22$
$t = 1$; $(3, 1)$

9. $m - 4n = 6$
$\underline{m - 2n = 18}$
$-2n = -12$
$n = 6$
$m - 4(6) = 6$
$m - 24 = 6$
$m = 30$; $(30, 6)$

10. $7g + h = 42$
$\underline{-3g + h = -8}$
$10g = 50$
$g = 5$
$7(5) + h = 42$
$35 + h = 42$
$h = 7$; $(5, 7)$

11. $3j - k = -10$
$\underline{-5j - k = 14}$
$8j = -24$
$j = -3$
$3(-3) - k = -10$
$-9 - k = -10$
$-k = -1$
$k = 1$; $(-3, 1)$

12. $-v + 5w = 12$
$\underline{-v - 3w = -4}$
$8w = 16$
$w = 2$
$-v + 5(2) = 12$
$-v + 10 = 12$
$-v = 2$
$v = -2$; $(-2, 2)$

13.
$$2x + 3y = -5$$
$$\underline{7x - 3y = 23}$$
$$9x = 18$$
$$x = 2$$
$$2(2) + 3y = -5$$
$$4 + 3y = -5$$
$$3y = -9$$
$$y = -3; (2, -3)$$

14.
$$-2m - n = 5$$
$$\underline{-2m - 3n = -7}$$
$$2n = 12$$
$$n = 6$$
$$-2m - 6 = 5$$
$$-2m = 11$$
$$m = -\frac{11}{2}; \left(-\frac{11}{2}, 6\right)$$

15.
$$2a - b = 1$$
$$\underline{-2a + b = -1}$$
$$0 = 0$$
$$\{(a,b): 2a - b = 1\}$$

16.
$$3x + 8y = -7$$
$$\underline{-3x - 8y = 7}$$
$$0 = 0$$
$$\{(x,y): 3x + 8y = -7\}$$

17.
$$3x - y = 8$$
$$\underline{-3x + \frac{1}{3}y = \frac{2}{3}}$$
$$-\frac{2}{3}y = 8\frac{2}{3}$$
$$-\frac{3}{2}\left(-\frac{2}{3}y\right) = -\frac{3}{2}\left(\frac{26}{3}\right)$$
$$y = -13$$
$$3x - (-13) = 8$$
$$3x + 13 = 8$$
$$3x = -5$$
$$x = -\frac{5}{3}; \left(-\frac{5}{3}, -13\right)$$

18.
$$2p - 3q = 11$$
$$\underline{-2p + \frac{1}{3}q = 1}$$
$$-2\frac{2}{3}q = 12$$
$$-\frac{3}{8}\left(-\frac{8}{3}q\right) = -\frac{3}{8}(12)$$
$$q = -\frac{9}{2}$$
$$2p - 3\left(-\frac{9}{2}\right) = 11$$
$$2p + \frac{27}{2} = 11$$
$$2p = -\frac{5}{2}$$
$$p = -\frac{5}{4}; \left(-\frac{5}{4}, -\frac{9}{2}\right)$$

19.
$$2c + 5d = 44$$
$$\underline{-6c + 5d = 8}$$
$$8c = 36$$
$$c = \frac{36}{8}$$
$$c = \frac{9}{2}$$
$$2\left(\frac{9}{2}\right) + 5d = 44$$
$$9 + 5d = 44$$
$$5d = 35$$
$$d = 7; \left(\frac{9}{2}, 7\right)$$

20.
$$4a - 7b = 3$$
$$\underline{16a - 7b = 12}$$
$$-12a = -9$$
$$a = \frac{9}{12}$$
$$a = \frac{3}{4}$$
$$4\left(\frac{3}{4}\right) - 7b = 3$$
$$3 - 7b = 3$$
$$-7b = 0$$
$$b = 0; \left(\frac{3}{4}, 0\right)$$

21.
$$6a - 5b = 6$$
$$\underline{7a - 5b = 7}$$
$$-a = -1$$
$$a = 1$$
$$6(1) - 5b = 6$$
$$6 - 5b = 6$$
$$-5b = 0$$
$$b = 0; (1, 0)$$

B 22.
$$4g - 2h = -14$$
$$\underline{-4g + 5h = 32}$$
$$3h = 18$$
$$h = 6$$
$$4g - 2(6) = -14$$
$$4g - 12 = -14$$
$$4g = -2$$
$$g = -\frac{1}{2}; \left(-\frac{1}{2}, 6\right)$$

23.
$$2m - 5n = 17$$
$$\underline{6m - 5n = 1}$$
$$-4m = 16$$
$$m = -4$$
$$2(-4) - 5n = 17$$
$$-8 - 5n = 17$$
$$-5n = 25$$
$$n = -5; (-4, -5)$$

24.
$$-7a + 4b = 13$$
$$\underline{-7a + 2b = 3}$$
$$2b = 10$$
$$b = 5$$
$$-7a + 4(5) = 13$$
$$-7a + 20 = 13$$
$$-7a = -7$$
$$a = 1; (1, 5)$$

25.
$$-7c + 2d = 31$$
$$\underline{-17c - 2d = 17}$$
$$-24c = 48$$
$$c = -2$$
$$-7(-2) + 2d = 31$$
$$14 + 2d = 31$$
$$2d = 17$$
$$d = \frac{17}{2}; \left(-2, \frac{17}{2}\right)$$

26.
$$3x + y = 1$$
$$\underline{5x + y = 2}$$
$$-2x = -1$$
$$x = \frac{1}{2}$$
$$3\left(\frac{1}{2}\right) + y = 1$$
$$\frac{3}{2} + y = 1$$
$$y = -\frac{1}{2}; \left(\frac{1}{2}, -\frac{1}{2}\right)$$

27.
$$2x - y = 1$$
$$\underline{-2x + 5y = 1}$$
$$4y = 2$$
$$y = \frac{1}{2}$$
$$2x - \frac{1}{2} = 1$$
$$2x = \frac{3}{2}$$
$$x = \frac{3}{4}; \left(\frac{3}{4}, \frac{1}{2}\right)$$

28.
$$4c + 3d = 3$$
$$\underline{c + 3d = 1}$$
$$3c = 2$$
$$c = \frac{2}{3}$$
$$4\left(\frac{2}{3}\right) + 3d = 3$$
$$\frac{8}{3} + 3d = 3$$
$$3d = \frac{1}{3}$$
$$d = \frac{1}{9}; \left(\frac{2}{3}, \frac{1}{9}\right)$$

29.
$$4e - 2f = 5$$
$$\underline{4e + 6f = 9}$$
$$-8f = -4$$
$$f = \frac{1}{2}$$
$$4e - 2\left(\frac{1}{2}\right) = 5$$
$$4e - 1 = 5$$
$$4e = 6$$
$$e = \frac{3}{2}; \left(\frac{3}{2}, \frac{1}{2}\right)$$

30.
$$0.4a - 0.2b = -1.4$$
$$\underline{0.4a - 0.5b = -3.2}$$
$$0.3b = 1.8$$
$$b = 6$$
$$0.4a - 0.2(6) = -1.4$$
$$0.4a - 1.2 = -1.4$$
$$0.4a = -0.2$$
$$a = -0.5; (-0.5, 6)$$

31.

$$-0.7p + 0.2q = 3.1$$
$$\underline{-1.7p - 0.2q = 1.7}$$
$$-2.4p = 4.8$$
$$p = -2$$
$$-0.7(-2) + 0.2q = 3.1$$
$$1.4 + 0.2q = 3.1$$
$$0.2q = 1.7$$
$$q = 8.5; \; (-2, 8.5)$$

32.

$$2v + 3w = 6$$
$$\underline{2v - 27w = 18}$$
$$30w = -12$$
$$w = -\frac{2}{5}$$
$$2v + 3\left(-\frac{2}{5}\right) = 6$$
$$2v - \frac{6}{5} = 6$$
$$2v = \frac{36}{5}$$
$$v = \frac{18}{5}; \; \left(\frac{18}{5}, -\frac{2}{5}\right)$$

33.

$$12j + 5k = 76$$
$$\underline{4j + 5k = 52}$$
$$8j = 24$$
$$j = 3$$
$$12(3) + 5k = 76$$
$$36 + 5k = 76$$
$$5k = 40$$
$$k = 8; \; (3, 8)$$

34.

$$0.12x - 1.2y = 3.024$$
$$\underline{1.34x - 1.2y = 6.928}$$
$$-1.22x = -3.904$$
$$x = 3.2$$
$$0.12(3.2) - 1.2y = 3.024$$
$$0.384 - 1.2y = 3.024$$
$$-1.2y = 2.64$$
$$y = -2.2; \; (3.2, -2.2)$$

35.

$$2.3x - 0.45y = 7.99$$
$$\underline{-2.3x + 1.6y = -4.31}$$
$$1.15y = 3.68$$
$$y = 3.2$$
$$2.3x - 0.45(3.2) = 7.99$$
$$2.3x - 1.44 = 7.99$$
$$2.3x = 9.43$$
$$x = 4.1; \; (4.1, 3.2)$$

36.

$$\frac{3}{2}x + \frac{5}{4}y = \frac{18}{13}$$
$$\underline{-\frac{3}{2}x + \frac{3}{4}y = \frac{8}{13}}$$
$$\frac{8}{4}y = \frac{26}{13}$$
$$2y = 2$$
$$y = 1$$
$$\frac{3}{2}x + \frac{5}{4}(1) = \frac{18}{13}$$
$$\frac{3}{2}x + \frac{5}{4} = \frac{18}{13}$$
$$\frac{3}{2}x = \frac{7}{52}$$
$$x = \frac{7}{78}; \; \left(\frac{7}{78}, 1\right)$$

37.

$$6r + \frac{3}{7}s = \frac{19}{3}$$
$$\underline{r + \frac{3}{7}s = \frac{4}{3}}$$
$$5r = \frac{15}{3}$$
$$5r = 5$$
$$r = 1$$
$$6(1) + \frac{3}{7}s = \frac{19}{3}$$
$$6 + \frac{3}{7}s = \frac{19}{3}$$
$$\frac{3}{7}s = \frac{1}{3}$$
$$s = \frac{7}{9}; \; \left(1, \frac{7}{9}\right)$$

C 38.

$$\frac{m + n}{2} - \frac{3n - 2}{2} = 0$$

$$\frac{-\left(\dfrac{m + n}{2}\right) - \dfrac{3n + 8}{2} = 0}{-\left(\dfrac{3n - 2}{2}\right) - \left(\dfrac{3n + 8}{2}\right) = 0}$$

$$\frac{-3n + 2 - 3n - 8}{2} = 0$$

$$\frac{-6n - 6}{2} = 0$$

$$-3n - 3 = 0$$

$$-3n = 3$$

$$n = -1$$

$$\frac{m + (-1)}{2} = \frac{3(-1) - 2}{2}$$

$$\frac{m - 1}{2} = \frac{-5}{2}$$

$$2m - 2 = -10$$

$$2m = -8$$

$$m = -4; \ (-4, -1)$$

39.

$$\frac{f + 3}{2} + g = 2$$

$$\frac{-\left(\dfrac{f + 3}{2}\right) + g = -5}{2g = -3}$$

$$g = -\frac{3}{2}$$

$$\frac{f + 3}{2} - \frac{3}{2} = 2$$

$$\frac{f + 3}{2} = \frac{7}{2}$$

$$f + 3 = 7$$

$$f = 4; \ \left(4, -\frac{3}{2}\right)$$

40.

$$\frac{p + 2}{3} - q = 3$$

$$\frac{\dfrac{p + 2}{3} - 4q = 0}{3q = 3}$$

$$q = 1$$

$$\frac{p + 2}{3} - 1 = 3$$

$$\frac{p + 2}{3} = 4$$

$$p + 2 = 12$$

$$p = 10; \ (10, 1)$$

41.

$$\frac{2j + 3}{9} - k = 6$$

$$\frac{\dfrac{2j + 3}{9} + 3k = 0}{-4k = 6}$$

$$k = -\frac{3}{2}$$

$$\frac{2j + 3}{9} - \left(-\frac{3}{2}\right) = 6$$

$$\frac{2j + 3}{9} + \frac{3}{2} = 6$$

$$\frac{2j + 3}{9} = \frac{9}{2}$$

$$4j + 6 = 81$$

$$4j = 75$$

$$j = \frac{75}{4}; \ \left(\frac{75}{4}, -\frac{3}{2}\right)$$

42.
$$6c + 4d = 16$$
$$\underline{3c - 4d = 2}$$
$$9c = 18$$
$$c = 2;$$
$$6(2) + 4d = 16$$
$$12 + 4d = 16$$
$$4d = 4$$
$$d = 1$$
$$c = 2 = \frac{1}{x}, x = \frac{1}{2};$$
$$d = 1 = \frac{1}{y}, y = 1; \left(\frac{1}{2}, 1\right)$$

43.
$$8c - 2d = -2$$
$$\underline{8c + 7d = 25}$$
$$-9d = -27$$
$$d = 3;$$
$$8c - 2(3) = -2$$
$$8c - 6 = -2$$
$$8c = 4$$
$$c = \frac{1}{2}$$
$$c = \frac{1}{2} = \frac{1}{x}, x = 2;$$
$$d = 3 = \frac{1}{y}, y = \frac{1}{3}; \left(2, \frac{1}{3}\right)$$

44.
$$c + 3d = 1$$
$$\underline{-4c + 3d = -3}$$
$$5c = 4$$
$$c = \frac{4}{5};$$
$$\frac{4}{5} + 3d = 1$$
$$3d = \frac{1}{5}$$
$$d = \frac{1}{15}$$
$$c = \frac{4}{5} = \frac{1}{x}, x = \frac{5}{4};$$
$$d = \frac{1}{15} = \frac{1}{y}, y = 15; \left(\frac{5}{4}, 15\right)$$

45.
$$-5c + 2d = 3$$
$$\underline{5c - 3d = -4}$$
$$-d = -1$$
$$d = 1;$$
$$-5c + 2(1) = 3$$
$$-5c + 2 = 3$$
$$-5c = 1$$
$$c = -\frac{1}{5}$$
$$c = -\frac{1}{5} = \frac{1}{x}, x = -5;$$
$$d = 1 = \frac{1}{y}, y = 1; (-5, 1)$$

46. Let c = the number of acres of corn, and s = the number of acres of soybeans.
$$\begin{cases} s + c = 240 \\ s = 80 + c \end{cases}$$

$$s + c = 240$$
$$\underline{s - c = 80}$$
$$2s = 320$$
$$s = 160;$$
$$160 + c = 240$$
$$c = 80$$

The Allens will grow 160 acres of soybeans and 80 acres of corn.

47. Let x = number of volts in one battery, and y = number of volts in the other battery.

$$\begin{cases} x + y = 4.5 \\ x - y = 1.5 \end{cases}$$

$$\begin{aligned} x + y &= 4.5 \\ \underline{x - y} &= \underline{1.5} \\ 2x &= 6.0 \\ x &= 3.0 \\ 3.0 + y &= 4.5 \\ y &= 1.5 \end{aligned}$$

The voltages of the two batteries are 3 v and 1.5 v.

page 463   Capsule Review

1. $3(2m + 5n) = 3(1)$, $6m + 15n = 3$ **2.** $-1(5p - q) = -1(2)$, $-5p + q = -2$

3. $-2(-3s + 6t) = -2(-8)$, $6s - 12t = 16$

4. $-5(-4x - 7y) = -5(-2)$, $20x + 35y = 10$

1. (a) $a + 2b = 4$ → $a + 2b = 4$ (b) $3(a + 2b) = 3(4)$ → $3a + 6b = 12$
   $-(a + 3b) = -(-2)$ → $-a - 3b = 2$ $-2(a + 3b) = -2(2)$ → $-2a - 6b = -4$

   $a + 2b = 4$
   $\underline{-a - 3b = 2}$
   $-b = 6$
   $b = -6$
   $a + 2(-6) = 4, a - 12 = 4, a = 16; (16, -6)$

2. (a) $2x - y = 4$ → $2x - y = 4$ (b) $3(2x - y) = 3(4)$ → $6x - 3y = 12$
   $-2(x + 3y) = -2(16)$ → $-2x - 6y = -32$ $x + 3y = 16$ → $x + 3y = 16$

   $2x - y = 4$
   $\underline{-2x - 6y = -32}$
   $-7y = -28$
   $y = 4$
   $2x - 4 = 4, 2x = 8, x = 4, (4, 4)$

3. (a) $3(2a - 3b) = 3(-4) →$ $6a - 9b = -12$ (b) $6a - 9b = -12$
   $-6a + 9b = -15$ $→ -6a + 9b = -15$ $-6a - 9b = -15$

   $6a - 9b = -12$
   $\underline{-6a + 9b = -15}$
   $0 \neq -27;$ no solution

4. (a) $-2(3s - 7t) = -2(13) → -6s + 14t = -26$ (b) $5(3s - 7t) = 5(13) → 15s - 35t = 65$
   $6s + 5t = 7$ → $6s + 5t = 7$ $7(6s + 5t) = 7(7)$ → $42s + 35t = 49$

   $-6s + 14t = -26$
   $\underline{6s + 5t = 7}$
   $19t = -19$
   $t = -1$
   $3s - 7(-1) = 13, 3s + 7 = 13, 3s = 6, s = 2; (2, -1)$

5. (a) $3(8x + 3y) = 3(13)$ → $24x + 9y = 39$ (b) $2(8x + 3y) = 2(13) → 16x + 6y = 26$
   $-8(3x + 2y) = -8(11) → -24x - 16y = -88$ $3(3x + 2y) = 3(11) →   9x + 6y = 33$

   $16x + 6y = 26$
   $\underline{9x + 6y = 33}$
   $7x = -7$
   $x = -1$
   $3(-1) + 2y = 11, -3 + 2y = 11, 2y = 14, y = 7; (-1, 7)$

6. (a) $3(4c - 15d) = 3(-13) \longrightarrow 12c - 45d = -39$

$\qquad 2(6c + 10d) = 2(13) \longrightarrow 12c + 20d = 26$

(b) $2(4c - 15d) = 2(-13) \longrightarrow 8c - 30d = -26$

$\qquad 3(6c + 10d) = 3(13) \longrightarrow 18c + 30d = 39$

$12c - 45d = -39$

$\underline{12c + 20d = 26}$

$\qquad -65d = -65$

$\qquad d = 1$

$4c - 15(1) = -13, \ 4c - 15 = -13, \ 4c = 2, \ c = \dfrac{1}{2}; \ \left(\dfrac{1}{2}, 1\right)$

7. Multiply the first equation by 5, and the second equation by 7. Subtract the second equation from the first. Solve for y. Substitute the value of y back into any equation and solve for x.

pages 465–466 Practice Exercises

A **1.** $3a - 4b = 1 \longrightarrow 3a - 4b = 1$

$\quad -4(12a - b) = -4(-11) \longrightarrow -48a + 4b = 44$

$\qquad 3a - 4b = 1$

$\qquad \underline{-48a + 4b = 44}$

$\qquad\quad -45a = 45$

$\qquad\qquad a = -1$

$\quad 3(-1) - 4b = 1, \ -3 - 4b = 1, \ -4b = 4, \ b = -1; \ (-1, -1)$

2. $-2(-5c + 3d) = -2(-16) \longrightarrow 10c - 6d = 32$

$\quad -10c + d = -22 \qquad\qquad \longrightarrow -10c + d = -22$

$\qquad 10c - 6d = 32$

$\qquad \underline{-10c + d = -22}$

$\qquad\quad -5d = 10$

$\qquad\qquad d = -2$

$\quad -5c + 3(-2) = -16, \ -5c - 6 = -16, \ -5c = -10, \ c = 2; \ (2, -2)$

3. $8u - 4v = 16 \longrightarrow 8u - 4v = 16$

$\quad -2(4u + 5v) = -2(22) \longrightarrow -8u - 10v = -44$

$\qquad 8u - 4v = 16$

$\qquad \underline{-8u - 10v = -44}$

$\qquad\quad -14v = -28$

$\qquad\qquad v = 2$

$\quad 4u + 5(2) = 22, \ 4u + 10 = 22, \ 4u = 12, \ u = 3; \ (3, 2)$

4. $2(-5m + 3n) = 2(-31) \longrightarrow -10m + 6n = -62$

$5(-2m - 5n) = 5(0) \longrightarrow -10m - 25n = 0$

$-10m + 6n = -62$

$\underline{-10m - 25n = 0}$

$31n = -62$

$n = -2$

$-2m - 5(-2) = 0, -2m + 10 = 0, -2m = -10, m = 5; (5, -2)$

5. $3(-3e + 4f) = 3(-6) \longrightarrow -9e + 12f = -18$

$2(5e - 6f) = 2(8) \longrightarrow 10e - 12f = 16$

$-9e + 12f = -18$

$\underline{10e - 12f = 16}$

$e = -2$

$-3(-2) + 4f = -6, 6 + 4f = -6, 4f = -12, f = -3; (-2, -3)$

6. $3(5k + 3l) = 3(17) \longrightarrow 15k + 9l = 51$

$2k - 9l = 17 \longrightarrow 2k - 9l = 17$

$15k + 9l = 51$

$\underline{2k - 9l = 17}$

$17k = 68$

$k = 4$

$2(4) - 9l = 17, 8 - 9l = 17, -9l = 9, l = -1; (4, -1)$

7. $8r - 5s = -11 \longrightarrow 8r - 5s = -11 \longrightarrow 8r - 5s = -11$

$-4r + 3s = -11 \longrightarrow 2(-4r + 3s) = 2(-11) \longrightarrow -8r + 6s = -22$

$8r - 5s = -11$

$\underline{-8r + 6s = -22}$

$s = -33$

$8r - 5(-33) = -11, 8r + 165 = -11, 8r = -176, r = -22; (-22, -33)$

8. $9p + 4q = -17 \longrightarrow 9p + 4q = -17 \longrightarrow 9p + 4q = -17$

$3p + 12q = -3 \longrightarrow 3(3p + 12q) = 3(-3) \longrightarrow 9p + 36q = -9$

$9p + 4q = -17$

$\underline{9p + 36q = -9}$

$-32q = -8$

$q = \dfrac{1}{4}$

$9p + 4\left(\dfrac{1}{4}\right) = -17, 9p + 1 = -17, 9p = -18, p = -2; \left(-2, \dfrac{1}{4}\right)$

9. $7(3x - 2y) = 7(6) \longrightarrow 21x - 14y = 42$

$2(5x + 7y) = 2(41) \longrightarrow 10x + 14y = 82$

$21x - 14y = 42$

$\underline{10x + 14y = 82}$

$\qquad 31x = 124$

$\qquad x = 4$

$3(4) - 2y = 6,\ 12 - 2y = 6,\ -2y = -6,\ y = 3;\ (4, 3)$

10. $-3(5m - 2n) = -3(8) \longrightarrow -15m + 6n = -24$

$5(3m - 5n) = 5(1) \qquad \longrightarrow 15m - 25n = 5$

$-15m + 6n = -24$

$\underline{15m - 25n = 5}$

$\qquad -19n = -19$

$\qquad n = 1$

$3m - 5(1) = 1,\ 3m - 5 = 1,\ 3m = 6,\ m = 2;\ (2, 1)$

11. $3(7a - 3b) = 3(-9) \longrightarrow 21a - 9b = -27$

$-4a + 9b = -24 \qquad \longrightarrow -4a + 9b = -24$

$21a - 9b = -27$

$\underline{-4a + 9b = -24}$

$\qquad 17a = -51$

$\qquad a = -3$

$7(-3) - 3b = -9,\ -21 - 3b = -9,\ -3b = 12,\ b = -4;\ (-3, -4)$

12. $-3x + 4y = -6 \longrightarrow 3(-3x + 4y) = 3(-6) \longrightarrow -9x + 12y = -18$

$5x - 6y = 8 \qquad \longrightarrow 2(5x - 6y) = 2(8) \qquad \longrightarrow 10x - 12y = 16$

$-9x + 12y = -18$

$\underline{10x - 12y = 16}$

$\qquad x = -2$

$-3(-2) + 4y = -6,\ 6 + 4y = -6,\ 4y = -12,\ y = -3;\ (-2, -3)$

13. $3(8c - 3d) = 3(5) \longrightarrow 24c - 9d = 15$

$16c + 9d = 5 \qquad \longrightarrow 16c + 9d = 5$

$24c - 9d = 15$

$\underline{16c + 9d = 5}$

$\qquad 40c = 20$

$\qquad c = \dfrac{1}{2}$

$8\left(\dfrac{1}{2}\right) - 3d = 5,\ 4 - 3d = 5,\ -3d = 1,\ d = -\dfrac{1}{3};\ \left(\dfrac{1}{2}, -\dfrac{1}{3}\right)$

14. $3s - 8t = 4 \longrightarrow 3s - 8t = 4$

$\quad 2(9s - 4t) = 2(5) \longrightarrow 18s - 8t = 10$

$\quad\quad 3s - 8t = 4$

$\quad\quad \underline{18s - 8t = 10}$

$\quad\quad\quad -15s = -6$

$\quad\quad\quad\quad s = \dfrac{2}{5}$

$3\left(\dfrac{2}{5}\right) - 8t = 4, \dfrac{6}{5} - 8t = 4, -8t = \dfrac{14}{5}, t = -\dfrac{7}{20}; \left(\dfrac{2}{5}, -\dfrac{7}{20}\right)$

15. $3x - y = 1 \longrightarrow 2(3x - y) = 2(1) \longrightarrow 6x - 2y = 2$

$\quad -9x + 2y = -5 \longrightarrow -9x + 2y = -5 \longrightarrow -9x + 2y = -5$

$\quad\quad 6x - 2y = 2$

$\quad\quad \underline{-9x + 2y = -5}$

$\quad\quad\quad -3x = -3$

$\quad\quad\quad\quad x = 1$

$\quad 3(1) = y + 1, 3 = y + 1, y = 2; (1, 2)$

B 16. $2a + 5b = 6 \longrightarrow 2(2a + 5b) = 2(6) \longrightarrow 4a + 10b = 12$

$\quad\quad 3a - 10b = 2 \longrightarrow 3a - 10b = 2 \longrightarrow 3a - 10b = 2$

$\quad\quad 4a + 10b = 12$

$\quad\quad \underline{3a - 10b = 2}$

$\quad\quad\quad\quad 7a = 14$

$\quad\quad\quad\quad\quad a = 2$

$\quad 2(2) + 5b = 6, 4 + 5b = 6, 5b = 2, b = \dfrac{2}{5}; \left(2, \dfrac{2}{5}\right)$

17. $u - 4v = 0 \longrightarrow u - 4v = 0 \longrightarrow u - 4v = 0$

$\quad\quad 3u + 2v = 7 \longrightarrow 2(3u + 2v) = 2(7) \longrightarrow 6u + 4v = 14$

$\quad\quad u - 4v = 0$

$\quad\quad \underline{6u + 4v = 14}$

$\quad\quad\quad 7u = 14; u = 2$

$\quad 2 = 4v, v = \dfrac{1}{2}; \left(2, \dfrac{1}{2}\right)$

18. $p - 2q = 0 \longrightarrow 2(p - 2q) = 2(0) \longrightarrow 2p - 4q = 0$

$\quad\quad 2p + 6q = 5 \longrightarrow 2p + 6q = 5 \longrightarrow 2p + 6q = 5$

$\quad\quad 2p - 4q = 0$

$\quad\quad \underline{2p + 6q = 5}$

$\quad\quad\quad -10q = -5; q = \dfrac{1}{2}$

$\quad p = 2\left(\dfrac{1}{2}\right), p = 1; \left(1, \dfrac{1}{2}\right)$

19. $4(6m + 12n) = 4(7) \longrightarrow 24m + 48n = 28$
$3(8m - 15n) = 3(-1) \longrightarrow 24m - 45n = -3$

$24m + 48n = 28$
$\underline{24m - 45n = -3}$
$93n = 31$
$n = \dfrac{1}{3}$

$6m + 12\left(\dfrac{1}{3}\right) = 7,\ 6m + 4 = 7,\ 6m = 3,\ m = \dfrac{1}{2};\ \left(\dfrac{1}{2}, \dfrac{1}{3}\right)$

20. $10r - 9s = 18 \longrightarrow \quad 10r - 9s = 18 \quad \longrightarrow \quad 10r - 9s = 18$
$2r + 6s = 1 \ \longrightarrow 5(2r + 6s) = 5(1) \longrightarrow 10r + 30s = 5$

$10r - 9s = 18$
$\underline{10r + 30s = 5}$
$-39s = 13$
$s = -\dfrac{1}{3}$

$6\left(-\dfrac{1}{3}\right) + 2r = 1,\ -2 + 2r = 1,\ 2r = 3,\ r = \dfrac{3}{2};\ \left(\dfrac{3}{2}, -\dfrac{1}{3}\right)$

21. $2p + 6 = 3 - q \longrightarrow 2p + q + 6 = 3 \quad\longrightarrow 2p + q = -3$
$3p - 3 = q - 4 \longrightarrow 3p - q - 3 = -4 \longrightarrow 3p - q = -1$

$2p + q = -3$
$\underline{3p - q = -1}$
$5p = -4$
$p = -\dfrac{4}{5}$

$2\left(-\dfrac{4}{5}\right) + 6 = 3 - q,\ -\dfrac{8}{5} + \dfrac{30}{5} = 3 - q,\ \dfrac{22}{5} = 3 - q,\ \dfrac{7}{5} = -q,\ q = -\dfrac{7}{5};\ \left(-\dfrac{4}{5}, -\dfrac{7}{5}\right)$

22. $4x + 12 = 3y + 7 \longrightarrow 4x - 3y + 12 = 7 \longrightarrow 4x - 3y = -5 \longrightarrow$
$2y - 10 = x + 5 \ \longrightarrow -x + 2y - 10 = 5 \longrightarrow -x + 2y = 15 \longrightarrow$

$4x - 3y = -5 \qquad \longrightarrow \quad 4x - 3y = -5$
$4(-x + 2y) = 4(15) \longrightarrow -4x + 8y = 60$

$4x - 3y = -5$
$\underline{-4x + 8y = 60}$
$5y = 55$
$y = 11$

$2(11) - 10 = x + 5,\ 22 - 10 = x + 5,\ 12 = x + 5,\ x = 7;\ (7, 11)$

23.

$$3(c - 2d) = 3(500) \longrightarrow 3c - 6d = 1500$$
$$100(0.030 + 0.02d) = 100(51) \longrightarrow 3c + 2d = 5100$$
$$3c - 6d = 1500$$
$$\underline{3c + 2d = 5100}$$
$$-8d = -3600$$
$$d = 450$$
$$c - 2(450) = 500, \; c - 900 = 500, \; c = 1400; \; (1400, 450)$$

24.

$$3e - 2f = -80 \longrightarrow 3e - 2f = -80$$
$$60(0.05e - 0.03f) = 60(2) \longrightarrow 3e - 1.8f = 120$$
$$3e - 2f = -80$$
$$\underline{3e - 1.8f = 120}$$
$$-0.2f = -200$$
$$f = 1000$$
$$3e - 2(1000) = -80, \; 3e - 2000 = -80, \; 3e = 1920, \; e = 640; \; (640, 1000)$$

25.

$$3m + 2n = 4(-5m + n) \longrightarrow 3m + 2n = -20m + 4n \longrightarrow -2n + 23m = 0$$
$$2n - 23m = 5 \longrightarrow 2n - 23m = 5 \longrightarrow 2n - 23m = 5$$
$$-2n + 23m = 0$$
$$\underline{2n - 23m = 5}$$
$$0 \neq 5; \text{ no solution}$$

26.

$$7u - 4v = 2(5v) \longrightarrow 7u - 4v = 10v \longrightarrow 7u - 14v = 0$$
$$7u - v = 5 \longrightarrow 7u - v = 5 \longrightarrow 7u - v = 5$$
$$7u - 14v = 0$$
$$\underline{7u - v = 5}$$
$$-13v = -5$$
$$v = \frac{5}{13}$$
$$5 + \frac{5}{13} = 7u, \; \frac{70}{13} = 7u, \; \frac{10}{13} = u; \; \left(\frac{10}{13}, \frac{5}{13}\right)$$

27.

$$3(-2a + b) = 5(4a - 2b) \longrightarrow -6a + 3b = 20a - 10b \longrightarrow$$
$$9(a + 2) = 3(1 - b) \longrightarrow 3a + 6 = 1 - b \longrightarrow$$
$$-26a + 13b = 0 \longrightarrow -26a + 13b = 0 \longrightarrow -26a + 13b = 0$$
$$3a + b = -5 \longrightarrow -13(3a + b) = -13(-5) \longrightarrow -39a - 13b = 65$$
$$-26a + 13b = 0$$
$$\underline{-39a - 13b = 65}$$
$$-65a = 65$$
$$a = -1$$
$$\frac{-1 + 2}{3} = \frac{1 - b}{9}, \; \frac{1}{3} = \frac{1 - b}{9}, \; 3 = 1 - b, \; 2 = -b, \; -2 = b; \; (-1, -2)$$

C 28. $-2\left(\dfrac{1}{x}+\dfrac{1}{y}\right) = -2(5) \longrightarrow -\dfrac{2}{x}-\dfrac{2}{y} = -10$ **29.** $\dfrac{1}{x}-\dfrac{2}{y}=8 \qquad\qquad \longrightarrow \dfrac{1}{x}-\dfrac{2}{y}=8$

$\qquad \dfrac{2}{x}+\dfrac{3}{y}=13 \qquad\qquad \longrightarrow \dfrac{2}{x}+\dfrac{3}{y}=13 \qquad\qquad 2\left(\dfrac{3}{x}+\dfrac{1}{y}\right)=2(10) \longrightarrow \dfrac{6}{x}+\dfrac{2}{y}=20$

$\qquad -\dfrac{2}{x}-\dfrac{2}{y}=-10 \qquad\qquad\qquad\qquad\qquad\qquad \dfrac{1}{x}-\dfrac{2}{y}=8$

$\qquad \dfrac{\dfrac{2}{x}+\dfrac{3}{y}=13}{\qquad\qquad} \qquad\qquad\qquad\qquad\qquad\qquad \dfrac{\dfrac{6}{x}+\dfrac{2}{y}=20}{\qquad\qquad}$

$\qquad\qquad \dfrac{1}{y}=3 \qquad\qquad\qquad\qquad\qquad\qquad\qquad\qquad \dfrac{7}{x}=28$

$\qquad\qquad y=\dfrac{1}{3} \qquad\qquad\qquad\qquad\qquad\qquad\qquad\qquad x=\dfrac{1}{4}$

$\qquad \dfrac{1}{x}+\dfrac{1}{\frac{1}{3}}=5, \dfrac{1}{x}+3=5, \dfrac{1}{x}=2, \qquad\qquad \dfrac{1}{\frac{1}{4}}-\dfrac{2}{y}=8, 4-\dfrac{2}{y}=8, -\dfrac{2}{y}=4,$

$\qquad x=\dfrac{1}{2}; \left(\dfrac{1}{2},\dfrac{1}{3}\right) \qquad\qquad\qquad\qquad\qquad -\dfrac{y}{2}=\dfrac{1}{4}, -y=\dfrac{1}{2}, y=-\dfrac{1}{2}; \left(\dfrac{1}{4},-\dfrac{1}{2}\right)$

30. $-2\left(\dfrac{1}{x}+\dfrac{3}{y}\right)=-2(10) \longrightarrow -\dfrac{2}{x}-\dfrac{6}{y}=-20$

$\qquad \dfrac{2}{x}+\dfrac{1}{y}=6 \qquad\qquad\qquad \longrightarrow \dfrac{2}{x}+\dfrac{1}{y}=6$

$\qquad -\dfrac{2}{x}-\dfrac{6}{y}=-20$

$\qquad \dfrac{\dfrac{2}{x}+\dfrac{1}{y}=6}{\qquad\qquad}$

$\qquad\qquad -\dfrac{5}{y}=-14$

$\qquad\qquad y=\dfrac{5}{14}$

$\qquad \dfrac{1}{x}+\dfrac{3}{\frac{5}{14}}=10, \dfrac{1}{x}+\dfrac{42}{5}=10, \dfrac{1}{x}=\dfrac{8}{5}, x=\dfrac{5}{8}; \left(\dfrac{5}{8},\dfrac{5}{14}\right)$

31.
$$ax + y = c \longrightarrow ax + y = c$$
$$-a(x + by) = -a(c) \longrightarrow -ax - aby = -ac$$

$$ax + y = c$$
$$\underline{-ax - aby = -ac}$$
$$y - aby = -ac + c$$
$$y(1 - ab) = c(1 - a)$$
$$y = \frac{c(1 - a)}{1 - ab}$$

$$ax + \frac{c(1 - a)}{1 - ab} = c, \ ax = c - \frac{c(1 - a)}{1 - ab}, \ ax = \frac{c(1 - ab) - c(1 - a)}{1 - ab},$$

$$ax = \frac{c[(1 - ab) - (1 - a)]}{1 - ab}, \ ax = \frac{c(-ab + a)}{1 - ab}, \ ax = \frac{ac(-b + 1)}{1 - ab}; \ x = \frac{c(1 - b)}{1 - ab}$$

32. $ax + y = c$
$$\underline{ax + by = 0}$$
$$y - by = c$$
$$y(1 - b) = c$$
$$y = \frac{c}{1 - b}$$

$$ax + \frac{c}{1 - b} = c, \ ax = c - \frac{c}{1 - b}, \ ax = \frac{c(1 - b) - c}{1 - b}, \ ax = \frac{c[(1 - b) - 1]}{1 - b}$$

$$ax = -\frac{bc}{1 - b}; \ x = -\frac{bc}{a(1 - b)}$$

33. $ax + y = c$
$$\underline{ax + y = 0}$$
$$0 = c \quad \{(x, y): ax + y = c \text{ and } c = 0\}$$

34.
$$1 = a(-1)^2 + b \longrightarrow a + b = 1 \longrightarrow a + b = 1$$
$$-15 = a(3)^2 + b \longrightarrow 9a + b = -15 \longrightarrow -9a - b = 15$$

$$a + b = 1$$
$$\underline{-9a - b = 15}$$
$$-8a = 16$$
$$a = -2$$
$$-2 + b = 1, \ b = 3; \ (-2, 3)$$

35.
$$20x + 25y = 1250 \longrightarrow 20x + 25y = 1250$$
$$25(x + y) = 25(55) \longrightarrow 25x + 25y = 1375$$

$$20x + 25y = 1250$$
$$\underline{25x + 25y = 1375}$$
$$-5x = -125$$
$$x = 25$$

36. Let x = number cups of orange juice used in orange-pineapple fruit punch.

Let y = number cups of pineapple juice used in orange-pineapple fruit punch.

$x + y = 112 \longrightarrow -2(x + y) = -2(112) \longrightarrow -2x - 2y = -224$

$3x + 2y = 274 \longrightarrow \quad 3x + 2y = 274 \quad\quad \longrightarrow \quad 3x + 2y = 274$

$-2x - 2y = -224$

$\underline{3x + 2y = 274}$

$x = 50$

$50 + y = 112; \; y = 62$

150 c of orange juice and 124 c of pineapple juice were used that Friday.

page 466   Test Yourself

1.

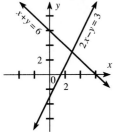

intersecting lines;
one solution

2.

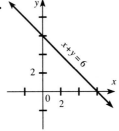

same line; infinitely
many solutions

3.

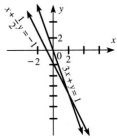

intersecting lines;
one solution

4. $\quad 2x + y = -10 \quad \longrightarrow \quad 2x + y = -10$

$-2(x + 3y) = -2(0) \longrightarrow -2x - 6y = 0$

$2x + y = -10$

$\underline{-2x - 6y = 0}$

$-5y = -10$

$y = 2$

$2x + 2 = -10, \; 2x = -12, \; x = -6; \; (-6, 2)$

5. $3x + 2y = 10$

$\underline{x - 2y = 14}$

$4x = 24$

$x = 6$

$6 - 2y = 14, \; -2y = 8, \; y = -4; \; (6, -4)$

6. $4x - 3y = 15 \quad \longrightarrow 4x - 3y = 15$

$4(x + y) = 4(9) \longrightarrow 4x + 4y = 36$

$4x - 3y = 15$

$\underline{4x + 4y = 36}$

$-7y = -21$

$y = 3$

$x + 3 = 9, \; x = 6; \; (6, 3)$

344   Chapter 11: Systems of Linear Equations

7. $3\left(\dfrac{1}{3}x\right) = 3\left(-\dfrac{1}{3}y + 4\right) \longrightarrow x = -y + 12$

$3\left(\dfrac{8}{3}\right) = 3\left(x + \dfrac{1}{3}y\right) \longrightarrow 8 = 3x + y$

$8 = 3(-y + 12) + y,\ 8 = -3y + 36 + y,\ 8 = -2y + 36,\ 2y = 28,\ y = 14;$

$\dfrac{1}{3}x = -\dfrac{1}{3}(14) + 4,\ \dfrac{1}{3}x = -\dfrac{2}{3},\ x = -2;\ (-2, 14)$

8. $3\left(-\dfrac{2}{3}y + 2x\right) = 3(-1) \longrightarrow -2y + 6x = -3$

$6\left(\dfrac{1}{3}y + x\right) = 6\left(\dfrac{1}{2}\right) \longrightarrow 2y + 6x = 3$

$\begin{array}{r} -2y + 6x = -3 \\ 2y + 6x = 3 \\ \hline 12x = 0 \\ x = 0 \end{array}$

$\dfrac{1}{3}y + 0 = \dfrac{1}{2},\ \dfrac{1}{3}y = \dfrac{1}{2},\ y = \dfrac{3}{2};\ \left(0, \dfrac{3}{2}\right)$

9. Let $x =$ the amount of money spent on advertising in newspapers and $y =$ the amount of money spent on advertising in magazines.

$\begin{cases} x + y = 33{,}000 \\ x = 2y \end{cases}$

$2y + y = 33{,}000,\ 3y = 33{,}000;\ y = 11{,}000$

$\$22{,}000$ is spent on advertising in newspapers and $\$11{,}000$ is spent on advertising in magazines.

page 468   Class Exercises

1. $t + u = 9$

2. $t + 4 = u$

3. $t = \dfrac{1}{2}u$

4. $10u + t = 10t + u + 9$

5. $\begin{cases} t + u = 9 \\ t = \dfrac{1}{2}u \end{cases}$

$\dfrac{1}{2}u + u = 9,\ \dfrac{3}{2}u = 9,\ u = 6;\ t + 6 = 9,\ t = 3;\ 36$

6. $\begin{cases} t = \dfrac{1}{2}u \longrightarrow t = \dfrac{1}{2}u \\ 9u = 9t + 9 \longrightarrow u = t + 1 \end{cases}$

$t = \dfrac{1}{2}(t + 1),\ t = \dfrac{1}{2}t + \dfrac{1}{2},\ \dfrac{1}{2}t = \dfrac{1}{2},\ t = 1;\ u = 1 + 1,\ u = 2;\ 12$

7. $\begin{cases} t + u = 9 \\ 9u = 9t + 9 \end{cases} \longrightarrow \begin{matrix} t + u = 9 \\ u = t + 1 \end{matrix}$

$t + t + 1 = 9,\ 2t + 1 = 9,\ 2t = 8,\ t = 4;\ 4 + u = 9,\ u = 5;\ 45$

8. $\begin{cases} t + 4 = u \\ t = \dfrac{1}{2}u \end{cases}$

$\dfrac{1}{2}u + 4 = u,\ 4 = \dfrac{1}{2}u,\ 8 = u;\ t = \dfrac{1}{2} \cdot 8,\ t = 4;\ 48$

pages 468–469   Practice Exercises

A   1. Let t = the tens digit and u = the units digit.

$\begin{cases} t + u = 6 \\ 10t + u = 6u \end{cases} \longrightarrow \begin{matrix} t + u = 6 \\ 10t = 5u \end{matrix} \longrightarrow \begin{matrix} t + u = 6 \\ 2t = u \end{matrix}$

$t + 2t = 6,\ 3t = 6,\ t = 2;\ 2 + u = 6,\ t = 4;$ The two-digit number is 24.

2. Let t = the tens digit and u = the units digit.

$\begin{cases} t + u = 6 \\ 10t + u = 10u + t + 18 \end{cases} \longrightarrow \begin{matrix} t + u = 6 \\ 9t - 9u = 18 \end{matrix} \longrightarrow \begin{matrix} t + u = 6 \\ t - u = 2 \end{matrix}$

$\begin{array}{r} t + u = 6 \\ t - u = 2 \\ \hline 2t = 8 \\ t = 4 \end{array}$

$4 + u = 6,\ u = 2;$ The two-digit number is 42.

3. Let t = the tens digit and u = the units digit.

$\begin{cases} u = 3t \\ 10t + u + 54 = 10u + t \end{cases} \longrightarrow \begin{matrix} u = 3t \\ 9t - 9u + 54 = 0 \end{matrix}$

$9t - 9(3t) + 54 = 0,\ 9t - 27t + 54 = 0,\ -18t = -54,\ t = 3;\ u = 3(3),$
$u = 9;$ The two-digit number is 39.

4. Let t = the tens digits and u = the units digit.

$\begin{cases} u = 3t \\ 10t + u = t + 12 \end{cases} \longrightarrow \begin{matrix} u = 3t \\ 9t + u = 12 \end{matrix}$

$9t + 3t = 12,\ 12t = 12,\ t = 1;\ u = 3(1),\ u = 3;$ The two-digit number is 13.

5. Let t = the tens digit and u = the units digit.

$\begin{cases} u = t + 6 \\ 10t + u = 9t + 10 \end{cases} \longrightarrow \begin{matrix} u = t + 6 \\ t + u = 10 \end{matrix}$

$t + t + 6 = 10,\ 2t + 6 = 10,\ 2t = 4,\ t = 2;\ u = 2 + 6,\ u = 8;$ The two-digit number is 28.

6. Let t = the tens digit and u = the units digit.

$\begin{cases} t + u = 12 \\ 10t + u = 8u + 1 \end{cases}$ $\rightarrow$ $\begin{aligned} t + u = 12 \\ 10t - 7u = 1 \end{aligned}$ $\rightarrow$ $\begin{aligned} -10(t + u) = -10(12) \\ 10t - 7u = 1 \end{aligned}$ $\rightarrow$ $\begin{aligned} -10t - 10u = -120 \\ 10t - 7u = 1 \end{aligned}$

$$\begin{aligned} -10t - 10u &= -120 \\ \underline{10t - 7u} &= \underline{1} \\ -17u &= -119 \\ u &= 7 \end{aligned}$$

$t + 7 = 12$, $t = 5$; The two-digit number is 57.

7. Let t = the tens digit and u = the units digit.

$\begin{cases} t + u = 10 \\ 10t + u - 54 = 10u + t \end{cases}$ $\rightarrow$ $\begin{aligned} t + u = 10 \\ 9t - 9u - 54 = 0 \end{aligned}$ $\rightarrow$ $\begin{aligned} t + u = 10 \\ t - u - 6 = 0 \end{aligned}$ $\rightarrow$ $\begin{aligned} t + u = 10 \\ t - u = 6 \end{aligned}$

$$\begin{aligned} t + u &= 10 \\ \underline{t - u} &= \underline{6} \\ 2t &= 16 \\ t &= 8 \end{aligned}$$

$8 + u = 10$, $u = 2$; The two-digit number is 82.

8. Let t = the tens digit and u = the units digit.

$\begin{cases} t + u = 12 \\ 10t + u + 36 = 10u + t \end{cases}$ $\rightarrow$ $\begin{aligned} t + u = 12 \\ 9t - 9u + 36 = 0 \end{aligned}$ $\rightarrow$ $\begin{aligned} t + u = 12 \\ t - u + 4 = 0 \end{aligned}$ $\rightarrow$ $\begin{aligned} t + u = 12 \\ t - u = -4 \end{aligned}$

$$\begin{aligned} t + u &= 12 \\ \underline{t - u} &= \underline{-4} \\ 2t &= 8 \\ t &= 4 \end{aligned}$$

$4 + u = 12$, $u = 8$; The two-digit number is 48.

9. Let t = the tens digit and u = the units digit.

$\begin{cases} 2t + u = 22 \\ 10t + u = 10u + t + 45 \end{cases}$ $\rightarrow$ $\begin{aligned} 2t + u = 22 \\ 9t - 9u = 45 \end{aligned}$ $\rightarrow$ $\begin{aligned} 2t + u = 22 \\ t - u = 5 \end{aligned}$

$$\begin{aligned} 2t + u &= 22 \\ \underline{t - u} &= \underline{5} \\ 3t &= 27 \\ t &= 9 \end{aligned}$$

$2(9) + u = 22$, $18 + u = 22$, $u = 4$; The two-digit number is 94.

B 10. Let t = the tens digit and u = the units digit.

$\begin{cases} 3u + 2 = t \\ 10t + u = 10u + t + 54 \end{cases}$ $\rightarrow$ $\begin{aligned} 3u + 2 = t \\ 9t - 9u = 54 \end{aligned}$ $\rightarrow$ $\begin{aligned} 3u + 2 = t \\ t - u = 6 \end{aligned}$

$(3u + 2) - u = 6$, $2u + 2 = 6$, $2u = 4$, $u = 2$; $3(2) + 2 = t$, $6 + 2 = t$, $8 = t$; The two-digit number is 82.

11. Let t = the tens digit and u = the units digit.
$$\begin{cases} t + u = 11 \\ 10t + u = 2(10u + t) + 7 \end{cases} \longrightarrow \begin{array}{l} t + u = 11 \\ 10t + u = 20u + 2t + 7 \end{array} \longrightarrow \begin{array}{l} t + u = 11 \longrightarrow \\ 8t - 19u = 7 \end{array} \longrightarrow$$

$$\begin{array}{l} -8(t + u) = -8(11) \longrightarrow -8t - 8u = -88 \\ 8t - 19u = 7 \qquad\quad \longrightarrow 8t - 19u = 7 \\ \hline -8t - 8u = -88 \\ \underline{8t - 19u = 7} \\ \qquad\quad -27u = -81 \\ \qquad\qquad u = 3 \qquad t + 3 = 11, t = 8; \text{ The two-digit number is 83.} \end{array}$$

12. Let t = the tens digit and u = the units digit.
$$\begin{cases} 10t + u = t + u + 9 \\ 4(10t + u) = 10u + t + 3 \end{cases} \longrightarrow \begin{array}{l} 9t = 9 \\ 40t + 4u = 10u + t + 3 \end{array} \longrightarrow \begin{array}{l} t = 1 \longrightarrow \\ 39t - 6u = 3 \end{array} \longrightarrow$$

$39(1) - 6u = 3,\ 39 - 6u = 3,\ -6u = -36,\ u = 6;$ The two-digit number is 16.

13. Let t = the tens digit and u = the units digit.
$$\begin{cases} 10t + u = 7(t + u) \\ 10u + t + 30 = 2(10t + u) \end{cases} \rightarrow \begin{array}{l} 10t + u = 7t + 7u \rightarrow \\ 10u + t + 30 = 20t + 2u \end{array} \rightarrow \begin{array}{l} 3t - 6u = 0 \rightarrow \\ 8u - 19t + 30 = 0 \rightarrow \end{array}$$

$$\begin{array}{l} 3t - 6u = 0 \qquad\quad \longrightarrow 4(3t - 6u) = 4(0) \qquad \longrightarrow 12t - 24u = 0 \\ -19t + 8u = -30 \longrightarrow 3(-19t + 8u) = 3(-30) \longrightarrow -57t + 24u = -90 \\ \\ 12t - 24u = 0 \\ \underline{-57t + 24u = -90} \\ \qquad\quad -45t = -90 \qquad 10(2) + u = 7(2 + u),\ 20 + u = 14 + 7u,\ 6 = 6u,\ 1 = u; \\ \qquad\qquad\quad t = 2 \qquad\qquad \text{The two-digit number is 21.} \end{array}$$

14. Let t = the tens digit and u = the units digit.
$$\begin{cases} \dfrac{1}{3}t = 2 \longrightarrow t = 6 \qquad\quad (6)u = 0,\ u = 0; \text{ The two-digit number is 60.} \\ tu = 0 \end{cases}$$

15. Let t = the tens digit and u = the units digit. $3t + u = 2(t + u)$, $3t + u = 2t + 2u$, $t = u$; The numbers are 11, 22, 33, 44, 55, 66, 77, 88 and 99.

16. $(-1)(4)\left(\dfrac{1}{2}\right)^2 = -1$ **17.** $0.5(4) - 4(-1)\left(\dfrac{1}{2}\right) = 2 + 2 = 4$ **18.** $\dfrac{2(4)^2}{(-1)\left(\frac{1}{2}\right)} = \dfrac{2(16)}{-\frac{1}{2}} = -64$

19. Let t represent the number of hours each plane has flown. $325t + 400t = 2900$, $725t = 2900$, $t = 4$; They will be 2900 mi apart in 4 h.

20. Let n represent the number of hours using both pipes together. $\dfrac{n}{8} + \dfrac{n}{12} = 1$,

$3n + 2n = 24$, $5n = 24$, $n = 4\dfrac{4}{5}$; It takes $4\dfrac{4}{5}$ h (4 h, 48 min) if pipes are used together.

21. $\dfrac{15}{12} = \dfrac{25}{y}$, $15y = 300$, $y = 20$ **22.** $(3)(15) = (a)(5)$, $45 = 5a$, $a = 9$

23. Let h = the hundreds digit, t = the tens digit and, u = the units digit.

$\begin{cases} h = t + u \\ u = 2t \\ (100h + 10t + u) - (100u + 10t + h) = 297 \end{cases}$ $\longrightarrow$ $h = t + u$ $\longrightarrow$
  $\longrightarrow$ $u = 2t$ $\longrightarrow$
$\longrightarrow$ $99h - 99u = 297$ $\longrightarrow$

$h = t + u \longrightarrow h = t + u$ $h = \dfrac{u}{2} + u$, $h = \dfrac{3}{2}u$; $\dfrac{3}{2}u - u = 3$, $\dfrac{1}{2}u = 3$,

$u = 2t \longrightarrow \dfrac{u}{2} = t$ $u = 6$; $6 = 2t$, $3 = t$; $h = 3 + 6$, $h = 9$;

$h - u = 3 \longrightarrow h - u = 3$ The three-digit number is 936.

24. Let h = the hundreds digit, t = the tens digit, and u = the units digit.

$\begin{cases} 2h = t + u \\ h - u = 3 \\ (100h + 10t + u) + (100u + 10t + h) = 867 \end{cases}$ $\longrightarrow$ $2h = t + u$
  $\longrightarrow$ $h = 3 + u$
$\longrightarrow$ $101h + 20t + 101u = 867$

$2(3 + u) = t + u$, $6 + 2u = t + u$, $t = u + 6$; $101(3 + u) + 20(u + 6) + 101u =$
867, $303 + 101u + 20u + 120 + 101u = 867$, $222u = 444$, $u = 2$; $h = 3 + 2$,
$h = 5$; $2(5) = t + 2$, $t = 8$; The three-digit number is 582.

page 471   Class Exercises

1.

	Age 8 years ago	Age 5 years ago	Age 2 years ago	Present age	Age in 1 year	Age in 4 years	Age in 7 years
Anne	$a - 8$	$a - 5$	$a - 2$	a	$a + 1$	$a + 4$	$a + 7$
Bill	$b - 8$	$b - 5$	$b - 2$	b	$b + 1$	$b + 4$	$b + 7$

2. $b - 3 = a$ **3.** $b - 8 = 2(a - 8)$ **4.** $a + 4 = \dfrac{5}{6}(b + 4)$

5. $b + 7 = \dfrac{7}{6}(a + 7)$ **6.** $a + b = 25$

7. $\begin{cases} b - 3 = a \\ a + b = 25 \end{cases}$ $(b - 3) + b = 25$, $2b - 3 = 25$, $2b = 28$, $b = 14$; $14 - 3 = a$,
       $11 = a$; Bill is 14 years old and Anne is 11 years old.

8. $\begin{cases} b - 3 = a \\ b - 8 = 2(a - 8) \end{cases}$ $\longrightarrow$ $b - 3 = a$ $\longrightarrow$ $b - 3 = a$
                      $\longrightarrow$ $b - 8 = 2a - 16$ $\longrightarrow$ $b - 2a = -8$

$b - 2(b - 3) = -8$, $b - 2b + 6 = -8$, $-b + 6 = -8$, $-b = -14$, $b = 14$;
$14 - 3 = a$, $11 = a$; Bill is 14 years old and Anne is 11 years old.

9. $\begin{cases} a + 4 = \dfrac{5}{6}(b + 4) \longrightarrow a + 4 = \dfrac{5}{6}b + \dfrac{20}{6} \longrightarrow 6a + 24 = 5b + 20 \longrightarrow \\[2mm] b + 7 = \dfrac{7}{6}(a + 7) \longrightarrow b + 7 = \dfrac{7}{6}a + \dfrac{49}{6} \longrightarrow 6b + 42 = 7a + 49 \longrightarrow \end{cases}$

$$6a - 5b = -4 \longrightarrow 7(6a - 5b) = 7(-4) \longrightarrow 42a - 35b = -28$$
$$-7a + 6b = 7 \longrightarrow 6(-7a + 6b) = 6(7) \longrightarrow -42a + 36b = 42$$

$$\begin{array}{r} 42a - 35b = -28 \\ -42a + 36b = 42 \\ \hline b = 14 \end{array}$$

$14 + 7 = \frac{7}{6}(a + 7)$, $21 = \frac{7}{6}(a + 7)$, $18 = a + 7$, $11 = a$; Bill is 14 years

old and Anne is 11 years old.

page 471–473 Practice Exercises

A 1. Let c = Cordell's present age, then $c - 8$ = Cordell's age eight years ago.
Let b = Beth's present age, then $b - 8$ = Beth's age eight years ago.
$$\begin{cases} c = 2b \\ c - 8 = 3(b - 8) \end{cases}$$
$2b - 8 = 3(b - 8)$, $2b - 8 = 3b - 24$, $-8 = b - 24$, $16 = b$; $c = 2(16)$,
$c = 32$; Beth is 16 years old and Cordell is 32 years old.

2. Let d = Dawn's present age, then $d - 4$ = Dawn's age four years ago.
Let l = Lois' present age, then $l - 4$ = Lois' age four years ago.
$$\begin{cases} d = l + 3 \\ d - 4 = 2(l - 4) \end{cases} \longrightarrow \begin{array}{l} d = l + 3 \\ d - 4 = 2l - 8 \end{array} \longrightarrow \begin{array}{l} d = l + 3 \\ d - 2l = -4 \end{array}$$
$(l + 3) - 2l = -4$, $-l + 3 = -4$, $-l = -7$, $l = 7$; $d = 7 + 3$, $d = 10$;
Dawn is 10 years old and Lois is 7 years old.

3. Let m = Mario's present age, then $m + 9$ = Mario's age in nine years.
Let n = Nadia's present age, then $n + 9$ = Nadia's age in nine years.
$$\begin{cases} m = 2n + 2 \\ n + 9 = \frac{2}{3}(m + 9) \end{cases} \longrightarrow \begin{array}{l} m = 2n + 2 \\ n + 9 = \frac{2}{3}m + 6 \end{array} \longrightarrow \begin{array}{l} m = 2n + 2 \\ n + 3 = \frac{2}{3}m \end{array}$$

$n + 3 = \frac{2}{3}(2n + 2)$, $n + 3 = \frac{4n}{3} + \frac{4}{3}$, $3(n + 3) = 3\left(\frac{4}{3}n + \frac{4}{3}\right)$, $3n + 9 = 4n + 4$,
$9 = n + 4$, $n = 5$; $m = 2(5) + 2$, $m = 10 + 2$, $m = 12$; Mario is
12 years old and Nadia is 5 years old.

4. Let t = Tanya's present age, then $t + 7$ = Tanya's age in seven years.
Let b = her brother's present age, then $b + 7$ = her brother's age in seven years.
$$\begin{cases} b = \frac{3}{4}t \\ (b + 7) + (t + 7) = 28 \end{cases} \longrightarrow \begin{array}{l} b = \frac{3}{4}t \\ b + t + 14 = 28 \end{array} \longrightarrow \begin{array}{l} b = \frac{3}{4}t \\ b + t = 14 \end{array} \longrightarrow$$

$\frac{3}{4}t + t = 14$, $\frac{7}{4}t = 14$, $t = 8$; $b = \frac{3}{4}(8)$, $b = 6$; Tanya is 8 years old and her

brother is 6 years old.

5. Let a = Adam's present age, then $a - 2$ = Adam's age two years ago, and $a + 3$ = Adam's age in three years.

Let s = his son's present age, then $s - 2$ = his son's age two years ago, and $s + 3$ = his son's age in three years.

$$\begin{cases} a - 2 = 6(s - 2) \longrightarrow a - 2 = 6s - 12 \longrightarrow a - 6s - 2 = -12 \longrightarrow a - 6s = -10 \\ a + 3 = 3\frac{1}{2}(s + 3) \longrightarrow a + 3 = \frac{7}{2}s + \frac{21}{2} \longrightarrow a - \frac{7}{2}s + 3 = \frac{21}{2} \longrightarrow a - \frac{7}{2}s = \frac{15}{2} \end{cases}$$

$$a - 6s = -10$$
$$\underline{a - \frac{7}{2}s = \frac{15}{2}}$$
$$-\frac{5}{2}s = -\frac{35}{2}; \ s = 7$$

$a - 2 = 6(7 - 2)$, $a - 2 = 6(5)$, $a - 2 = 30$, $a = 32$; Adam is 32 years old and his son is 7 years old.

6. Let b = Bob's present age, then $b - 6$ = Bob's age six years ago, and $b + 9$ = Bob's age in nine years.

Let p = Peri's present age, then $p - 6$ = Peri's age six years ago, and $p + 9$ = Peri's age in nine years.

$$\begin{cases} b - 6 = 4(p - 6) \longrightarrow b - 6 = 4p - 24 \longrightarrow b = 4p - 18 \\ b + 9 = \frac{3}{2}(p + 9) \longrightarrow b + 9 = \frac{3}{2}p + \frac{27}{2} \longrightarrow b = \frac{3}{2}p + \frac{9}{2} \end{cases}$$

$4p - 18 = \frac{3}{2}p + \frac{9}{2}, \frac{5}{2}p - 18 = \frac{9}{2}, \frac{5}{2}p = \frac{45}{2}, p = 9$; $b - 6 = 4(9 - 6)$,

$b - 6 = 4(3)$, $b - 6 = 12$, $b = 18$; Peri is 9 years old and Bob is 18 years old.

7. Let s = Sarat's present age, then $s - 14$ = Sarat's age fourteen years ago.

Let a = Akhil's present age, then $a - 14$ = Akhil's age fourteen years ago.

$$\begin{cases} s + 4 = a \qquad\qquad \longrightarrow s + 4 = a \qquad\qquad \longrightarrow s + 4 = a \\ a - 14 = 2(s - 14) \longrightarrow a - 14 = 2s - 28 \longrightarrow a = 2s - 14 \end{cases}$$

$s + 4 = 2s - 14$, $4 = s - 14$, $18 = s$; $18 + 4 = a$, $22 = a$; Akhil is 22 years old and Sarat is 18 years old.

8. Let p = Peter's present age, then $p + 2$ = Peter's age in two years.

Let e = Eve's present age, then $e + 2$ = Eve's age in two years.

$$\begin{cases} p + 2 = 2(e + 2) \longrightarrow p + 2 = 2e + 4 \longrightarrow p - 2e + 2 = 4 \longrightarrow p - 2e = 2 \\ p + e = 26 \qquad\qquad \longrightarrow p + e = 26 \qquad\qquad \longrightarrow \qquad p + e = 26 \longrightarrow p + e = 26 \end{cases}$$

$$p - 2e = 2$$
$$\underline{p + e = 26}$$
$$-3e = -24$$
$$e = 8$$

$p + 8 = 26$, $p = 18$; Peter is 18 years old and Eve is 8 years old.

9. Let l = Lauren's present age, then $l - 12$ = Lauren's age twelve years ago.
Let j = Joyce's present age, then $j - 12$ = Joyce's age twelve years ago.

$$\begin{cases} l + 18 = j \\ j - 12 = 4(l - 12) \end{cases} \longrightarrow \begin{array}{l} l + 18 = j \\ j - 12 = 4l - 48 \end{array} \longrightarrow \begin{array}{l} l + 18 = j \\ j = 4l - 36 \end{array}$$

$l + 18 = 4l - 36$, $18 = 3l - 36$, $54 = 3l$, $18 = l$; $18 + 18 = j$, $36 = j$;
Joyce is 36 years old and Lauren is 18 years old.

10. Let r = Rufus' present age, then $r + 3$ = Rufus' age in three years.
Let j = Jason's present age, then $j + 3$ = Jason's age in three years.

$$\begin{cases} r + 3 = \frac{2}{3}(j + 3) \\ r + j = 44 \end{cases} \longrightarrow \begin{array}{l} r + 3 = \frac{2}{3}j + 2 \\ r + j = 44 \end{array} \longrightarrow \begin{array}{l} r - \frac{2}{3}j = -1 \\ r + j = 44 \end{array}$$

$r - \frac{2}{3}j = -1$

$\underline{r + j = 44}$

$-\frac{5}{3}j = -45$

$j = 27$

$r + 27 = 44$, $r = 17$; Jason is 27 years old and Rufus is 17 years old.

11. Let h = Harvey's present age, then $h - 6$ = Harvey's age six years ago.
Let c = Carol's present age, then $c - 6$ = Carol's age six years ago.

$$\begin{cases} (h - 6) + (c - 6) = 71 \\ c + 27 = h \end{cases} \longrightarrow \begin{array}{l} h + c - 12 = 71 \\ c + 27 = h \end{array} \longrightarrow \begin{array}{l} h + c = 83 \\ c + 27 = h \end{array}$$

$(c + 27) + c = 83$, $2c + 27 = 83$, $2c = 56$, $c = 28$; $28 + 27 = h$, $55 = h$;
Harvey is 55 years old and Carol is 28 years old.

12. Let a = Alvin's present age, then $a - 21$ = Alvin's age twenty-one years ago.
Let t = Tomio's present age, then $t - 21$ = Tomio's age twenty-one years ago.

$$\begin{cases} a - 21 = 11(t - 21) \\ a + t = 54 \end{cases} \longrightarrow \begin{array}{l} a - 21 = 11t - 231 \\ a + t = 54 \end{array} \longrightarrow \begin{array}{l} a = 11t - 210 \\ a + t = 54 \end{array}$$

$(11t - 210) + t = 54$, $12t - 210 = 54$, $12t = 264$, $t = 22$; $a + 22 = 54$,
$a = 32$; Alvin is 32 years old and Tomio is 22 years old.

13. Let m = Mary's present age, then $m + 1$ = Mary's age in one year.
Let s = Seth's present age, then $s + 1$ = Seth's age in one year.

$$\begin{cases} m = s + 9 \\ m + 1 = 3(s + 1) \end{cases} \longrightarrow \begin{array}{l} m = s + 9 \\ m + 1 = 3s + 3 \end{array} \longrightarrow \begin{array}{l} m = s + 9 \\ m = 3s + 2 \end{array}$$

$s + 9 = 3s + 2$, $9 = 2s + 2$, $7 = 2s$, $3\frac{1}{2} = s$; $m = 3\frac{1}{2} + 9$, $m = 12\frac{1}{2}$;

Mary is $12\frac{1}{2}$ years old and Seth is $3\frac{1}{2}$ years old.

14. Let r = Roz's present age, then $r + 5$ = Roz's age in five years.
Let g = Gerry's present age, then $g + 5$ = Gerry's age in five years.
$$\begin{cases} r = 2g & \longrightarrow r = 2g \\ (r + 5) + (g + 5) = 28 & \longrightarrow r + g + 10 = 28 \end{cases}$$
$2g + g + 10 = 28$, $3g + 10 = 28$, $3g = 18$, $g = 6$; $r = 2(6)$, $r = 12$; Roz
is 12 years old and Gerry is 6 years old.

15. Let s = Susan's present age and r = Randi's present age.
$$\begin{cases} s = 3r \\ \dfrac{1}{4}(r + s) = 8 \end{cases}$$

$\dfrac{1}{4}(r + 3r) = 8$, $\dfrac{1}{4}(4r) = 8$, $r = 8$; $s = 3(8)$, $s = 24$; Randi is 8 years old and

Susan is 24 years old.

16. Let i = Ira's present age and s = Sofia's present age.
$$\begin{cases} i + s = 26 & \longrightarrow i = 26 - s \\ 4s - 2i = 8 & \longrightarrow 4s - 2i = 8 \end{cases}$$
$4s - 2(26 - s) = 8$, $4s - 52 + 2s = 8$, $6s - 52 = 8$, $6s = 60$, $s = 10$;
$i = 26 - 10$, $i = 16$; Ira is 16 years old and Sofia is 10 years old.

B 17. Let l = Liz's age in 1945 and b = Bill's age in 1945.
$$\begin{cases} l = 2b \\ l - b = 18 \end{cases}$$
$2b - b = 18$, $b = 18$; $l = 2(18)$, $l = 36$;
$1945 - 18 = 1927 \longrightarrow$ Bill was born in 1927.
$1945 - 36 = 1909 \longrightarrow$ Liz was born in 1909.

18. Let j = Joy's age in 1987 and a = Ali's age in 1987. ✳
$$\begin{cases} j = 3a \\ j = a + 8 \end{cases}$$
$3a = a + 8$, $2a = 8$, $a = 4$; $j = 3(4)$, $j = 12$; $1999 - 1987 = 12$,
Joy will be $12 + 12$ or 24 years old in 1999.
Ali will be $4 + 12$ or 16 years old in 1999.

19. Let b = Bill's present age, then $b - 6$ = Bill's age 6 years ago.
Let a = Ann's present age, then $a - 6$ = Ann's age 6 years ago.
$$\begin{cases} (b - 6) + 17 = 2(a - 6) \longrightarrow b + 11 = 2a - 12 \longrightarrow b + 11 = 2a - 12 \\ b + a = 100 \qquad\qquad \longrightarrow b + a = 100 \qquad \longrightarrow b = 100 - a \end{cases}$$
$(100 - a) + 11 = 2a - 12$, $111 - a = 2a - 12$, $111 = 3a - 12$,
$123 = 3a$, $41 = a$; $b = 100 - 41$, $b = 59$
Bill is 59 years old and Ann is 41 years old.

20. Let p = Patty's age in 1982, then $p + 4$ = Patty's age in 1986.
Let t = Tom's age in 1982, then $t + 4$ = Tom's age in 1986.

$$\begin{cases} p = 3t \\ (p + 4) - (t + 4) = 4 \end{cases} \longrightarrow \begin{array}{l} p = 3t \\ p - t = 4 \end{array}$$

$3t - t = 4$, $2t = 4$, $t = 2$; $p = 3(2)$, $p = 6$;
Patty was 6 years old in 1982. Tom was 2 years old in 1982. If the year is 1990, Patty would be $6 + 8$ or 14 years old and Tom would be $2 + 8$ or 10 years old.

21. Let s = Spero's present age, then $s + 2\frac{1}{2}$ = Spero's age in $2\frac{1}{2}$ years.

Let c = Chris' present age, then $c + 2\frac{1}{2}$ = Chris' age in $2\frac{1}{2}$ years.

$$\begin{cases} s + c = 50 \\ s + 2\frac{1}{2} = \frac{5}{6}\left(c + 2\frac{1}{2}\right) \end{cases} \longrightarrow \begin{array}{l} s + c = 50 \\ s + 2\frac{1}{2} = \frac{5}{6}c + \frac{25}{12} \end{array} \longrightarrow \begin{array}{l} s = 50 - c \\ s = \frac{5}{6}c - \frac{5}{12} \end{array}$$

$50 - c = \frac{5}{6}c - \frac{5}{12}$, $50 + \frac{5}{12} - c = \frac{5}{6}c$, $\frac{605}{12} = \frac{11}{6}c$, $605 = 22c$, $27\frac{1}{2} = c$;

$s + 27\frac{1}{2} = 50$, $s = 22\frac{1}{2}$; Chris is $27\frac{1}{2}$ years old and Spero is $22\frac{1}{2}$ years old.

㉒ Let s = Susie's present age, then $s + 4$ = Susie's age in 4 years.
Let w = Weiva's present age, then $w + 4$ = Weiva's age in 4 years.

$$\begin{cases} s = \frac{3}{4}w \\ s + 4 = \frac{4}{5}(w + 4) \end{cases} \longrightarrow \begin{array}{l} s = \frac{3}{4}w \\ s + 4 = \frac{4}{5}w + \frac{16}{5} \end{array} \longrightarrow \begin{array}{l} s = \frac{3}{4}w \\ s + \frac{4}{5} = \frac{4}{5}w \end{array}$$

$\frac{3}{4}w + \frac{4}{5} = \frac{4}{5}w$, $\frac{4}{5} = \frac{1}{20}w$, $16 = w$; $s = \frac{3}{4}(16)$, $s = 12$;

Weiva is 16 years old and Susie is 12 years old.

23. Let a = Armando's present age, then $a - 5$ = Armando's age five years ago.
Let z = Zelda's present age, then $z - 5$ = Zelda's age five years ago.

$$\begin{cases} a = 1\frac{1}{3}z \\ a - 5 = 1\frac{1}{2}(z - 5) \end{cases} \longrightarrow \begin{array}{l} a = 1\frac{1}{3}z \\ a - 5 = \frac{3}{2}z - \frac{15}{2} \end{array} \longrightarrow \begin{array}{l} a = 1\frac{1}{3}z \\ a + \frac{5}{2} = \frac{3}{2}z \end{array}$$

$1\frac{1}{3}z + \frac{5}{2} = \frac{3}{2}z$, $\frac{5}{2} = \frac{1}{6}z$, $15 = z$; $a = 1\frac{1}{3}(15)$, $a = 20$; Armando is 20 years old.

24. Let n = Nina's present age, then $n - 4$ = Nina's age four years ago.
Let l = Liu's present age, then $l - 4$ = Liu's age four years ago.

$$\begin{cases} n = \dfrac{2}{3}l \\ l - 4 = \dfrac{5}{3}(n - 4) \end{cases} \longrightarrow \begin{cases} n = \dfrac{2}{3}l \\ l - 4 = \dfrac{5}{3}n - \dfrac{20}{3} \end{cases} \longrightarrow \begin{cases} n = \dfrac{2}{3}l \\ l + \dfrac{8}{3} = \dfrac{5}{3}n \end{cases}$$

$l + \dfrac{8}{3} = \dfrac{5}{3}\left(\dfrac{2}{3}l\right), l + \dfrac{8}{3} = \dfrac{10}{9}l, \dfrac{8}{3} = \dfrac{1}{9}l, 24 = l; n = \dfrac{2}{3}(24), n = 16$; Nina is
16 years old.

C 25. Let e = Emma's age, v = Vangie's age, and k = Kim's age.

$$\begin{cases} e + v = 45 \\ \dfrac{1}{4}k = e - 12 \\ k = \dfrac{8}{5}e \end{cases}$$

$\dfrac{1}{4}\left(\dfrac{8}{5}e\right) = e - 12, \dfrac{8}{20}e = e - 12, -\dfrac{12}{20}e = -12, e = 20; k = \dfrac{8}{5}(20), k = 32;$
$20 + v = 45, v = 25$; Emma is 20 years old, Vangie is 25 years old, and
Kim is 32 years old.

26. Let m = Mary's present age, then $m - 13$ = Mary's age 13 years ago and
$m + 2$ = Mary's age in 2 years.
Let j = Jim's present age, then $j + 2$ = Jim's age in 2 years.
Let g = Grace's present age, then $g - 13$ = Grace's age 13 years ago.

$$\begin{cases} m - 13 = \dfrac{5}{17}(g - 13) \\ m + 2 = \dfrac{1}{2}(j + 2) \\ m + j + g = 118 \end{cases} \longrightarrow \begin{cases} \dfrac{17}{5}(m - 13) = g - 13 \\ 2(m + 2) = j + 2 \\ m + j + g = 118 \end{cases} \longrightarrow \begin{cases} \dfrac{17}{5}m - \dfrac{221}{5} = g - 13 \\ 2m + 4 = j + 2 \\ m + j + g = 118 \end{cases} \longrightarrow$$

$$\begin{cases} \dfrac{17}{5}m - \dfrac{221}{5} + \dfrac{65}{5} = g \\ 2m + 2 = j \\ m + j + g = 118 \end{cases} \longrightarrow \begin{cases} \dfrac{17}{5}m - \dfrac{156}{5} = g \\ 2m + 2 = j \\ m + j + g = 118 \end{cases}$$

$m + (2m + 2) + \left(\dfrac{17}{5}m - \dfrac{156}{5}\right) = 118, 5\left(m + 2m + 2 + \dfrac{17}{5}m - \dfrac{156}{5}\right) =$
$5(118), 5m + 10m + 10 + 17m - 156 = 590, 32m - 146 = 590,$
$32m = 736, m = 23; 23 + 2 = \dfrac{1}{2}(j + 2), 25 = \dfrac{1}{2}(j + 2), 50 = j + 2, 48 = j;$
$23 + 48 + g = 118, 71 + g = 118, g = 47$; Mary is 23 years old, Jim is

48 years old, and Grace is 47 years old. Five years from now Mary will be 28 years old, Jim will be 53 years old, and Grace will be 52 years old.

27. Let h = the number of grade points Diana received for History, e = the number of grade points she received for English, and m = the number of grade points she received for Mathematics.

$$\begin{cases} h + e + m = 255 \\ h = e + 10 \\ m + 95 = 2h \end{cases} \longrightarrow \begin{array}{l} h + e + m = 255 \\ h - 10 = e \\ m = 2h - 95 \end{array}$$

$h + (h - 10) + (2h - 95) = 255$, $4h - 105 = 255$, $4h = 360$, $h = 90$; $90 = e + 10$, $80 = e$; $90 + 80 + m = 255$, $170 + m = 255$, $m = 85$; Diana received 90 grade points for History, 80 for English, and 85 for Mathematics.

Diana's grade point average $= \dfrac{90 + 80 + 85}{3} = 85$.

28. **Project**

1. $x + (x - 4) - 6 = 0$, $2x - 4 - 6 = 0$, $2x - 10 = 0$, $2x = 10$, $x = 5$; $5 - 4 = y$, $1 = y$; $(5, 1) = A$

2. $(2y - 10) + y = 2$, $3y - 10 = 2$, $3y = 12$, $y = 4$; $2(4) - 10 = x$, $8 - 10 = x$, $-2 = x$; $(-2, 4) = M$

3. $\begin{array}{l} 2x + 3y = 8 \\ 3x + y = 5 \end{array} \longrightarrow \begin{array}{l} 2x + 3y = 8 \\ -3(3x + y) = -3(5) \end{array} \longrightarrow \begin{array}{l} 2x + 3y = 8 \\ -9x - 3y = -15 \end{array}$

$$\begin{array}{r} 2x + 3y = 8 \\ -9x - 3y = -15 \\ \hline -7x = -7 \\ x = 1 \end{array}$$

$3(1) + y = 5$, $3 + y = 5$, $y = 2$; $(1, 2) = N$

4. $2x + 3y = -5$
$\underline{5x + 3y = 1}$
$-3x = -6$
$x = 2$
$5(2) + 3y = 1, 10 + 3y = 1, 3y = -9, y = -3; (2, -3) = $ P

5. $4x + 3y = -2 \longrightarrow -2(4x + 3y) = -2(-2) \longrightarrow -8x - 6y = 4$
$8x - 2y = 12 \longrightarrow 8x - 2y = 12 \longrightarrow 8x - 2y = 12$
$-8x - 6y = 4$
$\underline{8x - 2y = 12}$
$-8y = 16$
$y = -2$
$4x + 3(-2) = -2, 4x - 6 = -2, 4x = 4, x = 1; (1, -2) = $ L

6. $-5x + 2y = -11 \longrightarrow 5(-5x + 2y) = 5(-11) \longrightarrow -25x + 10y = -55$
$3x + 5y = 19 \longrightarrow 2(3x + 5y) = 2(19) \longrightarrow 6x + 10y = 38$
$-25x + 10y = -55$
$\underline{6x + 10y = 38}$
$-31x = -93$
$x = 3$
$3(3) + 5y = 19, 9 + 5y = 19, 5y = 10, y = 2; (3, 2) = $ C

7. $2x + 5y = 18 \longrightarrow 2x + 5y = 18 \longrightarrow 2x + 5y = 18$
$-5x + y = -18 \longrightarrow 5(-5x + y) = 5(-18) \longrightarrow -25x + 5y = -90$
$2x + 5y = 18$
$\underline{-25x + 5y = -90}$
$27x = 108$
$x = 4$
$2(4) + 5y - 18 = 0, 8 + 5y - 18 = 0, 5y - 10 = 0, 5y = 10, y = 2; (4, 2) = $ F

8. $3x + y = 10$
$\underline{2x + y = 7}$
$x = 3$
$3(3) + y = 10, y = 1; (3, 1) = $ I

9. $4x - 2y = 20 \longrightarrow 4x - 2y = 20 \longrightarrow 4x - 2y = 20$
$x + 5y = -17 \longrightarrow -4(x + 5y) = -4(-17) \longrightarrow -4x - 20y = 68$
$4x - 2y = 20$
$\underline{-4x - 20y = 68}$
$-22y = 88$
$y = -4$
$x + 5(-4) = -17, x - 20 = -17, x = 3; (3, -4) = $ S

10. $4x + 3y = 7$

$\underline{4x + 4y = 12}$

$-y = -5$

$y = 5$

$4x = 7 - 3(5),\ 4x = 7 - 15,\ 4x = -8,\ x = -2;\ (-2, 5) = E$

Palindrome 1: 2-1 8-9 1-9 9-10-5-7-5-10-9-9 1-9 8 1-2

            MA IS AS   SELFLESS   AS I AM

Palindrome 2: 1 2-1-3, 1 4-5-1-3, 1 6-1-3-1-5, 4-1-3-1-2-1

            A MAN, A PLAN, A CANAL, PANAMA

pages 475–476 Class Exercises

1. $h + c$

2. $4.98h;\ 498h$

3. $6.50c;\ 650c$

4. $4.98h + 6.50c$ (in dollars)

5. $\begin{cases} h(4.98) + c(6.50) = (5.79)(15) \\ h + c = 15 \end{cases}$

6. No calculations with decimals.

7. $0.075x$

8. $0.075x + 0.08y$

9. $x + y$

10. $\begin{cases} x + y = 17{,}000 \\ 0.075x + 0.08y = 1347.50 \end{cases}$

pages 476–478 Practice Exercises

A **1.** Let $d =$ the number of dimes and $n =$ the number of nickels George saved.

$\begin{cases} n + d = 28 \\ 0.05n + 0.10d = 2.60 \end{cases} \rightarrow \begin{array}{l} n + d = 28 \\ 5n + 10d = 260 \end{array} \rightarrow \begin{array}{l} -5(n + d) = -5(28) \\ 5n + 10d = 260 \end{array} \rightarrow \begin{array}{l} -5n - 5d = -140 \\ 5n + 10d = 260 \end{array}$

$-5n - 5d = -140$

$\underline{5n + 10d = 260}$

$5d = 120$

$d = 24$

$n + 24 = 28,\ n = 4$; George saved 4 nickels and 24 dimes.

2. Let $n =$ the number of nickels and $q =$ the number of quarters Lorena has.

$\begin{cases} n + q = 26 \\ 0.05n + 0.25q = 3.10 \end{cases} \longrightarrow \begin{array}{l} n + q = 26 \\ 5n + 25q = 310 \end{array} \longrightarrow \begin{array}{l} -5(n + q) = -5(26) \\ 5n + 25q = 310 \end{array} \longrightarrow$

$-5n - 5q = -130$

$5n + 25q = 310$

$-5n - 5q = -130$

$\underline{5n + 25q = 310}$

$20q = 180$

$q = 9$

$n + 9 = 26,\ n = 17$; Lorena has 17 nickels and 9 quarters.

3. Let a = the number of adult tickets sold and c = number of children's tickets sold.

$$\begin{cases} a + c = 175 \\ 6a + 2c = 750 \end{cases} \longrightarrow \begin{array}{l} -6(a + c) = -6(175) \\ 6a + 2c = 750 \end{array} \longrightarrow \begin{array}{l} -6a - 6c = -1050 \\ 6a + 2c = 750 \end{array}$$

$$\begin{array}{r} -6a - 6c = -1050 \\ \underline{6a + 2c = 750} \\ -4c = -300 \\ c = 75 \end{array}$$

75 children's tickets were sold.

4. Let x = the number of \$6 tickets and y = the number of \$1.50 tickets sold.

$$\begin{cases} x + y = 371 \\ 6x + 1.5y = 822 \end{cases} \longrightarrow \begin{array}{l} x + y = 371 \\ 60x + 15y = 8220 \end{array} \longrightarrow \begin{array}{l} x = 371 - y \\ x + \frac{1}{4}y = 137 \end{array}$$

$371 - y + \frac{1}{4}y = 137$, $-\frac{3}{4}y = -234$, $y = -234\left(-\frac{4}{3}\right)$, $y = 312$;

312 were \$1.50 tickets.

5. Let p = the number of pounds of peanuts and r = the number of pounds of raisins put in the mixture.

$$\begin{cases} p + r = 50 \\ 1.20p + 2.10r = 50(1.47) \end{cases} \longrightarrow \begin{array}{l} p + r = 50 \\ 120p + 210r = 50(147) \end{array} \longrightarrow$$

$$\begin{array}{l} p + r = 50 \\ 120p + 210r = 7350 \end{array} \longrightarrow \begin{array}{l} p = 50 - r \\ 120p + 210r = 7350 \end{array}$$

$120(50 - r) + 210r = 7350$, $6000 - 120r + 210r = 7350$, $6000 + 90r = 7350$, $90r = 1350$, $r = 15$; $p + 15 = 50$, $p = 35$;

35 lb of peanuts and 15 lb of raisins are needed to make a 50 lb mixture.

6. Let p = the number of kilograms of pinto beans and k = the number of kilograms of kidney beans in the mixture.

$$\begin{cases} p + k = 100 \\ 0.69p + 0.89k = 81 \end{cases} \longrightarrow \begin{array}{l} p + k = 100 \\ 69p + 89k = 8100 \end{array} \longrightarrow \begin{array}{l} p = 100 - k \\ 69p + 89k = 8100 \end{array}$$

$69(100 - k) + 89k = 8100$, $6900 - 69k + 89k = 8100$, $6900 + 20k = 8100$, $20k = 1200$, $k = 60$; $p + 60 = 100$, $p = 40$; The mixture contains 40 kilograms of pinto beans and 60 kilograms of kidney beans.

7. Let x = the amount invested at 7.5% annual interest and y = the amount invested at 9% annual interest.

$$\begin{cases} x + y = 32{,}000 \\ 0.075x + 0.09y = 2670 \end{cases} \longrightarrow \begin{array}{l} x + y = 32{,}000 \\ 75x + 90y = 2{,}670{,}000 \end{array} \longrightarrow \begin{array}{l} y = 32{,}000 - x \\ 75x + 90y = 2{,}670{,}000 \end{array}$$

$75x + 90(32{,}000 - x) = 2{,}670{,}000$, $5x + 6(32{,}000 - x) = 178{,}000$,

$5x + 192{,}000 - 6x = 178{,}000$, $-x = -14{,}000$, $x = 14{,}000$; $14{,}000 + y = 32{,}000$,

$y = 18{,}000$; 18,000 is invested at the higher rate.

8. Let x = the amount invested at 5.9% annual interest and y = the amount invested at 6.75% annual interest.

$$\begin{cases} x + y = 7600 \\ 0.059x + 0.0675y = 481 \end{cases} \longrightarrow \begin{array}{l} y = 7600 - x \\ 0.059x + 0.0675y = 481 \end{array}$$

$0.059x + 0.0675(7600 - x) = 481,\ 0.059x + 513 - 0.0675x = 481,$
$-0.0085x + 513 = 481,\ -0.0085x = -32,\ x = 3764.71;\ \$3764.71$ is in the account at the lower interest.

9. Let f = the number of five dollar bills and t = the number of ten dollar bills.

$$\begin{cases} f + t = 124 \\ 5f + 10t = 840 \end{cases} \longrightarrow \begin{array}{l} -5(f + t) = -5(124) \\ 5f + 10t = 840 \end{array} \longrightarrow \begin{array}{l} -5f - 5t = -620 \\ 5f + 10t = 840 \end{array}$$

$$\begin{array}{r} -5f - 5t = -620 \\ \underline{5f + 10t = 840} \\ 5t = 220;\ t = 44 \end{array}$$

$f + 44 = 124,\ f = 80;$ The teller has 80 five dollar bills and 44 ten dollar bills.

10. Let f = the number of five dollar bills and o = the number of one dollar bills.

$$\begin{cases} f + o = 87 \\ 5f + o = 179 \end{cases}$$

$$\begin{array}{r} f + o = 87 \\ \underline{5f + o = 179} \\ -4f = -92;\ f = 23 \end{array}$$

$23 + o = 87,\ o = 64;$ There are 23 five dollar bills and 64 one dollar bills.

11. Let x = the amount invested at 7.5% and y = the amount invested at 6.0%.

$$\begin{cases} x = 2y \\ 0.075x + 0.06y = 840 \end{cases}$$

$0.075(2y) + 0.06y = 840,\ 0.15y + 0.06y = 840,\ 0.21y = 840,\ y = 4000;$
$x = 2(4000),\ x = 8000;$ Mrs. Chavis invested \$8000 at 7.5% and \$4000 at 6.0%.

B 12. Let x = the number of milliliters of 30% solution and y = the number of milliliters of 80% solution put in the mixture.

	Substance	No. mL solution	No. mL alcohol
Start with	30% alcohol solution	x	$0.3x$
Add	80% alcohol solution	y	$0.8y$
Finish with	50% alcohol solution	100	0.5(100)

$$\begin{cases} x + y = 100 \\ 0.3x + 0.8y = 0.5(100) \end{cases} \longrightarrow \begin{array}{l} y = 100 - x \\ 3x + 8y = 500 \end{array}$$

$3x + 8(100 - x) = 500,\ 3x + 800 - 8x = 500,\ -5x = -300,\ x = 60;$
$60 + y = 100,\ y = 40;$ 60 mL of the 30% alcohol solution and 40 mL of the 80% alcohol solution are put in mixture.

13. Let x = the number of gallons of milk with 4% butterfat and y = the number of gallons of cream with 40% butterfat added to the mixture.

	Substance	No. gal solution	No. gal butterfat
Start with	4% butterfat	x	$0.04x$
Add	40% butterfat	y	$0.4y$
Finish with	20% butterfat	36	$0.2(36)$

$$\begin{cases} x + y = 36 \\ 0.04x + 0.4y = 0.20(36) \end{cases} \longrightarrow \begin{array}{l} x + y = 36 \\ 4x + 40y = 720 \end{array} \longrightarrow \begin{array}{l} y = 36 - x \\ 4x + 40y = 720 \end{array}$$

$4x + 40(36 - x) = 720$, $x + 10(36 - x) = 180$, $x + 360 - 10x = 180$, $-9x = -180$, $x = 20$; $20 + y = 36$, $y = 16$; 20 gallons of milk and 16 gallons of cream are added to mixture.

14. Let f = the number of five dollar tickets and t = the number of two dollar tickets.

$$\begin{cases} t = 3\frac{1}{2}f \\ 5f + 500 = 2t \end{cases}$$

$5f + 500 = 2\left(3\frac{1}{2}f\right)$, $5f + 500 = 7f$, $500 = 2f$, $250 = f$;

250 five dollar tickets were printed.

15. Let a = the number of adults and c = the number of children who attended the football game.

$$\begin{cases} a + c = 350 \\ 2.25a + c = 600 \end{cases}$$

$$\begin{array}{l} a + c = 350 \\ \underline{2.25a + c = 600} \\ \quad -1.25a = -250 \\ \qquad a = 200 \end{array}$$

$200 + c = 350$, $c = 150$; 200 adults and 150 children attended the football game.

16. Let x = the number of liters of sauce with 70% tomato paste and y = the number of liters of sauce with 40% tomato paste added.

$$\begin{cases} x + y = 5 \\ 0.70x + 0.40y = 0.60(5) \end{cases} \longrightarrow \begin{array}{l} x + y = 5 \\ 70x + 40y = 300 \end{array} \longrightarrow \begin{array}{l} x = 5 - y \\ 70x + 40y = 300 \end{array}$$

$70(5 - y) + 40y = 300$, $7(5 - y) + 4y = 30$, $35 - 7y + 4y = 30$,

$-3y = -5$, $y = \dfrac{5}{3} = 1\frac{2}{3}$; $x + 1\frac{2}{3} = 5$, $x = 3\frac{1}{3}$; $3\frac{1}{3}$ liters of sauce with 70%

tomato paste and $1\frac{2}{3}$ liters of sauce with 40% tomato paste should be

used to make the new sauce.

17. Let $p =$ the number of pounds of peanuts and $w =$ number of pounds of walnuts in the box.

$$\begin{cases} p + 3 = w \\ 9.95w + 6.50p = 62.75 \end{cases} \longrightarrow \begin{array}{l} p + 3 = w \\ 995w + 650p = 6275 \end{array}$$

$995(p + 3) + 650p = 6275$, $199(p + 3) + 130p = 1255$, $199p + 597 + 130p = 1255$, $329p + 597 = 1255$, $329p = 658$, $p = 2$; $2 + 3 = w$, $5 = w$; 2 lbs of peanuts and 5 lbs of walnuts are in a box.

18. Let $n =$ the number of nickels and $d =$ the number of dimes.

$$\begin{cases} n + 10 = d \\ 5n + 10d = 310 \end{cases}$$

$5n + 10(n + 10) = 310$, $5n + 10n + 100 = 310$, $15n + 100 = 310$, $15n = 210$, $n = 14$; $14 + 10 = d$, $24 = d$; Rosa has 24 dimes and 14 nickels.

C 19. Let $p =$ the number of pennies, $n =$ the number of nickels, and $d =$ the number of dimes.

$$\begin{cases} p + n + d = 90 \\ 10d + 5n + p = 285 \\ 2(n + d) = p \end{cases} \longrightarrow \begin{array}{l} p + n + d = 90 \\ 10d + 5n + p = 285 \\ n + d = \dfrac{p}{2} \end{array}$$

$n = \dfrac{p}{2} - d$, $p + \dfrac{p}{2} - d + d = 90$, $p + \dfrac{p}{2} = 90$, $\dfrac{3}{2}p = 90$, $p = 60$;
$60 + n + d = 90$, $n + d = 30$, $d = 30 - n$; $10(30 - n) + 5n + 60 = 285$,
$300 - 10n + 5n + 60 = 285$, $-5n + 360 = 285$, $-5n = -75$, $n = 15$;
$60 + 15 + d = 90$, $75 + d = 90$, $d = 15$; There are 60 pennies, 15 nickels and 15 dimes.

20. Let $x =$ the amount invested in the first bank at 6% yearly interest, $y =$ the amount invested in the second bank at 6% yearly interest, and $z =$ the amount invested in the third bank at 7% yearly interest.

$$\begin{cases} x + y + z = 10{,}000 \\ 0.06x + 0.06y + 0.07z = 640 \\ x = y \end{cases}$$

$y + y + z = 10{,}000$, $2y + z = 10{,}000$, $z = 10{,}000 - 2y$;
$0.06y + 0.06y + 0.07(10{,}000 - 2y) = 640$,
$0.06y + 0.06y + 700 - 0.14y = 640$, $-0.02y + 700 = 640$, $-0.02y = -60$
$y = 3000$; $x = 3000$; $3000 + 3000 + z = 10{,}000$, $z = 4000$;
Miquel invested \$3000 in the first bank, \$3000 in the second bank and \$4000 in the third bank.

21. $n - 0.49$

22. $\dfrac{c}{-8}$

23. xy

24. $n + 0.5$

25. $n - 57 = -22$, $n = 35$; The number is 35.

26. $3n + 27 = 3$, $3n = -24$, $n = -8$; The number is -8.

27. 2.75×10^8 **28.** 5×10^3 **29.** 2.36×10^{-3} **30.** 5×10^{-6}

31. $\dfrac{j}{3^3 k^2} = \dfrac{j}{27k^2}$ **32.** $(x^9 y^{-6})(x^{-1} y) = x^8 y^{-5} = \dfrac{x^8}{y^5}$ **33.** $\dfrac{2y^{2\,+\,2}}{3m^{4\,+\,4}} = \dfrac{2y^4}{3m^8}$

page 478   Biography: Sonja Corvin-Kowalewski

Check students' work.

page 481   Class Exercises

1. Let t_1 = the time against the wind, t_2 = the time with the wind, and r = the average rate of speed for the round trip.

	rate	$\times$ time =	distance
Against wind	$255 - 15$	t_1	2160
With wind	$255 + 15$	t_2	2160

$\begin{cases} 240t_1 = 2160 \\ 270t_2 = 2160 \end{cases}$

$t_1 = \dfrac{2160}{270} = 9$ hours; $t_2 = \dfrac{2160}{270} = 8$ hours

Total time for the round trip = $9 + 8 = 17$ hours

$r \cdot 17 = 2(2160)$, $r \cdot 17 = 4320$, $r = 254.1$; The average rate of speed for the round trip is 254.1 mi/h.

2. Let c = the rate of the current and r = the rate of speed in still water.

	rate	$\times$ time =	distance
Downstream	$r + c$	3	36
Upstream	$r - c$	3	24

$\begin{cases} (r - c)3 = 24 \longrightarrow r - c = 8 \\ (r + c)3 = 36 \longrightarrow r + c = 12 \end{cases}$

$\begin{array}{l} r - c = 8 \\ \underline{r + c = 12} \\ \quad 2r = 20 \\ \quad\; r = 10 \end{array}$ $(10 + c)3 = 36$, $10 + c = 12$, $c = 2$; The rate of the current is 2 mi/h and the rate of speed of the boat in still water is 10 mi/h.

pages 481–483   Practice Exercises

A   **1.** Let t_1 = the time against the wind, t_2 = the time with the wind, and r = the average rate of speed.

	rate	$\times$ time =	distance
Against wind	$225 - 45$	t_1	1080
With wind	$225 + 45$	t_2	1080

$\begin{cases} 180t_1 = 1080 \\ 270t_2 = 1080 \end{cases}$ $t_1 = \dfrac{1080}{180} = 6$, $t_2 = \dfrac{1080}{270} = 4$;

Total time for the round trip = 6 + 4 = 10 hours.

$r \cdot 10 = 2160$, $r = 216$; The aircraft's average rate of speed is 216 mi/h.

2. Let t_1 = the time against the wind, t_2 = the time with the wind, and r = the average rate of speed.

	rate	× time =	distance
Against wind	12 − 3	t_1	45
With wind	12 + 3	t_2	45

$$\begin{cases} 9t_1 = 45 \\ 15t_2 = 45 \end{cases}$$

$t_1 = \dfrac{45}{9} = 5$, $t_2 \dfrac{45}{15} = 3$;

Total time for the round trip = 5 + 3 = 8 hours.

$r \cdot 8 = 90$; $r = 11.25$. The cyclists average rate of speed is 11.25 mi/h.

● 3. Let r = the rate of rowing in still water and c = the rate of the current.

	rate	× time =	distance
Downstream	$r + c$	2	16
Upstream	$r - c$	2	8

$$\begin{cases} (r + c)2 = 16 \longrightarrow r + c = 8 \\ (r - c)2 = 8 \longrightarrow r - c = 4 \end{cases}$$

$r + c = 8$
$\underline{r - c = 4}$
$\quad 2r = 12$
$\quad\;\; r = 6$

$(6 + c)2 = 16$, $6 + c = 8$, $c = 2$; The rate of rowing in still water is 6 mi/h and the rate of the current is 2 mi/h.

4. Let r = the rate of speed of the plane in calm air and w = the rate of the wind.

	rate	× time =	distance
Against wind	$r - w$	5	3000
With wind	$r + w$	5	4000

$$\begin{cases} (r - w)5 = 3000 \longrightarrow r - w = 600 \\ (r + w)5 = 4000 \longrightarrow r + w = 800 \end{cases}$$

$r - w = 600$
$\underline{r + w = 800}$
$\quad 2r = 1400$
$\quad\;\; r = 700$

$(700 + w)5 = 4000$, $700 + w = 800$, $w = 100$; The rate of speed of the plane in calm air is 700 km/h and the rate of the wind is 100 km/h.

5. Let r = the rate of speed of the boat in calm water and c = the rate of the current.

rate $\times$ time = distance

	rate	time	distance
Downstream	$r + c$	1	12
Upstream	$r - c$	2	12

$$\begin{cases} r + c = 12 & \longrightarrow r + c = 12 \\ (r - c)2 = 12 & \longrightarrow r - c = 6 \end{cases}$$

$r + c = 12$
$\underline{r - c = 6}$
$\quad 2r = 18$
$\quad\quad r = 9$

The rate of speed of the boat in calm water is 9 km/h.

6. Let r = the rate of speed of the plane in calm air and w = the rate of the wind.

rate $\times$ time = distance

	rate	time	distance
Against wind	$r - w$	3	1200
With wind	$r + w$	2	1200

$$\begin{cases} (r - w)3 = 1200 & \longrightarrow r - w = 400 \\ (r + w)2 = 1200 & \longrightarrow r + w = 600 \end{cases}$$

$r - w = 400$
$\underline{r + w = 600}$
$\quad 2r = 1000$
$\quad\quad r = 500$

The rate of speed of the plane in calm air is 500 mi/h.

7. Let r = the rate of speed of the plane in calm air and w = the rate of the wind.

rate $\times$ time = distance

	rate	time	distance
Against wind	$r - w$	5	2400
With wind	$r + w$	4	2400

$$\begin{cases} (r - w)5 = 2400 & \longrightarrow r - w = 480 \\ (r + w)4 = 2400 & \longrightarrow r + w = 600 \end{cases}$$

$r - w = 480$
$\underline{r + w = 600}$
$\quad 2r = 1080$
$\quad\quad r = 540$

$(540 + w)4 = 2400$, $540 + w = 600$, $w = 60$; The rate of speed of the plane in calm air is 540 km/h and the rate of the wind is 60 km/h.

8. Let r = the rate of speed of the plane in calm air and w = the rate of the wind.

rate $\times$ time = distance

	rate	time	distance
Against wind	$r - w$	5	700
With wind	$r + w$	$3\frac{1}{2}$	700

$$\begin{cases} (r-w)5 = 700 \longrightarrow r-w = 140 \\ (r+w)3\frac{1}{2} = 700 \longrightarrow r+w = 200 \end{cases}$$

$$\begin{array}{r} r - w = 140 \\ r + w = 200 \\ \hline 2r = 340 \\ r = 170 \end{array}$$

$(170 - w)5 = 700$, $170 - w = 140$, $-w = -30$, $w = 30$; The rate of the wind is 30 mi/h.

9. Let r = the rate of speed of the boat in still water and c = the rate of the current.

	rate	× time	= distance
Downstream	$r + c$	4	28
Upstream	$r - c$	7	28

$$\begin{cases} (r+c)4 = 28 \longrightarrow r+c = 7 \\ (r-c)7 = 28 \longrightarrow r-c = 4 \end{cases}$$

$$\begin{array}{r} r + c = 7 \\ r - c = 4 \\ \hline 2r = 11 \\ r = 5\frac{1}{2} \end{array}$$

Natasha's rate without the current was $5\frac{1}{2}$ mi/h.

10. Let r = Pam's swimming rate in still water and c = the rate of the current.

	rate	× time	= distance
Downstream	$r + c$	2	10
Upstream	$r - c$	8	10

$$\begin{cases} (r+c)2 = 10 \longrightarrow r+c = 5 \\ (r-c)8 = 10 \longrightarrow r-c = \dfrac{5}{4} \end{cases}$$

$$\begin{array}{r} r + c = 5 \\ r - c = \dfrac{5}{4} \\ \hline 2r = 6\dfrac{1}{4} \\ r = 3\dfrac{1}{8} \end{array}$$

$\left(3\frac{1}{8} + c\right)2 = 10$, $3\frac{1}{8} + c = 5$, $c = 1\frac{7}{8}$ or 1.875; The rate of the current is 1.875 mi/h.

11. Let t_1 = the time traveling against the wind and t_2 = the time traveling with the wind.

	rate	× time	= distance
Against wind	14	t_1	140
With wind	20	t_2	140

$$\begin{cases} 14t_1 = 140 \\ 20t_2 = 140 \end{cases}$$

$$t_1 = \frac{140}{14} = 10, \ t_2 = \frac{140}{20} = 7;$$

Total time = 10 + 7 = 17 hours. Noah traveled for 17 hours.

12. Let r = the rate of speed of the motorcycle in still air and w = the rate of the wind.

	rate	× time	= distance
Against wind	$r - w$	3	100
With wind	$r + w$	2	100

$$\begin{cases} (r - w)3 = 100 \longrightarrow r - w = \dfrac{100}{3} \\ (r + w)2 = 100 \longrightarrow r + w = 50 \end{cases}$$

$$r - w = \frac{100}{3}$$
$$\underline{r + w = 50}$$
$$2r = \frac{250}{3}$$
$$r = 41\frac{2}{3}$$

$$\left(41\frac{2}{3} + w\right)2 = 100, \ 41\frac{2}{3} + w = 50, \ w = 8\frac{1}{3}; \text{ The rate of the wind is } 8\frac{1}{3} \text{ mi/h.}$$

● 13. Let r = the rate of speed of the plane with no wind and w = the rate of the wind.

	rate	× time	= distance
Against wind	$r - w$	6	2100
With wind	$r + w$	5	2100

$$\begin{cases} (r - w)6 = 2100 \longrightarrow r - w = 350 \\ (r + w)5 = 2100 \longrightarrow r + w = 420 \end{cases}$$

$$r - w = 350$$
$$\underline{r + w = 420}$$
$$2r = 770$$
$$r = 385$$

The rate of speed of the plane with no wind is 385 mi/h.

14. Let r = the rate of speed of the plane with no wind and w = the rate of the wind.

	rate	× time	= distance
Against wind	$r - w$	$\frac{3}{4}$	300
With wind	$r + w$	$\frac{2}{3}$	300

$$\begin{cases} (r - w)\frac{3}{4} = 300 \longrightarrow r - w = 400 \\ (r + w)\frac{2}{3} = 300 \longrightarrow r + w = 450 \end{cases}$$

$$\begin{array}{r} r - w = 400 \\ \underline{r + w = 450} \\ 2r = 850 \\ r = 425 \end{array}$$

$(425 + w)\frac{2}{3} = 300$, $425 + w = 450$, $w = 25$; The rate of the wind is 25 mi/h.

B 15. Let r = the rate of speed of the boat in calm water and c = the rate of the current.

	rate	× time	= distance
Upstream	$r - c$	$\frac{5}{6}$	10
Downstream	$r + c$	$\frac{3}{4}$	15

$$\begin{cases} (r - c)\frac{5}{6} = 10 \longrightarrow r - c = 12 \\ (r + c)\frac{3}{4} = 15 \longrightarrow r + c = 20 \end{cases}$$

$$\begin{array}{r} r - c = 12 \\ \underline{r + c = 20} \\ 2r = 32 \\ r = 16 \end{array}$$

The rate of speed of the boat in calm water is 16 km/h.

16. Let r = the rate of swimming in still water and c = the rate of the current.

	rate	× time	= distance
Upstream	$r - c$	1	40
Downstream	$r + c$	1	80

$$\begin{cases} r - c = 40 \\ r + c = 80 \end{cases}$$

$$\begin{array}{r} r - c = 40 \\ r + c = 80 \\ \hline 2r = 120 \\ r = 60 \end{array}$$

The swimmer swims at 60 m/min in still water.

17. Let r = the rate of the steamboat in still water and c = the rate of the current.

rate × time = distance

Downstream	$r + c$	3	8.4
Upstream	$r - c$	3	$\frac{3}{7}(8.4)$

$$\begin{cases} (r + c)3 = 8.4 \\ (r - c)3 = \frac{3}{7}(8.4) \end{cases} \longrightarrow \begin{array}{l} (r + c)3 = 8.4 \\ (r - c)3 = 3.6 \end{array} \longrightarrow \begin{array}{l} r + c = 2.8 \\ r - c = 1.2 \end{array}$$

$$\begin{array}{r} r + c = 2.8 \\ r - c = 1.2 \\ \hline 2r = 4.0 \\ r = 2 \end{array}$$

$(2 + c)3 = 8.4$, $2 + c = 2.8$, $c = 0.8$; The rate of the current is 0.8 mi/h and the rate of the steamboat is 2 mi/h.

18. Let r = the rate of the canoe in still water and c = the rate of the current.

rate × time = distance

Downstream	$r + c$	6	40
Upstream	$r - c$	3(6)	$\frac{3}{5}(40)$

$$\begin{cases} (r + c)6 = 40 \\ (r - c)18 = \frac{3}{5}(40) \end{cases} \longrightarrow \begin{array}{l} (r + c)6 = 40 \\ (r - c)18 = 24 \end{array} \longrightarrow \begin{array}{l} r + c = \dfrac{40}{6} \\ r - c = \dfrac{24}{18} \end{array}$$

$$\begin{array}{r} r + c = \dfrac{40}{6} \\[2mm] r - c = \dfrac{24}{18} \\ \hline 2r = 8 \\ r = 4 \end{array}$$

$(4 + c)6 = 40$, $4 + c = 6\frac{2}{3}$, $c = 2\frac{2}{3}$;

The rate of the canoe in still water is 4 mi/h and the rate of the current is $2\frac{2}{3}$ mi/h.

19. Let r = the rate of the ultralight without wind and w = the rate of the wind.

rate $\times$ time = distance

Against wind	$r - w$	5	180
With wind	$r + w$	3	180

$$\begin{cases} (r - w)5 = 180 \longrightarrow r - w = 36 \\ (r + w)3 = 180 \longrightarrow r + w = 60 \end{cases}$$

$$\begin{aligned} r - w &= 36 \\ \underline{r + w} &= \underline{60} \\ 2r &= 96 \\ r &= 48 \end{aligned}$$

$(48 + w)3 = 180$, $48 + w = 60$, $w = 12$; The rate of the wind is 12 km/h.

20. Let r = the rate the eagle flies without wind and w = the rate of the wind.

rate $\times$ time = distance

Against wind	$r - w$	7	$\frac{1}{3}(300)$

$(r - w)7 = \frac{1}{3}(300)$, $(r - w)7 = 100$, $r - w = \frac{100}{7} = 14\frac{2}{7}$;

The rate of the eagle in the air against the wind is $14\frac{2}{7}$ mi/h.

C 21. Let r = the rate of the salmon, c = the rate of the current and d = the distance traveled upstream.

rate $\times$ time = distance

Upstream	$r - c$	t	d
Downstream	$r + c$	t	$2d$

$$\begin{cases} (r - c)t = d \longrightarrow r - c = \dfrac{d}{t} \longrightarrow r = \dfrac{d}{t} + c \\ (r + c)t = 2d \longrightarrow r + c = \dfrac{2d}{t} \longrightarrow r = \dfrac{2d}{t} - c \end{cases}$$

$$\frac{d}{t} + c = \frac{2d}{t} - c$$

22. Let w = rate of wind.

If r = the rate of speed of the plane in still air + the rate of the wind, then $r - w$ = the rate of the plane in still air.

If s = the rate of speed of the plane in still air − the rate of the wind, then $s + w$ = the rate of plane in still air.

$(r - w) + (s + w) = 2$ (the rate of speed of the plane in still air) then

$$\frac{(r - w) + (s + w)}{2} = \frac{r + s}{2}$$

$\dfrac{r + s}{2}$ is the rate of speed of the plane in still air.

Since r = the rate of speed of the plane in still air + the rate of the wind and $\frac{r+s}{2}$ is the rate of speed of the plane in still air, then $r - \frac{r+s}{2}$ = rate of the wind.

page 483 Career: Geophysicist

Let d = the distance the station is from the earthquake.
Let s = the time the primary wave travels, then $s + 16$ = the time the secondary wave travels.
$d = (5)(s)$
$d = (3)(s + 16)$
Since $d = d$, $5s = 3(s + 16)$, $5s = 3s + 48$, $2s = 48$, $s = 24$;
$d = 5s$, $d = 5(24)$, $d = 120$; The station is 120 miles from the earthquake.

page 484 Capsule Review

1.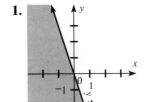

2.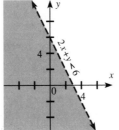

3.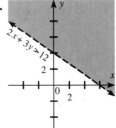

page 485 Class Exercises

1. C   **2.** A   **3.** D   **4.** B

5. $1 + -3 \overset{?}{\leq} 6$, $-2 \leq 6$; true
$1 - (-3) \overset{?}{<} 1$, $4 \nless 4$; false; no

6. $1 + 2 \overset{?}{\leq} 6$, $3 \leq 6$; true
$1 - 2 \overset{?}{<} 1$, $-1 < 1$; true; yes

7. $0 + 0 \overset{?}{\leq} 6$, $0 \leq 6$; true
$0 - 0 \overset{?}{<} 1$, $0 < 1$; true; yes

8. $6 + 0 \overset{?}{\leq} 6$, $6 \leq 6$; true
$6 - 0 \overset{?}{<} 1$, $6 \nless 1$; false; no

9. $-2 + 1 \overset{?}{\leq} 6$, $-1 \leq 6$; true
$-2 - 1 \overset{?}{<} 1$, $-3 < 1$; true; yes

10. $7 + 1 \overset{?}{\leq} 6$, $8 \nleq 6$; false
$7 - 1 \overset{?}{<} 1$, $6 \nless 1$; false; no

11. $8 + 6 \overset{?}{\leq} 6$, $14 \nleq 6$; false
$8 - 6 \overset{?}{<} 1$, $2 \nless 1$; false; no

12. $-3 + -1 \overset{?}{\leq} 6$, $-4 \leq 6$; true
$-3 - (-1) \overset{?}{<} 1$, $-2 < 1$; true; yes

A 1.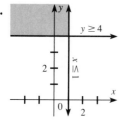

2.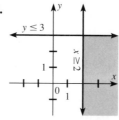

3.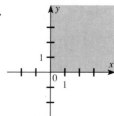

4.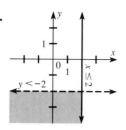

5.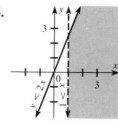

6.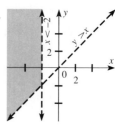

7.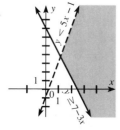

8.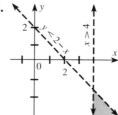

9.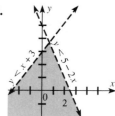

10.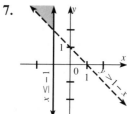

11.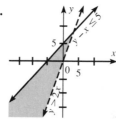

12.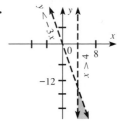

13.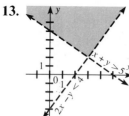

14.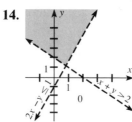

15.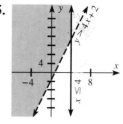

16.

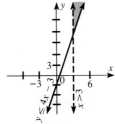

B 17.

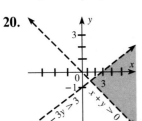

18.

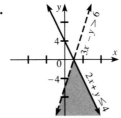

19.

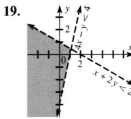

20.

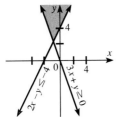

21.

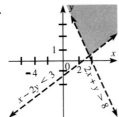

22.

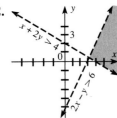

23.

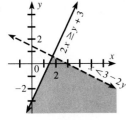

24.

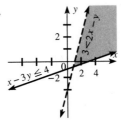

25.

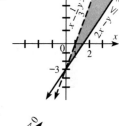

26.

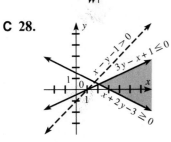

27.

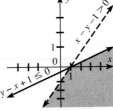

C 28.

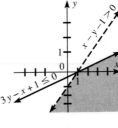

29.

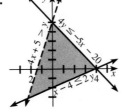

30.

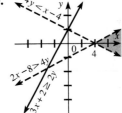

31.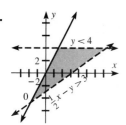

32.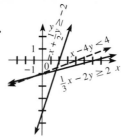

33.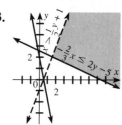

34. Let x = the number of pages typist A types and y = the number of pages typist B types.

$$\begin{cases} x \le \dfrac{2}{3}y \\ x + y > 20 \\ x + y < 30 \end{cases}$$

One possible solution is Typist A can type 10 pg/d and Typist B can type 15 pg/d.

35. Let l = the length of the lot and w = the width of the lot.

$$\begin{cases} l > 800 \\ 2l + 2w \le 2600 \end{cases}$$

$$\begin{array}{ccc} l = 900 \text{ ft} & & l = 1000 \text{ ft} \\ & \text{or} & \\ w = 400 \text{ ft} & & w = 300 \text{ ft} \end{array}$$

page 487   Test Yourself

1. Let t = the tens digit and u = the units digit.

$$\begin{cases} t + u = 11 \\ 10t + u + 9 = 10u + t \end{cases} \longrightarrow \begin{matrix} t + u = 11 \\ 9t - 9u = -9 \end{matrix} \longrightarrow \begin{matrix} 9(t + u) = 9(11) \\ 9t - 9u = -9 \end{matrix} \longrightarrow \begin{matrix} 9t + 9u = 99 \\ 9t - 9u = -9 \end{matrix}$$

$$\begin{aligned} 9t + 9u &= 99 \\ \underline{9t - 9u} &= \underline{-9} \\ 18t &= 90 \\ t &= 5 \end{aligned}$$

$5 + u = 11$, $u = 6$; The two-digit number is 56.

2. Let t = the tens digit and u = the units digit.

$$\begin{cases} t = 2u \\ 10t + u = 10u + t + 27 \end{cases} \longrightarrow \begin{matrix} t = 2u \\ 9t - 9u = 27 \end{matrix}$$

$9(2u) - 9u = 27$, $18u - 9u = 27$, $9u = 27$, $u = 3$; $t = 2(3)$, $t = 6$; The two-digit number is 63.

3. Let r = Robert's present age, then $r - 7$ = Robert's age seven years ago. Let d = Dawn's present age, then $d - 7$ = Dawn's age seven years ago.

$$\begin{cases} r = d + 4 \\ d - 7 = \dfrac{3}{4}(r - 7) \end{cases} \longrightarrow \begin{matrix} r = d + 4 \\ d - 7 = \dfrac{3}{4}r - \dfrac{21}{4} \end{matrix} \longrightarrow \begin{matrix} r = d + 4 \\ d - \dfrac{3}{4}r = \dfrac{7}{4}, \end{matrix}$$

$d - \dfrac{3}{4}(d + 4) = \dfrac{7}{4}$, $d - \dfrac{3}{4}d - 3 = \dfrac{7}{4}$, $\dfrac{1}{4}d - 3 = \dfrac{7}{4}$, $\dfrac{1}{4}d = \dfrac{19}{4}$, $d = 19$;

$r = 19 + 4$, $r = 23$; Robert is 23 years old and Dawn is 19 years old.

4. Let g = Greg's present age, then $g - 5$ = Greg's age five years ago.

Let j = Jane's present age, then $j - 5$ = Jane's age five years ago.

$$\begin{cases} g = 3j \\ g - 5 = 5(j - 5) \end{cases} \longrightarrow \begin{array}{l} g = 3j \\ g - 5 = 5j - 25 \end{array} \longrightarrow \begin{array}{l} g = 3j \\ g - 5 - 5j = -25 \end{array} \longrightarrow \begin{array}{l} g = 3j \\ g - 5j = -20 \end{array}$$

$3j - 5j = -20$, $-2j = -20$, $j = 10$; $g = 3(10)$; $g = 30$;

Jane is 10 years old and Greg is 30 years old.

5. Let d = the number of dimes and q = the number of quarters.

$$\begin{cases} d + g = 52 \\ 0.10d + 0.25q = 6.25 \end{cases} \longrightarrow \begin{array}{l} d + q = 52 \\ 10d + 25q = 625 \end{array} \longrightarrow$$

$$\begin{array}{l} -10(d + q) = -10(52) \\ 10d + 25q = 625 \end{array} \longrightarrow \begin{array}{l} -10d - 10q = -520 \\ 10d + 25q = 625 \end{array}$$

$$\begin{array}{r} -10d - 10q = -520 \\ \underline{10d + 25q = 625} \\ 15q = 105 \\ q = 7 \end{array}$$

$d + 7 = 52$, $d = 45$; Anne Marie has 45 dimes and 7 quarters.

6. Let x = the number of 25¢ stamps and y = the number of 2¢ stamps.

$$\begin{cases} x + y = 9 \\ 0.25x + 0.02y = 1.10 \end{cases} \longrightarrow \begin{array}{l} x + y = 9 \\ 25x + 2y = 110 \end{array} \longrightarrow \begin{array}{l} -2(x + y) = -2(9) \\ 25x + 2y = 110 \end{array} \longrightarrow$$

$$\begin{array}{l} -2x - 2y = -18 \\ 25x + 2y = 110 \end{array}$$

$$\begin{array}{r} -2x - 2y = -18 \\ \underline{25x + 2y = 110} \\ 23x = 92 \\ x = 4 \end{array}$$

$4 + y = 9$, $y = 5$; John has four 25¢ stamps and five 2¢ stamps.

7. Let t_1 = the time against the wind and t_2 = the time with the wind.

$$\begin{cases} 30t_1 = 120 \\ 60t_2 = 120 \end{cases}$$

	rate	× time =	distance
Against wind	45 − 15	t_1	120
With wind	45 + 15	t_2	120

$t_1 = \dfrac{120}{30} = 4$; $t_2 = \dfrac{120}{60} = 2$;

Total time = 4 + 2 = 6 hours.

$r \cdot 6 = 240$, $r = \dfrac{240}{6} = 40$; The average rate of speed for the trip is 40 mi/h.

8. Let r = the rate of speed of the boat in still water and c = the rate of the current.

	rate ×	time =	distance
Downstream	$r + c$	4	56
Upstream	$r - c$	6	36

$\begin{cases} (r + c)4 = 56 \longrightarrow r + c = 14 \\ (r - c)6 = 36 \longrightarrow r - c = 6 \end{cases}$

$\begin{aligned} r + c &= 14 \\ \underline{r - c} &= \underline{6} \\ 2r &= 20 \\ r &= 10 \end{aligned}$

$(10 + c)4 = 56$, $10 + c = 14$, $c = 4$; The rate of speed of the boat is 10 mi/h and rate of the current is 4 mi/h.

9.

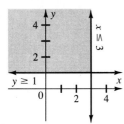

10.

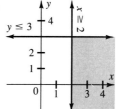

11.

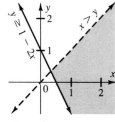

12.

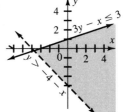

pages 488–489   Summary and Review

1.

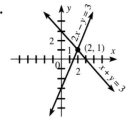

(2, 1); 1 solution;
lines intersect

2.

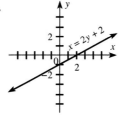

infinite number of
solutions; same line

3.

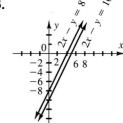

no solution;
parallel lines

4. $x + 2(2x) = 8$, $x + 4x = 8$, $5x = 8$, $x = \dfrac{8}{5}$; $y = 2\left(\dfrac{8}{5}\right)$, $y = \dfrac{16}{5}$; $\left(\dfrac{8}{5}, \dfrac{16}{5}\right)$

5. $2r + 3s = 27 \longrightarrow 2r + 3s = 27$
$4r = s + 9 \longrightarrow 4r - 9 = s$

$2r + 3(4r - 9) = 27$, $2r + 12r - 27 = 27$, $14r = 54$, $r = \dfrac{54}{14} = \dfrac{27}{7}$;

$4\left(\dfrac{27}{7}\right) - s = 9$, $\dfrac{108}{7} - s = 9$, $-s = -\dfrac{45}{7}$, $s = \dfrac{45}{7}$; $\left(\dfrac{27}{7}, \dfrac{45}{7}\right)$

6. $\dfrac{1}{4}c = d - 1 \longrightarrow c = 4d - 4$

$4c + 3d = 22 \longrightarrow 4c + 3d = 22$

$4(4d - 4) + 3d = 22$, $16d - 16 + 3d = 22$, $19d - 16 = 22$, $19d = 38$,
$d = 2$; $4c + 3(2) = 22$, $4c + 6 = 22$, $4c = 16$, $c = 4$; $(4, 2)$

7. Let $x =$ the length of the two equal sides and $y =$ the length of the base.
$\begin{cases} 2x + y = 21 \\ y = x + 3 \end{cases}$

$2x + (x + 3) = 21$, $3x + 3 = 21$, $3x = 18$, $x = 6$; $6 + 3 = y$, $9 = y$;
The lengths of the sides of the triangle are 6 ft, 6 ft, and 9 ft.

8. $2a + b = 8$
$\underline{a - b = 4}$
$3a = 12$
$a = 4$
$4 - b = 4$
$-b = 0$
$b = 0$; $(4, 0)$

9. $3p + q = 13$
$\underline{2p - q = 2}$
$5p = 15$
$p = 3$
$2(3) - q = 2$
$6 - q = 2$
$-q = -4$
$q = 4$; $(3, 4)$

10. $\dfrac{1}{2}m - \dfrac{1}{2}n = 10$

$\underline{\dfrac{3}{4}m + \dfrac{1}{2}n = 20}$

$\dfrac{5}{4}m = 30$

$m = 24$

$\dfrac{1}{2}(24) - \dfrac{1}{2}n = 10$

$12 - \dfrac{1}{2}n = 10$

$-\dfrac{1}{2}n = -2$

$n = 4$; $(24, 4)$

11. $x + 4y = 17 \longrightarrow x + 4y = 17$
$-2(3x + 2y) = -2(6) \longrightarrow -6x - 4y = -12$
$x + 4y = 17$
$\underline{-6x - 4y = -12}$
$-5x = 5$
$x = -1$

$-1 + 4y = 17$, $4y = 18$, $y = \dfrac{9}{2}$; $\left(-1, \dfrac{9}{2}\right)$

12. $4(6a + 7b) = 4(-33) \longrightarrow 24a + 28b = -132$

$7(5a - 4b) = 7(2) \longrightarrow 35a - 28b = 14$

$24a + 28b = -132$

$\underline{35a - 28b = 14}$

$ 59a = -118$

$ a = -2$

$6(-2) + 7b = -33, -12 + 7b = -33, 7b = -21, b = -3; (-2, -3)$

13. $2\left(\dfrac{1}{4}c + d\right) = 2\left(\dfrac{7}{2}\right) \longrightarrow \dfrac{1}{2}c + 2d = 7$

$\dfrac{1}{2}c - \dfrac{1}{4}d = 1 \qquad \longrightarrow \dfrac{1}{2}c - \dfrac{1}{4}d = 1$

$\dfrac{1}{2}c + 2d = 7$

$\underline{\dfrac{1}{2}c - \dfrac{1}{4}d = 1}$

$\phantom{\dfrac{1}{2}c} \dfrac{9}{4}d = 6$

$\phantom{\dfrac{1}{2}c} d = \dfrac{8}{3}$

$\dfrac{1}{4}c + \dfrac{8}{3} = \dfrac{7}{2}, \dfrac{1}{4}c = \dfrac{5}{6}, c = \dfrac{10}{3}; \left(\dfrac{10}{3}, \dfrac{8}{3}\right)$

14. Let t = the tens digit and u = the units digit.

$\begin{cases} t + u = 5 \\ 2(10t + u) + 13 = 10u + t \end{cases} \longrightarrow \begin{matrix} t + u = 5 \\ 20t + 2u + 13 = 10u + t \end{matrix} \longrightarrow$

$\begin{matrix} t + u = 5 \longrightarrow \\ 19t - 8u + 13 = 0 \longrightarrow \end{matrix} \begin{matrix} t + u = 5 \\ 19t - 8u = -13 \end{matrix} \longrightarrow \begin{matrix} u = 5 - t \\ 19t - 8u = -13 \end{matrix}$

$19t - 8(5 - t) = -13, 19t - 40 + 8t = -13, 27t - 40 = -13, 27t = 27$

$t = 1; 1 + u = 5, u = 4;$ The two-digit number is 14.

15. Let s = Sue's present age, then $s + 5$ = Sue's age in five years.

Let a = Amy's present age, then $a + 5$ = Amy's age in five years.

$\begin{cases} s + a = 35 \\ 2(a + 5) = s + 5 \end{cases} \longrightarrow \begin{matrix} s + a = 35 \\ 2a + 10 = s + 5 \end{matrix} \longrightarrow \begin{matrix} s + a = 35 \\ 2a + 5 = s \end{matrix}$

$(2a + 5) + a = 35, 3a + 5 = 35, 3a = 30, a = 10; s + 10 = 35, s = 25;$

Sue is 25 years old and Amy is 10 years old.

16. Let x = the number of 25¢ stamps and y = the number of 10¢ stamps.

$\begin{cases} x + y = 22 \\ 0.25x + 0.10y = 3.40 \end{cases}$

$\begin{matrix} x + y = 22 \\ 25x + 10y = 340 \end{matrix} \longrightarrow \begin{matrix} y = 22 - x \\ 25x + 10y = 340 \end{matrix}$

$25x + 10(22 - x) = 340, 25x + 220 - 10x = 340, 15x + 220 = 340,$

$15x = 120$, $x = 8$; $8 + y = 22$, $y = 14$; Ed bought eight 25¢ stamps and fourteen 10¢ stamps.

17. Let r = the rate of speed of the boat in still water and c = the rate of the current.

$$\text{rate} \times \text{time} = \text{distance}$$

Downstream	$r + c$	$1\frac{1}{2}$	12
Upstream	$r - c$	6	12

$$\begin{cases} (r + c)1\frac{1}{2} = 12 \longrightarrow r + c = 8 \\ (r - c)6 = 12 \longrightarrow r - c = 2 \end{cases}$$

$\quad r + c = 8$
$\quad r - c = 2$
$\quad \overline{\quad 2r = 10}$
$\quad\quad r = 5$

$(5 - c)6 = 12$, $5 - c = 2$, $-c = -3$, $c = 3$; The rate of speed of the boat in still water is 5 mi/h and the rate of the current is 3 mi/h.

18.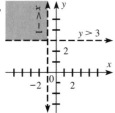

19.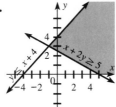

20.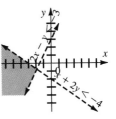

page 490   Chapter Test

1.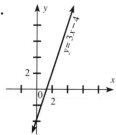

$\{(x, y): 3x - y = 4\}$

2.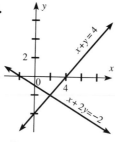

$(2, -2)$

3.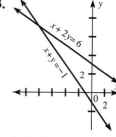

$(-8, 7)$

4. $x - (2x - 3) = 1$, $x - 2x + 3 = 1$, $-x + 3 = 1$, $-x = -2$, $x = 2$; $y = 2(2) - 3$, $y = 1$; $(2, 1)$

5. $2(x + y) = 2(2) \longrightarrow 2x + 2y = 4$
$3x - 2y = -9 \longrightarrow 3x - 2y = -9$

$2x + 2y = 4$
$\underline{3x - 2y = -9}$
$5x = -5$
$x = -1$
$-1 + y = 2, y = 3; (-1, 3)$

6. $2x + \dfrac{x}{3} = 0, \dfrac{7}{3}x = 0, x = 0, 2(0) + y = 0, y = 0; (0, 0)$

7. $2x + 1 = y$
$3x = y + 2$
$3x = (2x + 1) + 2, 3x = 2x + 3, x = 3; 2(3) + 1 = y, y = 7; (3, 7)$

8. $6\left(\dfrac{x}{3} - \dfrac{y}{2}\right) = 6(6) \longrightarrow 2x - 3y = 36 \longrightarrow 2x - 3y = 36 \longrightarrow 2x - 3y = 36$

$8\left(\dfrac{x}{2} + \dfrac{y}{8}\right) = 8(2) \longrightarrow 4x + y = 16 \longrightarrow 3(4x + y) = 3(16) \longrightarrow 12x + 3y = 48$

$2x - 3y = 36$
$\underline{12x + 3y = 48}$
$14x = 84$
$x = 6$

$\dfrac{6}{3} - \dfrac{y}{2} = 6, 2 - \dfrac{y}{2} = 6, 4 - y = 12, y = -8; (6, -8)$

9. $5x - 4y = -7$
$\underline{-3x + 4y = 1}$
$2x = -6$
$x = -3$
$5(-3) = 4y - 7, -15 = 4y - 7, -8 = 4y, y = -2; (-3, -2)$

10.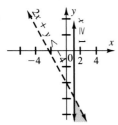

11.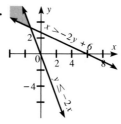

12.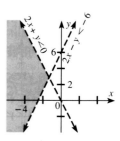

13. Let t = the tens digit and u = the units digit.
$\begin{cases} t + u = 9 \\ 10t + u + 45 = 10u + t \end{cases} \longrightarrow \begin{array}{l} t + u = 9 \\ 9t + 9u = -45 \end{array} \longrightarrow \begin{array}{l} t = 9 - u \\ 9t - 9u = -45 \end{array}$

$9(9 - u) - 9u = -45, 81 - 9u - 9u = -45, -18u = -126, u = 7;$
$t + 7 = 9; t = 2;$ The two-digit number is 27.

14. Let p = the number of pounds of peanuts and c = the number of pounds of cashews in a mixture.

$$\begin{cases} 4.40c + 1.20p = 6 \longrightarrow 44c + 12p = 60 \\ c + p = 3 \qquad\qquad \longrightarrow \qquad\quad p = 3 - c \end{cases}$$

$44c + 12(3 - c) = 60,\ 44c + 36 - 12c = 60,\ 32c + 36 = 60,\ 32c = 24,$

$c = \dfrac{24}{32} = \dfrac{3}{4} = 0.75;\ 0.75 + p = 3,\ p = 2.25;$

0.75 lb of cashews and 2.25 lb of peanuts can be bought to get exactly 3 lb of nuts for $6.

15. Let r = the rate of speed of the barge in still water and c = the rate of the stream's current.

	rate	× time =	distance
Upstream	$r - c$	4	16
Downstream	$r + c$	2	16

$$\begin{cases} (r - c)4 = 16 \longrightarrow r - c = 4 \\ (r + c)2 = 16 \longrightarrow r + c = 8 \end{cases}$$

$\begin{array}{r} r - c = 4 \\ r + c = 8 \\ \hline 2r = 12 \\ r = 6 \end{array}$

$(6 - c)4 = 16,\ 6 - c = 4,\ -c = -2;\ c = 2;$

The rate of speed of the barge in still water is 6 mi/h and the rate of the stream's current is 2 mi/h.

16. Let a = Allen's present age, then $a - 3$ = Allen's age 3 years ago.
Let r = Randy's present age, then $r - 3$ = Randy's age 3 years ago.

$$\begin{cases} r + 4 = a \qquad\longrightarrow\quad r + 4 = a \qquad\longrightarrow\quad r + 4 = a \\ 2(r - 3) = a - 3 \longrightarrow 2r - 6 = a - 3 \longrightarrow 2r - 3 = a \end{cases}$$

$2r - 3 = r + 4,\ r - 3 = 4,\ r = 7;\ 7 + 4 = a,\ 11 = a;$

Allen is 11 years old and Randy is 7 years old.

Challenge Let r = the rate of speed of the plane in still air and c = the rate of the jetstream.

	rate	× time =	distance
Against jetstream	$r - c$	2	800
With jetstream	$r + c$	$1\dfrac{3}{5}$	800

$$\begin{cases} (r - c)2 = 800 \longrightarrow r - c = 400 \\ (r + c)1\dfrac{3}{5} = 800 \longrightarrow r + c = 500 \end{cases}$$

$$r - c = 400$$
$$\underline{r + c = 500}$$
$$2r = 900$$
$$r = 450$$

$(450 - c)2 = 800$, $450 - c = 400$, $-c = -50$, $c = 50$;

The rate of the jetstream is 50 mi/h.

page 491 Preparing for Standardized Tests

1. B; $2ab(3a^2 - 5b)^2 = 2(-3)(5)[3(-3)^2 - 5(5)]^2 = -30(27 - 25)^2 = -120$

2. C; $2x - 7y = 8 \longrightarrow 2x - 7y = 8$
$2(x - 4y) = 2(3) \longrightarrow \underline{2x - 8y = 6}$
$$y = 2$$
$2x - 8(2) = 6$, $2x - 16 = 6$, $2x = 22$, $x = 11$

3. E; Let x = the rate of the boat and y = the rate of the river.

	rate	× time	= distance
Up the river	$x - y$	5	30
Down the river	$x + y$	3	30

$$\begin{cases} (x + y)3 = 30 \longrightarrow x + y = 10 \\ (x - y)5 = 30 \longrightarrow \underline{x - y = 6} \end{cases}$$
$$2x = 16$$
$$x = 8$$

$8 + y = 10$, $y = 2$; The rate of the river is 2 mi/h.

4. A; $3x + 21 + 10x - 20 = 11x - 33$, $13x + 1 = 11x - 33$, $2x = -34$, $x = -17$

5. C; Let x = Martin's present age and y = his sister's present age.

$$\begin{cases} x - 3 = \frac{1}{2}(y - 3) \longrightarrow 2x - 6 = y - 3 \longrightarrow 2x - y = 3 \\ x + y = 21 \longrightarrow x + y = 21 \longrightarrow x + y = 21 \end{cases}$$
$$2x - y = 3$$
$$\underline{x + y = 21}$$
$$3x = 24; \ x = 8$$

6. B; $\dfrac{54.81}{2.7} = 20.3$

7. C; $m = \dfrac{5 - 3}{5 - 4} = 2$; $y - 3 = 2(x - 4)$, $y - 3 = 2x - 8$, $y = 2x - 5$

8. C; $3 @ \$ \ 7.90 = \$23.70$
$1 @ \$22.50 = 22.50$
$4 @ \$ \ 1.95 = \underline{7.80}$
$\$54.00$ 6% of $\$54.00 = 0.06(54) = \$3.24$

9. E; Let t = the tens digit and u = the units digit.

$$\begin{cases} t + u = 11 \\ 10t + u + 27 = 10u + t \end{cases} \longrightarrow \begin{array}{l} t + u = 11 \\ 9t - 9u = -27 \end{array} \longrightarrow \begin{array}{l} t + u = 11 \\ t - u = -3 \end{array}$$

$$\begin{array}{r} t + u = 11 \\ \underline{t - u = -3} \\ 2t = 8 \\ t = 4 \end{array}$$

$$4 + u = 11, \; u = 7$$

10. E; $2' \times 12''$ gives two shelves, $20'' \times 6''$ gives 1, and $38'' \times 8''$ gives 2, for a total of 5 shelves.

11. D; $2x + (3x - 1) = 9$, $5x = 10$, $x = 2$. Then $y = 3(2) - 1 = 5$ so the point of intersection is $(2, 5)$.

page 492   Mixed Review

1. 733.04   **2.** 351.13   **3.** 215.91   **4.** 729.3

5. $10.48 \div 2.62 + 4 = 4 + 4 = 8$ **6.** $3.2 \cdot 3.2 + 2 = 10.4 + 2 = 12.24$

7. $50.5 - 4.8 \div 6 + 3 = 50.5 - 0.8 + 3 = 52.7$

8. $15 \cdot 15 = 225$ **9.** $13 \cdot 13 = 169$

10. $(-4)(-4)(-4) = -64$ **11.** $(-2)(-2)(-2)(-2) = 16$

12. $-(5 \cdot 5) = -25$ **13.** $(-5)(-5) = 25$

14. $0.3 \cdot 0.3 \cdot 0.3 = 0.027$ **15.** $0.01 \cdot 0.01 \cdot 0.01 \cdot 0.01 = 0.00000001$

16. $a^2 + 5a - 5a - 25 = a^2 - 25$ **17.** $y^2 - 4y - 3y + 12 = y^2 - 7y + 12$

18. $2x^2 - 6x + 6x - 18 = 2x^2 - 18$ **19.** $3b^2 - 6b + 5b - 10 = 3b^2 - b - 10$

20. $6m^2 - 4m + 12m - 8 = 6m^2 + 8m - 8$

21. $12x^2 - 15x + 8x - 10 = 12x^2 - 7x - 10$

22. 1.5 m = 150 cm, 18.5 cm − 150 cm = 35 cm

23. 3 lb 8 oz + 4 lb 9 oz = 7 lb 17 oz = 8 lb 1 oz

Chapter 12 Radicals

page 494 Capsule Review

1. $4 \cdot 4 = 16$ **2.** $-2 \cdot -2 = 4$ **3.** $\dfrac{1}{2} \cdot \dfrac{1}{2} = \dfrac{1}{4}$ **4.** $12 \cdot 12 = 144$

5. $\left(-\dfrac{1}{3}\right)^2$ **6.** $(-4)^2$ **7.** s^2 **8.** $a^2 b^2$

page 496 Class Exercises

1. $\sqrt{7 \cdot 7} = 7$ **2.** $-\sqrt{11 \cdot 11} = -11$ **3.** $\pm\sqrt{12 \cdot 12} = \pm 12$

4. $\pm\sqrt{17 \cdot 17} = \pm\sqrt{17}$ **5.** $\sqrt{5 \cdot 5} = 5$ **6.** $\sqrt{9 \cdot 9} = 9$

7. $\dfrac{\sqrt{1}}{\sqrt{49}} = \dfrac{\sqrt{1}}{\sqrt{7^2}} = \dfrac{1}{7}$ **8.** $-\dfrac{\sqrt{1}}{\sqrt{16}} = -\dfrac{\sqrt{1}}{\sqrt{4^2}} = -\dfrac{1}{4}$ **9.** $\pm\sqrt{40 \cdot 40} = \pm 40$

10. $\pm\sqrt{24 \cdot 24} = \pm 24$

page 496 Practice Exercises

A **1.** $\sqrt{5 \cdot 5} = 5$ **2.** $-\sqrt{9 \cdot 9} = -9$ **3.** $-\sqrt{16 \cdot 16} = -16$

4. $\sqrt{19 \cdot 19} = 19$ **5.** $\dfrac{\sqrt{4}}{\sqrt{9}} = \dfrac{\sqrt{2^2}}{\sqrt{3^2}} = \dfrac{2}{3}$ **6.** $\pm\dfrac{\sqrt{16}}{\sqrt{25}} = \pm\dfrac{\sqrt{4^2}}{\sqrt{5^2}} = \pm\dfrac{4}{5}$

7. $-\dfrac{\sqrt{1}}{\sqrt{144}} = -\dfrac{\sqrt{1}}{\sqrt{12^2}} = -\dfrac{1}{12}$ **8.** $\dfrac{\sqrt{1}}{\sqrt{121}} = \dfrac{\sqrt{1}}{\sqrt{11^2}} = \dfrac{1}{11}$

9. $\pm\dfrac{\sqrt{144}}{\sqrt{225}} = \pm\dfrac{\sqrt{12^2}}{\sqrt{15^2}} = \pm\dfrac{12}{15} = \pm\dfrac{4}{5}$ **10.** $\pm\dfrac{\sqrt{196}}{\sqrt{361}} = \pm\dfrac{\sqrt{14^2}}{\sqrt{19^2}} = \pm\dfrac{14}{19}$

11. $-\sqrt{100 \cdot 100} = -100$ **12.** $-\sqrt{300 \cdot 300} = -300$

13. $\sqrt{33 \cdot 33} = 33$ **14.** $\sqrt{28 \cdot 28} = 28$

B **15.** $\sqrt{\dfrac{1}{100}} = \dfrac{\sqrt{1}}{\sqrt{100}} = \dfrac{1}{10} = 0.1$ **16.** $\sqrt{\dfrac{4}{100}} = \dfrac{\sqrt{4}}{\sqrt{100}} = \dfrac{2}{10} = 0.2$

17. $\sqrt{\dfrac{64}{100}} = \dfrac{\sqrt{64}}{\sqrt{100}} = \dfrac{8}{10} = 0.8$ **18.** $\sqrt{\dfrac{25}{100}} = \dfrac{\sqrt{25}}{\sqrt{100}} = \dfrac{5}{10} = 0.5$

19. $-\sqrt{\dfrac{196}{100}} = -\dfrac{\sqrt{196}}{\sqrt{100}} = -\dfrac{14}{10} = -1.4$ **20.** $\sqrt{\dfrac{289}{100}} = \dfrac{\sqrt{289}}{\sqrt{100}} = \dfrac{17}{10} = 1.7$

21. $\sqrt{\dfrac{9}{10,000}} = \dfrac{\sqrt{9}}{\sqrt{10,000}} = \dfrac{3}{100} = 0.03$ **22.** $\sqrt{\dfrac{25}{10,000}} = \dfrac{\sqrt{25}}{\sqrt{10,000}} = \dfrac{5}{100} = 0.05$

23. $\pm\sqrt{\dfrac{625}{10,000}} = \pm\dfrac{\sqrt{625}}{\sqrt{10,000}} = \pm\dfrac{25}{100} = \pm 0.25$ **24.** $\pm\sqrt{\dfrac{196}{10,000}} = \pm\dfrac{\sqrt{196}}{\sqrt{10,000}} = \pm\dfrac{14}{100} = \pm 0.14$

25. $\sqrt{\dfrac{1296}{10,000}} = \dfrac{\sqrt{1296}}{\sqrt{10,000}} = \dfrac{36}{100} = 0.36$ **26.** $\sqrt{\dfrac{1089}{10,000}} = \dfrac{\sqrt{1089}}{\sqrt{10,000}} = \dfrac{33}{100} = 0.33$

C 27. $\sqrt{\dfrac{25}{9}} = \dfrac{\sqrt{25}}{\sqrt{9}} = \dfrac{5}{3}$ **28.** $\sqrt{\dfrac{1}{81}} = \dfrac{\sqrt{1}}{\sqrt{81}} = \dfrac{1}{9}$

29. $\sqrt{\dfrac{144}{225}} = \dfrac{\sqrt{144}}{\sqrt{225}} = \dfrac{12}{15} = \dfrac{4}{5}$ **30.** $\sqrt{\dfrac{121}{361}} = \dfrac{\sqrt{121}}{\sqrt{361}} = \dfrac{11}{19}$

31. $\sqrt[4]{3 \cdot 3 \cdot 3 \cdot 3} = 3$ **32.** $\sqrt[4]{10 \cdot 10 \cdot 10 \cdot 10} = 10$

33. $\sqrt[3]{5 \cdot 5 \cdot 5} = 5$ **34.** $\sqrt[3]{4 \cdot 4 \cdot 4} = 4$

35. $t = \sqrt{\dfrac{144}{16}}, \; t = \sqrt{\dfrac{12 \cdot 12}{4 \cdot 4}}, \; t = \dfrac{12}{4} = 3$; It takes an object 3s to fall 144 ft.

page 496   Historical Note

1.

15   21   28

2.

25   36   49

3.

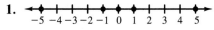

page 497   Capsule Review

1.

2.

3.

4.

5.

page 498 Class Exercises

    1. 1.732

    3. 5.196

    5. 5.568

    7. 9.220

    9. 6.708

    2. 1.414

    4. 7.616

    6. 9.849

    8. 4.359

    10. 8.718

11–15. Answers may vary. One possible answer is given.

11. $\sqrt{41}$ (since $6 = \sqrt{36}$ and $7 = \sqrt{49}$)

12. '$\sqrt{101}$ (since $10 = \sqrt{100}$ and $11 = \sqrt{121}$)

13. $-\sqrt{19}$ (since $-5 = -\sqrt{25}$ and $-4 = -\sqrt{16}$)

14. $\sqrt{13.5}$ (since $13 < 13.5 < 14$)

15. $\sqrt{1.23}$ (since $1.1 = \sqrt{1.21}$ and $1.2 = \sqrt{1.44}$)

pages 498–499 Practice Exercises

A   **1.** $\sqrt{5} \approx 2.24$

    3. $\sqrt{22} \approx 4.69$

    5. $\sqrt{83} \approx 9.11$

    7. $\sqrt{95} \approx 9.75$

    9. $\sqrt{27} \approx 5.20$

    11. $\sqrt{15} \approx 3.87$

    13. $\sqrt{120} \approx 10.95$

    15. $-\sqrt{99} \approx -9.95$

    2. $\sqrt{8} \approx 2.83$

    4. $\sqrt{17} \approx 4.12$

    6. $\sqrt{67} \approx 8.19$

    8. $\sqrt{59} \approx 7.68$

    10. $\sqrt{32} \approx 5.66$

    12. $\sqrt{12} \approx 3.46$

    14. $\sqrt{149} \approx 12.21$

    16. $-\sqrt{123} \approx -11.09$

17–28. Answers may vary. Possible answers are given.

B **17.** $\sqrt{27}$, $\sqrt{28}$, $\sqrt{29}$ (since $5 = \sqrt{25}$ and $6 = \sqrt{36}$)

18. $\sqrt{90}$, $\sqrt{91}$, $\sqrt{92}$ (since $9 = \sqrt{81}$ and $10 = \sqrt{100}$)

19. $-\sqrt{10}$, $-\sqrt{11}$, $-\sqrt{12}$ (since $-4 = -\sqrt{16}$ and $-3 = -\sqrt{9}$)

20. $-\sqrt{70}$, $-\sqrt{71}$, $-\sqrt{72}$ (since $-8 = -\sqrt{64}$ and $-9 = -\sqrt{81}$)

21. $\sqrt{150}$, $\sqrt{151}$, $\sqrt{152}$ (since $12 = \sqrt{144}$ and $13 = \sqrt{169}$)

22. $\sqrt{430}$, $\sqrt{431}$, $\sqrt{432}$ (since $20 = \sqrt{400}$ and $21 = \sqrt{441}$)

23. $\sqrt{15.4}$, $\sqrt{15.5}$, $\sqrt{15.6}$ (since these radicands are all between 15 and 16)

24. $\sqrt{27.5}$, $\sqrt{27.6}$, $\sqrt{27.7}$ (since these radicands are all between 27 and 28)

C **25.** $\sqrt{0.51}$, $\sqrt{0.52}$, $\sqrt{0.53}$ (since $0.7 = \sqrt{0.49}$ and $0.8 = \sqrt{0.64}$)

26. $\sqrt{0.0002}$, $\sqrt{0.0003}$, $\sqrt{0.00035}$ (since $0.01 = \sqrt{0.0001}$ and $0.02 = \sqrt{0.0004}$)

27. $\sqrt{3}$, $\sqrt{3.1}$, $\sqrt{3.2}$ (since $2 = \sqrt{4}$)

28. $\sqrt{9.1}$, $\sqrt{9.2}$, $\sqrt{9.3}$ (since $3 = \sqrt{9}$ and $\pi \approx \sqrt{9.8696}$)

29. $6 < \sqrt{48} < 7$, $48.0 \div 7 \approx 6.8$, $(7 + 6.8) \div 2 \approx 6.9$, $48.00 \div 6.9 \approx 6.95$, $(6.9 + 6.95) \div 2 \approx 6.93$, $\sqrt{48} \approx 6.93$;
The approximate length of one side of the rug is 6.93 ft.

30. Let $w =$ the width of the poster, then $3w =$ the length of the poster.
$w(3w) = 192$, $3w^2 = 192$, $w^2 = 64$, $w = 8$; then $3w = 3(8) = 24$.
The dimensions of the poster are 24 in. × 8 in.

31. $2^3 \cdot 3^2$ **32.** $2 \cdot 3 \cdot 5^2$ **33.** $2^2 \cdot 5 \cdot 11$ **34.** $3^2 \cdot 5 \cdot 7$

35. $7 < \sqrt{50} < 8$, $50 \div 7 \approx 7.1$, $(7 + 7.1) \div 2 \approx 7.1$, $50 \div 7.1 \approx 7.04$, $(7.1 + 7.04) \div 2 = 7.07$; $\sqrt{50} \approx 7.07$

36. $5 < \sqrt{32} < 6$, $32 \div 6 \approx 5.3$, $(6 + 5.3) \div 2 \approx 5.7$, $32 \div 5.7 \approx 5.61$, $(5.7 + 5.61) \div 2 \approx 5.66$; $\sqrt{32} \approx 5.66$

37. $3 < \sqrt{10.5} < 4$, $10.5 \div 3 = 3.5$, $(3 + 3.5) \div 2 \approx 3.3$, $10.5 \div 3.3 \approx 3.18$, $(3.3 + 3.18) \div 2 = 3.24$; $\sqrt{10.5} \approx 3.24$

38. $22 < \sqrt{500} < 23$, $500 \div 22 \approx 22.7$, $(22 + 22.7) \div 2 \approx 22.4$, $500 \div 22.4$ ≈ 22.32, $(22.4 + 22.32) \div 2 = 22.36$; $\sqrt{500} \approx 22.36$

page 500　Capsule Review

1. $8.3 = 8\dfrac{3}{10} = \dfrac{83}{10}$

2. $1.5 = 1\dfrac{5}{10} = \dfrac{15}{10} = \dfrac{3}{2}$

3. $4 = \dfrac{4}{1}$

4. $0 = \dfrac{0}{1}$ or $\dfrac{0}{\text{any nonzero integer}}$

5. $-2\dfrac{1}{2} = \dfrac{-5}{2}$ or $\dfrac{5}{-2}$

6. $-1\dfrac{3}{4} = \dfrac{-7}{4}$ or $\dfrac{7}{-4}$

7. $6.01 = 6\dfrac{1}{100} = \dfrac{601}{100}$

8. $5.99 = 5\dfrac{99}{100} = \dfrac{599}{100}$

page 501　Class Exercises

1. $5 \div 8 = 0.625$

2. $-5 \div 2 = -2.5$

3. $-1 \div 6 = -0.1\overline{6}$

4. $2 \div 9 = 0.\overline{2}$

5. $0.75 = \dfrac{75}{100} = \dfrac{3}{4}$

6. $1.8 = 1\dfrac{8}{10} = \dfrac{18}{10} = \dfrac{9}{5}$

7. Let $x = 0.\overline{6}$.　$(10x = 6.\overline{6}) - (x = 0.\overline{6})$, $9x = 6$; $x = \dfrac{6}{9} = \dfrac{2}{3}$

8. Let $x = 0.8\overline{3}$.　$(10x = 8.3\overline{3}) - (x = 0.8\overline{3})$, $9x = 7.5$; $x = \dfrac{7.5}{9} = \dfrac{75}{90} = \dfrac{5}{6}$

pages 501–502　Practice Exercises

A **1.** $1 \div 8 = 0.125$

2. $7 \div 2 = 3.5$

3. $7 \div 3 = 2.\overline{3}$

4. $2 \div 3 = 0.\overline{6}$

5. $-5 \div 4 = -1.25$

6. $8 \div 5 = 1.6$

7. $4 \div 9 = 0.\overline{4}$

8. $-5 \div 11 = -0.\overline{45}$

9. $-51 \div 4 = -12.75$

10. $37 \div 8 = 4.625$

11. $5 \div 33 = 0.\overline{15}$

12. $-8 \div 11 = -0.\overline{72}$

13. $2.33 = 2\dfrac{33}{100} = \dfrac{233}{100}$

14. $-1.6 = -1\dfrac{6}{10} = \dfrac{-16}{10} = \dfrac{-8}{5}$

15. Let $x = 0.\overline{3}$. $(10x = 3.\overline{3}) - (x = 0.\overline{3})$, $9x = 3$; $x = \dfrac{3}{9} = \dfrac{1}{3}$

16. Let $x = 0.\overline{1}$. $(10x = 1.\overline{1}) - (x = 0.\overline{1})$, $9x = 1$; $x = \dfrac{1}{9}$

B 17. Let $x = 0.\overline{27}$. $(100x = 27.\overline{27}) - (x = 0.\overline{27})$, $99x = 27$; $x = \dfrac{27}{99} = \dfrac{3}{11}$

18. Let $x = 0.\overline{45}$. $(100x = 45.\overline{45}) - (x = 0.\overline{45})$, $99x = 45$; $x = \dfrac{45}{99} = \dfrac{5}{11}$

19. Let $x = 1.\overline{1213}$. $(10{,}000x = 11213.\overline{1213}) - (x = 1.\overline{1213})$, $9999x = 11212$; $x = \dfrac{11212}{9999}$

20. Let $x = -5.2\overline{3}$. $(100x = -523.\overline{3}) - (x = -5.2\overline{3})$, $99x = -518.1$; $x = \dfrac{-518.1}{99} = \dfrac{-157}{30}$

21. Let $x = 0.\overline{325}$. $(1000x = 325.\overline{325}) - (x = 0.\overline{325})$, $999x = 325$; $x = \dfrac{325}{999}$

22. Let $x = 0.\overline{125}$. $(1000x = 125.\overline{125}) - (x = 0.\overline{125})$, $999x = 125$; $x = \dfrac{125}{999}$

23. Let $x = 0.\overline{142857}$. $(1{,}000{,}000x = 142857.\overline{142857}) - (x = 0.\overline{142857})$, $999{,}999x = 142857$; $x = \dfrac{142857}{999{,}999} = \dfrac{1}{7}$

24. Let $x = 0.\overline{857142}$. $(1{,}000{,}000x = 857142.\overline{857142}) - (x = 0.\overline{857142})$, $999{,}999x = 857142$; $x = \dfrac{857142}{999{,}999} = \dfrac{6}{7}$

25. $\dfrac{2}{3}$; Let $x = 0.\overline{3}$. $(10x = 3.\overline{3}) - (x = 0.\overline{3})$, $9x = 3$, $x = \dfrac{1}{3}$; $\dfrac{2}{3} + \dfrac{1}{3} = \dfrac{3}{3} = 1$

26. Let $x = 0.\overline{36}$. $(100x = 36.\overline{36}) - (x = 0.\overline{36})$, $99x = 36$, $x = \dfrac{36}{99} = \dfrac{4}{11}$;

$\dfrac{4}{11} + \dfrac{7}{11} = \dfrac{11}{11} = 1$

27. $\dfrac{1}{18}$; Let $x = 0.3\overline{8}$. $(10x = 3.8\overline{8}) - (x = 0.3\overline{8})$, $9x = 3.5$, $x = \dfrac{3.5}{9} = \dfrac{7}{18}$;

$\dfrac{1}{18} + \dfrac{7}{18} = \dfrac{8}{18} = \dfrac{4}{9}$

28. $\frac{5}{7}$; Let $x = 0.\overline{285714}$. $(1,000,000x = 285714.\overline{285714}) - (x = 0.\overline{285714})$,

$999,999x = 285714$; $x = \dfrac{285714}{999,999} = \dfrac{2}{7}; \dfrac{5}{7} + \dfrac{2}{7} = \dfrac{7}{7} = 1$

C 29. $20 = 2 \cdot 2 \cdot 5$ or $2^2 \cdot 5$; $80 = 2 \cdot 2 \cdot 2 \cdot 2 \cdot 5$ or $2^4 \cdot 5$; $125 = 5 \cdot 5 \cdot 5$ or 5^3; The prime factors are all 2s or 5s.

30. $6 = 2 \cdot 3$; $12 = 2 \cdot 2 \cdot 3$ or $2^2 \cdot 3$; 11 is prime; yes

31. A rational number represents a terminating decimal if the prime factors of its denominator consist of only the factors 2 and/or 5. A rational number represents a repeating decimal if the prime factors of its denominator include factors other than 2 and/or 5.

32. $\dfrac{20.2 \text{ amp}}{0.12 \text{ amp/s}} = 168.\overline{3}\text{s}$

33. Answers may vary, depending upon the initial approximation. ($\sqrt{10} \approx 3.162277660$)

34. The closer the initial approximation is to the actual square root, the sooner the process yields the square root.

page 503   Capsule Review

1. 30

2. 60

3. 10

4. $\dfrac{7}{12}$

5. $-\dfrac{1}{20}$

6. 3.74

7. 9.49

8. 11.62

9. -6.71

10. -1.41

page 505   Class Exercises

1. $\sqrt{4 \cdot 5} = \sqrt{4} \cdot \sqrt{5} = 2\sqrt{5}$; 4 is a factor.

2. $\sqrt{9 \cdot 3} = \sqrt{9} \cdot \sqrt{3} = 3\sqrt{3}$; 9 is a factor.

3. $\sqrt{9x^2 \cdot 2x} = \sqrt{9x^2} \cdot \sqrt{2x} = 3|x|\sqrt{2x}$; $9x^2$ is a factor.

4. $\sqrt{9b^4 \cdot 5} = \sqrt{9b^4} \cdot \sqrt{5} = 3b^2\sqrt{5}$; $9b^4$ is a factor.

5. $\sqrt{3 + 9} = \sqrt{12} = \sqrt{4 \cdot 3} = \sqrt{4} \cdot \sqrt{3} = 2\sqrt{3}$

6. $\sqrt{6 - 3} = \sqrt{3}$

7. $\sqrt{-15 - 10} = \sqrt{-25}$; $\sqrt{-25}$ is not a real number.

8. $\sqrt{9 - 1} = \sqrt{8} = \sqrt{4} \cdot \sqrt{2} = 2\sqrt{2}$

9. $3x - 9 \geq 0$, $3x \geq 9$, $x \geq 3$

10. $2x + 5 \geq 0$, $2x \geq -5$, $x \geq -\dfrac{5}{2}$

11. $-2x - 7 \geq 0$, $-2x \geq 7$, $x \leq -\dfrac{7}{2}$

12. $-x + 4 \geq 0$, $-x \geq -4$, $x \leq 4$

pages 505–506   Practice Exercises

A   1. $\sqrt{15 - 7} = \sqrt{8} = \sqrt{4} \cdot \sqrt{2} = 2\sqrt{2}$ 2. $\sqrt{-30 - 6} = \sqrt{-36}$;
                                                       $\sqrt{-36}$ is not a real number.

    3. $\sqrt{21 + 6} = \sqrt{27} = \sqrt{9} \cdot \sqrt{3} = 3\sqrt{3}$ 4. $5x - 10 \geq 0, \; 5x \geq 10, \; x \geq 2$

    5. $2x + 7 \geq 0, \; 2x \geq -7, \; x \geq -\dfrac{7}{2}$ 6. $-4x + 2 \geq 0, \; -4x \geq -2, \; x \leq \dfrac{2}{4}, \; x \leq \dfrac{1}{2}$

    7. $-3x - 9 \geq 0, \; -3x \geq 9, \; x \leq -3$ 8. $\sqrt{8} \cdot \sqrt{3} = \sqrt{4} \cdot \sqrt{2} \cdot \sqrt{3} = 2\sqrt{6}$

    9. $\sqrt{16} \cdot \sqrt{3} = \sqrt{4} \cdot \sqrt{4} \cdot \sqrt{3} = 4\sqrt{3}$ 10. $\sqrt{49} \cdot \sqrt{2} = 7\sqrt{2}$

    11. $\sqrt{9} \cdot \sqrt{6} = 3\sqrt{6}$ 12. $\sqrt{144} \cdot \sqrt{2} = 12\sqrt{2}$

    13. $\sqrt{25 \cdot 6} = \sqrt{25} \cdot \sqrt{6} = 5\sqrt{6}$ 14. $\sqrt{28 \cdot 28} = 28$

    15. $\sqrt{100 \cdot 10} = \sqrt{100} \cdot \sqrt{10} = 10\sqrt{10}$ 16. $\sqrt{100x^2 \cdot 2} = \sqrt{100x^2} \cdot \sqrt{2} = 10x\sqrt{2}$

    17. $\sqrt{25x^4 \cdot 3} = \sqrt{25x^4} \cdot \sqrt{3} = 5x^2\sqrt{3}$ 18. $\sqrt{9x^2 \cdot 7x} = \sqrt{9x^2} \cdot \sqrt{7x} = 3x\sqrt{7x}$

    19. $\sqrt{100x^6 \cdot 3} = \sqrt{100x^6} \cdot \sqrt{3} = 10x^3\sqrt{3}$ 20. $\sqrt{4b^8 \cdot 2} = \sqrt{4b^8} \cdot \sqrt{2} = 2b^4\sqrt{2}$

    21. $\sqrt{4y^{12} \cdot 3} = \sqrt{4y^{12}} \cdot \sqrt{3} = 2y^6\sqrt{3}$ 22. $\sqrt{c^4 \cdot 5c} = \sqrt{c^4} \cdot \sqrt{5c} = c^2\sqrt{5c}$

    23. $\sqrt{d^6 \cdot 15d} = \sqrt{d^6} \cdot \sqrt{15d} = d^3\sqrt{15d}$

    24. $-\sqrt{4m^6 \cdot 30} = -\sqrt{4m^6} \cdot \sqrt{30} = -2m^3\sqrt{30}$

    25. $-\sqrt{25x^{16} \cdot 5} = -\sqrt{25x^{16}} \cdot \sqrt{5} = -5x^8\sqrt{5}$

    26. $\sqrt{100b^8 \cdot b} = \sqrt{100b^8} \cdot \sqrt{b} = 10b^4\sqrt{b}$

    27. $\sqrt{12p^4 \cdot 12p^4} = 12p^4$

B   28. $\sqrt{a^2 \cdot b} = \sqrt{a^2} \cdot \sqrt{b} = a\sqrt{b}$ 29. $\sqrt{x^4 \cdot y} = \sqrt{x^4} \cdot \sqrt{y} = x^2\sqrt{y}$

    30. $\sqrt{m^4 \cdot n^6} = \sqrt{m^4} \cdot \sqrt{n^6} = m^2n^3$ 31. $\sqrt{p^{12} \cdot q^{10}} = \sqrt{p^{12}} \cdot \sqrt{q^{10}} = p^6 q^5$

    32. $\sqrt{64x^4 \cdot 5x} = \sqrt{64x^4} \cdot \sqrt{5x} = 8x^2\sqrt{5x}$

    33. $\sqrt{81x^{12} \cdot 3x} = \sqrt{81x^{12}} \cdot \sqrt{3x} = 9x^6\sqrt{3x}$

    34. $-\sqrt{64n^8 \cdot 2n} = -\sqrt{64n^8} \cdot \sqrt{2n} = -8n^4\sqrt{2n}$

    35. $-\sqrt{25a^{10} \cdot 6a} = -\sqrt{25a^{10}} \cdot \sqrt{6a} = -5a^5\sqrt{6a}$

    36. $\sqrt{4a^4 \cdot 22} = \sqrt{4a^4} \cdot \sqrt{22} = 2a^2\sqrt{22}$

    37. $\sqrt{100r^6 \cdot 5r} = \sqrt{100r^6} \cdot \sqrt{5r} = 10r^3\sqrt{5r}$

    38. $\sqrt{144a^{12} \cdot 5} \; \sqrt{144a^{12}} \cdot \sqrt{5} = 12a^6\sqrt{5}$

    39. $\dfrac{\sqrt{36x^4 \cdot x \cdot y}}{\sqrt{49}} = \dfrac{6}{7}x^2\sqrt{xy}$ 40. $\dfrac{-\sqrt{100a^2 \cdot a}}{\sqrt{64}} = -\dfrac{10a}{8}\sqrt{a} = -\dfrac{5}{4}a\sqrt{a}$

    41. $\sqrt{0.09c^4 \cdot c} = \sqrt{0.09c^4} \cdot \sqrt{c} = 0.3c^2\sqrt{c}$

42. $\sqrt{0.04p^8 \cdot p} = \sqrt{0.04p^8} \cdot \sqrt{p} = 0.2p^4\sqrt{p}$

43. $\sqrt{0.16m^2 \cdot m} = \sqrt{0.16m^2} \cdot \sqrt{m} = 0.4m\sqrt{m}$

C 44. $\sqrt{\sqrt{144}} = \sqrt{12} = \sqrt{4 \cdot 3} = \sqrt{4} \cdot \sqrt{3} = 2\sqrt{3}$

45. $\sqrt{\sqrt{324}} = \sqrt{18} = \sqrt{9 \cdot 2} = \sqrt{9} \cdot \sqrt{2} = 3\sqrt{2}$

46. $\sqrt{\sqrt{625}} = \sqrt{25} = 5$

47. $\sqrt{\sqrt{10,000}} = \sqrt{100} = 10$ **48.** $\sqrt{(x+2)(x+2)} = \sqrt{(x+2)^2} = x + 2$

49. $\sqrt{(a-3)(a-3)} = \sqrt{(a-3)^2} = a - 3$ **50.** $\sqrt{(3x+1)(3x+1)} = \sqrt{(3x+1)^2} = 3x + 1$

51. $v = \sqrt{64 \cdot 128}$, $v = \sqrt{64} \cdot \sqrt{128}$, $v = \sqrt{64} \cdot \sqrt{64} \cdot \sqrt{2}$, $v = 64\sqrt{2}$ ft/sec

52. $s^2 = 125$, $s = \sqrt{125}$, $s = \sqrt{25 \cdot 5}$, $s = \sqrt{25} \cdot \sqrt{5}$, $s = 5\sqrt{5}$ in.

53. $T = 2\pi\sqrt{\dfrac{8}{32}}$, $T = 2\pi\sqrt{\dfrac{1}{4}}$, $T = 2\pi\left(\dfrac{1}{2}\right)$, $T = \pi$ sec

page 506 Algebra in Police Science

1. $s = 2\sqrt{5 \cdot 25}$, $s = 2\sqrt{125}$, $s \approx 22$ mi/h

2. $s = 2\sqrt{5 \cdot 80}$, $s = 2\sqrt{400}$, $s = 40$ mi/h

3. $s = 2\sqrt{5 \cdot 125}$, $s = 2\sqrt{625}$, $s = 50$ mi/h

4. $s = 2\sqrt{5 \cdot 175}$, $s = 2\sqrt{875}$, $s \approx 59$ mi/h

5. $s = 2\sqrt{5 \cdot 200}$, $s = 2\sqrt{1000}$, $s \approx 63$ mi/h

page 507 Integrating Algebra: Image Formation

1. $c + 2c = 6$, $3c = 6$, $c = 2$ **2.** $2b + 30 = 5b$, $3b = 30$, $b = 10$

3. $18 + 2a = 3a$, $a = 18$

4. $\dfrac{1}{70} + \dfrac{1}{5} = \dfrac{1}{f}$, $f + 14f = 70$, $15f = 70$, $f = \dfrac{70}{15} = \dfrac{14}{3}$ cm

5. $\dfrac{1}{66} + \dfrac{1}{q} = \dfrac{1}{\frac{66}{23}}$, $\dfrac{1}{66} + \dfrac{1}{q} = \dfrac{23}{66}$, $q + 66 = 23q$, $22q = 66$, $q = 3$ cm

page 509 Capsule Review

1. $3m^3$ **2.** $9a^2b$ **3.** unlike terms

4. $3x^2 + 3x$ **5.** $5y^3 + 7y$ **6.** unlike terms

1. $6\sqrt{2}$ 2. $-3\sqrt{3}$ 3. $-3\sqrt{a}$ 4. $-5\sqrt{x}$ 5. $-3\sqrt{6}$

6. $-\sqrt{10} - 2\sqrt{7}$ 7. $4\sqrt{y} - \sqrt{y} = 3\sqrt{y}$ 8. $-5\sqrt{b} + 2\sqrt{b} = -3\sqrt{b}$

9. $\sqrt{25 \cdot 3} + 2\sqrt{9 \cdot 3} - \sqrt{4 \cdot 3} = 5\sqrt{3} + 6\sqrt{3} - 2\sqrt{3} = 9\sqrt{3}$

10. $-\sqrt{16 \cdot 5} - 3\sqrt{9 \cdot 5} + \sqrt{4 \cdot 5} = -4\sqrt{5} - 9\sqrt{5} + 2\sqrt{5} = -11\sqrt{5}$

11. $\sqrt{4x^2 \cdot 3x} + 2\sqrt{9x^2 \cdot 3x} = 2x\sqrt{3x} + 6x\sqrt{3x} = 8x\sqrt{3x}$

12. $2\sqrt{4y^2 \cdot 7y} + \sqrt{9y^2 \cdot 7y} = 4y\sqrt{7y} + 3y\sqrt{7y} = 7y\sqrt{7y}$

pages 510–511   Practice Exercises

A   1. $5\sqrt{5}$ 2. $-3\sqrt{2}$ 3. $7\sqrt{10}$ 4. $-\sqrt{7}$

5. $2\sqrt{9 \cdot 3} + 3\sqrt{4 \cdot 4} = 6\sqrt{3} + 12$ 6. $3\sqrt{9 \cdot 2} - \sqrt{25} = 9\sqrt{2} - 5$

7. $(6\sqrt{3} - \sqrt{3}) - 4\sqrt{5} = 5\sqrt{3} - 4\sqrt{5}$ 8. $(-\sqrt{7} + 4\sqrt{7}) - 7\sqrt{3} = 3\sqrt{7} - 7\sqrt{3}$

9. $\sqrt{4 \cdot 2} + 2\sqrt{25 \cdot 2} - \sqrt{9 \cdot 2} = 2\sqrt{2} + 10\sqrt{2} - 3\sqrt{2} = 9\sqrt{2}$

10. $2\sqrt{4 \cdot 3} - \sqrt{16 \cdot 3} + 3\sqrt{9 \cdot 3} = 4\sqrt{3} - 4\sqrt{3} + 9\sqrt{3} = 9\sqrt{3}$

11. $8\sqrt{36 \cdot 2} - 3\sqrt{4 \cdot 2} - \sqrt{49 \cdot 2} = 48\sqrt{2} - 6\sqrt{2} - 7\sqrt{2} = 35\sqrt{2}$

12. $-5\sqrt{16 \cdot 5} + \sqrt{25 \cdot 5} + 10\sqrt{9 \cdot 5} = -20\sqrt{5} + 5\sqrt{5} + 30\sqrt{5} = 15\sqrt{5}$

13. $3\sqrt{x} - 4\sqrt{x} + 5\sqrt{x} = 4\sqrt{x}$ 14. $18\sqrt{a} + 10\sqrt{a} - 16\sqrt{a} = 12\sqrt{a}$

15. $\sqrt{4 \cdot 3x} - \sqrt{9 \cdot 3x} + \sqrt{16 \cdot 3x} = 2\sqrt{3x} - 3\sqrt{3x} + 4\sqrt{3x} = 3\sqrt{3x}$

16. $\sqrt{9 \cdot 2y} - 2\sqrt{16 \cdot 2y} - \sqrt{25 \cdot 2y} = 3\sqrt{2y} - 8\sqrt{2y} - 5\sqrt{2y} = -10\sqrt{2y}$

17. $4\sqrt{9 \cdot 6} - \sqrt{6} + 5\sqrt{4 \cdot 6} = 12\sqrt{6} - \sqrt{6} + 10\sqrt{6} = 21\sqrt{6}$

18. $2\sqrt{25 \cdot 10} - 3\sqrt{64 \cdot 10} - 5\sqrt{10} = 10\sqrt{10} - 24\sqrt{10} - 5\sqrt{10} = -19\sqrt{10}$

B 19. $10x\sqrt{y} - 12x\sqrt{y} - 5x\sqrt{y} = -7x\sqrt{y}$

20. $-27b\sqrt{a} + 2b\sqrt{a} - 15b\sqrt{a} = -40b\sqrt{a}$

21. $\sqrt{36y^2 \cdot 2x} - 2\sqrt{49y^2 \cdot 2x} = 6y\sqrt{2x} - 14y\sqrt{2x} = -8y\sqrt{2x}$

22. $-3\sqrt{25c^4 \cdot 5d} + \sqrt{16c^4 \cdot 5d} = -15c^2\sqrt{5d} + 4c^2\sqrt{5d} = -11c^2\sqrt{5d}$

23. $4\sqrt{64a^2 \cdot 5a} - 2\sqrt{36a^2 \cdot 5a} = 32a\sqrt{5a} - 12a\sqrt{5a} = 20a\sqrt{5a}$

24. $\sqrt{4x^4 \cdot 11x} + \sqrt{9x^4 \cdot 11x} = 2x^2\sqrt{11x} + 3x^2\sqrt{11x} = 5x^2\sqrt{11x}$

25. $\sqrt{25a^2 \cdot 7b} - 3\sqrt{16a^2 \cdot 7b} = 5a\sqrt{7b} - 12a\sqrt{7b} = -7a\sqrt{7b}$

26. $-5\sqrt{64s^2 \cdot 3rs} + 2\sqrt{100s^2 \cdot 3rs} = -40s\sqrt{3rs} + 20s\sqrt{3rs} = -20s\sqrt{3rs}$

27. $5\sqrt{16 \cdot 11} - 3\sqrt{9 \cdot 11} + 4\sqrt{81 \cdot 2} = 20\sqrt{11} - 9\sqrt{11} + 36\sqrt{2} = 11\sqrt{11} + 36\sqrt{2}$

28. $-3\sqrt{100 \cdot 2} + 2\sqrt{49 \cdot 3} - \sqrt{169 \cdot 2} = -30\sqrt{2} + 14\sqrt{3} - 13\sqrt{2} =$
$-43\sqrt{2} + 14\sqrt{3}$

29. $\dfrac{1}{2}\sqrt{3} - \dfrac{2}{2}\sqrt{3} = \dfrac{1}{2}\sqrt{3} - \sqrt{3} = -\dfrac{1}{2}\sqrt{3}$ **30.** $-\dfrac{3}{4}\sqrt{5} - \dfrac{5}{4}\sqrt{5} = -\dfrac{8}{4}\sqrt{5} = -2\sqrt{5}$

31. $\sqrt{\dfrac{y^2}{9} \cdot 5x} - \sqrt{\dfrac{y^2}{4} \cdot 5x} = \dfrac{y}{3}\sqrt{5x} - \dfrac{y}{2}\sqrt{5x} = \dfrac{2y}{6}\sqrt{5x} - \dfrac{3y}{6}\sqrt{5x} = -\dfrac{y}{6}\sqrt{5x}$

32. $-\sqrt{\dfrac{t^2}{16} \cdot 5t} + \sqrt{\dfrac{t^2}{36} \cdot 5t} = -\dfrac{t}{4}\sqrt{5t} + \dfrac{t}{6}\sqrt{5t} = -\dfrac{3t}{12}\sqrt{5t} + \dfrac{2t}{12}\sqrt{5t} = -\dfrac{t}{12}\sqrt{5t}$

C 33. $a\sqrt{b^2 \cdot ab} + ab\sqrt{ab} - b\sqrt{a^2 \cdot ab} = ab\sqrt{ab} + ab\sqrt{ab} - ab\sqrt{ab} = ab\sqrt{ab}$

34. $3x\sqrt{25x^2 \cdot x} - x\sqrt{4x^2 \cdot 2} + 3x\sqrt{9x^2 \cdot x} = 15x^2\sqrt{x} - 2x^2\sqrt{x} + 9x^2\sqrt{x} = 22x^2\sqrt{x}$

35. $y\sqrt{4x^4 \cdot 2x} + xy\sqrt{9x^2 \cdot 2x} - x\sqrt{25x^2y^2 \cdot 2x} = 2x^2y\sqrt{2x} + 3x^2y\sqrt{2x} -$
$5x^2y\sqrt{2x} = 0$

36. $a\sqrt{9a^2 \cdot 3a} - b\sqrt{16b^2 \cdot 3b} - a^2\sqrt{25 \cdot 3a} = 3a^2\sqrt{3a} - 4b^2\sqrt{3b} - 5a^2\sqrt{3a} =$
$-2a^2\sqrt{3a} - 4b^2\sqrt{3b}$

37. $\sqrt{125} + \sqrt{125} + \sqrt{125} + \sqrt{125} = 4\sqrt{125} = 4\sqrt{25 \cdot 5} = 20\sqrt{5}$ cm

38. $4\sqrt{2} + 4\sqrt{6} + 8\sqrt{2} = (4\sqrt{2} + 8\sqrt{2}) + 4\sqrt{6} = 12\sqrt{2} + 4\sqrt{6}$ cm

page 511   Biography

Answers may vary.

page 512   Capsule Review

1. 25   **2.** 30   **3.** $\sqrt{9 \cdot 6} = 3\sqrt{6}$ **4.** $\sqrt{9 \cdot 7} = 3\sqrt{7}$

5. $\sqrt{r^4s^4 \cdot s} = r^2s^2\sqrt{s}$ **6.** $\sqrt{a^2b^{10} \cdot a} = ab^5\sqrt{a}$

7. $\sqrt{4x^2y^6 \cdot 2xy} = 2xy^3\sqrt{2xy}$ **8.** $\sqrt{9c^2 \cdot 3d} = 3c\sqrt{3d}$

page 514   Class Exercises

1. $\sqrt{50} = \sqrt{25 \cdot 2} = 5\sqrt{2}$ **2.** $-\sqrt{12} = -\sqrt{4 \cdot 3} = -2\sqrt{3}$

3. $9\sqrt{4} = 18$ **4.** $25\sqrt{9} = 75$

5. $-6\sqrt{56y^2} = -6\sqrt{4y^2 \cdot 14} = -12y\sqrt{14}$ **6.** $\sqrt{48x^2} = \sqrt{16x^2 \cdot 3} = 4x\sqrt{3}$

7. $-5\sqrt{3} - 2\sqrt{9} = -5\sqrt{3} - 6$

8. $2\sqrt{12} - 3\sqrt{2} = 2\sqrt{4 \cdot 3} - 3\sqrt{2} = 4\sqrt{3} - 3\sqrt{2}$

9. $10 + 15\sqrt{6} - 4\sqrt{6} - 6\sqrt{36} = 10 + 15\sqrt{6} - 4\sqrt{6} - 36 = -26 + 11\sqrt{6}$

10. $12 - 16\sqrt{7} + 9\sqrt{7} - 12\sqrt{49} = 12 - 16\sqrt{7} + 9\sqrt{7} - 84 = -72 - 7\sqrt{7}$

11. $9 + 3\sqrt{5} - 3\sqrt{5} - \sqrt{25} = 9 + 3\sqrt{5} - 3\sqrt{5} - 5 = 4$

12. $\sqrt{9} + 7\sqrt{3} - 7\sqrt{3} - 49 = 3 + 7\sqrt{3} - 7\sqrt{3} - 49 = -46$

13. The sum of the two terms which contain irrational numbers, $-a\sqrt{b}$ and $a\sqrt{b}$, will always be zero.

14. Multiplying radical expressions is like multiplying a binomial. The FOIL method may be used when multiplying radical expressions. For example: $(2 + \sqrt{5})(2 - \sqrt{5}) = 4 - 2\sqrt{5} + 2\sqrt{5} - 5 = 4 - 5 = -1$.

pages 514–515 Practice Exercises

A 1. $\sqrt{16} = 4$ **2.** $\sqrt{36} = 6$ **3.** $\sqrt{12} = \sqrt{4 \cdot 3} = 2\sqrt{3}$

4. $\sqrt{50} = \sqrt{25 \cdot 2} = 5\sqrt{2}$ **5.** $\sqrt{49} = 7$ **6.** $\sqrt{25} = 5$

7. $3\sqrt{100} = 30$ **8.** $2\sqrt{90} = 2\sqrt{9 \cdot 10} = 6\sqrt{10}$

9. $-10\sqrt{18} = -10\sqrt{9 \cdot 2} = -30\sqrt{2}$ **10.** $15\sqrt{45} = 15\sqrt{9 \cdot 5} = 45\sqrt{5}$

11. $6\sqrt{21}$ **12.** $-20\sqrt{30}$

13. $12\sqrt{350} = 12\sqrt{25 \cdot 14} = 60\sqrt{14}$ **14.** $14\sqrt{98} = 14\sqrt{49 \cdot 2} = 98\sqrt{2}$

15. $\sqrt{\dfrac{15}{15}} = \sqrt{1} = 1$ **16.** $\sqrt{\dfrac{18}{6}} = \sqrt{3}$

17. $6\sqrt{24x^2} = 6\sqrt{4x^2 \cdot 6} = 12x\sqrt{6}$ **18.** $24\sqrt{50a^2} = 24\sqrt{25a^2 \cdot 2} = 120a\sqrt{2}$

19. $9\sqrt{n^2} = 9n$ **20.** $4\sqrt{m^2} = 4m$

B 21. $4\sqrt{3} + \sqrt{9} = 4\sqrt{3} + 3$ **22.** $\sqrt{25} - 3\sqrt{5} = 5 - 3\sqrt{5}$

23. $\sqrt{36} - 2\sqrt{6} = 6 - 2\sqrt{6}$ **24.** $2\sqrt{2} + \sqrt{4} = 2\sqrt{2} + 2$

25. $6\sqrt{4} - 9\sqrt{2} + 2\sqrt{2} - 3 = 12 - 9\sqrt{2} + 2\sqrt{2} - 3 = 9 - 7\sqrt{2}$

26. $20 + 25\sqrt{3} - 16\sqrt{3} - 20\sqrt{9} = 20 + 25\sqrt{3} - 16\sqrt{3} - 60 = -40 + 9\sqrt{3}$

27. $36 + 6\sqrt{7} - 6\sqrt{7} - \sqrt{49} = 36 + 6\sqrt{7} - 6\sqrt{7} - 7 = 29$

28. $\sqrt{9} + 4\sqrt{3} - 4\sqrt{3} - 16 = 3 + 4\sqrt{3} - 4\sqrt{3} - 16 = -13$

29. $3\sqrt{12a^2b^2} = 3\sqrt{4a^2b^2 \cdot 3} = 6ab\sqrt{3}$ **30.** $-2\sqrt{24x^2y^2} = -2\sqrt{4x^2y^2 \cdot 6} = -4xy\sqrt{6}$

31. $\sqrt{c^2d^2e^3} = \sqrt{c^2d^2e^2 \cdot e} = cde\sqrt{e}$ **32.** $\sqrt{a^5b^2c^5} = \sqrt{a^4b^2c^4 \cdot ac} = a^2bc^2\sqrt{ac}$

33. $5a\sqrt{\dfrac{a^2}{b^2}} = \dfrac{5a^2}{b}$ **34.** $3r\sqrt{\dfrac{4r^2}{s^2}} = \dfrac{6r^2}{s}$

35. $(3\sqrt{7} + 2)(3\sqrt{7} + 2) = 9\sqrt{49} + 6\sqrt{7} + 6\sqrt{7} + 4 = 67 + 12\sqrt{7}$

36. $(5 - 2\sqrt{6})(5 - 2\sqrt{6}) = 25 - 10\sqrt{6} - 10\sqrt{6} + 4\sqrt{36} = 49 - 20\sqrt{6}$

C 37. $12 + 3\sqrt{2} - 4\sqrt{3} - \sqrt{6}$ **38.** $\sqrt{30} - 4\sqrt{5} + 2\sqrt{6} - 8$

39. $(\sqrt{x} + 2\sqrt{y})(\sqrt{x} + 2\sqrt{y}) = \sqrt{x^2} + 2\sqrt{xy} + 2\sqrt{xy} + 4\sqrt{y^2} =$

$$x + 2\sqrt{xy} + 2\sqrt{xy} + 4y = x + 4\sqrt{xy} + 4y$$

40. $(5\sqrt{u} + 2\sqrt{v})(5\sqrt{u} + 2\sqrt{v}) = 25\sqrt{u^2} + 10\sqrt{uv} + 10\sqrt{uv} + 4\sqrt{v^2} = 25u + 10\sqrt{uv} + 10\sqrt{uv} + 4v = 25u + 20\sqrt{uv} + 4v$

41. $8\sqrt{m^2} + 2\sqrt{mn} - 12\sqrt{mn} - 3\sqrt{n^2} = 8m + 2\sqrt{mn} - 12\sqrt{mn} - 3n = 8m - 10\sqrt{mn} - 3n$

42. $5\sqrt{x^2} - 2\sqrt{xy} + 20\sqrt{xy} - 8\sqrt{y^2} = 5x + 18\sqrt{xy} - 8y$

43. $A = \frac{1}{2}bh$, $A = \frac{1}{2} \cdot 2\sqrt{3} \cdot 2\sqrt{2}$; $A = 2\sqrt{6}m^2$

44. $A = bh$, $A = \dfrac{5\sqrt{2} + 3}{3} \cdot \dfrac{3\sqrt{2} - 2}{5}$, $A = \dfrac{15\sqrt{4} - 10\sqrt{2} + 9\sqrt{2} - 6}{15}$; $A = \dfrac{24 - \sqrt{2}}{15}$ cm²

45. $A = s^2$, $A = (4\sqrt{x})^2$, $A = 16\sqrt{x^2}$, $A = 16x$ ft²

46. $A = \frac{1}{2}h(b_1 + b_2)$, $A = \frac{1}{2} \cdot \sqrt{x} \cdot (\sqrt{x} + 2\sqrt{y})$, $A = \frac{1}{2}(\sqrt{x^2} \cdot 2\sqrt{xy})$, $A = \frac{1}{2}\sqrt{x^2} + \sqrt{xy}$; $A = (\frac{1}{2}x + \sqrt{xy})$ m²

47. $(\sqrt{2} + \sqrt{3})(\sqrt{2} + \sqrt{3}) = \sqrt{4} + \sqrt{6} + \sqrt{6} + \sqrt{9} = 2 + \sqrt{6} + \sqrt{6} + 3 = 5 + 2\sqrt{6}$

48. $(\sqrt{2} + \sqrt{3})(\sqrt{2} + \sqrt{3})(\sqrt{2} + \sqrt{3}) = (\sqrt{2} + \sqrt{3})(5 + 2\sqrt{6}) = 5\sqrt{2} + 2\sqrt{12} + 5\sqrt{3} + 2\sqrt{18} = 5\sqrt{2} + 4\sqrt{3} + 5\sqrt{3} + 6\sqrt{2} = 11\sqrt{2} + 9\sqrt{3}$

49. $(\sqrt{2} + \sqrt{3})(\sqrt{2} + \sqrt{3})(\sqrt{2} + \sqrt{3})(\sqrt{2} + \sqrt{3}) = (\sqrt{2} + \sqrt{3})(11\sqrt{2} + 9\sqrt{3}) = 11\sqrt{4} + 9\sqrt{6} + 11\sqrt{6} + 9\sqrt{9} = 22 + 9\sqrt{6} + 11\sqrt{6} + 27 = 49 + 20\sqrt{6}$

50. $(\sqrt{2} + \sqrt{3} + \sqrt{5})(\sqrt{2} + \sqrt{3} + \sqrt{5}) = \sqrt{4} + \sqrt{6} + \sqrt{10} + \sqrt{6} + \sqrt{9} + \sqrt{15} + \sqrt{10} + \sqrt{15} + \sqrt{25} = 2 + \sqrt{6} + \sqrt{10} + \sqrt{6} + 3 + \sqrt{15} + \sqrt{10} + \sqrt{15} + 5 = 10 + 2\sqrt{6} + 2\sqrt{10} + 2\sqrt{15}$

51. $(\sqrt{x} + \sqrt{y} + \sqrt{z})^2 = x + y + z + 2\sqrt{xy} + 2\sqrt{xz} + 2\sqrt{yz}$

52. $(\sqrt{w} + \sqrt{x} + \sqrt{y} + \sqrt{z})^2 = w + x + y + z + 2\sqrt{wx} + 2\sqrt{wy} + 2\sqrt{wz} + 2\sqrt{xy} + 2\sqrt{xz} + 2\sqrt{yz}$

page 515 Test Yourself

1. $\sqrt{45} \approx 6.71$ **2.** $\sqrt{80} \approx 8.94$ **3.** Let $x = 0.\overline{3}$. $(10x = 3.\overline{3}) - (x = 0.\overline{3})$, $9x = 3$, $x = \dfrac{3}{9} = \dfrac{1}{3}$ **4.** Let $x = 0.\overline{12}$. $(100x = 12.\overline{12}) - (x = 0.\overline{12})$, $99x = 12$, $x = \dfrac{12}{99} = \dfrac{4}{33}$ **5.** $\sqrt{14 \cdot 14} = 14$ **6.** $-\sqrt{300 \cdot 300} = -300$

7. $\dfrac{\sqrt{25}}{\sqrt{16}} = \dfrac{5}{4}$ **8.** $\pm\dfrac{\sqrt{4}}{\sqrt{81}} = \pm\dfrac{2}{9}$ **9.** $-\sqrt{16 \cdot 3} = -4\sqrt{3}$

10. $\sqrt{9y^2 \cdot 7} = 3y\sqrt{7}$ **11.** $\sqrt{25x^2 \cdot x} = 5x\sqrt{x}$ **12.** $\dfrac{\sqrt{64}}{\sqrt{9}} = \dfrac{8}{3}$

13. $(6\sqrt{2} - \sqrt{2}) - 4\sqrt{7} = 5\sqrt{2} - 4\sqrt{7}$

14. $\sqrt{4 \cdot 2} - 2\sqrt{25 \cdot 2} + \sqrt{9 \cdot 2} = 2\sqrt{2} - 10\sqrt{2} + 3\sqrt{2} = -5\sqrt{2}$

15. $12\sqrt{12} = 12\sqrt{4 \cdot 3} = 24\sqrt{3}$

16. $2\sqrt{3} - 2\sqrt{18} = 2\sqrt{3} - 2\sqrt{9 \cdot 2} = 2\sqrt{3} - 6\sqrt{2}$

page 516   Capsule Review

1. $\dfrac{\sqrt{4}}{\sqrt{9}} = \dfrac{2}{3}$ **2.** $-\dfrac{\sqrt{16}}{\sqrt{25}} = -\dfrac{4}{5}$ **3.** $\dfrac{\sqrt{64}}{\sqrt{49}} = \dfrac{8}{7}$ **4.** $\dfrac{\sqrt{1}}{\sqrt{144}} = \dfrac{1}{12}$

5. $-\dfrac{\sqrt{1}}{\sqrt{100}} = -\dfrac{1}{10}$ **6.** $\dfrac{\sqrt{36}}{\sqrt{121}} = \dfrac{6}{11}$ **7.** $\dfrac{\sqrt{225}}{\sqrt{81}} = \dfrac{15}{9} = \dfrac{5}{3}$ **8.** $\dfrac{2z^4}{10} = 0.2z^4$

page 518   Class Exercises

1. $\dfrac{\sqrt{1}}{\sqrt{5}} \cdot \dfrac{\sqrt{5}}{\sqrt{5}} = \dfrac{\sqrt{5}}{\sqrt{25}} = \dfrac{\sqrt{5}}{5}$ or $\dfrac{1}{5}\sqrt{5}$ **2.** $\dfrac{\sqrt{1}}{\sqrt{2}} \cdot \dfrac{\sqrt{2}}{\sqrt{2}} = \dfrac{\sqrt{2}}{\sqrt{4}} = \dfrac{\sqrt{2}}{2}$ or $\dfrac{1}{2}\sqrt{2}$

3. $\dfrac{\sqrt{2}}{\sqrt{5}} \cdot \dfrac{\sqrt{5}}{\sqrt{5}} = \dfrac{\sqrt{10}}{\sqrt{25}} = \dfrac{\sqrt{10}}{5}$ or $\dfrac{1}{5}\sqrt{10}$ **4.** $\dfrac{\sqrt{3}}{\sqrt{7}} \cdot \dfrac{\sqrt{7}}{\sqrt{7}} = \dfrac{\sqrt{21}}{\sqrt{49}} = \dfrac{\sqrt{21}}{7}$ or $\dfrac{1}{7}\sqrt{21}$

5. $\dfrac{2\sqrt{2}}{\sqrt{18}} \cdot \dfrac{\sqrt{2}}{\sqrt{2}} = \dfrac{2\sqrt{4}}{\sqrt{36}} = \dfrac{4}{6} = \dfrac{2}{3}$ **6.** $\dfrac{3\sqrt{2}}{4\sqrt{32}} \cdot \dfrac{\sqrt{2}}{\sqrt{2}} = \dfrac{3\sqrt{4}}{4\sqrt{64}} = \dfrac{6}{32} = \dfrac{3}{16}$

7. $\dfrac{\sqrt{4}}{\sqrt{x}} \cdot \dfrac{\sqrt{x}}{\sqrt{x}} = \dfrac{\sqrt{4x}}{\sqrt{x^2}} = \dfrac{2\sqrt{x}}{x}$ or $\dfrac{2}{x}\sqrt{x}$ **8.** $\dfrac{\sqrt{9}}{\sqrt{b^3}} \cdot \dfrac{\sqrt{b}}{\sqrt{b}} = \dfrac{\sqrt{9b}}{\sqrt{b^4}} = \dfrac{3\sqrt{b}}{b^2}$ or $\dfrac{3}{b^2}\sqrt{b}$

9. $1 + \sqrt{2}$ **10.** $2 - \sqrt{3}$ **11.** $5 - 3\sqrt{5}$ **12.** $4 + 2\sqrt{7}$

pages 518–519   Practice Exercises

A 1. $\dfrac{\sqrt{2}}{\sqrt{11}} \cdot \dfrac{\sqrt{11}}{\sqrt{11}} = \dfrac{\sqrt{22}}{\sqrt{121}} = \dfrac{\sqrt{22}}{11}$ or $\dfrac{1}{11}\sqrt{22}$ **2.** $\dfrac{\sqrt{3}}{\sqrt{10}} \cdot \dfrac{\sqrt{10}}{\sqrt{10}} = \dfrac{\sqrt{30}}{\sqrt{100}} = \dfrac{\sqrt{30}}{10}$ or $\dfrac{1}{10}\sqrt{30}$

3. $-2\dfrac{\sqrt{1}}{\sqrt{2}} \cdot \dfrac{\sqrt{2}}{\sqrt{2}} = -2\dfrac{\sqrt{2}}{\sqrt{4}} = -\sqrt{2}$ **4.** $-3\dfrac{\sqrt{1}}{\sqrt{3}} \cdot \dfrac{\sqrt{3}}{\sqrt{3}} = -3\dfrac{\sqrt{3}}{\sqrt{9}} = -\sqrt{3}$

5. $\dfrac{3\sqrt{8}}{\sqrt{3}} \cdot \dfrac{\sqrt{3}}{\sqrt{3}} = \dfrac{3\sqrt{24}}{\sqrt{9}} = \dfrac{6\sqrt{6}}{3} = 2\sqrt{6}$ **6.** $\dfrac{2\sqrt{12}}{-\sqrt{5}} \cdot \dfrac{\sqrt{5}}{\sqrt{5}} = \dfrac{2\sqrt{60}}{-\sqrt{25}} = \dfrac{4\sqrt{15}}{-5}$ or $-\dfrac{4}{5}\sqrt{15}$

7. $-\sqrt{\dfrac{90}{10}} = -\sqrt{9} = -3$ **8.** $\dfrac{1}{2}\sqrt{\dfrac{75}{3}} = \dfrac{1}{2}\sqrt{25} = \dfrac{1}{2} \cdot 5 = \dfrac{5}{2}$

9. $\dfrac{5\sqrt{2}}{3\sqrt{6}} \cdot \dfrac{\sqrt{6}}{\sqrt{6}} = \dfrac{5\sqrt{12}}{3\sqrt{36}} = \dfrac{10\sqrt{3}}{18} = \dfrac{5\sqrt{3}}{9}$ or $\dfrac{5}{9}\sqrt{3}$ **10.** $2\dfrac{\sqrt{3}}{\sqrt{8}} \cdot \dfrac{\sqrt{2}}{\sqrt{2}} = \dfrac{2\sqrt{6}}{\sqrt{16}} = \dfrac{2\sqrt{6}}{4} = \dfrac{\sqrt{6}}{2}$ or $\dfrac{1}{2}\sqrt{6}$

11. $-\dfrac{3}{4}\sqrt{\dfrac{5}{15}} = -\dfrac{3}{4}\sqrt{\dfrac{1}{3}} = -\dfrac{3}{4}\dfrac{\sqrt{1}}{\sqrt{3}}\cdot\dfrac{\sqrt{3}}{\sqrt{3}} = -\dfrac{3}{4}\dfrac{\sqrt{3}}{\sqrt{9}} = -\dfrac{1}{4}\sqrt{3}$

12. $-\dfrac{6}{2}\sqrt{\dfrac{6}{30}} = -3\sqrt{\dfrac{1}{5}} = -3\dfrac{\sqrt{1}}{\sqrt{5}}\cdot\dfrac{\sqrt{5}}{\sqrt{5}} = -3\dfrac{\sqrt{5}}{\sqrt{25}} = -\dfrac{3}{5}\sqrt{5}$

13. $\dfrac{\sqrt{16}}{\sqrt{y}}\cdot\dfrac{\sqrt{y}}{\sqrt{y}} = \dfrac{4\sqrt{y}}{y}$ or $\dfrac{4}{y}\sqrt{y}$

14. $\dfrac{\sqrt{25}}{\sqrt{a}}\cdot\dfrac{\sqrt{a}}{\sqrt{a}} = \dfrac{5\sqrt{a}}{a}$ or $\dfrac{5}{a}\sqrt{a}$

15. $\dfrac{\sqrt{8}}{\sqrt{x}}\cdot\dfrac{\sqrt{x}}{\sqrt{x}} = \dfrac{\sqrt{8x}}{\sqrt{x^2}} = \dfrac{2\sqrt{2x}}{x}$ or $\dfrac{2}{x}\sqrt{2x}$

16. $\dfrac{\sqrt{12}}{\sqrt{b}}\cdot\dfrac{\sqrt{b}}{\sqrt{b}} = \dfrac{\sqrt{12b}}{\sqrt{b^2}} = \dfrac{2\sqrt{3b}}{b}$ or $\dfrac{2}{b}\sqrt{3b}$

17. $\dfrac{3}{\sqrt{a^3}}\cdot\dfrac{\sqrt{a}}{\sqrt{a}} = \dfrac{3\sqrt{a}}{\sqrt{a^4}} = \dfrac{3\sqrt{a}}{a^2}$ or $\dfrac{3}{a^2}\sqrt{a}$

18. $\dfrac{-5}{\sqrt{x^5}}\cdot\dfrac{\sqrt{x}}{\sqrt{x}} = \dfrac{-5\sqrt{x}}{\sqrt{x^6}} = \dfrac{-5\sqrt{x}}{x^3}$ or $-\dfrac{5}{x^3}\sqrt{x}$

19. $\dfrac{-2}{3\sqrt{y^3}}\cdot\dfrac{\sqrt{y}}{\sqrt{y}} = \dfrac{-2\sqrt{y}}{3\sqrt{y^4}} = \dfrac{-2\sqrt{y}}{3y^2}$ or $-\dfrac{2}{3y^2}\sqrt{y}$

20. $\sqrt{\dfrac{3}{12m}} = \sqrt{\dfrac{1}{4m}} = \dfrac{\sqrt{1}}{\sqrt{4m}}\cdot\dfrac{\sqrt{m}}{\sqrt{m}} = \dfrac{\sqrt{m}}{2m}$ or $\dfrac{1}{2m}\sqrt{m}$

21. $\sqrt{4c^2} = 2c$

22. $\sqrt{9x^4} = 3x^2$

23. $\sqrt{\dfrac{a^3b^4}{ab}} = \sqrt{a^2b^3} = \sqrt{a^2b^2\cdot b} = ab\sqrt{b}$

24. $\sqrt{\dfrac{cd}{c^3d^3}} = \sqrt{\dfrac{1}{c^2d^2}} = \dfrac{1}{cd}$

B 25. $\dfrac{5}{3-\sqrt{2}}\cdot\dfrac{3+\sqrt{2}}{3+\sqrt{2}} = \dfrac{15+5\sqrt{2}}{9-\sqrt{4}} = \dfrac{15+5\sqrt{2}}{7}$

26. $\dfrac{-6}{5+\sqrt{7}}\cdot\dfrac{5-\sqrt{7}}{5-\sqrt{7}} = \dfrac{-30+6\sqrt{7}}{25-\sqrt{49}} = \dfrac{-30+6\sqrt{7}}{18} = \dfrac{-5+\sqrt{7}}{3}$

27. $\dfrac{-3}{\sqrt{5}-1}\cdot\dfrac{\sqrt{5}+1}{\sqrt{5}+1} = \dfrac{-3\sqrt{5}-3}{\sqrt{25}-1} = \dfrac{-3\sqrt{5}-3}{4}$

28. $\dfrac{2}{\sqrt{6}+3}\cdot\dfrac{\sqrt{6}-3}{\sqrt{6}-3} = \dfrac{2\sqrt{6}-6}{\sqrt{36}-9} = \dfrac{2\sqrt{6}-6}{-3}$

29. $\dfrac{\sqrt{5}-1}{\sqrt{5}+3}\cdot\dfrac{\sqrt{5}-3}{\sqrt{5}-3} = \dfrac{\sqrt{25}-3\sqrt{5}-\sqrt{5}+3}{\sqrt{25}-9} = \dfrac{8-4\sqrt{5}}{-4} = -2+\sqrt{5}$

30. $\dfrac{\sqrt{6}+3}{\sqrt{6}-4}\cdot\dfrac{\sqrt{6}+4}{\sqrt{6}+4} = \dfrac{\sqrt{36}+4\sqrt{6}+3\sqrt{6}+12}{\sqrt{36}-16} = \dfrac{18+7\sqrt{6}}{-10}$

31. $\dfrac{2+\sqrt{3}}{1-\sqrt{5}}\cdot\dfrac{1+\sqrt{5}}{1+\sqrt{5}} = \dfrac{2+2\sqrt{5}+\sqrt{3}+\sqrt{15}}{1-\sqrt{25}} = \dfrac{2+2\sqrt{5}+\sqrt{3}+\sqrt{15}}{-4}$

32. $\dfrac{4-\sqrt{3}}{2+\sqrt{7}}\cdot\dfrac{2-\sqrt{7}}{2-\sqrt{7}} = \dfrac{8-4\sqrt{7}-2\sqrt{3}+\sqrt{21}}{4-\sqrt{49}} = \dfrac{8-4\sqrt{7}-2\sqrt{3}+\sqrt{21}}{-3}$

33. $\dfrac{-5}{2\sqrt{11}+2}\cdot\dfrac{2\sqrt{11}-2}{2\sqrt{11}-2} = \dfrac{-10\sqrt{11}+10}{4\sqrt{121}-4} = \dfrac{-10\sqrt{11}+10}{40} = \dfrac{-\sqrt{11}+1}{4}$

34. $\dfrac{-7}{3\sqrt{10} - 5} \cdot \dfrac{3\sqrt{10} + 5}{3\sqrt{10} + 5} = \dfrac{-21\sqrt{10} - 35}{9\sqrt{100} - 25} = \dfrac{-21\sqrt{10} - 35}{65}$

35. $\dfrac{1 - 3\sqrt{7}}{3\sqrt{3} + 2} \cdot \dfrac{3\sqrt{3} - 2}{3\sqrt{3} - 2} = \dfrac{3\sqrt{3} - 2 - 9\sqrt{21} + 6\sqrt{7}}{9\sqrt{9} - 4} = \dfrac{3\sqrt{3} - 2 - 9\sqrt{21} + 6\sqrt{7}}{23}$

36. $\dfrac{3\sqrt{2} + 2}{3 - 2\sqrt{5}} \cdot \dfrac{3 + 2\sqrt{5}}{3 + 2\sqrt{5}} = \dfrac{9\sqrt{2} + 6\sqrt{10} + 6 + 4\sqrt{5}}{9 - 4\sqrt{25}} = \dfrac{9\sqrt{2} + 6\sqrt{10} + 6 + 4\sqrt{5}}{-11}$

C 37. $\dfrac{2\sqrt{7} + 3\sqrt{5}}{3\sqrt{7} - 2\sqrt{5}} \cdot \dfrac{3\sqrt{7} + 2\sqrt{5}}{3\sqrt{7} + 2\sqrt{5}} = \dfrac{6\sqrt{49} + 4\sqrt{35} + 9\sqrt{35} + 6\sqrt{25}}{9\sqrt{49} - 4\sqrt{25}} = \dfrac{72 + 13\sqrt{35}}{43}$

38. $\dfrac{5\sqrt{2} - \sqrt{3}}{\sqrt{2} + 3\sqrt{3}} \cdot \dfrac{\sqrt{2} - 3\sqrt{3}}{\sqrt{2} - 3\sqrt{3}} = \dfrac{5\sqrt{4} - 15\sqrt{6} - \sqrt{6} + 3\sqrt{9}}{\sqrt{4} - 9\sqrt{9}} = \dfrac{19 - 16\sqrt{6}}{-25}$

39. $\dfrac{\sqrt{x} - \sqrt{y}}{\sqrt{x}} \cdot \dfrac{\sqrt{x}}{\sqrt{x}} = \dfrac{\sqrt{x^2} - \sqrt{xy}}{\sqrt{x^2}} = \dfrac{x - \sqrt{xy}}{x}$

40. $\dfrac{\sqrt{x} + 2\sqrt{y}}{5\sqrt{x}} \cdot \dfrac{5\sqrt{x}}{5\sqrt{x}} = \dfrac{5\sqrt{x^2} + 10\sqrt{xy}}{25\sqrt{x^2}} = \dfrac{5x + 10\sqrt{xy}}{25x} = \dfrac{5(x + 2\sqrt{xy})}{25x} = \dfrac{x + 2\sqrt{xy}}{5x}$

41. $\dfrac{2\sqrt{x} - 3}{\sqrt{x} + 1} \cdot \dfrac{\sqrt{x} - 1}{\sqrt{x} - 1} = \dfrac{2\sqrt{x^2} - 2\sqrt{x} - 3\sqrt{x} + 3}{\sqrt{x^2} + \sqrt{x} - \sqrt{x} - 1} = \dfrac{2x - 5\sqrt{x} + 3}{x - 1}$

42. $\dfrac{3\sqrt{x} + 4\sqrt{y}}{3\sqrt{x} - 4\sqrt{y}} \cdot \dfrac{3\sqrt{x} + 4\sqrt{y}}{3\sqrt{x} + 4\sqrt{y}} = \dfrac{9\sqrt{x^2} + 12\sqrt{xy} + 12\sqrt{xy} + 16\sqrt{y^2}}{9\sqrt{x^2} - 12\sqrt{xy} + 12\sqrt{xy} - 16\sqrt{y^2}} = \dfrac{9x + 24\sqrt{xy} + 16y}{9x - 16y}$

43. $\dfrac{r_1}{r_2} = \dfrac{\sqrt{25}}{\sqrt{80}}, \dfrac{r_1}{r_2} = \dfrac{\sqrt{25}}{\sqrt{80}} \cdot \dfrac{\sqrt{5}}{\sqrt{5}}, \dfrac{r_1}{r_2} = \dfrac{\sqrt{125}}{\sqrt{400}}, \dfrac{r_1}{r_2} = \dfrac{5\sqrt{5}}{20}, \dfrac{r_1}{r_2} = \dfrac{\sqrt{5}}{4}$

44. $T = 2\pi\sqrt{\dfrac{4}{32}}, T = 2\pi\sqrt{\dfrac{1}{8}}, T = 2\pi\dfrac{\sqrt{1}}{\sqrt{8}} \cdot \dfrac{\sqrt{2}}{\sqrt{2}}, T = 2\pi\dfrac{\sqrt{2}}{4}, T = \dfrac{\sqrt{2}}{2}\pi$ s

45. Reasons: 1. Commutative property of addition; 2. Associative property of addition; 3. Inverse property of addition; 4. Identity property of addition

46. $2y = -3x + 4$, $y = -\dfrac{3}{2}x + 2$; slope: $-\dfrac{3}{2}$; y-int.: 2

47. $y = 2x$; slope: 2; y-int.: 0   **48.** $y = x - 9$; slope: 1; y-int.: -9

49. $-6y = -x - 18$, $y = \dfrac{1}{6}x + 3$; slope: $\dfrac{1}{6}$; y-int.: 3

50. $5y = 3x$, $y = \dfrac{3}{5}x$; slope: $\dfrac{3}{5}$; y-int.: 0

51. $-2y = -2x - 1$, $y = x + \dfrac{1}{2}$; slope: 1; y-int.: $\dfrac{1}{2}$

52. Let t represent the tens digit and u represent the units digit.
$$\begin{cases} 5u + 15 = 7t \\ t + u = 9 \longrightarrow t = 9 - u \end{cases}$$
$5u + 15 = 7(9 - u)$, $5u + 15 = 63 - 7u$, $12u = 48$, $u = 4$; $t + 4 = 9$, $t = 5$; The number is 54.

page 520   Capsule Review

1. $3x - 9 \geq 0$, $3x \geq 9$, $x \geq \dfrac{9}{3}$; $x \geq 3$

2. $-2x + 5 \geq 0$, $-2x \geq -5$; $x \leq \dfrac{5}{2}$

3. $18 - 5x \geq 0$, $-5x \geq -18$; $x \leq \dfrac{18}{5}$

4. $6 + 3x \geq 0$, $3x \geq -6$; $x \geq -2$

page 521   Class Exercises

1. $(\sqrt{x})^2 = 5^2$; $x = 25$

2. $(\sqrt{q})^2 = 12^2$; $q = 144$

3. $(\sqrt{2b})^2 = 16^2$, $2b = 256$; $b = 128$

4. $(\sqrt{3y})^2 = 9^2$, $3y = 81$; $y = 27$

5. $\sqrt{3t} = 2$, $(\sqrt{3t})^2 = 2^2$, $3t = 4$; $t = \dfrac{4}{3}$

6. $\sqrt{5p} = 2$, $(\sqrt{5p})^2 = 2^2$, $5p = 4$; $p = \dfrac{4}{5}$

7. $\sqrt{2x + 1} = 3$; $2x + 1 = 9$

8. $\sqrt{3a - 1} = 2$; $3a - 1 = 4$

9. $\sqrt{4y + 1} = -1$; $4y + 1 = 1$

10. $\sqrt{3c + 1} = 5$; $3c + 1 = 25$

11. $-3 = -2\sqrt{y}$; $4y = 9$

12. $-3\sqrt{2x} = -6$; $2x = 4$

pages 522–523   Practice Exercises

A **1.** $(\sqrt{y})^2 = 8^2$; $y = 64$

2. $(\sqrt{a})^2 = 9^2$; $a = 81$

3. $(\sqrt{3m})^2 = 18^2$, $3m = 324$; $m = 108$

4. $(\sqrt{4x})^2 = 8^2$, $4x = 64$; $x = 16$

5. $(\sqrt{n})^2 = \left(\dfrac{1}{4}\right)^2$; $n = \dfrac{1}{16}$

6. $(\sqrt{m})^2 = \left(\dfrac{2}{3}\right)^2$; $m = \dfrac{4}{9}$

7. $\left(\sqrt{\dfrac{x}{3}}\right)^2 = 2^2$, $\dfrac{x}{3} = 4$; $x = 12$

8. $\left(\sqrt{\dfrac{b}{7}}\right)^2 = 3^2$, $\dfrac{b}{7} = 9$; $b = 63$

9. $(\sqrt{y - 2})^2 = 3^2$, $y - 2 = 9$; $y = 11$

10. $(\sqrt{a + 4})^2 = 6^2$, $a + 4 = 36$; $a = 32$

11. $\sqrt{4x} = 4$, $(\sqrt{4x})^2 = 4^2$, $4x = 16$; $x = 4$

12. $\sqrt{2z} = 9$, $(\sqrt{2z})^2 = 9^2$, $2z = 81$; $z = \dfrac{81}{2}$

13. $\sqrt{m} = -1$; no solution

14. $2\sqrt{b} = -10$; no solution

15. $-3\sqrt{y} = -6$, $\sqrt{y} = 2$, $(\sqrt{y})^2 = 2^2$; $y = 4$

16. $-\sqrt{a} = -7$, $\sqrt{a} = 7$, $(\sqrt{a})^2 = 7^2$; $a = 49$

17. $\sqrt{5x - 1} = 4$, $(\sqrt{5x - 1})^2 = 4^2$, $5x - 1 = 16$, $5x = 17$; $x = \dfrac{17}{5}$

18. $\sqrt{3x + 4} = 3$, $(\sqrt{3x + 4})^2 = 3^2$, $3x + 4 = 9$, $3x = 5$; $x = \dfrac{5}{3}$

19. $\sqrt{6m - 1} = 3$, $(\sqrt{6m - 1})^2 = 3^2$, $6m - 1 = 9$, $6m = 10$; $m = \dfrac{5}{3}$

20. $\sqrt{2x - 3} = 6$, $(\sqrt{2x - 3})^2 = 6^2$, $2x - 3 = 36$, $2x = 39$; $x = \dfrac{39}{2}$

21. $-3\sqrt{2b} = -6$, $\sqrt{2b} = 2$, $(\sqrt{2b})^2 = 2^2$, $2b = 4$; $b = 2$

22. $-2\sqrt{y} = -3$, $(-2\sqrt{y})^2 = (-3)^2$, $4y = 9$; $y = \dfrac{9}{4}$

23. $\sqrt{\dfrac{2m}{3}} = 2$, $\left(\sqrt{\dfrac{2m}{3}}\right)^2 = 2^2$, $\dfrac{2m}{3} = 4$, $2m = 12$; $m = 6$

24. $\sqrt{\dfrac{5n}{4}} = 10$, $\left(\sqrt{\dfrac{5n}{4}}\right)^2 = 10^2$, $\dfrac{5n}{4} = 100$, $5n = 400$; $n = 80$

B 25. $(\sqrt{x})^2 = (3\sqrt{2})^2$; $x = 18$ **26.** $(\sqrt{t})^2 = (2\sqrt{5})^2$; $t = 20$

27. $(\sqrt{5y^2 - 7})^2 = (2y)^2$, $5y^2 - 7 = 4y^2$, $y^2 = 7$; $y = \sqrt{7}$

28. $(\sqrt{3n^2 + 12})^2 = (3n)^2$, $3n^2 + 12 = 9n^2$, $6n^2 = 12$, $n^2 = 2$; $n = \sqrt{2}$

29. $(\sqrt{y^2 + 3})^2 = (y + 1)^2$, $y^2 + 3 = y^2 + 2y + 1$, $2y + 1 = 3$, $2y = 2$; $y = 1$

30. $(t - 5)^2 = (\sqrt{t^2 - 35})^2$, $t^2 - 10t + 25 = t^2 - 35$, $-10t + 25 = -35$, $-10t = -60$; $t = 6$

31. $(\sqrt{2x - 3})^2 = (\sqrt{3x - 2})^2$, $2x - 3 = 3x - 2$, $x = -1$; no solution

32. $(\sqrt{3y - 5})^2 = (\sqrt{5y - 3})^2$, $3y - 5 = 5y - 3$, $2y = -2$, $y = -1$; no solution

33. $(\sqrt{2b + 1})^2 = (\sqrt{4b + 7})^2$, $2b + 1 = 4b + 7$, $2b = -6$, $b = -3$; no solution

34. $(2\sqrt{5a - 1})^2 = (3\sqrt{2a + 4})^2$, $4(5a - 1) = 9(2a + 4)$, $20a - 4 = 18a + 36$, $2a = 40$; $a = 20$

35. $(\sqrt{3y + 4})^2 = (y - 2)^2$, $3y + 4 = y^2 - 4y + 4$, $y^2 - 7y = 0$, $y(y - 7) = 0$, $y = 0$, $y = 7$; 0 is an extraneous solution.

36. $(m - 4)^2 = (\sqrt{2m - 5})^2$, $m^2 - 8m + 16 = 2m - 5$, $m^2 - 10m + 21 = 0$, $(m - 7)(m - 3) = 0$, $m = 7$, $m = 3$; 3 is an extraneous solution.

C 37. $(\sqrt{y} + 2)^2 = (\sqrt{y + 16})^2$, $y + 4\sqrt{y} + 4 = y + 16$, $4\sqrt{y} = 12$, $\sqrt{y} = 3$, $(\sqrt{y})^2 = 3^2$; $y = 9$

38. $(\sqrt{x - 1})^2 = (2 - \sqrt{x})^2$, $x - 1 = 4 - 4\sqrt{x} + x$, $4\sqrt{x} = 5$, $(4\sqrt{x})^2 = 5^2$, $16x = 25$; $x = \dfrac{25}{16}$

39. $(\sqrt[3]{2x + 3})^3 = 7^3$, $2x + 3 = 343$, $2x = 340$, $x = 170$

40. $(\sqrt[3]{3y + 5})^3 = (\sqrt[3]{5 - 2y})^3$, $3y + 5 = 5 - 2y$, $5y = 0$, $y = 0$

41. Cubing a number does not change its sign.

42. $3 = \sqrt{\dfrac{2d}{32}}$, $9 = \dfrac{2d}{32}$, $288 = 2d$; $d = 144$ ft

43. $1 = 2(3.14)\sqrt{\dfrac{l}{980}}$, $1 = 6.28\sqrt{\dfrac{l}{980}}$, $1 = 39.4\left(\dfrac{l}{980}\right)$ $\dfrac{1}{39.4} = \dfrac{l}{980}$, $l = \dfrac{980}{39.4}$; $l \approx 25$ cm

44. $-3t^2 - 12t - t^2 + 5t = -4t^2 - 7t$ **45.** $-4h^2 - 3h - 2h^2 + 8h = -6h^2 + 5h$

46. $4b^2 + 2b - 5 - 8b^2 - 6 = -4b^2 + 2b - 11$

47. $-6a^5 - 8a^4 + 10a^3$

48. $10d^2 + 5d - 6d - 3 = 10d^2 - d - 3$ **49.** $(3x + 5)(3x + 5) = 9x^2 + 30x + 25$

50. $\begin{cases} 5a + 8b = 20 \longrightarrow 5a + 8b = 20 \\ 3a - 4b = 12 \longrightarrow 6a - 8b = 24 \end{cases}$
$11a = 44$, $a = 4$; $5(4) + 8b = 20$, $20 + 8b = 20$, $8b = 0$, $b = 0$; $(4, 0)$

51. $\begin{cases} 2x + 5y = 21 \longrightarrow 4x + 10y = 42 \\ 3x - 2y = -16 \longrightarrow 15x - 10y = -80 \end{cases}$
$19x = -38$, $x = -2$; $2(-2) + 5y = 21$, $-4 + 5y = 21$, $5y = 25$, $y = 5$; $(-2, 5)$

52. $V = 3.5\sqrt{7}$, $V = 9.26$; $V \approx 9$ km   **53.** $21 = 3.5\sqrt{h}$, $6 = \sqrt{h}$; $h = 36$ m

54. $55 = 3.5\sqrt{h}$, $15.71 = \sqrt{h}$; $h \approx 247$ m

55. 8 km = 8000 m; $V = 3.5\sqrt{8000}$; $V \approx 313$ km

56. $200 = 3.5\sqrt{h}$, $\sqrt{h} = 57.14$, $h = 3265$, $8000 - 3265 \approx 4735$; 4735 m

page 524 Capsule Review

1. 36   **2.** $9 + 4 = 13$ **3.** $100 - 64 = 36$

4. 324   **5.** 6   **6.** 9

7. $\sqrt{169 - 25} = \sqrt{144} = 12$ **8.** $\sqrt{9 + 16} = \sqrt{25} = 5$ **9.** $\sqrt{3} \cdot \sqrt{3} = 3$

10. $\sqrt{11} \cdot \sqrt{11} = 11$

11. $(x + 1)(x + 1) = x^2 + 2x + 1$ **12.** $(y - 2)(y - 2) = y^2 - 4y + 4$

page 526 Class Exercises

1. $c^2 = 6^2 + 8^2$, $c^2 = 36 + 64$, $c^2 = 100$, $c = 10$

2. $c^2 = 10^2 + 24^2$, $c^2 = 100 + 576$, $c^2 = 676$, $c = 26$

3. $13^2 = 5^2 + b^2$, $169 = 25 + b^2$, $b^2 = 144$; $b = 12$

4. $5^2 = a^2 + 4^2$, $25 = a^2 + 16$, $a^2 = 9$; $a = 3$

5. $(\sqrt{13})^2 = 2^2 + b^2$, $13 = 4 + b^2$, $b^2 = 9$; $b = 3$

6. $c^2 = (\sqrt{5})^2 + (\sqrt{11})^2$, $c^2 = 5 + 11$, $c^2 = 16$; $c = 4$

A 1. $c^2 = 3^2 + 4^2$, $c^2 = 9 + 16$, $c^2 = 25$; $c = 5$

2. $c^2 = 5^2 + 12^2$, $c^2 = 25 + 144$, $c^2 = 169$; $c = 13$

3. $25^2 = 24^2 + b^2$, $625 = 576 + b^2$, $b^2 = 49$, $b = 7$

4. $10^2 = a^2 + 8^2$, $100 = a^2 + 64$, $a^2 = 36$, $a = 6$

5. $c^2 = 12^2 + 9^2$, $c^2 = 144 + 81$, $c^2 = 225$, $c = 15$

6. $c^2 = 7^2 + (\sqrt{15})^2$, $c^2 = 49 + 15$, $c^2 = 64$, $c = 8$

7. $26^2 = a^2 + 24^2$, $676 = a^2 + 576$, $a^2 = 100$, $a = 10$

8. $41^2 = 40^2 + b^2$, $1681 = 1600 + b^2$, $b^2 = 81$; $b = 9$

9. $3^2 + 4^2 \stackrel{?}{=} 6^2$, $9 + 16 \stackrel{?}{=} 36$, $25 \neq 36$; no

10. $18^2 + 15^2 \stackrel{?}{=} 23^2$, $324 + 225 \stackrel{?}{=} 529$, $549 \neq 529$; no

11. $1^2 + 1^2 \stackrel{?}{=} (\sqrt{2})^2$, $1 + 1 \stackrel{?}{=} 2$, $2 = 2$; yes

12. $1^2 + (\sqrt{3})^2 \stackrel{?}{=} 2^2$, $1 + 3 \stackrel{?}{=} 4$, $4 = 4$; yes

B 13. $c^2 = \left(\dfrac{2}{5}\right)^2 + \left(\dfrac{3}{10}\right)^2$, $c^2 = \dfrac{4}{25} + \dfrac{9}{100}$, $c^2 = \dfrac{16}{100} + \dfrac{9}{100}$, $c^2 = \dfrac{25}{100}$, $c = \dfrac{5}{10}$,

$c = \dfrac{1}{2}$ m

14. $(1.0)^2 = a^2 + (0.6)^2$, $1 = a^2 + 0.36$, $a^2 = 0.64$, $a = 0.8$ m

15. $c^2 = (\sqrt{13})^2 + 6^2$, $c^2 = 13 + 36$, $c^2 = 49$, $c = 7$ cm

16. $(5\sqrt{2})^2 = a^2 + 5^2$, $50 = a^2 + 25$, $a^2 = 25$, $a = 5$ in.

C 17. Diagram 1: Area of large square = Areas of 2 small squares + Areas of 4 small triangles, so $(a + b)^2 = a^2 + b^2 + 4\left(\dfrac{1}{2}ab\right)$.

Diagram 2: Area of blue square = Area of red square + Areas of 4 small triangles, so $(a + b)^2 = c^2 + 4\left(\dfrac{1}{2}ab\right)$.

18. Since $(a + b)^2 = a^2 + b^2 + 4\left(\dfrac{1}{2}ab\right)$, and $(a + b)^2 = c^2 + 4\left(\dfrac{1}{2}ab\right)$, then, by

the transitive property of equality, $a^2 + b^2 + 4(\frac{1}{2}ab) = c^2 + 4(\frac{1}{2}ab)$.

Therefore, by the subtraction property for equations, $a^2 + b^2 = c^2$.

19. $c^2 = 8^2 + 15^2$, $c^2 = 64 + 225$, $c^2 = 289$, $c = 17$; 17 ft

20. $c^2 = 600^2 + 800^2$, $c^2 = 360{,}000 + 640{,}000$, $c^2 = 1{,}000{,}000$, $c = 1{,}000$; 1,000 lb

21. $(8\sqrt{2})^2 = a^2 + a^2$, $2a^2 = (64)(2)$, $2a^2 = 128$, $a^2 = 64$, $a = 8$; 8 cm

22. $c^2 = 9^2 + 12^2$, $c^2 = 81 + 144$, $c^2 = 225$, $c = 15$; 15 mi

23. Let x = the first integer, then $x + 1$ = the second integer and $x + 2$ = the third integer. The equation is: $(x + 2)^2 = x^2 + (x + 1)^2$, solve $x^2 + 4x + 4 = x^2 + x^2 + 2x + 1$, $x^2 + 4x + 4 = 2x^2 + 2x + 1$, $x^2 - 2x - 3 = 0$, $(x - 3)(x + 1) = 0$, $x - 3 = 0$, $x = 3$, Then $x + 1 = 4$, and $x + 2 = 5$. The lengths of the sides of a right triangle are 3, 4, 5.

24. 9 ft $-$ 4 ft = 5 ft, $c^2 = 12^2 + 5^2$, $c^2 = 144 + 25$, $c^2 = 169$, $c = 13$; 13 ft

page 527   Algebra in Civil Engineering

1. $(120.5)^2 = 120^2 + h^2$, $14,520.25 = 14,400 + h^2$, $h^2 = 120.25$, $h = 10.97$; $h \approx 11$ in.

2. $(300.5)^2 = (300)^2 + h^2$, $90,300.25 = 90,000 + h^2$, $h^2 = 300.25$, $h \approx 17$ in.

page 528   Capsule Review

1. $|2 - 5| = 3$ **2.** $|7 - 3| = 4$ **3.** $|8 - (-1)| = 9$

4. $|-5 - 1| = 6$ **5.** $|-5 - (-4)| = 1$ **6.** $|-2 - (-13)| = 11$

page 529   Class Exercises

1. $x_1 = 0$, $x_2 = -3$, $y_1 = 0$, $y_2 = 4$; $\sqrt{(-3 - 0)^2 + (4 - 0)^2} = \sqrt{(-3)^2 + 4^2} = \sqrt{9 + 16} = \sqrt{25} = 5$

2. $x_1 = 5$, $x_2 = 0$, $y_1 = 12$, $y_2 = 0$; $\sqrt{(0 - 5)^2 + (0 - 12)^2} = \sqrt{(-5)^2 + (-12)^2} = \sqrt{25 + 144} = \sqrt{169} = 13$

3. $x_1 = 2$, $x_2 = 5$, $y_1 = 2$, $y_2 = -2$; $\sqrt{(5 - 2)^2 + (-2 - 2)^2} = \sqrt{3^2 + (-4)^2} = \sqrt{9 + 16} = \sqrt{25} = 5$

4. $x_1 = 1$, $x_2 = 6$, $y_1 = 4$, $y_2 = 9$; $\sqrt{(6 - 1)^2 + (9 - 4)^2} = \sqrt{5^2 + 5^2} = \sqrt{25 + 25} = \sqrt{50} = 5\sqrt{2}$

5. $x_1 = 3$, $x_2 = -1$, $y_1 = 1$, $y_2 = 6$; $\sqrt{(-1 - 3)^2 + (6 - 1)^2} = \sqrt{(-4)^2 + 5^2} = \sqrt{16 + 25} = \sqrt{41}$

6. $x_1 = 5$, $x_2 = 4$, $y_1 = 6$, $y_2 = -1$; $\sqrt{(4 - 5)^2 + (-1 - 6)^2} = \sqrt{(-1)^2 + (-7)^2} = \sqrt{1 + 49} = \sqrt{50} = 5\sqrt{2}$

7. No, squaring the difference makes the signs irrelevant.

page 530   Practice Exercises

A   **1.** $\sqrt{(4 - 1)^2 + (-2 - 2)^2} = \sqrt{3^2 + (-4)^2} = \sqrt{9 + 16} = \sqrt{25} = 5$

    2. $\sqrt{(-2 - 4)^2 + (10 - 2)^2} = \sqrt{(-6)^2 + 8^2} = \sqrt{36 + 64} = \sqrt{100} = 10$

3. $\sqrt{(-2-4)^2+[4-(-2)]^2}=\sqrt{(-6)^2+6^2}=\sqrt{36+36}=\sqrt{72}=6\sqrt{2}$

4. $\sqrt{[-5-(-6)]^2+(4-(-2)^2}=\sqrt{1^2+6^2}=\sqrt{1+36}=\sqrt{37}$

5. $\sqrt{[7-(-5)]^2+(6-1)^2}=\sqrt{12^2+5^2}=\sqrt{144+25}=\sqrt{169}=13$

6. $\sqrt{(5-5)^2+(-1-13)^2}=\sqrt{0^2+(-14)^2}=\sqrt{0+196}=\sqrt{196}=14$

7. $\sqrt{(-2-3)^2+(-1-1)^2}=\sqrt{(-5)^2+(-2)^2}=\sqrt{25+4}=\sqrt{29}$

8. $\sqrt{[-9-(-5)]^2+(6-0)^2}=\sqrt{(-4)^2+6^2}=\sqrt{16+36}=\sqrt{52}=2\sqrt{13}$

9. $\sqrt{(2-10)^2+(-3-8)^2}=\sqrt{(-8)^2+(-11)^2}=\sqrt{64+121}=\sqrt{185}$

10. $\sqrt{(4-7)^2+[-3-(-9)]^2}=\sqrt{(-3)^2+6^2}=\sqrt{9+36}=\sqrt{45}=3\sqrt{5}$

11. $\sqrt{(3-8)^2+[2-(-10)]^2}=\sqrt{(-5)^2+12^2}=\sqrt{25+144}=\sqrt{169}=13$

12. $\sqrt{[7-(-2)]^2+(-8-4)^2}=\sqrt{9^2+(-12)^2}=\sqrt{81+144}=\sqrt{225}=15$

B 13. AB: $\sqrt{(8-0)^2+(0-0)^2}=\sqrt{8^2+0}=\sqrt{64}=8;$
BC: $\sqrt{(4-8)^2+(3-0)^2}=\sqrt{(-4)^2+3^2}=\sqrt{16+9}=5;$
AC: $\sqrt{(4-0)^2+(3-0)^2}=\sqrt{4^2+3^2}=\sqrt{16+9}=5$

14. RS: $\sqrt{[0-(-1)]^2+(0-7)^2}=\sqrt{1^2+(-7)^2}=\sqrt{1+49}=5\sqrt{2};$
ST: $\sqrt{(8-0)^2+(4-0)^2}=\sqrt{8^2+4^2}=\sqrt{64+16}=4\sqrt{5};$
RT: $\sqrt{[8-(-1)]^2+(4-7)^2}=\sqrt{9^2+(-3)^2}=\sqrt{81+9}=3\sqrt{10}$

15. XY: $\sqrt{[-1-(-4)]^2+(6-2)^2}=\sqrt{3^2+4^2}=\sqrt{9+16}=5;$
YZ: $\sqrt{[5-(-1)]^2+(4-6)^2}=\sqrt{6^2+(-2)^2}=\sqrt{36+4}=2\sqrt{10};$
XZ: $\sqrt{[5-(-4)]^2+(4-2)^2}=\sqrt{9^2+2^2}=\sqrt{81+4}=\sqrt{85}$

16. DE: $\sqrt{(5-1)^2+(5-5)^2}=\sqrt{4^2+0^2}=\sqrt{16+0}=4;$
EF: $\sqrt{(5-5)^2+(1-5)^2}=\sqrt{0^2+(-4)^2}=\sqrt{0+16}=4;$
DF: $\sqrt{(5-1)^2+(1-5)^2}=\sqrt{4^2+(-4)^2}=\sqrt{16+16}=4\sqrt{2}$

17. AB: $\sqrt{[-1-3)]^2+(3-6)^2}=\sqrt{(-4)^2+(-3)^2}=\sqrt{16+9}=5;$
BC: $\sqrt{[5-(-1)]^2+(-5-3)^2}=\sqrt{6^2+(-8)^2}=\sqrt{36+64}=10;$
AC: $\sqrt{(5-3)^2+(-5-6)^2}=\sqrt{2^2+(-11)^2}=\sqrt{4+121}=5\sqrt{5}$

18. RS: $\sqrt{(0-7)^2+[4-(-3)]^2}=\sqrt{(-7)^2+7^2}=\sqrt{49+49}=7\sqrt{2};$
ST: $\sqrt{(8-0)^2+(-1-4)^2}=\sqrt{8^2+(-5)^2}=\sqrt{64+25}=\sqrt{89};$
RT: $\sqrt{(8-7)^2+[-1-(-3)]^2}=\sqrt{1^2+2^2}=\sqrt{1+4}=\sqrt{5}$

19. MN: $\sqrt{[-7-(-7)]^2+(-7-5)^2}=\sqrt{0^2+(-12)^2}=\sqrt{0+144}=12;$
NP: $\sqrt{[2-(-7)]^2+[-7-(-7)]^2}=\sqrt{9^2+0^2}=\sqrt{81+0}=9;$
MP: $\sqrt{[2-(-7)]^2+(-7-5)^2}=\sqrt{9^2+(-12)^2}=\sqrt{81+144}=15$

20. KJ: $\sqrt{(9-6)^2+(7-3)^2}=\sqrt{3^2+4^2}=\sqrt{9+16}=5;$
KL: $\sqrt{(10-9)^2+(0-7)^2}=\sqrt{1^2+(-7)^2}=\sqrt{1+49}=5\sqrt{2};$
JL: $\sqrt{(10-6)^2+(0-3)^2}=\sqrt{4^2+(-3)^2}=\sqrt{16+9}=5$

C 21. $\sqrt{(4-1)^2 + (y-1)^2} = 5$, $\sqrt{9 + (y-1)^2} = 5$, $9 + (y-1)^2 = 25$,
$(y-1)^2 = 16$, $y - 1 = \pm 4$, $y = 1 \pm 4$; $y = 5$ or -3

22. $\sqrt{(x-2)^2 + [3-(-1)]^2} = 5$, $\sqrt{(x-2)^2 + 16} = 5$, $(x-2)^2 + 16 = 25$,
$(x-2)^2 = 9$, $x - 2 = \pm 3$, $x = 2 \pm 3$; $x = 5$ or -1

23. distance from $(-2, 4)$ to $(1, 0)$: $\sqrt{[1-(-2)]^2 + (0-4)^2} = \sqrt{9 + 16} = 5$;
distance from $(-2, 4)$ to $(-5, 0)$: $\sqrt{[-5-(-2)]^2 + (0-4)^2} = \sqrt{9 + 16} = 5$;
distance from $(-2, 4)$ to $(1, 8)$: $\sqrt{[1-(-2)]^2 + (8-4)^2} = \sqrt{9 + 16} = 5$; yes

24. The quadrilateral $ABCD$ is a rhombus, since $AB = BC = CD = DA$.

25. $\begin{cases} x + 2y = 5 \longrightarrow x = 5 - 2y \\ 2x + y = 1 \end{cases}$
$2(5 - 2y) + y = 1$, $10 - 4y + y = 1$, $10 - 3y = 1$, $-3y = -9$, $y = 3$;
$x = 5 - 2(3)$, $x = 5 - 6$, $x = -1$; $(-1, 3)$

26. $\begin{cases} x + 3y = 10 \longrightarrow x = 10 - 3y \\ 2x + y = -5 \end{cases}$
$2(10 - 3y) + y = -5$, $20 - 6y + y = -5$, $20 - 5y = -5$, $-5y = -25$,
$y = 5$; $x = 10 - 3(5)$, $x = 10 - 15$, $x = -5$; $(-5, 5)$

page 532   Class Exercises

1. $\left(\dfrac{2+8}{2}, \dfrac{4+(-4)}{2}\right)$, $\left(\dfrac{10}{2}, \dfrac{0}{2}\right)$, $(5, 0)$

2. $\left(\dfrac{-4+6}{2}, \dfrac{3+(-11)}{2}\right)$, $\left(\dfrac{2}{2}, -\dfrac{8}{2}\right)$, $(1, -4)$

3. $\left(\dfrac{-5+7}{2}, \dfrac{2+7}{2}\right)$, $\left(\dfrac{2}{2}, \dfrac{9}{2}\right)$, $\left(1, 4\dfrac{1}{2}\right)$

4. $\left(\dfrac{-6+0}{2}, \dfrac{0+6}{2}\right)$, $\left(-\dfrac{6}{2}, \dfrac{6}{2}\right)$, $(-3, 3)$

5. $\left(\dfrac{x+0}{2}, \dfrac{0+y}{2}\right)$, $\left(\dfrac{x}{2}, \dfrac{y}{2}\right)$

6. $\left(\dfrac{2a+6a}{2}, \dfrac{b-4b}{2}\right)$, $\left(\dfrac{8a}{2}, -\dfrac{3b}{2}\right)$, $\left(4a, -\dfrac{3b}{2}\right)$

7. Find the coordinates of the midpoint, then use the distance formula to find the distances to the vertices.

pages 532–535   Practice Exercises

A 1. $\left(\dfrac{-6+(-4)}{2}, \dfrac{4+6}{2}\right)$, $\left(-\dfrac{10}{2}, \dfrac{10}{2}\right)$, $(-5, 5)$

2. $\left(\dfrac{-4+(-1)}{2}, \dfrac{-1+(-4)}{2}\right)$, $\left(-\dfrac{5}{2}, -\dfrac{5}{2}\right)$, $\left(-2\dfrac{1}{2}, -2\dfrac{1}{2}\right)$

3. $\left(\dfrac{-7 + (-5)}{2}, \dfrac{3 + 3}{2}\right), \left(-\dfrac{12}{2}, \dfrac{6}{2}\right), (-6, 3)$

4. $\left(\dfrac{5 + 8}{2}, \dfrac{-2 + (-2)}{2}\right), \left(\dfrac{13}{2}, -\dfrac{4}{2}\right), \left(6\dfrac{1}{2}, -2\right)$

5. $\left(\dfrac{1 + (-8)}{2}, \dfrac{7 + (-3)}{2}\right), \left(-\dfrac{7}{2}, \dfrac{4}{2}\right), \left(-3\dfrac{1}{2}, 2\right)$

6. $\left(\dfrac{-5 + (-7)}{2}, \dfrac{2 + 5}{2}\right), \left(-\dfrac{12}{2}, \dfrac{7}{2}\right), \left(-6, 3\dfrac{1}{2}\right)$

7. $\left(\dfrac{2a + a}{2}, \dfrac{b + 2b}{2}\right), \left(\dfrac{3a}{2}, \dfrac{3b}{2}\right)$

8. $\left(\dfrac{3a + (-2a)}{2}, \dfrac{-3b + (-3b)}{2}\right), \left(\dfrac{a}{2}, -\dfrac{6b}{2}\right), \left(\dfrac{a}{2}, -3b\right)$

9. $\left(\dfrac{0 + 10}{2}, \dfrac{0 + 0}{2}\right), \left(\dfrac{10}{2}, \dfrac{0}{2}\right)$, $(5, 0)$; $C(5, 0)$, $B(5, 5)$, $D(5, -5)$

10. $A(2a, 0)$, $D(0, -2a)$, $E(-2a, 0)$

B 11. a. $(6, 6)$

   b. $\sqrt{(6 - 0)^2 + (6 - 0)^2} = \sqrt{36 + 36} = \sqrt{72} = 6\sqrt{2}$
   c. $\sqrt{(6 - 0)^2 + (0 - 6)^2} = \sqrt{36 + 36} = \sqrt{72} = 6\sqrt{2}$
   d. The length of the diagonals of a square are the same.

12. a. $(0, 4)$ **b.** $(7, 0)$
   c. $\sqrt{(7 - 0)^2 + (4 - 0)^2} = \sqrt{49 + 16} = \sqrt{65}$
   d. $\sqrt{(7 - 0)^2 + (0 - 4)^2} = \sqrt{49 + 16} = \sqrt{65}$
   e. The lengths of the diagonals of a rectangle are the same.

13. $\left(\dfrac{0 + 6}{2}, \dfrac{6 + 0}{2}\right); \left(\dfrac{6}{2}, \dfrac{6}{2}\right); (3, 3); \left(\dfrac{0 + 6}{2}, \dfrac{0 + 6}{2}\right), \left(\dfrac{6}{2}, \dfrac{6}{2}\right), (3, 3)$

14. $\left(\dfrac{0 + 7}{2}, \dfrac{0 + 4}{2}\right), \left(\dfrac{7}{2}, \dfrac{4}{2}\right), \left(\dfrac{7}{2}, 2\right); \left(\dfrac{0 + 7}{2}, \dfrac{4 + 0}{2}\right), \left(\dfrac{7}{2}, \dfrac{4}{2}\right), \left(\dfrac{7}{2}, 2\right)$

15. $\sqrt{[5 - (-2)]^2 + (2 - 1)^2} = \sqrt{7^2 + 1^2} = \sqrt{49 + 1} = \sqrt{50} = 5\sqrt{2}$; $AB = 5\sqrt{2}$, $\sqrt{(1 - 5)^2 + (-2 - 2)^2} = \sqrt{(-4)^2 + (-4)^2} = \sqrt{16 + 16} = \sqrt{32} = 4\sqrt{2}$; $BC = 4\sqrt{2}$, $\sqrt{[1 - (-2)]^2 + (-2 - 1)^2} = \sqrt{3^2 + (-3)^2} = \sqrt{9 + 9} = \sqrt{18} = 3\sqrt{2}$; $AC = 3\sqrt{2}$; Yes, triangle ABC is a right triangle. $(AB)^2 = (AC)^2 + (BC)^2$, $(5\sqrt{2})^2 = (3\sqrt{2})^2 + (4\sqrt{2})^2$, $50 = 18 + 32$, $50 = 50$

16. a. $\left(\dfrac{0 + 3}{2}, \dfrac{0 + 4}{2}\right), \left(\dfrac{3}{2}, \dfrac{4}{2}\right), \left(1\dfrac{1}{2}, 2\right)$

   b. $\left(\dfrac{9 + 6}{2}, \dfrac{0 + 4}{2}\right), \left(\dfrac{15}{2}, \dfrac{4}{2}\right), \left(7\dfrac{1}{2}, 2\right)$

   c. $\sqrt{\left(\dfrac{15}{2} - \dfrac{3}{2}\right)^2 + (2 - 2)^2} = \sqrt{\left(\dfrac{12}{2}\right)^2 + (0)^2} = \sqrt{36} = 6$

   d. $\sqrt{(3 - 6)^2 + (4 - 4)^2} = \sqrt{(-3)^2 + 0^2} = \sqrt{9} = 3$
   e. $\sqrt{(9 - 0)^2 + (0 - 0)^2} = \sqrt{9^2 + 0^2} = \sqrt{81} = 9$

   f. $MN = \dfrac{1}{2}(AB + CD)$, $6 = \dfrac{1}{2}(9 + 3)$, $6 = \dfrac{1}{2}(12)$, $6 = 6$

17. a. $\left(\dfrac{0+8}{2}, \dfrac{0+0}{2}\right)$, $\left(\dfrac{8}{2}, \dfrac{0}{2}\right)$, $(4, 0)$

b. $\left(\dfrac{8+6}{2}, \dfrac{0+6}{2}\right)$, $\left(\dfrac{14}{2}, \dfrac{6}{2}\right)$, $(7, 3)$

c. $\sqrt{(4-7)^2 + [0-(-3)]^2} = \sqrt{(-3)^2 + 3^2} = \sqrt{9+9} = \sqrt{18} = 3\sqrt{2}$; $JK = 3\sqrt{2}$

d. $\sqrt{(6-0)^2 + (6-0)^2} = \sqrt{6^2 + 6^2} = \sqrt{36+36} = \sqrt{72} = 6\sqrt{2}$; $PR = 6\sqrt{2}$

e. $JK = \dfrac{1}{2}PR$, $3\sqrt{2} = \dfrac{1}{2}(6\sqrt{2})$, $3\sqrt{2} = 3\sqrt{2}$

C 18. a. $AP = \sqrt{(x-0)^2 + (y-3)^2}$, $AP = \sqrt{x^2 + (y-3)^2}$

b. $BP = \sqrt{(x-4)^2 + (y-1)^2}$

c. $\sqrt{x^2 + (y-3)^2} = \sqrt{(x-4)^2 + (y-1)^2}$

d. $x^2 + (y-3)^2 = (x-4)^2 + (y-1)^2$, $x^2 + y^2 - 6y + 9 =$

$x^2 - 8x + 16 + y^2 - 2y + 1$, $4y = 8x - 8$, $y = \dfrac{8(x-1)}{4}$, $y = 2(x-1)$

pages 534–535   Mixed Problem Solving Review

1. Let $x =$ the kilograms of garbanzo beans, then $27 - x =$ the kilograms of pinto beans. $1.59(x) + (27 - x)(1.29) = 38.43$, $1.59x + 34.83 - 1.29x = 38.43$, $0.3x + 34.83 = 38.43$, $0.3x = 3.6$, $x = 12$
12 kg of garbanzo beans and 15 kg of pinto beans are needed to form a mixture of 27 kg at $38.43.

2. Let $s =$ the age of Suki, then $s - 7 =$ the age of Suki seven years ago. Let $b =$ the age of her brother, then $b - 7 =$ the age of her brother seven years ago.

$\begin{cases} s = 0.75(b) \\ (s - 7) + (0.5)(s - 7) = (b - 7) \end{cases}$

$0.75b - 7 + 0.5(0.75b - 7) = b - 7$, $0.75b + 0.375b - 3.5 = b$,
$1.125b - 3.5 = b$, $0.125b = 3.5$, $b = 28$; Then $s = 0.75(28)$, $s = 21$.
Suki is 21 years old and her brother is 28 years old.

3. Let $r =$ the rate of Josh and $c =$ the current of the water.

$\begin{cases} 2(r + c) = 11 \\ 2(r - c) = 7 \end{cases}$

$4r = 18$, $r = \dfrac{18}{4}$, $r = 4.5$, 4.5 mi/h

4. Let $x =$ how many milliliters of water.
$25(15\%) = (25 + x)(10\%)$, $3.75 = 2.5 + 0.1(x)$, $0.1x = 1.25$, $x = 12.5$
12.5 mL of water must be added to obtain a 10% solution.

page 535   Project

Check students' work.

page 535 Test Yourself

1. $\dfrac{\sqrt{4}}{\sqrt{8}} = \dfrac{2}{2\sqrt{2}} = \dfrac{1}{\sqrt{2}} \cdot \dfrac{\sqrt{2}}{\sqrt{2}} = \dfrac{\sqrt{2}}{2}$

2. $\dfrac{5}{3 - \sqrt{2}} \cdot \dfrac{3 + \sqrt{2}}{3 + \sqrt{2}} = \dfrac{5(3 + \sqrt{2})}{9 - 2} = \dfrac{15 + 5\sqrt{2}}{7}$

3. $\sqrt{t^2 + 6t + 9} = \sqrt{(t + 3)(t + 3)} = (t + 3)$

4. $\dfrac{\sqrt{48x^3y^2}}{\sqrt{3xy}} \cdot \dfrac{\sqrt{3xy}}{\sqrt{3xy}} = \dfrac{\sqrt{144x^4y^3}}{3xy} = \dfrac{12x^2y\sqrt{y}}{3xy} = 4x\sqrt{y}$

5. $\dfrac{-3\sqrt{5}}{4 - \sqrt{15}} \cdot \dfrac{4 + \sqrt{15}}{4 + \sqrt{15}} = \dfrac{-3\sqrt{5}(4 + \sqrt{15})}{16 - 15} = -12\sqrt{5} - 3\sqrt{75} = -12\sqrt{5} - 15\sqrt{3}$

6. $\dfrac{5}{2 - \sqrt{3}} \cdot \dfrac{2 + \sqrt{3}}{2 + \sqrt{3}} = \dfrac{5(2 + \sqrt{3})}{4 - 3} = 10 + 5\sqrt{3}$

7. $2t + 1 = 25,\ 2t = 24,\ t = 12$

8. $-2\sqrt{a} = -14,\ \sqrt{a} = 7,\ a = 49$

9. $\sqrt{7x + 8} = 8,\ 7x + 8 = 64,\ 7x = 56,\ x = 8$

10. $-5h + 18 = 4h,\ 18 = 9h,\ h = 2$

11. $x^2 = 5^2 + 12^2,\ x^2 = 25 + 144,\ x^2 = 169;\ x = 13$

12. $41^2 = x^2 + 9^2,\ 1681 = x^2 + 81,\ x^2 = 1600,\ x = 40$

13. $(5\sqrt{2})^2 = x^2 + 5^2,\ 50 = x^2 + 25,\ x^2 = 25;\ x = 5$

14. $d = \sqrt{(-3 - 10)^2 + (4 - 7)^2},\ d = \sqrt{(-13)^2 + (-3)^2},\ d = \sqrt{169 + 9},$

$d = \sqrt{178};\ M\left(\dfrac{10 + (-3)}{2}, \dfrac{7 + 4}{2}\right),\ M\left(\dfrac{7}{2}, \dfrac{11}{2}\right),\ M\left(3\dfrac{1}{2}, 5\dfrac{1}{2}\right)$

15. $d = \sqrt{(3 - 8)^2 + [5 - (-10)]^2},\ d = \sqrt{(-5)^2 + 15^2},\ d = \sqrt{25 + 225},$

$d = \sqrt{250},\ d = 5\sqrt{10};\ M\left(\dfrac{8 + 3}{2}, \dfrac{-10 + 5}{2}\right),\ M\left(\dfrac{11}{2}, -\dfrac{5}{2}\right),\ M\left(5\dfrac{1}{2}, -2\dfrac{1}{2}\right)$

pages 536–537 Summary and Review

1. 5

2. $\dfrac{2}{3}$

3. $\dfrac{9}{6} = \dfrac{3}{2}$

4. 30

5. 2.24

6. 3.16

7. 9.43

8. 7.75

9–12. Answers may vary. Possible answers are given.

9. $\sqrt{55}, \sqrt{56}, \sqrt{57}$ (since $7 = \sqrt{49}$ and $8 = \sqrt{64}$)

10. $\sqrt{226}, \sqrt{227}, \sqrt{228}$ (since $15 = \sqrt{225}$ and $16 = \sqrt{256}$)

11. $\sqrt{21.5}, \sqrt{21.6}, \sqrt{21.7}$ (since these radicands are between 21 and 22)

12. $\sqrt{5}, \sqrt{6}, \sqrt{7}$ (since $3 = \sqrt{9}$)

13. $-7 \div 2 = -3.5$ **14.** $1 \div 9 = 0.\overline{1}$ **15.** $5 \div 6 = 0.8\overline{3}$ **16.** $0.25 = \dfrac{25}{100} = \dfrac{1}{4}$

17. Let $x = 0.\overline{6}$. $(10x = 6.\overline{6}) - (x = 0.\overline{6})$, $9x = 6$; $x = \dfrac{6}{9} = \dfrac{2}{3}$

18. Let $x = 0.\overline{54}$. $(100x = 54.\overline{54}) - (x = 0.\overline{54})$, $99x = 54$; $x = \dfrac{54}{99} = \dfrac{6}{11}$

19. $\sqrt{4 \cdot 5} = 2\sqrt{5}$ **20.** $\sqrt{9 \cdot 5} = 3\sqrt{5}$

21. $\sqrt{49x^2 \cdot 2} = 7x\sqrt{2}$ **22.** $\sqrt{4y^2 \cdot 3y} = 2y\sqrt{3y}$

23. $(\sqrt{5} - 4\sqrt{5}) + 2\sqrt{6} = -3\sqrt{5} + 2\sqrt{6}$ **24.** $-10\sqrt{18x^2} = -30x\sqrt{2}$

25. $2\sqrt{6} - 5\sqrt{3}$ **26.** $6 + 4\sqrt{5} - 9\sqrt{5} - 30 = -5\sqrt{5} - 24$
                                         or $-24 - 5\sqrt{5}$

27. $\dfrac{\sqrt{3}}{\sqrt{5}} \cdot \dfrac{\sqrt{5}}{\sqrt{5}} = \dfrac{\sqrt{15}}{5}$

28. $\dfrac{1 + \sqrt{5}}{2 - \sqrt{5}} \cdot \dfrac{2 + \sqrt{5}}{2 + \sqrt{5}} = \dfrac{(1 + \sqrt{5})(2 + \sqrt{5})}{4 - 5} = \dfrac{2 + \sqrt{5} + 2\sqrt{5} + 5}{-1} = -1(7 + 3\sqrt{5}) = -7 - 3\sqrt{5}$

29. $(\sqrt{3x - 1})^2 = 4^2$, $3x - 1 = 16$, $3x = 17$; $x = \dfrac{17}{3}$

30. $2\sqrt{x} = -6$, $\sqrt{x} = -3$; no solution

31. $\sqrt{2y + 1} = 3$, $(\sqrt{2y + 1})^2 = 3^2$, $2y + 1 = 9$, $2y = 8$; $y = 4$

32. $c^2 = 10^2 + 24^2$, $c^2 = 100 + 576$, $c^2 = 676$; $c = 26$

33. $(\sqrt{2})^2 = 1^2 + b^2$, $2 = 1 + b^2$, $b^2 = 1$; $b = 1$

34. $d = \sqrt{(0 - 0)^2 + (-12 - 9)^2}$, $d = \sqrt{0^2 + (-21)^2}$, $d = \sqrt{441}$, $d = 21$

35. $d = \sqrt{[-9 - (-3)]^2 + (-6 - 7)^2}$, $d = \sqrt{(-6)^2 + (-13)^2}$, $d = \sqrt{36 + 169}$, $d = \sqrt{205}$

36. $d = \sqrt{(3 - 0)^2 + (9 - 0)^2}$, $d = \sqrt{3^2 + 9^2}$, $d = \sqrt{9 + 81}$, $d = \sqrt{90}$, $d = 3\sqrt{10}$,
$M\left(\dfrac{0 + 3}{2}, \dfrac{0 + 9}{2}\right)$, $M\left(\dfrac{3}{2}, \dfrac{9}{2}\right)$, $M\left(1\frac{1}{2}, 4\frac{1}{2}\right)$

1. $\dfrac{10}{7}$

2. 30

3. $\sqrt{225 \cdot 2} = 15\sqrt{2}$

4. $\sqrt{9y^2 \cdot 3y} = 3y\sqrt{3y}$

5. Answers may vary. Possible answers: $\sqrt{19}$, $\sqrt{20}$, $\sqrt{21}$ (since $4 = \sqrt{16}$ and $5 = \sqrt{25}$)

6. Let $x = 0.\overline{3}$. $(10x = 3.\overline{3}) - (x = 0.\overline{3})$, $9x = 3$, $x = \dfrac{3}{9} = \dfrac{1}{3}$

7. Let $x = 1.\overline{45}$. $(100x = 145.\overline{45}) - (x = 1.\overline{45})$; $99x = 144$; $x = \dfrac{144}{99} = \dfrac{16}{11}$

8. $(2\sqrt{3} - 4\sqrt{3}) + 5\sqrt{4} = -2\sqrt{3} + 10$

9. $\sqrt{4b^2 \cdot 3} - \sqrt{9b^2 \cdot 3} = 2b\sqrt{3} - 3b\sqrt{3} = -b\sqrt{3}$

10. $6\sqrt{18} = 6\sqrt{9 \cdot 2} = 18\sqrt{2}$

11. $2\sqrt{10} - 3\sqrt{2}$

12. $\dfrac{\sqrt{5}}{\sqrt{8}} \cdot \dfrac{\sqrt{2}}{\sqrt{2}} = \dfrac{\sqrt{10}}{\sqrt{16}} = \dfrac{\sqrt{10}}{4}$

13. $\dfrac{5}{\sqrt{10} - 2} \cdot \dfrac{\sqrt{10} + 2}{\sqrt{10} + 2} = \dfrac{5\sqrt{10} + 10}{10 - 4} = \dfrac{5\sqrt{10} + 10}{6}$ or $\dfrac{10 + 5\sqrt{10}}{6}$

14. $\sqrt{2x + 1} = 4$, $(\sqrt{2x + 1})^2 = 4^2$; $2x + 1 = 16$, $2x = 15$; $x = \dfrac{15}{2}$

15. $\sqrt{3x - 1} = 5$, $(\sqrt{3x - 1})^2 = 5^2$, $3x - 1 = 25$, $3x = 26$; $x = \dfrac{26}{3}$

16. $4^2 = (\sqrt{7})^2 + b^2$, $16 = 7 + b^2$, $b^2 = 9$, $b = 3$

17. $\sqrt{[7 - (-2)]^2 + (-8 - 4)^2} = \sqrt{9^2 + (-12)^2} = \sqrt{81 + 144} = \sqrt{225} = 15$

18. $\left(\dfrac{-4 + 4}{2}, \dfrac{0 - 0}{2}\right)$, $\left(\dfrac{0}{2}, \dfrac{0}{2}\right)$, $(0, 0)$; $\left(\dfrac{4 + 0}{2}, \dfrac{0 + 8}{2}\right)$, $\left(\dfrac{4}{2}, \dfrac{8}{2}\right)$, $(2, 4)$; $\left(\dfrac{0 - 4}{2}, \dfrac{8 + 0}{2}\right)$, $\left(-\dfrac{4}{2}, \dfrac{8}{2}\right)$, $(-2, 4)$

19. $c^2 = 12^2 + 9^2$, $c^2 = 144 + 81$, $c^2 = 225$, $c = 15$

20. perimeter: $5\sqrt{y} + 5\sqrt{y} + 2\sqrt{y} + 2\sqrt{y} = 14\sqrt{y}$ ft
    area: $5\sqrt{y} \cdot 2\sqrt{y} = 10\sqrt{y^2} = 10y$ ft^2

21. Let n = the number. $\sqrt{2n} - 5 = 61$ $\sqrt{2n} = 66$, $(\sqrt{2n})^2 = 66^2$, $2n = 4,356$; $n = 2178$

22. $c^2 = a^2 + b^2$, $c^2 = 24^2 + 7^2$, $c^2 = 576 + 49$, $c^2 = 625$, $c = 25$; 25 m

Challenge

$(l + \sqrt{l^2 - k^2})^2 - 2l(l + \sqrt{l^2 - k^2}) + k^2 = l^2 + 2l\sqrt{l^2 - k^2} + l^2 - k^2 - 2l^2 - 2l\sqrt{l^2 - k^2} + k^2 = 0$ and $(l - \sqrt{l^2 - k^2})^2 - 2l(l - \sqrt{l^2 - k^2}) + k^2 = l^2 - 2l\sqrt{l^2 - k^2} + l^2 - k^2 - 2l^2 + 2l\sqrt{l^2 - k^2} + k^2 = 0$

page 539   Preparing for Standardized Tests

1. Using rounding, the amounts become: 6.47 million + 1.47 million + 3.46 million + 4.99 million + 12.35 million + 0.95 million = 29.69 million; 30 million

2. Since $50^2 = 2500$, try 53. Since 53^2 is 2809, $\sqrt{2809} = 53$.

3. 30% off \$348 is 70% of \$348 = 0.7(348) = \$243.60
   25% off \$295 is 75% of \$295 = 0.75(295) = \$221.25, which is the lower of the two prices.

4. $\sqrt{2x - 5} = 3$, $2x - 5 = 9$, $2x = 14$; $x = 7$

5. $1.125 = 1\frac{1}{8} = \frac{9}{8}$

6. 5 pages left out of 20 is $\frac{1}{4}$ of the 20 or 25%

7. $\frac{3}{x + 1} - \frac{2}{x} = \frac{2}{x + 1}$, $3x - 2(x + 1) = 2x$, $3x - 2x - 2 = 2x$, $-2 = x$, $x = -2$

8. $24^2 + b^2 = 25^2$, $b^2 = 25^2 - 24^2$, $b^2 = 625 - 576$, $b^2 = 49$; $b = 7$

9. If the second equation is subtracted from the first, then $2x + 2y = 1$.
   Dividing by 2, $x + y = \frac{1}{2}$.

10. The shaded area = area of the circle − area of the square =
    $\pi(14)^2 - 16^2 = \frac{22}{7} \cdot 196 - 256 = 22 \cdot 28 - 256 = 616 - 256 = 360$ sq cm

11. slope = $\frac{\text{rise}}{\text{run}} = -\frac{2}{3}$ 12. The x-intercept is 3.

13. $\frac{8}{\frac{1}{2}} = \frac{28}{x}$, $8x = 14$, $x = \frac{14}{8}$, $x = \frac{7}{4}$; $1\frac{3}{4}$ L

14. $1 - \left(\frac{2}{5} + 3 \cdot \frac{1}{20} + \frac{1}{10}\right) = 1 - \left(\frac{2}{5} + \frac{3}{20} + \frac{1}{10}\right) = 1 - \left(\frac{8}{20} + \frac{3}{20} + \frac{2}{20}\right) = 1 - \frac{13}{20} = \frac{7}{20}$

pages 540–542   Cumulative Review (Chapters 1–12)

1. $-2, \sqrt{4}, 3, 6$

2. $-\sqrt{\frac{9}{16}}, -\frac{1}{3}, 0, \frac{1}{2}, \sqrt{\frac{4}{9}}$

3. b; $\dfrac{1}{x^2} - \dfrac{2}{3x} - \dfrac{5}{3} = 0$, $\dfrac{3}{3x^2} - \dfrac{2x}{3x^2} - \dfrac{5x^2}{3x^2} = 0$, $3 - 2x - 5x^2 = 0$, $5x^2 + 2x - 3 = 0$,

$(5x - 3)(x + 1) = 0$, $5x - 3 = 0$, $x = \dfrac{3}{5}$; $(x + 1) = 0$, $x = -1$

4. e; $|x| - 3 = -2$, $|x| = 1$, $x = \pm 1$, $\{-1, 1\}$

5. a; $-3 < -3x < 3$, $-1 < x < 1$

6. c; $2x - 3 > -1$, $2x > 2$, $x > 1$

7. d; $\begin{cases} -x - 7y = -6 \rightarrow -3x - 21y = -18 \\ 3x + y = -2 \rightarrow \quad\ \ 3x + y = -2 \end{cases}$

$\quad -20y = -20$, $y = 1$; $3x + 1 = -2$, $3x = -3$, $x = -1$, $\{(-1, 1)\}$

8. $15x = 5x - 20$, $10x = -20$, $x = -2$ $\qquad$ **9.** $-\dfrac{3}{4}a > -3$, $a < 4$

10. $|2y - 1| = 5$, $2y - 1 = 5$, $2y = 6$, $y = 3$, or $2y - 1 = -5$, $y = -2$

11. $\dfrac{8m}{8} - \dfrac{6}{8} - \dfrac{5m}{8} = 0$, $\dfrac{3m}{8} - \dfrac{6}{8} = 0$, $3m - 6 = 0$, $3m = 6$, $m = 2$

12. $\dfrac{2}{3}t - 1 = 7$, $\dfrac{2}{3}t = 8$, $t = 12$

13. $6c = 24$, $c = 4$, $2(4) + 3d = 14$, $8 + 3d = 14$, $3d = 6$, $d = 2$, $(4, 2)$

14. $z(z - 9) = 0$, $z = 0$, $z - 9 = 0$, $z = 9$

15. $\dfrac{3}{b + 2} - \dfrac{1}{b - 2} = 0$, $\dfrac{3(b - 2) - (b + 2)}{(b + 2)(b - 2)} = 0$, $3b - 6 - b - 2 = 0$, $2b - 8 = $

0, $2b = 8$, $b = 4$

16. $\sqrt{n - 3} = -4$; no solution

17. $4r - 5r - 5 = -3r + 2 + 5$, $-r - 5 = -3r + 7$, $2r = 12$, $r = 6$

18. $n \le 3$ and $n > -2$, $-2 < n \le 3$

19. $(g + 5)(g - 3) = 0$, $g + 5 = 0$, $g = -5$, $g - 3 = 0$, $g = 3$

20. $3w + 1 = 9(2w - 1)$, $3w + 1 = 18w - 9$, $15w = 10$, $w = \dfrac{10}{15} = \dfrac{2}{3}$

21. $55 - 40 = 15$; \$15 per hour $\qquad$ **22.** $40 - 20 = 20$; \$20 per hour

23. $40 + 4(15) = 100$, $20 + 4(20) = 100$, 5 h

24. $40 + 2(15) = 70$; $40 + 3(15) = 85$; $20 + 2(20) = 60$; $20 + 3(20) = 80$

25. $|-6| = 6$ $\qquad$ **26.** $-\left[\dfrac{1}{3}\right] = -\dfrac{1}{3}$ $\qquad$ **27.** $\dfrac{\sqrt{9}}{\sqrt{144}} = \dfrac{3}{12} = \dfrac{1}{4}$ $\qquad$ **28.** 1.44×10^{-6}

29. $6p^{2+1}q^{1+3} = 6p^3q^4$

30. $c^2 - 2cd + d^2 - d^2 = c^2 - 2cd$

31. $-\dfrac{2}{3}a^{5-3}b^{2-3} = -\dfrac{2}{3}a^2b^{-1} = -\dfrac{2a^2}{3b}$

32. $\dfrac{-2}{3\sqrt{2x}} \cdot \dfrac{\sqrt{2x}}{\sqrt{2x}} = \dfrac{-2\sqrt{2x}}{3(2x)} = -\dfrac{\sqrt{2x}}{3x}$

33. $\dfrac{3 - m}{(m - 3)(m + 2)} = -\dfrac{1}{m + 2}$

34. $\dfrac{(z + 2)(z - 2)}{3(z + 2)} = \dfrac{z - 2}{3}$

35. $\dfrac{s + t}{\dfrac{3t}{st} + \dfrac{3s}{st}} = \dfrac{s + t}{\dfrac{3(t + s)}{st}} = \dfrac{st}{3}$

36. $\sqrt{4x^2 \cdot 3xy} = 2x\sqrt{3xy}$

37. $2(3^2 + 1) - 3\sqrt{2^2 - 1} = 2(10) - 3\sqrt{2^2 - 1} = 20 - 3\sqrt{3}$

38. $2[(h^2 - 6h + 9)] + 2[h - 9] = 2h^2 - 12h + 18 + 2h - 18 = 2h^2 - 10h$

39. $2y = x - 6,\ y = \dfrac{x}{2} - 3,\ y = \dfrac{1}{2}x - 3$

40. $y - 1 = -2(x - 1),\ y - 1 = -2x + 2,\ y = -2x + 3$

41. $m = \dfrac{4 - (-2)}{1 - (-1)} = \dfrac{4 + 2}{1 + 1} = \dfrac{6}{2} = 3;\ y - 4 = 3(x - 1),\ y - 4 = 3x - 3,\ y = 3x + 1$

42. $(m - 2)(m - 9)$

43. $3xy^2(4x) - 3xy^2(3) = 3xy^2(4x - 3)$

44. $(2a^3 + 3)(2a^3 + 3) = (2a^3 + 3)^2$

45. prime

46. $(2b + 3)(b - 6)$

47. $3p(p^2) + 3p(3p) + 3p(1) = 3p(p^2 + 3p + 1)$

48. $(y - 2)(x - 3)$

49. $(m + 2n)(3 - 2m)$

50. $45 + 2(25) = 45 + 50,\ \$95;$
$45 + 3(25) = 45 + 75,\ \$120$

51.

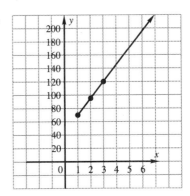

52. \$170

53. $220 = 45 + x(25),\ 25x = 175,\ x = 7;$ 7 days

54. $\left[\left(-\dfrac{1}{2} + \left(-\dfrac{3}{4}\right)\right)4\right]^2 - 1 = \left[\left(-\dfrac{1}{2} - \dfrac{3}{4}\right)4\right]^2 - 1 = \left[\left(-\dfrac{2}{4} - \dfrac{3}{4}\right)4\right]^2 - 1 =$
$\left[\left(-\dfrac{5}{4}\right)4\right]^2 - 1 = [-5]^2 - 1 = 25 - 1 = 24$

55. $\sqrt{4} \div \left(-\frac{1}{2}\right) = 2 \div -\frac{1}{2} = 2(-2) = -4$

56. $4 + \left(-\frac{3}{4}\right) \div -\frac{1}{2} = 4 + -\frac{3}{4}(-2) = 4 + \frac{3}{2} = \frac{8}{2} + \frac{3}{2} = \frac{11}{2} = 5\frac{1}{2}$

57. $\sqrt{-4} \div \left(-\frac{1}{2}\right) = \sqrt{-4}(-2) = \sqrt{8} = 2\sqrt{2}$

58. $\left[\left(-\frac{1}{2}\right)^2 + \left(-\frac{3}{4}\right)\right]\left[\left(-\frac{1}{2}\right)^2 - \left(-\frac{3}{4}\right)\right] = \left(\frac{1}{4} - \frac{3}{4}\right)\left(\frac{1}{4} + \frac{3}{4}\right) = \left(-\frac{2}{4}\right)\left(\frac{4}{4}\right) = \left(-\frac{2}{4}\right) = -\frac{1}{2}$

59. $\sqrt{\left[-\frac{1}{2} - \left(-\frac{3}{4}\right)\right]}4 = \sqrt{\left(-\frac{1}{2} + \frac{3}{4}\right)}4 = \sqrt{\left(-\frac{2}{4} + \frac{3}{4}\right)}4 = \sqrt{\frac{1}{4}}(4) = \sqrt{1} = 1$

60. $\dfrac{\left(-\frac{1}{2}\right)^2(4)}{-\frac{1}{2} + 4} = \dfrac{\frac{1}{4}(4)}{-\frac{1}{2} + \frac{8}{2}} = \dfrac{1}{\frac{7}{2}} = \frac{2}{7}$

61. $\sqrt{\dfrac{-\left(-\frac{1}{2}\right)(\sqrt{4})}{\left[-\frac{1}{2} - \left(-\frac{3}{4}\right)\right]^2}} = \sqrt{\dfrac{\frac{1}{2}(2)}{\left[-\frac{1}{2} - \left(-\frac{3}{4}\right)\right]^2}} = \sqrt{\dfrac{1}{\left(-\frac{2}{4} + \frac{3}{4}\right)^2}} = \sqrt{\dfrac{1}{\left(\frac{1}{4}\right)^2}} = \sqrt{\dfrac{1}{\frac{1}{16}}} = \sqrt{16} = 4$

62. $\sqrt{4} + \dfrac{-\frac{3}{4} - \left(-\frac{1}{2}\right)^2}{4} = 2 + \dfrac{-\frac{3}{4} - \frac{1}{4}}{4} = 2 + \dfrac{-\frac{4}{4}}{4} = 2 + \dfrac{-1}{4} = \frac{8}{4} - \frac{1}{4} = \frac{7}{4} = 1\frac{3}{4}$

63. 4

64. 0.46

65. -0.2

66. $-10a^5d^3$

67. $9m^4n^6$

68. $(3t - 2)^2$

69. $\dfrac{p - 1}{p - 2}$

70. $\dfrac{3y}{x^2z}$

71. $\dfrac{c + 3}{c - 1}$

72. $\dfrac{1}{h - 1}$

73. $3\sqrt{2} + 4\sqrt{5}$

74. $-\sqrt{2}$

75. $2r^2 + 6r - 1$

76. $3b^5 - 9b^4 + 6b^2$

77. $xy^2 + 2xy$

78. $8x\sqrt{y}$

79. $\{-7, -3 -1\}$

80. $\{5, 1, 2\}$

81. $\{5, 6, 6\frac{1}{2}\}$

82. $\{-5\frac{1}{2}, \frac{1}{2}, 3\frac{1}{2}\}$

83. $2.15

84. 3, 5, 7 or $-3, -1, 1$

85. crew: 7 mi/h; current: 2 mi/h

86. 8.5%

87. $w = 10$ ft; $l = 16$ ft

88. 300

89. $l = \dfrac{2}{w + 4}$

90. $\dfrac{3}{2}, \dfrac{9}{2}$

Chapter 13   Quadratic Equations and Functions

page 544   Capsule Review

1. $\pm\, 2x$ 2. $\pm\, 4y$ 3. $\pm\, 5mn$ 4. $\pm\, (3y - 1)$

page 545   Class Exercises

1. $5x^2 = 30$, $x^2 = 6$; $x = \pm\, \sqrt{6}$ 2. $n^2 = \pm\, \sqrt{\dfrac{9}{16}}$, $n = \dfrac{\sqrt{9}}{\sqrt{16}}$; $n = \pm\, \dfrac{3}{4}$

3. $3(x - 5)^2 = 75$, $(x - 5)^2 = 25$, $x - 5 = \pm 5$; $x = 10$ and $x = 0$

4. $(x - 2)^2 = 7$, $x - 2 = \pm\sqrt{7}$; $x = 2 \pm \sqrt{7}$

5. $5(m + 3)^2 = 25$, $(m + 3)^2 = 5$, $m + 3 = \pm\sqrt{5}$; $m = -3 \pm \sqrt{5}$

6. $4(x + 4)^2 = 256$, $(x + 4)^2 = 64$, $x + 4 = \pm 8$; $x = -12$ and 4

pages 545–546   Practice Exercises

A 1. $p^2 = \dfrac{4}{25}$, $p = \pm\, \dfrac{\sqrt{4}}{\sqrt{25}}$; $p = \pm\, \dfrac{2}{5}$ 2. $n^2 = \dfrac{1}{100}$, $n = \pm\, \dfrac{\sqrt{1}}{\sqrt{100}}$; $n = \pm\, \dfrac{1}{10}$

3. $x^2 = 49$, $x = \pm\, \sqrt{49}$; $x = \pm 7$ 4. $y^2 = 64$, $y = \pm\, \sqrt{64}$; $y = \pm 8$

5. $3x^2 = 18$, $x^2 = \sqrt{6}$; $x = \pm 6$ 6. $5m^2 = 35$, $m^2 = 7$; $m = \pm\, \sqrt{7}$

7. $2n^2 = 16$, $n^2 = 8$, $n = \pm\, \sqrt{8}$; $n = \pm\, 2\sqrt{2}$

8. $4y^2 = 48$, $y^2 = 12$; $y = \pm 2\sqrt{3}$ 9. $x^2 - 15 = 0$, $x^2 = 15$; $x = \pm\sqrt{15}$

10. $y^2 - 10 = 0$, $y^2 = 10$; $y = \pm\, \sqrt{10}$

11. $2n^2 - 6 = 0$, $2n^2 = 6$, $n^2 = 3$; $n = \pm\, \sqrt{3}$

12. $3m^2 - 15 = 0$, $3m^2 = 15$, $m^2 = 5$; $m = \pm\, \sqrt{5}$

13. $4x^2 + 9 = 0$, $4x^2 = -9$, $x^2 = -\dfrac{9}{4}$; no real solution

14. $6y^2 + 24 = 0$, $6y^2 = -24$, $y^2 = -4$; no real solution

15. $3p^2 - 5 = 7$, $3p^2 = 12$, $p^2 = 4$; $p = \pm 2$

16. $2n^2 - 13 = -1$, $2n^2 = 12$, $n^2 = 6$; $n = \pm\, \sqrt{6}$

17. $8m^2 - 40 = 0$, $8m^2 = 40$, $m^2 = 5$; $m = \pm\, \sqrt{5}$

18. $6z^2 = 42$, $z^2 = 7$; $z = \pm\, \sqrt{7}$

19. $(x - 1)^2 = 9$, $x - 1 = \pm\, \sqrt{9}$, $x - 1 = \pm\, 3$; $x = 4$ or $x = -2$

20. $(y - 3)^2 = 36$, $(y - 3) = \pm\sqrt{36}$, $y - 3 = \pm\, 6$, $y = \pm\, 6 + 3$; $y = 9$ or $y = -3$

21. $(2x + 1)^2 = 16$, $2x + 1 = \pm \sqrt{16}$, $2x + 1 = \pm 4$, $2x = 1 \pm 4$, $x = \dfrac{-1 \pm 4}{2}$;

$x = \dfrac{3}{2}$ or $x = -\dfrac{5}{2}$

22. $(3z + 2)^2 = 4$, $3z + 2 = \pm \sqrt{4}$, $3z + 2 = \pm 2$, $3z = -2 \pm 2$, $z = \dfrac{-2 \pm 2}{3}$;

$z = 0$ or $z = -\dfrac{4}{3}$

23. $5(2x - 1)^2 = 45$, $(2x - 1)^2 = 9$, $2x - 1 = \pm \sqrt{9}$, $2x - 1 = \pm 3$, $2x = 1 \pm 3$, $x = \dfrac{1 \pm 3}{2}$; $x = 2$ or $x = -1$

24. $6(3y + 2)^2 = 24$, $(3y + 2)^2 = 4$, $3y + 2 = \pm \sqrt{4}$, $3y + 2 = \pm 2$, $3y = -2 \pm 2$, $y = \dfrac{-2 \pm 2}{3}$; $y = 0$ or $y = -\dfrac{4}{3}$

25. $\dfrac{1}{9}p^2 - 1 = -\dfrac{3}{4}$, $\dfrac{1}{9}p^2 = -\dfrac{3}{4} + 1$, $\dfrac{1}{9}p^2 = \dfrac{1}{4}$, $p^2 = \dfrac{1}{4}(9)$, $p^2 = \dfrac{9}{4}$, $p = \pm \dfrac{\sqrt{9}}{\sqrt{4}}$; $p = \pm \dfrac{3}{2}$

26. $\dfrac{4}{25}n^2 + \dfrac{6}{16} = \dfrac{15}{16}$, $\dfrac{4}{25}n^2 = \dfrac{15}{16} - \dfrac{6}{16}$, $\dfrac{4}{25}n^2 = \dfrac{9}{16}$, $n^2 = \dfrac{9}{16}\left(\dfrac{25}{4}\right)$, $n^2 = \dfrac{225}{64}$,

$n = \pm \dfrac{\sqrt{225}}{\sqrt{64}}$; $n = \pm \dfrac{15}{8}$

27. $(x + 1)^2 = 72$, $x + 1 = \pm \sqrt{72}$, $x + 1 = \pm 6\sqrt{2}$; $x = -1 \pm 6\sqrt{2}$

28. $(n - 9)^2 = 27$, $n - 9 = \pm \sqrt{27}$, $n - 9 = \pm 3\sqrt{3}$; $n = 9 \pm 3\sqrt{3}$

29. $\left(m + \dfrac{2}{3}\right)^2 = \dfrac{1}{9}$, $m + \dfrac{2}{3} = \pm \sqrt{\dfrac{1}{9}}$, $m + \dfrac{2}{3} = \pm \dfrac{1}{3}$, $m = -\dfrac{2}{3} \pm \dfrac{1}{3}$; $m = -\dfrac{1}{3}$ or m $= -1$

30. $\left(n + \dfrac{3}{4}\right)^2 = \dfrac{1}{16}$, $n + \dfrac{3}{4} = \pm\sqrt{\dfrac{1}{16}}$, $n + \dfrac{3}{4} = \pm \dfrac{1}{4}$, $n = -\dfrac{3}{4} \pm \dfrac{1}{4}$; $n = -1$ or $n = -\dfrac{1}{2}$

31. $\left(t + \dfrac{2}{3}\right)^2 = \dfrac{5}{9}$, $t + \dfrac{2}{3} = \pm \sqrt{\dfrac{5}{9}}$, $t + \dfrac{2}{3} = \pm\dfrac{\sqrt{5}}{3}$, $t = -\dfrac{2}{3} \pm \dfrac{\sqrt{5}}{3}$; $t = \dfrac{-2 \pm \sqrt{5}}{3}$

32. $\left(y + \dfrac{3}{4}\right)^2 = \dfrac{3}{16}$, $y + \dfrac{3}{4} = \pm \sqrt{\dfrac{3}{16}}$, $y + \dfrac{3}{4} = \pm \dfrac{\sqrt{3}}{4}$, $y = -\dfrac{3}{4} \pm \dfrac{\sqrt{3}}{4}$; $y = \dfrac{-3 \pm \sqrt{3}}{4}$

33. $2(5x + 1)^2 = 50$, $(5x + 1)^2 = 25$, $5x + 1 = \pm \sqrt{25}$, $5x + 1 = \pm 5$, $5x = -1 \pm 5$,

$x = \dfrac{-1 \pm 5}{5}$; $x = \dfrac{4}{5}$ or $x = -\dfrac{6}{5}$

34. $3(2s + 4)^2 = 36$, $(2s + 4)^2 = 12$, $2s + 4 = \pm \sqrt{12}$, $2s + 4 = \pm 2\sqrt{3}$,

$2s = -4 \pm 2\sqrt{3}$, $s = \dfrac{-4 \pm 2\sqrt{3}}{2}$; $s = -2 \pm \sqrt{3}$

35. $5(z - 3)^2 = 35$, $(z - 3)^2 = 7$, $z - 3 = \pm \sqrt{7}$; $z = 3 \pm \sqrt{7}$

36. $7(m + 4)^2 = 70$, $(m + 4)^2 = 10$, $m + 4 = \pm\sqrt{10}$; $m = -4 \pm \sqrt{10}$

37. $(x - 5)^2 + 3 = 27$, $(x - 5)^2 = 24$, $x - 5 = \pm\sqrt{24}$, $x = 5 \pm \sqrt{24}$;
$x = 5 \pm 2\sqrt{6}$

38. $(z + 3)^2 - 5 = 49$, $(z + 3)^2 = 54$, $z + 3 = \pm\sqrt{54}$, $z + 3 = \pm 3\sqrt{6}$;
$z = -3 \pm 3\sqrt{6}$

39. $2(3x - 5)^2 + 3 = 7$, $2(3x - 5)^2 = 4$, $(3x - 5)^2 = 2$, $3x - 5 = \pm\sqrt{2}$,
$3x = 5 \pm \sqrt{2}$; $x = \dfrac{5 \pm \sqrt{2}}{3}$

40. $3(2t + 7)^2 + 1 = 37$, $3(2t + 7)^2 = 36$, $(2t + 7)^2 = 12$, $2t + 7 = \pm\sqrt{12}$,
$2t + 7 = \pm 2\sqrt{3}$, $2t = -7 \pm 2\sqrt{3}$; $t = \dfrac{-7 \pm 2\sqrt{3}}{2}$

41. $3(2p + 1)^2 + 5 = 59$, $3(2p + 1)^2 = 54$, $(2p + 1)^2 = 18$, $2p + 1 = \pm\sqrt{18}$,
$2p + 1 = \pm 3\sqrt{2}$, $2p = -1 \pm 3\sqrt{2}$; $p = \dfrac{-1 \pm 3\sqrt{2}}{2}$

42. $5(6m - 5)^2 - 3 = 57$, $5(6m - 5)^2 = 60$, $(6m - 5)^2 = 12$, $6m - 5 = \pm\sqrt{12}$,
$6m - 5 = \pm 2\sqrt{3}$, $6m = 5 \pm 2\sqrt{3}$; $m = \dfrac{5 \pm 2\sqrt{3}}{6}$

43. $6(2r + 3)^2 - 5 = 85$, $6(2r + 3)^2 = 90$, $(2r + 3)^2 = 15$, $2r + 3 = \pm\sqrt{15}$,
$2r = -3 \pm \sqrt{15}$; $r = \dfrac{-3 \pm \sqrt{15}}{2}$

44. $\left(x - \dfrac{3}{5}\right)^2 - \dfrac{7}{9} = -\dfrac{1}{3}$, $\left(x - \dfrac{3}{5}\right)^2 = -\dfrac{1}{3} + \dfrac{7}{9}$, $\left(x - \dfrac{3}{5}\right)^2 = -\dfrac{3}{9} + \dfrac{7}{9}$, $\left(x - \dfrac{3}{5}\right)^2 = \dfrac{4}{9}$,
$x - \dfrac{3}{5} = \pm\sqrt{\dfrac{4}{9}}$, $x - \dfrac{3}{5} = \pm\dfrac{2}{3}$, $x = \dfrac{3}{5} \pm \dfrac{2}{3}$, $x = \dfrac{9}{15} \pm \dfrac{10}{15}$; $x = -\dfrac{1}{15}$ or $x = \dfrac{19}{15}$

45. $\left(y - \dfrac{1}{3}\right)^2 - \dfrac{5}{8} = -\dfrac{1}{2}$, $\left(y - \dfrac{1}{3}\right)^2 = -\dfrac{1}{2} + \dfrac{5}{8}$, $\left(y - \dfrac{1}{3}\right)^2 = -\dfrac{4}{8} + \dfrac{5}{8}$, $\left(y - \dfrac{1}{3}\right)^2 = \dfrac{1}{8}$,
$y - \dfrac{1}{3} = \pm\sqrt{\dfrac{1}{8}}$, $y - \dfrac{1}{3} = \pm\dfrac{1}{2\sqrt{2}}$, $y = \dfrac{1}{3} \pm \dfrac{1}{2\sqrt{2}} \cdot \dfrac{\sqrt{2}}{\sqrt{2}}$; $y = \dfrac{1}{3} \pm \dfrac{\sqrt{2}}{4}$

46. $5\left(x - \dfrac{1}{3}\right)^2 + \dfrac{1}{2} = \dfrac{3}{5}$, $5\left(x - \dfrac{1}{3}\right)^2 = \dfrac{3}{5} - \dfrac{1}{2}$, $5\left(x - \dfrac{1}{3}\right)^2 = \dfrac{6}{10} - \dfrac{5}{10}$, $5\left(x - \dfrac{1}{3}\right)^2 = \dfrac{1}{10}$,
$\left(x - \dfrac{1}{3}\right)^2 = \dfrac{1}{50}$, $x - \dfrac{1}{3} = \pm\sqrt{\dfrac{1}{50}}$, $x - \dfrac{1}{3} = \pm\dfrac{1}{5\sqrt{2}}$, $x = \dfrac{1}{3} \pm \dfrac{1}{5\sqrt{2}} \cdot \dfrac{\sqrt{2}}{\sqrt{2}}$; $x = \dfrac{1}{3} \pm \dfrac{\sqrt{2}}{10}$

47. $5\left(z - \dfrac{1}{2}\right)^2 + \dfrac{1}{3} = \dfrac{3}{4}$, $5\left(z - \dfrac{1}{2}\right)^2 = \dfrac{3}{4} - \dfrac{1}{3}$, $5\left(z - \dfrac{1}{2}\right)^2 = \dfrac{9}{12} - \dfrac{4}{12}$, $5\left(z - \dfrac{1}{2}\right)^2 = \dfrac{5}{12}$,
$\left(z - \dfrac{1}{2}\right)^2 = \dfrac{1}{12}$, $z - \dfrac{1}{2} = \pm\sqrt{\dfrac{1}{12}}$, $z - \dfrac{1}{2} = \pm\dfrac{1}{2\sqrt{3}}$, $z = \dfrac{1}{2} \pm \dfrac{1}{2\sqrt{3}}$,
$z = \dfrac{1}{2} \pm \dfrac{1}{2\sqrt{3}} \cdot \dfrac{\sqrt{3}}{\sqrt{3}}$; $z = \dfrac{1}{2} \pm \dfrac{\sqrt{3}}{6}$

48. $x^2 + 2x + 1 = 4$, $(x + 1)^2 = 4$, $x + 1 = \pm \sqrt{4}$, $x + 1 = \pm 2$; $x = -1 \pm 2$; $x = 1$ or $x = -3$

49. $y^2 - 4y + 4 = 25$, $(y - 2)^2 = 25$, $y - 2 = \pm\sqrt{25}$, $y - 2 = \pm 5$, $y = 2 \pm 5$; $y = 7$ or $y = -3$

50. $25x^2 - 10x + 1 = 16$, $(5x - 1)^2 = 16$, $5x - 1 = \pm\sqrt{16}$, $5x - 1 = \pm 4$, $5x = 1 \pm 4$; $x = 1$ or $x = -\dfrac{3}{5}$

51. $16m^2 + 24m + 9 = 9$, $(4m + 3)^2 = 9$, $4m + 3 = \pm\sqrt{9}$, $4m + 3 = \pm 3$, $4m = -3 \pm 3$, $m = \dfrac{-3 \pm 3}{4}$; $m = -\dfrac{3}{2}$ or $m = 0$

52. $x^2 + x + \dfrac{1}{4} = 4$, $\left(x + \dfrac{1}{2}\right)^2 = 4$, $x + \dfrac{1}{2} = \pm\sqrt{4}$, $x + \dfrac{1}{2} = \pm 2$, $x = -\dfrac{1}{2} \pm 2$; $x = \dfrac{3}{2}$ or $x = -\dfrac{5}{2}$

53. $y^2 - 18y + 81 = 25$, $(y - 9)^2 = 25$, $y - 9 = \pm\sqrt{25}$, $y - 9 = \pm 5$, $y = 9 \pm 5$; $y = 14$ or $y = 4$

54. $y^2 - 12y + 36 = 84$, $(y - 6)^2 = 84$, $y - 6 = \pm\sqrt{84}$, $y = 6 \pm \sqrt{84}$; $y = 6 \pm 2\sqrt{21}$

55. $b^2 + 14b + 42 = -7$, $b^2 + 14b + 49 = 0$, $(b + 7)^2 = 0$, $b + 7 = 0$; $b = -7$

56. $x^2 + y^2 = r^2$, $x^2 = r^2 - y^2$; $x = \pm \sqrt{r^2 - y^2}$

57. $\dfrac{x^2}{16} + \dfrac{y^2}{4} = 1$, $x^2 + 4y^2 = 16$, $x^2 = 16 - 4y^2$, $x = \pm\sqrt{16 - 4y^2}$

58. The ellipse would appear rotated, passing through the points $(2, 0)$, $(0, 4)$, $(-2, 0)$, and $(0, -4)$. This could be verified by solving for x or y, substituting values, and plotting the points.

page 547   Capsule Review

1. yes; $(x + 2)^2$ **2.** yes; $(2m + 3)^2$ **3.** no   **4.** yes; $\left(z - \dfrac{1}{5}\right)^2$ **5.** yes; $\left(n - \dfrac{1}{2}\right)^2$ **6.** yes; $(7p - 2)^2$ **7.** no   **8.** yes; $\left(n - \dfrac{2}{5}\right)^2$ **9.** yes; $\left(c - \dfrac{3}{4}\right)^2$

page 549   Class Exercises

1. $\dfrac{1}{2}(10) = 5$, $5^2 = 25$, $y^2 + 10y + 25$; $c = 25$

2. $\dfrac{1}{2}(-6) = -3$, $(-3)^2 = 9$, $x^2 - 6x + 9$; $c = 9$

Chapter 13: Quadratic Equations and Functions   **419**

3. $\frac{1}{2}\left(-\frac{2}{3}\right) = -\frac{1}{3}, \left(-\frac{1}{3}\right)^2 = \frac{1}{9}, p^2 - \frac{2}{3}p + \frac{1}{9}; c = \frac{1}{9}$

4. $\frac{1}{2}(-8) = -4, (-4)^2 = 16, x^2 - 8x + 16 = -15 + 16, (x - 4)^2 = 1, x - 4 = \pm 1, x = 5, 3$

5. $\frac{1}{2}(18) = 9, (9)^2 = 81, z^2 + 18z + 81 = -56 + 81, (z + 9)^2 = 25, z + 9 = \pm 5, z = -14, -4$

6. $\frac{1}{2}(-1) = -\frac{1}{2}, \left(-\frac{1}{2}\right)^2 = \frac{1}{4}, n^2 - n + \frac{1}{4} = \frac{1}{4} + \frac{1}{4}, \left(n - \frac{1}{2}\right)^2 = \frac{1}{2}, n - \frac{1}{2} = \pm\sqrt{\frac{1}{2}}, n = \frac{1 \pm \sqrt{2}}{2}$

pages 549–550   Practice Exercises

A 1. $\frac{1}{2}(8) = 4, 4^2 = 16, m^2 + 8m + 16; c = 16$

2. $\frac{1}{2}(12) = 6, 6^2 = 36, t^2 + 12t + 36; c = 36$

3. $\frac{1}{2}(-10) = -5, (-5)^2 = 25, y^2 - 10y + 25; c = 25$

4. $\frac{1}{2}(3) = \frac{3}{2}, \left(\frac{3}{2}\right)^2 = \frac{9}{4}, y^2 + 3y + \frac{9}{4} = 3 + \frac{9}{4}, y^2 + 3y + \frac{9}{4} = \frac{21}{4}, \left(y + \frac{3}{2}\right)^2 = \frac{21}{4},$ $y + \frac{3}{2} = \pm\frac{\sqrt{21}}{2}, y = -\frac{3}{2} \pm \frac{\sqrt{21}}{2}; y = \frac{-3 \pm \sqrt{21}}{2}$

5. $\frac{1}{2}(-4) = -2, (-2)^2 = 4, b^2 - 4b + 4 = 5 + 4, b^2 - 4b + 4 = 9, (b - 2)^2 = 9,$ $b - 2 = \pm\sqrt{9}, b - 2 = \pm 3, b = 2 \pm 3; b = 5 \text{ or } b = -1$

6. $\frac{1}{2}(-2) = -1, (-1)^2 = 1, x^2 - 2x + 1 = 15 + 1, x^2 - 2x + 1 = 16,$ $(x - 1)^2 = 16, x - 1 = \pm\sqrt{16}, x - 1 = \pm 4, x = 1 \pm 4; x = 5 \text{ or } x = -3$

7. $\frac{1}{2}(12) = 6, 6^2 = 36, z^2 + 12z + 36 = 45 + 36, z^2 + 12z + 36 = 81,$ $(z + 6)^2 = 81, z + 6 = \pm\sqrt{81}, z + 6 = \pm 9, z = -6 \pm 9; z = 3 \text{ or } z = -15$

8. $\frac{1}{2}(12) = 6, 6^2 = 36, h^2 + 12h + 36 = 28 + 36, h^2 + 12h + 36 = 64,$ $(h + 6)^2 = 64, h + 6 = \pm\sqrt{64}, h + 6 = \pm 8, h = -6 \pm 8; h = -14 \text{ or } h = 2$

9. $\frac{1}{2}(4) = 2$, $2^2 = 4$, $t^2 + 4t + 4 = 12 + 4$, $t^2 + 4t + 4 = 16$, $(t + 2)^2 = 16$,

$t + 2 = \pm\sqrt{16}$, $t + 2 = \pm 4$, $t = -2 \pm 4$; $t = -6$ or $t = 2$

10. $\frac{1}{2}(-5) = -\frac{5}{2}$, $\left(-\frac{5}{2}\right)^2 = \frac{25}{4}$, $t^2 - 5t + \frac{25}{4} = 2\frac{3}{4} + \frac{25}{4}$, $t^2 - 5t + \frac{25}{4} = \frac{36}{4}$,

$\left(t - \frac{5}{2}\right)^2 = \frac{36}{4}$, $t - \frac{5}{2} = \pm\sqrt{\frac{36}{4}}$, $t - \frac{5}{2} = \pm\frac{6}{2}$, $t = \frac{5}{2} \pm \frac{6}{2}$; $t = \frac{11}{2}$ or $t = -\frac{1}{2}$

11. $\frac{1}{2}(7) = \frac{7}{2}$, $\left(\frac{7}{2}\right)^2 = \frac{49}{4}$, $y^2 + 7y + \frac{49}{4} = 7\frac{3}{4} + \frac{49}{4}$, $y^2 + 7y + \frac{49}{4} = \frac{31}{4} + \frac{49}{4}$,

$y^2 + 7y + \frac{49}{4} = \frac{80}{4}$, $\left(y + \frac{7}{2}\right)^2 = \frac{80}{4}$, $y + \frac{7}{2} = \pm\sqrt{\frac{80}{4}}$, $y + \frac{7}{2} = \pm\frac{4\sqrt{5}}{2}$,

$y = -\frac{7}{2} \pm \frac{4\sqrt{5}}{2}$, $y = \frac{-7 \pm 4\sqrt{5}}{2}$; $y = -\frac{7}{2} \pm 2\sqrt{5}$

12. $\frac{1}{2}(3) = \frac{3}{2}$, $\left(\frac{3}{2}\right)^2 = \frac{9}{4}$, $x^2 + 3x + \frac{9}{4} = -1 + \frac{9}{4}$, $x^2 + 3x + \frac{9}{4} = \frac{5}{4}$, $\left(x + \frac{3}{2}\right)^2 = \frac{5}{4}$,

$x + \frac{3}{2} = \pm\sqrt{\frac{5}{4}}$, $x + \frac{3}{2} = \pm\frac{\sqrt{5}}{2}$, $x = -\frac{3}{2} \pm \frac{\sqrt{5}}{2}$; $x = \frac{-3 \pm \sqrt{5}}{2}$

13. $\frac{1}{2}(-3) = -\frac{3}{2}$, $\left(-\frac{3}{2}\right)^2 = \frac{9}{4}$, $m^2 - 3m + \frac{9}{4} = -1 + \frac{9}{4}$, $m^2 - 3m + \frac{9}{4} = \frac{5}{4}$,

$\left(m - \frac{3}{2}\right)^2 = \frac{5}{4}$, $m - \frac{3}{2} = \pm\frac{\sqrt{5}}{2}$, $m = \frac{3}{2} \pm \frac{\sqrt{5}}{2}$; $m = \frac{3 \pm \sqrt{5}}{2}$

14. $\frac{1}{2}(3) = \frac{3}{2}$, $\left(\frac{3}{2}\right)^2 = \frac{9}{4}$, $x^2 + 3x + \frac{9}{4} = \frac{9}{4}$, $\left(x + \frac{3}{2}\right)^2 = \frac{9}{4}$, $\left(x + \frac{3}{2}\right) = \pm\sqrt{\frac{9}{4}}$,

$x + \frac{3}{2} = \pm\frac{3}{2}$, $x = -\frac{3}{2} \pm \frac{3}{2}$; $x = 0$ or $x = -3$

15. $\frac{1}{2}(5) = \frac{5}{2}$, $\left(\frac{5}{2}\right)^2 = \frac{25}{4}$, $c^2 + 5c + \frac{25}{4} = \frac{25}{4}$, $\left(c + \frac{5}{2}\right)^2 = \frac{25}{4}$, $c + \frac{5}{2} = \pm\sqrt{\frac{25}{4}}$,

$c + \frac{5}{2} = \pm\frac{5}{2}$, $c = -\frac{5}{2} \pm \frac{5}{2}$; $c = 0$ or $c = -5$

16. $\frac{1}{2}(-6) = -3$, $(-3)^2 = 9$, $r^2 - 6r + 9 = -5 + 9$, $r^2 - 6r + 9 = 4$,

$(r - 3)^2 = 4$, $r - 3 = \pm\sqrt{4}$, $r - 3 = \pm 2$; $r = 5$ or $r = 1$

17. $\frac{1}{2}(8) = 4$, $4^2 = 16$, $p^2 + 8p + 16 = 9 + 16$, $p^2 + 8p + 16 = 25$, $(p + 4)^2 = 25$,

$p + 4 = \pm\sqrt{25}$, $p + 4 = \pm 5$, $p = -4 \pm 5$; $p = -9$ or $p = 1$

18. $\frac{1}{2}(-3) = -\frac{3}{2}$, $\left(-\frac{3}{2}\right)^2 = \frac{9}{4}$, $t^2 - 3t + \frac{9}{4} = 10 + \frac{9}{4}$, $t^2 - 3t + \frac{9}{4} = \frac{49}{4}$,

$\left(t - \frac{3}{2}\right)^2 = \frac{49}{4}$, $t - \frac{3}{2} = \pm\sqrt{\frac{49}{4}}$, $t - \frac{3}{2} = \pm\frac{7}{2}$, $t = \frac{3}{2} \pm \frac{7}{2}$; $t = 5$ or $t = -2$

B 19. $\frac{1}{2}(5) = \frac{5}{2}, \left(\frac{5}{2}\right)^2 = \frac{25}{4}, p^2 + 5p + \frac{25}{4} = -6 + \frac{25}{4}, p^2 + 5p + \frac{25}{4} = \frac{1}{4},$

$\left(p + \frac{5}{2}\right)^2 = \frac{1}{4}, p + \frac{5}{2} = \pm \sqrt{\frac{1}{4}}, p + \frac{5}{2} = \pm \frac{1}{2}, p = -\frac{5}{2} \pm \frac{1}{2}; p = -3 \text{ or } p = -2$

20. $\frac{1}{2}(-3) = -\frac{3}{2}, \left(-\frac{3}{2}\right)^2 = \frac{9}{4}, n^2 - 3n + \frac{9}{4} = -\frac{5}{4} + \frac{9}{4}, n^2 - 3n + \frac{9}{4} = 1,$

$\left(n - \frac{3}{2}\right)^2 = 1, n - \frac{3}{2} = \pm \sqrt{1}, n - \frac{3}{2} = \pm 1, n = \frac{3}{2} \pm 1; n = \frac{1}{2} \text{ or } n = \frac{5}{2}$

21. $\frac{1}{2}(-1) = -\frac{1}{2}, \left(-\frac{1}{2}\right)^2 = \frac{1}{4}, r^2 - r + \frac{1}{4} = 3\frac{3}{4} + \frac{1}{4}, r^2 - r + \frac{1}{4} = \frac{16}{4}, r^2 - r + \frac{1}{4} = 4,$

$\left(r - \frac{1}{2}\right)^2 = 4, r - \frac{1}{2} = \pm \sqrt{4}, r - \frac{1}{2} = \pm 2, r = \frac{1}{2} \pm 2; r = \frac{5}{2} \text{ or } r = -\frac{3}{2}$

22. $\frac{1}{2}(-4) = -2, (-2)^2 = 4, x^2 - 4x + 4 = -2 + 4, x^2 - 4x + 4 = 2,$

$(x - 2)^2 = 2, x - 2 = \pm \sqrt{2}; x = 2 \pm \sqrt{2}$

23. $4y^2 - 6y = \frac{1}{2}, y^2 - \frac{3}{2}y = \frac{1}{8}, \frac{1}{2}\left(-\frac{3}{2}\right) = -\frac{3}{4}, \left(-\frac{3}{4}\right)^2 = \frac{9}{16}, y^2 - \frac{3}{2}y + \frac{9}{16} = \frac{1}{8} + \frac{9}{16},$

$y^2 - \frac{3}{2}y + \frac{9}{16} = \frac{11}{16}, \left(y - \frac{3}{4}\right)^2 = \frac{11}{16}, y - \frac{3}{4} = \pm \sqrt{\frac{11}{16}}, y - \frac{3}{4} = \pm \frac{\sqrt{11}}{4},$

$y = \frac{3}{4} \pm \frac{\sqrt{11}}{4}; y = \frac{3 \pm \sqrt{11}}{4}$

24. $3b^2 - 12b = 1, b^2 - 4b = \frac{1}{3}, \frac{1}{2}(-4) = -2, (-2)^2 = 4, b^2 - 4b + 4 = \frac{1}{3} + 4,$

$b^2 - 4b + 4 = \frac{13}{3}, (b - 2)^2 = \frac{13}{3}, b - 2 = \pm \sqrt{\frac{13}{3}}, b = 2 \pm \sqrt{\frac{13}{3}},$

$b = 2 \pm \frac{\sqrt{13}}{\sqrt{3}} \cdot \frac{\sqrt{3}}{\sqrt{3}}; b = 2 \pm \frac{\sqrt{39}}{3}$

25. $p^2 - 2p = -\frac{4}{5}, \frac{1}{2}(-2) = -1, (-1)^2 = 1, p^2 - 2p + 1 = -\frac{4}{5} + 1,$

$p^2 - 2p + 1 = \frac{1}{5}, (p - 1)^2 = \frac{1}{5}, p - 1 = \pm \sqrt{\frac{1}{5}}, p = 1 \pm \sqrt{\frac{1}{5}},$

$p = 1 \pm \frac{1 \cdot \sqrt{5}}{\sqrt{5} \cdot \sqrt{5}}; p = 1 \pm \frac{\sqrt{5}}{5}$

26. $t^2 - \frac{5}{2}t = 2, \frac{1}{2}\left(-\frac{5}{2}\right) = -\frac{5}{4}, \left(-\frac{5}{4}\right)^2 = \frac{25}{16}, t^2 - \frac{5}{2}t + \frac{25}{16} = 2 + \frac{25}{16}, t^2 - \frac{5}{2}t + \frac{25}{16} = \frac{57}{16},$

$\left(t - \frac{5}{4}\right)^2 = \frac{57}{16}, t - \frac{5}{4} = \pm \sqrt{\frac{57}{16}}, t - \frac{5}{4} = \pm \frac{\sqrt{57}}{4}, t = \frac{5}{4} \pm \frac{\sqrt{57}}{4}; t = \frac{5 \pm \sqrt{57}}{4}$

27. $c^2 + \dfrac{c}{2} = \dfrac{5}{2}, \dfrac{1}{2}\left(\dfrac{1}{2}\right) = \dfrac{1}{4}, \left(\dfrac{1}{4}\right)^2 = \dfrac{1}{16}, c^2 + \dfrac{c}{2} + \dfrac{1}{16} = \dfrac{5}{2} + \dfrac{1}{16}, c^2 + \dfrac{c}{2} + \dfrac{1}{16} = \dfrac{41}{16},$

$\left(c + \dfrac{1}{4}\right)^2 = \dfrac{41}{16}, c + \dfrac{1}{4} = \pm\sqrt{\dfrac{41}{16}}, c + \dfrac{1}{4} = \pm\dfrac{\sqrt{41}}{4}, c = -\dfrac{1}{4} \pm \dfrac{\sqrt{41}}{4}; c = \dfrac{-1 \pm \sqrt{41}}{4}$

28. $z^2 - 5z = \dfrac{3}{2}, \dfrac{1}{2}(-5) = -\dfrac{5}{2}, \left(-\dfrac{5}{2}\right)^2 = \dfrac{25}{4}, z^2 - 5z + \dfrac{25}{4} = \dfrac{3}{2} + \dfrac{25}{4},$

$z^2 - 5z + \dfrac{25}{4} = \dfrac{31}{4}, \left(z - \dfrac{5}{2}\right)^2 = \dfrac{31}{4}, z - \dfrac{5}{2} = \pm\sqrt{\dfrac{31}{4}}, z - \dfrac{5}{2} = \pm\dfrac{\sqrt{31}}{2},$

$z = \dfrac{5}{2} \pm \dfrac{\sqrt{31}}{2}; z = \dfrac{5 \pm \sqrt{31}}{2}$

29. $2n^2 + 8n = 5, n^2 + 4n = \dfrac{5}{2}, \dfrac{1}{2}(4) = 2, 2^2 = 4, n^2 + 4n + 4 = \dfrac{5}{2} + 4,$

$n^2 + 4n + 4 = \dfrac{13}{2}, (n + 2)^2 = \dfrac{13}{2}, n + 2 = \pm\sqrt{\dfrac{13}{2}}, n + 2 = \pm\dfrac{\sqrt{26}}{2},$

$n = -2 \pm \dfrac{\sqrt{26}}{2}, n = \dfrac{-4 \pm \sqrt{26}}{2}$

30. $2m^2 - m = 9, m^2 - \dfrac{m}{2} = \dfrac{9}{2}, \dfrac{1}{2}\left(-\dfrac{1}{2}\right) = -\dfrac{1}{4}, \left(-\dfrac{1}{4}\right)^2 = \dfrac{1}{16}, m^2 - \dfrac{m}{2} + \dfrac{1}{16} =$

$\dfrac{9}{2} + \dfrac{1}{16}, m^2 - \dfrac{m}{2} + \dfrac{1}{16} = \dfrac{73}{16}, \left(m - \dfrac{1}{4}\right)^2 = \dfrac{73}{16}, m - \dfrac{1}{4} = \pm\sqrt{\dfrac{73}{16}}, m - \dfrac{1}{4} =$

$\pm\dfrac{\sqrt{73}}{4}, m = \dfrac{1}{4} \pm \dfrac{\sqrt{73}}{4}, m = \dfrac{1 \pm \sqrt{73}}{4}$

C 31. $3n^2 - 8n = -4, n^2 - \dfrac{8}{3}n = -\dfrac{4}{3}, \dfrac{1}{2}\left(-\dfrac{8}{3}\right) = -\dfrac{4}{3}, \left(-\dfrac{4}{3}\right)^2 = \dfrac{16}{9}, n^2 - \dfrac{8}{3}n + \dfrac{16}{9} = -\dfrac{4}{3} + \dfrac{16}{9},$

$n^2 - \dfrac{8}{3}n + \dfrac{16}{9} = \dfrac{4}{9}, \left(n - \dfrac{4}{3}\right)^2 = \dfrac{4}{9}, n - \dfrac{4}{3} = \pm\sqrt{\dfrac{4}{9}}, n - \dfrac{4}{3} = \pm\dfrac{2}{3},$

$n = \dfrac{4}{3} \pm \dfrac{2}{3}; n = 2 \text{ or } n = \dfrac{2}{3}$

32. $2x^2 - 5x = 12, x^2 - \dfrac{5}{2}x = 6, \dfrac{1}{2}\left(-\dfrac{5}{2}\right) = -\dfrac{5}{4}, \left(-\dfrac{5}{4}\right)^2 = \dfrac{25}{16}, x^2 - \dfrac{5}{2}x + \dfrac{25}{16} = 6 + \dfrac{25}{16},$

$x^2 - \dfrac{5}{2}x + \dfrac{25}{16} = \dfrac{121}{16}, \left(x - \dfrac{5}{4}\right)^2 = \dfrac{121}{16}, x - \dfrac{5}{4} = \pm\sqrt{\dfrac{121}{16}}, x - \dfrac{5}{4} = \pm\dfrac{11}{4},$

$x = \dfrac{5}{4} \pm \dfrac{11}{4}; x = 4 \text{ or } x = -\dfrac{3}{2}$

33. $2y^2 + 5y = -1, y^2 + \dfrac{5}{2}y = -\dfrac{1}{2}, \dfrac{1}{2}\left(\dfrac{5}{2}\right) = \dfrac{5}{4}, \left(\dfrac{5}{4}\right)^2 = \dfrac{25}{16}, y^2 + \dfrac{5}{2}y + \dfrac{25}{16} = -\dfrac{1}{2} + \dfrac{25}{16},$

$y^2 + \dfrac{5}{2}y + \dfrac{25}{16} = \dfrac{17}{16}, \left(y + \dfrac{5}{4}\right)^2 = \dfrac{17}{16}, y + \dfrac{5}{4} = \pm\sqrt{\dfrac{17}{16}}, y + \dfrac{5}{4} = \pm\dfrac{\sqrt{17}}{4},$

$y = -\dfrac{5}{4} \pm \dfrac{\sqrt{17}}{4}; y = \dfrac{-5 \pm \sqrt{17}}{4}$

34. $3z^2 - 4z = 1$, $z^2 - \dfrac{4}{3}z = \dfrac{1}{3}$, $\dfrac{1}{2}\left(-\dfrac{4}{3}\right) = -\dfrac{2}{3}$, $\left(-\dfrac{2}{3}\right)^2 = \dfrac{4}{9}$, $z^2 - \dfrac{4}{3}z + \dfrac{4}{9} = \dfrac{1}{3} + \dfrac{4}{9}$,

$z^2 - \dfrac{4}{3}z + \dfrac{4}{9} = \dfrac{7}{9}$, $\left(z - \dfrac{2}{3}\right)^2 = \dfrac{7}{9}$, $z - \dfrac{2}{3} = \pm\sqrt{\dfrac{7}{9}}$, $z = \dfrac{2}{3} \pm \dfrac{\sqrt{7}}{3}$; $z = \dfrac{2 \pm \sqrt{7}}{3}$

35. $3y^2 - 5y = 10$, $y^2 - \dfrac{5}{3}y = \dfrac{10}{3}$, $\dfrac{1}{2}\left(-\dfrac{5}{3}\right) = -\dfrac{5}{6}$, $\left(-\dfrac{5}{6}\right)^2 = \dfrac{25}{36}$, $y^2 - \dfrac{5}{3}y + \dfrac{25}{36} = \dfrac{10}{3} + \dfrac{25}{36}$,

$y^2 - \dfrac{5}{3}y + \dfrac{25}{36} = \dfrac{145}{36}$, $\left(y - \dfrac{5}{6}\right)^2 = \dfrac{145}{36}$, $y - \dfrac{5}{6} = \pm\sqrt{\dfrac{145}{36}}$, $y - \dfrac{5}{6} = \pm\dfrac{\sqrt{145}}{6}$,

$y = \dfrac{5}{6} \pm \dfrac{\sqrt{145}}{6}$; $y = \dfrac{5 \pm \sqrt{145}}{6}$

36. $6b^2 - 10b = -3$, $b^2 - \dfrac{5}{3}b = -\dfrac{1}{2}$, $\dfrac{1}{2}\left(-\dfrac{5}{3}\right) = -\dfrac{5}{6}$, $\left(-\dfrac{5}{6}\right)^2 = \dfrac{25}{36}$, $b^2 - \dfrac{5}{3}b + \dfrac{25}{36} =$

$-\dfrac{1}{2} + \dfrac{25}{36}$, $b^2 - \dfrac{5}{3}b + \dfrac{25}{36} = \dfrac{7}{36}$, $\left(b - \dfrac{5}{6}\right)^2 = \dfrac{7}{36}$, $b - \dfrac{5}{6} = \pm\sqrt{\dfrac{7}{36}}$, $b - \dfrac{5}{6} = \pm\dfrac{\sqrt{7}}{6}$,

$b = \dfrac{5}{6} \pm \dfrac{\sqrt{7}}{6}$; $b = \dfrac{5 \pm \sqrt{7}}{6}$

37. $y^2 + 2 = \dfrac{7}{2}y$, $y^2 - \dfrac{7}{2}y = -2$, $\dfrac{1}{2}\left(-\dfrac{7}{2}\right) = -\dfrac{7}{4}$, $\left(-\dfrac{7}{4}\right)^2 = \dfrac{49}{16}$,

$y^2 - \dfrac{7}{2}y + \dfrac{49}{16} = -2 + \dfrac{49}{16}$, $y^2 - \dfrac{7}{2}y + \dfrac{49}{16} = \dfrac{17}{16}$, $\left(y - \dfrac{7}{4}\right)^2 = \dfrac{17}{16}$, $y - \dfrac{7}{4} = \pm\sqrt{\dfrac{17}{16}}$,

$y - \dfrac{7}{4} = \pm\dfrac{\sqrt{17}}{4}$, $y = \dfrac{7}{4} \pm \dfrac{\sqrt{17}}{4}$; $y = \dfrac{7 \pm \sqrt{17}}{4}$

38. $3(y - 1) - (y - 2) = 2(y^2 - 3y + 2)$, $3y - 3 - y + 2 = 2y^2 - 6y + 4$,

$2y - 1 = 2y^2 - 6y + 4$, $2y^2 - 8y + 5 = 0$, $2y^2 - 8y = -5$, $y^2 - 4y = -\dfrac{5}{2}$,

$\dfrac{1}{2}(-4) = -2$, $(-2)^2 = 4$, $y^2 - 4y + 4 = -\dfrac{5}{2} + 4$, $y^2 - 4y + 4 = \dfrac{3}{2}$, $(y - 2)^2 = \dfrac{3}{2}$,

$y - 2 = \sqrt{\dfrac{3}{2}}$, $y - 2 = \dfrac{\sqrt{3} \cdot \sqrt{2}}{\sqrt{2} \cdot \sqrt{2}}$, $y - 2 = \pm\dfrac{\sqrt{6}}{2}$; $y = 2 \pm \dfrac{\sqrt{6}}{2}$

39. $\dfrac{1}{2}(b) = \dfrac{b}{2}$, $\left(\dfrac{b}{2}\right)^2 = \dfrac{b^2}{4}$, $y^2 + by + \dfrac{b^2}{4} = -1 + \dfrac{b^2}{4}$, $y^2 + by + \dfrac{b^2}{4} = \dfrac{-4 + b^2}{4}$, $\left(y + \dfrac{b}{2}\right)^2 =$

$\dfrac{-4 + b^2}{4}$, $y + \dfrac{b}{2} = \pm\sqrt{\dfrac{b^2 - 4}{4}}$, $y = -\dfrac{b}{2} \pm \sqrt{\dfrac{b^2 - 4}{4}}$; $y = \dfrac{-b \pm \sqrt{b^2 - 4}}{2}$

40. $\dfrac{1}{2}(b) = \dfrac{b}{2}$, $\left(\dfrac{b}{2}\right)^2 = \dfrac{b^2}{4}$, $x^2 + bx + \dfrac{b^2}{4} = \dfrac{b^2}{4} - 2$, $x^2 + bx + \dfrac{b^2}{4} = \dfrac{b^2}{4} - \dfrac{8}{4}$, $x^2 + bx + \dfrac{b^2}{4} =$

$\dfrac{b^2 - 8}{4}$, $\left(x + \dfrac{b}{2}\right)^2 = \dfrac{b^2 - 8}{4}$, $x + \dfrac{b}{2} = \pm\sqrt{\dfrac{b^2 - 8}{4}}$, $x = -\dfrac{b}{2} \pm \dfrac{\sqrt{b^2 - 8}}{2}$;

$x = \dfrac{-b \pm \sqrt{b^2 - 8}}{2}$

41. $\frac{1}{2}(b) = \frac{b}{2}, \left(\frac{b}{2}\right)^2 = \frac{b^2}{4}, x^2 + bx + \frac{b^2}{4} = \frac{b^2}{4} - c, x^2 + bx + \frac{b^2}{4} = \frac{b^2 - 4c}{4}, \left(x + \frac{b}{2}\right)^2 = \frac{b^2 - 4c}{4},$

$x + \frac{b}{2} = \pm \sqrt{\frac{b^2 - 4c}{4}}, x + \frac{b}{2} = \pm \frac{\sqrt{b^2 - 4c}}{2}, x = -\frac{b}{2} \pm \frac{\sqrt{b^2 - 4c}}{2}, x = \frac{-b \pm \sqrt{b^2 - 4c}}{2}$

42. $x^2 + \frac{b}{a}x + \frac{c}{a} = 0, \frac{1}{2}\left(\frac{b}{a}\right) = \frac{b}{2a}, \left(\frac{b}{2a}\right)^2 = \frac{b^2}{4a^2}, x^2 + \frac{b}{a}x + \frac{b^2}{4a^2} = \frac{b^2}{4a^2} - \frac{c}{a},$

$x^2 + \frac{b}{a}x + \frac{b^2}{4a^2} = \frac{b^2 - 4ac}{4a^2}, \left(x + \frac{b}{2a}\right)^2 = \frac{b^2 - 4ac}{4a^2}, x + \frac{b}{2a} = \pm \sqrt{\frac{b^2 - 4ac}{4a^2}},$

$x = -\frac{b}{2a} \pm \frac{\sqrt{b^2 - 4ac}}{2a}; x = \frac{-b \pm \sqrt{b^2 - 4ac}}{2a}$

43. $t^2 - 3.7t = 10, \frac{1}{2}(-3.7) = -1.85, (-1.85)^2 = 3.4225, t^2 - 3.7t + 3.4225 =$

$10 + 3.4225, (t - 1.85)^2 = 13.4225, t - 1.85 = \pm \sqrt{13.4225}, t - 1.85 \approx$

$\pm 3.6637, t \approx 1.85 \pm 3.66, t \approx -1.81 \text{ or } t \approx 5.51$

44. $p^2 - 5.2p = -6, \frac{1}{2}(-5.2) = -2.6, (-2.6)^2 = 6.76, p^2 - 5.2p + 6.76 =$

$-6 + 6.76, (p - 2.6)^2 = 0.76, p - 2.6 = \pm \sqrt{0.76}, p - 2.6 \approx \pm 0.8718,$

$p \approx 2.6 \pm 0.87, p \approx 1.73 \text{ or } p \approx 3.47$

45. $y^2 - 5.4y = 10, \frac{1}{2}(-5.4) = -2.7, (-2.7)^2 = 7.29, y^2 - 5.4y + 7.29 =$

$10 + 7.29, (y - 2.7)^2 = 17.29, y - 2.7 = \pm \sqrt{17.29}, y - 2.7 \approx \pm 4.1581,$

$y \approx 2.7 \pm 4.16, y \approx -1.46 \text{ or } y \approx 6.86$

46. $b^2 - 2\sqrt{2}b = 2, \frac{1}{2}(-2\sqrt{2}) = -\sqrt{2}, (-\sqrt{2})^2 = 2, b^2 - 2\sqrt{2}b + 2 = 2 + 2,$

$(b - \sqrt{2})^2 = 4, b - \sqrt{2} = \pm 2, b = \sqrt{2} \pm 2, b \approx -0.59 \text{ or } b \approx 3.41$

47. $16 - 4u = 2t, t = 8 - 2u$

48. $14a + 7c = 3b, 7c = 3b - 14a, c = \frac{3}{7}b - 2a$

49. $f(-3) = 2(-3) - 5 = -6 - 5 = -11$

50. $g(f(3)) = g(2(3) - 5) = g(6 - 5) = g(1) = (1)^2 - (1) + 2 = 1 - 1 + 2 = 2$

51. $g(4) - f(7) = [(4)^2 - (4) + 2] - [2(7) - 5] = (16 - 4 + 2) - (14 - 5) =$
$14 - 9 = 5$

52. $\dfrac{9b^{4-2}}{5a^{5-3}} = \dfrac{9b^2}{5a^2}$

53. $\dfrac{(2x+1)(2x+1)}{(3x-2)(2x+1)} = \dfrac{2x+1}{3x-2}$

54. $\sqrt{25x^4y^4} \cdot \sqrt{6y} = 5x^2y^2\sqrt{6y}$

55. $-\sqrt{4a^6b^2} \cdot \sqrt{22a} = -2|a^3b|\sqrt{22a}$

56. $2\sqrt{2} - 8\sqrt{2} = -6\sqrt{2}$

57. $9\sqrt{5} + 4\sqrt{5} = 13\sqrt{5}$

page 551   Capsule Review

1. $3x^2 - 7x - 5 = 0$ **2.** $4x^2 - x = 0$ **3.** $x^2 + 7 = 0$

4. $-2(x^2 + 10x + 25) = 3x$, $-2x^2 - 20x - 50 = 3x$; $-2x^2 - 23x - 50 = 0$

page 552   Class Exercises

1. $a = 1, b = 6, c = 1$

$x = \dfrac{-6 \pm \sqrt{6^2 - 4(1)(1)}}{2(1)}$, $x = \dfrac{-6 \pm \sqrt{36 - 4}}{2}$, $x = \dfrac{-6 \pm \sqrt{32}}{2}$, $x = \dfrac{-6 \pm 4\sqrt{2}}{2}$,

$x = -3 \pm 2\sqrt{2}$; The solutions are $-3 + 2\sqrt{2}$ and $-3 - 2\sqrt{2}$.

2. $a = 3, b = -4, c = -7$

$x = \dfrac{4 \pm \sqrt{(-4)^2 - 4(3)(-7)}}{2(3)}$, $x = \dfrac{4 \pm \sqrt{16 + 84}}{6}$, $x = \dfrac{4 \pm \sqrt{100}}{6}$, $x = \dfrac{4 \pm 10}{6}$;

The solutions are $\dfrac{4 + 10}{6} = \dfrac{7}{3}$ and $\dfrac{4 - 10}{6} = -1$.

3. $a = 1, b = -5, c = 0$

$y = \dfrac{5 \pm \sqrt{(-5)^2 - 4(1)(0)}}{2(1)}$, $y = \dfrac{5 \pm \sqrt{25 - 0}}{2}$, $y = \dfrac{5 \pm \sqrt{25}}{2}$, $y = \dfrac{5 \pm 5}{2}$; The

solutions are $\dfrac{5 + 5}{2} = 5$ and $\dfrac{5 - 5}{2} = 0$.

4. A negative radicand will result and there will be no real solution.

pages 552–553   Practice Exercises

A   **1.** $x = \dfrac{-5 \pm \sqrt{5^2 - 4(1)(6)}}{2(1)}$, $x = \dfrac{-5 \pm \sqrt{25 - 24}}{2}$, $x = \dfrac{-5 \pm \sqrt{1}}{2}$, $x = \dfrac{-5 \pm 1}{2}$;

The solutions are $\dfrac{-5 + 1}{2} = -2$ and $\dfrac{-5 - 1}{2} = -3$.

2. $y = \dfrac{1 \pm \sqrt{(-1)^2 - 4(1)(-6)}}{2(1)}$, $y = \dfrac{1 \pm \sqrt{1 + 24}}{2}$, $y = \dfrac{1 \pm \sqrt{25}}{2}$, $y = \dfrac{1 \pm 5}{2}$; The

solutions are $\dfrac{1 + 5}{2} = 3$ and $\dfrac{1 - 5}{2} = -2$.

3. $c = \dfrac{3 \pm \sqrt{(-3)^2 - 4(1)(-10)}}{2(1)}$, $c = \dfrac{3 \pm \sqrt{9 + 40}}{2}$, $c = \dfrac{3 \pm \sqrt{49}}{2}$, $c = \dfrac{3 \pm 7}{2}$; The

solutions are $\dfrac{3 + 7}{2} = 5$ and $\dfrac{3 - 7}{2} = -2$.

4. $m = \dfrac{-6 \pm \sqrt{6^2 - 4(1)(8)}}{2(1)}$, $m = \dfrac{-6 \pm \sqrt{36 - 32}}{2}$, $m = \dfrac{-6 \pm \sqrt{4}}{2}$, $m = -\dfrac{6 \pm 2}{2}$;

The solutions are $-\dfrac{6 + 2}{2} = -2$ and $-\dfrac{6 - 2}{2} = -4$.

5. $p^2 - 9p + 18 = 0$; $p = \dfrac{9 \pm \sqrt{(-9)^2 - 4(1)(18)}}{2(1)}$, $p = \dfrac{9 \pm \sqrt{81 - 72}}{2}$,

$p = \dfrac{9 \pm \sqrt{9}}{2}$, $p = \dfrac{9 \pm 3}{2}$; The solutions are $\dfrac{9 + 3}{2} = 6$ and $\dfrac{9 - 3}{2} = 3$.

6. $t^2 - 3t + 2 = 0$; $t = \dfrac{3 \pm \sqrt{(-3)^2 - 4(1)(2)}}{2(1)}$, $t = \dfrac{3 \pm \sqrt{9 - 8}}{2}$, $t = \dfrac{3 \pm \sqrt{1}}{2}$,

$t = \dfrac{3 \pm 1}{2}$; The solutions are $\dfrac{3 + 1}{2} = 2$ and $\dfrac{3 - 1}{2} = 1$.

7. $y = \dfrac{3 \pm \sqrt{(-3)^2 - 4(3)(-1)}}{2(3)}$, $y = \dfrac{3 \pm \sqrt{9 + 12}}{6}$, $y = \dfrac{3 \pm \sqrt{21}}{6}$; The solutions

are $\dfrac{3 + \sqrt{21}}{6}$ and $\dfrac{3 - \sqrt{21}}{6}$.

8. $n = \dfrac{5 \pm \sqrt{(-5)^2 - 4(2)(-12)}}{2(2)}$, $n = \dfrac{5 \pm \sqrt{25 + 96}}{4}$, $n = \dfrac{5 \pm \sqrt{121}}{4}$, $n = \dfrac{5 \pm 11}{4}$;

The solutions are $\dfrac{5 + 11}{4} = 4$ and $\dfrac{5 - 11}{4} = -\dfrac{3}{2}$.

9. $4x^2 - 12x + 9 = 0$; $x = \dfrac{12 \pm \sqrt{(-12)^2 - 4(4)(9)}}{2(4)}$, $x = \dfrac{12 \pm \sqrt{144 - 144}}{8}$,

$x = \dfrac{12 \pm 0}{8}$; The solution is $\dfrac{3}{2}$.

10. $5b^2 - 4b - 33 = 0$; $b = \dfrac{4 \pm \sqrt{(-4)^2 - 4(5)(-33)}}{2(5)}$, $b = \dfrac{4 \pm \sqrt{16 + 660}}{10}$, $b =$

$\dfrac{4 \pm \sqrt{676}}{10}$, $b = \dfrac{4 \pm 26}{10}$; The solutions are $\dfrac{4 + 26}{10} = 3$ and $\dfrac{4 - 26}{10} = -\dfrac{11}{5}$.

11. $x^2 - 2x - 10 = 0$; $x = \dfrac{2 \pm \sqrt{(-2)^2 - 4(1)(-10)}}{2(1)}$, $x = \dfrac{2 \pm \sqrt{4 + 40}}{2}$, $x = \dfrac{2 \pm \sqrt{44}}{2}$,

$x = \dfrac{2 \pm 2\sqrt{11}}{2}$, $x = 1 \pm \sqrt{11}$; The solutions are $1 + \sqrt{11}$ and $1 - \sqrt{11}$.

12. $3x^2 - 8x + 2 = 0$; $x = \dfrac{8 \pm \sqrt{(-8)^2 - 4(3)(2)}}{2(3)}$, $x = \dfrac{8 \pm \sqrt{64 - 24}}{6}$, $x = \dfrac{8 \pm \sqrt{40}}{6}$,

$x = \dfrac{8 \pm 2\sqrt{10}}{6}$, $x = \dfrac{4 \pm \sqrt{10}}{3}$; The solutions are $\dfrac{4 + \sqrt{10}}{3}$ and $\dfrac{4 - \sqrt{10}}{3}$.

13. $x = \dfrac{-7 \pm \sqrt{7^2 - 4(6)(-5)}}{2(6)}$, $x = \dfrac{-7 \pm \sqrt{49 + 120}}{12}$, $x = \dfrac{-7 \pm \sqrt{169}}{12}$, $x = \dfrac{-7 \pm 13}{12}$;

The solutions are $\dfrac{-7 + 13}{12} = \dfrac{1}{2}$ and $\dfrac{-7 - 13}{12} = -\dfrac{5}{3}$.

14. $p = \dfrac{-5 \pm \sqrt{5^2 - 4(2)(3)}}{2(2)}$, $p = \dfrac{-5 \pm \sqrt{25 - 24}}{4}$, $p = \dfrac{-5 \pm \sqrt{1}}{4}$, $p = \dfrac{-5 \pm 1}{4}$;

The solutions are $\dfrac{-5 + 1}{4} = -1$ and $\dfrac{-5 - 1}{4} = -\dfrac{3}{2}$.

15. $y = \dfrac{-3 \pm \sqrt{3^2 - 4(2)(-1)}}{2(2)}$, $y = \dfrac{-3 \pm \sqrt{9 + 8}}{4}$, $y = \dfrac{-3 \pm \sqrt{17}}{4}$; The solutions

are $\dfrac{-3 + \sqrt{17}}{4}$ and $\dfrac{-3 - \sqrt{17}}{4}$.

16. $n = \dfrac{4 + \sqrt{(-4)^2 - 4(3)(-2)}}{2(3)}$, $n = \dfrac{4 \pm \sqrt{16 + 24}}{6}$, $n = \dfrac{4 \pm \sqrt{40}}{6}$, $n = \dfrac{4 \pm 2\sqrt{10}}{6}$,

$n = \dfrac{2 \pm \sqrt{10}}{3}$; The solutions are $\dfrac{2 + \sqrt{10}}{3}$ and $\dfrac{2 - \sqrt{10}}{3}$.

17. $3z^2 - 8z + 4 = 0$; $z = \dfrac{8 \pm \sqrt{8^2 - 4(3)(4)}}{2(3)}$, $z = \dfrac{8 \pm \sqrt{64 - 48}}{6}$, $z = \dfrac{8 \pm \sqrt{16}}{6}$,

$z = \dfrac{8 \pm 4}{6}$; The solutions are $\dfrac{8 + 4}{6} = 2$ and $\dfrac{8 - 4}{6} = \dfrac{2}{3}$.

18. $6y^2 - y - 2 = 0$; $y = \dfrac{1 \pm \sqrt{(-1)^2 - 4(6)(-2)}}{2(6)}$, $y = \dfrac{1 \pm \sqrt{1 + 48}}{12}$, $y = \dfrac{1 \pm \sqrt{49}}{12}$,

$y = \dfrac{1 \pm 7}{12}$; The solutions are $\dfrac{1 + 7}{12} = \dfrac{2}{3}$ and $\dfrac{1 - 7}{12} = -\dfrac{1}{2}$.

19. $m = \dfrac{-6 \pm \sqrt{6^2 - 4(1)(-10)}}{2(1)}$, $m = \dfrac{-6 \pm \sqrt{36 + 40}}{2}$, $m = \dfrac{-6 \pm \sqrt{76}}{2}$, $m = \dfrac{-6 \pm 2\sqrt{19}}{2}$,

$m = -3 \pm \sqrt{19}$; The solutions are $-3 + \sqrt{19}$ and $-3 - \sqrt{19}$.

20. $m = \dfrac{-4 \pm \sqrt{4^2 - 4(1)(-6)}}{2(1)}$, $m = \dfrac{-4 \pm \sqrt{16 + 24}}{2}$, $m = \dfrac{-4 \pm \sqrt{40}}{2}$, $m = \dfrac{-4 \pm 2\sqrt{10}}{2}$,

$m = -2 \pm \sqrt{10}$; The solutions are $-2 + \sqrt{10}$ and $-2 - \sqrt{10}$.

21. $c^2 + 7c + 5 = 0$; $c = \dfrac{-7 \pm \sqrt{7^2 - 4(1)(5)}}{2(1)}$, $c = \dfrac{-7 \pm \sqrt{49 - 20}}{2}$, $c = \dfrac{-7 \pm \sqrt{29}}{2}$;

The solutions are $\dfrac{-7 + \sqrt{29}}{2}$ and $\dfrac{-7 - \sqrt{29}}{2}$.

22. $2p^2 + 4p - 5 = 0$; $p = \dfrac{-4 \pm \sqrt{4^2 - 4(2)(-5)}}{2(2)}$, $p = \dfrac{-4 \pm \sqrt{16 + 40}}{4}$, $p = \dfrac{-4 \pm \sqrt{56}}{4}$,

$p = \dfrac{-4 \pm 2\sqrt{14}}{4}$, $p = \dfrac{-2 \pm \sqrt{14}}{2}$; The solutions are $\dfrac{-2 + \sqrt{14}}{2}$ and $\dfrac{-2 - \sqrt{14}}{2}$.

23. $2b^2 - 2b - 5 = 0$; $b = \dfrac{2 \pm \sqrt{(-2)^2 - 4(2)(-5)}}{2(2)}$, $b = \dfrac{2 \pm \sqrt{4 + 40}}{4}$, $b = \dfrac{2 \pm \sqrt{44}}{4}$,

$b = \dfrac{2 \pm 2\sqrt{11}}{4}$, $b = \dfrac{1 \pm \sqrt{11}}{2}$; The solutions are $\dfrac{1 + \sqrt{11}}{2}$ and $\dfrac{1 - \sqrt{11}}{2}$.

24. $5n^2 - 8n - 2 = 0$; $n = \dfrac{8 \pm \sqrt{(-8)^2 - 4(5)(-2)}}{2(5)}$, $n = \dfrac{8 \pm \sqrt{64 + 40}}{10}$, $n = \dfrac{8 \pm \sqrt{104}}{10}$,

$n = \dfrac{8 \pm 2\sqrt{26}}{10}$, $n = \dfrac{4 \pm \sqrt{26}}{5}$; The solutions are $\dfrac{4 + \sqrt{26}}{5}$ and $\dfrac{4 - \sqrt{26}}{5}$.

B 25. $2t^2 + 8t + 3t + 12 = 1$, $2t^2 + 11t + 12 = 1$, $2t^2 + 11t + 11 = 0$;

$t = \dfrac{-11 \pm \sqrt{11^2 - 4(2)(11)}}{2(2)}$, $t = \dfrac{-11 \pm \sqrt{121 - 88}}{4}$, $t = \dfrac{-11 \pm \sqrt{33}}{4}$; The

solutions are $\dfrac{-11 + \sqrt{33}}{4}$ and $\dfrac{-11 - \sqrt{33}}{4}$.

26. $2x^2 + 2x - 5x - 5 = 2$, $2x^2 - 3x - 5 = 2$, $2x^2 - 3x - 7 = 0$;

$x = \dfrac{3 \pm \sqrt{(-3)^2 - 4(2)(-7)}}{2(2)}$, $x = \dfrac{3 \pm \sqrt{9 + 56}}{4}$, $x = \dfrac{3 \pm \sqrt{65}}{4}$; The solutions are

$\dfrac{3 + \sqrt{65}}{4}$ and $\dfrac{3 - \sqrt{65}}{4}$.

27. $\dfrac{2y^2}{3} - y + \dfrac{1}{6} = 0$, $4y^2 - 6y + 1 = 0$; $y = \dfrac{6 \pm \sqrt{(-6)^2 - 4(4)(1)}}{2(4)}$, $y = \dfrac{6 \pm \sqrt{36 - 16}}{8}$,

$y = \dfrac{6 \pm \sqrt{20}}{8}$, $y = \dfrac{6 \pm 2\sqrt{5}}{8}$, $y = \dfrac{3 \pm \sqrt{5}}{4}$; The solutions are $\dfrac{3 + \sqrt{5}}{4}$ and $\dfrac{3 - \sqrt{5}}{4}$.

28. $\dfrac{m^2}{3} - m + \dfrac{1}{2} = 0$, $2m^2 - 6m + 3 = 0$; $m = \dfrac{6 \pm \sqrt{(-6)^2 - 4(2)(3)}}{2(2)}$,

$m = \dfrac{6 \pm \sqrt{36 - 24}}{4}$, $m = \dfrac{6 \pm \sqrt{12}}{4}$, $m = \dfrac{6 \pm 2\sqrt{3}}{4}$, $m = \dfrac{3 \pm \sqrt{3}}{2}$; The solutions are

$\dfrac{3 + \sqrt{3}}{2}$ and $\dfrac{3 - \sqrt{3}}{2}$.

29. $\dfrac{x^2}{3} - \dfrac{x}{2} - \dfrac{3}{2} = 0$, $2x^2 - 3x - 9 = 0$; $x = \dfrac{3 \pm \sqrt{(-3)^2 - 4(2)(-9)}}{2(2)}$, $x = \dfrac{3 \pm \sqrt{9 + 72}}{4}$,

$x = \dfrac{3 \pm \sqrt{81}}{4}$, $x = \dfrac{3 \pm 9}{4}$; The solutions are $\dfrac{3 + 9}{4} = 3$ and $\dfrac{3 - 9}{4} = -\dfrac{3}{2}$.

30. $\dfrac{c^2}{3} - \dfrac{5c}{6} - \dfrac{1}{2} = 0$, $2c^2 - 5c - 3 = 0$; $c = \dfrac{5 \pm \sqrt{(-5)^2 - 4(2)(-3)}}{2(2)}$, $c = \dfrac{5 \pm \sqrt{25 + 24}}{4}$,

$c = \dfrac{5 \pm \sqrt{49}}{4}$, $c = \dfrac{5 \pm 7}{4}$; The solutions are $\dfrac{5 + 7}{4} = 3$ and $\dfrac{5 - 7}{4} = -\dfrac{1}{2}$.

31. $(x - 1)(x - 2) - 2x(x + 1) = 0$, $x^2 - 2x - x + 2 - 2x^2 - 2x = 0$,

$-x^2 - 5x + 2 = 0$, $x^2 + 5x - 2 = 0$; $x = \dfrac{-5 \pm \sqrt{5^2 - 4(1)(-2)}}{2(1)}$, $x = \dfrac{-5 \pm \sqrt{25 + 8}}{2}$,

$x = \dfrac{-5 \pm \sqrt{33}}{2}$; The solutions are $\dfrac{-5 + \sqrt{33}}{2}$ and $\dfrac{-5 - \sqrt{33}}{2}$.

32. $(x + 2)(x - 4) - 3x = 0$, $x^2 - 4x + 2x - 8 - 3x = 0$, $x^2 - 5x - 8 = 0$;

$x = \dfrac{5 \pm \sqrt{(-5)^2 - 4(1)(-8)}}{2(1)}$, $x = \dfrac{5 \pm \sqrt{25 + 32}}{2}$, $x = \dfrac{5 \pm \sqrt{57}}{2}$; The solutions

are $\dfrac{5 + \sqrt{57}}{2}$ and $\dfrac{5 - \sqrt{57}}{2}$.

33. $1 + \dfrac{2}{c^2} - \dfrac{7}{2c} = 0$, $2c^2 + 4 - 7c = 0$, $2c^2 - 7c + 4 = 0$; $c = \dfrac{7 \pm \sqrt{(-7)^2 - 4(2)(4)}}{2(2)}$,

$c = \dfrac{7 \pm \sqrt{49 - 32}}{4}$, $c = \dfrac{7 \pm \sqrt{17}}{4}$; The solutions are $\dfrac{7 + \sqrt{17}}{4}$ and $\dfrac{7 - \sqrt{17}}{4}$.

34. $\dfrac{1}{y} - \dfrac{2}{y^2} + 6 = 0$, $y - 2 + 6y^2 = 0$, $6y^2 + y - 2 = 0$; $y = \dfrac{-1 \pm \sqrt{1^2 - 4(6)(-2)}}{2(6)}$,

$y = \dfrac{-1 \pm \sqrt{1 + 48}}{12}$, $y = \dfrac{-1 \pm \sqrt{49}}{12}$, $y = \dfrac{-1 \pm 7}{12}$; The solutions are $\dfrac{-1 + 7}{12} = \dfrac{1}{2}$

and $\dfrac{-1 - 7}{12} = -\dfrac{2}{3}$.

35. $\dfrac{x^2}{x - 2} - \dfrac{2x}{x - 2} + 3 = 0$, $x^2 - 2x + 3(x - 2) = 0$, $x^2 + x - 6 = 0$;

$x = \dfrac{-1 \pm \sqrt{1^2 - 4(1)(-6)}}{2(1)}$, $x = \dfrac{-1 \pm \sqrt{1 + 24}}{2}$, $x = \dfrac{-1 \pm \sqrt{25}}{2}$, $x = \dfrac{-1 \pm 5}{2}$;

The solutions are $\dfrac{-1 + 5}{2} = 2$ and $\dfrac{-1 - 5}{2} = -3$.

36. $\dfrac{3}{m-2} - \dfrac{1}{m-1} - 2 = 0$, $3(m-1) - (m-2) - 2(m-2)(m-1) = 0$,

$3m - 3 - m + 2 - 2(m^2 - m - 2m + 2) = 0$, $-3 + 2m + 2 - 2m^2 +$

$2m + 4m - 4 = 0$, $-2m^2 + 8m - 5 = 0$, $2m^2 - 8m + 5 = 0$;

$m = \dfrac{8 \pm \sqrt{(-8)^2 - 4(2)(5)}}{2(2)}$, $m = \dfrac{8 \pm \sqrt{64 - 40}}{4}$, $m = \dfrac{8 \pm \sqrt{24}}{4}$, $m = \dfrac{8 \pm 2\sqrt{6}}{4}$,

$m = \dfrac{4 \pm \sqrt{6}}{2}$; The solutions are $\dfrac{4 + \sqrt{6}}{2}$ and $\dfrac{4 - \sqrt{6}}{2}$.

C 37. $-x^2 + 4x + k^2 = 0$; $x = \dfrac{-4 \pm \sqrt{4^2 - 4(-1)(k^2)}}{2(-1)}$, $x = \dfrac{-4 \pm \sqrt{16 + 4k^2}}{-2}$,

$x = \dfrac{-4 \pm \sqrt{4(4 + k^2)}}{-2}$, $x = \dfrac{-4 \pm 2\sqrt{4 + k^2}}{-2}$, $x = 2 \pm \sqrt{4 + k^2}$; The solutions

are $2 + \sqrt{4 + k^2}$ and $2 - \sqrt{4 + k^2}$.

38. $x = \dfrac{-2k \pm \sqrt{(2k)^2 - 4(1)(k^2)}}{2}$, $x = \dfrac{-2k \pm \sqrt{4k^2 - 4k^2}}{2}$, $x = \dfrac{-2k}{2}$, $x = -k$; The

solution is $-k$.

39. $\dfrac{x^2}{a} + \dfrac{x}{b} + \dfrac{1}{c} = 0$, $bcx^2 + acx + ab = 0$; $x = \dfrac{-ac \pm \sqrt{(ac)^2 - 4(bc)(ab)}}{2(bc)}$,

$x = \dfrac{-ac \pm \sqrt{a^2c^2 - 4ab^2c}}{2bc}$; The solutions are $\dfrac{-ac + \sqrt{a^2c^2 - 4ab^2c}}{2bc}$ and

$\dfrac{-ac - \sqrt{a^2c^2 - 4ab^2c}}{2bc}$.

40. $\dfrac{x^2}{a} - \dfrac{5x}{b} - \dfrac{1}{c} = 0$, $bcx^2 - 5acx - ab = 0$; $x = \dfrac{5ac \pm \sqrt{(-5ac)^2 - 4(bc)(-ab)}}{2(bc)}$,

$x = \dfrac{5ac \pm \sqrt{25a^2c^2 + 4ab^2c}}{2bc}$; The solutions are $\dfrac{5ac + \sqrt{25a^2c^2 + 4ab^2c}}{2bc}$ and

$\dfrac{5ac - \sqrt{25a^2c^2 + 4ab^2c}}{2bc}$.

41. $(2x + a)^2 - (x + a) - 6 = 0$, $4x^2 + 4xa + a^2 - x - a - 6 = 0$,

$4x^2 + 4xa - x + a^2 - a - 6 = 0$, $4x^2 + (4a - 1)x + (a^2 - a - 6) = 0$;

$x = \dfrac{-4a + 1 \pm \sqrt{(4a - 1)^2 - 4(4)(a^2 - a - 6)}}{2(4)}$,

$x = \dfrac{-4a + 1 \pm \sqrt{16a^2 - 8a + 1 - 16a^2 + 16a + 96}}{8}$, $x = \dfrac{-4a + 1 \pm \sqrt{8a + 97}}{8}$;

The solutions are $\dfrac{-4a + 1 + \sqrt{8a + 97}}{8}$ and $\dfrac{-4a + 1 - \sqrt{8a + 97}}{8}$.

42. $(x - b)^2 - (x - b) - 4 = 0$, $x^2 - 2xb + b^2 - x + b - 4 = 0$,

$x^2 - 2xb - x + b^2 + b - 4 = 0$, $x^2 + (-2b - 1)x + (b^2 + b - 4) = 0$,

$$x = \frac{2b + 1 \pm \sqrt{(-2b - 1)^2 - 4(1)(b^2 + b - 4)}}{2(1)},$$

$$x = \frac{2b + 1 \pm \sqrt{4b^2 + 4b + 1 - 4b^2 - 4b + 16}}{2}, \; x = \frac{2b + 1 \pm \sqrt{17}}{2}; \text{ The}$$

solutions are $\dfrac{2b + 1 + \sqrt{17}}{2}$ and $\dfrac{2b + 1 - \sqrt{17}}{2}$.

43. Let w = the width of the rectangle, then $w - 1$ = the length of the
rectangle. $12 = w(w - 1)$, $12 = w^2 - w$, $w^2 - w - 12 = 0$

$$w = \frac{1 \pm \sqrt{(-1)^2 - 4(1)(-12)}}{2(1)}, \; w = \frac{1 \pm \sqrt{1 + 48}}{2}, \; w = \frac{1 \pm \sqrt{49}}{2}, \; w = \frac{1 \pm 7}{2};$$

Since the width is a positive number, the solution is $\dfrac{1 + 7}{2} = 4$; The

width = 4 ft and the length = $4 - 1 = 3$ ft.

44. Let x = the first consecutive integer, then $x + 1$ = the second consecutive
integer. The equation is: $3x^2 = 5(x + 1) + 7$, solve $3x^2 = 5x + 5 + 7$,
$3x^2 = 5x + 12$, $3x^2 - 5x - 12 = 0$

$$x = \frac{5 \pm \sqrt{(-5)^2 - 4(3)(-12)}}{2(3)}, \; x = \frac{5 \pm \sqrt{25 + 144}}{6}, \; x = \frac{5 \pm \sqrt{169}}{6},$$

$x = \dfrac{5 \pm 13}{6}$; Since the square of a number is positive, the solution is $\dfrac{5 + 13}{6} = 3$;

Then $x + 1 = 3 + 1 = 4$. The two consecutive integers are 3, 4.

45. For $x = \dfrac{-b + \sqrt{b^2 - 4ac}}{2a}$: $a\left(\dfrac{-b + \sqrt{b^2 - 4ac}}{2a}\right)^2 + b\left(\dfrac{-b + \sqrt{b^2 - 4ac}}{2a}\right) +$

$$c = a\left(\frac{b^2 - 2b\sqrt{b^2 - 4ac} + b^2 - 4ac}{4a^2}\right) + \left(\frac{-b^2 + b\sqrt{b^2 - 4ac}}{2a}\right) + c =$$

$$\frac{2b^2 - 2b\sqrt{b^2 - 4ac} - 4ac}{4a} + \frac{-b^2 + b\sqrt{b^2 - 4ac}}{2a} + c = \frac{b^2 - b\sqrt{b^2 - 4ac} - 2ac}{2a} +$$

$$\frac{-b^2 + b\sqrt{b^2 - 4ac}}{2a} + \frac{2ac}{2a} = \frac{0}{2a} = 0; \text{ For } x = \frac{-b - \sqrt{b^2 - 4ac}}{2a}:$$

$$a\left(\frac{-b - \sqrt{b^2 - 4ac}}{2a}\right)^2 + b\left(\frac{-b - \sqrt{b^2 - 4ac}}{2a}\right) + c =$$

$$a\left(\frac{b^2 + 2b\sqrt{b^2 - 4ac} + b^2 - 4ac}{4a^2}\right) + \left(\frac{-b^2 - b\sqrt{b^2 - 4ac}}{2a}\right) + c =$$

$$\frac{2b^2 + 2b\sqrt{b^2 - 4ac} - 4ac}{4a} + \frac{-b^2 - b\sqrt{b^2 - 4ac}}{2a} + c = \frac{b^2 + b\sqrt{b^2 - 4ac} - 2ac}{2a} +$$

$$\frac{-b^2 - b\sqrt{b^2 - 4ac}}{2a} + \frac{2ac}{2a} = \frac{0}{2a} = 0$$

46. $(x + 5)(x - 8) = 0$, $x^2 - 3x - 40 = 0$

47. $(x + 6)(x + 2) = 0$, $x^2 + 8x + 12 = 0$

48. $(3x - 1)(5x - 2) = 0$, $15x^2 - 11x + 2 = 0$

49. $(2x + 1)(x - 5) = 0$, $2x^2 - 9x - 5 = 0$

50. $(x + \sqrt{3})(x - \sqrt{3}) = 0$, $x^2 - 3 = 0$

51. $[x - (\sqrt{2} + 1)] \cdot [x - (\sqrt{2} - 1)] = x^2 - x(\sqrt{2} - 1) - x(\sqrt{2} + 1) + (\sqrt{2} + 1)(\sqrt{2} - 1) = x^2 - \sqrt{2}x + x - \sqrt{2}x - x + 2 - 1 = x^2 - 2\sqrt{2}x + 1 = 0$

page 555   Class Exercises

1. Square-Root Property; The equation is in the form $x^2 = k$.

2. Square-Root Property; The equation is in the form $(ax + b)^2 = k$.

3. Factoring; The equation is in the form $ax^2 + bx + c$.

4. Quadratic Formula; The equation is in the form $ax^2 + bx + c = 0$.

5. Completing the Square; The equation is in the form $x^2 + bx + c = 0$, where b is even.

6. Quadratic Formula; The equation is in the form $ax^2 + bx + c = 0$.

7. Quadratic Formula; The equation is in the form $ax^2 + bx + c = 0$.

8. Quadratic Formula; The equation is in the form $ax^2 + bx + c = 0$.

9. Factoring; The constant term is zero.

10. Factoring; The equation is in the form $ax^2 + bx + c$.

11. Square-Root Property; The equation is in the form $x^2 = k$.

12. Square-Root Property; The equation is in the form $(ax + b)^2 = k$.

13. $x = \dfrac{250 \pm \sqrt{250^2 - 4(175)(-225)}}{2(175)}$, $x = -0.626$, $x = 2.054$

14. $x^2 + \dfrac{b}{a}x$ can always be transformed into a perfect square by adding the number, $\left(\dfrac{1}{2} \cdot \dfrac{b}{a}\right)^2$ and the quadratic formula is derived by completing the square.

15. Not all quadratics are in the form $x^2 = k$ or $(ax + b)^2 = k$.

A **1.** $x^2 - x - 2 = 0$, $(x - 2)(x + 1)$, $x - 2 = 0$, $x = 2$, $x + 1 = 0$, $x = -1$

2. $y^2 - 6y + 9 = 0$, $(y - 3)(y - 3)$, $y - 3 = 0$; $y = 3$

3. $6m^2 = 72$, $m^2 = \dfrac{72}{6}$, $m^2 = 12$, $m = \sqrt{12}$; $m = \pm 2\sqrt{3}$

4. $4x^2 = 80$, $x^2 = \dfrac{80}{4}$, $x^2 = 20$, $x = \sqrt{20}$; $x = \pm 2\sqrt{5}$

5. $z^2 - 4z + 3 = 0$, $(z - 3)(z - 1) = 0$, $z - 3 = 0$, $z = 3$, $z - 1 = 0$, $z = 1$

6. $x^2 + 8x - 20 = 0$, $(x + 10)(x - 2) = 0$, $x + 10 = 0$, $x = -10$, $x - 2 = 0$, $x = 2$

7. $x = \dfrac{-3 \pm \sqrt{(3)^2 - 4(4)(-1)}}{2(4)}$, $x = \dfrac{-3 \pm \sqrt{9 + 16}}{8}$, $x = \dfrac{-3 \pm \sqrt{25}}{8}$, $x = \dfrac{-3 \pm 5}{8}$;

The solutions are $\dfrac{-3 + 5}{8} = \dfrac{1}{4}$ and $\dfrac{-3 - 5}{8} = -1$.

8. $(2x + 3)(x + 1) = 0$, $2x + 3 = 0$, $2x = -3$, $x = -\dfrac{3}{2}$; $x + 1 = 0$, $x = -1$

9. $t(3t - 2) = 0$, $t = 0$, $3t - 2 = 0$, $3t = 2$; $t = \dfrac{2}{3}$

10. $4x^2 = 20$, $x^2 = \dfrac{20}{4}$, $x^2 = 5$; $x = \pm \sqrt{5}$

11. $x^2 + \dfrac{1}{6}x - \dfrac{1}{6} = 0$, $6x^2 + x - 1 = 0$, $x = \dfrac{-1 \pm \sqrt{(1)^2 - 4(6)(-1)}}{2(6)}$,

$x = \dfrac{-1 \pm \sqrt{1 + 24}}{12}$, $x = \dfrac{-1 \pm \sqrt{25}}{12}$, $x = \dfrac{-1 \pm 5}{12}$; The solutions are $\dfrac{-1 + 5}{12} = \dfrac{1}{3}$ and

$\dfrac{-1 - 5}{12} = -\dfrac{1}{2}$.

12. $m^2 - \dfrac{7}{6}m - \dfrac{1}{2} = 0$, $6m^2 - 7m - 3 = 0$, $(3m + 1)(2m - 3) = 0$, $3m + 1 = 0$,

$3m = -1$, $m = -\dfrac{1}{3}$; $2m - 3 = 0$, $2m = 3$, $m = \dfrac{3}{2}$

13. $n^2 - 12n - 45 = 0$; $n = \dfrac{12 \pm \sqrt{(-12)^2 - 4(1)(-45)}}{2(1)}$, $n = \dfrac{12 \pm \sqrt{144 + 180}}{2}$,

$n = \dfrac{12 \pm \sqrt{324}}{2}$, $n = \dfrac{12 \pm 18}{2}$; The solutions are $\dfrac{12 + 18}{2} = 15$ and $\dfrac{12 - 18}{2} = -3$.

14. $t^2 + 14t - 15 = 0$, $(t + 15)(t - 1) = 0$, $t + 15 = 0$, $t = -15$, $t - 1 = 0$, $t = 1$

15. $2x^2 - 9x + 8 = 0$, $x = \dfrac{9 \pm \sqrt{(-9)^2 - 4(2)(8)}}{2(2)}$, $x = \dfrac{9 \pm \sqrt{81 - 64}}{4}$,

$x = \dfrac{9 \pm \sqrt{17}}{4}$; The solutions are $\dfrac{9 + \sqrt{17}}{4}$ and $\dfrac{9 - \sqrt{17}}{4}$.

16. $2x^2 + 7x - 9 = 0$, $(2x + 9)(x - 1) = 0$, $2x + 9 = 0$, $2x = -9$, $x = -\dfrac{9}{2}$;

$x - 1 = 0$, $x = 1$

17. $3x - 2 = \pm\sqrt{10}$, $3x = \pm\sqrt{10} + 2$, $x = \dfrac{\pm\sqrt{10} + 2}{3}$

18. $4t - 1 = \pm\sqrt{15}$, $4t = \pm\sqrt{15} + 1$, $t = \dfrac{\pm\sqrt{15} + 1}{4}$

19. $4x^2 - 4x - 44 = 0$, $x^2 - x - 11 = 0$, $x = \dfrac{1 \pm \sqrt{(-1)^2 - 4(1)(-11)}}{2(1)}$,

$x = \dfrac{1 \pm \sqrt{1 + 44}}{2}$, $x = \dfrac{1 \pm \sqrt{45}}{2}$, $x = \dfrac{1 \pm 3\sqrt{5}}{2}$; The solutions are $\dfrac{1 + 3\sqrt{5}}{2}$ and $\dfrac{1 - 3\sqrt{5}}{2}$.

20. $9y^2 + 42y + 17 = 0$; $y = \dfrac{-42 \pm \sqrt{42^2 - 4(9)(17)}}{2(9)}$, $y = \dfrac{-42 \pm \sqrt{1764 - 612}}{18}$,

$y = \dfrac{-42 \pm \sqrt{1152}}{18}$, $y = \dfrac{-42 \pm \sqrt{36 \cdot 32}}{18}$, $y = \dfrac{-42 \pm 24\sqrt{2}}{18}$, $y = \dfrac{-7 \pm 4\sqrt{2}}{3}$;

The solutions are $\dfrac{-7 + 4\sqrt{2}}{3}$ and $\dfrac{-7 - 4\sqrt{2}}{3}$.

21. $2x^2 - 6x + 3 = 0$; $x = \dfrac{6 \pm \sqrt{(-6)^2 - 4(2)(3)}}{2(2)}$, $x = \dfrac{6 \pm \sqrt{36 - 24}}{4}$,

$x = \dfrac{6 \pm \sqrt{12}}{4}$, $x = \dfrac{6 \pm 2\sqrt{3}}{4}$, $x = \dfrac{3 \pm \sqrt{3}}{2}$; The solutions are $\dfrac{3 + \sqrt{3}}{2}$ and $\dfrac{3 - \sqrt{3}}{2}$.

22. $3m^2 + 2m - 3 = 0$; $m = \dfrac{-2 \pm \sqrt{2^2 - 4(3)(-3)}}{2(3)}$, $m = \dfrac{-2 \pm \sqrt{4 + 36}}{6}$,

$m = \dfrac{-2 \pm \sqrt{40}}{6}$, $m = \dfrac{-2 \pm 2\sqrt{10}}{6}$, $m = \dfrac{-1 \pm \sqrt{10}}{3}$; The solutions are

$\dfrac{-1 + \sqrt{10}}{3}$ and $\dfrac{-1 - \sqrt{10}}{3}$.

23. $\dfrac{2}{x^2} + \dfrac{3}{x} - 5 = 0$, $2 + 3x - 5x^2 = 0$, $5x^2 - 3x - 2 = 0$, $(5x + 2)(x - 1) = 0$,

$5x + 2 = 0$, $5x = -2$, $x = -\dfrac{2}{5}$; $x - 1 = 0$, $x = 1$

24. $\dfrac{4}{m^2} - \dfrac{2}{m} - 3 = 0$, $4 - 2m - 3m^2 = 0$, $3m^2 + 2m - 4 = 0$, $m = \dfrac{-2 \pm \sqrt{(2)^2 - 4(3)(-4)}}{2(3)}$,

$m = \dfrac{-2 \pm \sqrt{4 + 48}}{6}$, $m = \dfrac{-2 \pm \sqrt{52}}{6}$, $m = \dfrac{-2 \pm 2\sqrt{52}}{6}$, $m = \dfrac{-1 \pm \sqrt{13}}{3}$; The solutions

are $\dfrac{-1 + \sqrt{13}}{3}$ and $\dfrac{-1 - \sqrt{13}}{3}$.

25. $n^2 - 8n + 16 - 3n + 12 - 10 = 0$, $n^2 - 11n + 18 = 0$, $(n - 9)(n - 2) = 0$,
$n - 9 = 0$, $n = 9$; $n - 2 = 0$, $n = 2$

26. $(2y + 5)^2 + 7(2y + 5) + 6 = 0$, $4y^2 + 20y + 25 + 14y + 35 + 6 = 0$,
$4y^2 + 34y + 66 = 0$, $(2y + 6)(2y + 11) = 0$, $2y + 6 = 0$, $2y = -6$,

$y = -\dfrac{6}{2} = -3$; $2y + 11 = 0$, $2y = -11$, $y = -\dfrac{11}{2}$

27. $(3x - 8)^2 - (2x - 5)^2 = 0$, $9x^2 - 48x + 64 - 4x^2 + 20x - 25 = 0$, $5x^2 - 28x +$

$39 = 0$, $(5x - 13)(x - 3) = 0$, $5x - 13 = 0$, $5x = 13$, $x = \dfrac{13}{5}$; $x - 3 = 0$, $x = 3$

28. $(4x + 6)^2 - (2x + 4)^2 = 0$, $16x^2 + 48x + 36 - 4x^2 - 16x - 16 = 0$,
$12x^2 + 32x + 20 = 0$, $3x^2 + 8x + 5 = 0$, $(3x + 5)(x + 1) = 0$, $3x + 5 = 0$,

$3x = -5$, $x = -\dfrac{5}{3}$; $x + 1 = 0$, $x = -1$

C 29. $2b^2x^2 - 3bx + 1 = 0$, $(2bx - 1)(bx - 1) = 0$, $2bx - 1 = 0$, $2bx = 1$,

$x = \dfrac{1}{2b}$; $bx - 1 = 0$, $bx = 1$, $x = \dfrac{1}{b}$

30. $x^2 - cx - 2c^2 = 0$, $(x + c)(x - 2c) = 0$, $x + c = 0$, $x = -c$; $x - 2c = 0$, $x = 2c$

31. $(x - b)^2 + 5(x - b) + 4 = 0$, $[(x - b) + 4][(x - b) + 1] = 0$, $x - b + 4 = 0$,
$x = b - 4$; $x - b + 1 = 0$, $x = b - 1$

32. $x^2 - 2ax + a^2 + x^2 + 2ax + a^2 = 5a$, $2x^2 + 2a^2 = 5a$, $2x^2 = 5a - 2a^2$,

$x^2 = \dfrac{5a - 2a^2}{2}$, $x = \dfrac{\pm\sqrt{5a - 2a^2}}{\sqrt{2}} \cdot \dfrac{\sqrt{2}}{\sqrt{2}} = \dfrac{\pm\sqrt{10a - 4a^2}}{2}$

33. $\dfrac{1}{4}x - \dfrac{1}{16}b = \pm\dfrac{1}{6}$, $\dfrac{1}{4}x = \pm\dfrac{1}{6} + \dfrac{1}{16}b$, $x = \pm\dfrac{4}{6} + \dfrac{4}{16}b$, $x = \pm\dfrac{2}{3} + \dfrac{1}{4}b$, $x = \dfrac{3b \pm 8}{12}$

34. $\dfrac{1}{3}x + \dfrac{1}{4}a = \pm\dfrac{1}{3}$, $\dfrac{1}{3}x = \pm\dfrac{1}{3} - \dfrac{1}{4}a$, $x = \pm 1 - \dfrac{3}{4}a$, $x = \dfrac{-3a \pm 4}{4}$

35. $s = \dfrac{1}{2}at^2$, $64 = \dfrac{1}{2}(32)t^2$, $64 = 16t^2$, $4 = t^2$, $t = \sqrt{4}$, $t = 2$; 2s

36. $s = 96t - 16t^2$, $96 = 96t - 16t^2$, $16t^2 - 96t + 96 = 0$, $t^2 - 6t + 6 = 0$;
$t = \dfrac{6 \pm \sqrt{(-6)^2 - 4(1)(6)}}{2(1)}$, $t = \dfrac{6 \pm \sqrt{36 - 24}}{2}$, $t = \dfrac{6 \pm \sqrt{12}}{2}$, $t = \dfrac{6 \pm 2\sqrt{3}}{2}$,
$t = 3 \pm \sqrt{3}$; The solutions are $3 + \sqrt{3} = 4.73$ s and $3 - \sqrt{3} = 1.27$ s.

37. Let w = the width, then $2w$ = the length of the rectangular floor. $32 = 2w(w)$, $32 = 2w^2$, $w^2 = 16$, $w = 4$, then $2w = 8$; length = 8 ft and width = 4 ft

38. Let x = the first even integer, then $x + 2$ = the second even integer. The equation is: $x(x + 2) = 224$, solve $x^2 + 2x = 224$, $x^2 + 2x - 224 = 0$, $(x + 16)(x - 14) = 0$, $x + 16 = 0$, $x = -16$; $x - 14 = 0$, $x = 14$. Then $x + 2 = -16 + 2 = -14$ or $x + 2 = 14 + 2 = 16$. The two consecutive even integers are -16, -14 or 16, 14.

39. Let x = the integer. $x^2 = 18x - 81$, $x^2 - 18x + 81 = 0$, $(x - 9)(x - 9) = 0$, $x - 9 = 0$, $x = 9$; The integer is 9.

40. Let w = the width, then $3w + 4$ = the length of a rectangle.

$24 = (3w + 4)(w)$, $24 = 3w^2 + 4w$, $3w^2 + 4w - 24 = 0$, $w = \dfrac{-4 \pm \sqrt{4^2 - 4(3)(-24)}}{2(3)}$,

$w = \dfrac{-4 \pm \sqrt{16 + 288}}{6}$, $w = \dfrac{-4 \pm \sqrt{304}}{6}$, $w = \dfrac{-4 \pm 4\sqrt{19}}{6}$, $w = \dfrac{-2 \pm 2\sqrt{19}}{3}$;

Since the width is a positive number, the width is $\dfrac{-2 + 2\sqrt{19}}{3} = 2.24$ and

$3w + 4 = 3(2.24) + 4 = 10.72$ m. The length = 10.72 m and the width = 2.24 m.

41. Let x = the number. $3x^2 = 2x$, $3x^2 - 2x = 0$, $x(3x - 2) = 0$, $x = 0$,

$3x - 2 = 0$, $3x = 2$, $x = \dfrac{2}{3}$; The number is $\dfrac{2}{3}$ or 0.

page 556   Test Yourself

1. $5x^2 = 40$, $x^2 = 8$, $x = \pm\sqrt{8}$, $x = \pm 2\sqrt{2}$

2. $6(3z - 2)^2 = 24$, $(3z - 2)^2 = 4$, $3z - 2 = \pm\sqrt{4}$, $3z - 2 = \pm 2$, $3z = \pm 2 + 2$,

$z = \dfrac{\pm 2 + 2}{3}$; $z = \dfrac{2 + 2}{3} = \dfrac{4}{3}$; $z = \dfrac{-2 + 2}{3} = 0$

3. $m^2 - 2m - 3 = 0$, $m^2 - 2m = 3$, $\dfrac{1}{2}(-2) = -1$, $(-1)^2 = 1$, $m^2 - 2m + 1 = 3 + 1$, $(m - 1)(m - 1) = 4$, $(m - 1)^2 = 4$, $m - 1 = \pm\sqrt{4}$, $m = \pm 2 + 1$, $m = 2 + 1 = 3$; $m = -2 + 1 = -1$

4. $2x^2 - 4x = -1$, $x^2 - 2x = -\dfrac{1}{2}$, $\dfrac{1}{2}(-2) = -1$, $(-1)^2 = 1$, $x^2 - 2x + 1 = -\dfrac{1}{2} + 1$,

$x^2 - 2x + 1 = \dfrac{1}{2}$, $(x - 1)(x - 1) = \dfrac{1}{2}$, $(x - 1)^2 = \dfrac{1}{2}$, $x - 1 = \pm\sqrt{\dfrac{1}{2}}$,

$x - 1 = \pm\dfrac{1}{\sqrt{2}}$, $x = \pm\dfrac{1}{\sqrt{2}} + 1$, $x = \dfrac{\pm\sqrt{2}}{2} + 1$, $x = \dfrac{\pm\sqrt{2} + 2}{2}$

5. $x^2 - 2x - 5 = 0$; $x = \dfrac{2 \pm \sqrt{(-2)^2 - 4(1)(-5)}}{2(1)}$, $x = \dfrac{2 \pm \sqrt{4 + 20}}{2}$, $x = \dfrac{2 \pm \sqrt{24}}{2}$,

$x = \dfrac{2 \pm 2\sqrt{6}}{2}$, $x = 1 \pm \sqrt{6}$; The solutions are $1 + \sqrt{6}$ and $1 - \sqrt{6}$.

6. $2n^2 - 8n + 5 = 0$; $n = \dfrac{8 \pm \sqrt{(-8)^2 - 4(2)(5)}}{2(2)}$, $n = \dfrac{8 \pm \sqrt{64 - 40}}{4}$, $n = \dfrac{8 \pm \sqrt{24}}{4}$,

$n = \dfrac{8 \pm 2\sqrt{6}}{4}$, $n = \dfrac{4 \pm \sqrt{6}}{2}$; The solutions are $\dfrac{4 + \sqrt{6}}{2}$ and $\dfrac{4 - \sqrt{6}}{2}$.

7. $4t^2 - 9 = 0$, $4t^2 = 9$, $t^2 = \dfrac{9}{4}$, $t = \pm \sqrt{\dfrac{9}{4}}$, $t = \pm \dfrac{3}{2}$

8. $y^2 - 2y = 3$, $y^2 - 2y - 3 = 0$, $(y - 3)(y + 1) = 0$, $y - 3 = 0$, $y = 3$; $y + 1 = 0$, $y = -1$

9. $3x^2 + 5x - 2 = 0$, $(3x - 1)(x + 2)$, $3x - 1 = 0$, $3x = 1$, $x = \dfrac{1}{3}$; $x + 2 = 0$, $x = -2$

10. $m^2 + 4m - 9 = 0$; $m = \dfrac{-4 \pm \sqrt{4^2 - 4(1)(-9)}}{2(1)}$, $m = \dfrac{-4 \pm \sqrt{16 + 36}}{2}$,

$m = \dfrac{-4 \pm \sqrt{52}}{2}$, $m = \dfrac{-4 \pm 2\sqrt{13}}{2}$, $m = -2 \pm \sqrt{13}$; The solutions are

$-2 + \sqrt{13}$ and $-2 - \sqrt{13}$.

page 557   Capsule Review

1. $f(x) = 2x + 1$

x	-2	-1	0	1	2
$f(x)$	-3	-1	1	3	5

$f(-2) = 2(-2) + 1$, $f(-2) = -4 + 1$, $f(-2) = -3$;
$f(-1) = 2(-1) + 1$, $f(-1) = -2 + 1$, $f(-1) = -1$;
$f(0) = 2(0) + 1$, $f(0) = 0 + 1$, $f(0) = 1$;
$f(1) = 2(1) + 1$, $f(1) = 2 + 1$, $f(1) = 3$;
$f(2) = 2(2) + 1$, $f(2) = 4 + 1$, $f(2) = 5$;
$m = \dfrac{-1 - (-3)}{-1 - (-2)}$, $m = \dfrac{-1 + 3}{-1 + 2}$, $m = \dfrac{2}{1}$, $m = 2$; y-intercept $= 1$

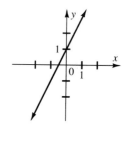

2. $g(x) = -x + 1$

x	-4	-2	0	2	4
$g(x)$	5	3	1	-1	-3

$g(-4) = -(-4) + 1$, $g(-4) = 4 + 1$, $g(-4) = 5$;
$g(-2) = -(-2) + 1$, $g(-2) = 2 + 1$, $g(-2) = 3$;
$g(0) = -(0) + 1$, $g(0) = 0 + 1$, $g(0) = 1$;
$g(2) = -(2) + 1$, $g(2) = -2 + 1$, $g(2) = -1$;
$g(4) = -(4) + 1$, $g(4) = -4 + 1$, $g(4) = -3$;
$m = \dfrac{3 - 5}{-2 - (-4)}$, $m = \dfrac{3 - 5}{-2 + 4}$, $m = \dfrac{-2}{2}$, $m = -1$; y-intercept $= 1$

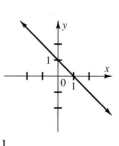

3. $h(x) = 5x$; $m = 5$
  y-intercept $= 0$

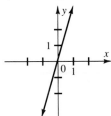

4. $f(x) = -x$; $m = -1$
  y-intercept $= 0$

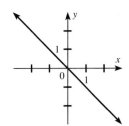

5. $h(x) = 3$; $m = 0$

  y-intercept $= 3$

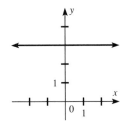

6. $g(x) = -\dfrac{1}{2}$; $m = 0$

  y-intercept $= -\dfrac{1}{2}$

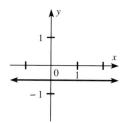

page 560 Class Exercises

1. $g(x) = 2x^2$, $a = 2$; Since 2 is a positive number, the function opens upward and has a minimum point.

2. $h(x) = 2x - x^2$, $a = -1$; Since -1 is a negative number, the function opens downward and has a maximum point.

3. $g(x) = x^2 + 3x + 4$, $a = 1$; Since 1 is a positive number, the function opens upward and has a minimum point.

4. $g(x) = x^2 - x - 12$; $x = -\dfrac{b}{2a} = \dfrac{1}{2(1)} = \dfrac{1}{2}$; $y = \left(\dfrac{1}{2}\right)^2 - \dfrac{1}{2} - 12$, $y = \dfrac{1}{4} - \dfrac{1}{2} - 12$,

$y = \dfrac{1}{4} - \dfrac{2}{4} - \dfrac{48}{4}$, $y = -\dfrac{49}{4}$; The vertex is $\left(\dfrac{1}{2}, -\dfrac{49}{4}\right)$. $x = -\dfrac{b}{2a} = \dfrac{1}{2}$;

Therefore, $x = \dfrac{1}{2}$ is the axis of symmetry.

5. $f(x) = 3 - 2x - x^2$, $x = -\dfrac{b}{2a} = \dfrac{2}{2(-1)} = -1$, $y = 3 - 2(-1) - (-1)^2$,

$y = 3 + 2 - 1$, $y = 4$; The vertex is $(-1, 4)$. $x = -\dfrac{b}{2a} = -1$;

Therefore, $x = -1$ is the axis of symmetry.

6. $h(x) = 3x^2$, $x = -\dfrac{b}{2a} = \dfrac{0}{2(3)} = 0$, $y = 3(0)^2$, $y = 0$;   The vertex is $(0, 0)$.

$x = -\dfrac{b}{2a} = 0$; Therefore, $x = 0$ is the axis of symmetry.

7. $h(x) = 4 - x^2$, $x = -\dfrac{b}{2a} = -\dfrac{0}{2(-1)} = 0$, $y = 4 - 0^2$, $y = 4$; The vertex is $(0, 4)$.

$x = -\dfrac{b}{2a} = 0$; Therefore, $x = 0$ is the axis of symmetry.

8. $g(x) = x^2 - 9$, $x = -\dfrac{b}{2a} = \dfrac{0}{2(1)} = 0$, $y = 0^2 - 9$, $y = -9$; The vertex is $(0, -9)$.

$x = -\dfrac{b}{2a} = 0$; Therefore, $x = 0$ is the axis of symmetry.

9. $f(x) = x^2 + x$, $x = -\dfrac{b}{2a} = -\dfrac{1}{2(1)} = -\dfrac{1}{2}$, $y = \left(-\dfrac{1}{2}\right)^2 - \dfrac{1}{2}$, $y = \dfrac{1}{4} - \dfrac{1}{2}$, $y = \dfrac{1}{4} - \dfrac{2}{4}$,

$y = -\dfrac{1}{4}$; The vertex is $\left(-\dfrac{1}{2}, -\dfrac{1}{4}\right)$. $x = -\dfrac{b}{2a} = -\dfrac{1}{2}$; Therefore, $x = -\dfrac{1}{2}$

is the axis of symmetry.

10. If c is positive, each point is c units above the corresponding point on $y = ax^2 + bx$. If c is negative, each point is c units below the corresponding point.

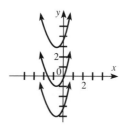

pages 560–561   Practice Exercises

A 1.

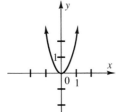

upward; minimum point

2.

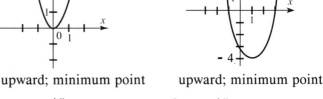

upward; minimum point

3.

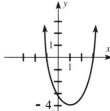

upward; minimum point

4.

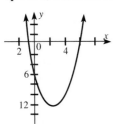

upward; minimum point

5.

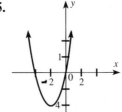

upward; minimum point

6.

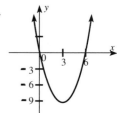

upward; minimum point

7.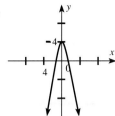

downward, maximum point

8.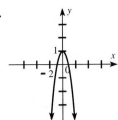

downward, maximum point

9.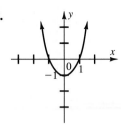

upward; minimum point

10.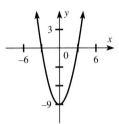

upward; minimum point

11.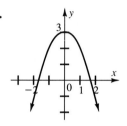

downward, maximum point

12.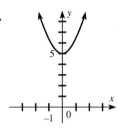

upward; minimum point

13. $s(x) = x^2 + 3x + 2$;

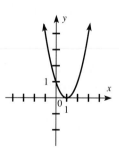

$x = -\dfrac{3}{2(1)} = -\dfrac{3}{2}$;

$y = \left(-\dfrac{3}{2}\right)^2 + 3\left(-\dfrac{3}{2}\right) + 2$,

$y = \dfrac{9}{4} - \dfrac{9}{2} + 2$, $y = \dfrac{9}{4} - \dfrac{18}{4} + \dfrac{8}{4}$,

$y = -\dfrac{1}{4}$; The vertex is $\left(-\dfrac{3}{2}, -\dfrac{1}{4}\right)$.

The axis of symmetry is $x = -\dfrac{b}{2a}$, $x = -\dfrac{3}{2}$. $x^2 + 3x + 2 = 0$, $(x + 1)(x + 2) = 0$,

$x + 1 = 0$, $x = -1$; $x + 2 = 0$, $x = -2$; The x-intercepts are $-1, -2$.

14. $g(x) = x^2 - 2x + 1$;

$x = -\dfrac{2}{2(1)} = \dfrac{2}{2} = 1$; $y = 1^2 - 2(1) + 1$,

$y = 1 - 2 + 1$, $y = 0$; The vertex is $(1, 0)$.

The axis of symmetry is $x = -\dfrac{b}{2a}$, $x = 1$.

$x^2 - 2x + 1 = 0$, $(x - 1)(x - 1) = 0$, $x - 1 = 0$, $x = 1$;
The x-intercept is 1.

15. $r(x) = x^2 + 5x + 8$;

$x = -\dfrac{5}{2(1)} = -\dfrac{5}{2}$;

$y = \left(-\dfrac{5}{2}\right)^2 + 5\left(-\dfrac{5}{2}\right) + 8$

$y = \dfrac{25}{4} - \dfrac{25}{2} + 8$

$y = \dfrac{25}{4} - \dfrac{50}{4} + \dfrac{32}{4}, y = \dfrac{7}{4}$;

The vertex is $\left(-\dfrac{5}{2}, \dfrac{7}{4}\right)$.

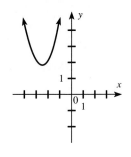

The axis of symmetry is $x = -\dfrac{b}{2a}$, $x = -\dfrac{5}{2}$. $x^2 + 5x + 8 = 0$; $a = 1$, $b = 5$,

$c = 8$; $\dfrac{-5 \pm \sqrt{5^2 - 4(1)(8)}}{2(1)}$, $\dfrac{-5 \pm \sqrt{25 - 32}}{2}$, $\dfrac{-5 \pm \sqrt{-7}}{2}$; Since $\sqrt{-7}$ is not

a real number, there are no x-intercepts.

16. $f(x) = x^2 - 4x + 5$; $x = -\dfrac{-4}{2(1)} = \dfrac{4}{2} = 2$;

$y = 2^2 - 4(2) + 5$, $y = 4 - 8 + 5$, $y = 1$;
The vertex is $(2, 1)$.
The axis of symmetry is $x = -\dfrac{b}{2a}$, $x = 2$.

$x^2 - 4x + 5 = 0$; $a = 1$, $b = -4$, $c = 5$;

$\dfrac{4 \pm \sqrt{(-4)^2 - 4(1)(5)}}{2(1)}$, $\dfrac{4 \pm \sqrt{16 - 20}}{2}$, $\dfrac{4 \pm \sqrt{-4}}{2}$; Since $\sqrt{-4}$ is

not a real number, there are no x-intercepts.

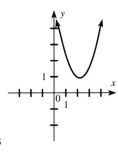

B 17. $h(x) = 7 - 6x - x^2$;

$x = -\dfrac{-6}{2(-1)} = -\dfrac{6}{2} = -3$; $y = 7 - 6(-3) - (-3)^2$,

$y = 7 + 18 - 9$, $y = 16$; The vertex is $(-3, 16)$.

The axis of symmetry is $x = -\dfrac{b}{2a}$, $x = -3$.

$x^2 + 6x - 7 = 0$, $(x + 7)(x - 1) = 0$, $x + 7 = 0$, $x = -7$;
$x - 1 = 0$, $x = 1$; The x-intercepts are -7, 1.

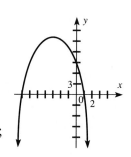

18. $f(x) = 5 - 4x - x^2$;

$x = -\dfrac{-4}{2(-1)} = \dfrac{4}{-2} = -2; \ y = 5 - 4(-2) - (-2)^2,$

$y = 5 + 8 - 4, \ y = 9;$ The vertex is $(-2, 9)$.

The axis of symmetry is $x = -\dfrac{b}{2a}, \ x = -2.$

$5 - 4x - x^2 = 0, \ x^2 + 4 - 5 = 0, \ (x + 5)(x - 1) = 0,$

$x + 5 = 0, \ x = -5; \ x - 1 = 0, \ x = 1;$ The x-intercepts are $-5, 1.$

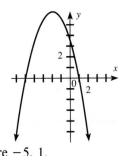

19. $g(x) = 5 + 3x - 2x^2; \ x = -\dfrac{3}{2(-2)} = \dfrac{3}{4},$

$y = 5 + 3\left(\dfrac{3}{4}\right) - 2\left(\dfrac{3}{4}\right)^2, \ y = 5 + \dfrac{9}{4} - 2\left(\dfrac{9}{16}\right),$

$y = 5 + \dfrac{9}{4} - \dfrac{9}{8}, \ y = \dfrac{40}{8} + \dfrac{18}{8} - \dfrac{9}{8}, \ y = \dfrac{49}{8};$

The vertex is $\left(\dfrac{3}{4}, \dfrac{49}{8}\right).$

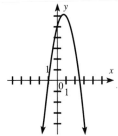

The axis of symmetry is $x = -\dfrac{b}{2a}, \ x = \dfrac{3}{4}. \quad 5 + 3x - 2x^2 = 0, \ 2x^2 - 3x - 5 = 0,$

$(2x - 5)(x + 1) = 0, \ 2x - 5 = 0, \ 2x = 5, \ x = \dfrac{5}{2}; \ x + 1 = 0, \ x = -1;$

The x-intercepts are $\dfrac{5}{2}, -1.$

20. $h(x) = 10 - x - 3x^2; \ x = -\dfrac{-1}{2(-3)} = -\dfrac{1}{6};$

$y = 10 - \left(-\dfrac{1}{6}\right) - 3\left(-\dfrac{1}{6}\right)^2,$

$y = 10 + \dfrac{1}{6} - 3\left(\dfrac{1}{36}\right), \ y = 10 + \dfrac{1}{6} - \dfrac{1}{12},$

$y = \dfrac{120}{12} + \dfrac{2}{12} - \dfrac{1}{12}, \ y = \dfrac{121}{12};$ The vertex

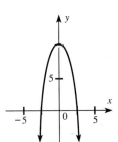

is $\left(-\dfrac{1}{6}, \dfrac{121}{12}\right).$ The axis of symmetry is $x = -\dfrac{b}{2a}, \ x = -\dfrac{1}{6}. \quad 10 - x - 3x^2 = 0,$

$3x^2 + x - 10 = 0, \ (3x - 5)(x + 2) = 0, \ 3x - 5 = 0, \ 3x = 5, \ x = \dfrac{5}{3}, \ x + 2 = 0,$

$x = -2;$ The x-intercepts are $\dfrac{5}{3}, -2.$

21. $f(x) = -\frac{1}{2}x^2 - 5$; $x = -\dfrac{0}{2\left(-\frac{1}{2}\right)} = 0$;

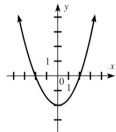

$y = -\frac{1}{2}(0) - 5$, $y = -5$; The vertex is $(0, -5)$.

The axis of symmetry is $x = -\dfrac{b}{2a}$, $x = 0$.

$-\frac{1}{2}x^2 - 5 = 0$, $-\frac{1}{2}x^2 = 5$, $x^2 = -10$, $x = \sqrt{-10}$;

Since $\sqrt{-10}$ is not a real number, there are no x-intercepts.

22. $g(x) = -2 + \frac{1}{2}x^2$; $x = -\dfrac{0}{2\left(\frac{1}{2}\right)} = 0$;

$y = -2 + \frac{1}{2}(0)^2$, $y = -2$; The vertex is $(0, -2)$.

The axis of symmetry is $x = -\dfrac{b}{2a}$, $x = 0$.

$-2 + \frac{1}{2}x^2 = 0$, $\frac{1}{2}x^2 = 2$, $x^2 = 4$, $x = \pm 2$; The x-intercepts are ± 2.

23. $m(x) = \frac{3}{4}x^2 + 2$; $x = -\dfrac{0}{2\left(\frac{3}{4}\right)} = 0$;

$y = \frac{3}{4}(0)^2 + 2$, $y = 2$; The vertex is $(0, 2)$.

The axis of symmetry is $x = -\dfrac{b}{2a}$, $x = 0$.

$\frac{3}{4}x^2 + 2 = 0$, $\frac{3}{4}x^2 = -2$, $x^2 = -2\left(\frac{4}{3}\right)$, $x^2 = -\frac{8}{3}$, $x = \sqrt{-\frac{8}{3}}$;

Since $\sqrt{-\frac{8}{3}}$ is not a real number, there are no x-intercepts.

24. $r(x) = \frac{1}{5}x^2 + 3$; $x = -\dfrac{0}{2\left(\frac{1}{5}\right)} = 0$;

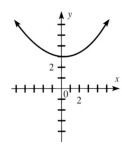

$y = \frac{1}{5}(0)^2 + 3$, $y = 3$; The vertex is $(0, 3)$.

The axis of symmetry is $x = -\dfrac{b}{2a}$, $x = 0$.

$\frac{1}{5}x^2 + 3 = 0$, $\frac{1}{5}x^2 = -3$, $x^2 = -15$, $x = \sqrt{-15}$; Since $\sqrt{-15}$ is not a real

number, there are no x-intercepts.

25. $g(x) = (x + 2)^2$, $g(x) = x^2 + 4x + 4$;

$x = -\dfrac{4}{2(1)} = -2$; $y = (-2 + 2)^2$, $y = 0$;

The vertex is $(-2, 0)$. The axis of symmetry is

$x = -\dfrac{b}{2a}$, $x = -2$. The x-intercept is -2.

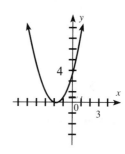

26. $f(x) = (x + 3)^2$, $f(x) = x^2 + 6x + 9$;

$x = -\dfrac{6}{2(1)} = -3$; $y = (-3 + 3)^2$, $y = 0$; The vertex is $(-3, 0)$.

The axis of symmetry is $x = -\dfrac{b}{2a}$, $x = -3$.

$(x + 3)^2 = 0$, $x + 3 = 0$, $x = -3$; The x-intercept is -3.

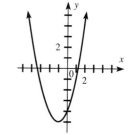

27. $h(x) = (x + 1)^2 - 5$, $h(x) = x^2 + 2x + 1 - 5$,

$h(x) = x^2 + 2x - 4$; $x = -\dfrac{2}{2(1)} = -1$;

$y = (-1 + 1)^2 - 5$, $y = 0 - 5$, $y = -5$;
The vertex is $(-1, -5)$.

The axis of symmetry is $x = -\dfrac{b}{2a}$, $x = -1$. $(x + 1)^2 - 5 = 0$,

$(x + 1)^2 = 5$, $x + 1 = \sqrt{5}$, $x = -1 \pm\sqrt{5}$; The x-intercepts
are $-1 \pm \sqrt{5}$.

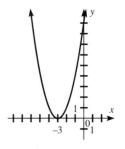

28. $r(x) = (x + 4)^2 - 3$, $r(x) = x^2 + 8x + 16 - 3$,

$r(x) = x^2 + 8x + 13$; $x = -\dfrac{8}{2(1)} = -4$;

$y = (-4 + 4)^2 - 3$, $y = 0 - 3$, $y = -3$;
The vertex is $(-4, -3)$.

The axis of symmetry is $x = -\dfrac{b}{2a}$, $x = -4$. $(x + 4)^2 - 3 = 0$,

$(x + 4)^2 = 3$, $x + 4 = \pm \sqrt{3}$, $x = -4 \pm \sqrt{3}$; The x-intercepts
are $-4 \pm \sqrt{3}$.

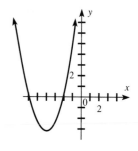

C 29. $g(x) = -2(x + 3)^2 - 4$,

$g(x) = -2(x^2 + 6x + 9) - 4$,

$g(x) = -2x^2 - 12x - 18 - 4$,

$g(x) = -2x^2 - 12x - 22; \ x = -\dfrac{-12}{2(-2)} = -3;$

$y = -2(-3 + 3)^2 - 4, \ y = 0 - 4, \ y = -4;$

The vertex point is $(-3, -4)$.

The axis of symmetry is $x = -\dfrac{b}{2a}, \ x = -3. \quad -2(x + 3)^2 - 4 = 0,$

$-2(x + 3)^2 = 4, \ (x + 3)^2 = -2, \ x + 3 = \pm\sqrt{-2}, \ x = \pm\sqrt{-2} - 3;$ Since $\sqrt{-2}$ is not a real number, there are no x-intercepts.

30. $h(x) = -4\left(x + \dfrac{1}{2}\right)^2 - 4$,

$h(x) = -4\left(x^2 + x + \dfrac{1}{4}\right) - 4$,

$h(x) = -4x^2 - 4x - 1 - 4$,

$h(x) = -4x^2 - 4x - 5; \ x = -\dfrac{-4}{2(-4)} = -\dfrac{1}{2};$

$y = -4\left(-\dfrac{1}{2} + \dfrac{1}{2}\right)^2 - 4, \ y = 0 - 4, \ y = -4;$

The vertex point is $\left(-\dfrac{1}{2}, -4\right)$. The axis of symmetry is $x = -\dfrac{b}{2a}, \ x = -\dfrac{1}{2}.$

$-4\left(x + \dfrac{1}{2}\right)^2 - 4 = 0, \ -4\left(x + \dfrac{1}{2}\right)^2 = 4, \ \left(x + \dfrac{1}{2}\right)^2 = -1, \ x + \dfrac{1}{2} = \pm\sqrt{-1},$

$x = \pm\sqrt{-1} - \dfrac{1}{2};$ Since the $\sqrt{-1}$ is not a real number, there are no x-intercepts.

31.

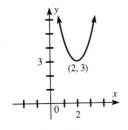

upward; minimum point

32.

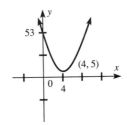

upward; minimum point

33.

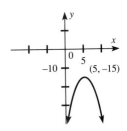

downward; maximum point

34.

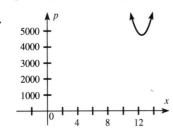

35.

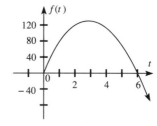

$x = -\dfrac{-25}{2(1)} = 12.5$; minimum profit $=$ $\qquad$ $t = -\dfrac{96}{2(-16)} = 3$; maximum height $=$

$(12.5)^2 - 25(12.5) + 5000 = \$4843.75$ $\qquad$ $96(3) - 16(3)^2 = 144$ ft

36. $-6 < 3x \le 6,\ -2 < x \le 2$ $\qquad$ **37.** $5t > 20$ or $-2t > 12,\ t > 4$ or $t < -6$

38. $2x + 3 \le -11$ or $2x + 3 \ge 11,\ x \le -7$ or $x \ge 4$

39. $4t - 12 = -20$ or $4t - 12 = 20,\ t = -2$ or $t = 8$

40. $3b - 2 = 64,\ 3b = 66,\ b = 22$ $\qquad$ **41.** $\dfrac{a}{8} = 9,\ a = 72$

42. $5y + 1 = 4y^2 - 32y + 64,\ 4y^2 - 37y + 63 = 0,\ (4y - 9)(y - 7) = 0,$

$\quad y = 7;\ \left(\dfrac{9}{4} \text{ is an extraneous solution.}\right)$

43. $d + 2\sqrt{d} + 1 = d + 7,\ 2\sqrt{d} = 6,\ \sqrt{d} = 3,\ d = 9$

pages 562–563   Integrating Algebra: Physics

1. The x-axis represents time; the y-axis represents height.

2. Approximately 6.25 s $\qquad$ **3.** Approximately 156.25 ft $\qquad$ **4.** Approximately 3.12 s

5. Graph $H(t) = -16t^2 + 120t$.
   a. Approximately 224.98 ft
   b. Approximately 7.50 s; Approximately 0.96 s
   c. The rocket with the greater initial speed flies higher and is in flight for a longer period of time.

6. Graph $H(t) = -16t^2 = 100t$ and $H(t) = -2.66t^2 + 100t$.
   a. Earth: Approximately 156.25 ft; moon: Approximately 939.79 ft
   b. Earth: lands in approximately 6.25 s; is at 100 ft in approximately 1.25 s; moon: lands in approximately 37.59 s; is at 100 ft in approximately 1.03 s

7.

	Venus	Mars	Jupiter	Saturn
a.	173.61 ft	409.82 ft	60.49 ft	140.82 ft
b.	6.94 s	16.39 s	2.42 s	5.63 s
c.	90 ft/s	38 ft/s	255 ft/s	110 ft/s
d.	94 ft/s	61 ft/s	160 ft/s	105 ft/s

page 564   Capsule Review

1. $t = \dfrac{-(-3) \pm \sqrt{(-3)^2 - 4(1)(-10)}}{2(1)} = \dfrac{3 \pm \sqrt{49}}{2} = \dfrac{3 \pm 7}{2};\ t = 5,\ -2$

2. $2y^2 + 4y - 3 = 0;\ y = \dfrac{-4 \pm \sqrt{4^2 - 4(2)(-3)}}{2(2)} = \dfrac{-4 \pm \sqrt{40}}{4} = \dfrac{-4 \pm 2\sqrt{10}}{4};\ y = \dfrac{-2 \pm \sqrt{10}}{2}$

3. $x = \dfrac{-(-4) \pm \sqrt{(-4)^2 - 4(12)(-3)}}{2(12)} = \dfrac{4 \pm \sqrt{160}}{24} = \dfrac{4 \pm 4\sqrt{10}}{24}; x = \dfrac{1 \pm \sqrt{10}}{6}$

4. $m = \dfrac{-(-10) \pm \sqrt{(-10)^2 - 4(1)(25)}}{2(1)} = \dfrac{10 \pm \sqrt{0}}{2} = \dfrac{10}{2}; m = 5$

5. $t = \dfrac{-0 \pm \sqrt{0^2 - 4(6)(-1)}}{2(6)} = \dfrac{\pm\sqrt{24}}{12} = \dfrac{\pm 2\sqrt{6}}{12}; t = \dfrac{\pm\sqrt{6}}{6}$

6. $z = \dfrac{-0 \pm \sqrt{0^2 - 4(1)(4)}}{2(1)} = \dfrac{\pm\sqrt{-16}}{2}$; no real solutions

page 566　Class Exercises

1. $0^2 - 4(1)(-1) = 4$; 2 real solutions

2. $0^2 - 4(1)(4) = -16$; no real solutions

3. $6^2 - 4(1)(9) = 0$; 1 real solution

4. $4^2 - 4(1)(3) = 4$; 2 real solutions

5. $3^2 - 4(-2)(-2) = -7$; no real solutions

6. $(-20)^2 - 4(4)(25) = 0$; 1 real solution

page 566　Practice Exercises

A　**1.** $2^2 - 4(1)(-15) = 64$, 2 real solutions

2. $(-5)^2 - 4(1)(4) = 9$; 2 real solutions

3. $(-8)^2 - 4(1)(16) = 0$; 1 real solution

4. $(-10)^2 - 4(1)(25) = 0$; 1 real solution

5. $4^2 - 4(1)(-6) = 40$; 2 real solutions

6. $(-1)^2 - 4(1)(-3) = 13$; 2 real solutions

7. $(-3)^2 - 4(2)(2) = -7$; no real solutions

8. $(-4)^2 - 4(3)(5) = -44$; no real solutions

9. $7^2 - 4(-2)(-3) = 25$; 2 real solutions

10. $(-3)^2 - 4(-2)(2) = 25$; 2 real solutions

11. $5^2 - 4(-6)(-4) = -71$; no real solutions

12. $(-3)^2 - 4(4)(3) = -39$; no real solutions

B　**13.** $4x^2 + 12x + 9 = 0$; $12^2 - 4(4)(9) = 0$; 1 real solution

14. $9x^2 - 30x + 25 = 0$; $(-30)^2 - 4(9)(25) = 0$; 1 real solution

15. $\dfrac{3}{2}x^2 - 4x + \dfrac{1}{2} = 0$; $(-4)^2 - 4\left(\dfrac{3}{2}\right)\left(\dfrac{1}{2}\right) = 13$; 2 real solutions

16. $x^2 + 3x - 3 = 0$; $3^2 - 4(1)(-3) = 21$; 2 real solutions

17. $\left(\dfrac{1}{2}\right)^2 - 4(2)\left(\dfrac{2}{3}\right) = -\dfrac{61}{12}$; no real solutions

18. $\left(-\dfrac{5}{4}\right)^2 - 4\left(\dfrac{1}{5}\right)(-1) = \dfrac{189}{80}$; 2 real solutions

19. $(-12)^2 - 4(-4)(9) = 288$; 2

20. $(-40)^2 - 4(25)(-16) = 3200$; 2

C 21. $\left(\frac{1}{2}\right)^2 - 4(1)\left(-\frac{17}{4}\right) = \frac{69}{4}; 2$ 　　　　　　**22.** $(-1)^2 - 4\left(-\frac{1}{2}\right)\left(-\frac{11}{6}\right) = -\frac{8}{3}; 0$

23. $(-k)^2 - 4(4)(625) = k^2 - 10{,}000$; Since $|k| > 100$, $k^2 - 10{,}000 > 0$; 2

24. $(-8)^2 - 4(k)(3) = 64 - 12k$; Since $k > 5\frac{1}{3}$, $64 - 12k = 0$; 0

25. $(-60)^2 - 4(2)(480) = -240$; no

26. $7000 = 5400 + 300x - 50x^2$, $0 = -1600 + 300x - 50x^2$; $a = -50$, $b = 300$, $c = -1600$; $300^2 - 4(-50)(-1600) = -230{,}000$; no

27. There are 2 real solutions. The given points indicate that the graph crosses the x-axis once, and if it does so once, it must do so twice.

page 569　Class Exercises

1. Let i represent income and s represent sales.　$i = 0.15s + 12{,}000$

2. $m = \dfrac{12.2 - 8.3}{9 - 6} = \dfrac{20.0 - 12.2}{15 - 9} = 1.3$

3. $20.0 = 1.3(15) + b$, $0.5 = b$; $c = 1.3n + 0.5$

4. The slope 1.3 is the increase in cost per 1 mile. The intercept 0.5 is the cost at zero miles.

pages 569–571　Practice Exercises

A 1. linear function; $\dfrac{70 - 58}{18 - 3} = 0.8$; $58 = 0.8(3) + b$, $55.6 = b$; $c = 0.8w + 55.6$;
The slope is a \$0.80 increase for each ounce and the c-intercept is the constant cost of production.

2. not a linear function

3. linear function; $\dfrac{3.02 - 1.88}{6.3 - 1.5} = 0.24$; $1.88 = 0.24(1.5) + b$, $1.52 = b$; $c = 0.24h + 1.52$;
The slope is a \$0.24 increase for each hour and the c-intercept is a basic cost of \$1.52.

4. linear function; $\dfrac{6.12 - 4.56}{2.1 - 0.5} = 0.98$; $4.56 = 0.98(0.5) + b$, $4.07 = b$;
$p = 0.98 g + 4.07$; The slope is a \$0.98 increase for each gram and the p-intercept \$4.07 is the basic price.

5. Let m represent the monthly charge and c represent the total cost. $c = 24.5m + 75$

6. Let h represent the hourly charge and c represent the consulting cost. $c = 55h + 35$

7. Since there are 4 quarts in 1 gallon, the leak escapes at a rate of $\frac{1}{4}$ h. Let h represent the number of hours and g represent the number of gallons left. $g = -\frac{1}{4}h + 55$.

8. Let x represent the rate it is filling and v represent the volume. $v = 2.5x + 3000$

9. $\frac{325 - 150}{6 - 2.5} = 50$; Let h represent the number of hours and c represent the total cost. $325 = -50(6) + b$, $25 = b$; $c = 50h + 25$; $200 = 50h + 25$, $175 = 50h$, $3.5 = h$

10. $\frac{144 - 80}{76 - 60} = 4$; Let c represent the number of chirps and t represent the temperature. $144 = 4(76) + b$, $-160 = b$; $c = 4t - 160$; $180 = 4t - 160$, $340 = 46$, $85 = t$; $85°$

B 11. $f(30) = -0.01(30)^2 + 0.8(30) = 15$ bushels per tree

12. $16 = -0.01t^2 + 0.8t$, $0.01t^2 - 0.8t + 16 = 0$; $a = 0.01$, $b = -0.8$, $c = 16$;
$$t = \frac{-(-0.8) \pm \sqrt{(-0.8)^2 - 4(0.01)(16)}}{2(0.01)} = \frac{0.8 \pm \sqrt{0}}{0.02} = \frac{0.8}{0.02}; \ t = 40 \text{ trees}$$

13. $a = -0.01$, $b = 0.8$, $c = 0$; $t = \frac{-0.8 \pm \sqrt{(0.8)^2 - 4(-0.01)(0)}}{2(-0.01)} = \frac{-0.8 \pm \sqrt{0}}{-0.02} = \frac{0.8}{0.02}$;
$t = 40$ trees; $f(40) = -0.01(40)^2 + 0.8(40) = 16$ bushels per tree.

14. The t-intercept represents the number of trees when there is no yield, and the $f(t)$-intercept represents the yield when no trees are planted.

15. Let $R(x)$ represent the annual receipts. $R(x) = (6 - 0.25x)(500 + 50x)$

16. $(6 - 25x)(500 + 50x) = 3000 + 175x - 12.5x^2$; Since $a < 0$, the maximum value occurs at the vertex. $x = -\frac{175}{2(-12.5)} = 7$; $R(7) = [6 - 25(7)][500 + 50(7)] = 3612.50$; The maximum receipts of $3612.50 occur after 7 decreases.

17. Let $s(x)$ represent ticket sales. $s(x) = (6 + 0.5x)(600 - 25x)$

18. $(6 + 0.5x)(600 - 25x) = 3600 + 150x - 12.5x^2$; Since $a < 0$, the maximum value occurs at the vertex. $x = -\dfrac{150}{2(-12.5)} = 6$; $s(6) = [6 + 0.5(6)][600 - 25(6)] = 4050$; The maximum ticket sales of $4050 occur after 6 increases.

page 571   Mixed Problem Solving Review

1. Let x represent the length of a side of the square, then $1 + 10x$ represents the longer sides of the resulting rectangle and $1 - 15x$ represents the shorter sides. The perimeter p of the resulting rectangle is $2l + 2w = 2(1 + 10x) + 2(1 - 15x)$. So $234 = 2.2x + 1.7x$, $234 = 3.9x$, $60 = x$; Therefore, the perimeter of the original square is $60(4) = 240$.

2. Let x represent the rate, in hours, of the second printer working alone.

Work rate · time worked = part of job done

	Work rate	time worked	part of job done
printer 1	$\dfrac{1}{3}$	2	$\dfrac{2}{3}$
printer 2	$\dfrac{1}{x}$	2	$\dfrac{2}{x}$

$\dfrac{2}{3} + \dfrac{2}{x} = 1$, $\dfrac{2x + 6}{3x} = 1$, $2x + 6 = 3x$, $6 = x$; it would take the second printer 6h to do a job alone.

3. Let m represent the number of months and c represent the total cost. $c = 15m + 35$; $125 = 15m + 35$, $90 = 15m$, $6 = m$; Cable has been connected for 6 months.

4. Let x represent the units digit, $x + 2$ represent the tens digit, and y represent the original number.

$$\begin{cases} x + 10(x + 2) = y \longrightarrow 11x + 20 = y \\ (x + 2) + 10x = \dfrac{4}{7}y \longrightarrow 11x + 2 = \dfrac{4}{7}y \end{cases}$$

$$\begin{aligned} 11x + 20 &= y \\ \underline{11x + 2 = \dfrac{4}{7}y} \\ -18 &= -\dfrac{3}{7}y \\ y &= -18\left(-\dfrac{7}{3}\right) \\ y &= 42 \end{aligned}$$

The two digit number is 42.

Check students' work.

1. $x^2 - 5x + 6 = 0$, $(x - 3)(x - 2) = 0$, $x - 3 = 0$, $x = 3$; $x - 2 = 0$, $x = 2$

2. $x^2 + 9x + 14 = 0$, $(x + 7)(x + 2) = 0$, $x + 7 = 0$, $x = -7$; $x + 2 = 0$, $x = -2$

3. $x^2 + 11x = 0$, $x(x + 11) = 0$, $x = 0$; $x + 11 = 0$, $x = -11$

pages 573–574   Class Exercises

1. $-(4) = -4$; -12

2. $-(-6) = 6$; 8

3. $p^2 - 3p - 10 = 0$; $-(-3) = 3$; -10

4. $m^2 + 3m + 2 = 0$; $-(3) = -3$; 2

5. -5; 0

6. $n^2 - 5n - 6 = 0$; $-(-5) = 5$; -6

7. $\dfrac{b}{a} = -[1 + (-2)] = 1$; $\dfrac{c}{a} = (1)(-2) = -2$; $x^2 + x - 2 = 0$

8. $\dfrac{b}{a} = -[-2 + (-3)] = 5$; $\dfrac{c}{a} = (-2)(-3) = 6$; $x^2 + 5x + 6 = 0$

9. $\dfrac{b}{a} = -\left(-\dfrac{1}{3} + 1\right) = -\dfrac{2}{3}$; $\dfrac{c}{a} = \left(-\dfrac{1}{3}\right)(1) = -\dfrac{1}{3}$; $x^2 - \dfrac{2}{3}x - \dfrac{1}{3} = 0$; $3x^2 - 2x - 1 = 0$

10. $\dfrac{b}{a} = -\left[3 + \left(-\dfrac{3}{2}\right)\right] = -\dfrac{3}{2}$; $\dfrac{c}{a} = (3)\left(-\dfrac{3}{2}\right) = -\dfrac{9}{2}$; $x^2 - \dfrac{3}{2}x - \dfrac{9}{2} = 0$; $2x^2 - 3x - 9 = 0$

11. Since -4 is the only solution, the equation is $(x + 4)^2$, or $x^2 + 8x + 16 = 0$. The sum of the solutions is -8 and the product is 16.

pages 574–575   Practice Exercises

A 1. $\dfrac{b}{a} = -(1 + 3) = -4$; $\dfrac{c}{a} = (1)(3) = 3$; $x^2 - 4x + 3 = 0$

2. $\dfrac{b}{a} = -(3 + 6) = -9$; $\dfrac{c}{a} = (3)(6) = 18$; $x^2 - 9x + 18 = 0$

3. $\dfrac{b}{a} = -(-2 + 6) = -4$; $\dfrac{c}{a} = (-2)(6) = -12$; $x^2 - 4x - 12 = 0$

4. $\dfrac{b}{a} = -(-3 + 2) = 1$; $\dfrac{c}{a} = (-3)(2) = -6$; $x^2 + x - 6 = 0$

5. $\dfrac{b}{a} = -[-5 + (-1)] = 6$; $\dfrac{c}{a} = (-5)(-1) = 5$; $x^2 + 6x + 5 = 0$

6. $\dfrac{b}{a} = -[-1 + (-2)] = 3$; $\dfrac{c}{a} = (-1)(-2) = 2$; $x^2 + 3x + 2 = 0$

7. $\dfrac{b}{a} = -(-5 + 0) = 5$; $\dfrac{c}{a} = (-5)(0) = 0$; $x^2 + 5x = 0$

8. $\dfrac{b}{a} = -(0 + 3) = -3$; $\dfrac{c}{a} = (0)(3) = 0$; $x^2 - 3x = 0$

9. $\dfrac{b}{a} = -\left(-3 + \dfrac{4}{5}\right) = \dfrac{11}{5}$; $\dfrac{c}{a} = (-3)\left(\dfrac{4}{5}\right) = -\dfrac{12}{5}$; $x^2 + \dfrac{11}{5}x - \dfrac{12}{5} = 0$; $5x^2 + 11x - 12 = 0$

10. $\dfrac{b}{a} = -\left(-\dfrac{1}{2} + 2\right) = -\dfrac{3}{2}$; $\dfrac{c}{a} = \left(-\dfrac{1}{2}\right)(2) = -1$; $x^2 - \dfrac{3}{2}x - 1 = 0$; $2x^2 - 3x - 2 = 0$

11. $\dfrac{b}{a} = -\left(\dfrac{2}{3} + 1\right) = -\dfrac{5}{3}$; $\dfrac{c}{a} = \left(\dfrac{2}{3}\right)(1) = \dfrac{2}{3}$; $x^2 - \dfrac{5}{3}x + \dfrac{2}{3} = 0$; $3x^2 - 5x + 2 = 0$

12. $\dfrac{b}{a} = -\left(-\dfrac{3}{2} + 3\right) = -\dfrac{3}{2}$; $\dfrac{c}{a} = \left(-\dfrac{3}{2}\right)(3) = -\dfrac{9}{2}$; $x^2 - \dfrac{3}{2}x - \dfrac{9}{2} = 0$; $2x^2 - 3x - 9 = 0$

13. $\dfrac{b}{a} = 0$, $\dfrac{c}{a} = -16$; $-(-4 + 4) = 0$; $(-4)(4) = -16$; yes

14. $\dfrac{b}{a} = 0$, $\dfrac{c}{a} = 9$; $-[3 + (-3)] = 0$; $(3)(-3) \neq 9$; no

15. $\dfrac{b}{a} = -3$, $\dfrac{c}{a} = 2$; $-[-2 + (-1)] \neq -3$, no

16. $\dfrac{b}{a} = -7$, $\dfrac{c}{a} = 12$; $-[-3 + (-4)] = 7$; $(-3)(-4) = 12$; yes

B 17. $\dfrac{b}{a} = -\left(-\dfrac{5}{2} + \dfrac{5}{2}\right) = 0$; $\dfrac{c}{a} = \left(-\dfrac{5}{2}\right)\left(\dfrac{5}{2}\right) = -\dfrac{25}{4}$; $x^2 - \dfrac{25}{4} = 0$; $4x^2 - 25 = 0$

18. $\dfrac{b}{a} = -\left[\dfrac{2}{3} + \left(-\dfrac{2}{3}\right)\right] = 0$; $\dfrac{c}{a} = \left(\dfrac{2}{3}\right)\left(-\dfrac{2}{3}\right) = -\dfrac{4}{9}$; $x^2 - \dfrac{4}{9} = 0$; $9x^2 - 4 = 0$

19. $\dfrac{b}{a} = -\left(-\dfrac{3}{2} + \dfrac{4}{3}\right) = \dfrac{1}{6}$; $\dfrac{c}{a} = \left(-\dfrac{3}{2}\right)\left(\dfrac{4}{3}\right) = -2$; $x^2 + \dfrac{1}{6}x - 2 = 0$; $6x^2 + x - 12 = 0$

20. $\dfrac{b}{a} = -\left[\dfrac{3}{2} + \left(-\dfrac{1}{2}\right)\right] = -1$; $\dfrac{c}{a} = \left(\dfrac{3}{2}\right)\left(-\dfrac{1}{2}\right) = -\dfrac{3}{4}$; $x^2 - x - \dfrac{3}{4} = 0$; $4x^2 - 4x - 3 = 0$

21. $\dfrac{b}{a} = -[1 + \sqrt{6} + (1 - \sqrt{6})] = -2$; $\dfrac{c}{a} = (1 + \sqrt{6})(1 - \sqrt{6}) = 1 - 6 = -5$; $x^2 - 2x - 5 = 0$

22. $\dfrac{b}{a} = -[-2 + \sqrt{3} + (-2 - \sqrt{3})] = 4$; $\dfrac{c}{a} = (-2 + \sqrt{3})(-2 - \sqrt{3}) = 4 - 3 = 1$; $x^2 + 4x + 1 = 0$

23. $\dfrac{b}{a} = -[5\sqrt{2} + 1 + (5\sqrt{2} - 1)] = -10\sqrt{2}$; $\dfrac{c}{a} = (5\sqrt{2} + 1)(5\sqrt{2} - 1) =$ $50 - 1 = 49$; $x^2 - 10\sqrt{2}x + 49 = 0$

24. $\dfrac{b}{a} = -[2\sqrt{3} + 5 + (2\sqrt{3} - 5)] = -4\sqrt{3}; \dfrac{c}{a} = (2\sqrt{3} + 5)(2\sqrt{3} - 5) =$

$12 - 25 = -13; x^2 - 4\sqrt{3}x - 13 = 0$

25. $2x^2 - 3x - 35 = 0; \dfrac{b}{a} = -\dfrac{3}{2}, \dfrac{c}{a} = -\dfrac{35}{2}; -\left(-\dfrac{3}{2} + 5\right) = -\left(-\dfrac{3}{2} + \dfrac{10}{2}\right) = -\dfrac{3}{2};$

$\left(-\dfrac{7}{2}\right)(5) = -\dfrac{35}{2};$ yes

26. $6x^2 - x - 3 = 0; \dfrac{b}{a} = -\dfrac{1}{6}, \dfrac{c}{a} = -3; -\left[1 + \left(-\dfrac{2}{3}\right)\right] = -\dfrac{1}{3};$ no

C 27. $\dfrac{b}{a} = -1, \dfrac{c}{a} = -1; -[1 + \sqrt{2} + (1 - \sqrt{2})] = -2;$ no

28. $\dfrac{b}{a} = -10, \dfrac{c}{a} = 20; -[5 + \sqrt{5} + (5 - \sqrt{5})] = -10; (5 + \sqrt{5})(5 - \sqrt{5}) =$

$25 - 5 = 20;$ yes

29. $p^2 + 3p + 1 = 0; \dfrac{b}{a} = 3, \dfrac{c}{a} = 1; -\left[\dfrac{-3 + \sqrt{5}}{2} + \left(\dfrac{-3 - \sqrt{5}}{2}\right)\right] = 3;$

$\left(\dfrac{-3 + \sqrt{5}}{2}\right)\left(\dfrac{-3 - \sqrt{5}}{2}\right) = \dfrac{9 - 5}{4} = 1;$ yes

30. $3r^2 - 2r - 2 = 0; \dfrac{b}{a} = -\dfrac{2}{3}, \dfrac{c}{a} = -\dfrac{2}{3}; -\left[\dfrac{1 + \sqrt{7}}{3} + \left(\dfrac{1 - \sqrt{7}}{3}\right)\right] = -\dfrac{2}{3};$

$\left(\dfrac{1 + \sqrt{7}}{3}\right)\left(\dfrac{1 - \sqrt{7}}{3}\right) = \dfrac{1 - 7}{9} = -\dfrac{2}{3};$ yes

31. $\dfrac{-b + \sqrt{b^2 - 4ac}}{2a} + \dfrac{-b - \sqrt{b^2 - 4ac}}{2a} = \dfrac{-2b}{2a} = \dfrac{-b}{a} = -\dfrac{b}{a}$

32. $\left(\dfrac{-b + \sqrt{b^2 - 4ac}}{2a}\right)\left(\dfrac{-b - \sqrt{b^2 - 4ac}}{2a}\right) = \dfrac{b^2 - (b^2 - 4ac)}{4a^2} = \dfrac{4ac}{4a^2} = \dfrac{c}{a}$

33. $\dfrac{b}{a} = -22; \dfrac{c}{a} = 120; x^2 - 22x + 120 = 0, (x - 10)(x - 12) = 0, x = 10 \text{ or } x = 12$

34. $\dfrac{b}{a} = 18; \dfrac{c}{a} = 72; x^2 + 18x + 72 = 0, (x + 12)(x + 6) = 0, x = -12 \text{ or } x = -6$

35. $\dfrac{b}{a} = -\dfrac{7}{6}, \dfrac{c}{a} = \dfrac{1}{3}; x^2 - \dfrac{7}{6}x + \dfrac{1}{3} = 0, 6x^2 - 7x + 2 = 0, (3x - 2)(2x - 1) = 0,$

$x = \dfrac{2}{3} \text{ or } x = \dfrac{1}{2}$

36. $\dfrac{b}{a} = \dfrac{23}{20}, \dfrac{c}{a} = \dfrac{3}{10}; x^2 + \dfrac{23}{20}x + \dfrac{3}{10} = 0, 20x^2 + 23x + 6 = 0, (5x + 2)(4x + 3) =$

$0, x = -\dfrac{2}{5} \text{ or } x = -\dfrac{3}{4}$

1. $x = -\dfrac{0}{2(1)}$, $x = 0$; $y = 0^2$, $y = 0$; The vertex is $(0, 0)$.

The axis of symmetry is $x = \dfrac{0}{2(1)}$, $x = 0$.

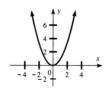

2. $x = -\dfrac{0}{2(-1)}$, $x = 0$; $y = 5 - 0^2 = 5$; The vertex is $(0, 5)$.

The axis of symmetry is $x = -\dfrac{0}{2(-1)}$, $x = 0$.

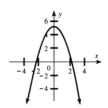

3. $x = -\left(\dfrac{-2}{2(1)}\right) = 1$; $y = 1^2 - 2(1) = -1$; The vertex is

$(1, -1)$. The axis of symmetry is $x = -\left(\dfrac{-2}{2(1)}\right)$, $x = 1$.

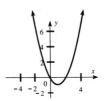

4. Since $a > 0$, the graph opens upward and has a minimum point.

5. Since $a < 0$, the graph opens downward and has a maximum point.

6. Since $a > 0$, the graph opens upward and has a minimum point.

7. $(-3)^2 - 4(1)(4) = -7$; no real solutions   **8.** $12^2 - 4(4)(9) = 0$; 1 real solution

9. $2z^2 - 5z - 12 = 0$; $(-5)^2 - 4(2)(-12) = 121$; 2 real solutions

10. $f(15) = 40(15) - 15^2 = 600 - 225 = 375$; $375

11. $40x - x^2 = 351$, $x^2 - 40x + 351 = 0$, $(x - 27)(x - 13) = 0$, $x = 27$ people or $x = 13$ people

12. $40x - x^2 = 0$; Since $a < 0$, the maximum profit/person occurs at the vertex. $x = -\dfrac{40}{2(-1)} = 20$; $40(20) - 20^2 = 400$; The maximum profit of $400 occurs with 20 people.

13. $40x - x^2 = 0$, $x(40 - x) = 0$, $x = 0$ people or $x = 40$ people

14. $\dfrac{b}{a} = -(2 + 5) = -7$; $\dfrac{c}{a} = (2)(5) = 10$; $x^2 - 7x + 10 = 0$

15. $\frac{b}{a} = -\left(-\frac{3}{2} + 2\right) = -\frac{1}{2}; \frac{c}{a} = \left(-\frac{3}{2}\right)(2) = -3; x^2 - \frac{1}{2}x - 3 = 0,$
$2x^2 - x - 6 = 0$

16. $\frac{b}{a} = -[1 + \sqrt{3} + (1 - \sqrt{3})] = -2; \frac{c}{a} = (1 + \sqrt{3})(1 - \sqrt{3}) = -2;$
$x^2 - 2x - 2 = 0$

pages 576–577   Summary and Review

1. $4p^2 = 80, p^2 = 20, p = \pm\sqrt{20}, p = \pm 2\sqrt{5}$

2. $(y - 1)^2 = 9, y - 1 = \pm\sqrt{9}, y - 1 = \pm 3, y = 4 \text{ or } y = -2$

3. $(m + 3)^2 - 4 = 32, (m + 3)^2 = 36, m + 3 = \pm\sqrt{36}, m + 3 = \pm 6, m = 3 \text{ or } m = -9$

4. $\frac{1}{2}(4) = 2, 2^2 = 4; x^2 + 4x + 4 = 12 + 4, (x + 2)^2 = 16, x + 2 = \pm\sqrt{16},$
$x + 2 = \pm 4, x = 2 \text{ or } x = -6$

5. $\frac{1}{2}(-6) = -3, (-3)^2 = 9; y^2 - 6y + 9 = 5 + 9, (y - 3)^2 = 14, y - 3 = \pm\sqrt{14},$
$y = \pm\sqrt{14} + 3$

6. $\frac{1}{2}(0) = 0, 0^2 = 0; z^2 = \frac{9}{2}, z = \pm\frac{3}{\sqrt{2}} \cdot \frac{\sqrt{2}}{\sqrt{2}}, z = \pm\frac{3\sqrt{2}}{2}$

7. $a = 3, b = 7, c = 3; x = \frac{-7 \pm \sqrt{7^2 - 4(3)(3)}}{2(3)}, x = \frac{-7 \pm \sqrt{49 - 36}}{6}; x = \frac{-7 \pm \sqrt{13}}{6}$

8. $4z^2 - 20z + 5 = 0; a = 4, b = -20, c = 5; z = \frac{-(-20) \pm \sqrt{(-20)^2 - 4(4)(5)}}{2(4)},$
$z = \frac{20 \pm \sqrt{400 - 80}}{8}, z = \frac{20 \pm \sqrt{320}}{8}, z = \frac{20 \pm 8\sqrt{5}}{8}; z = \frac{5 \pm 2\sqrt{5}}{5}$

9. $4p^2 - 6p + 9 = 0; a = 4, b = -6, c = 9; p = \frac{-(-6) \pm \sqrt{(-6)^2 - 4(4)(9)}}{2(4)},$
$p = \frac{6 \pm \sqrt{36 - 144}}{8}, p = \frac{6 \pm \sqrt{-108}}{8}; \text{ no real solutions}$

10. $4t^2 - 8t = 0, 4t(t - 2) = 0, 4t = 0, t = 0; t - 2 = 0, t = 2$

11. $5x^2 = 30, x^2 = 6, x = \pm\sqrt{6}$

12. $(y + 3)^2 = 16, y + 3 = \pm\sqrt{16}, y + 3 = \pm 4, y = 1 \text{ or } y = -7$

13. $3p^2 - 10p + 1 = 0; a = 3, b = -10, c = 1; p = \frac{-(-10) \pm \sqrt{(-10)^2 - 4(3)(1)}}{2(3)},$
$p = \frac{10 \pm \sqrt{100 - 12}}{6}, p = \frac{10 \pm \sqrt{88}}{6}, p = \frac{10 \pm 2\sqrt{22}}{6}; p = \frac{5 \pm \sqrt{22}}{3}$

14. $k^2 + 10k + 30 = 0$; $a = 1$, $b = 10$, $c = 30$; $k = \dfrac{-10 \pm \sqrt{10^2 - 4(1)(30)}}{2(1)}$, $k =$

$\dfrac{-10 \pm \sqrt{100 - 120}}{2}$, $k = \dfrac{-10 \pm \sqrt{-20}}{2}$; no real solutions

15. $m^2 + 5m - 24 = 0$, $(m + 8)(m - 3) = 0$, $m + 8 = 0$, $m = -8$ or $m - 3 = 0$, $m = 3$

16. 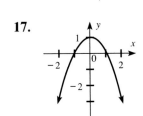 $x^2 = 0$, $x = 0$; The x-intercept is 0; minimum point;

The axis of symmetry is $x = -\dfrac{0}{2(1)}$, $x = 0$.

17. 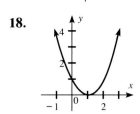 $1 - x^2 = 0$, $x^2 = 1$, $x = \pm 1$; The x-intercepts are 1 and -1; maximum point; The axis of symmetry is

$x = -\dfrac{0}{2(-1)}$, $x = 0$.

18. $x^2 - 2x + 1 = 0$, $(x - 1)^2 = 0$, $x - 1 = 0$, $x = 1$; The x-intercept is 1; minimum point; The axis of

symmetry is $x = -\dfrac{(-2)}{2(1)}$, $x = 1$.

19. $4^2 - 4(1)(-12) = 64$; 2 real solutions

20. $x^2 - 14x + 49 = 0$; $(-14)^2 - 4(1)(49) = 0$; 1 real solution

21. $(-2)^2 - 4(1)(25) = -96$; no real solutions

22. $\dfrac{b}{a} = -[8 + (-1)] = -7$; $\dfrac{c}{a} = (8)(-1) = -8$; $x^2 - 7x - 8 = 0$

23. $\dfrac{b}{a} = -[-6 + (-5)] = 11$; $\dfrac{c}{a} = (-6)(-5) = 30$; $x^2 + 11x + 30 = 0$

24. $\dfrac{b}{a} = -\left[\dfrac{4}{5} + (-1)\right] = \dfrac{1}{5}$; $\dfrac{c}{a} = \left(\dfrac{4}{5}\right)(-1) = -\dfrac{4}{5}$; $x^2 + \dfrac{1}{5}x - \dfrac{4}{5} = 0$; $5x^2 + x - 4 = 0$

25. $\dfrac{b}{a} = -\left[\dfrac{10}{3} + \left(-\dfrac{5}{2}\right)\right] = -\dfrac{5}{6}$; $\dfrac{c}{a} = \left(\dfrac{10}{3}\right)\left(-\dfrac{5}{2}\right) = -\dfrac{25}{3}$; $x^2 - \dfrac{5}{6}x - \dfrac{25}{3} = 0$; $6x^2 - 5x - 50 = 0$

26. $-0.6x^2 + 15x - 4 = 0$; Since $a < 0$, the maximum profit/sale occurs at

the vertex. $x = -\dfrac{15}{2(-0.6)} = 12.5$; $-0.6(12.5)^2 + 15(12.5) - 4 = 89.75$;

The maximum profit/sale of $89.75 occurs after 12.5 sales.

1. $2x^2 = 24$, $x^2 = 12$, $x = \pm\sqrt{12}$, $x = \pm 2\sqrt{3}$

2. $2(y - 1)^2 = 50$, $(y - 1)^2 = 25$, $y - 1 = \pm\sqrt{25}$, $y - 1 = \pm 5$, $y = 6$ or $y = -4$

3. $\frac{1}{2}(-2) = -1$, $(-1)^2 = 1$; $c^2 - 2c + 1 = 15 + 1$, $(c - 1)^2 = 16$, $c - 1 = \pm\sqrt{16}$,

$c - 1 = \pm 4$, $c = 1 \pm 4$; $c = 5$ or $c = -3$

4. $m^2 + 5m = \frac{3}{2}$; $\frac{1}{2}(5) = \frac{5}{2}$, $\left(\frac{5}{2}\right)^2 = \frac{25}{4}$; $m^2 + 5m + \frac{25}{4} = \frac{3}{2} + \frac{25}{4}$, $\left(m + \frac{5}{2}\right)^2 = \frac{31}{4}$,

$m + \frac{5}{2} = \pm\sqrt{\frac{31}{4}}$, $m + \frac{5}{2} = \pm\frac{\sqrt{31}}{2}$, $m = \frac{-5 \pm \sqrt{31}}{2}$

5. $6x^2 + 7x - 5 = 0$; $a = 6$, $b = 7$, $c = -5$; $\dfrac{-7 \pm \sqrt{7^2 - 4(6)(-5)}}{2(6)}$,

$\dfrac{-7 \pm \sqrt{49 + 120}}{12}$, $\dfrac{-7 \pm \sqrt{169}}{12}$, $\dfrac{-7 \pm 13}{12}$; $x = \dfrac{1}{2}$ or $x = -\dfrac{5}{3}$

6. $2n^2 - 2n - 5 = 0$; $a = 2$, $b = -2$, $c = -5$; $\dfrac{-(-2) \pm \sqrt{(-2)^2 - 4(2)(-5)}}{2(2)}$,

$\dfrac{2 \pm \sqrt{4 + 40}}{4}$, $\dfrac{2 \pm \sqrt{44}}{4}$, $\dfrac{2 \pm 2\sqrt{11}}{4}$; $n = \dfrac{1 \pm \sqrt{11}}{2}$

7. $m^2 - 6m - 7 = 0$, $(m - 7)(m + 1) = 0$, $m = 7$ or $m = -1$

8. $p^2 + \dfrac{p}{6} - \dfrac{1}{6} = 0$, $6p^2 + p - 1 = 0$; $(3p - 1)(2p + 1) = 0$, $p = \dfrac{1}{3}$ or $p = -\dfrac{1}{2}$

9. 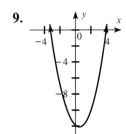 $x^2 - x - 12 = 0$, $(x - 4)(x + 3) = 0$, $x = 4$ or $x = -3$;
The x-intercepts are 4 and -3; minimum point;

The axis of symmetry is $x = -\dfrac{(-1)}{2(1)}$, $x = \dfrac{1}{2}$.

10. $-x^2 - 2 = 0$, $x^2 = -2$, $x = \pm\sqrt{-2}$; Since $\sqrt{-2}$ is
not a real number, there are no x-intercepts;

maximum point; The axis of symmetry is $x = -\dfrac{0}{2(-1)}$, $x = 0$.

11. $(-8)^2 - 4(1)(16) = 0$; 1 real solution

12. $7^2 - 4(2)(-3) = 73$; 2 real solutions

13. $3x^2 - 2x + 5 = 0$; $(-2)^2 - 4(3)(5) = -56$; no real solutions

14. $\dfrac{b}{a} = -(3 + 2) = -5, \dfrac{c}{a} = (3)(2) = 6$; $x^2 - 5x + 6 = 0$

15. $\dfrac{b}{a} = -\left[\dfrac{5}{2} + \left(-\dfrac{9}{4}\right)\right] = -\dfrac{1}{4}$; $\dfrac{c}{a} = \left(\dfrac{5}{2}\right)\left(-\dfrac{9}{4}\right) = -\dfrac{45}{8}$; $x^2 - \dfrac{1}{4}x - \dfrac{45}{8} = 0$; $8x^2 - 2x - 45 = 0$

16. $\dfrac{b}{a} = -\left(-\dfrac{2}{3} + 0\right) = \dfrac{2}{3}$; $\dfrac{c}{a} = \left(-\dfrac{2}{3}\right)(0) = 0$; $x^2 + \dfrac{2}{3}x = 0$; $3x^2 + 2x = 0$

17. Let $w =$ the width, then $2w + 5 =$ the length of the rectangle.
$(2w + 5)(w) = 20$, $2w^2 + 5w = 20$, $2w^2 + 5w - 20 = 0$,
$w = \dfrac{-5 \pm \sqrt{5^2 - 4(2)(-20)}}{2(2)} = \dfrac{-5 \pm \sqrt{25 + 160}}{4} = \dfrac{-5 \pm \sqrt{185}}{4}$; Since width
is always positive, $w = \dfrac{-5 + \sqrt{185}}{4}$, $w = 2.2$ ft; $2w + 5 = 2(2.2) + 5 = 9.4$ ft

18. $3300 = 115x - x^2$, $x^2 - 115x + 3300 = 0$, $(x - 60)(x - 55) = 0$, $x = 60$
sales or $x = 55$ sales

Challenge

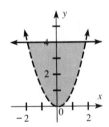

page 579   **Preparing for Standardized Tests**

1. D; $(2x + 4)(2x + 4) = 4x^2 + 16x + 16$; $k = 16$

2. B; $\dfrac{x - 64}{108 - 144} = \dfrac{12 - x}{27 - 108}$, $\dfrac{x - 64}{-36} = \dfrac{12 - x}{-81}$, $-81x + 5184 = -432 + 36x$,
$-117x = -5616$, $x = 48$

3. A; $\begin{array}{r} 10.08 \\ \times\ 0.05 \\ \hline 0.5040 \end{array}$

4. A; $2x - 3 = 5x + 18$, $-3x = 21$, $x = -7$

5. B; $|3(-4) + 2(5)| = |-12 + 10| = |-2| = 2$

6. E; $2x + 5 = \sqrt{7}$, $2x = \sqrt{7} - 5$, $x = \dfrac{\sqrt{7} - 5}{2}$

7. D; $5x - 7 < 3x + 11$, $2x < 18$, $x < 9$

8. B; $180 = 2 \cdot 90 = 2 \cdot 2 \cdot 45 = 2 \cdot 2 \cdot 9 \cdot 5 = 2 \cdot 2 \cdot 3 \cdot 3 \cdot 5 = 2^2 \cdot 3^2 \cdot 5$

9. B; $x^2 - 7x - 18 = 0$, $(x - 9)(x + 2) = 0$, $x = 9$ or $x = -2$

10. C; $5(-1)^3(2) + 2(-1)^2(2)^2 - 3(-1)(2)^3 = -10 + 8 + 24 = 22$

11. B; 10:15 P.M. could be written as 9:75 P.M.; $9:75 - 6:48 = 3:27$

12. E; Let x = the first odd integer, then $x + 2$ = the second. $x(x + 2) = 195$, $x^2 + 2x = 195$, $x^2 + 2x - 195 = 0$, $(x + 15)(x - 13) = 0$; Since x is positive, $x = 13$.

page 580   Mixed Review

1. $-4, -3, -1, 2, 4$

2. $-2, -1.75, -1.5, -1, 0$

3. $-3.1, -3.0, 2.08, 2.8, 2.88$

4. $-\frac{3}{4}, -\frac{2}{3}, 0, \frac{1}{4}, \frac{1}{3}$

5. $-\frac{2}{3}, -\frac{3}{5}, -\frac{4}{7}, \frac{5}{6}, \frac{7}{8}$

6. $-\frac{2}{3}, -0.6, -\frac{1}{2}, 0.7, \frac{3}{4}$

7. 450

8. Jan.

9. $300 - 100 = 200$

10. $300 + 450 + 100 + 150 + 300 = 1300$

11. $(1.8)^2 = (1.8)(1.8) = 3.24$

12. $(0.7)^2 = (0.7)(0.7) = 0.49$

13. $(-2.6)^2 = (-2.6)(-2.6) = 6.76$

14. $-(3.2)^2 = -(3.2)(3.2) = -10.24$

15. $-(4.9)^2 = -(-4.9)(-4.9) = -24.01$

16. $(1.2)^2 = (1.2)(1.2) = 144$

17. $(-5.1)^2 = (-5.1)(-5.1) = 26.01$

18. $-(6.5)^2 = -(6.5)(6.5) = -42.25$

19. $-(-8.3)^2 = -(-8.3)(-8.3) = -68.89$

20. $(7.4)^2 = (7.4)(7.4) = 54.76$

21. $P(\text{blue}) = \dfrac{\text{number of blue marbles}}{\text{total number of marbles}} = \dfrac{3}{10}$

22. $P(\text{green}) = \dfrac{\text{number of green marbles}}{\text{total number of marbles}} = \dfrac{5}{10} = \dfrac{1}{2}$

23. $P(\text{not blue}) = 1 - P(\text{blue}) = 1 - \dfrac{3}{10} = \dfrac{7}{10}$

24. $P(\text{not green}) = 1 - P(\text{green}) = 1 - \dfrac{1}{2} = \dfrac{1}{2}$

Chapter 14   Statistics and Probability

page 582   Capsule Review

   1. $-11, -5, -2, -1, 0, 4, 6, 7, 9$

   2. $-4, -3.2, 1.2, 3, 5.8$

   3. $-3\frac{1}{2}, -3, -\frac{1}{2}, \sqrt{3}, 3, 3\frac{1}{2}$

   4. $-6.10, -6.08, 6.0, 6.1, 2\pi$

page 584   Class Exercises

   1. 7, 8, 8, 9, 9, 9, 10, 10, 11; mode = 9; median = 9;

     mean $= \dfrac{7 + 8 + 8 + 9 + 9 + 9 + 10 + 10 + 11}{9} = 9$

   2. 24 and 26

   3. $\dfrac{25 + 26}{2} = 25.5$

   4. $\dfrac{254}{10} = 25.4$

pages 584–585   Practice Exercises

A **1.** 65, 67, 67, 68, 69, 69, 70, 71, 71, 71; mode = 71; median $= \dfrac{69 + 69}{2} = 69$;

     mean $= \dfrac{65 + 67 + 67 + 68 + 69 + 69 + 70 + 71 + 71 + 71}{10} = \dfrac{688}{10} = 68.8$

   2. 21, 22, 23, 23, 24, 24, 25, 26, 26, 27, 28, 29, 30, 33;

     mode = 23, 24 and 26; median $= \dfrac{25 + 26}{2} = 25.5$; mean $=$

     $\dfrac{21 + 22 + 23 + 23 + 24 + 24 + 25 + 26 + 26 + 27 + 28 + 29 + 30 + 33}{14} = \dfrac{361}{14} = 25.8$

   3. 148, 153, 155, 159, 162, 164, 167, 168, 171, 183;

     mode = none; median $= \dfrac{162 + 164}{2} = 163$;

     mean $= \dfrac{148 + 153 + 155 + 159 + 162 + 164 + 167 + 168 + 171 + 183}{10} = \dfrac{1630}{10} = 163$

   4. 400, 478, 495, 517, 525, 539, 560; mode = none; median = 517;

     mean $= \dfrac{400 + 478 + 495 + 517 + 525 + 539 + 560}{7} = \dfrac{3514}{7} = 502$

   5. 1.1, 1.2, 1.2, 1.3, 1.4, 1.4, 1.5, 1.5, 1.7, 1.8, 2.1;

     mode = 1.2, 1.4 and 1.5; median = 1.4;

     mean $= \dfrac{1.1 + 1.2 + 1.2 + 1.3 + 1.4 + 1.4 + 1.5 + 1.5 + 1.7 + 1.8 + 2.1}{11} = \dfrac{16.2}{11} = 1.5$

6. 34.5, 35.2, 36.8, 37.2, 37.9, 41.0, 41.0, 41.5, 42.1, 42.6

mode = 41.0; median = $\dfrac{37.9 + 41.0}{2} = \dfrac{78.9}{2} = 39.5$; mean =

$\dfrac{34.5 + 35.2 + 36.8 + 37.2 + 37.9 + 41.0 + 41.0 + 41.5 + 42.1 + 42.6}{10} = \dfrac{389.8}{10} = 39.0$

7. mode = 2; median = 2; mean = $\dfrac{100}{50} = 2$ **8.** mode = 1; median = 2; mean = $\dfrac{75}{25} = 3$

B **9.** Change 16 to 12 or 19 to 12 or 12 to 8. **10.** $\dfrac{9 + 11}{2} = 10$; insert 10

11. $\dfrac{6 + 8 + n + 10 + 16}{5} = 13, \dfrac{n + 40}{5} = 13, n + 40 = 65, n = 25$

12. $\dfrac{18 + 35 + n + 9 + 15}{5} = 17, \dfrac{n + 77}{5} = 17, n + 77 = 85, n = 8$

C **13.** All three measures will decrease by 8. **14.** All three measures will increase by 4.

15. All three measures will be halved.

16. The mode is squared, the mean is not squared, and the median may or may not be squared.

17. Let x_1 = the first bowling score, x_2 = the second bowling score, and so on. $\dfrac{x_1 + x_2 + \cdots + x_{30}}{30} = 145, x_1 + x_2 + \cdots + x_{30} = 145(30), x_1 + x_2 + \cdots + x_{30} =$

$4350; \dfrac{x_5 + \cdots + x_{26}}{22} = 148, x_5 + \cdots + x_{26} = 148(22), x_5 + \cdots + x_{26} =$

$3256, \dfrac{x_1 + x_2 + x_3 + x_4 + 3256 + x_{27} + x_{28} + x_{29} + x_{30}}{30} = 145$

$\dfrac{x_1 + x_2 + x_3 + x_4 + x_{27} + x_{28} + x_{29} + x_{30} + 3256}{30} = 145,$

$x_1 + x_2 + x_3 + x_4 + x_{27} + x_{28} + x_{29} + x_{30} + 3256 = 145(30),$
$x_1 + x_2 + x_3 + x_4 + x_{27} + x_{28} + x_{29} + x_{30} + 3256 = 4350,$
$x_1 + x_2 + x_3 + x_4 + x_{27} + x_{28} + x_{29} + x_{30} = 1094,$

mean of the removed scores = $\dfrac{x_1 + x_2 + x_3 + x_4 + x_{27} + x_{28} + x_{29} + x_{30}}{8} = \dfrac{1094}{8} = 136.75$

18. median **19.** mode **20.** mean

page 585 Algebra in Demography

1. The population increased by a factor of about 15 since 1900.

2. $\dfrac{408{,}442 + 587{,}718 + 654{,}153 + 785{,}880}{4} = \dfrac{2{,}436{,}193}{4} = 609{,}048.25$

3. Increase   **4.** Answers may vary.   **5.** Answers may vary.

page 586   Capsule Review

1. 72   **2.** 73

3. $\frac{1}{18}$(70 + 70 + 71 + 71 + 72 + 72 + 72 + 72 + 73 + 73 + 73 + 74 + 74 + 75 +

75 + 76 + 77 + 78) = $\frac{1318}{18}$ = 73.22

page 587–588   Class Exercises

1.

Intervals	Tally	Frequency
60–64	||	2
65–69	|||	3
70–74	₪ ||	7
75–79	₪	5
80–84	₪	5
85–89	|||	3
90–94	||	2
95–99	||	2
100–104			1

2.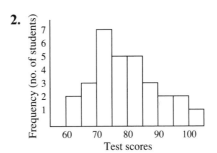

3. $\frac{25}{30}$ = $83\frac{1}{3}$%

4.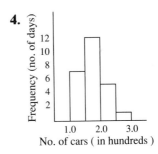

5. mode = 150–190; median = 150–190;

mean = $\frac{1}{25}$(1.5 + 1.0 + 1.2 + 1.5 + 1.7 + 2.3 + 1.8 + 1.9 + 1.3 + 1.5 + 1.6 +

2.2 + 1.8 + 1.9 + 1.9 + 1.7 + 1.4 + 1.3 + 2.0 + 2.2 + 2.5 + 2.0 + 1.9 +

1.4 + 1.4) = $\frac{42.9}{25}$ = 1.7 hundreds, or 170

6. Pie charts, bar graphs, or line graphs

A 1.

Interval	Tally	Frequency
45-49	|	1
50-54	|	1
55-59	|||	3
60-64	|||	3
65-69	⧸⧸⧸⧸ |	6
70-74	||||	4
75-79	⧸⧸⧸⧸ |	6
80-84	||	2
85-89	||	2
90-94	||	2

2. 70-74

3. 70.97

4.

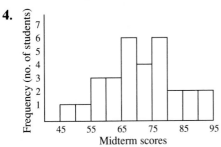

Midterm scores

5. 16 students

6.

Interval	Tally	Frequency
10-19	|||	3
20-29	⧸⧸⧸⧸ ||	7
30-39	⧸⧸⧸⧸ ||	7
40-49	⧸⧸⧸⧸ ⧸⧸⧸⧸	10
50-59	||||	4
60-69	||||	4

7. 40-49

8. 38.9 yr

9.

10. $\dfrac{14}{35} = 0.40 = 40\%$

B 11. $\dfrac{74}{98} = 0.755 = 75.5\%$

12. $20-29

13. 24.5%; 49%

14. $0-9; $20-29

15.

Interval	Tally	Frequency
0-4	ⅢⅠ ⅢⅠ	9
5-9	ⅢⅠ ⅢⅠ ⅠⅠ	12
10-14	Ⅲ	3
15-20	Ⅰ	1

mode = 4, 5, and 6;
median = 6; mean = 6.1

16.

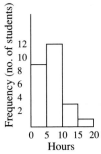

C 17.

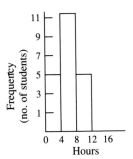

18.

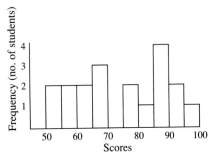

19.

Interval	Tally	Frequency
10,000-14,999	ⅢⅠ	4
15,000-19,999	ⅢⅠ	4
20,000-24,999	Ⅲ	3
25,000-29,999		0
30,000-34,999	Ⅰ	1
35,000-39,999	Ⅰ	1

20.

Interval	Tally	Frequency
101-200	ⅢⅠ ⅢⅠ Ⅰ	11
201-300	ⅠⅠ	2
301-400	ⅢⅠ	4
401-500		0
501-600	ⅠⅠ	2
601-700	Ⅰ	1

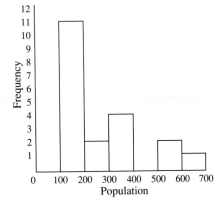

21. $5x + 4$ **22.** $4x - 7$ **23.** $2(x - 12)$

24. Check students' work.

page 590   Capsule Review

1. 10 **2.** $\dfrac{1}{8}$ **3.** 1.1 **4.** 144 **5.** $\dfrac{9}{16}$

page 592   Class Exercises

1. mode $= 1$; median $= \dfrac{1 + 2}{2} = 1.5$; mean $= \dfrac{0 + 0 + 1 + 1 + 1 + 2 + 2 + 3 + 4 + 6}{10} = 2$

2. range $= 6 - 0 = 6$;

No. of Pets	Deviation from mean	Freq. × (Dev.)²
0	$0 - 2 = -2$	2(4)
1	$1 - 2 = -1$	3(1)
2	$2 - 2 = 0$	2(0)
3	$3 - 2 = 1$	1(1)
4	$4 - 2 = 2$	1(4)
6	$6 - 2 = 4$	1(16)

Sum of squares = 32

variance $= \dfrac{32}{10} = 3.2$; standard deviation $= \sqrt{3.2} = 1.8$

pages 592–593   Practice Exercises

A 1. mean $= 150$;

Wages	Deviation from mean	Freq. × (Dev.)²
100	$100 - 150 = -50$	5(2500)
150	$150 - 150 = 0$	2(0)
200	$200 - 150 = 50$	1(2500)
250	$250 - 150 = 100$	2(10,000)

Sum of squares = 35,000

variance $= \dfrac{35,000}{10} = \$3,500$

standard deviation $= \sqrt{3500} = \$59.16$

2. mean $= 140$;

Wages	Deviation from mean	Freq. × (Dev.)²
100	$100 - 140 = -40$	4(1600)
150	$150 - 140 = 10$	5(100)
200	$200 - 140 = 60$	0(3600)
250	$250 - 140 = 110$	1(12,100)

Sum of squares = 19,000

variance $= \dfrac{19,000}{10} = \$1,900$

standard deviation $= \sqrt{1900} = \$43.59$

3. 5, 8, 8, 9, 10, 11, 12;

mode = 8; median = 9; mean = $\dfrac{5 + 8 + 8 + 9 + 10 + 11 + 12}{7} = \dfrac{63}{7} = 9$;

Temp. Readings	Deviation from mean	Freq. × (Dev.)²
5	5 − 9 = −4	1(16)
8	8 − 9 = −1	2(1)
9	9 − 9 = 0	1(0)
10	10 − 9 = 1	1(1)
11	11 − 9 = 2	1(4)
12	12 − 9 = 3	1(9)
	Sum of squares = 32	

range = 12 − 5 = 7

variance = $\dfrac{32}{7} = 4.6$

standard deviation = $\sqrt{4.6} = 2.1$

4. mode = \$10 and \$12; median = $\dfrac{11 + 12}{2} = \$11.50$;

mean = $\dfrac{7 + 10 + 10 + 11 + 12 + 12 + 14 + 20}{8} = \dfrac{96}{8} = 12; \$12$

Wages	Deviation from mean	Freq. × (Dev.)²
7	7 − 12 = −5	1(25)
10	10 − 12 = −2	2(4)
11	11 − 12 = −1	1(1)
12	12 − 12 = 0	2(0)
14	14 − 12 = 2	1(4)
20	20 − 12 = 8	1(64)
	Sum of squares = 102	

range = 20 − 7 = \$13

variance = $\dfrac{102}{8} = \$12.8$

standard deviation = $\sqrt{12.8} = \$3.6$

B 5. 29.0, 29.0, 29.4, 29.4, 29.6, 29.9, 30.1;

mode = 29.0 and 29.4; median = 29.4;

mean = $\dfrac{29.0 + 29.0 + 29.4 + 29.4 + 29.6 + 29.9 + 30.1}{7} = 29.5$;

Readings	Deviation from mean	Freq. × (Dev.)²
29.0	29.0 − 29.5 = −0.5	2(0.25)
29.4	29.4 − 29.5 = −0.1	2(0.01)
29.6	29.6 − 29.5 = 0.1	1(0.01)
29.9	29.9 − 29.5 = 0.4	1(0.16)
30.1	30.1 − 29.5 = 0.6	1(0.36)
	Sum of squares = 1.05	

range = 30.1 − 29.0 = 1.1

variance = $\dfrac{1.05}{7} = 0.2$

standard deviation = $\sqrt{0.2} = 0.4$

6. 23,000, 26,000, 32,000, 37,000, 42,000, 45,000, 46,000, 47,000, 50,000, 52,000;

mode = none; median = $\dfrac{42,000 + 45,000}{2} = 43,500$;

$$\text{mean} = \frac{1}{10}(23{,}000 + 26{,}000 + 32{,}000 + 37{,}000 + 42{,}000 + 45{,}000 + 46{,}000 +$$

$$47{,}000 + 50{,}000 + 52{,}000) = \frac{400{,}000}{10} = 40{,}000;$$

Population	Deviation from mean	Freq. × (Dev.)²
23,000	23,000 − 40,000 = −17,000	1(289,000,000)
26,000	26,000 − 40,000 = −14,000	1(196,000,000)
32,000	32,000 − 40,000 = −8,000	1(64,000,000)
37,000	37,000 − 40,000 = −3,000	1(9,000,000)
42,000	42,000 − 40,000 = 2,000	1(4,000,000)
45,000	45,000 − 40,000 = 5,000	1(25,000,000)
46,000	46,000 − 40,000 = 6,000	1(36,000,000)
47,000	47,000 − 40,000 = 7,000	1(49,000,000)
50,000	50,000 − 40,000 = 10,000	1(100,000,000)
52,000	52,000 − 40,000 = 12,000	1(144,000,000)

Sum of squares = 916,000,000

range = 52,000 − 23,000 = 29,000

$$\text{variance} = \frac{916{,}000{,}000}{10} = 91{,}600{,}000$$

standard deviation = $\sqrt{91{,}600{,}000}$ = 9570.8

C 7. $1.76, $2.08, $2.86, $3.25, $4.29, $4.78, $5.06, $5.12, $6.32, $6.45;

mode = none; median = $\dfrac{4.29 + 4.78}{2}$ = $4.54;

$$\text{mean} = \frac{1.76 + 2.08 + 2.86 + 3.25 + 4.29 + 4.78 + 5.06 + 5.12 + 6.32 + 6.45}{10} = \$4.20;$$

Earnings per share	Deviation from mean	Freq. × (Dev.)²
$1.76	1.76 − 4.20 = −2.44	1(5.95)
$2.08	2.08 − 4.20 = −2.12	1(4.49)
$2.86	2.86 − 4.20 = −1.34	1(1.80)
$3.25	3.25 − 4.20 = −0.95	1(0.90)
$4.29	4.29 − 4.20 = 0.09	1(0.01)
$4.78	4.78 − 4.20 = 0.58	1(0.34)
$5.06	5.06 − 4.20 = 0.86	1(0.74)
$5.12	5.12 − 4.20 = 0.92	1(0.85)
$6.32	6.32 − 4.20 = 2.12	1(4.49)
$6.45	6.45 − 4.20 = 2.25	1(5.06)

Sum of squares = 24.63

range = 6.45 − 1.76 = $4.69

$$\text{variance} = \frac{24.63}{10} = \$2.46$$

standard deviation = $\sqrt{2.46}$ = $1.57

8. 11, 17, 17, 20, 22, 24, 24;

$$\text{mean} = \frac{11 + 17 + 17 + 20 + 22 + 24 + 24}{7} = 19.3;$$

Grams	Deviation from mean	Freq. × (Dev.)²
11	$11 - 19.3 = -8.3$	$1(68.89)$
17	$17 - 19.3 = -2.3$	$2(5.29)$
20	$20 - 19.3 = 0.7$	$1(0.49)$
22	$22 - 19.3 = 0.7$	$1(0.49)$
24	$24 - 19.3 = 4.7$	$2(22.09)$
	Sum of squares = 124.63	

$$\text{variance} = \frac{124.63}{7} = 17.80$$

$$\text{standard deviation} = \sqrt{17.80} = 4.2$$

9. $\text{mean} = \dfrac{\$21{,}000 + \$24{,}000 + \$30{,}000 + \$25{,}000 + \$30{,}000}{5} = \$26{,}000;$

Wages	Deviation from mean	Freq. × (Dev.)²
21,000	$21{,}000 - 26{,}000 = -5{,}000$	$1(25{,}000{,}000)$
24,000	$24{,}000 - 26{,}000 = -2{,}000$	$1(4{,}000{,}000)$
25,000	$25{,}000 - 26{,}000 = -1{,}000$	$1(1{,}000{,}000)$
30,000	$30{,}000 - 26{,}000 = 4{,}000$	$2(16{,}000{,}000)$
	Sum of squares = 62,000,000	

$\text{variance} = \dfrac{62{,}000{,}000}{5} = 12{,}400{,}000;$ standard deviation $= \sqrt{12{,}400{,}000} = \$3521.36$

page 593 Test Yourself

1. Let $x_3 =$ the third lowest score, let $x_4 =$ the fourth lowest score and so on up to x_{22} which is the twenty-second lowest score.

$$\frac{60 + 62 + x_3 + x_4 \cdots + x_{22} + 92 + 97}{24} = 74, \quad \frac{x_3 + x_4 + \cdots + x_{22} + 311}{24} = 74,$$

$x_3 + x_4 + \cdots + x_{22} + 311 = 1776,\ x_3 + x_4 + \cdots + x_{22} = 1465;\ \dfrac{x_3 + x_4 + \cdots + x_{22}}{20} =$

$\dfrac{1465}{20} = 73.25;$ The mean of the twenty remaining scores is 73.25.

2. 18, 19, 19, 19, 20, 21, 21, 22, 22, 24, 25, 25, 25, 28, 28, 29, 30, 33, 35,

37, 38, 38, 40, 41; mode = 19 and 25; median = 25; mean $= \dfrac{657}{24} = 27$

3.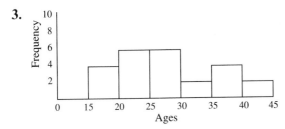

$\dfrac{10}{24} \times 100 = 41.\overline{6};$ 42% of the employees are under 25.

4.

Ages	Deviation from mean	Freq. × (Dev.)^2
18	$18 - 27 = -9$	$1(81)$
19	$19 - 27 = -8$	$3(64)$
20	$20 - 27 = -7$	$1(49)$
21	$21 - 27 = -6$	$2(36)$
22	$22 - 27 = -5$	$2(25)$
24	$24 - 27 = -3$	$1(9)$
25	$25 - 27 = -2$	$3(4)$
28	$28 - 27 = 1$	$2(1)$
29	$29 - 27 = 2$	$1(4)$
30	$30 - 27 = 3$	$1(9)$
33	$33 - 27 = 6$	$1(36)$
35	$35 - 27 = 8$	$1(64)$
37	$37 - 27 = 10$	$1(100)$
38	$38 - 27 = 11$	$2(121)$
40	$40 - 27 = 13$	$1(169)$
41	$41 - 27 = 14$	$1(196)$

Sum of squares = 1287

range $= 41 - 18 = 23$

variance $= \dfrac{1287}{24} = 53.6$

standard deviation $= \sqrt{53.6} = 7.3$

5. mean $= \dfrac{\text{Sum of all the cars that pass over a 24 h period}}{24} = \dfrac{1725}{24} = 71.9;$

No. of cars	Deviation from mean	Freq. × (Dev.)^2
10	$10 - 71.9 = -61.9$	$2(3831.61)$
20	$20 - 71.9 = -51.9$	$1(2693.61)$
40	$40 - 71.9 = -31.9$	$2(1017.61)$
50	$50 - 71.9 = -21.9$	$1(479.61)$
60	$60 - 71.9 = -10.9$	$2(118.81)$
70	$70 - 71.9 = -1.9$	$5(3.61)$
75	$75 - 71.9 = 3.1$	$1(9.61)$
80	$80 - 71.9 = 8.1$	$3(65.61)$
85	$85 - 71.9 = 13.1$	$1(171.61)$
95	$95 - 71.9 = 23.1$	$1(533.61)$
100	$100 - 71.9 = 28.1$	$2(789.61)$
120	$120 - 71.9 = 48.1$	$1(2313.61)$
130	$130 - 71.9 = 58.1$	$1(3375.61)$
140	$140 - 71.9 = 68.1$	$1(4637.61)$

Sum of squares = 25,945.04

variance $= \dfrac{25{,}945.04}{24} = 1081.04$

standard deviation $= \sqrt{1081.04} = 32.9$

page 595 Integrating Algebra: Scattergrams

1. As the age increases the height also increases. The upward slant of the best fit line indicates that there is a positive correlation. The slope is $\frac{3}{4}$.

2.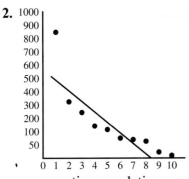

yes; negative correlation

3.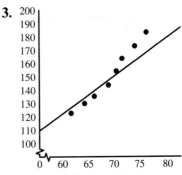

yes; positive correlation

4.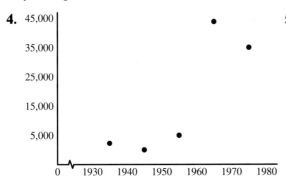

No correlation, cannot predict the certainty of immigration for 1981–1990.

5.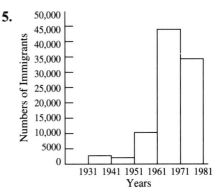

The bar graph more clearly illustrates the interval nature of the given data.

6. Answers may vary.

page 596 Capsule Review

1. $\frac{3}{5}$ or 60%　　**2.** $\frac{3}{8}$ or 37.5%　　**3.** $\frac{4}{7}$ or 57.1%　　**4.** $\frac{3}{4}$ or 75%

page 598 Class Exercises

1. $\frac{5}{10} = \frac{1}{2}$　　　　**2.** $\frac{2}{10} = \frac{1}{5}$　　　　**3.** $\frac{1}{10}$

4. $\frac{10}{10} = 1$　　　　**5.** $\frac{4}{10} = \frac{2}{5}$　　　　**6.** $\frac{4}{10} = \frac{2}{5}$

7. $\frac{0}{10} = 0$ **8.** $\frac{2}{10} = \frac{1}{5}$ **9.** $\frac{9}{10}$

10. Yes, because $P(E)$ and $P(\overline{E})$ are the only two possible situations. Therefore, their sum is 1. $P(E) + P(\overline{E}) = 1$, $P(\overline{E}) = 1 - P(E)$

11. Yes, because odds are defined as $\frac{P(E)}{P(\overline{E})}$. Therefore, if $P(\overline{E}) = 1 - P(E)$, then

Odds $(E) = \dfrac{P(E)}{1 - P(E)}$.

pages 598–599   Practice Exercises

A 1. $\frac{8}{24} = \frac{1}{3}$ **2.** $\frac{8}{24} = \frac{1}{3}$ **3.** $\frac{3}{24} = \frac{1}{8}$

4. $\frac{3}{24} = \frac{1}{8}$ **5.** $\frac{16}{24} = \frac{2}{3}$ **6.** $\frac{16}{24} = \frac{2}{3}$

7. $\frac{6}{24} = \frac{1}{4}$ **8.** $\frac{0}{24} = 0$ **9.** $\frac{6}{24} = \frac{1}{4}$

10. $\frac{15}{24} = \frac{5}{8}$ **11.** $\frac{21}{24} = \frac{7}{8}$ **12.** $\frac{21}{24} = \frac{7}{8}$

13. $\frac{5}{31}$ **14.** $\frac{10}{31}$

15. $\dfrac{\frac{14}{30}}{\frac{16}{30}} = \dfrac{14}{16} = \dfrac{7}{8}$, odds for rain in April is 7 to 8.

16. $\dfrac{\frac{21}{30}}{\frac{9}{30}} = \dfrac{21}{9} = \dfrac{7}{3}$; The odds against rain in September is 7 to 3.

17. $\frac{24}{31}$ **18.** $\dfrac{\frac{21}{30}}{\frac{9}{30}} = \dfrac{21}{9} = \dfrac{7}{3}$; The odds against rain in June is 7 to 3.

B 19. $\frac{5}{9}$ **20.** $\frac{4}{9}$ **21.** $\frac{2}{9}$ **22.** $\frac{1}{9}$

23. $\frac{3}{9} = \frac{1}{3}$ **24.** $\frac{2}{9}$ **25.** $\frac{0}{9} = 0$ **26.** $\frac{0}{9} = 0$

C **27.** $\dfrac{12}{24} = \dfrac{1}{2}$ **28.** $\dfrac{20}{24} = \dfrac{5}{6}$ **29.** $\dfrac{16}{24} = \dfrac{2}{3}$ **30.** $\dfrac{12}{24} = \dfrac{1}{2}$

31. $\dfrac{\frac{3}{4}}{\frac{1}{4}} = \dfrac{3}{1}$; The odds against guessing the correct answer is 3 to 1.

32. $\dfrac{2}{7}$ is the probability of taking a \$5 bill; $\dfrac{1}{7}$ is the probability of taking a \$10 bill.

33. distance from (2, 6) to (5, 2) = $\sqrt{(5-2)^2 + (2-6)^2} = \sqrt{3^2 + (-4)^2} = \sqrt{9+16} = \sqrt{25} = 5$; distance from (5, 2) to (−3, −4) = $\sqrt{(-3-5)^2 + (-4-2)^2} = \sqrt{(-8)^2 + (-6)^2} = \sqrt{64+36} = \sqrt{100} = 10$; distance from (−3, −4) to (2, 6) = $\sqrt{(2+3)^2 + (6+4)^2} = \sqrt{5^2 + 10^2} = \sqrt{25+100} = \sqrt{125}$; $(5)^2 + (10)^2 = (\sqrt{125})^2$; Yes, the points can be connected to form a right triangle.

34. distance from (−3, −2) to (0, 4) = $\sqrt{(0+3)^2 + (4+2)^2} = \sqrt{9+36} = \sqrt{45}$; distance from (0, 4) to (7, 3) = $\sqrt{(7-0)^2 + (3-4)^2} = \sqrt{49+1} = \sqrt{50}$; distance from (7, 3) to (−3, −2) = $\sqrt{(-3-7)^2 + (-2-3)^2} = \sqrt{100+25} = \sqrt{125}$; $(\sqrt{45})^2 + (\sqrt{50})^2 \neq (\sqrt{125})^2$; No, the points cannot be connected to form a right triangle.

35. Answers may vary.

page 601 Class Exercises

1.

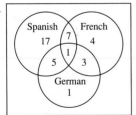

$\dfrac{7}{28} = \dfrac{1}{4}$; $\dfrac{1}{4}$ is the probability of selecting a student at random who only studies Spanish and French.

2.

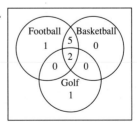

$\dfrac{1}{9}$ is the probability of choosing a parent who prefers only golf.

pages 601–602 Practice Exercises

A 1.

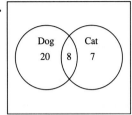

$\frac{7}{35} = \frac{1}{5}$; The probabality of randomly selecting a student who will only have a cat is $\frac{1}{5}$.

2.

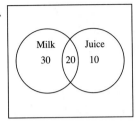

$\frac{30}{60} = \frac{1}{2}$; The probability of randomly selecting a student who will drink only milk is $\frac{1}{2}$.

B 3.

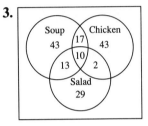

$\frac{10}{157}$ is the probability of randomly selecting a person who has had soup, chicken and salad.

4.

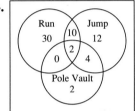

$\frac{4}{60} = \frac{1}{15}$; The probability of randomly selecting a member who only jumped and pole vaulted is $\frac{1}{15}$.

C 5.

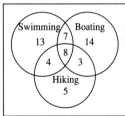

$\frac{7 + 3 + 8}{54} = \frac{18}{54} = \frac{1}{3}$

The probability of randomly selecting a camper who chose swimming and boating or boating and hiking is $\frac{1}{3}$.

page 602 Mixed Problem Solving Review

1. Let x = the width of room, then $x + 3$ = the length of room.
$x(x + 3) = 108$, $x^2 + 3x - 108 = 0$, $(x + 12)(x - 9) = 0$
$x + 12 = 0 \quad | \quad x - 9 = 0$
$\qquad x = -12 \quad | \qquad x = 9$
The width of the room is 9 ft and the length of the room is $9 + 3 = 12$ ft.

2. Let a = Al's present age, then $a - 5$ = his age five years ago.
Let f = Al's father's present age, then $f - 5$ = his age five years ago.
$$\begin{cases} f = 3a \\ f - 5 = 4(a - 5) \end{cases}$$
$3a - 5 = 4(a - 5)$, $3a - 5 = 4a - 20$, $-5 = a - 20$, $15 = a$; $f = 3(15)$, $f = 45$;
Al is 15 years old and his father is 45 years old.

3. If masses m_1 and m_2 are placed on a lever d_1 and d_2 distances respectively from fulcrum, then $m_1d_1 = m_2d_2$. $20 \times 15 = 30 \times d$, $300 = 30d$, $d = 10$;
The 30 kg object is 10 ft from fulcrum.

4. Let n = the number of hours working together.

	work rate ·	time worked =	part of job done
George	$\dfrac{1}{12}$	n	$\dfrac{n}{12}$
Joe	$\dfrac{1}{15}$	n	$\dfrac{n}{15}$

$\dfrac{n}{12} + \dfrac{n}{15} = 1$, $60\left(\dfrac{n}{12} + \dfrac{n}{15}\right) = 60(1)$, $5n + 4n = 60$, $9n = 60$, $n = \dfrac{60}{9} = 6\dfrac{2}{3}$.

It will take George and Joe $6\dfrac{2}{3}$ h to paint the room together.

5. Let n = the number of nickels and q = the number of quarters.
$$\begin{cases} n + q = 45 \\ 5n + 25q = 965 \end{cases} \longrightarrow \begin{array}{l} -5(n + q) = (-5)45 \\ 5n + 25q = 965 \end{array} \longrightarrow \begin{array}{l} -5n - 5q = -225 \\ 5n + 25q = 965 \end{array}$$

$$\begin{array}{l} -5n - 5q = -225 \\ \underline{5n + 25q = 965} \\ \quad\quad 20q = 740 \\ \quad\quad\quad q = 37 \end{array}$$

$n + 37 = 45$, $n = 8$;
Eric has 37 quarters and 8 nickels.

6. Let r = the rate of the boat in still water and c = the rate of the current.

	rate ·	time =	distance
downstream	$r + c$	4	64
upstream	$r - c$	5	35

$$\begin{cases} (r + c)4 = 64 \longrightarrow r + c = 16 \\ (r - c)5 = 35 \longrightarrow r - c = 7 \end{cases}$$

$$\begin{array}{l} r + c = 16 \\ \underline{r - c = 7} \\ 2r = 23, \; r = \dfrac{23}{2}; \end{array}$$ The rate of boat in still water is $\dfrac{23}{2}$ mi/h.

page 602 Project

Check students' diagrams.

page 603 Capsule Review

1. $P(R) = \dfrac{4}{10} = \dfrac{2}{5}$

2. $P(G) = \dfrac{2}{10} = \dfrac{1}{5}$

3. $P(Y) = \dfrac{3}{10}$

4. $P(\overline{Y}) = 1 - \dfrac{3}{10} = \dfrac{7}{10}$

5. $P(\text{B or G}) = \dfrac{1}{10} + \dfrac{2}{10} = \dfrac{3}{10}$

6. $P(\text{R or G}) = \dfrac{2}{5} + \dfrac{1}{5} = \dfrac{3}{5}$

7. $P(\text{a pencil}) = 1$

8. $P(\text{a pen}) = 0$

page 606 Class Exercises

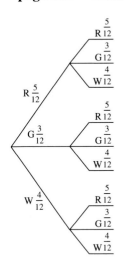

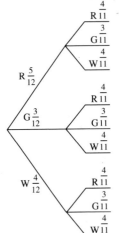

1. $P(\text{1st R and 2nd W}) =$
$$\left(\dfrac{5}{12}\right)\left(\dfrac{4}{12}\right) = \dfrac{20}{144} = \dfrac{5}{36}$$

2. $P(\text{1st R and 2nd R}) =$
$$\left(\dfrac{5}{12}\right)\left(\dfrac{5}{12}\right) = \dfrac{25}{144}$$

3. $P(\text{1st R and 2nd }\overline{R}) =$
$$\left(\dfrac{5}{12}\right)\left(\dfrac{7}{12}\right) = \dfrac{35}{144}$$

4. $P(\text{1st R and 2nd W}) =$
$$\left(\dfrac{5}{12}\right)\left(\dfrac{4}{11}\right) = \dfrac{20}{132} = \dfrac{5}{33}$$

5. $P(\text{1st R and 2nd R}) =$
$$\left(\dfrac{5}{12}\right)\left(\dfrac{4}{11}\right) = \dfrac{20}{132} = \dfrac{5}{33}$$

6. $P(\text{1st R and 2nd }\overline{R}) =$
$$\left(\dfrac{5}{12}\right)\left(\dfrac{7}{11}\right) = \dfrac{35}{132}$$

7. $P(s) = \dfrac{5}{15}$, $P(t) = \dfrac{7}{15}$, $P(s \text{ and } t) = \dfrac{2}{15}$; $P(s \text{ or } t) = P(s) + P(t) - P(s \text{ and } t) =$
$$\dfrac{5}{15} + \dfrac{7}{15} - \dfrac{2}{15} = \dfrac{10}{15} = \dfrac{2}{3}$$

pages 606–607 Practice Exercises

A **1.** $P(\text{1st 5 and 2nd 3}) = \left(\dfrac{3}{8}\right)\left(\dfrac{2}{8}\right) = \dfrac{6}{64} = \dfrac{3}{32}$ **2.** $P(\text{1st 4 and 2nd 2}) = \left(\dfrac{1}{8}\right)\left(\dfrac{2}{8}\right) = \dfrac{2}{64} = \dfrac{1}{32}$

3. $P(\text{1st 5 and 2nd 5}) = \left(\dfrac{3}{8}\right)\left(\dfrac{3}{8}\right) = \dfrac{9}{64}$ **4.** $P(\text{1st 3 and 2nd 3}) = \left(\dfrac{2}{8}\right)\left(\dfrac{2}{8}\right) = \dfrac{4}{64} = \dfrac{1}{16}$

5. $P(\text{1st even and 2nd odd}) = \left(\dfrac{3}{8}\right)\left(\dfrac{5}{8}\right) = \dfrac{15}{64}$ **6.** $P(\text{1st odd and 2nd odd}) = \left(\dfrac{5}{8}\right)\left(\dfrac{5}{8}\right) = \dfrac{25}{64}$

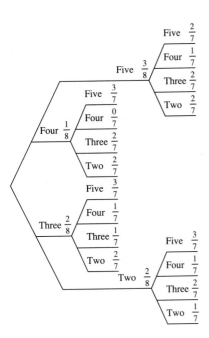

Five $\frac{2}{7}$
Four $\frac{1}{7}$
Five $\frac{3}{8}$
Three $\frac{2}{7}$
Two $\frac{2}{7}$

Five $\frac{3}{7}$
Four $\frac{0}{7}$
Four $\frac{1}{8}$
Three $\frac{2}{7}$
Two $\frac{2}{7}$

Five $\frac{3}{7}$
Four $\frac{1}{7}$
Three $\frac{2}{8}$
Three $\frac{1}{7}$
Two $\frac{2}{7}$

Five $\frac{3}{7}$
Four $\frac{1}{7}$
Two $\frac{2}{8}$
Three $\frac{2}{7}$
Two $\frac{1}{7}$

7. $P(\text{1st 5 and 2nd 3}) = \left(\frac{3}{8}\right)\left(\frac{2}{7}\right) = \frac{6}{56} = \frac{3}{28}$

8. $P(\text{1st 4 and 2nd 2}) = \left(\frac{1}{8}\right)\left(\frac{2}{7}\right) = \frac{2}{56} = \frac{1}{28}$

9. $P(\text{1st 4 and 2nd 4}) = \left(\frac{1}{8}\right)\left(\frac{0}{7}\right) = 0$

10. $P(\text{1st 2 and 2nd 2}) = \left(\frac{2}{8}\right)\left(\frac{1}{7}\right) = \frac{2}{56} = \frac{1}{28}$

11. $P(\text{1st even and 2nd even}) = \left(\frac{1}{8}\right)\left(\frac{0}{7}\right) + \left(\frac{1}{8}\right)\left(\frac{2}{7}\right) +$
$\left(\frac{2}{8}\right)\left(\frac{1}{7}\right) + \left(\frac{2}{8}\right)\left(\frac{1}{7}\right) = 0 + \frac{2}{56} + \frac{2}{56} + \frac{2}{56} = \frac{6}{56} = \frac{3}{28}$

12. $P(\text{1st odd and 2nd even}) = \left(\frac{3}{8}\right)\left(\frac{1}{7}\right) + \left(\frac{3}{8}\right)\left(\frac{2}{7}\right) +$
$\left(\frac{2}{8}\right)\left(\frac{1}{7}\right) + \left(\frac{2}{8}\right)\left(\frac{2}{7}\right) = \frac{3}{56} + \frac{6}{56} + \frac{2}{56} + \frac{4}{56} = \frac{15}{56}$

13. $P(\text{4 or 5}) = P(4) + P(5) = \frac{1}{8} + \frac{3}{8} = \frac{1}{2}$ **14.** $P(\text{2 or 3}) = P(2) + P(3) = \frac{2}{8} + \frac{2}{8} = \frac{1}{2}$

15. $P(\text{3 or 5}) = P(3) + P(5) = \frac{2}{8} + \frac{3}{8} = \frac{5}{8}$ **16.** $P(\text{2 or 4}) = P(2) + P(4) = \frac{2}{8} + \frac{1}{8} = \frac{3}{8}$

17. $P(\text{2 or 5}) = P(2) + P(5) = \frac{2}{8} + \frac{3}{8} = \frac{5}{8}$ **18.** $P(\text{3 or 4}) = P(3) + P(4) = \frac{2}{8} + \frac{1}{8} = \frac{3}{8}$

B 19. $P(a, e, i, o, u) = \frac{5}{26}$

20. $P(\text{The letter is a consonant.}) = \frac{21}{26}$

21. $P(p, r, o, b, a, i, l, t, y) = \frac{9}{26}$

22. $P(\text{The letter is contained in the word } probability \text{ and the word } statistics.) = \frac{3}{26}$

23. $P(\text{The letter is contained in the word } probability \text{ or the word } statistics.) =$
$\frac{9}{26} + \frac{5}{26} - \frac{3}{26} = \frac{11}{26}$

24. $P(\text{ace or heart}) = \frac{4}{52} + \frac{13}{52} - \frac{1}{52} = \frac{16}{52} = \frac{4}{13}$

25. $P(\text{mysteries or science fiction}) = \frac{6}{20} + \frac{10}{20} - \frac{2}{20} = \frac{14}{20} = \frac{7}{10}$

26. $P(\text{girl or an Algebra student}) = \dfrac{32}{50} + \dfrac{40}{50} - \dfrac{25}{50} = \dfrac{47}{50}$

C 27. If 8 eat salads, $8 - 5 = 3$, 3 eat only salads. If 12 eat sandwiches, $12 - 5 = 7$, 7 eat only sandwiches. $P(\text{only sandwich or only salad}) = \left(\dfrac{3}{28}\right)\left(\dfrac{2}{27}\right) +$

$\left(\dfrac{3}{28}\right)\left(\dfrac{7}{27}\right) + \left(\dfrac{7}{28}\right)\left(\dfrac{3}{27}\right) + \left(\dfrac{7}{28}\right)\left(\dfrac{6}{27}\right) = \dfrac{90}{756} = \dfrac{5}{42}$

28. Let $x =$ the number of students who take both, $y =$ the number of students who take only Biology, and $z =$ the number of students who take only Geometry.

$\begin{cases} x + y = 12 \\ x + z = 10 \\ x + y + z = 19 \end{cases}$ $\quad \begin{aligned} 2x + y + z &= 22 \\ x + y + z &= 19 \\ \hline x &= 3 \end{aligned}$

Since $x = 3$, $y = 12 - 3 = 9$, and $z = 10 - 3 = 7$.

$P \text{ (only Biology and only Biology)} = \left(\dfrac{9}{30}\right)\left(\dfrac{8}{29}\right) = \dfrac{72}{870} = \dfrac{12}{145}$

29. Let $x =$ the percent of people who purchased only a microwave, then $2x =$ the percent of people who purchased only a dishwasher.
$x + 2x = 60$, $3x = 60$, $x = 20$; $2x = 2(20) = 40$; 40%

30. $\left(\dfrac{15}{27}\right)\left(\dfrac{14}{26}\right)\left(\dfrac{13}{25}\right)\left(\dfrac{12}{24}\right)\left(\dfrac{11}{23}\right) = 0.04$; $\left(\dfrac{12}{27}\right)\left(\dfrac{11}{26}\right)\left(\dfrac{10}{25}\right)\left(\dfrac{9}{24}\right)\left(\dfrac{8}{23}\right) = 0.01$; $0.04 + 0.01 = 0.05$

31. $P(\$2) = \left(\dfrac{3}{7}\right)\left(\dfrac{2}{6}\right) = \dfrac{1}{7}$, $P(\$10) = \left(\dfrac{4}{7}\right)\left(\dfrac{3}{6}\right) = \dfrac{2}{7}$; $P(\$2 \text{ or } \$10) = \dfrac{1}{7} + \dfrac{2}{7} = \dfrac{3}{7}$

page 607 Test Yourself

1. $P(\text{red}) = \dfrac{3}{8}$

2. $P(2) = \dfrac{4}{10} = \dfrac{2}{5}$

3. $P(\text{blue}) = 0$

4. $P(2 \text{ or } 3) = 1$

5. $P(\text{1st green and 2nd 2}) = \left(\dfrac{5}{8}\right)\left(\dfrac{4}{10}\right) = \dfrac{1}{4}$

6. $P(\text{1st red and 2nd 3}) = \left(\dfrac{3}{8}\right)\left(\dfrac{6}{10}\right) = \dfrac{18}{80} = \dfrac{9}{40}$

7. $P(\text{1st red and 2nd red}) = \left(\dfrac{3}{8}\right)\left(\dfrac{2}{7}\right) = \dfrac{6}{56} = \dfrac{3}{28}$

8. If 20 people drive to work, $20 - 8 = 12$ only drive to work. $P(\text{only drive}) = \dfrac{12}{30} = \dfrac{2}{5}$

1. 7, 7, 10, 12, 12, 24; mode = 7 and 12; median = $\dfrac{10 + 12}{2}$ = 11;

 mean = $\dfrac{7 + 7 + 10 + 12 + 12 + 24}{6}$ = 12

2. −5.5, −2, 0, 0.5, 1.5, 3.5, 3.5, 3.8; mode = 3.5; median = $\dfrac{0.5 + 1.5}{2}$ = 1;

 mean = $\dfrac{-5.5 + (-2) + 0 + 0.5 + 1.5 + 3.5 + 3.5 + 3.8}{8}$ = 0.66

3.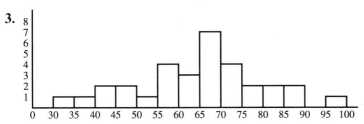

4. 6.0 − 4.0 = 2.0

5. mean = $\dfrac{4.0 + 5.0 + 6.0 + 4.5 + 4.0 + 6.0 + 5.5}{7}$ = 5.0

Data	Deviation from Mean	Frequency × (Dev.)²
4.0	4.0 − 5.0 = −1.0	2(1)
4.5	4.5 − 5.0 = −0.5	1(0.25)
5.0	5.0 − 5.0 = 0	1(0)
5.5	5.5 − 5.0 = 0.5	1(0.25)
6.0	6.0 − 5.0 = 1.0	2(1)

Sum of squares = 4.5

variance = $\dfrac{4.5}{7}$ = 0.64

standard deviation = $\sqrt{0.64}$ = 0.80

6. $P(8) = \dfrac{4}{80} = \dfrac{1}{20}$ 7. $P(\text{red}) = \dfrac{20}{80} = \dfrac{1}{4}$ 8. $P(\text{blue}) = \dfrac{20}{80} = \dfrac{1}{4}$ 9. $P(\text{blue 5}) = \dfrac{1}{80}$

10. odds in favor of rain = $\dfrac{\frac{20}{30}}{\frac{10}{30}} = \dfrac{2}{1}$ = 2 to 1 for rain

11. $P(\text{1st gold and 2nd red}) = \left(\dfrac{3}{18}\right)\left(\dfrac{6}{18}\right) = \dfrac{18}{324} = \dfrac{1}{18}$

12. $P(\text{1st gold and 2nd gold}) = \left(\dfrac{3}{18}\right)\left(\dfrac{3}{18}\right) = \dfrac{9}{324} = \dfrac{1}{36}$

13. $P(\text{red or white}) = \dfrac{6}{18} + \dfrac{4}{18} = \dfrac{10}{18} = \dfrac{5}{9}$ 14. $P(\text{white or blue}) = \dfrac{4}{18} + \dfrac{5}{18} = \dfrac{9}{18} = \dfrac{1}{2}$

15. $P(\text{1st gold and 2nd red}) = \left(\dfrac{3}{18}\right)\left(\dfrac{6}{17}\right) = \dfrac{18}{306} = \dfrac{1}{17}$

16. $P(\text{1st white and 2nd white}) = \left(\dfrac{4}{18}\right)\left(\dfrac{3}{17}\right) = \dfrac{12}{306} = \dfrac{2}{51}$

17. $P(\text{1st blue and 2nd gold}) = \left(\dfrac{5}{18}\right)\left(\dfrac{3}{17}\right) = \dfrac{15}{306} = \dfrac{5}{102}$

18. $P(\text{1st gold and 2nd gold}) = \left(\dfrac{3}{18}\right)\left(\dfrac{2}{17}\right) = \dfrac{6}{306} = \dfrac{1}{51}$

19. Since the events are inclusive, $P(\text{science or math}) = \dfrac{8}{20} + \dfrac{5}{20} - \dfrac{3}{10} = \dfrac{10}{20} = \dfrac{1}{2}$.

page 610   Chapter Test

1. 53, 56, 58, 70, 70, 75, 90, 93;
mode = 70; median = 70; mean =
$$\dfrac{53 + 56 + 58 + 70 + 70 + 75 + 90 + 93}{8} = 70.6$$

2.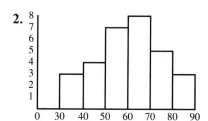

3. mean $= \dfrac{8 + 10 + 10 + 11 + 12 + 14 + 14 + 16 + 16 + 19}{10} = 13$

Data	Deviation from Mean	Frequency × (Dev.)²
8	8 − 13 = −5	1(25)
10	10 − 13 = −3	2(9)
11	11 − 13 = −2	1(4)
12	12 − 13 = −1	1(1)
14	14 − 13 = 1	2(1)
16	16 − 13 = 3	2(9)
19	19 − 13 = 6	1(36)
	Sum of squares = 104	

range = 19 − 8 = 11

variance $= \dfrac{104}{10} = 10.4$

standard deviation $= \sqrt{10.4} = 3.2$

4. $P(B) = \dfrac{5}{20} = \dfrac{1}{4}$

5. $P(T) = \dfrac{4}{20} = \dfrac{1}{5}$

6. $P(\overline{Y}) = 1 - P(Y) = 1 - \dfrac{5}{20} = \dfrac{15}{20} = \dfrac{3}{4}$

7. $P(\text{any model}) = 1$

8. $P(\text{1st G and 2nd G}) = \left(\dfrac{5}{20}\right)\left(\dfrac{5}{20}\right) = \dfrac{25}{400} = \dfrac{1}{16}$

9. $P(\text{1st S and 2nd T}) = \left(\dfrac{4}{20}\right)\left(\dfrac{4}{20}\right) = \dfrac{16}{400} = \dfrac{1}{25}$

10. $P(R \text{ or } B) = \dfrac{5}{20} + \dfrac{5}{20} = \dfrac{10}{20} = \dfrac{1}{2}$

11. $P(\text{M or T}) = \dfrac{4}{20} + \dfrac{4}{20} = \dfrac{8}{20} = \dfrac{2}{5}$

12. Odds in favor of rain $= \dfrac{\frac{12}{30}}{\frac{18}{30}} = \dfrac{12}{18} = 2$ to 3 for rain

13. $P(\text{W and W}) = \left(\dfrac{2}{5}\right)\left(\dfrac{2}{5}\right) = \dfrac{4}{25}$

14. Since the events are inclusive, $P(\text{car or public transportation}) =$
$\dfrac{20}{60} + \dfrac{37}{60} - \dfrac{12}{60} = \dfrac{45}{60} = \dfrac{3}{4}$

Challenge $P(\text{at least 1}) = 1 - P(\text{none}) = 1 - \left(\dfrac{9}{12}\right)\left(\dfrac{8}{11}\right)\left(\dfrac{7}{10}\right) = 1 - \dfrac{21}{55} = \dfrac{34}{55}$

page 611 Preparing for Standardized Tests

1. B; $\dfrac{7}{20} = 0.35$; $0.35 > 0.035$

2. C; 80% of 45 $= 0.8 \times 45 = 36.0$

3. A; $a^2b + ab^2 = 0(-1) + 0(-1)^2 = 0$; $5b + 4a = 5(-1) + 4(0) = -5$

4. A; $7x = 4 \cdot 21$, $x = 12$; $12 > 10$

5. C; $|5 - 4(3 - 6)| = |5 - 4(-3)| = |5 + 12| = 17$

6. B; $\sqrt{16 + 9} = \sqrt{25} = 5$; $4 + 3 = 7$; $5 < 7$

7. A; $9x + 1 \geq 2x - 6$, $7x \geq -7$, $x \geq -1$; $x > -2$

8. A; $b^2 - 4ac = 5^2 - 4 \cdot 2 \cdot (-5) = 25 + 40 = 65$; $65 > 50$

9. C; sum of the roots: $-\dfrac{b}{a} = -\dfrac{5}{2}$; product of the roots: $\dfrac{c}{a} = -\dfrac{5}{2}$

10. A; $\dfrac{11}{12} \cdot \dfrac{18}{7} = \dfrac{33}{14} = 2\,\dfrac{5}{14}$ $\sqrt{14} > 2$ $2\sqrt{7}$

11. D; The triangle could be isosceles if $x = 10$ or if $x = 15$.

12. A; $\dfrac{2}{3} + \dfrac{1}{2} = \dfrac{4 + 3}{6} = \dfrac{7}{6} > \dfrac{3}{5}$

13. C; $f(-1) = 12(-1)^2 - 3 = 12 - 3 = 9$; $f(1) = 12(1)^2 - 3 = 9$

14. B; $x + 28 + 105 = 180$, $x + 133 = 180$, $x = 47$, $57 > 47$

15. B; $x + y = 90$; $90 < 180$

16. D; Either x or y can be the larger or they can be equal.

page 612   Cumulative Review Chapters (1–14)

1. $<$ **2.** $=$ **3.** $>$ **4.** $=$

5. $=$ **6.** $>$ **7.** $<$ **8.** $>$

9. $<$ **10.** $=$

11.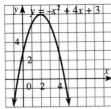

12.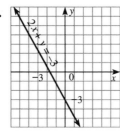

13.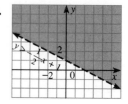

14. $\sqrt{81} - (-\sqrt{36}) = \sqrt{81} + \sqrt{36} = 9 + 6 = 15$

15. $-\sqrt{18} + \sqrt{50} = -\sqrt{9 \cdot 2} + \sqrt{25 \cdot 2} = -3\sqrt{2} + 5\sqrt{2} = 2\sqrt{2}$

16. $3\sqrt{8} \cdot 2\sqrt{3} = (3 \cdot 2) \cdot (\sqrt{8} \cdot \sqrt{3}) = 6\sqrt{24} = 6\sqrt{4 \cdot 6} = 12\sqrt{6}$

17. $\dfrac{2\sqrt{3}}{1 + \sqrt{2}} \left(\dfrac{1 - \sqrt{2}}{1 - \sqrt{2}}\right) = \dfrac{2\sqrt{3}\,(1 - \sqrt{2})}{1 - 2} = \dfrac{2\sqrt{3}\,(1 - \sqrt{2})}{-1} = 2\sqrt{3}(\sqrt{2} - 1) = 2\sqrt{6} - 2\sqrt{3}$

18. $\dfrac{2\sqrt{6}}{6\sqrt{3}}\left(\dfrac{\sqrt{3}}{\sqrt{3}}\right) = \dfrac{2\sqrt{18}}{6\sqrt{9}} = \dfrac{2\sqrt{9 \cdot 2}}{6(3)} = \dfrac{6\sqrt{2}}{18} = \dfrac{\sqrt{2}}{3}$

19. $\dfrac{(-2\sqrt{2})^2}{\sqrt{6}\sqrt{2}} = \dfrac{(-2)^2(\sqrt{2})^2}{\sqrt{12}} = \dfrac{4(2)}{\sqrt{4 \cdot 3}} = \dfrac{8}{2\sqrt{3}} = \dfrac{4}{\sqrt{3}}\left(\dfrac{\sqrt{3}}{\sqrt{3}}\right) = \dfrac{4\sqrt{3}}{3}$

20. Let s = the length of a side of the square. Use the Pythagorean theorem with two sides of the square and the diagonal. $s^2 + s^2 = (7\sqrt{2})^2$, $2s^2 = (7 \cdot 7) \cdot (\sqrt{2} \cdot \sqrt{2})$, $2s^2 = 49 \cdot 2$, $s^2 = 49$, $s = 7$ or -7. Since length cannot be negative, the length is 7 cm.

21. Let a = Al's present age, then $a - 9$ = Al's age nine years ago.
Let f = his father's present age, then $f - 9$ = his father's age nine years ago.

$$\begin{cases} 2\frac{1}{4}a = f \\ 3(a - 9) = (f - 9) \end{cases} \longrightarrow \begin{array}{l} 2\frac{1}{4}a = f \\ 3(a - 9) + 9 = f \end{array}$$

$2\frac{1}{4}a = 3(a - 9) + 9$, $2\frac{1}{4}a = 3a - 27 + 9$, $2\frac{1}{4}a = 3a - 18$, $\frac{3}{4}a = 18$, $a = 24$;

$2\frac{1}{4}(24) = f, f = 54$ yr. Al is 24 years old and his father is 54 years old.

22. Let x = the number of people needed. $12 \cdot 4 = 8 \cdot x, 48 = 8x, 6 = x$

23. 2.7, 3.2, 3.3, 3.3, 3.4, 3.7, 3.7, 3.7, 4.0, 4.2, 4.7, 4.9

$$\text{mean} = \frac{2.7 + 3.2 + 3.3 + 3.3 + 3.4 + 3.7 + 3.7 + 3.7 + 4.0 + 4.2 + 4.7 + 4.9}{12} = 3.7;$$

mode = 3.7; median = 3.7

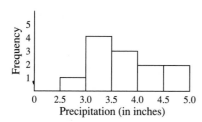

Data	Deviation from Mean	Frequency $\times$ (Dev.)^2
2.7	$2.7 - 3.7 = -1.0$	1(1)
3.2	$3.2 - 3.7 = -0.5$	1(0.25)
3.3	$3.3 - 3.7 = -0.4$	2(0.16)
3.4	$3.4 - 3.7 = -0.3$	1(0.09)
3.7	$3.7 - 3.7 = 0$	3(0)
4.0	$4.0 - 3.7 = 0.3$	1(0.09)
4.2	$4.2 - 3.7 = 0.5$	1(0.25)
4.7	$4.7 - 3.7 = 1.0$	1(1)
4.9	$4.9 - 3.7 = 1.2$	1(1.44)

Sum of squares = 4.44

$$\text{variance} = \frac{4.44}{12} = 0.37; \text{ standard deviation} = \sqrt{0.37} = 0.61$$

Chapter 15   Right Triangle Relationships

page 614   Capsule Review

1.

2.

3.

4.

page 616   Class Exercises

1. $\overline{RO}, \overline{OP}, \overline{YO}, \overline{TO}, \overline{YP}$

2. $\angle ROY, \angle TOY, \angle ROT, \angle ROP, \angle POT$

3. acute

4. obtuse

5. right

6. obtuse

7. $90° - 15° = 75°$

8. $90° - 75° = 15°$

9. $90° - 45° = 45°$

10. $90° - 89° = 1°$

11. $180° - 140° = 40°$

12. $180° - 10° = 170°$

13. $180° - 90° = 90°$

14. $180° - 150° = 30°$

pages 616–617   Practice Exercises

A 1.
ray

2.
ray

3.
point

4.
point

5.
line segment

6.
line segment

7.
line

8.
line

9. $47° + 43° = 90°$; complementary

10. $12° + 78° = 90°$; complementary

11. $2° + 178° = 180°$; supplementary

12. $154° + 26° = 180°$; supplementary

B 13. $m\angle N = 90 - 30 = 60$; acute

14. $m\angle M = 90 - 75 = 15$; acute

15. $m\angle B = 180 - 78 = 102$; obtuse

16. $m\angle A = 180 - 25 = 155$; obtuse

17. Let x = the measure of the angle, then $90 - x$ = the measure of the complement. $x = 90 - x + 40$, $2x = 130$, $x = 65$; $90 - x = 25$; $25°$, $60°$

18. Let x = the measure of the angle, then $90 - x$ = the measure of the complement. $x = 90 - x - 30$, $2x = 60$, $x = 30$; $90 - x = 60$; $30°$, $60°$

19. Let x = the measure of the angle, then $90 - x$ = the measure of the complement, and $180 - x$ = the measure of the supplement. $x = 90 - x - 50$, $2x = 40$, $x = 20$; $180 - x = 160$; $20°$, $160°$

20. Let x = the measure of the angle, then $180 - x$ = the measure of the supplement. $x = 180 - x + 10$, $2x = 190$, $x = 95$; $180 - x = 85$; $95°$, $85°$

C 21. Let x = the measure of the angle, then $90 - x$ = the measure of the complement, and $180 - x$ = the measure of the supplement. $3x - 10 = 90 - x$, $4x = 100$, $x = 25$; $180 - x = 155$; $25°$, $155°$

22. Let x = the measure of the angle, then $180 - x$ = the measure of the supplement. $4x - 45 = 180 - x$, $5x = 225$, $x = 45$; $45°$

23. Let x = the measure of the angle, then $90 - x$ = the measure of the complement, and $180 - x$ = the measure of the supplement. $2(180 - x) = 5(90 - x)$, $360 - 2x = 450 - 5x$, $3x = 90$, $x = 30$; $30°$

24. $90° - 28° = 62°$; $180° - 28° = 152°$ **25.** $90° - 35° = 55°$

26. Let $x = 0.\overline{6}$. $(10x = 6.\overline{6}) - (x = 0.\overline{6})$, $9x = 6$, $x = \dfrac{6}{9} = \dfrac{2}{3}$

27. $0.32 = \dfrac{32}{100} = \dfrac{8}{25}$

28. Let $x = 0.\overline{14}$. $(100x = 14.\overline{14}) - (x = 0.\overline{14})$, $99x = 14$, $x = \dfrac{14}{99}$

29. Let $x = 2.\overline{36}$. $(100x = 236.\overline{36}) - (x = 2.\overline{36})$, $99x = 234$, $x = \dfrac{234}{99} = \dfrac{26}{11}$

30. mode: 6; median: $\dfrac{7 + 10}{2} = \dfrac{17}{2} = 8.5$; mean: $\dfrac{3 + 6 + 6 + 7 + 10 + 15 + 17 + 21}{8} = 10.625$

31. mode: -8.5; median: $\dfrac{-3 + (-1.4)}{2} = \dfrac{-4.4}{2} = -2.2$; mean: $\dfrac{-20}{10} = -2$

page 618   Capsule Review

1. $9b - 13 = 26 - 4b$, $13b = 39$, $b = 3$ **2.** $2x - 48 = 3x - 144$, $96 = x$

3. $9w + 5 = w + 16$, $8w = 11$, $w = \dfrac{11}{8}$ **4.** $3x - 15 = 18 - 3x$, $6x = 33$, $x = \dfrac{11}{2}$

pages 620–621   Class Exercises

1. acute triangle; equilateral triangle   **2.** right triangle; scalene triangle

3. obtuse triangle; isosceles triangle   **4.** $45 + 30 + m\angle O = 180$, $m\angle O = 105$

5. $90 + 13 + m\angle O = 180$, $m\angle O = 77$ **6.** Check students' work.

7. The sum of the measures of the angles in a triangle is 180.

pages 621–622   Practice Exercises

A 1. obtuse; isosceles   **2.** acute; equilateral   **3.** right; scalene

4. right; scalene   **5.** right; isosceles   **6.** obtuse; scalene

7. $70 + 40 + m\angle L = 180$, $m\angle L = 70$ **8.** $30 + 40 + m\angle L = 180$, $m\angle L = 110$

9. $33 + 80 + m\angle L = 180$, $113 + m\angle L = 180$, $m\angle L = 67$

10. $50 + 72 + m\angle L = 180$, $122 + m\angle L = 180$, $m\angle L = 58$

11. $82 + 73 + m\angle L = 180$, $m\angle L = 25$ **12.** $42 + 64 + m\angle L = 180$, $m\angle L = 74$

13. Since $MN = NO$, $m\angle M = m\angle O$, and $m\angle M + m\angle O + 80 = 180$. So $m\angle M + m\angle M + 80 = 180$, $2(m\angle M) = 100$, $m\angle M = 50$.

14. Since $WY = XY$, $m\angle W = m\angle X$, and $m\angle W + m\angle X + 24 = 180$. So $m\angle X + m\angle X + 24 = 180$, $2(m\angle X) = 156$, $m\angle X = 78$

15. Since $\triangle ABC$ is isosceles, $AB = AC$, $m\angle B = m\angle C$, and $m\angle B + m\angle C + 90 = 180$. So $m\angle B + m\angle B + 90 = 180$, $2(m\angle B) = 90$, $m\angle B = 45$

16. Since $\triangle DEF$ is equilateral, $m\angle D = m\angle E = m\angle F$. So $m\angle E + m\angle E + m\angle E = 180$, $3(m\angle E) = 180$, $m\angle E = 60$

B 17. Since the triangle is isosceles, one of the other angles measures 33. Let $x =$ the measure of the third angle. $33 + 33 + x = 180$, $66 + x = 180$, $x = 114$

18. Since the triangle is isosceles, one of the other angles measures 47. Let $x =$ the measure of the third angle. $47 + 47 + x = 180$, $94 + x = 180$, $x = 86$

19. Let $x =$ the measure of the first angle, then $x + 1 =$ the measure of the second angle, and $x + 2 =$ the measure of the third angle. $x + (x + 1) + (x + 2) = 180$, $3x + 3 = 180$, $3x = 177$, $x = 59$; $x + 1 = 60$; $x + 2 = 61$

20. Let $x =$ the measure of the first angle, then $x + 2 =$ the measure of the second angle, and $x + 4 =$ the measure of the third angle. $x + (x + 2) + (x + 4) = 180$, $3x + 6 = 180$, $3x = 174$, $x = 58$; $x + 2 = 60$; $x + 4 = 62$

21. Let $x =$ the measure of the third angle, then $2.5x =$ the measure of each of the two equal angles. $2.5x + 2.5x + x = 180$, $6x = 180$, $x = 30$; $2.5x = 75$; the angles measure $30°$, $75°$, and $75°$.

22. Let $x =$ the measure of the third angle, then $\frac{1}{2}x =$ the measure of each of the two equal angles. $\frac{1}{2}x + \frac{1}{2}x + x = 180$, $2x = 180$, $x = 90$; $\frac{1}{2}x = 45$; The angles measure $45°$, $45°$, and $90°$.

23. $m\angle Q = 3(m\angle R)$, $m\angle S = 5(m\angle R) + 9$; $m\angle R + 3(m\angle R) + 5(m\angle R) + 9 = 180$, $9(m\angle R) + 9 = 180$, $9(m\angle R) = 171$, $m\angle R = 19$; $m\angle Q = 57$; $m\angle S = 104$

24. $m\angle C = 4(m\angle A)$, $m\angle B = 2(m\angle A) - 2$; $m\angle A + 4(m\angle A) + 2(m\angle A) - 2 = 180$, $7(m\angle A) - 2 = 180$, $7(m\angle A) = 182$, $m\angle A = 26$; $m\angle C = 104$; $m\angle B = 50$

25. $m\angle N = 7(m\angle M) + 11$, $m\angle O = 5(m\angle M)$; $m\angle M + 7(m\angle M) + 11 + 5(m\angle M) = 180$, $13(m\angle M) + 11 = 180$, $13(m\angle M) = 169$, $m\angle M = 13$; $m\angle N = 102$; $m\angle O = 65$

26. $m\angle Z = 6(m\angle Y) + 22$; $53 + 6(m\angle Y) + 22 + m\angle Y = 180$, $7(m\angle Y) + 75 = 180$, $7(m\angle Y) = 105$, $m\angle Y = 15$; $m\angle Z = 112$

C 27. $m\angle M = (180 - m\angle U) - 4$, $m\angle S = \frac{1}{2}(m\angle U) - 30$; $m\angle U + (180 - m\angle U) -$

$4 + \frac{1}{2}(m\angle U) - 30 = 180$, $\frac{1}{2}(m\angle U) + 146 = 180$, $\frac{1}{2}(m\angle U) = 34$, $m\angle U = 68$;

$m\angle M = 108$; $m\angle S = 4$

28. $m\angle U = \frac{1}{3}(180 - m\angle F) + 8$, $m\angle N = (90 - m\angle F) + 31$; $m\angle F + \frac{1}{3}(180 - m\angle F) +$

$8 + (90 - m\angle F) + 31, = 180$, $-\frac{1}{3}(m\angle F) + 189 = 180$, $-\frac{1}{3}(m\angle F) = -9$,

$m\angle F = 27$; $m\angle U = 59$; $m\angle N = 94$

29. $x^2 + 40^2 = 50^2$
$x^2 = 900$
$x = 30$
$180° - (90° + 53°) = 37°$

30. $60^2 + 130^2 = x^2$
$20{,}500 = x^2$
$x \approx 143.18$
$180° - (90° + 65°) = 25°$

31. $x^2 + x^2 = 14.15^2$
$2x^2 = 200.22$
$x^2 = 100.11$
$x \approx 10$
$90° \div 2 = 45°$

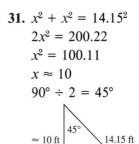

32. A triangle cannot have more than one obtuse angle because the measures of two or more obtuse angles total more than 180°.

33. 33

page 623   Capsule Review

1. symmetric property   **2.** reflexive property   **3.** transitive property

page 625   Class Exercises

1. $AE = 4$ and $EI = 4$; congruent   **2.** $DF = 2$ and $JL = 2$; congruent

3. $CG = 4$ and $EJ = 5$; not congruent   **4.** $\overline{AC}$

5. 90   **6.** $m\angle M$ **7.** $\overline{ZA}$ **8.** $\overline{AB}$

9. $A\leftrightarrow T$; $B\leftrightarrow O$; $C\leftrightarrow D$ **10.** $\overline{AB}\leftrightarrow\overline{TO}$; $\overline{BC}\leftrightarrow\overline{OD}$; $\overline{AC}\leftrightarrow\overline{TD}$

11. $\angle A\leftrightarrow\angle T$; $\angle B\leftrightarrow\angle O$; $\angle C\leftrightarrow\angle D$

pages 625–627   Practice Exercises

A 1. $BF = 4$ and $KO = 4$; congruent   **2.** $CH = 5$ and $IN = 5$; congruent

3. $EI = 4$ and $KM = 2$; not congruent   **4.** $AF = 5$ and $CE = 2$; not congruent

5. $JM = 3$ and $DG = 3$; congruent

6. $BE = 3$ and $KN = 3$; congruent

7. reflexive

8. transitive

9. symmetric

10. reflexive

11. transitive

12. symmetric

B 13. $\angle T \leftrightarrow \angle C$; $\angle O \leftrightarrow \angle A$; $\angle P \leftrightarrow \angle R$

14. $\overline{TO} \leftrightarrow \overline{CA}$; $\overline{OP} \leftrightarrow \overline{AR}$; $\overline{TP} \leftrightarrow \overline{CR}$

15. true

16. false

17. $\overline{PA} \leftrightarrow \overline{DA}$; $\overline{AL} \leftrightarrow \overline{AN}$; $\overline{PL} \leftrightarrow \overline{DN}$

18. $\angle PAL \leftrightarrow \angle DAN$; $\angle L \leftrightarrow \angle N$; $\angle D \leftrightarrow \angle P$

19. false

20. true

C 21. $\triangle ETC$; $\triangle CRE$; $\triangle TER$

22. $\triangle QUR$; $\triangle URS$; $\triangle SQU$

23. $\triangle AUD \cong \triangle FUD$; $\angle A \leftrightarrow \angle F$; $\angle AUD \leftrightarrow \angle FUD$; $\angle ADU \leftrightarrow \angle FDU$; $\overline{AU} \leftrightarrow \overline{FU}$; $\overline{UD} \leftrightarrow \overline{UD}$; $\overline{AD} \leftrightarrow \overline{FD}$

24. $\triangle BFT \cong \triangle DFT \cong \triangle TUD \cong \triangle SUD \cong \triangle DGS \cong \triangle EGS$; $\triangle BTD \cong \triangle STD \cong \triangle DSE$

25. $\triangle PST \cong \triangle QRU$

page 627 Test Yourself

1.
ray

2.
line segment

3.
line

4. Let $90 - m\angle T$ = the measure of the complement. $= m\angle T + 10$, $2(90 - m\angle T)$
$= m\angle T + 10$, $180 - 2(m\angle T)$, $3(m\angle T) = 170$, $m\angle T = \dfrac{170}{3} = 56\dfrac{2}{3}$

5. Let $90 - m\angle X$ = the measure of the complement, then $180 - m\angle X =$
the measure of the supplement. $90 - m\angle X = \dfrac{1}{4}(180 - m\angle X)$, $90 - m\angle X$
$= 45 - \dfrac{1}{4}(m\angle X)$, $\dfrac{3}{4}(m\angle X) = 45$, $m\angle X = 60$

6. right triangle; scalene triangle

7. acute triangle; equilateral triangle

8. obtuse triangle; isosceles triangle

9. Since $\overline{HI} \cong \overline{JI}$, $HI = JI$, $m\angle H = m\angle J$, and $m\angle H + m\angle J + 48 = 180$. So
$m\angle H + m\angle H + 48 = 180$, $2(m\angle H) = 132$, $m\angle H = 66$, and $m\angle J = 66$.

10. true **11.** false **12.** false **13.** true **14.** true

page 628 Capsule Review

1. $5x = 60$, $x = \dfrac{60}{5} = 12$

2. $3x = 36$, $x = \dfrac{36}{3} = 12$

3. $9n = 765$, $n = 85$ **4.** $22x = 77$, $x = \dfrac{7}{2}$ **5.** $4x = 36$, $x = 9$

6. $5x = 105$, $x = 21$ **7.** $21x = 21$, $x = 1$ **8.** $6x = 3$, $x = \dfrac{1}{2}$

page 630   Class Exercises

1. $\angle G \leftrightarrow \angle M$; $\angle F \leftrightarrow \angle N$; $\angle E \leftrightarrow \angle O$ **2.** $\overline{GF} \leftrightarrow \overline{MN}$; $\overline{FE} \leftrightarrow \overline{NO}$; $\overline{GE} \leftrightarrow \overline{MO}$

3. false   **4.** true   **5.** NM

6. $\dfrac{EG}{OM} = \dfrac{GF}{MN}$ **7.** $\dfrac{NO}{FE} = \dfrac{MN}{GF}$ **8.** similar

9. similar   **10.** impossible to determine   **11.** similar

pages 630–632   Practice Exercises

A 1. similar; $\overline{AB} \leftrightarrow \overline{DE}$; $\overline{BC} \leftrightarrow \overline{EF}$; $\overline{AC} \leftrightarrow \overline{DF}$ **2.** not similar

3. similar; $\overline{EQ} \leftrightarrow \overline{ED}$; $\overline{EP} \leftrightarrow \overline{EL}$; $\overline{PQ} \leftrightarrow \overline{LD}$ **4.** similar; $\overline{PO} \leftrightarrow \overline{AR}$; $\overline{PH} \leftrightarrow \overline{AT}$; $\overline{OH} \leftrightarrow \overline{RT}$

5. $\dfrac{7}{YZ} = \dfrac{4}{8}$, $4(YZ) = 56$, $YZ = \dfrac{56}{4} = 14$; $\dfrac{5}{XZ} = \dfrac{4}{8}$, $4(XZ) = 40$, $XZ = \dfrac{40}{4} = 10$

6. $\dfrac{36}{YZ} = \dfrac{18}{6}$, $18(YZ) = 216$, $YZ = \dfrac{216}{18} = 12$; $\dfrac{21}{XZ} = \dfrac{18}{6}$, $18(XZ) = 126$, $XZ = \dfrac{126}{18} = 7$

7. $\dfrac{4}{AB} = \dfrac{10}{7.5}$, $10(AB) = 30$, $AB = \dfrac{30}{10} = 3$; $\dfrac{8}{BC} = \dfrac{10}{7.5}$, $10(BC) = 60$, $BC = \dfrac{60}{10} = 6$

8. $\dfrac{3}{AB} = \dfrac{2.5}{7.5}$, $2.5(AB) = 22.5$, $AB = \dfrac{22.5}{2.5} = 9$; $\dfrac{4}{AC} = \dfrac{2.5}{7.5}$, $2.5(AC) = 30$, $AC = \dfrac{30}{2.5} = 12$

9. $\dfrac{7}{MN} = \dfrac{4}{12}$, $4(MN) = 84$, $MN = \dfrac{84}{4} = 21$; $\dfrac{18}{DF} = \dfrac{12}{4}$, $12(DF) = 72$, $DF = \dfrac{72}{12} = 6$

10. $\dfrac{54}{MO} = \dfrac{18}{15}$, $18(MO) = 810$, $MO = \dfrac{810}{18} = 45$; $\dfrac{45}{DE} = \dfrac{15}{18}$, $15(DE) = 810$, $DE = \dfrac{810}{15} = 54$

11. $\dfrac{4}{MO} = \dfrac{2.4}{6}$, $2.4(MO) = 24$, $MO = \dfrac{24}{2.4} = 10$; $\dfrac{5}{EF} = \dfrac{6}{2.4}$, $6(EF) = 12$, $EF = \dfrac{12}{6} = 2$

12. $\dfrac{8}{NO} = \dfrac{12}{7.2}$, $12(NO) = 57.6$, $NO = \dfrac{57.6}{12} = 4.8$; $\dfrac{3.6}{DF} = \dfrac{7.2}{12}$, $7.2(DF) = 43.2$, $DF = \dfrac{43.2}{7.2} = 6$

B 13. $\dfrac{21}{PR} = \dfrac{12}{20}$, $12(PR) = 420$, $PR = \dfrac{420}{12} = 35$; $\dfrac{15}{AR} = \dfrac{12}{20}$, $12(AR) = 300$, $AR = \dfrac{300}{12} = 25$

14. $\dfrac{50}{BC} = \dfrac{30}{20}$, $30(BC) = 1000$, $BC = \dfrac{1000}{30} = 33\dfrac{1}{3}$; $\dfrac{40}{AC} = \dfrac{30}{20}$, $30(AC) = 800$, $AC = \dfrac{800}{30} = 26\dfrac{2}{3}$

15. $\dfrac{16}{AB} = \dfrac{20}{8}$, $20(AB) = 128$, $AB = \dfrac{128}{20} = 6.4$; $\dfrac{12}{PR} = \dfrac{8}{20}$, $8(PR) = 240$, $PR = \dfrac{240}{8} = 30$

16. $\dfrac{63}{AC} = \dfrac{81}{54}$, $81(AC) = 3402$, $AC = \dfrac{3402}{81} = 42$; $\dfrac{28}{AP} = \dfrac{54}{81}$, $54(AP) = 2268$, $AP = \dfrac{2268}{54} = 42$

17. $\dfrac{3.6}{TP} = \dfrac{3}{4}$, $3(TP) = 14.4$, $TP = \dfrac{14.4}{3} = 4.8$; $\dfrac{5.6}{AR} = \dfrac{4}{3}$, $4(AR) = 16.8$, $AR = \dfrac{16.8}{4} = 4.2$,
$AR = AJ = 4.2$

18. $\dfrac{11}{JR} = \dfrac{22}{3}$; $22(JR) = 33$, $JR = \dfrac{33}{22} = 1.5$; $\dfrac{33}{OP} = \dfrac{3}{22}$, $3(OP) = 726$, $OP = \dfrac{726}{3} = 242$,
$OP = TO = 242$

C 19. $\dfrac{x + 6}{x} = \dfrac{12}{10}$, $12x = 10(x + 6)$, $12x = 10x + 60$, $2x = 60$, $x = \dfrac{60}{2} = 30$

20. $\dfrac{40}{x} = \dfrac{20}{12}$, $20x = 480$, $x = \dfrac{480}{20} = 24$

21. $\dfrac{2.8}{x + 2.8} = \dfrac{0.8}{1.0}$, $0.8(x + 2.8) = 2.8$, $0.8x + 2.24 = 2.8$, $0.8x = 0.56$, $x = \dfrac{0.56}{0.8} = 0.7$

22. $\dfrac{5.6}{5.6 - x} = \dfrac{8.4}{2.4}$, $8.4(5.6 - x) = 13.44$, $47.04 - 8.4x = 13.44$, $8.4x = 33.60$, $x = 4$

23. $\dfrac{18}{h} = \dfrac{3}{2}$, $3h = 36$, $h = \dfrac{36}{3} = 12$ m   **24.** $\dfrac{2}{h} = \dfrac{0.3}{1.8}$, $0.3h = 3.6$, $h = \dfrac{3.6}{0.3} = 12$ m

25. $\dfrac{12}{h} = \dfrac{2}{100}$, $2h = 1200$, $h = \dfrac{1200}{2} = 600$ m   **26.** $\dfrac{30}{h} = \dfrac{3}{2.1}$, $3h = 63$, $h = \dfrac{63}{3} = 21$ m

27. Let x = the length of the side of the garden alongside the 60-ft side of the
sidewalk and y = the length of the side of the garden alongside the 45-ft
side of the sidewalk. $\dfrac{60}{x} = \dfrac{20}{14.5}$; $20x = 870$, $x = \dfrac{870}{20} = 43.5$; $\dfrac{45}{y} = \dfrac{20}{14.5}$,

$20y = 652.5$, $y = \dfrac{652.5}{20} = 32.625$; The perimeter of the garden is 14.5 ft +

43.5 ft + 32.625 = 90.625 ft; $90.625 \text{ ft} \cdot \dfrac{1 \text{ yd}}{3 \text{ ft}} \cdot \dfrac{\$15}{1 \text{ yd}} = \$453.13$; It would

cost \$453.13 to put a fence around the garden.

28. $4x(x^2 + 3x - 4) = 4x(x + 4)(x - 1)$ **29.** $3(25a^2 + 20ab + 4b^2) = 3(5a + 2b)^2$

30. $2c(c^4 - 16) = 2c(c^2 + 4)(c^2 - 4) = 2c(c^2 + 4)(c + 2)(c - 2)$

31. $p^2(4q + 5) - 4r^2(4q + 5) = (p^2 - 4r^2)(4q + 5) = (p + 2r)(p - 2r)(4q + 5)$

32. $8s = 7(s + 2)$, $8s = 7s + 14$, $s = 14$

33. $5(p - 2) = 3(p + 4)$, $5p - 10 = 3p + 12$, $2p = 22$, $p = 11$

34. $(m + 3)(2m - 1) = m(m + 7)$, $2m^2 + 5m - 3 = m^2 + 7m$,
$m^2 - 2m - 3 = 0$, $(m - 3)(m + 1) = 0$, $m = 3$ or $m = -1$

page 633   Capsule Review

1. $\dfrac{1}{\sqrt{5}} = \dfrac{1}{\sqrt{5}} \cdot \dfrac{\sqrt{5}}{\sqrt{5}} = \dfrac{\sqrt{5}}{5}$

2. $\dfrac{3}{\sqrt{2}} = \dfrac{3}{\sqrt{2}} \cdot \dfrac{\sqrt{2}}{\sqrt{2}} = \dfrac{3\sqrt{2}}{2}$

3. $\dfrac{1}{3\sqrt{2}} = \dfrac{1}{3\sqrt{2}} \cdot \dfrac{\sqrt{2}}{\sqrt{2}} = \dfrac{\sqrt{2}}{6}$

4. $\dfrac{5}{\sqrt{10}} = \dfrac{5}{\sqrt{10}} \cdot \dfrac{\sqrt{10}}{\sqrt{10}} = \dfrac{5\sqrt{10}}{10} = \dfrac{\sqrt{10}}{2}$

5. $\dfrac{\sqrt{3}}{\sqrt{6}} = \dfrac{\sqrt{3}}{\sqrt{6}} \cdot \dfrac{\sqrt{6}}{\sqrt{6}} = \dfrac{\sqrt{18}}{6} = \dfrac{3\sqrt{2}}{6} = \dfrac{\sqrt{2}}{2}$

6. $\dfrac{\sqrt{2}}{\sqrt{5}} = \dfrac{\sqrt{2}}{\sqrt{5}} \cdot \dfrac{\sqrt{5}}{\sqrt{5}} = \dfrac{\sqrt{10}}{5}$

7. $\dfrac{\sqrt{3}}{\sqrt{50}} = \dfrac{\sqrt{3}}{\sqrt{50}} \cdot \dfrac{\sqrt{50}}{\sqrt{50}} = \dfrac{\sqrt{150}}{50} = \dfrac{5\sqrt{6}}{50} = \dfrac{\sqrt{6}}{10}$

8. $\dfrac{\sqrt{7}}{\sqrt{21}} = \dfrac{\sqrt{7}}{\sqrt{21}} \cdot \dfrac{\sqrt{21}}{\sqrt{21}} = \dfrac{\sqrt{147}}{21} = \dfrac{7\sqrt{3}}{21} = \dfrac{\sqrt{3}}{3}$

page 634   Class Exercises

1. $\dfrac{5}{13}$
2. $\dfrac{12}{13}$
3. $\dfrac{5}{12}$
4. $\dfrac{12}{13}$
5. $\dfrac{5}{13}$
6. $\dfrac{12}{5}$

7. $\dfrac{1}{\sqrt{5}} = \dfrac{1}{\sqrt{5}} \cdot \dfrac{\sqrt{5}}{\sqrt{5}} = \dfrac{\sqrt{5}}{5}$
8. $\dfrac{2}{\sqrt{5}} = \dfrac{2}{\sqrt{5}} \cdot \dfrac{\sqrt{5}}{\sqrt{5}} = \dfrac{2\sqrt{5}}{5}$
9. $\dfrac{1}{2}$

10. $\dfrac{2}{\sqrt{5}} = \dfrac{2}{\sqrt{5}} \cdot \dfrac{\sqrt{5}}{\sqrt{5}} = \dfrac{2\sqrt{5}}{5}$
11. $\dfrac{1}{\sqrt{5}} = \dfrac{1}{\sqrt{5}} \cdot \dfrac{\sqrt{5}}{\sqrt{5}} = \dfrac{\sqrt{5}}{5}$
12. $\dfrac{2}{1} = 2$

pages 635–636   Practice Exercises

A 1. $\dfrac{3}{5}$
2. $\dfrac{4}{5}$
3. $\dfrac{3}{4}$
4. $\dfrac{4}{5}$
5. $\dfrac{3}{5}$
6. $\dfrac{4}{3}$

7. $\sin A = \dfrac{5}{13}$; $\cos A = \dfrac{12}{13}$; $\tan A = \dfrac{5}{12}$; $\sin B = \dfrac{12}{13}$; $\cos B = \dfrac{5}{13}$; $\tan B = \dfrac{12}{5}$

8. $\sin A = \dfrac{24}{25}$; $\cos A = \dfrac{7}{25}$; $\tan A = \dfrac{24}{7}$; $\sin B = \dfrac{7}{25}$; $\cos B = \dfrac{24}{25}$; $\tan B = \dfrac{7}{24}$

9. $\sin A = \dfrac{12}{37}$; $\cos A = \dfrac{35}{37}$; $\tan A = \dfrac{12}{35}$; $\sin B = \dfrac{35}{37}$; $\cos B = \dfrac{12}{37}$; $\tan B = \dfrac{35}{12}$

10. $\sin A = \dfrac{2\sqrt{3}}{7}$; $\cos A = \dfrac{\sqrt{37}}{7}$; $\tan A = \dfrac{2\sqrt{3}}{\sqrt{37}} = \dfrac{2\sqrt{3}}{\sqrt{37}} \cdot \dfrac{\sqrt{37}}{\sqrt{37}} = \dfrac{2\sqrt{111}}{37}$;

$\sin B = \dfrac{\sqrt{37}}{7}$; $\cos B = \dfrac{2\sqrt{3}}{7}$; $\tan B = \dfrac{\sqrt{37}}{2\sqrt{3}} = \dfrac{\sqrt{37}}{2\sqrt{3}} \cdot \dfrac{\sqrt{3}}{\sqrt{3}} = \dfrac{\sqrt{111}}{6}$

11. $\sin A = \dfrac{2}{2\sqrt{2}} = \dfrac{1}{\sqrt{2}} \cdot \dfrac{\sqrt{2}}{\sqrt{2}} = \dfrac{\sqrt{2}}{2}$; $\cos A = \dfrac{2}{2\sqrt{2}} = \dfrac{1}{\sqrt{2}} \cdot \dfrac{\sqrt{2}}{\sqrt{2}} = \dfrac{\sqrt{2}}{2}$; $\tan A = \dfrac{2}{2} = 1$

$\sin B = \dfrac{2}{2\sqrt{2}} = \dfrac{1}{\sqrt{2}} \cdot \dfrac{\sqrt{2}}{\sqrt{2}} = \dfrac{\sqrt{2}}{2}$; $\cos B = \dfrac{2}{2\sqrt{2}} = \dfrac{1}{\sqrt{2}} \cdot \dfrac{\sqrt{2}}{\sqrt{2}} = \dfrac{\sqrt{2}}{2}$; $\tan B = \dfrac{2}{2} = 1$

12. $\sin A = \dfrac{6}{4\sqrt{3}} = \dfrac{3}{2\sqrt{3}} \cdot \dfrac{\sqrt{3}}{\sqrt{3}} = \dfrac{3\sqrt{3}}{6} = \dfrac{\sqrt{3}}{2}$; $\cos A = \dfrac{2\sqrt{3}}{4\sqrt{3}} = \dfrac{1}{2}$; $\tan A = \dfrac{6}{2\sqrt{3}} =$

$\dfrac{3}{\sqrt{3}} \cdot \dfrac{\sqrt{3}}{\sqrt{3}} = \dfrac{3\sqrt{3}}{3} = \sqrt{3}$; $\sin B = \dfrac{2\sqrt{3}}{4\sqrt{3}} = \dfrac{1}{2}$; $\cos B = \dfrac{6}{4\sqrt{3}} = \dfrac{3}{2\sqrt{3}} \cdot \dfrac{\sqrt{3}}{\sqrt{3}} = \dfrac{3\sqrt{3}}{6} = \dfrac{\sqrt{3}}{2}$;

$\tan B = \dfrac{2\sqrt{3}}{6} = \dfrac{\sqrt{3}}{3}$

B 13. $(TQ)^2 = 5^2 + 12^2$, $(TQ)^2 = 169$, $TQ = 13$; $\sin Q = \dfrac{5}{13}$; $\cos Q = \dfrac{12}{13}$; $\tan Q = \dfrac{5}{12}$;

$\sin T = \dfrac{12}{13}$; $\cos T = \dfrac{5}{13}$; $\tan T = \dfrac{12}{5}$

14. $(RT)^2 + 2^2 = 13^2$, $(RT)^2 = 165$, $RT = \sqrt{165}$; $\sin Q = \dfrac{\sqrt{165}}{13}$; $\cos Q = \dfrac{2}{13}$;

$\tan Q = \dfrac{\sqrt{165}}{2}$; $\sin T = \dfrac{2}{13}$; $\cos T = \dfrac{\sqrt{165}}{13}$; $\tan T = \dfrac{2}{\sqrt{165}} = \dfrac{2}{\sqrt{165}} \cdot \dfrac{\sqrt{165}}{\sqrt{165}} = \dfrac{2\sqrt{165}}{165}$

15. $(QT)^2 = (2\sqrt{61})^2 + 10^2$, $(QT)^2 = 344$, $QT = \sqrt{344} = 2\sqrt{86}$; $\sin Q = \dfrac{2\sqrt{61}}{2\sqrt{86}} =$

$\dfrac{\sqrt{61}}{\sqrt{86}} \cdot \dfrac{\sqrt{86}}{\sqrt{86}} = \dfrac{\sqrt{5246}}{86}$; $\cos Q = \dfrac{10}{2\sqrt{86}} = \dfrac{5}{\sqrt{86}} \cdot \dfrac{\sqrt{86}}{\sqrt{86}} = \dfrac{5\sqrt{86}}{86}$; $\tan Q =$

$\dfrac{2\sqrt{61}}{10} = \dfrac{\sqrt{61}}{5}$; $\sin T = \dfrac{10}{2\sqrt{86}} = \dfrac{5}{\sqrt{86}} \cdot \dfrac{\sqrt{86}}{\sqrt{86}} = \dfrac{5\sqrt{86}}{86}$; $\cos T = \dfrac{2\sqrt{61}}{2\sqrt{86}} = \dfrac{\sqrt{61}}{\sqrt{86}} \cdot \dfrac{\sqrt{86}}{\sqrt{86}} =$

$\dfrac{\sqrt{5246}}{86}$; $\tan T = \dfrac{10}{2\sqrt{61}} = \dfrac{5}{\sqrt{61}} \cdot \dfrac{\sqrt{61}}{\sqrt{61}} = \dfrac{5\sqrt{61}}{61}$

16. $(QR)^2 + (3\sqrt{3})^2 = 6^2$, $(QR)^2 = 9$, $QR = 3$; $\sin Q = \dfrac{3\sqrt{3}}{6} = \dfrac{\sqrt{3}}{2}$; $\cos Q = \dfrac{3}{6} = \dfrac{1}{2}$;

$\tan Q = \dfrac{3\sqrt{3}}{3} = \sqrt{3}$; $\sin T = \dfrac{3}{6} = \dfrac{1}{2}$; $\cos T = \dfrac{3\sqrt{3}}{6} = \dfrac{\sqrt{3}}{2}$; $\tan T = \dfrac{3}{3\sqrt{3}} =$

$\dfrac{1}{\sqrt{3}} \cdot \dfrac{\sqrt{3}}{\sqrt{3}} = \dfrac{\sqrt{3}}{3}$

17. $(TR)^2 + 3^2 = (\sqrt{30})^2$, $(TR)^2 = 21$, $TR = \sqrt{21}$; $\sin Q = \dfrac{\sqrt{21}}{\sqrt{30}} = \dfrac{\sqrt{21}}{\sqrt{30}} \cdot \dfrac{\sqrt{30}}{\sqrt{30}} =$

$\dfrac{\sqrt{630}}{30} = \dfrac{3\sqrt{70}}{30} = \dfrac{\sqrt{70}}{10}$; $\cos Q = \dfrac{3}{\sqrt{30}} = \dfrac{3}{\sqrt{30}} \cdot \dfrac{\sqrt{30}}{\sqrt{30}} = \dfrac{3\sqrt{30}}{30} = \dfrac{\sqrt{30}}{10}$; $\tan Q = \dfrac{\sqrt{21}}{3}$;

$\sin T = \dfrac{3}{\sqrt{30}} = \dfrac{3}{\sqrt{30}} \cdot \dfrac{\sqrt{30}}{\sqrt{30}} = \dfrac{3\sqrt{30}}{30} = \dfrac{\sqrt{30}}{10}$; $\cos T = \dfrac{\sqrt{21}}{\sqrt{30}} = \dfrac{\sqrt{21}}{\sqrt{30}} \cdot \dfrac{\sqrt{30}}{\sqrt{30}} =$

$\dfrac{\sqrt{630}}{30} = \dfrac{3\sqrt{70}}{30} = \dfrac{\sqrt{70}}{10}$; $\tan T = \dfrac{3}{\sqrt{21}} = \dfrac{3}{\sqrt{21}} \cdot \dfrac{\sqrt{21}}{\sqrt{21}} = \dfrac{3\sqrt{21}}{21} = \dfrac{\sqrt{21}}{7}$

18. $(TR)^2 + (5\sqrt{3})^2 = 10^2$, $(TR)^2 = 25$, $TR = 5$; $\sin Q = \dfrac{5}{10} = \dfrac{1}{2}$; $\cos Q = \dfrac{5\sqrt{3}}{10} = \dfrac{\sqrt{3}}{2}$;

$\tan Q = \dfrac{5}{5\sqrt{3}} = \dfrac{1}{\sqrt{3}} \cdot \dfrac{\sqrt{3}}{\sqrt{3}} = \dfrac{\sqrt{3}}{3}$; $\sin T = \dfrac{5\sqrt{3}}{10} = \dfrac{\sqrt{3}}{2}$; $\cos T = \dfrac{5}{10} = \dfrac{1}{2}$;

$\tan T = \dfrac{5\sqrt{3}}{5} = \sqrt{3}$

C 19. $\cos J = \dfrac{k}{l}$; $\sin K = \dfrac{k}{l}$

20. $\cos K = \dfrac{j}{l}$; $\sin J = \dfrac{j}{l}$

21. $(\sin J)^2 + (\cos J)^2 = \left(\dfrac{j}{l}\right)^2 + \left(\dfrac{k}{l}\right)^2 = \dfrac{l^2}{l^2} = 1$

22. $\dfrac{\sin K}{\cos K} = \dfrac{\frac{k}{l}}{\frac{j}{l}} = \dfrac{k}{j} = \tan K$

23. Let d = diagonal length. $d^2 = 8^2 + 6^2$, $d^2 = 100$, $d = 10$ m; sine: $\dfrac{3}{5}$; cosine: $\dfrac{4}{5}$; tangent: $\dfrac{3}{4}$

24. Let x = the distance from the boat to its starting point.
$x^2 = 15^2 + 10^2$, $x^2 = 225 + 100$, $x^2 = 325$, $x = \sqrt{325} = 5\sqrt{13}$ mi;
sine: $\dfrac{10}{5\sqrt{13}} = \dfrac{2\sqrt{13}}{13}$; cosine: $\dfrac{15}{5\sqrt{13}} = \dfrac{3\sqrt{13}}{13}$; tangent: $\dfrac{10}{15} = \dfrac{2}{3}$

25. Let h = the height of the lodge. $30^2 + h^2 = 80^2$, $900 + h^2 = 6400$, $h^2 = 5500$,
$h = 10\sqrt{55}$ ft; sine: $\dfrac{10\sqrt{55}}{80} = \dfrac{\sqrt{55}}{8}$; cosine: $\dfrac{30}{80} = \dfrac{3}{8}$; tangent: $\dfrac{10\sqrt{55}}{30} = \dfrac{\sqrt{55}}{3}$

26. $324 = 2l + 2(32)$, $324 = 2l + 64$, $260 = 2l$, $l = 130$

27. $168 = \dfrac{1}{3}(7)(4)h$, $168 = \dfrac{28}{3}h$, $h = 18$

28. $(-2a^2)(9a^8) = -18a^{10}$

29. $-\sqrt{24a^4b^4} = -\sqrt{4a^4b^4} \cdot \sqrt{6} = -2a^2b^2\sqrt{6}$

30. $\sqrt{180x^6y^7} = \sqrt{36x^6y^6} \cdot \sqrt{5y} = 6|x^3|y^3\sqrt{5y}$

31. $2z^3 - 8z^2 + 2z + 3z^2 - 12z + 3 = 2z^3 - 5z^2 - 10z + 3$

32. $\dfrac{-(x-4)}{(x+8)(x-3)} \cdot \dfrac{(x+4)(x+8)}{(x+4)(x-4)} = -\dfrac{1}{x-3}$

33. $\dfrac{2(m+5)}{(3m-1)(2m+5)} \cdot \dfrac{(3m-1)(m+5)}{(m+5)(m+5)} = \dfrac{2}{2m+5}$

34. $\dfrac{2(t-3)}{(t+3)(t-3)} - \dfrac{3(t+3)}{(t+3)(t-3)} + \dfrac{18}{(t+3)(t-3)} =$

$\dfrac{2t-6-3t-9+18}{(t+3)(t-3)} = \dfrac{-t+3}{(t+3)(t-3)} = -\dfrac{1}{t+3}$

35. $\dfrac{5}{15} + \dfrac{6}{15} = \dfrac{11}{15}$

36. $\dfrac{6}{15} \cdot \dfrac{4}{14} = \dfrac{4}{35}$

37. $\dfrac{5}{15} \cdot \dfrac{6}{15} = \dfrac{2}{15}$

38. In a right triangle, the leg opposite one of the acute angles is the leg adjacent to the other acute angle.

39. If tan $A = 1$, the legs of the triangle ($\overline{BC}$ and $\overline{AC}$) have the same length, and $\triangle ABC$ is isosceles. In isosceles triangles, the angles opposite the equal sides are equal in measure. Since $\angle A$ and $\angle B$ are equal in measure and complementary, each has a measure of 45°.

pages 637–638 Integrating Algebra: Networks

Check students' graphs for Exercises 1–6. Exercise 4 cannot be done. Check students' work for Exercises 7–12.

7. inside   **8.** outside   **9.** outside

10. not possible   **11.** outside   **12.** not possible

13. Answers may vary.   **14.** Check students' work.

page 639 Capsule Review

1. $c^2 = 4^2 + 3^2$, $c^2 = 25$, $c = 5$; $\sin A = \dfrac{4}{5}$; $\cos A = \dfrac{3}{5}$; $\tan A = \dfrac{4}{3}$; $\sin B = \dfrac{3}{5}$;

$\cos B = \dfrac{4}{5}$; $\tan B = \dfrac{3}{4}$

2. $3^2 = a^2 + 2^2$, $5 = a^2$, $\sqrt{5} = a$; $\sin A = \dfrac{\sqrt{5}}{3}$; $\cos A = \dfrac{2}{3}$; $\tan A = \dfrac{\sqrt{5}}{2}$;

$\sin B = \dfrac{2}{3}$; $\cos B = \dfrac{\sqrt{5}}{3}$; $\tan B = \dfrac{2}{\sqrt{5}} = \dfrac{2}{\sqrt{5}} \cdot \dfrac{\sqrt{5}}{\sqrt{5}} = \dfrac{2\sqrt{5}}{5}$

3. $c^2 = (2\sqrt{5})^2 + (2\sqrt{6})^2$, $c^2 = 44$, $c = \sqrt{44}$, $c = 2\sqrt{11}$; $\sin A = \dfrac{2\sqrt{5}}{2\sqrt{11}} = \dfrac{\sqrt{5}}{\sqrt{11}} \cdot \dfrac{\sqrt{11}}{\sqrt{11}} =$

$\dfrac{\sqrt{55}}{11}$; $\cos A = \dfrac{2\sqrt{6}}{2\sqrt{11}} = \dfrac{\sqrt{6}}{\sqrt{11}} \cdot \dfrac{\sqrt{11}}{\sqrt{11}} = \dfrac{\sqrt{66}}{11}$; $\tan A = \dfrac{2\sqrt{5}}{2\sqrt{6}} = \dfrac{\sqrt{5}}{\sqrt{6}} \cdot \dfrac{\sqrt{6}}{\sqrt{6}} = \dfrac{\sqrt{30}}{6}$;

$\sin B = \dfrac{2\sqrt{6}}{2\sqrt{11}} = \dfrac{\sqrt{6}}{\sqrt{11}} \cdot \dfrac{\sqrt{11}}{\sqrt{11}} = \dfrac{\sqrt{66}}{11}$; $\cos B = \dfrac{2\sqrt{5}}{2\sqrt{11}} = \dfrac{\sqrt{5}}{\sqrt{11}} \cdot \dfrac{\sqrt{11}}{\sqrt{11}} = \dfrac{\sqrt{55}}{11}$;

$\tan B = \dfrac{2\sqrt{6}}{2\sqrt{5}} = \dfrac{\sqrt{6}}{\sqrt{5}} \cdot \dfrac{\sqrt{5}}{\sqrt{5}} = \dfrac{\sqrt{30}}{5}$

page 641 Class Exercises

1. 0.7071   **2.** 0.2588   **3.** 0.5774   **4.** 0.6018

5. 0.6561   **6.** 0.3057   **7.** 0.9877   **8.** 0.3256

9. 60°   **10.** 45°   **11.** 25°   **12.** 69°   **13.** 62°   **14.** 33°

15. $\sin 30° = \dfrac{x}{26}$

16. $\cos 45° = \dfrac{6}{x}$

17. $\tan x° = \dfrac{9}{4}$

pages 641–642 Practice Exercises

A 1. 0.9962

2. 0.3907

3. 1.5399

4. 0.2867

5. 0.9563

6. 0.1392

7. 0.4067

8. 0.0349

9. 28.6363

10. 2.1445

11. 0.8192

12. 0.7314

13. 30°

14. 20°

15. 45°

16. 75°

17. 55°

18. 73°

19. 18°

20. 41°

21. 2°

B 22. $\tan 40° = \dfrac{5}{AC}$, $0.8391 = \dfrac{5}{AC}$, $AC = \dfrac{5}{0.8391} = 6$; $\sin 40° = \dfrac{5}{AB}$, $AB = \dfrac{5}{0.6428} = 8$

23. $\sin 32° = \dfrac{BC}{42}$, $0.5299 = \dfrac{BC}{42}$, $BC = 0.5299(42) = 22$; $\cos 32° = \dfrac{AC}{42}$, $0.8480 = \dfrac{AC}{42}$, $AC = 0.8480(42) = 36$

24. $\tan 71° = \dfrac{17}{BC}$, $2.9042 = \dfrac{17}{BC}$, $BC = \dfrac{17}{2.9042} = 6$; $\sin 71° = \dfrac{17}{AB}$, $AB = \dfrac{17}{0.9455} = 18$

25. $\tan 5° = \dfrac{AC}{50}$, $0.0875 = \dfrac{AC}{50}$, $AC = 0.0875(50) = 4$; $\cos 5° = \dfrac{50}{AB}$, $AB = \dfrac{50}{0.9962} = 50$

26. $\tan 50° = \dfrac{YZ}{50}$, $1.1918 = \dfrac{YZ}{50}$, $YZ = 1.1918(50) = 60$; $\cos 50° = \dfrac{50}{XZ}$, $XZ = \dfrac{50}{0.6428} = 78$

27. $\tan 22° = \dfrac{32}{XY}$, $0.4040 = \dfrac{32}{XY}$, $XY = \dfrac{32}{0.4040} = 79$; $\sin 22° = \dfrac{32}{XZ}$, $XZ = \dfrac{32}{0.3746} = 85$

28. $\sin 71° = \dfrac{XY}{34}$, $0.9455 = \dfrac{XY}{34}$, $XY = 0.9455(34) = 32$; $\cos 71° = \dfrac{YZ}{34}$, $0.3256 = \dfrac{YZ}{34}$, $YZ = 0.3256(34) = 11$

29. $\tan 14° = \dfrac{15}{YZ}$, $0.2493 = \dfrac{15}{YZ}$, $YZ = \dfrac{15}{0.2493} = 60$; $\sin 14° = \dfrac{15}{XZ}$, $XZ = \dfrac{15}{0.2419} = 62$

30. $\sin K = \dfrac{12}{13} = 0.9231$; By the table, $m\angle K \approx 67$. So $m\angle J = 90 - 67 = 23$.

31. $\cos K = \dfrac{3}{5} = 0.6$; By the table, $m\angle K \approx 53$. So $m\angle J = 90 - 53 = 37$.

32. $\tan K = \dfrac{12}{7} = 1.7143$; By the table, $m\angle K \approx 60$. So $m\angle J = 90 - 60 = 30$.

33. $\tan K = \dfrac{3}{9} = 0.3333$; By the table, $m\angle K \approx 18$. So $m\angle J = 90 - 18 = 72$.

C 34. Since $\cos B = 0.4226$, $m\angle B = 65$. $\sin 65° = \dfrac{AC}{18}$, $0.9063 = \dfrac{AC}{18}$,

$AC = 0.9063(18) = 16.3$

35. Since $\sin B = 0.6157$, $m\angle B = 38$. $\cos 38° = \dfrac{AB}{25}$, $0.7880 = \dfrac{AB}{25}$,

$AB = 0.7880(25) = 19.7$

36. Since $\tan C = 0.2867$, $m\angle C = 16$. $\sin 16° = \dfrac{12}{BC}$, $0.2756 = \dfrac{12}{BC}$,

$BC = \dfrac{12}{0.2756} = 43.5$

37. Since $\tan C = 4.7046$, $m\angle C = 78$. $\cos 78° = \dfrac{22}{BC}$, $0.2079 = \dfrac{22}{BC}$,

$BC = \dfrac{22}{0.2079} = 105.8$

38. $\tan 37° = \dfrac{AB}{200}$, $0.7536 = \dfrac{AB}{200}$, $AB = 0.7536(200) = 151$ m

39. Let x = the length of the ramp. $\sin 5° = \dfrac{5}{x}$, $0.0872 = \dfrac{5}{x}$, $x = \dfrac{5}{0.0872} = 57$ ft

40. Let h = the height of the pole. $\tan 47° = \dfrac{h}{4}$, $1.0724 = \dfrac{h}{4}$, $h = 1.0724(4) = 4$ m

41. Let h = the height of the kite. $\sin 65° = \dfrac{h}{60}$, $0.9063 = \dfrac{h}{60}$, $h = 0.9063(60) = 54$ m

42. a. 0.8660, 0.8660
  b. 0.3420, 0.3420
  c. 0.9962, 0.9962
  Conjecture: The sine of an acute angle has the same value as the cosine of its complement.

page 645   Class Exercises

For Exercises 1–5, answers may vary.

1. A fact that is taken for granted.   **2.** probably both

3. yes   **4.** They can lead to incorrect solutions.

5. Drawings may be based on hidden assumptions.

pages 646–647   Practice Exercises

A 1. yes   **2.** yes   **3.** yes   **4.** no

5. $\angle C = 90°$ **6.** The figure is square.

B 7.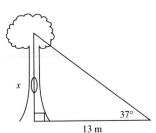

13 m

Assume a right triangle (level land).

$\tan 37° = \dfrac{x}{13}$, $x = 10$ m

8.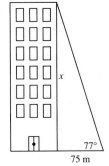

77°
75 m

Assume a right triangle (level land).

$\tan 77° = \dfrac{x}{75}$, $x = 325$ m

9.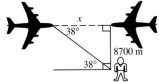

8700 m

Assume a right triangle (level land).

$\tan 38° = \dfrac{8700}{x}$, $x = 11{,}135$ m

10.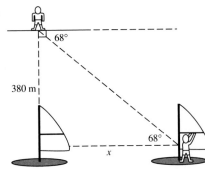

Assume a right triangle (level land).

$\tan 68° = \dfrac{380}{x}$, $x = 154$ m

11.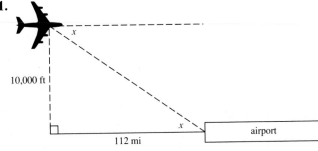

Assume a flat earth and 112 m ahead means horizontal distance.

112 mi = 591, 360 ft

$\tan x = \dfrac{10{,}000}{591{,}360}$, $x = 1°$

C 12.

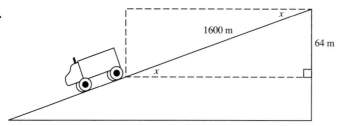

Assume a constant slope for the hill.

$\sin x = \dfrac{64}{1600}$, $x = 2°$

13.

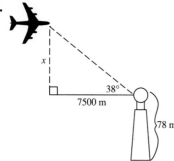

Assume a flat earth.

$\tan 38° = \dfrac{x}{7500}$, $x = 5860$ m

altitude $= 5860 + 78 = 5938$ m

page 647 Mixed Problem Solving Review

1. Let w = the width, then $4w$ = the length of the rectangle.
$w(4w) = 4[2(4w) + 2w]$, $4w^2 = 32w + 8w$, $4w^2 - 40w = 0$, $4w(w - 10) = 0$;
Since width is positive, $w = 10$ ft and $4w = 4(10) = 40$ ft.

2. Let x = the first integer and y = the second integer.
$$\begin{cases} x - y = 12 \\ 2x + y = 84 \end{cases}$$
$3x = 96$
$x = 32$
$32 - y = 12$, $y = 20$

3. Let x = the score received on the fifth exam.
$\dfrac{85 + 78 + 80 + 94 + x}{5} > 85$, $\dfrac{337 + x}{5} > 85$, $337 + x > 425$, $x > 88$; 89

Check students' work.

1. $\frac{10}{AN} = \frac{12}{4}$, $12(AN) = 40$, $AN = \frac{40}{12} = \frac{10}{3} = 3\frac{1}{3}$

2. $(XZ)^2 = 3^2 + 5^2$, $(XZ)^2 = 34$, $XZ = \sqrt{34}$; $\sin X = \frac{5}{\sqrt{34}} \cdot \frac{\sqrt{34}}{\sqrt{34}} = \frac{5\sqrt{34}}{34}$;

$\cos X = \frac{3}{\sqrt{34}} \cdot \frac{\sqrt{34}}{\sqrt{34}} = \frac{3\sqrt{34}}{34}$; $\tan X = \frac{5}{3}$

3. $\tan A = \frac{4}{12} = \frac{1}{3}$; By the table, $m\angle A \approx 18°$; So $m\angle B = 90 - 18 = 72$.

4. Let x = the angle of depression. 3 mi = 15,840 ft; $\tan x = \frac{246}{15,840}$, $x = 1°$

1. $m\angle N = 180 - 67 = 113$

2. Let x = the measure of the angle, then $90 - x$ = the measure of the complement. $x = (90 - x) + 20$, $x = 110 - x$, $2x = 110$, $x = 55$; $90 - x = 35$

3. right triangle; isosceles triangle

4. $m\angle M + 55 + 63 = 180$, $m\angle M + 118 = 180$, $m\angle M = 62$

5. $\angle P \leftrightarrow \angle C$; $\angle PAD \leftrightarrow \angle CAR$; $\angle D \leftrightarrow \angle R$ **6.** $\overline{PA} \leftrightarrow \overline{CA}$; $\overline{AD} \leftrightarrow \overline{AR}$; $\overline{PD} \leftrightarrow \overline{RC}$

7. false

8. $\frac{15}{DG} = \frac{10}{4}$, $10(DG) = 60$, $DG = \frac{60}{10} = 6$; $\frac{3}{AT} = \frac{4}{10}$, $4(AT) = 30$, $AT = \frac{30}{4} = 7.5$

9. $\frac{50}{BC} = \frac{25}{15}$, $25(BC) = 750$, $BC = \frac{750}{25} = 30$; $\frac{32}{AE} = \frac{15}{25}$, $15(AE) = 800$, $AE = \frac{800}{15} =$

$\frac{160}{3} = 53\frac{1}{3}$

10. $(AB)^2 = 18^2 + 24^2$, $(AB)^2 = 900$, $AB = 30$; $\sin A = \frac{18}{30} = \frac{3}{5}$; $\cos A = \frac{24}{30} = \frac{4}{5}$;

$\tan A = \frac{18}{24} = \frac{3}{4}$; $\sin B = \frac{24}{30} = \frac{4}{5}$; $\cos B = \frac{18}{30} = \frac{3}{5}$; $\tan B = \frac{24}{18} = \frac{4}{3}$

11. 0.8746   **12.** 30°

13. $\tan 34° = \frac{YZ}{40}$, $0.6745(40) = YZ$, $YZ = 27$; $\cos 34° = \frac{40}{XZ}$, $\frac{40}{0.8290} = XZ$, $XZ = 48$

In Exercises 14–15, assume right triangles (level land).

14.
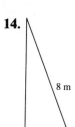
2 m, 8 m, A

$\cos A = \dfrac{2}{8}$; $m\angle A \approx 76$

15.
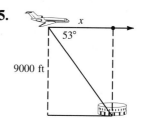
9000 ft, 53°, x

$\tan 53° = \dfrac{9000}{x}$;

$x \approx 6782$ ft

page 650   Chapter Test

1. ⟵————•———⟶   ray
       0   2

2. true

3. $m\angle N = 90 - 71 = 19$

4. Let x = the measure of the angle, then $180 - x$ = the measure of the supplement. $x = (180 - x) + 18$, $x = 198 - x$, $2x = 198$, $x = 99$; $180 - x = 81$

5. right triangle; scalene triangle

6. $m\angle B + 35 + 21 = 180$, $m\angle B = 124$

7. Since $MN = NO$, $m\angle M = m\angle O$, and $m\angle M + m\angle O + 100 = 180$.
   So $m\angle M + m\angle M + 100 = 180$, $2(m\angle M) = 80$, $m\angle M = 40$

8. $\angle R \leftrightarrow \angle T$; $\angle RIG \leftrightarrow \angle TIN$; $\angle G \leftrightarrow \angle N$; $\overline{RI} \leftrightarrow \overline{TI}$; $\overline{IG} \leftrightarrow \overline{IN}$; $\overline{RG} \leftrightarrow \overline{TN}$

9. true

10. $\dfrac{15}{IT} = \dfrac{9}{6}$, $9(IT) = 90$, $IT = \dfrac{90}{9} = 10$; $\dfrac{5}{PS} = \dfrac{6}{9}$, $6(PS) = 45$, $PS = \dfrac{45}{6} = 7.5$

11. $\dfrac{15}{FT} = \dfrac{12}{18}$, $12(FT) = 270$, $FT = \dfrac{270}{12} = 22.5$; $\dfrac{10}{JM} = \dfrac{18}{12}$, $18(JM) = 120$, $JM = \dfrac{120}{18} = 6\dfrac{2}{3}$

12. $(AB)^2 = 16^2 + 30^2$, $(AB)^2 = 1156$, $AB = 34$; $\sin A = \dfrac{16}{34} = \dfrac{8}{17}$; $\sin B = \dfrac{30}{34} = \dfrac{15}{17}$

13. 0.3256

14. 70°

15. $\tan 68° = \dfrac{YZ}{10}$, $YZ = (2.475)(10)$, $YZ = 25$; $\cos 68° = \dfrac{10}{XZ}$, $XZ = \dfrac{10}{0.3746}$, $XZ = 27$

Challenge   Since the slope of the line is $\dfrac{1}{4}$, the acute angle the line makes with the x-axis is the angle whose tangent is $\dfrac{1}{4}$. So $\tan x = \dfrac{1}{4}$, $x \approx 14°$.

page 651   Preparing for Standardized Tests

1. $f(-2) = 3(-2)^2 + 5(-2) + 1 = 3(4) - 10 + 1 = 3$

2. $\dfrac{57 + 52 + 61 + 57 + 49 + 54 + 61 + 57 + 53 + 51 + 53}{11} = \dfrac{605}{11} = 55$

3. median: 54   **4.** 57   **5.** $61 - 49 = 12$

6. $0.0075(64) = 0.48$

7. $2(1.66) + 1.89 + 2.99 = 8.20$; total amount paid $= 8.20(1.05) = \$8.61$

8. $P(2, 4, \text{ or } 6) = P(2) + P(4) + P(6) = \dfrac{1}{6} + \dfrac{1}{6} + \dfrac{1}{6} = \dfrac{3}{6} = \dfrac{1}{2}$

9. $5x - 2y = 8$, $-2y = -5x + 8$, $y = \dfrac{5}{2}x - 4$; slope $= \dfrac{5}{2}$

10. $0.012 + 0.5 + 0.04 = 0.552 = \dfrac{552}{1000} = \dfrac{69}{125}$

11. $x^2 + 3x - 2x - 6 - 5 = x^2 + x - x - 1$, $x - 11 = -1$, $x = 10$

12. $x^2 - x + 5 = 0$; $b^2 - 4ac = 1 - 4(1)(5) = -19 < 0$; 0 real solutions

13. Let x represent the first odd integer, then $x + 2$ and $x + 4$ represent the next two consecutive odd integers. $x(x + 2) = 13(x + 4) + 8$, $x^2 + 2x = 13x + 52 + 8$, $x^2 + 2x = 13x + 60$, $x^2 - 11x - 60 = 0$, $(x - 15)(x + 4) = 0$, $x = 15$

14. Let t represent the tens digit, and u represent the units digit.
$$\begin{cases} t + u = 11 \\ 10u + t = 10t + u - 9 \end{cases} \longrightarrow \begin{aligned} t + u &= 11 \\ 9t - 9u &= 9 \end{aligned} \longrightarrow \begin{aligned} t + u &= 11 \\ t - u &= 1 \end{aligned}$$
$2t = 12$, $t = 6$; $6 + u = 11$, $u = 5$; The number is 65.

15. $\dfrac{JA}{JO} = \dfrac{JN}{JB}$, $\dfrac{12}{18} = \dfrac{JN}{9}$, $\dfrac{2}{3} = \dfrac{JN}{9}$, $3(JN) = 18$, $JN = 6$

16. $\dfrac{JA}{JO} = \dfrac{AN}{OB}$, $\dfrac{12}{18} = \dfrac{15}{OB}$, $\dfrac{2}{3} = \dfrac{15}{OB}$, $2(OB) = 45$, $OB = 22.5$

17. $\begin{cases} s + m + l = 96 \\ s + l = m \end{cases} \longrightarrow \begin{aligned} m + (s + l) &= 96 \\ s + l &= m \end{aligned}$; $2m = 96$, $m = 48$

18. $A = B + 6.29$, $17.3 - 6.29 = 11.01$ grams

page 652–654 Cumulative Review Chapters (1–15)

1. $d = \sqrt{(5 - 2)^2 + (7 - 3)^2} = \sqrt{3^2 + 4^2} = \sqrt{9 + 16} = \sqrt{25} = 5$

2. $d = \sqrt{[7 - (-1)]^2 + [4 - (-2)]^2} = \sqrt{8^2 + 6^2} = \sqrt{64 + 36} = \sqrt{100} = 10$

3. $d = \sqrt{(6 - 3)^2 + [1 - (-2)]^2} = \sqrt{3^2 + 3^2} = \sqrt{9 + 9} = \sqrt{18} = 3\sqrt{2}$

4. $d = \sqrt{(-3 - 5)^2 + [1 - (-4)]^2} = \sqrt{(-8)^2 + 5^2} = \sqrt{64 + 25} = \sqrt{89}$

5. $d = \sqrt{[3 - (-1)]^2 + (6 - 2)^2} = \sqrt{4^2 + 4^2} = \sqrt{16 + 16} = \sqrt{32} = 4\sqrt{2}$

6. $d = \sqrt{[2 - (-2)]^2 + [1 - (-5)]^2} = \sqrt{4^2 + 6^2} = \sqrt{16 + 36} + \sqrt{52} = 2\sqrt{13}$

7. $|2 - (3\sqrt{2})^2| = |2 - 18| = |-16| = 16$

8. $\sqrt{3}(2 + \sqrt{6}) = 2\sqrt{3} + \sqrt{18} = 2\sqrt{3} + 3\sqrt{2}$

9. $3\sqrt{12} + 3\sqrt{18} - 2\sqrt{27} = 6\sqrt{3} + 9\sqrt{2} - 6\sqrt{3} = 9\sqrt{2}$

10. $a - 3(a - 2) = a - 3a + 6 = -2a + 6$

11. $\sqrt{20c^2d^3} = 2cd\sqrt{5d}$

12. $(2b^3 - 1)(2b^3 - 1) = 4b^6 - 2b^3 - 2b^3 + 1 = 4b^6 - 4b^3 + 1$

13. $(2z - 6)(3z + 2) = 6z^2 + 4z - 18z - 12 = 6z^2 - 14z - 12$

14. $(2\sqrt{3} - 1)(2\sqrt{3} - 1) = 12 - 2\sqrt{3} - 2\sqrt{3} + 1 = 13 - 4\sqrt{3}$

15. $(2r^2t)^3(3rt^3) = (8r^6t^3)(3rt^3) = 24r^7t^6$

16. $\dfrac{(8m^3n^5)^2}{(-4m^4n^3)^2} = \dfrac{64m^6n^{10}}{16m^8n^6} = \dfrac{4n^4}{m^2}$

17. $\dfrac{2\sqrt{3}}{3 + \sqrt{2}} = \dfrac{2\sqrt{3}}{3 + \sqrt{2}} \cdot \dfrac{3 - \sqrt{2}}{3 - \sqrt{2}} = \dfrac{6\sqrt{3} - 2\sqrt{6}}{9 - 2} = \dfrac{6\sqrt{3} - 2\sqrt{6}}{7}$

18. $\dfrac{3x - 7}{5} - \dfrac{3 - 2x}{5} = \dfrac{3x - 7 - 3 + 2x}{5} = \dfrac{5x - 10}{5} = \dfrac{5(x - 2)}{5} = x - 2$

19. $\dfrac{24d^6}{\sqrt{5}} \cdot \dfrac{45}{\sqrt{6d^2}} = \dfrac{1080d^6}{\sqrt{30d^2}} \cdot \dfrac{\sqrt{30}}{\sqrt{30}} = \dfrac{1080d^6\sqrt{30}}{30d^2} = 36d^4\sqrt{30}$

20. $\dfrac{4x^2 - 9}{x + 2} \div \dfrac{2x + 3}{x^2 - 4} = \dfrac{(2x + 3)(2x - 3)}{x + 2} \cdot \dfrac{(x + 2)(x - 2)}{2x + 3} = (2x - 3)(x - 2) = 2x^2 - 7x + 6$

21. $\dfrac{2x^2 + 2x}{x^2 - 9} + \dfrac{x + 1}{3 - x} = \dfrac{-2x^2 - 2x}{(3 - x)(3 + x)} + \left(\dfrac{x + 1}{3 - x}\right)\left(\dfrac{3 + x}{3 + x}\right) =$

$\dfrac{-2x^2 - 2x + 3x + x^2 + 3 + x}{(3 - x)(3 + x)} = \dfrac{-x^2 + 2x + 3}{(3 - x)(3 + x)} = \dfrac{x^2 - 2x - 3}{(x - 3)(x + 3)} = \dfrac{(x - 3)(x + 1)}{(x - 3)(x + 3)} = \dfrac{x + 1}{x + 3}$

22. $(2.5 \times 10^6)(0.3 \times 10^{-2}) = 0.75 \times 10^4 = 7.5 \times 10^3$

23. $7(3c - 2) + 6 - 5c = 21c - 14 + 6 - 5c = 16c - 8 = 8(2c - 1) = 16c - 8$

24. $(2 \div 3)2^2 + 8 \div 3 = \left(\dfrac{2}{3}\right)(4) + \dfrac{8}{3} = \dfrac{8}{3} + \dfrac{8}{3} = \dfrac{16}{3}$

25. $(2\sqrt{3} + 3\sqrt{2})(3\sqrt{3} - 2\sqrt{2}) = 18 - 4\sqrt{6} + 9\sqrt{6} - 12 = 6 + 5\sqrt{6}$

26. $-\dfrac{2}{3}g^2h(12gh^3 - 3h) = -8g^3h^4 + 2g^2h^2$

27. $(8x^3 + 27) \div (2x + 3) = \dfrac{(2x + 3)(4x^2 - 6x + 9)}{2x + 3} = 4x^2 - 6x + 9$

28. $(9p^2 - 3p + 1)(3p + 1) = 27p^3 + 9p^2 - 9p^2 - 3p + 3p + 1 = 27p^3 + 1$

29. $(2a^2b - 3c)(2ab^2 + 3c) = 4a^3b^3 + 6a^2bc - 6ab^2c - 9c^2$

30. $\dfrac{36w^5 - 48w^3 + 12w^2}{-12w^2} = \dfrac{-12w^2(-3w^3 + 4w - 1)}{-12w^2} = -3w^3 + 4w - 1$

31. $\dfrac{8m^2 - 6mn - 9n^2}{12m^2 + mn - 6n^2} = \dfrac{(4m + 3n)(2m - 3n)}{(4m + 3n)(3m - 2n)} = \dfrac{2m - 3n}{3m - 2n}$

32. Let $t =$ the tens digit and $u =$ the ones digit.
$\begin{cases} t + u = 13 \\ 10u + t = 10t + u + 27 \end{cases} \longrightarrow \begin{array}{l} u = 13 - t \\ 9u - 9t = 27 \end{array}$
$9(13 - t) - 9t = 27, \; 117 - 9t - 9t = 27, \; -18t = -90, \; t = 5; \; u = 13 - 5, \; u = 8; \; 58$

33. Let $x =$ the first integer, then $x + 1 =$ the second integer, and $x + 2 =$ the third integer. $x(x + 1) = (x + 2)^2 - 34, \; x^2 + x = x^2 + 4x + 4 - 34,$
$x = 4x - 30, \; 30 = 3x, \; x = 10, \; x + 1 = 11, \; x + 2 = 12$

34. Let $x =$ the amount of water to be added. $30(0.15) = (30 + x)(0.10), \; 4.5 = 0.10(30 + x); \; 4.5 = 3 + 0.10x, \; 1.5 = 0.10x, \; x = 15; \; 15$ mL

35. $\dfrac{81 + 92 + 86 + x}{4} > 85; \; \dfrac{259 + x}{4} > 85, \; 259 + x > 340, \; x > 81; \; 82$

36. $16x^2 - 9y^2 = (4x - 3y)(4x + 3y)$ **37.** $t^2 + 2t - 35 = (t + 7)(t - 5)$

38. $c^2 - 13c + 36 = (c - 9)(c - 4)$ **39.** $2m^2 + m - 15 = (2m - 5)(m + 3)$

40. $6m^2 + 5m - 6 = (3m - 2)(2m + 3)$

41. $6a^3 - 24ab^2 = 6a(a^2 - 4b^2) = 6a(a - 2b)(a + 2b)$

42. $8a^3b + 12a^2b^2 - 4a^2b = 4a^2b(2a + 3b - 1)$

43. $rt^2 - 9r + t^3 - 9t = r(t^2 - 9) + t(t^2 - 9) = (r + t)(t^2 - 9) = (r + t)(t + 3)(t - 3)$

44. $0.2x + 3.9 = 1.5, \; 0.2x = -2.4, \; x = -12$ **45.** $\dfrac{3}{5}r + 10 = 19, \; \dfrac{3}{5}r = 9, \; r = 15$

46. $2 - 3(a - 4) = 2a - 1, \; 2 - 3a + 12 = 2a - 1, \; -5a = -15, \; a = 3$

47. $0.18m = 90, \; m = \dfrac{90}{0.18}, \; m = 500$

48. $2y^2 - 3y - 27 = 0, \; (2y - 9)(y + 3) = 0, \; 2y - 9 = 0, \; y = \dfrac{9}{2} \text{ or } y + 3 = 0, \; y = -3$

49. $c^2 + 5c - 5c = 4, \; c^2 = 4, \; c = \pm\sqrt{4}, \; c = \pm 2$

50. $8 - 5t > 13, \; -5t > 5, \; t < -1$

51. $|d - 7| = 3, \; d - 7 = 3 \quad \text{or} \quad d - 7 = -3$
$\qquad\qquad\qquad\qquad d = 10 \qquad\qquad\qquad d = 4$

52. $\sqrt{3n} + 1 = 10, \; \sqrt{3n} = 9, \; 3n = 81, \; n = 27$

53. $|p| + 6 = 5, \; |p| = -1;$ no solution

54. $|4 - 3x| < 10$, $4 - 3x < 10$ and $4 - 3x > -10$

$-3x < 6$ $\quad\vdots\quad$ $-3x > -14$

$x > -2$ $\quad\vdots\quad$ $x < \dfrac{14}{3}$

$-2 < x < \dfrac{14}{3}$

55. $\sqrt{b^2} + 2 = b - 2$, $\sqrt{b^2} = b - 4$, $b^2 = b^2 - 8b + 16$, $-8b + 16 = 0$, $-8b = -16$, $b = 2$; Substituting into the original equation gives $\sqrt{2^2} + 2 \stackrel{?}{=} 2 - 2$, $2 + 2 \neq 2 - 2$. Therefore, there is no solution.

56. $\dfrac{u}{u + 1} = 2 - \dfrac{3}{u + 1}$, $\dfrac{u}{u + 1} = \dfrac{2(u + 1)}{u + 1} - \dfrac{3}{u + 1}$, $u = 2(u + 1) - 3$, $u = 2u + 2 - 3$, $u = 2u - 1$, $-u = -1$, $u = 1$

57. $2x + 3y = 5 \longrightarrow \quad 2x + 3y = 5$

$5x + 3y = 11 \longrightarrow \quad \underline{-5x - 3y = -11}$

$-3x = -6$

$x = 2$

$2(2) + 3y = 5$, $4 + 3y = 5$, $3y = 1$, $y = \dfrac{1}{3}$; $\left(2, \dfrac{1}{3}\right)$

58. $\dfrac{2h}{h^2 - 4} + 3 = \dfrac{3h}{h + 2}$, $\dfrac{2h}{(h + 2)(h - 2)} + \dfrac{3(h + 2)(h - 2)}{(h + 2)(h - 2)} = \dfrac{3h(h - 2)}{(h + 2)(h - 2)}$,

$2h + 3h^2 - 12 = 3h^2 - 6h$, $2h - 12 = -6h$, $8h = 12$, $h = \dfrac{12}{8} = \dfrac{3}{2}$

59. $|w - 5| > 1$, $w - 5 > 1$ or $w - 5 < -1$

$w > 6$ $\quad\vdots\quad$ $w < 4$

$w > 6$ or $w < 4$

60. $9x^2 = 18$, $x^2 = 2$, $x = \pm\sqrt{2}$

61. $|a - 2| \leq 3$, $a - 2 \leq 3$ and $a - 2 \geq -3$

$a \leq 5$ $\quad\vdots\quad$ $a \geq -1$

$-1 \leq a \leq 5$

62. $a + 3(2a + 1) = 30$, $a + 6a + 3 = 30$, $7a = 27$, $a = \dfrac{27}{7}$; $b = 2\left(\dfrac{27}{7}\right) + 1 = \dfrac{61}{7}$; $\left(\dfrac{27}{7}, \dfrac{61}{7}\right)$

63. $(m - 2)^2 = 36$, $m - 2 = \pm\sqrt{36}$, $m - 2 = \pm 6$, $m = 2 \pm 6$; $m = 8, -4$

64. $6r + 7t = 24 \longrightarrow \quad 6r + 7t = 24$

$2r - 3t = -8 \longrightarrow \quad \underline{-6r + 9t = 24}$

$16t = 48$

$t = 3$

$6r + 7(3) = 24$, $6r + 21 = 24$, $6r = 3$, $r = \dfrac{1}{2}$; $\left(\dfrac{1}{2}, 3\right)$

65. $d^2 - 6d - 16 = 0$, $(d - 8)(d + 2) = 0$, $d - 8 = 0$, $d = 8$ or $d + 2 = 0$, $d = -2$

66. $3x^2 - 2x - 5 = 0$, $(3x - 5)(x + 1) = 0$, $3x - 5 = 0$, $x = \dfrac{5}{3}$ or $x + 1 = 0$, $x = -1$

67. $5a^2 = 20a + 60$, $5a^2 - 20a - 60 = 0$, $a^2 - 4a - 12 = 0$, $(a - 6)(a + 2) = 0$,
$a - 6 = 0$, $a = 6$ or $a + 2 = 0$, $a = -2$

68. $p - 5p + 10 + p - 3 = 2$, $-3p + 7 = 2$, $-3p = -5$, $p = \dfrac{5}{3}$

69. $5 + \sqrt{3a} - 2 = 9$, $3 + \sqrt{3a} = 9$, $\sqrt{3a} = 6$, $3a = 36$, $a = 12$

70. $(-1)^2 - 4(2)(1) = 1 - 8 = -7$; no real solutions

71. $3^2 - 4(1)(-2) = 9 + 8 = 17$; 2 real solutions

72. $(-2)^2 - 4(1)(-1) = 4 + 4 = 8$; 2 real solutions

73. $\dfrac{b}{a} = -(2 + 5) = -7$; $\dfrac{c}{a} = (2)(5) = 10$; $x^2 - 7x + 10 = 0$

74. $\dfrac{b}{a} = -\left(\dfrac{1}{3} + 2\right) = -\dfrac{7}{3}$; $\dfrac{c}{a} = \left(\dfrac{1}{3}\right)(2) = \dfrac{2}{3}$; $x^2 - \dfrac{7}{3}x + \dfrac{2}{3} = 0$; $3x^2 - 7x + 2 = 0$

75. $\dfrac{b}{a} = -\left(-3 + \dfrac{1}{2}\right) = \dfrac{5}{2}$; $\dfrac{c}{a} = (-3)\left(\dfrac{1}{2}\right) = -\dfrac{3}{2}$; $x^2 + \dfrac{5}{2}x - \dfrac{3}{2} = 0$; $2x^2 + 5x - 3 = 0$

76. mode: 8; median: $\dfrac{8 + 9}{2} = 8.5$; mean: $\dfrac{6 + 8 + 8 + 9 + 14 + 15}{6} = 10$; range: $15 - 6 = 9$

77. 21, 21, 32, 32, 35, 63; mode: 21, 32; median: 32
mean: $\dfrac{21 + 21 + 32 + 32 + 35 + 63}{6} = 34$; range: $63 - 21 = 42$

78. mean $= \dfrac{0 + 2 + 4 + 4 + 5}{5} = 3$

Data	Deviations from Mean	Frequency $\times$ (Dev.)^2
0	$0 - 3 = -3$	$1(9) = 9$
2	$2 - 3 = -1$	$1(1) = 1$
4	$4 - 3 = 1$	$2(1) = 2$
5	$5 - 3 = 2$	$1(4) = 4$
		Sum of squares $= 16$

range: $5 - 0 = 5$; variance: $\dfrac{16}{5} = 3.2$; standard deviation $= \sqrt{3.2} = 1.8$

79. $P(\text{green}) = \dfrac{4}{12} = \dfrac{1}{3}$

80. $P(\text{green or red}) = P(\text{green}) + P(\text{red}) = \dfrac{4}{12} + \dfrac{5}{12} = \dfrac{9}{12} = \dfrac{3}{4}$

81. $P(\text{blue}) = 0$

82. $P(\text{1st green and 2nd green}) = \dfrac{4}{12}\left(\dfrac{4}{12}\right) = \dfrac{16}{144} = \dfrac{1}{9}$

83. $P(\text{1st green and 2nd not green}) = \dfrac{4}{12}\left(\dfrac{8}{12}\right) = \dfrac{32}{144} = \dfrac{2}{9}$

84. $P(\text{1st green and 2nd green}) = \dfrac{4}{12}\left(\dfrac{3}{11}\right) = \dfrac{12}{132} = \dfrac{1}{11}$

85. $P(\text{1st red and 2nd green}) = \dfrac{5}{12}\left(\dfrac{4}{11}\right) = \dfrac{20}{132} = \dfrac{5}{33}$

86. Let x = the measure of the angle, then $90 - x$ = the measure of the complement. $x = (90 - x) - 20$, $x = 70 - x$, $2x = 70$, $x = 35$; $90 - x = 55$

87. $m\angle B = 2(m\angle A)$, $m\angle C = m\angle A + 20$; $m\angle A + 2(m\angle A) + (m\angle A + 20) = 180$, $4(m\angle A) + 20 = 180$, $4(m\angle A) = 160$, $m\angle A = 40$; $m\angle B = 2(40) = 80$; $m\angle C = 40 + 20 = 60$

88. Let h = the height of the tree. $\tan 33° = \dfrac{h}{8}$, $h = 5$ ft

89. $A = lw = 3\sqrt{x}(2\sqrt{x}) = 6x$ cm²

90. Area $= s^2$, $625 = s^2$, $25 = s$; Perimeter $= 4s = 4(25) = 100$; Therefore, 100 ft is needed.

91. $m = \dfrac{6 - 2}{4 - 3} = \dfrac{4}{1} = 4$

92. $m = \dfrac{1 - 0}{\dfrac{3}{4} - \left(-\dfrac{1}{2}\right)} = \dfrac{1}{\dfrac{5}{4}} = \dfrac{4}{5}$

93.

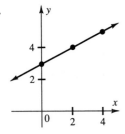

94.

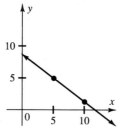

95. $\{3\}$ **96.** no solution   **97.** $\{4\}$

98. $\{-2, 0, 1, 3, 4\}$ **99.** $\{3, 4\}$ **100.** $\{-2, 0\}$

101. $x^2 - 30x + 125 = 0$, $(x - 25)(x - 5) = 0$, $x - 25 = 0$, $x = 25$ or $x - 5 = 0$, $x = 5$

102. Let w = the width, then $w + 5$ = the length of the rectangle. $w(w + 5) = 36$, $w^2 + 5w - 36 = 0$, $(w + 9)(w - 4) = 0$, $w + 9 = 0$ or $w - 4 = 0$; Since width is positive, $w - 4 = 0$, $w = 4$ m and $w + 5 = 4 + 5 = 9$ m.

Extra Practice

page 655

1. $w + 6 = \dfrac{4}{3} + 6 = \dfrac{22}{3}$

2. $(wx) - 1 = \dfrac{4}{3}(3) - 1 = 3$

3. $(9w) \div 4 = 9\left(\dfrac{4}{3}\right) \div 4 = 3$

4. $3 + z = 3 + (-4.1) = -1.1$

5. $y + z = -\dfrac{1}{2} + (-4.1) = -4.6$

6. $|x + y| = \left|3 + \left(-\dfrac{1}{2}\right)\right| = 2\dfrac{1}{2}$

7. $y - x = -\dfrac{1}{2} - 3 = -3\dfrac{1}{2}$

8. $5 - (y - z) = 5 - \left[-\dfrac{1}{2} - (-4.1)\right] = 5 - (3.6) = 1.4$

9. $|z| - |y| = |-4.1| - \left|-\dfrac{1}{2}\right| = 4.1 - 0.5 = 3.6$

10. $(xyw) + z = (3)\left(-\dfrac{1}{2}\right)\left(\dfrac{4}{3}\right) + (-4.1) = -2 + (-4.1) = -6.1$

11. $w[x + (6y)] = \dfrac{4}{3}\left[3 + 6\left(-\dfrac{1}{2}\right)\right] = \dfrac{4}{3}(0) = 0$

12. $\left(\dfrac{y}{w}\right) - (2z) = \dfrac{-\dfrac{1}{2}}{\dfrac{4}{3}} - 2(-4.1) = -\dfrac{3}{8} + 8.2 = -0.375 + 8.2 = 7.825$

13. $u + z - 11.3 = -5 + (-4.1) - 11.3 = -20.4$

14. $(uw) - z + 0.3 = (-5)\left(\dfrac{4}{3}\right) - (-4.1) + 0.3 = -\dfrac{20}{3} + 4.4 = -6.\overline{6} + 4.4 = -2.2\overline{6}$

15. $\dfrac{|(6w) - (4y)|}{-0.5} = \dfrac{\left|6\left(\dfrac{4}{3}\right) - 4\left(-\dfrac{1}{2}\right)\right|}{-0.5} = \dfrac{|8 + 2|}{-0.5} = \dfrac{10}{-0.5} = -20$

16. $i + 5$ **17.** $55 - n$ **18.** $\dfrac{m}{h}$ **19.** $-3\dfrac{1}{2}$ **20.** $-\dfrac{1}{2}$ **21.** 2

22. $1\dfrac{1}{2}$ **23.** E **24.** C **25.** F **26.** H

27. $-7 = \dfrac{-7}{1}$ **28.** $0 = \dfrac{0}{1}$ **29.** $3\dfrac{2}{3} = \dfrac{11}{3}$ **30.** $-1.26 = \dfrac{-126}{100}$

31. $-(-3) = 3$ **32.** $-[-(-8)] = -8$ **33.** $(-7)(-9)\left(\dfrac{1}{3}\right) = 21$

34. $\dfrac{2}{5}(10)(-x) = -4x$ **35.** $-2 < 0$ **36.** $-4 < -3$

37. $-\dfrac{1}{2} < -\dfrac{1}{3}$ **38.** $-2.1 = -2.1$

39. $-11.1 + (-4.6) + 2.3 < 11.1 + (-4.6) - (+2.3)$; $-13.4 < 4.2$

40. $0.56 + 2.12 + (-2.12) > -0.56 + (-2.21) + (2.21)$; $0.56 > -0.56$

41. Let x represent the average temperature at sunrise.

$$x = \frac{-1.5° + (-3.1°) + 2.2° + 0.4° + (-1.1°) + 3.6° + 0.5°}{7} = \frac{1°}{7} \approx 0.1°$$

The average temperature was $0.1°$F.

page 656

1. $2(-2) - 8 = -4 - 8 = -12$

2. $8 - \dfrac{4}{0}$; no solution

3. $3\{4 - 2[4(-2) - 3(8)] + 3(0)\} = 3[4 - 2(-32) + 0] = 3(4 + 64) = 3(68) = 204$

4. $\dfrac{-0 + (-5)}{2(4)(8)} = -\dfrac{5}{64}$

5. $5(-2)^3 - 2(-5)^2 = 5(-8) - 2(25) = -40 - 50 = -90$

6. $[2 - (4)(-2)](8 - 0) \div [2(-5) + 4] = (10)(8) \div (-6) = -\dfrac{80}{6} = -\dfrac{40}{3}$

7. $\dfrac{8^2 - 4^2}{(8 - 4)^4} = \dfrac{48}{256} = \dfrac{3}{16}$

8. $4(4)(0)[2(8) + (-2)] = 0(14) = 0$

9. $-4[2(-5)^2 + (-2)] \div (0^2 - 8^2) = -4[50 + (-2)] \div (-64) = -192 \div (-64) = 3$

10. Distributive property

11. Associative property of addition

12. Commutative property of addition

13. Commutative property of multiplication

14. $-r^3 + 6r^2 - 3r$

15. $5j^2k + jk + 3jk^2$

16. $5m^2 - (2m^2 - 2m^2 - m) = 5m^2 + m$

17. $3x - 2(-x + 3x^2 + 6x) = 3x + 2x - 6x^2 - 12x = -6x^2 - 7x$

18. $2g - (-h + g - g^2 + h - g^2 + h + g) = 2g - (-2g^2 + 2g + h) = 2g^2 - h$

19. $n - 3$ **20.** $5 \cdot n^2$ **21.** $8(n^2 + 1)$

22. a. $s - 5$ **b.** $s - 6$ **c.** $s - 11$

23. $13 - n = -24$ **24.** $-6n - 2 = 10$ **25.** $38 + \dfrac{7}{n} = 50$

26. $10a + 7.5c = 1520$

27. $-2 > 3y + 5; -2 > 3(-3) + 5;$ true; $-2 > 3(1) + 5;$ false; $-2 > 3(0) + 5;$ false; $-2 > 3\left(\dfrac{1}{3}\right) + 5;$ false; $\{-3\}$

28. $-2y = 1; -2(-3) = 1;$ false; $-2(1) = 1;$ false; $-2(0) = 1;$ false; $-2\left(\dfrac{1}{3}\right) = 1;$ false; $\varnothing$

29. $y + 2 > -4; (-3) + 2 > -4;$ true; $1 + 2 > -4;$ true; $(0) + 2 > -4;$ true; $\left(\dfrac{1}{3}\right) + 2 > -4;$ true; $\left\{-3, 1, 0, \dfrac{1}{3}\right\}$

page 657

1. $a + 26 - 26 = 31 - 26, a = 5$

2. $\dfrac{98}{14} = \dfrac{14m}{14}, m = 7$

3. $2\left(\dfrac{1}{2}y\right) = 2(-17), y = -34$

4. $4.5 + b - 4.5 = 11.2 - 4.5, b = 6.7$

5. $8\left(\dfrac{f}{8}\right) = 8(-3), f = -24$

6. $-\dfrac{2}{3} - \dfrac{5}{3} = \dfrac{5}{3} + c - \dfrac{5}{3}, c = -\dfrac{7}{3}$

7. $-2.23 + s + 2.23 = 0 + 2.23, s = 2.23$

8. $-\dfrac{4}{9}(0) = -\dfrac{4}{9}\left(-\dfrac{9}{4}r\right), r = 0$

9. $3.45 + t - 3.45 = 2.82 - 3.45, t = -0.63$

10. $d - 4.2 + 4.2 = -3.1 + 4.2, d = 1.1$

11. $-1(-9.6) = -1(-h), h = 9.6$

12. $-8.21 + e + 8.21 = -8.21 + 8.21, e = 0$

13. $4k + 3 - 3 = 11 - 3, 4k = 8, \dfrac{4k}{4} = \dfrac{8}{4}, k = 2$

14. $-4 + \dfrac{m}{6} + 4 = 2 + 4, \dfrac{m}{6} = 6, 6\left(\dfrac{m}{6}\right) = 6(6), m = 36$

15. $6.4p - 1.5 + 1.5 = -0.22 + 1.5, 6.4p = 1.28, \dfrac{6.4p}{6.4} = \dfrac{1.28}{6.4}, p = 0.2$

16. $n - 11 = -15, n - 11 + 11 = -15 + 11, n = -4$

17. $\dfrac{7}{8}n = 49, \dfrac{8}{7}\left(\dfrac{7}{8}n\right) = \dfrac{8}{7}(49), n = 56$

18. $0.25n - 7 = -9$, $0.25n - 7 + 7 = -9 + 7$, $0.25n = -2$, $\dfrac{0.25n}{0.25} = -\dfrac{2}{0.25}$, $n = -8$

19. Let x represent Jan's age. $3\frac{1}{2}x = 56$, $\frac{7}{2}x = 56$, $\frac{2}{7}\left(\frac{7}{2}x\right) = \frac{2}{7}(56)$, $x = 16$; Jan is 16 years old.

20. Reflexive property; Subtraction property; Given; Substitution property

21. Given; Definition of division; Substitution property; Multiplication property for equations; Identity property for multiplication

22. $48 = 2l + 2(11)$, $48 = 2l + 22$, $48 - 22 = 2l + 22 - 22$, $26 = 2l$, $\frac{1}{2}(26) = \frac{1}{2}(2l)$, $l = 13$

23. $251.2 = 3.14(4^2)h$, $251.2 = 50.24h$, $\dfrac{251.2}{50.24} = \dfrac{50.24h}{50.24}$, $h = 5$

24. $60.4 = \frac{1}{2}(2.0)(3.8 + b)$, $60.4 = 3.8 + b$, $60.4 - 3.8 = 3.8 + b - 3.8$, $b = 56.6$

page 658

1. $2a = 14$; $a = 7$

2. $-15 = -3b + 3$, $-18 = -3b$; $6 = b$

3. $3c + 2c + 4 = -6$, $5c + 4 = -6$, $5c = -10$; $c = -2$

4. $-20d - 24 = 2$, $-20d = 26$; $d = -\dfrac{13}{10}$

5. $5e = 7e + 6$, $-2e = 6$; $e = -3$

6. $12f + 18 = -2f + 18$, $14f + 18 = 18$, $14f = 0$; $f = 0$

7. $2g + 12 - 4g = 5g + 19$, $-2g + 12 = 5g + 19$, $-7g + 12 = 19$, $-7g = 7$; $g = -1$

8. $-h + 2h + 16 = -2 + 6h + 3$, $h + 16 = 1 + 6h$, $-5h + 16 = 1$, $-5h = -15$; $h = 3$

9. $-6a - 15 = 9$, $-6a = 24$; $a = -4$

10. $9k + 4 = 15 - 2k$, $11k + 4 = 15$, $11k = 11$; $k = 1$

11. $-0.1l + 0.7 = 1.4$, $-0.1l = 0.7$; $l = -7$

12. $0.06m - 0.09 + 0.05 = 5$, $0.06m - 0.04 = 5$, $0.06m = 5.04$; $m = 84$

13. i, $i + 2$, $i + 4$

14. $c - 1$, $c - 2$

15. Let x = the first integer, then $x + 1$ = the second integer, $x + 2$ = the third integer, $x + 3$ = the fourth integer, and $x + 4$ = the fifth integer. The equation is: $x + (x + 2) = -28$; solve $2x + 2 = -28$, $2x = -30$, $x = -15$. Then $x + 1 = -14$, $x + 2 = -13$, $x + 3 = -12$, and $x + 4 = -11$. The five consecutive integers are -15, -14, -13, -12, -11.

16. Let n = the number.

$n = 0.45(350)$;

$n = 157.5$

17. Let r = the percent.

24 is what percent of 108?

$24 = r(108)$;

$$r = \frac{24}{108} = \frac{2}{9} = 0.\overline{2} = 22.\overline{2}\%$$

18. Let t = the number.

0.15 is 30% of t.

$0.15 = (0.30)t$;

$0.5 = t$

19. Let r = the percent.

5 is what percent of 50,000?

$5 = r(50,000)$;

$$r = \frac{5}{50,000} = \frac{1}{10,000} = 0.0001 = 0.01\%$$

20. Change in price: $0.30(12.50) = \$3.75$

New price: $12.50 - 3.75 = 8.75$; new price: \$8.75

21. Change: $15.60 - 12 = 3.60$

Let r = percent of increase. 3.60 is what percent of 12?

$$360 = r(12), r = \frac{3.60}{12} = 0.30 = 30\%$$

22. Let w = original weight. Then $0.80w$ = new weight.

$$0.80w = 124, w = \frac{124}{0.80} = 155 \text{ lb}$$

23. Let x = the number of mL of water added.

	Substance	Total No. mL	No. mL Pure Acid
Start with	20% acid solution	15	$0.20(15)$
Add	Water	x	0
Finish with	5% acid solution	$x + 15$	$0.05(x + 15)$

$0.20(15) = 0.05(x + 15)$, $3 = 0.05x + 0.75$, $2.25 = 0.05x$, $x = 45$;
45 mL of water must be added to obtain a 5% acid solution.

24. Let x = the number of mL of ester added.

	Substance	Total No. g	Grams pure Ester
Start with	10% ester solution	25	$0.10(25)$
Add	100% pure ester	x	x
Finish with	25% ester solution	$x + 25$	$0.25(x + 25)$

$0.10(25) + x = 0.25(x + 25)$, $2.5 + x = 0.25x + 6.25$, $0.75x = 3.75$, $x = 5$;
5 g of ester must be added to obtain a 25% ester solution.

25. $p - r - t = r + s - t - r + t$, $p - r - t = s + 0$, $s = p - r - t$

26. $\dfrac{5ca}{5c} = \dfrac{3d}{5c}, \ a = \dfrac{3d}{5c}$

27. $2p - 6q = 5p, \ -6q = 3p, \ -2q = p$

28.

rate	× time	= distance
plane 1	540	t	540t
plane 2	630	t	630t

$540t + 630t = 1755, \ 1170t = 1755, \ t = 1.5$; It will take the two planes 1.5 h to be 1755 mi apart.

29.

rate	× time	= distance
Robin	25	t	25t
2nd cyclist	20	$t + 1$	$20(t + 1)$

$25t = 20(t + 1), \ 25t = 20t + 20, \ 5t = 20, \ t = 4$;
It will take 4 h for Robin to overtake the first cyclist.

30.

rate	× time	= distance
truck	55	t	55t
car	40	t	40t

$55t = 40t + 10, \ 15t = 10, \ t = \dfrac{2}{3}$; It will take 40 min before they are 10 mi apart.

page 659

1.

2.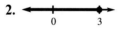

3.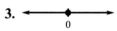

4.

5.

6. $a + 11 - 11 > -6 - 11; \ a > -17$

7. $\dfrac{4b}{4} \leq -\dfrac{15}{4}; \ b \leq -\dfrac{15}{4}$

8. $c + 5 < 7, \ c + 5 - 5 < 7 - 5; \ c < 2$

9. $6\left(-\dfrac{2}{3}\right) \geq 6\left(d + \dfrac{1}{6}\right), \ -4 \geq 6d + 1, \ -5 \leq 6d; \ -\dfrac{5}{6} \geq d$

10. $12\left(-\dfrac{5}{3}\right) > 12\left(-\dfrac{5}{4}e\right), \ -20 > -15e, \ \dfrac{-20}{-15} < \dfrac{-15e}{-15}, \ \dfrac{4}{3} < e$

11. $6f - 3 - 5f \leq 7, \ f - 3 \leq 7; \ f \leq 10$

12. $\dfrac{-2g}{-2} \leq \dfrac{0}{-2}; \ g \leq 0$

13. $j \leq -1.002 + 0.12; j \leq -0.882$

(number line: closed dot at −0.882, shaded left)
$-0.882 \quad 0$

14. $-h < -1.5; h > 1.5$

(number line: open dot at 1.5, shaded right)
$0 \quad 1.5$

15. $13l - 10 > 16, 13l > 26; l > 2$

(number line: open dot at 2, shaded right)
$0 \quad 2$

16. $4 - 1 + t \geq 2t - 3, 3 + t \geq 2t - 3, -t \geq -6; t \leq 6$

(number line: closed dot at 6, shaded left)
$0 \quad 6$

17. $3 - 3m > -9m, 3 > -6m; -\dfrac{1}{2} < m$

(number line: open dot at $-\frac{1}{2}$, shaded right)
$-\dfrac{1}{2} \quad 0$

18. $\dfrac{-6n}{-6} < \dfrac{3}{-6}$ or $-25 \geq -5n$

  $n < -\dfrac{1}{2}$ or $5 \leq n$

  The solution set is $\{n: n < -\dfrac{1}{2}$ or $n \geq 5\}$.

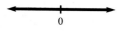
$-\dfrac{1}{2} \quad 0 \qquad 5$

19. $16 - 8p > 0$ and $p \leq -2$
  $-8p > -16$ and
  $p < 2$ and $p \leq -2$
  The solution set is $\{p: p \leq -2\}$.

$-2 \quad 0$

20. $3q \geq 9$ or $-5q > -15$
  $q \geq 3$ or $q < 3$
  The solution set is {all real numbers}.

(number line: shaded all, tick at 0)
0

21. $-12 < -3r < -3$
  The solution set is $\{r: 1 < r < 4\}$.

(number line: open dots at 1 and 4, shaded between)
$0 \; 1 \qquad 4$

22. $x = 1.1$ or $x = -1.1$
  The solution set is $\{-1.1, 1.1\}$.

(number line: closed dots at −1.1 and 1.1)
$-1.1 \quad 0 \quad 1.1$

23. $z - 2 = 3$ or $z - 2 = -3$
  $z = 5$ or $z = -1$
  The solution set is $\{-1, 5\}$.

(number line: closed dots at −1 and 5)
$-1 \; 0 \qquad 5$

24. $3v + 6 = 9$ or $3v + 6 = -9$
  $3v = 3$ or $3v = -15$
  $v = 1$ or $v = -5$
  The solution set is $\{-5, 1\}$.

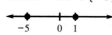
$-5 \qquad 0 \; 1$

25. $|2w - 1| = 4$;
  $2w - 1 = 4$ or $2w - 1 = -4$
  $2w = 5$ or $2w = -3$
  $w = \dfrac{5}{2}$ or $w = -\dfrac{3}{2}$
  The solution set is $\left\{-\dfrac{3}{2}, \dfrac{5}{2}\right\}$.

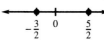
$-\dfrac{3}{2} \quad 0 \qquad \dfrac{5}{2}$

26. $|1 - 2p| = \frac{1}{2}$;

$$1 - 2p = \frac{1}{2} \quad or \quad 1 - 2p = -\frac{1}{2}$$

$$-2p = -\frac{1}{2} \quad or \quad -2p = -\frac{3}{2}$$

$$p = \frac{1}{4} \quad or \quad p = \frac{3}{4}$$

The solution set is $\left\{\frac{1}{4}, \frac{3}{4}\right\}$.

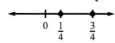

27. $\left|\frac{2g}{3}\right| = -19$;

No solution, the absolute value of a number cannot be negative.

28. $m < 0.5$ *and* $m > -0.5$
The solution set is $\{m: -0.5 < m < 0.5\}$.

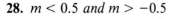

29. $n + 8 \geq 14 \quad or \quad n + 8 \leq -14$
$\qquad n \geq 6 \quad or \qquad n \leq -22$
The solution set is $\{n: n \geq 6$
or $n \leq -22\}$.

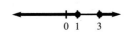

30. $3l - 5 = 0$; $l = \frac{5}{3}$

The solution set is $\left\{\frac{5}{3}\right\}$.

31. $|2h + 1| > 0$;
$$2h + 1 > 0 \quad and \quad 2h + 1 < 0$$
$$2h > -1 \quad and \qquad 2h < -1$$
$$h > -\frac{1}{2} \quad and \qquad h < -\frac{1}{2}$$
The solution set is $\Big\{$all real numbers

except $-\frac{1}{2}\Big\}$.

32. $-6 + 2|1 - 2e| < 3$, $2|1 - 2e| < 9$,

$|1 - 2e| < 4\frac{1}{2}$;

$$1 - 2e < 4\frac{1}{2} \quad and \quad 1 - 2e > -4\frac{1}{2}$$

$$-2e < 3\frac{1}{2} \quad and \qquad -2e > -5\frac{1}{2}$$

$$e > -\frac{7}{4} \quad and \qquad e < \frac{11}{4}$$

The solution set is $\left\{e: -\frac{7}{4} < e < \frac{11}{4}\right\}$.

33. $|2 - g| \geq 1$;
$$2 - g \geq 1 \quad or \quad 2 - g \leq -1$$
$$-g \geq -1 \quad or \qquad -g \leq -3$$
$$g \leq 1 \quad or \qquad g \geq 3$$
The solution set is $\{g: g \leq 1$ or $g \geq 3\}$.

34. $x > -1$ **35.** $x \le \dfrac{1}{2}$ **36.** $x \ne 1$

37. $x < -\dfrac{1}{2}$ or $x > 1$ **38.** $-2 \le x \le 3$ **39.** $|x| < 2$ or $-2 < x < 2$

40. Let $x =$ the number of stamps. $0.35x \le 15$, $x \le 42.8$; Mark can buy at most 42 stamps.

41. Let $r =$ the plate's radius. $2\pi r < 36$, $2(3.14)r < 36$, $r < \dfrac{36}{6.28}$, $r < 5.73$;
The radius is less than 5.73 inches.

42. Let $l =$ the length of the photograph. $2(19) + 2l \ge 150$, $38 + 2l \ge 150$, $2l \ge 112$, $l \ge 56$; The length must be at least 56 cm.

page 660

1. $(-2) \cdot 11 \cdot h^4 \cdot h^3 = -22h^7$ **2.** $(1.5 \times 2.0) \times (10^8 \times 10) = 3.0 \times 10^9$

3. $5 \cdot (-7) \cdot j \cdot j^2 \cdot k^2 \cdot k^2 = -35j^3k^4$

4. $\dfrac{1}{b^{4-1}} = \dfrac{1}{b^3}$ **5.** $\dfrac{-14}{7} \cdot \dfrac{c^3}{c^2} \cdot \dfrac{d^2}{d^3} = \dfrac{-2c^{3-2}}{d^{3-2}} = -\dfrac{2c}{d}$

6. $\dfrac{1.8}{0.3} \times \dfrac{10^4}{10} = 6 \times 10^3$ **7.** -1

8. $a^{2 \cdot 4} = a^8$ **9.** $(-3)^2 b^2 (c^4)^2 = 9b^2c^8$

10. $\dfrac{5}{-15} \cdot \dfrac{e}{e^4} \cdot \dfrac{f^7}{f^2} \cdot \dfrac{1}{g} = \dfrac{f^{7-2}}{-3e^{4-1}g} = \dfrac{f^5}{-3e^3g}$ **11.** $\dfrac{(2c^3)^3}{d^3} = \dfrac{8c^6}{d^3}$

12. $\dfrac{(-5)^2}{20} \cdot \dfrac{g^2}{g^2} \cdot \dfrac{h^{3 \cdot 2}}{h} = \dfrac{25h^{6-1}}{20} = \dfrac{5h^5}{4}$

13. $5(3)^3 \left[(-2)\left(-\dfrac{1}{2}\right)^5 \right] = 5(27)\left[-2\left(\dfrac{1}{-32}\right) \right] = 5(27)\left(\dfrac{1}{16}\right) = 8\dfrac{7}{16}$ or 8.4375

14. $\dfrac{-7(3)^2 \left[4\left(-\dfrac{1}{2}\right)^3 \right]^2}{9(-2)^2} = \dfrac{-63\left[4\left(-\dfrac{1}{8}\right) \right]^2}{9(4)} = \dfrac{-63\left(\dfrac{1}{4}\right)}{9(4)} = -\dfrac{7}{16}$ or -0.4375

15. $\dfrac{-5(-2)^5}{\left[-\left(-\dfrac{1}{2}\right)^7 \right]^2} = \dfrac{-5(-32)}{\left(\dfrac{1}{128}\right)^2} = 128^2(-5)(-32) = 2{,}621{,}440$

16. 8.14×10^{-3} **17.** 2.31×10^7 **18.** 1.2×10^{-5}

19. $(0.7)^2 \times 10^4 = 0.49 \times 10^4 = 4.9 \times 10^3$

20. $(3.4 \times 4.2) \times (10^{-2} \times 10^1) = 14.28 \times 10^{-1} = 1.428 \times 10^0$ or 1.428

21. $\dfrac{1.9}{38} \times \dfrac{10^2}{10^2} = 0.05 \times 10^{2-2} = 0.05 \times 10^0 = 5 \times 10^{-2}$

22. $2x^2y^2 - xy^2 + 3xy^2 - 7 = 2x^2y^2 + 2xy^2 - 7$

23. $-2ax^2 + 6a^2x^3 + 5ax^2 = 6a^2x^3 + 5ax^2 - 2ax^2 = 6a^2x^3 + 3ax^2$

24. $-3x^3 + 6x^2 - 5$ **25.** $8x^4 - 16x^3 - 4x^2 - x - 18$

26. $2d + d - 7x - 5x = -12x + 3d$

27. $x^4 + 2x^3 - x + 5 - 3x^3 - 2x^2 = x^4 - x^3 - 2x^2 - x + 5$

28. $-6y^3 + 3y^2 - 9y$ **29.** $-14z^6 + 2z^5 - 20z^3$

30. $-16a^3b^3 + 14a^5b - 2a^2b^2$

31. $6w^3 - 5w^2 + 14w^3 - 16w^2 = 20w^3 - 21w^2$

32. $2x^2 + 10x + 3x + 15 = 2x^2 + 13x + 15$

33. $c^2 + 11c - 7c - 77 = c^2 + 4c - 77$

34. $10t^2 - 25t - 4t + 10 = 10t^2 - 29t + 10$

35. $12p^3 - 4p^2 + 8p + 21p^2 - 7p + 14 = 12p^3 + 17p^2 + p + 14$

36. $6a^2 - 14ab + 3ab - 7b^2 = 6a^2 - 11ab - 7b^2$

37. $15d^5 - 6d^4 - 3d^3 - 20d^3 + 8d^2 + 4d = 15d^5 - 6d^4 - 23d^3 + 8d^2 + 4d$

38. Let l = the length, then $l - 6$ = the width. So the area = $l(l - 6) = l^2 - 6l$.

39. $(12x + 7)(12x + 7) = 144x^2 + 84x + 84x + 49 = 144x^2 + 168x + 49$

40. $16a^2 + 4a - 4a - 1 = 16a^2 - 1$

41. $(-2b + c)(-2b + c) = 4b^2 - 2bc - 2bc + c^2 = 4b^2 - 4bc + c^2$

42. $(5c^2 - 2d^3)(5c^2 - 2d^3) = 25c^4 - 10c^2d^3 - 10c^2d^3 + 4d^6 = 25c^4 - 20c^2d^3 + 4d^6$

43. $169x^4 + 13x^2y - 13x^2y - y^2 = 169x^4 - y^2$

44. $(11 - 3e^3)(11 - 3e^3) = 121 - 33e^3 - 33e^3 + 9e^6 = 121 - 66e^3 + 9e^6$

45. First 2 positive even integers: $2 + 4 = 6 = (2 \cdot 3)$; First 3 positive even integers: $2 + 4 + 6 = 12 = (3 \cdot 4)$; First 4 positive even integers: $2 + 4 + 6 + 8 = 20 = (4 \cdot 5)$; First n positive even integers: $2 + 4 + \ldots + 2n = n(n + 1)$.

page 661

1. $48 = 2 \cdot 24 = 2 \cdot 2 \cdot 12 = 2 \cdot 2 \cdot 2 \cdot 6 = 2 \cdot 2 \cdot 2 \cdot 2 \cdot 3 = 2^4 \cdot 3$

2. $720 = 2 \cdot 360 = 2 \cdot 2 \cdot 180 = 2 \cdot 2 \cdot 2 \cdot 90 = 2 \cdot 2 \cdot 2 \cdot 2 \cdot 45 = 2^4 \cdot 3 \cdot 15 = 2^4 \cdot 3 \cdot 3 \cdot 5 = 2^4 \cdot 3^2 \cdot 5$

3. $41 = 1 \cdot 41$; prime

4. $960 = 2 \cdot 480 = 2 \cdot 2 \cdot 240 = 2 \cdot 2 \cdot 2 \cdot 120 = 2 \cdot 2 \cdot 2 \cdot 2 \cdot 60 = 2 \cdot 2 \cdot 2 \cdot 2 \cdot 2 \cdot 30 = 2 \cdot 2 \cdot 2 \cdot 2 \cdot 2 \cdot 2 \cdot 15 = 2^6 \cdot 3 \cdot 5$

5. $4375 = 5 \cdot 875 = 5 \cdot 5 \cdot 175 = 5 \cdot 5 \cdot 5 \cdot 35 = 5 \cdot 5 \cdot 5 \cdot 5 \cdot 7 = 5^4 \cdot 7$

6. $1119 = 3 \cdot 373$ 　　　　　　　　**7.** $3(4m^2 + 9m^3 - m - 2)$

8. $2r(4r^4 - 2r^3 + 16r^2 + 3)$ 　　　　**9.** $-3ab(a - 5a^2b + 6b^2)$

10. $7(2xy^2 - 3x^4 + 5y^2 - 4)$ 　　　　**11.** $(y + 4)(y + 3)$

12. $(b - 4)(b - 2)$ 　　**13.** $(x - 16)(x - 9)$ 　　**14.** $(n + 8)(n + 9)$

15. $(j - 11)^2$ 　　**16.** $(s + 4k)(s + 3k)$ 　　**17.** $(t + 2)(t - 1)$

18. $(r + 10)(r - 15)$ 　　**19.** $(m + 14)(m - 3)$ 　　**20.** not factorable

21. $(x - 2y)(x - 3y)$ 　　**22.** $(m + 18n)(m - 4n)$ 　　**23.** $(2h + 1)(h + 1)$

24. $(5m - 1)(m - 1)$ 　　**25.** $(2t - 1)(t - 2)$ 　　**26.** $(3n - 1)(n + 1)$

27. $(4k - 1)(k + 1)$ 　　**28.** $(7l + 5)(l + 1)$ 　　**29.** $(2x - 3)(x + 4)$

30. $(2y - 1)(3y - 5)$ 　　**31.** $(4z + 3)(z + 2)$ 　　**32.** not factorable

33. $(16q - 15)(q + 1)$ 　　**34.** $(4g - 3h)(g + 4h)$ 　　**35.** $(13k + 8)(3k + 1)$

36. $(3ab - 8)(2ab + 5)$ 　　**37.** not factorable 　　**38.** $(5y - 3)(5y + 3)$

39. $(2w + 1)^2$ 　　**40.** $(3u - r)^2$ 　　**41.** $(1 - 7a)(1 + 7a)$

42. $2(2m^2 + mn - n^2) = 2(2m - n)(m + n)$

43. $(ab^2)^2 - (c^3)^2 = (ab^2 - c^3)(ab^2 + c^3)$ 　　**44.** $(14 - 11b)(14 + 11b)$

45. $25v^2 - 60v + 36 = (5v - 6)(5v - 6) = (5v - 6)^2$

46. $(9x - 13)^2$ 　　　　　　　　**47.** $16(q^2 + 3)$

48. $3a^2(a^2 - 1) = 3a^2(a - 1)(a + 1)$ 　　　**49.** $7(x^2 + 2xy + 4y^2)$

50. $(1 - x) - y(1 - x) = (1 - y)(1 - x)$

51. $p(2p^2 + 3p - 20) = p(2p - 5)(p + 4)$

52. $5(1 - 8w^2 + 16w^4) = 5(1 - 4w^2)^2 = 5(1 - 2w^2)(1 + 2w^2)$

53. $(a + 1)^2 - b^2 = [(a + 1) - b][(a + 1) + b] = (a + 1 - b)(a + 1 + b)$

54. $[(3x - 2) - 2y][(3x - 2) + 2y] = (3x - 2 - 2y)(3x - 2 + 2y)$

55. $2(81 - d^4) = 2(-1)(d^4 - 81) = -2(d^2 + 9)(d^2 - 9) = -2(d^2 + 9)(d - 3)(d + 3)$

56.
$$2y + 3 = 0 \quad \vdots \quad 3y - 2 = 0$$
$$2y = -3 \quad \vdots \quad 3y = 2$$
$$y = -\frac{3}{2} \quad \vdots \quad y = \frac{2}{3}$$

57.
$$m(5 - 6m) = 0$$
$$m = 0 \quad \vdots \quad 5 - 6m = 0$$
$$ \quad \vdots \quad -6m = -5$$
$$ \quad \vdots \quad m = \frac{5}{6}$$

58. $(v + 5)^2 = 0,\ v + 5 = 0;\ v = -5$

59.
$$j^2 - 6j - 7 = 0;\ (j - 7)(j + 1) = 0$$
$$j - 7 = 0 \quad \vdots \quad j + 1 = 0$$
$$j = 7 \quad \vdots \quad j = -1$$

60.
$$(3x + 1)(x - 1) = 0$$
$$3x + 1 = 0 \quad \vdots \quad x - 1 = 0$$
$$3x = -1 \quad \vdots \quad x = 1$$
$$x = -\frac{1}{3} \quad \vdots$$

61.
$$2a^2 + 2a - 4 = 20,\ 2a^2 + 2a - 24 = 0,$$
$$2(a^2 + a - 12) = 0,$$
$$2(a + 4)(a - 3) = 0$$
$$a + 4 = 0 \quad \vdots \quad a - 3 = 0$$
$$a = -4 \quad \vdots \quad a = 3$$

62.
$$3a(a^2 + 2a + 1) = 0,\ 3a(a + 1)^2 = 0$$
$$a = 0 \quad \vdots \quad a + 1 = 0$$
$$ \quad \vdots \quad a = -1$$

63.
$$(6x - 7)(6x + 7) = 0$$
$$6x - 7 = 0 \quad \vdots \quad 6x + 7 = 0$$
$$6x = 7 \quad \vdots \quad 6x = -7$$
$$x = \frac{7}{6} \quad \vdots \quad x = -\frac{7}{6}$$

64. Let n = the number. $n^2 - 3n = 28,\ n^2 - 3n - 28 = 0,\ (n - 7)(n + 4) = 0$
$$n - 7 = 0 \quad \vdots \quad n + 4 = 0$$
$$n = 7 \quad \vdots \quad n = -4$$

65. Let b = the base, then $b - 5$ = the height of the triangle.
$$33 = \frac{1}{2}b(b - 5),\ 66 = b^2 - 5b,\ 0 = b^2 - 5b - 66,\ 0 = (b - 11)(b + 6)$$
$$b - 11 = 0 \quad \vdots \quad b + 6 = 0$$
$$b = 11 \quad \vdots \quad b = -6$$
-6 is not a solution, therefore, the base is 11 m and the height is $b - 5 = 6$ m.

66. Let m = the first integer, then $m + 1$ = the second integer and $m + 2$ = the third integer. The equation is: $(2m)^2 = 30 + 3(m + 1)(m + 2)$, solve
$m^2 - 9m - 36 = 0,\ (m - 12)(m + 3) = 0$
$$m - 12 = 0 \quad \vdots \quad m + 3 = 0$$
$$m = 12 \quad \vdots \quad m = -3$$
The three consecutive integers are 12, 13, 14 or $-3, -2, -1$.

page 662

1. $4x = 0,\ x = 0$

2. $y + 1 = 0,\ y = -1$

3. $a^2 + 3a = 0,\ a(a + 3) = 0,\ a = 0$ or $a + 3 = 0,\ a = -3$

4. $z^2 - 2z + 1 = 0,\ (z - 1)^2 = 0,\ z - 1 = 0;\ z = 1$

5. $\dfrac{5 \cdot 3v}{9v^2 \cdot 3v} = \dfrac{5}{9v^2}$

6. Fraction is in simplest form.

7. $\dfrac{-(-a + b)}{b - a} = \dfrac{-(b - a)}{b - a} = -1$

8. $\dfrac{(2x + 1)(x - 3)}{x - 3} = 2x + 1$

9. $-\dfrac{5m \cdot m}{2n} \cdot \dfrac{2n \cdot 3n^3}{5m \cdot 2} = -\dfrac{3mn^3}{2}$

10. $\dfrac{x - 3}{x + 5} \cdot \dfrac{2(x + 5)}{-1(-3 + x)} = -2$

11. $\dfrac{2v + 3}{2v \cdot v^2} \cdot \dfrac{v^2(v + 4)}{2v + 3} = \dfrac{v + 4}{2v}$

12. $\dfrac{(2y + 1)(y - 2)}{3y \cdot y} \cdot \dfrac{6y(2y - 1)}{y - 2} = \dfrac{2(4y^2 - 1)}{y} = \dfrac{8y^2 - 2}{y}$

13. $\dfrac{-14c^2}{d} \cdot \dfrac{d^2}{21c^3} = \dfrac{-2 \cdot 7c^2}{d} \cdot \dfrac{d \cdot d}{3c \cdot 7c^2} = -\dfrac{2d}{3c}$

14. $\dfrac{9h^2 + 3h}{h + 1} \cdot \dfrac{1}{6h + 2} = \dfrac{3h(3h + 1)}{h + 1} \cdot \dfrac{1}{2(3h + 1)} = \dfrac{3h}{2h + 2}$

15. $\dfrac{5j - 15}{6j} \cdot \dfrac{9j^2}{6 - 2j} = \dfrac{5(j - 3)}{3j \cdot 2} \cdot \dfrac{3j \cdot 3j}{-2(j - 3)} = -\dfrac{15j}{4}$

16. $\dfrac{x^3 - 4x}{3x} \cdot \dfrac{x - 5}{x + 2} \cdot \dfrac{x + 5}{x - 2} = \dfrac{x(x - 2)(x + 2)}{3x} \cdot \dfrac{x - 5}{x + 2} \cdot \dfrac{x + 5}{x - 2} = \dfrac{x^2 - 25}{3}$

17. LCD is $3 \cdot 2 \cdot a^2 \cdot b^2 = 6a^2b^2$; $\dfrac{2}{3ab} \cdot \dfrac{2ab}{2ab} - \dfrac{5}{2a^2b} \cdot \dfrac{3b}{3b} + \dfrac{6}{a^2b^2} \cdot \dfrac{6}{6} = \dfrac{4ab - 15b + 36}{6a^2b^2}$

18. LCD is $x - y$; $\dfrac{y}{x - y} - \dfrac{x}{y - x} \cdot \dfrac{(-1)}{(-1)} = \dfrac{y}{x - y} - \dfrac{-x}{x - y} = \dfrac{y + x}{x - y}$

19. LCD is $(d - 2)(d + 2) = d^2 - 4$; $\dfrac{(d + 1)(d + 2)}{(d - 2)(d + 2)} - \dfrac{1}{-1(d - 2)} = \dfrac{d + 1}{d - 2} - \dfrac{-1}{d - 2} =$

$\dfrac{d + 1 + 1}{d - 2} = \dfrac{d + 2}{d - 2}$

20. LCD is $m^2 - n^2$; $\dfrac{2n(n + m)}{(m - n)(m + n)} + 1 \cdot \dfrac{m - n}{m - n} = \dfrac{2n + m - n}{m - n} = \dfrac{m + n}{m - n}$

21. LCD is x^3; $\dfrac{\left(\dfrac{1}{x} - \dfrac{3}{x^2}\right)}{\left(2x - \dfrac{1}{x^3} + 1\right)} \cdot \dfrac{x^3}{x^3} = \dfrac{x^2 - 3x}{2x^4 - 1 + x^3} = \dfrac{x^2 - 3x}{2x^4 + x^3 - 1}$

22. LCD is c^2; $\dfrac{\left(1 - \dfrac{2}{c} - \dfrac{3}{c^2}\right)}{\left(\dfrac{2}{c} + \dfrac{2}{c^2}\right)} \cdot \dfrac{c^2}{c^2} = \dfrac{c^2 - 2c - 3}{2c + 2} = \dfrac{(c - 3)(c + 1)}{2(c + 1)} = \dfrac{c - 3}{2}$

23. $\dfrac{5j^2}{20jkl} - \dfrac{10jk^2l}{20jkl} + \dfrac{25j^2kl^2}{20jkl} - \dfrac{30kl}{20jkl} = \dfrac{j}{4kl} - \dfrac{k}{2} + \dfrac{5jl}{4} - \dfrac{3}{2j}$

24.
$$
\begin{array}{r}
3x^2 + 4x + 11 + \dfrac{17}{x-2} \\
x - 2\overline{)3x^3 - 2x^2 + 3x - 5} \\
\underline{-(3x^3 - 6x^2)} \\
4x^2 + 3x \\
\underline{-(4x^2 - 8x)} \\
11x - 5 \\
\underline{-(11x - 22)} \\
17
\end{array}
$$

25.
$$
\begin{array}{r}
y^2 + 3y + 9 \\
y - 3\overline{)y^3 + 0y^2 + 0y - 27} \\
\underline{-(y^3 - 3y^2)} \\
3y^2 + 0y \\
\underline{-(3y^2 - 9y)} \\
9y - 27 \\
\underline{-(9y - 27)} \\
0
\end{array}
$$

26. $\dfrac{3 \cdot 13}{13 \cdot 13} = \dfrac{3}{13}$; 3:13

27. $\dfrac{300 \text{ cm}}{20 \text{ cm}} = 15{:}1$

28. $\dfrac{3.25 \times 60 \text{ min}}{20 \text{ min}} = \dfrac{3.25 \times 3}{1} = \dfrac{9.75}{1} = \dfrac{975}{100} = \dfrac{39}{4}$ or 39:4

29. LCD is 6; $\left[\dfrac{1}{6}x^2 + \dfrac{1}{3}x\right]6 = \left(\dfrac{5}{2}\right)6$, $x^2 + 2x = 15$, $x^2 + 2x - 15 = 0$, $(x + 5)(x - 3) = 0$,
$x = -5$; $x = 3$

30. LCD is $2t(t - 1)$; $\left[\dfrac{8}{t^2 - t} + \dfrac{t}{1 - t}\right] \cdot 2t(t - 1) = \dfrac{7}{2} \cdot 2t(t - 1)$, $2 \cdot 8 + 2(-1)t^2 =$

$7t^2 - 7t$, $16 - 2t^2 = 7t^2 - 7t$, $0 = 9t^2 - 7t - 16$, $0 = (9t - 16)(t + 1)$, $t = \dfrac{16}{9}$; $t = -1$

31. Let $r =$ Sue's present rate, then $r + 2 =$ Sue's faster rate.

LCD is $r(r + 2)$ $\dfrac{12}{r} \cdot r(r + 2) = \dfrac{16}{r + 2} \cdot r(r + 2)$, $12r + 24 = 16r$; $r = 6$; Sue
is walking 6 km/h.

page 663

1. 1st quadrant

2. 2nd quadrant

3. x-axis

4. 3rd quadrant

5. y-axis

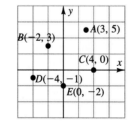

6. $3(2) - 2(3) \overset{?}{=} 0$, $6 - 6 = 0$; yes

$3\left(-\dfrac{2}{3}\right) - 2\left(-\dfrac{3}{2}\right) \overset{?}{=} 0$, $-2 + 3 \neq 0$; no

$3(-3) - 2(-2) \overset{?}{=} 0$, $-9 + 4 \neq 0$; no

$3\left(\dfrac{5}{6}\right) - 2\left(\dfrac{5}{4}\right) \overset{?}{=} 0$, $\dfrac{5}{2} - \dfrac{5}{2} = 0$; yes

$3\left(-\dfrac{8}{3}\right) - 2(-4) \overset{?}{=} 0$, $-8 + 8 = 0$; yes

7. no; $2 + (-1) \neq -1$

8. yes; $-1 + 0 = -1$

9. yes; $0 + (-1) = -1$

10. no; $1 + 1 \neq -1$

11. $0.56

12. 120 oz

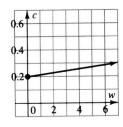

13. $\dfrac{y_2 - y_1}{x_2 - x_1} = \dfrac{-5 - 3}{2 - (-4)} = \dfrac{-8}{6} = -\dfrac{4}{3}$

14. $\dfrac{y_2 - y_1}{x_2 - x_1} = \dfrac{1 - (-3)}{7 - 7} = \dfrac{4}{0}$; no slope

15. $\dfrac{y_2 - y_1}{x_2 - x_1} = \dfrac{-\dfrac{1}{5} - (-1)}{\dfrac{4}{3} - \left(-\dfrac{3}{2}\right)} = \dfrac{-\dfrac{1}{5} + 1}{\dfrac{4}{3} + \dfrac{3}{2}} \cdot \dfrac{2 \cdot 3 \cdot 5}{2 \cdot 3 \cdot 5} = \dfrac{-6 + 30}{40 + 45} = \dfrac{24}{85}$

16.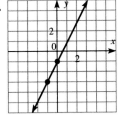

17.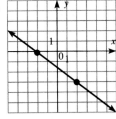

18.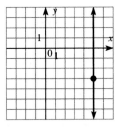

19.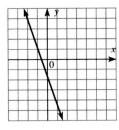

$y = -3x - 2$

$m = -3;\ b = -2$

20.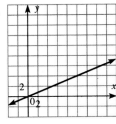

$-5y = -2x;\ y = \dfrac{2}{5}x$

$m = \dfrac{2}{5};\ b = 0$

21.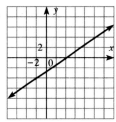

$\dfrac{1}{2}y = \dfrac{1}{3}x - \dfrac{3}{2}$

$3y = 2x - 9$

$y = \dfrac{2}{3}x - 3$

$m = \dfrac{2}{3};\ b = -3$

22. $y = -2x + 4$, $m = -2$; $4x + 2y = 0$, $y = 2x$; $m = -2$;
Equal slopes, Lines are parallel.

23. $y = -\frac{4}{3}x + \frac{3}{5}$, $m = -\frac{4}{3}$; $15y - 20x = 0$, $y = \frac{4}{3}x$; $m = \frac{4}{3}$;

Unequal slopes, Lines are not parallel.

24. $m = \dfrac{-4 - 6}{3 - (-2)} = -2$, $y = -2x + b$, $6 = -2(-2) + b$, $b = 2$;

$y = -2x + 2$; $2x + y = 2$

25. $y = 0x + b$, $4 = 0(-6) + b$, $b = 4$; $y = 0x + 4$, $y = 4$

26. 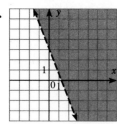 **27.** 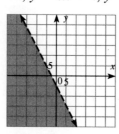 **28.**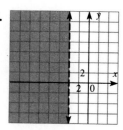

page 664

1. a. $(-3, 4)$, $(-2, 4)$, $(-1, 4)$, $(0, 4)$, $(0, 2)$, $(1, 2)$, $(2, 0)$

b. Domain: $\{-3, -2, -1, 0, 1, 2\}$; Range: $\{0, 2, 4\}$

c. 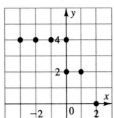 **d.** No

2. a. $(-1, -1)$, $\left(0, -\frac{1}{2}\right)$, $(1, 0)$, $\left(2, \frac{1}{2}\right)$, $(3, 1)$

b. Domain: $\{-1, 0, 1, 2, 3\}$; Range: $\left\{-1, -\frac{1}{2}, 0, \frac{1}{2}, 1\right\}$

c. Yes

3. a. 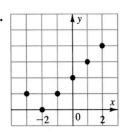 **b.** Yes

4. a. Answers may vary. Sample answer: $(0, 2)$, $(1, 2)$, $(-2, 2)$

b. Domain: {Set of all real numbers}; Range: {2}

c. This relation is a constant function.

d. $y = 2$

5. $f(0) = -3(0) + 5 = 0 + 5 = 5$

6. $f\left(-\frac{1}{3}\right) = -3\left(-\frac{1}{3}\right) + 5 = 1 + 5 = 6$

7. $g(-20) = \frac{1}{10}(-20) = -2$

8. $g(0.2) = \frac{1}{10}(0.2) = 0.02$

9. $g(9) = \frac{1}{10}(9) = \frac{9}{10}; f\left(\frac{9}{10}\right) = -3\left(\frac{9}{10}\right) + 5 = -\frac{27}{10} + 5 = \frac{23}{10}$

10. $f(0) = g(5) = \frac{1}{10}(5) = \frac{1}{2}$

11. $f(1) \cdot g\left(\frac{1}{2}\right) = [-3(1) + 5] \cdot \left[\frac{1}{10}\left(\frac{1}{2}\right)\right] = 2 \cdot \frac{1}{20} = \frac{1}{10}$

12. $g(5) + f\left(-\frac{5}{3}\right) = \frac{1}{10}(5) + \left[-3\left(-\frac{5}{3}\right) + 5\right] = \frac{1}{2} + (5 + 5) = 10\frac{1}{2} \text{ or } \frac{21}{2}$

13. direct variation; $k = \frac{3}{2}, y = \frac{3}{2}x$

14. indirect variation; $k = 48$, $xy = 48$ or $y = \frac{48}{x}$

15. $69 = k(3), k = 23; y = 23x, y = 23(5) = 115$

16. $4(12) = k, k = 48; y = \frac{48}{x} = \frac{48}{6} = 8$

17. $C = \frac{5}{9}(5 - 32), C = \frac{5}{9}(-27), C = -15°$

18. $100 = \frac{5}{9}(F - 32), 9(100) = 5F - 160, 900 + 160 = 5F, 1060 = 5F, F = 212°$

page 665

1.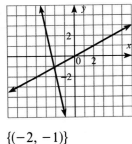

$\{(-2, -1)\}$

2.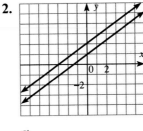

$\emptyset$

3.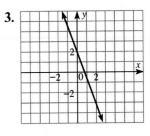

$\{(x, y): 5x + 2y = 4\}$

4. Solve first equation for x: $x = -5y - 4$; substitute $-3(-5y - 4) + 2y = 12$, $15y + 12 + 2y = 12, 17y + 12 = 12, 17y = 0, y = 0; x + 5(0) = -4, x = -4; (-4, 0)$

5.
$$2a - 3b = 18$$
$$\underline{-4a + 3b = -24}$$
$$-2a = -6$$
$$a = 3$$
$2(3) - 3b = 18, 6 - 3b = 18, -3b = 12; b = -4; (3, -4)$

6. Solve the first equation for c: $c = -10 + 5d$; substitute $-2(-10 + 5d) + 5d$ $= -5, 20 - 10d + 5d = -5, 20 - 5d = -5, -5d = -25, d = 5; -c = 10 - 5(5), -c = 10 - 25, -c = -15, c = 15; (15, 5)$

7.
$$5x + 7y = 11 \quad \rightarrow \quad -2(5x + 7y) = -2(11) \quad \rightarrow \quad -10x - 14y = -22$$
$$10x + 14y = 13 \quad \rightarrow \quad 10x + 14y = 13 \quad \rightarrow \quad 10x + 14y = 13$$
$$-10x - 14y = -22$$
$$\underline{10x + 14y = 13}$$
$$0 = -9; \text{ no solution}$$

8.
$$-3y + 4x = -4 \rightarrow \quad -3y + 4x = -4 \rightarrow -3y + 4x = -4$$
$$5y - 2x = 9 \quad \rightarrow 2(5y - 2x) = 2(9) \rightarrow \quad 10y - 4x = 18$$
$$-3y + 4x = -4$$
$$\underline{10y - 4x = 18}$$
$$7y = 14$$
$$y = 2$$
$5(2) - 2x = 9, 10 - 2x = 9, -2x = -1; x = \dfrac{1}{2}; \left(\dfrac{1}{2}, 2\right)$

9.
$$4j - 9k = 3 \quad \rightarrow \quad -5(4j - 9k) = -5(3) \rightarrow -20j + 45k = -15$$
$$5j + 12k = -4 \quad \rightarrow \quad 4(5j + 12k) = 4(-4) \rightarrow \quad 20j + 48k = -16$$
$$-20j + 45k = -15$$
$$\underline{20j + 48k = -16}$$
$$93k = -31$$
$$k = -\dfrac{1}{3}$$
$5j + 12\left(-\dfrac{1}{3}\right) = -4, 5j - 4 = -4, 5j = 0, j = 0; \left(0, -\dfrac{1}{3}\right)$

10. $3\left(2 + \dfrac{4}{3}v\right) - 4v = 6, 6 + 4v - 4v = 6, 6 = 6, 0 = 0; \{(u, v): 3u - 4v = 6\}$

11. $8[r - (-0.5)] = \dfrac{7}{2} - 5(-0.5), 16(r + 0.5) = 7 - 10(-0.5), 16r + 8 = 7 + 5,$

$16r = 4, r = \dfrac{1}{4}; 8\left(\dfrac{1}{4} - t\right) = \dfrac{7}{2} - 5t, 2 - 8t = \dfrac{7}{2} - 5t, 4 - 16t = 7 - 10t,$

$-6t = 3, t = -\dfrac{1}{2}; \left(\dfrac{1}{4}, -\dfrac{1}{2}\right)$

12. $-3q + 2p = 6p + 12 \rightarrow -3q - 4p = 12 \rightarrow -1(-3q - 4p) = -1(12) \rightarrow 3q + 4p = -12$

$\quad\quad p - q = 2p + 2 \quad \rightarrow \quad -q - p = 2 \quad \rightarrow \quad 3(-q - p) = 3(2) \quad \rightarrow -3q - 3p = 6$

$\quad\quad 3q + 4p = -12$

$\quad\quad \underline{-3q - 3p = 6}$

$\quad\quad\quad\quad\quad p = -6$

$\quad\quad \dfrac{-6 - q}{2} = -6 + 1, \; -6 - q = -12 + 2, \; -q = -4, \; q = 4; \; (-6, 4)$

13. Let x = the smaller number and y = the larger number.

$\begin{cases} x + y = 22 \\ y = 2x + 1 \end{cases}$

$\quad x + (2x + 1) = 22, \; 3x + 1 = 22, \; 3x = 21, \; x = 7; \; y = 2(7) + 1, \; y = 15;$
The numbers are 7 and 15.

14. Let x = the amount invested at 7.5% and y = the amount invested at 8.5%.

$\begin{cases} x + y = 1200 & \rightarrow & x = 1200 - y \\ 0.075x + 0.085y = 96.50 & \rightarrow 0.075x + 0.085y = 96.50 \end{cases}$

$\quad 0.075(1200 - y) + 0.085y = 96.50, \; 90 - 0.075y + 0.085y = 96.50,$
$\quad 0.01y = 6.50, \; y = 650; \; x + 650 = 1200, \; x = 550;$ The amount invested
at 7.5% is \$550 and the amount invested at 8.5% is \$650.

15. Let u = the units digit and t = the tens digit.

$\begin{cases} u + t = 10 \\ 10u + t = 10t + u - 18 \end{cases}$

$\quad\quad u + t = 10 \quad \rightarrow 9(u + t) = 9(10) \rightarrow 9u + 9t = 90$

$\quad 9u - 9t = -18 \rightarrow 9u - 9t = -18 \quad \rightarrow 9u - 9t = -18$

$\quad\quad 9u + 9t = 90$

$\quad\quad \underline{9u - 9t = -18}$

$\quad\quad\quad 18u = 72$

$\quad\quad\quad\quad u = 4$

$\quad 4 + t = 10, \; t = 6;$ The original number is 64.

16. Let l = Lonnie's present age, then $l + 13$ = Lonnie's age in thirteen years.
Let s = his sister's present age, then $s + 13$ = his sister's age in thirteen years.

$\begin{cases} l = s - 6 & \rightarrow l = s - 6 & \rightarrow l = s - 6 \\ l + 13 = \frac{4}{5}(s + 13) \rightarrow l + 13 = \frac{4}{5}s + \frac{52}{5} \rightarrow 5l + 65 = 4s + 52 \end{cases}$

$\quad 5(s - 6) + 65 = 4s + 52, \; 5s - 30 + 65 = 4s + 52, \; s = 17; \; l = 17 - 6, \; l = 11$
Lonnie is 17 years old and his sister is 11 years old.

17. Let d = number of dollar bills and f = the number of five dollar bills.

$$d + f = 40 \quad \rightarrow \quad -1(d + f) = -1(40) \rightarrow -d - f = -40$$
$$1d + 5f = 144 \rightarrow \quad d + 5f = 144 \quad \rightarrow \quad d + 5f = 144$$

$$\begin{array}{r} -d - f = -40 \\ \underline{d + 5f = 144} \\ 4f = 104 \\ f = 26 \end{array}$$

$d + 26 = 40$, $d = 14$; There are 14 one dollar bills and 26 five dollar bills.

18. Let s = the number of pounds of sunflower seeds and p = the number of pounds of peanuts.

$$\begin{cases} s + p = 16 \\ 2.35s + 1.85p = 32.40 \end{cases}$$

$s = 16 - p$; $2.35(16 - p) + 1.85p = 32.40$, $37.60 - 2.35p + 1.85p = 32.40$, $-0.50p = -5.20$, $p = 10.4$; $s + 10.4 = 16$, $s = 5.6$; 10.4 pounds of peanuts and 5.6 pounds of sunflower seeds are used in the mixture.

19. Let c = the rate of the current and r = the rate of the boat in still water.

	rate	× time =	distance
downstream	$r + c$	3	15
upstream	$r - c$	$3\frac{3}{4}$	15

$$(r + c)3 = 15 \rightarrow r + c = 5 \qquad \rightarrow r + c = 5$$
$$(r - c)3\frac{3}{4} = 15 \rightarrow (r - c)\frac{15}{4} = 15 \rightarrow r - c = 4$$

$$\begin{array}{r} r + c = 5 \\ \underline{r - c = 4} \\ 2r = 9 \\ r = \dfrac{9}{2} \end{array}$$

$\dfrac{9}{2} + c = 5$, $c = \dfrac{1}{2}$; The speed of the current is $\dfrac{1}{2}$ mi/h.

20.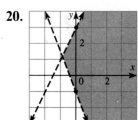

21.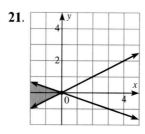

22.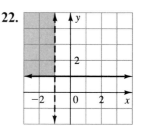

page 666

1. $-\sqrt{15 \cdot 15} = -15$

2. $\dfrac{\sqrt{5}}{\sqrt{3 \cdot 3}} = \dfrac{\sqrt{5}}{3}$

3. $\sqrt{16 \cdot 6} = \sqrt{16} \cdot \sqrt{6} = 4\sqrt{6}$ **4.** $\sqrt{25x^2 \cdot 2x} = \sqrt{25x^2} \cdot \sqrt{2x} = 5x\sqrt{2x}$

5. $3\sqrt{4 \cdot 3} - \sqrt{3} + 2\sqrt{5} = 6\sqrt{3} - \sqrt{3} + 2\sqrt{5} = 5\sqrt{3} + 2\sqrt{5}$

6. $15\sqrt{3a^2} = 15\sqrt{3 \cdot a^2} = 15(\sqrt{3} \cdot \sqrt{a^2}) = 15a\sqrt{3}$

7. $\sqrt{(0.01b)^2(0.01b)^2} = 0.01b^2$ **8.** $\dfrac{1}{\sqrt{27}} \cdot \dfrac{\sqrt{3}}{\sqrt{3}} = \dfrac{\sqrt{3}}{\sqrt{81}} = \dfrac{\sqrt{3}}{9}$

9. $\dfrac{\sqrt{2}}{2} + 3\sqrt{4 \cdot 2} - \sqrt{16 \cdot 2} = \dfrac{1}{2}\sqrt{2} + 6\sqrt{2} - 4\sqrt{2} = \dfrac{5}{2}\sqrt{2}$

10. $\sqrt{\dfrac{40c^3d}{10cd}} = \sqrt{4c^2} = 2c$ **11.** $\sqrt{144 \cdot 2 \cdot x \cdot y^2 \cdot z^2 \cdot z} = 12yz\sqrt{2xz}$

12. $2(4\sqrt{5}) + (3\sqrt{5})(4\sqrt{5}) = 8\sqrt{5} + 12\sqrt{25} = 8\sqrt{5} + 12 \cdot 5 = 8\sqrt{5} + 60$

13. $(5\sqrt{3} + 3\sqrt{2})(5\sqrt{3} + 3\sqrt{2}) = 25\sqrt{9} + 15\sqrt{6} + 15\sqrt{6} + 9\sqrt{4} = $
$75 + 15\sqrt{6} + 15\sqrt{6} + 18 = 93 + 30\sqrt{6}$

14. $49\sqrt{25} + 7\sqrt{5} - 7\sqrt{5} - 1 = 245 + 7\sqrt{5} - 7\sqrt{5} - 1 = 244$

15. $\dfrac{-6}{3\sqrt{3} - 2} \cdot \dfrac{3\sqrt{3} + 2}{3\sqrt{3} + 2} = \dfrac{-18\sqrt{3} - 12}{9\sqrt{9} + 6\sqrt{3} - 6\sqrt{3} - 4} = \dfrac{-18\sqrt{3} - 12}{27 - 4} = \dfrac{-18\sqrt{3} - 12}{23}$

16. 7.141 **17.** 9.695 **18.** 2.646 **19.** 12.042

20. $3 < \sqrt{11} < 4$, $11 \div 3 \approx 3.7$, $(3 + 3.7) \div 2 \approx 3.3$; $\sqrt{11} \approx 3.3$

21. $15 < \sqrt{241} < 16$, $241 \div 15 \approx 16.1$, $(15 + 16.1) \div 2 \approx 15.5$; $\sqrt{241} \approx 15.5$

22. $9 < \sqrt{87.9} < 10$, $87.9 \div 9 \approx 9.7$, $(9 + 9.7) \div 2 \approx 9.4$; $\sqrt{87.9} \approx 9.4$

23. $21 < \sqrt{459.6} < 22$, $459.6 \div 21 \approx 21.9$, $(21 + 21.9) \div 2 \approx 21.4$, $\sqrt{459.6} \approx 21.4$

24. $-3 \div 8 = -0.375$ **25.** $25 \div 3 = 8.\overline{3}$ **26.** $-7 \div 25 = -0.28$

27. $5 \div 7 = 0.\overline{714285}$ **28.** $0.13 = \dfrac{13}{100}$

29. Let $x = -0.\overline{4}$. $(10x = -4.\overline{4}) - (x = -0.\overline{4})$, $9x = -4$; $x = -\dfrac{4}{9}$

30. $9.01 = 9\dfrac{1}{100} = \dfrac{901}{100}$

31. Let $x = 0.\overline{12}$. $(100x = 12.\overline{12}) - (x = 0.\overline{12})$, $99x = 12$; $x = \dfrac{12}{99} = \dfrac{4}{33}$;

$3.12 = 3\dfrac{4}{33} = \dfrac{103}{33}$

32. $3\sqrt{x} = 2$, $(3\sqrt{x})^2 = 2^2$, $9x = 4$, $x = \dfrac{4}{9}$

33. No solution since $\sqrt{5 - 2x} \not> 0$.

34. $(2\sqrt{1 - 2v})^2 = (3\sqrt{v})^2$, $4(1 - 2v) = 9v$, $4 - 8v = 9v$, $-17v = -4$, $v = \dfrac{4}{17}$

35. $c^2 = 12^2 + 16^2$, $c^2 = 144 + 256$, $c^2 = 400$, $c = 20$

36. $25^2 = 7^2 + b^2$, $b^2 = 25^2 - 7^2$, $b^2 = 625 - 49$, $b^2 = 576$, $b = 24$

37. $14.5^2 = a^2 + 11.4^2$, $a^2 = 14.5^2 - 11.4^2$, $a^2 = 210.25 - 129.96$, $a^2 = 80.29$, $a = 9.0$

38. $(1.0)^2 + (2.4)^2 \stackrel{?}{=} (2.6)^2$, $1 + 5.76 \stackrel{?}{=} 6.76$, $6.76 = 6.76$; yes

39. $(3\sqrt{3})^2 + 9^2 \stackrel{?}{=} (6\sqrt{3})^2$, $27 + 81 \stackrel{?}{=} 108$, $108 = 108$; yes

40. $7^2 + 12^2 \stackrel{?}{=} 13^2$, $49 + 144 \stackrel{?}{=} 169$, $193 \neq 169$; no

41. $\sqrt{[2 - (-1)]^2 + (7 - 3)^2} = \sqrt{3^2 + 4^2} = \sqrt{9 + 16} = \sqrt{25} = 5$

42. $\sqrt{(-3 - 2)^2 + [-10 - (-5)]^2} = \sqrt{(-5)^2 + (-5)^2} = \sqrt{25 + 25} = \sqrt{50} = \sqrt{25 \cdot 2} = 5\sqrt{2}$

43. $PQ = \sqrt{[-2 - (-2)]^2 + (13 - 1)^2} = \sqrt{0^2 + 12^2} = \sqrt{144} = 12$
$QR = \sqrt{[-7 - (-2)]^2 + (1 - 13)^2} = \sqrt{(-5)^2 + (-12)^2} = \sqrt{25 + 144} = \sqrt{169} = 13$
$PR = \sqrt{[-7 - (-2)]^2 + (1 - 1)^2} = \sqrt{(-5)^2 + 0^2} = \sqrt{25} = 5$

44. $HK = \sqrt{(0 - 5)^2 + (0 - 12)^2} = \sqrt{(-5)^2 + (-12)^2} = \sqrt{25 + 144} = \sqrt{169} = 13$
$KL = \sqrt{(-3 - 0)^2 + (-4 - 0)^2} = \sqrt{(-3)^2 + (-4)^2} = \sqrt{9 + 16} = \sqrt{25} = 5$
$HL = \sqrt{(-3 - 5)^2 + (-4 - 12)^2} = \sqrt{(-8)^2 + (-16)^2} = \sqrt{64 + 256} = \sqrt{320} = \sqrt{64 \cdot 5} = 8\sqrt{5}$

page 667

1. $z = \pm\sqrt{225}$, $z = \pm\sqrt{15 \cdot 15}$, $z = \pm 15$

2. $y^2 = 27$, $y = \pm\sqrt{27}$, $y = \pm\sqrt{9 \cdot 3}$, $y = \pm 3\sqrt{3}$

3. $\sqrt{(x - 5)^2} = \sqrt{98}$, $x - 5 = \pm\sqrt{98}$, $x = 5 \pm \sqrt{98}$, $x = 5 \pm \sqrt{49 \cdot 2}$, $x = 5 \pm 7\sqrt{2}$

4. $\sqrt{(2b + 1)^2} = \sqrt{4}$, $2b + 1 = \pm 2$; $2b + 1 = 2$, $b = \dfrac{1}{2}$; or $2b + 1 = -2$, $b = -\dfrac{3}{2}$.

5. $\dfrac{1}{2}(-4) = -2$, $(-2)^2 = 4$, $r^2 - 4r = 96$, $r^2 - 4r + 4 = 96 + 4$, $r^2 - 4r + 4 = 100$, $r^2 - 4r - 96 = 0$, $(r + 8)(r - 12) = 0$; $r = -8$ and $r = 12$

6. $\dfrac{1}{2}\left(-1\right)^2 = -\dfrac{1}{2}$, $\left(-\dfrac{1}{2}\right)^2 = \dfrac{1}{4}$, $q^2 - q + \dfrac{1}{4}$, $= 2 + \dfrac{1}{4}$, $q^2 - q + \dfrac{1}{4} = \dfrac{9}{4}$, $\left(q - \dfrac{1}{2}\right)^2 = \dfrac{9}{4}$,
$q - \dfrac{1}{2} = \pm\dfrac{3}{2}$, $q = \dfrac{3}{2} + \dfrac{1}{2} = 2$ or $q = -\dfrac{3}{2} + \dfrac{1}{2} = -1$

7. $\frac{1}{2}(2) = 1$, $1^2 = 1$, $p^2 + 2p = \frac{7}{2}$, $p^2 + 2p + 1 = \frac{7}{2} + 1$, $p^2 + 2p + 1 = \frac{9}{2}$, $(t + 1)^2 = \frac{9}{2}$,

$\sqrt{(t + 1)^2} = \sqrt{\frac{9}{2}}$, $t + 1 = \pm\sqrt{\frac{9}{2}}$, $t + 1 = \pm\frac{\sqrt{9}}{\sqrt{2}}$, $t + 1 = \pm\frac{\sqrt{9}}{\sqrt{2}} \cdot \frac{\sqrt{2}}{\sqrt{2}}$,

$t + 1 = \pm\frac{\sqrt{9 \cdot 2}}{2}$, $t + 1 = \pm\frac{3\sqrt{2}}{2}$, $t = -1 \pm \frac{3\sqrt{2}}{2}$

8. $x = \frac{-(-6) \pm \sqrt{(-6)^2 - 4(1)(4)}}{2(1)}$, $x = \frac{6 \pm \sqrt{36 - 16}}{2}$, $x = \frac{6 \pm \sqrt{20}}{2}$, $x = \frac{6 \pm \sqrt{4 \cdot 5}}{2}$,

$x = \frac{6 \pm 2\sqrt{5}}{2}$, $x = 3 \pm \sqrt{5}$

9. $6y^2 + 11y - 10 = 0$; $y = \frac{-11 \pm \sqrt{(11)^2 - 4(6)(-10)}}{2(6)}$, $y = \frac{-11 \pm \sqrt{121 - (-240)}}{12}$,

$y = \frac{-11 \pm \sqrt{361}}{12}$, $y = \frac{-11 \pm 19}{12}$, $y = \frac{8}{12} = \frac{2}{3}$ and $y = -\frac{30}{12} = -\frac{5}{2}$

10. $9z^2 - 12z - 1 = 0$; $z = \frac{-(-12) \pm \sqrt{(-12)^2 - 4(9)(-1)}}{2(9)}$, $z = \frac{12 \pm \sqrt{144 + 36}}{18}$,

$z = \frac{12 \pm \sqrt{180}}{18}$, $z = \frac{12 \pm \sqrt{36 \cdot 5}}{18}$, $z = \frac{12 \pm 6\sqrt{5}}{18}$, $z = \frac{2 \pm \sqrt{5}}{3}$

11. $(t + 10)(t - 11) = 0$; $t = -10$ and $t = 11$

12. $\sqrt{(3m - 4)^2} = \pm\sqrt{50}$, $3m - 4 = \pm\sqrt{50}$, $3m = 4 \pm\sqrt{50}$, $m = \frac{4 \pm \sqrt{50}}{3}$,

$m = \frac{4 \pm \sqrt{25 \cdot 2}}{3}$, $m = \frac{4 \pm 5\sqrt{2}}{3}$

13. $m^2 = \frac{49}{16}$, $m = \pm\sqrt{\frac{49}{16}}$, $m = \frac{\pm\sqrt{49}}{\pm\sqrt{16}}$, $m = \pm\frac{7}{4}$

14. $10b^2 + 21b - 10 = 0$, $(2b + 5)(5b - 2) = 0$; $b = -\frac{5}{2}$ and $b = \frac{2}{5}$

15. $c = \frac{-2 \pm \sqrt{2^2 - 4(1)(-1)}}{2(1)}$, $c = \frac{-2 \pm \sqrt{4 + 4}}{2}$, $c = \frac{-2 \pm \sqrt{8}}{2}$, $c = \frac{-2 \pm \sqrt{4 \cdot 2}}{2}$,

$c = \frac{-2 \pm 2\sqrt{2}}{2}$, $c = -1 \pm \sqrt{2}$

16. When using the quadratic formula a negative radicand will result, therefore no real solution.

17. Let $x =$ the first integer and $x + 2 =$ the second integer. $x(x + 2) = 288$, $x^2 + 2x = 288$, $x^2 + 2x - 288 = 0$, $(x - 16)(x + 18) = 0$; $x = 16$ and $x = -18$; If $x = 16$, then $x + 2 = 16 + 2 = 18$. If $x = -18$, then $x + 2 = -18 + 2 = -16$.

18. $x = -\dfrac{0}{2(-1)}$, $x = -\dfrac{0}{-2}$, $x = 0$

$y = -(0)^2$, $y = 0$

The vertex is $(0, 0)$

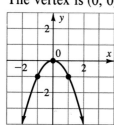

19. $x = -\dfrac{0}{2(2)}$, $x = -\dfrac{0}{4}$, $x = 0$

$y = 2(0)^2 + 3$, $y = 0 + 3$, $y = 3$

The vertex is $(0, 3)$

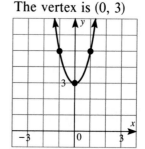

20. $f(x) = x^2 + 4x + 4$

$x = -\dfrac{4}{2(1)}$, $x = -\dfrac{4}{2}$, $x = -2$

$y = (-2)^2 + 4(-2) + 4$,

$y = 4 - 8 + 4$, $y = 0$

The vertex is $(-2, 0)$

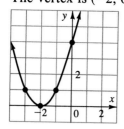

21. $x = -\dfrac{(-6)}{2(1)}$, $x = \dfrac{6}{2}$, $x = 3$

$y = 3^2 - 6(3) + 6$, $y = 9 - 18 + 6$,

$y = -3$; The vertex is $(3, -3)$

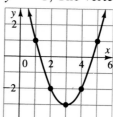

22. $x = -\dfrac{12}{2(-3)}$, $x = \dfrac{12}{6}$, $x = 2$

$y = -3(2)^2 + 12(2) + 1$,

$y = -12 + 24 + 1$, $y = 13$

The vertex is $(2, 13)$

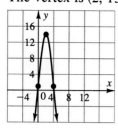

23. $x = -\dfrac{8}{2(2)}$, $x = -\dfrac{8}{4}$, $x = -2$

$y = 2(-2)^2 + 8(-2)$, $y = 8 - 16$,

$y = -8$; The vertex is $(-2, -8)$

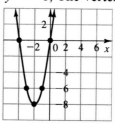

24. $a = 1$, $b = 12$, $c = 36$; $12^2 - 4(1)(36) = 144 - 144 = 0$; 1 real solution

25. $a = 4$, $b = -8$, $c = 1$; $(-8)^2 - 4(4)(1) = 64 - 16 = 48$; 2 real solutions

26. $a = 3$, $b = -2$, $c = 5$; $(-2)^2 - 4(3)(5) = 4 - 60 = -56$; no real solutions

27. $a = 1$, $b = 2$, $c = 1$; $2^2 - 4(1)(1) = 4 - 4 = 0$; one intercept

28. $a = -2$, $b = 1$, $c = -2$; $1^2 - 4(-2)(-2) = 1 - 16 = -15$; no intercept

29. $a = 3$, $b = -1$, $c = -1$; $(-1)^2 - 4(3)(-1) = 1 + 12 = 13$; two intercepts

30. $\dfrac{b}{a} = -(-7 + 11) = -4$; $\dfrac{c}{a} = (-7)(11) = -77$; $x^2 - 4x - 77 = 0$

31. $\dfrac{b}{a} = -\left(-\dfrac{2}{3} + 5\right) = -\dfrac{13}{3}$; $\dfrac{c}{a} = -\left(\dfrac{2}{3}\right)(5) = -\dfrac{10}{3}$; $x^2 - \dfrac{13}{3}x - \dfrac{10}{3} = 0$,
$3x^2 - 13x - 10 = 0$

32. $\dfrac{b}{a} = -\left(\dfrac{4}{5} + -\dfrac{1}{2}\right) = -\left[\dfrac{8}{10} + \left(-\dfrac{5}{10}\right)\right] = -\dfrac{3}{10}$; $\dfrac{c}{a} = \dfrac{4}{5} - \dfrac{1}{2} = -\dfrac{4}{10}$;
$x^2 - \dfrac{3}{10}x - \dfrac{4}{10} = 0$, $10x^2 - 3x - 4 = 0$

33. $\dfrac{b}{a} = -[(-3 + \sqrt{2}) + (-3 - \sqrt{2})] = 6$; $\dfrac{c}{a} = (-3 + \sqrt{2})(-3 - \sqrt{2}) =$
$9 + 3\sqrt{2} - 3\sqrt{2} - 2 = 7$; $x^2 + 6x + 7 = 0$

34. a. For (12, 4.8) and (20, 8): $m = \dfrac{8 - 4.8}{20 - 12} = \dfrac{3.2}{8} = \dfrac{32}{80} = \dfrac{2}{5}$

For (20, 8) and (35, 14): $m = \dfrac{14 - 8}{35 - 20} = \dfrac{6}{15} = \dfrac{2}{5}$

Since the slope is constant, it is a linear function.
Let c = the number of cups oatmeal and n = the number of people.

$c = \dfrac{2}{5}n + b$; Use (20, 8) to find b: $8 = \dfrac{2}{5}(20) + b$, $8 = 8 + b$, $b = 0$;

$c = \dfrac{2}{5}n + 0$, $c = \dfrac{2}{5}n$

b. Slope $\dfrac{2}{5}$ means the average person uses $\dfrac{2}{5}$ c of oatmeal per breakfast.

The n- and c-intercepts, both zero, mean that no oatmeal is used if no one is at breakfast.

page 668

1. mode: -3; median: 0

mean: $\dfrac{-3 - 3 - 2 + 0 + 1 + 3 + 4}{7} = \dfrac{-8 + 8}{7} = 0$

2. mode: none; median: $\dfrac{-1.1 + 2.7}{2} = \dfrac{1.6}{2} = 0.8$

mean: $\dfrac{-4.0 - 2.3 - 1.1 + 2.7 + 3.8 + 4.2}{6} = \dfrac{-7.4 + 10.7}{6} = \dfrac{3.3}{6} = 0.55$

3. $\dfrac{6 + 3 + 1}{30} = \dfrac{10}{30} = \dfrac{1}{3}$ or $33\frac{1}{3}\%$

4. If the student distances are arranged in order, the first 8 student distances will be in the interval $0 - 1$. The average of the 15th and 16th student distances is the median distance. These are both in the interval $1 - 2$.

5. $\dfrac{8 \times 0.5 + 12 \times 1.5 + 6 \times 2.5 + 3 \times 3.5 + 1 \times 4.5}{30} = \dfrac{52}{30} = 1.7$ mi

6. $\dfrac{3 + 2 + 2 + 4 + 3 + 6 + 9 + 13 + 11 + 5 + 0 + 2}{12} = \dfrac{60}{12} = 5$

7. $13 - 0 = 13$

8. $\dfrac{1}{12} \cdot [1(5 - 0)^2 + 3(5 - 2)^2 + 2(5 - 3)^2 + 1(5 - 4)^2 + 1(5 - 5)^2 + 1(5 - 6)^2$

$+ 1(5 - 9)^2 + 1(5 - 11)^2 + 1(5 - 13)^2] = \dfrac{1}{12} \cdot [25 + 27 + 8 + 1 +$

$0 + 1 + 16 + 36 + 64] = \dfrac{178}{12} = 14.8\overline{3}$

9. $\sqrt{14.83} \approx 3.85$

10. $\dfrac{4}{11}$

11. $\dfrac{2}{11}$

12. $\dfrac{0}{11} = 0$

13. $\dfrac{4}{11}$

14. $\dfrac{1}{30}$

15. $\dfrac{1 + 4 + 15}{30} = \dfrac{20}{30} = \dfrac{2}{3}$

16. $\dfrac{10}{30} = \dfrac{1}{3}$

17. $\dfrac{3}{14} \cdot \dfrac{3}{14} = \dfrac{9}{196}$

18. $\dfrac{6}{14} \cdot \dfrac{6}{14} = \left(\dfrac{3}{7}\right)^2 = \dfrac{9}{49}$

19. $\dfrac{9}{14} \cdot \dfrac{9}{14} + \dfrac{5}{14} \cdot \dfrac{5}{14} = \dfrac{81 + 25}{196} = \dfrac{106}{196} = \dfrac{53}{98}$

page 669

1. $25 + 65 + 25 = 115$

2. 65

3. $65 + 25 + 65 = 155$

4. Since $25 + 65 = 90$, $\angle CFE$, $\angle BFD$, and $\angle AFC$ are right angles.

5. $m\angle BFE = 115$, $90 < 115 < 180$. Thus, $\angle BFE$ is obtuse. $m\angle AFD = 155$, $90 < 155 < 180$. Thus $\angle AFD$ is obtuse.

6. Answers may vary. Sample answers are: $\angle EFD$ and $\angle AFD$; $\angle AFB$ and $\angle BFE$; $\angle AFC$ and $\angle CFE$.

7. Let $x =$ the measure of the angle, then $180 - x =$ the measure of its supplement. $x = 15 + \dfrac{1}{2}(180 - x)$. $x = 15 + 90 - \dfrac{1}{2}x$, $1\frac{1}{2}x = 105$, $x = 70$, then $180 - x = 110$

8. Let x = the measure of each angle opposite one of the congruent sides. $2x + 74 = 180$, $x = 53$; Since $53 < 90$ and $74 < 90$, the triangle is acute.

9. Let $a = m\angle A$, then $m\angle B = \frac{1}{5}a$. $a + \frac{1}{5}a + 90 = 180$, $a = 75$; $\frac{1}{5}a = 15$

10. $\angle J$ and $\angle P$; $\angle K$ and $\angle Q$; $\angle L$ and $\angle R$

11. $\overline{JK}$ and $\overline{PQ}$; $\overline{JL}$ and $\overline{PR}$; $\overline{KL}$ and $\overline{QR}$ **12.** True

13. In $\triangle LMN$, $m\angle L = 180 - (75 + 33) = 72$. In $\triangle RST$, $m\angle R = 180 - (72 + 33) = 75$. Thus, the three angles of $\triangle LMN$ are congruent to the corresponding angles of $\triangle RST$. Therefore, the two triangles are similar: $\triangle LMN \sim \triangle SRT$.

14. $\dfrac{BC}{EF} = \dfrac{AB}{DE}$; $\dfrac{16}{12} = \dfrac{12}{DE}$; $144 = 16(DE)$; $9 = DE$

15. $\dfrac{c}{b}$ **16.** $\dfrac{a}{b}$ **17.** $\dfrac{a}{c}$

18. $(GK)^2 + (KH)^2 = (GH)^2$, $(GK)^2 + (2\sqrt{3})^2 = 4^2$, $(GK)^2 + 12 = 16$, $(GK)^2 = 4$, $GK = 2$

19. $\sin G = \dfrac{2\sqrt{3}}{4} = \dfrac{\sqrt{3}}{2}$ **20.** $\tan H = \dfrac{2}{2\sqrt{3}} = \dfrac{2}{2\sqrt{3}} \cdot \dfrac{\sqrt{3}}{\sqrt{3}} = \dfrac{\sqrt{3}}{3}$

21. a. $\sin 23° = 0.3907$ **b.** $\tan 23° = 0.4245$ **c.** $\cos 23° = 0.9205$

22. a. $m\angle B = 8$ **b.** $\tan 8° = 0.1405$ **c.** $\sin 8° = 0.1392$

23. a. $\sin A = \dfrac{15}{25}$ **b.** $\cos B = \dfrac{15}{25}$
$= 0.6000$ $= 0.6000$
$m\angle A = 37$ $m\angle B = 53$
c. $(BC)^2 + (AC)^2 = (AB)^2$; $15^2 + (AC)^2 = 25^2$, $(AC)^2 = 400$, $AC = 20$

24. Hidden assumption: The lighthouse stands at the edge of the water.
$\tan 12° = \dfrac{350}{BC}$, $0.2126 = \dfrac{350}{BC}$, $BC = \dfrac{350}{0.2126} = 1647$ ft to the nearest foot, using a calculator; (1646 ft using the table on page 671).